NANOMATERIALS, POLYMERS, AND DEVICES

NANOMATERIALS, POLYMERS, AND DEVICES

Materials Functionalization and Device Fabrication

Edited by

ERIC S. W. KONG
Research Professor, Shanghai Jiao Tong University
Research Institute of Micro/Nanometer Science and Technology
Shanghai, China

Published by John Wiley & Sons, Inc., Hoboken, New Jersey
Published simultaneously in Canada

For general information on our other products and services or for technical support, please contact our Customer Care Department within the United States at (800) 762-2974, outside the United States at (317) 572-3993 or fax (317) 572-4002.

Wiley also publishes its books in a variety of electronic formats. Some content that appears in print may not be available in electronic formats. For more information about Wiley products, visit our web site at www.wiley.com.

Library of Congress Cataloging-in-Publication Data:

Nanomaterials, polymers, and devices : materials functionalization and device fabrication / edited by Eric S. W. Kong.
    pages cm
  Includes index.
  ISBN 978-0-470-04806-1 (cloth)
1. Nanostructured materials. I. Kong, Eric S. W.
  TA418.9.N35N3293 2015
  620.1′92–dc23

                                                                    2014042896

Printed in the United States of America

10 9 8 7 6 5 4 3 2 1

1   2015

CONTENTS

CONTRIBUTORS

Sehee Ahn, School of Nano-Bioscience and Chemical Engineering, Ulsan National Institute Science and Technology (UNIST), Ulsan, Republic of Korea

Sukang Bae, Soft Innovative Materials Research Center, Institute of Advanced Composite Materials, Korea Institute of Science and Technology, Eunha-ri, Bongdong-eup, Wanju-gun, Jeollabuk-do, South Korea

Jaehoon Bang, Department of Mechanical Science and Engineering, Department of Materials Science and Engineering, University of Illinois, Urbana-Champaign, USA

Dylan J. Boday, International Business Machines, Corporation, Tucson, AZ, USA

Qing Cao, IBM T.J. Watson Research Center, Yorktown Heights, NY, USA

Jonghyun Choi, Department of Mechanical Science and Engineering, Department of Materials Science and Engineering, University of Illinois, Urbana-Champaign, USA

Won-Kook Choi, Interface Control Research Center, Korea Institute of Science and Technology

Woong Choi, School of Advanced Materials Engineering, Kookmin University, Seoul, South Korea

SungGyu Chun, Department of Mechanical Science and Engineering, Department of Materials Science and Engineering, University of Illinois, Urbana-Champaign, USA

Manfred Eich, Institute of Optical and Electronic Materials, Hamburg University of Technology, Hamburg, Germany

Zhiyong Fan, Department of Electronic and Computer Engineering, Hong Kong University of Science and Technology, Kowloon, Hong Kong SAR, China

Chao Gao, MOE Key Laboratory of Macromolecular Synthesis and Functionalization, Department of Polymer Science and Engineering, Zhejiang University, Hangzhou, People's Republic of China

Karthikeyan Gopalsamy, MOE Key Laboratory of Macromolecular Synthesis and Functionalization, Department of Polymer Science and Engineering, Zhejiang University, Hangzhou, People's Republic of China

Tingyi Gu, Optical Nanostructures Laboratory, Center for Integrated Science and Engineering, Solid-State Science and Engineering, and Mechanical Engineering, Columbia University, New York, NY, USA

Shu-jen Han, IBM T.J. Watson Research Center, Yorktown Heights, NY, USA

Helmut Heidrich, Fraunhofer Institute for Telecommunications, Heinrich Hertz Institute, Berlin, Germany

Cheuk-Lam Ho, Department of Chemistry and Institute of Advanced Materials, Hong Kong Baptist University, Hong Kong, People's Republic of China

Johnny C. Ho, Department of Physics and Materials Science, City University of Hong Kong, Kowloon, Hong Kong SAR, China

Diana L. Huffaker, University of California, Los Angeles, CA, USA

Jacek J. Jasieniak, Materials Science and Engineering, CSIRO, Clayton, Victoria, Australia

Han-Ik Joh, Carbon Convergence Materials Research Center

Norbert Keil, Fraunhofer Heinrich Hertz Institute (HHI), Berlin, Germany

Sunkook Kim, Department of Electronic and Radio Engineering, Kyung Hee University, Yongin, South Korea

Hyunhyub Ko, School of Nano-Bioscience and Chemical Engineering, Ulsan National Institute Science and Technology (UNIST), Ulsan, Republic of Korea

Eric S.W. Kong, Research Institute of Micro/Nanometer Science & Technology, Shanghai Jiao Tong University, Shanghai, People's Republic of China

Frederik C. Krebs, Department of Energy Conversion and Storage, Technical University of Denmark, Roskilde, Denmark

Joseph P. Kuczynski, International Business Machines, Corporation, Tucson, AZ, USA

Sungho Lee, Carbon Convergence Materials Research Center

Lei Liao, Department of Physics and Key Laboratory of Artificial Micro- and Nano-Structures of Ministry of Education, Wuhan UniversityWuhan, China

Xi Liu, Department of Electronic and Computer Engineering, Hong Kong University of Science and Technology, Kowloon, Hong Kong SAR, China

Peter Lützow, Fraunhofer Institute for Telecommunications, Heinrich Hertz Institute, Berlin, Germany

Brandon I. MacDonald, QD Vision, Lexington, MA, USA

Giacomo Mariani, University of California, Los Angeles, CA, USA

Paul Mulvaney, School of Chemistry and Bio21 Institute, University of Melbourne, Parkville, Victoria, Australia

SungWoo Nam, Department of Mechanical Science and Engineering, Department of Materials Science and Engineering, University of Illinois, Urbana-Champaign, USA

Jonghwa Park, School of Nano-Bioscience and Chemical Engineering, Ulsan National Institute Science and Technology (UNIST), Ulsan, Republic of Korea

Daniel Pergande, Fraunhofer Institute for Telecommunications, Heinrich Hertz Institute, Berlin, Germany

Xiaoling Shi, Department of Physics and Materials Science, City University of Hong Kong, Kowloon, Hong Kong SAR, China

Brandon Smith, Department of Mechanical Science and Engineering, Department of Materials Science and Engineering, University of Illinois, Urbana-Champaign, USA

Dong-Ick Son, Interface Control Research Center, Korea Institute of Science and Technology

Haiyan Sun, MOE Key Laboratory of Macromolecular Synthesis and Functionalization, Department of Polymer Science and Engineering, Zhejiang University, Hangzhou, People's Republic of China

Thomas Tromholt, Department of Energy Conversion and Storage, Technical University of Denmark, Roskilde, Denmark

Michael C. Wang, Department of Mechanical Science and Engineering, Department of Materials Science and Engineering, University of Illinois, Urbana-Champaign, USA

Xiao-Dong Wen, State Key Laboratory of Coal Conversion, Institute of Coal Chemistry, Chinese Academy of Sciences, Taiyuan, People's Republic of China; Synfuels China Co Ltd, Huairou, Beijing, People's Republic of China

Jason T. Wertz, International Business Machines, Corporation, Tucson, AZ, USA

Chee W. Wong, Optical Nanostructures Laboratory, Center for Integrated Science and Engineering, Solid-State Science and Engineering, and Mechanical Engineering, Columbia University, New York, NY, USA

Wai-Yeung Wong, Department of Chemistry and Institute of Advanced Materials, Hong Kong Baptist University, Hong Kong, People's Republic of China

Jan H. Wülbern, Tomographic Imaging Systems, Philips Technologie GmbH, Innovative Technologies, Research Laboratories, Hamburg, Germany

Zhen Xu, MOE Key Laboratory of Macromolecular Synthesis and Functionalization, Department of Polymer Science and Engineering, Zhejiang University, Hangzhou, People's Republic of China

Tao Yang, Baker Laboratory, Department of Chemistry and Chemical Biology, Cornell University, Ithaca, NY, USA

Vanessa Zamora, Fraunhofer Institute for Telecommunications, Heinrich Hertz Institute, Berlin, Germany

Ziyang Zhang, Fraunhofer Heinrich Hertz Institute (HHI), Berlin, Germany

FOREWORD

For ages, the classical disciplines of physics, chemistry, biology, engineering, and the materials science developed and grew in a rather isolated mode, independently of each other, in separate silos. But with the advent of the nanosciences, we have seen a remarkable change in the scientific landscape. Originally, the new trend was identified as a separate and distinctly different subdiscipline of, for example, the physics of materials, where it was observed that the mere size of an object matters and largely defines the different properties of materials. Examples are the change of the optoelectronic properties of nanometer-sized semiconducting particles, the quantum dots. Whenever the dimensions of such an object reach the limit of intrinsic length scales, in this case the exciton radius of an electron–hole pair, fundamental properties like the energy bandgap begin to change and become size dependent. The consequence is well known: the emission wavelength of the recombination radiation after photoexcitation changes and quantum dots of different radii – although chemically identical – emit fluorescence light of different colors, a direct consequence of the quantum confinement.

Other examples are the size- and shape-dependent optical properties of nanoscopic metallic particles. In the bulk, the collective oscillation of the nearly free electron gas of the conduction band electron cloud leads to the well-defined plasma frequency of the material, characteristic for each metal. However, if the metallic particle, owing to its small dimensions in the nanometer range, "senses" the boundary to its dielectric surrounding specific resonances occur, observable as localized surface plasmon modes. Their frequency and polarization are strongly dependent on the size and the shape of the metallic nanoparticle.

A different yet also remarkable trend could be observed in recent years in chemistry of materials. Especially polymers for a long time were considered to be typical bulk materials. Even though it was known that many details of the properties and functions of structural bulk polymers could be understood only at the level of the molecular and supramolecular organization of individual chain molecules – their chemical nature, their linearity versus branching, their entanglement with other chains, the presence of soft and hard segments, etc. – those individual polymer molecules were only vaguely defined and typically characterized by rather fuzzy quantities like an average molecular mass and, at best, a molecular mass distribution, that is, their polydispersity. This changed dramatically with the appearance of certain novel polymeric molecular architectures, in particular, the dendrimers. These macromolecules could be synthesized in a strictly monodisperse fashion, all molecules with the absolutely same molecular mass of up to several hundred thousand Daltons. With some of them being also shape persistent all of a sudden, it became obvious that these molecules can be seen as well-defined objects, as building blocks for supramolecular architectures, and as individual units for the assembly of nanomaterials.

Of course, materials scientists saw the enormous potential for the buildup of nanomaterials in a truly bottom-up approach by assembling materials, object by object, using building blocks of different materials. This paradigm shift offered fascinating novel strategies for the fabrication of synthetic materials that were never made or observed before because their structures and properties were completely defined by the man-made protocol of how to assemble them into 2D or 3D architectures.

This brought physics and chemistry again much closer to each other: the mutual interest of understanding the properties of nanoscopic objects from classical materials like metals and semiconductors and the novel synthetic efforts in generating organic and/or polymeric objects with property profiles that were defined – almost at will – by the chemical nature of the building blocks containing chromophores, redox centers, charges, etc., resulted in a convergence of the classical scientific disciplines that was not observed for the past decades.

In passing, we note that naturally, almost inevitably, objects of biological origin came into the game, too. Whether these were (large) proteins, micellar or vesicular structures from lipid molecules, or simple or highly complex supramolecular assemblies from DNA molecules – their use as building blocks in nanomaterials widened the spectrum of properties that could be tailored into these materials by another dimension.

As a result of this merging of ideas and concepts from different classical research areas in materials science, we see a very exciting cross-disciplinary effort in designing and fabricating nanoscopic materials with tailor-made properties and functionalities. Of course, this new horizon in materials properties triggered the phantasy of those engineers that design new functional devices. Their approach was governed for a long time by a top-down strategy, increasing the functionality and the integration density of sensors, logic gates, storage devices, etc., by making the individual functional unit smaller and smaller. Looking at Moore's law, for example, in microelectronics, this concept has been remarkably successful for a remarkably long time. However, the limit of this scaling down concept is now more serious than ever: if structural features in a single transistor reach 22 nm, if the current that switches the gate from on to off is carried by a few electrons only (and if the costs for the development of the next generation of storage device cannot be borne by a single player anymore ...), new strategies are needed urgently.

The bottom-up approach that the nanosciences offer, using nanomaterials that have never been available, designed by multidisciplinary concepts, using the principle of self-organization of the individual building blocks, for the construction of hierarchical systems, following examples from nature – all this has inspired the imagination of materials scientists and is currently in the center of the interest of materials science and engineering groups around the world.

Many of them have contributed to this book describing in numerous chapters the fabrication, the assembly, and the structural characterization of nanomaterials and refer to their functional potential that they offer for the device engineers. Most of them focus, for good reasons, on the currently most promising forms of nanocarbon, that is, single- or multiwalled carbon nanotubes and graphene, the monomolecular thin layer of graphite. Awarded with Nobel Prizes, both novel nanomaterials are currently the most widely studied nanomaterials worldwide. The set of articles assembled in this book represent a unique snapshot of where the field is and where it is heading toward. This book starts off with the first five chapters on the synthesis and functionalization of nanomaterials. The second part of this book covers various devices, with emphasis on nanofabrication. Light-emitting diodes are discussed in Chapters 6 and 7. Nanophotonics and nano-optics are covered between Chapters 8 and 9. Between Chapters 12 and 14, photovoltaic devices are discussed in depth. In the last section of the book, between Chapters 15 and 19, nanoelectronic devices are covered in detail. Enjoy this unique journey into the world of nanomaterials, polymers, and the devices made from them.

WOLFGANG KNOLL
AIT Austrian Institute of Technology, Vienna, Austria

1

THE FUNCTIONALIZATION OF CARBON NANOTUBES AND NANO-ONIONS

KARTHIKEYAN GOPALSAMY[1], ZHEN XU[1], CHAO GAO[1], AND ERIC S.W. KONG[2]

[1]*MOE Key Laboratory of Macromolecular Synthesis and Functionalization, Department of Polymer Science and Engineering, Zhejiang University, Hangzhou, People's Republic of China*
[2]*Research Institute of Micro/Nanometer Science & Technology, Shanghai Jiao Tong University, Shanghai, People's Republic of China*

1.1 FUNCTIONALIZATION OF CARBON NANOTUBES

The unique structure and morphology of carbon nanotubes have kept attracting researchers to explore their novel properties and applications since their discovery by Iijima in 1991 (1). Carbon nanotube (CNT) is a tubular structure made of carbon atoms and denoted as single-walled or multiwalled CNTs, having diameter of nanometer scale but length in micrometers. As in graphite, the carbon atoms in CNTs are in sp^2 hybridization and these sp^2 carbon sets give a great mechanical strength to CNTs. Single-walled carbon nanotubes (SWCNTs) are narrow and possess the simplest geometry and have been observed with diameters ranging from 0.4 to 3 nm. Multiwalled carbon nanotubes (MWCNTs) possess diameters of up to 100 nm. In general, pure CNTs have all carbons bonded in a hexagonal pattern except at their ends, and other defects in the sidewalls and the formation of various patterns resulting from the mass production generally humiliate desired properties.

Starting from diamond and fullerene to related nanostructures (Fig. 1.1) (2), CNTs have progressive properties, and during the past decade, CNTs had fascinated much attention for their electronic and mechanical properties. One of the main challenges in implementing serious applications of CNTs is functionalizing them. The lack of solubility and the difficult processing of CNTs in solvents impose great limitations to the practical applications of CNTs. Therefore, surface functionalization of CNTs has become the focus of research in recent years in order to improve their compatibility with matrices. Several methods, such as covalent or noncovalent functionalization, have been developed to achieve effective dispersion and bonding without the loss of properties of CNTs. The scope of applicability of CNTs has been expanded by extensive research particularly on tuning structural and electronic properties. According to the reports in recent times, CNTs have been used for many purposes and still implementation has been done day to day. Among those diverse applications, CNTs alone find some limitations in thermal, electrical, and mechanical properties, and this can be improved by combining them with certain polymers. Grafting polymers on CNTs is likely to improve the thermal and electrical properties (3, 4). Nanotube reinforcement into the polymer matrix has been considered as the primary need in order to obtain strongly enough composites (5). In general, CNTs have hollow structure and a very high aspect ratio (length-to-diameter ratio) which help them to form a network of conductive tubes. The surface properties of CNT-grafted polymers have been reported early by Downs and Baker (6) and Thostenson et al. (7). In particular, surface area tends to increase by the growth of nanofibers when compared to conventional process as predicted by Downs et al. It was observed that the surface area increased around 300 times and the interfacial shear strength by 4.75 times. From then, the feasibility of grafting CNT–polymers has been investigated to a larger extent and the nanotube composites have been designed according to the requirement of specific applications. The development of polymer–CNT composite methodology is based on improving the dispersion and controlling the orientation of CNTs in the polymer matrix (8). In general, high resistance at the nanotube–matrix interface limits thermal transfer along percolating networks of CNTs. However, viscoelasticity of the nanocomposites has been greatly enhanced by the nanotube network interpenetrating the polymer template. The small cross-sectional dimensions and extreme length of the CNTs allow bending to a large extent to have good intertube interactions under processing conditions. It has been clearly explained that the CNTs not only influence the electrical or thermal properties but they also eliminate the die swell, a problem faced during polymer processing (9).

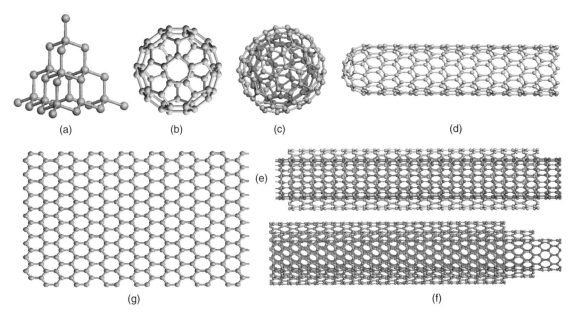

Figure 1.1   Model structures of carbon: (a) diamond, (b) fullerene, (c) multilayer fullerene, (d) single-walled carbon nanotubes, (e) double-walled carbon nanotubes, (f) multiwalled carbon nanotubes, and (g) graphene. (*Source*: Reproduced with permission from Reference 2).

1.1.1   Surface Chemistry of CNTs (Small Molecules)

After the CNTs came into existence, the structure, surface properties, and formation mechanism took the prime place to prove one or the other. Surface modification of pristine CNTs with no damage to CNT sidewalls is indispensable in current materials world when concerned with the mass production. Much of the applications involve the interaction of the CNT with the surrounding medium, and in particular, for example, molecules encapsulated or bonded/grafted in or to the CNTs. In these systems, the hydrophobic–hydrophilic behavior of the CNT is considered to be important to achieve well-designed surface-modified CNT materials. In general, CNTs line up into bundle of ropes held together by van der Waals (VDW) forces, more distinctively, pi-stacking. Ruoff et al. suggested that adjacent nanotubes which are present within the multiple cylindrical layers could be deformed by VDW forces, thereby destroying cylindrical symmetry (10). It is also important to note that the interaction between a CNT and the environment mainly consists of VDW forces (either attractive or repulsive force). It is these forces that are strong enough to insert molecules inside the CNT channel. The next step of explanation could be well understood in a simple CNT–molecule–water system. The molecule remains encapsulated inside the CNT channels, and based on the energy change during this process, it was proposed that the largest energy is contributed by the creation of cavities in the water to contain the nanotubes (11). In this section, it is more confined to polymer-grafted CNTs and thereafter will be discussed briefly about various grafting methods, type of polymerization, properties, and applications. Grafting of polymers to CNTs is of much interest since the first polymer-grafted MWCNTs were reported by Jin et al. (12). The solubility of polymers/CNT composites in organic solvents is considerably good, and various methods were proposed with the aim of dispersing both CNTs and polymer-grafted CNTs (13, 14).

Studies on inner surface of CNTs and nanoprotrusion processes have been reported by Chamberlain et al. (15). The convex surface is reactive and the inner concave surface is inert and can hold or encapsulate the molecules in it. But metal particles (Re complex) could deform the sidewalls at atomic scale as stated by this work. This sidewall tube opening would be similar to the nanobud-like structure. Such a negative curvature, a characteristic of nanobuds, is considered important in electronic properties of CNTs (15). Synthesis, properties, and commercial applications of CNTs make them a great deal to be considered in future (16, 17).

1.1.2   Linear Polymer-Grafted CNTs

Highly soluble linear polymers were used for grafting them as covalently attached moieties to solubilize CNTs by Riggs et al. (18). The resulted linear polymer-grafted CNTs were found to be strongly luminescent, showing a new opening for the grafted CNTs served as electron acceptors. The functionalized nanotubes were synthesized to contain both short and long nanotubes and mixed with poly-(propionylethylenimine-co-ethylenimine) (MW 200 000, EI mole fraction 15%). Luminescence of the polymer-bound MWCNTs and SWCNTs were in the same order of magnitude, although the long SWCNTs bound to the polymer displayed less luminescence. Starting from this report, many researchers reported the covalently grafted CNTs associated with good optical and photoluminescence properties.

Apart from MWCNTs, SWCNTs are also grafted and the possibility of dissolving them in various solvents took priority. In such a way, a variety of linear polymers such as polyvinyl pyrrolidone (PVP) and polystyrene sulfonate (PSS) have been chosen for grafting to SWCNTs by Smalley et al. (19). A general phenomenon called wrapping has been employed for grafting water-soluble polymers onto SWCNTs to

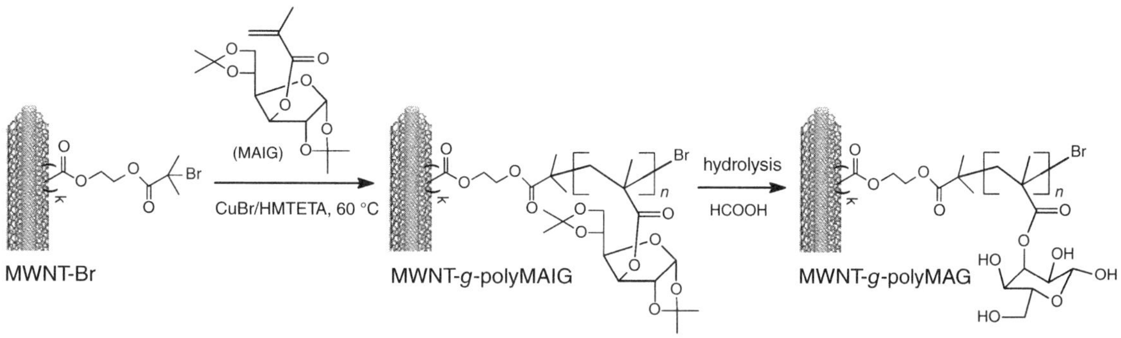

Figure 1.2 Synthetic strategy for grafting linear glycopolymer from surfaces of MWNTs by ATRP. (*Source*: Reproduced with permission from Reference 23).

avoid hydrophobic interface between CNTs and aqueous medium depending on thermodynamic factors. The stability of the polymer–CNT materials obtained by this method endured high ionic strength as they open up an idea for biological applications. Also, a simple method was introduced by Iijima et al. to chemically react the linear polymer poly(methyl methacrylate) (PMMA) with SWCNTs by ultrasonication (20). The aforementioned method originated while purifying the CNTs by ultrasonic irradiation and the functionalization mainly depended on frequency and duration of sonication. The organic molecules/polymer PMMA with SWCNTs could be synthesized by controlling ultrasonic irradiation, where the polymer molecules get dissociated at the hot spots produced by waves and also CNTs sidewalls get damaged to form bonding with the polymer. In another early work, SWNTs and MWNTs have been functionalized with polyvinyl alcohol (PVA) and PVA–CNT composite thin films were prepared (21). The thin films were achieved due to the highly soluble nature in polar solvents such as DMSO and water. The linear polymer PVA is considerably a good hosting polymer matrix for providing composites with high optical properties. Homogeneous dispersion of CNTs in the solvent and high optical quality of the CNT–PVA displays the effective functionalization by this method.

An *in situ* polycondensation approach has been introduced by Gao et al. to bond linear polyurea and polyurethane on amino-functionalized MWNTs. Polyurea-grafted MWNTs presented self-assembly behaviors, forming flat- or flowerlike morphologies in the solid state. Moreover, Raman spectra indicated that the signals of polymer-functionalized CNTs are dependent on the polymer species involved and not on the content of the polymer. Structure and morphology of the MWNT/polyurea molecular nanocomposites were studied by TEM and SEM. The grafted polymers onto the tubes were uniform, and it was determined that the higher the quantity of the grafted polymer, the thicker is the polymer shell. Covalent linkage between the core and the shell referred to as molecular nanocomposites could well explain the difference between grafted NTs and crude CNTs as the surface remains smooth in the latter. The main objective involved the synthesis of various flowerlike structures which would pave the way for its scope in supramolecular chemistry (22).

Also, linear glycopolymer-grafted MWNTs by surface-initiated atom transfer radical polymerization (ATRP) were reported by Gao et al. (Fig. 1.2) (23). In this work, linear biocompatible and water-soluble polymers were grafted from the surface of MWC-NTs with 3-*O*-methacryloyl-1,2:5,6-di-*O*-isopropylidene-D-glucofuranose (MAIG). The structural elucidation of the resulting grafted MWNT-*g*-polyMAIG was predicted by NMR and FTIR. MWNT-*g*-polyMAIG was deprotected in 80% formic acid to get multihydroxy MWNT-*g*-polyMAIG. The solubility of deprotected MWNT-*g*-polyMAIG was studied, and the material was highly soluble in polar solvents such as water, methanol, DMSO, and DMF. SEM, TEM, and SFM observations on the morphology and nanostructures of the resulting grafted MWNTs showed that a nanowire-like morphology could be observed and the space among nanowires becomes smaller for MWNT-*g*-polyMAIG. A diffused nanowire-like morphology was observed for the lower polymer content which results from the cleavage of the protected units. From the TEM images, for the sample obtained after 10.5 h, the core–shell structure of polymer-grafted CNTs can be distinctly observed under high magnification. On deprotecting the material, as obtained after 29 h, a core–shell structure (4–6 nm) of polymer shell was observed which evidenced grafting of polymer on nanotube surface. The grafting method of linear glycopolymers is biocompatible as the deprotected glycopolymers with hydroxyl groups are water soluble, leading to effective grafting over CNTs. These grafting techniques would pave the way for their focus on bionanotechnology in the future.

Wu et al. reported the grafting of bromo-ended polystyrene (Br–PS–Br and PS–Br) to MWNTs over CuBr/bipy catalytic system (24). ATRP technique was used to prepare the linear polymer grafting with distinct optical limiting effect. Also, solubility of the linear polymer PS/MWNTs was observed to be good in DCB, THF, and $CHCl_3$. The PS-modified MWCNTs were found to bond together as polymer chains were covalently connected to the convex walls of nanotubes. Optical limiting properties was tested for the PS-MWCNTs and found to be preserved by the covalent attachment of PS to CNTs. The optical limiting results from the open-aperture z-scan of MWCNT-PS and MWCNTs in chloroform showed a reduction in the transmittance. In general, the mechanism of the optical performance of the linear polymer–CNT solution could be regarded to both nonlinear scattering and nonlinear absorption, one or the other and the proof are under progress. Solvation of CNTs and functionalizing them remain a challenge till now that has to be boosted up to be available for biology and medicine where the solubility factor is much more critical. To abide this, water solubility of CNTs has been improved via noncovalent interactions in poly(ethylene glycol) (PEG)-based ABA linear copolymers as reported by Adeli's group (25). The water-soluble CNT nanoparticles were obtained by the grafting method using linear-dendritic copolymer, as the method finds better observations than that observed for other methods. The hybrid materials were found to be a good system for loading the drugs and deliver to kill cancer cells.

Recently, CNTs modified with phenylacetylene groups were found to enhance the reaction of porphyrin monomers with CuAAC to give rise to linear polymers on the CNT surfaces (26). These novel nanotube–porphyrin polymer hybrids are photoactive and found to possess good optical and electrochemical properties. It was attributed that covalent bonding of the polymer to the CNT surface enhances the stability of CNTs, and moreover, the attached polymer shields a large area of the CNT surface from VDW attraction to the neighbor CNT. This was supported by previous work by Qin et al. when a linear polymer, poly(sodium 4-styrenesulfonate) (PSS), was grafted onto SWCNTs in aqueous solution by *in situ* method (27). The method was applicable to functionalize, solubilize, and purify the pristine SWCNTs in lab scale.

1.1.2.1 "Grafting to" Approach The "grafting to" technique comprises a method of attachment of as-prepared or preformed end-functionalized polymer on to the CNT reactive surface groups. In general "grafting to" method involves the generation of carboxylic acid groups on the surface of CNTs and then covalently linking the polymers with these carboxylic acid groups. Also, the polymers should possess functional groups for grafting with CNTs in this approach. A controlled synthesis and comprehensible characterization of polymer structure can be offered prior to the grafting process by this technique. In particular, the method provides synthesis path of definite polymer structures and also enables the prepared polymers to attach consequently to the CNT surface (28).

Initially, the functionalization of CNTs by "grafting to" method was reported by Fu et al. This study involves the preparation of CNTs with functional COOH groups and then converting them to acyl chlorides by refluxing with thionyl chloride. The resulting CNTs with acyl chloride moieties were further treated with the hydroxyl groups of the dendritic polymer (29). Qin et al. used "grafting to" approach for functionalizing SWNTs with polystyrene. The azide end-functional polystyrene (PS–N_3) was synthesized by ATRP of styrene through end-group transformation and then grafted to SWCNTs via cycloaddition reaction. In this case, it was suggested that the grafting density was not high due to the low reactivity and high steric hindrance of the macromolecules (30). Thus, this method has some limitations as the grafted polymer chains will not allow a further high extent of grafting. To overcome these limitations, existing methods need to be modified and implemented to achieve higher degree of polymer grafting. As an alternative to "grafting to" technique, Lou et al. (31) reported the direct covalent bonding of polymeric radicals, released by thermolysis of 2,6,6-tetramethylpiperidinyl-1-oxyl-(TEMPO)-end-capped polymers to pristine nanotubes. The resulted polymer-modified MWCNTs were observed to be good in dispersion in various solvents. A hydrophobic polymer, poly2VP, which is soluble in apolar solvents and also is polyelectrolyte in water at low pH, was chosen to graft CNTs. The end-capped poly2VP type of polymer forms a shell-like layer over the CNTs, which help CNTs to dissolve in organic solvents and in acid solution. More clearly, poly2VP with positive charge plays an important role in depositing CNTs over oppositely charged layers.

1.1.2.2 "Grafting from" Approach Until now, several strategies have been proposed based on the noncovalent and covalent bonding of organic molecules to the surface of CNTs. Grafting of polymer chain on CNTs can be carried out by either "grafting from" or "grafting to" technique (28, 32). In order to obtain specific functionalities over CNTs, polymer molecules in different solvents has been preformed for grafting CNTs by grafting from technique and the grafted material as a whole is found to possess good dispersibility. "Grafting from" approach relies on the immobilization of initiators at the surface of CNTs, followed by *in situ* surface-initiated polymerization to generate tethered polymer chains. Polymer molecules grafted on the surface of CNTs in the presence of active functional groups (–COOH, –NH_2, –OH) seemed to be an effective method for nanocomposites. "Grafting from" approach was first used for the fabrication of CNT–polymer composites by an *in situ* radical polymerization process (33). The "grafting from" approach is based on the initial immobilization of initiators onto the nanotube surface, followed by *in situ* polymerization with the formation of polymer molecules attached to CNTs. The double bonds of the CNTs were opened by the monomer molecules and more exactly, the surface of the nanotube was considered to be important for this method of grafting. Early works were more focused on this phenomena and PSS was first grafted on the surface of CNTs through this method by Ajayan et al. (34). Guldi et al. reported the use of grafting from method for the grafting of PSS from the CNT surface with positively charged porphyrin, and in the following work, CNT–porphyrin hybrids were incorporated onto indium tin oxide (ITO) electrodes, which find importance in high efficient energy conversion devices (35, 36). The advantage of this approach is that polymer-functionalized nanotubes with high grafting density can be prepared. However, this process needs a strict control on the amounts of initiator and substrate.

1.1.2.3 In Situ *Polymerization Approach* *In situ* polymerization is a common method that has been widely explored for the synthesis of polymer-grafted nanotubes and other polymer processing techniques. In this method, polymer macromolecules can be grafted onto the walls of CNTs and more suitable for the preparation of insoluble and thermally unstable polymers, which cannot be processed by melt or solution processing. Also, this processing technique enables high nanotube loading and high miscibility with the polymer molecules. It has been practically shown that *in situ* polymerization provided stable dispersion of CNTs even after thermal processing. Moreover, CNTs can induce the local ordering of certain polymers during *in situ* polymerization. Mechanical properties of the composites can be improved due to ordered polymer packing on the CNTs that enhance stress transfer from the polymer to CNTs across the interface (37, 38).

To design novel nanomaterials like CNTs with progressive properties, *in situ* method provides an easy way to graft CNTs with polymer by various strategies depending on the interaction between the polymer and CNTs (Fig. 1.3).

The first *in situ* radical polymerization was introduced by Jia et al. in 1999 to graft CNTs to produce homogeneous composites. The method includes the use of PMMA grafted to high content of CNTs (33). In order to attach the carbon network of CNTs, initiator and monomer were mixed via *in situ* method to get the longer polymer chains grafted to CNTs. Following this, "*in situ*" bulk polymerization provided an easy way to predict the synthesis and dispersion characteristics of MWCNT composites with polymers. Owing to the grafting nature of PMMA, a uniform dispersion of MWCNTs in PMMA matrix could be achieved better than that prepared from composites by post mixing methods (40).

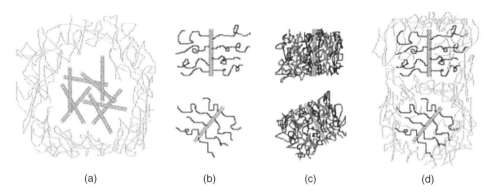

(a)   (b)   (c)   (d)

Figure 1.3 Schematic illustration of (a) CNTs dispersed mechanically in polymer matrix, (b) polymer-bonded CNTs, (c) polymer-coated CNTs by layer-by-layer self-assembly approach, and (d) polymer-functionalized CNTs dispersed in free polymer matrix. (*Source*: Reproduced with permission from Reference 39).

TABLE 1.1 Young's Modulus and Elongation at Break of MWNTs/PA1010 Composites Made by *In Situ* Polymerization and Melt-Mixing

Sample[a]	MWNTs Content (wt%)[b]	Young's Modulus (MPa)	Increment (%)[c]	Increment (%)[d]	Elongation at Break (%)
LPA1010	0	1021	-	-	37
NTPA-1	1	1301	27.4	8.1	35
NTPA-2	2.5	1302	27.5	8.2	31
NTPA-5	5	1449	41.9	20.4	23
NTPA-10	10	1748	71.2	45.3	17
NTPA-20	20	1813	77.6	50.7	9
NTPA-30	30	1912	87.3	58.9	5
NTPA-5B	5	1226	20.1	1.9	27
NTPA-10B	10	1694	65.9	40.8	19
NTPA-30B	30	1769	73.3	47.0	8

Source: Reproduced with permission from Reference 39.
[a]Samples of NTPA-1 to NTPA-30 were made by the *in situ* polymerization method, and samples of NTPA-5B to NTPA-30B were made by the melt-mixing method.
[b]The content represents the feed weight percentage.
[c]Increment with LPA1010 as the reference, calculated by the equation: Young's modulus of (composites-LPA1010)/Young's modulus of LPA1010.
[d]Increment with CPA1010 as the reference, calculated by the equation: Young's modulus of (composites-CPA1010)/Young's modulus of CPA1010.

As a suitable method to graft polymers over CNTs, *in situ* method was used along with other techniques to achieve a strong interfacial adhesion between the CNTs and the polymer matrix. Other early works (41–43) continued the interest for *in situ* method, and now most composites involve a combination along with "*in situ*" polymerization technique. Kong et al. employed a novel *in situ* ATRP "grafting from" approach to functionalize MWCNTs and provided a new idea for extending it further to copolymerization systems to synthesize nanocomposites based on CNTs (44).

Homogeneous surface coating of long carbon nanotubes (MWCNTs) has been performed by *in situ* polymerization of ethylene to achieve high-performance polyolefinic nanocomposites. The method used here is a form of polymerization filling technique (PFT) and applied to polymerize ethylene directly onto the CNT surface. Higher tensile properties with nanofiller loading of 1 wt% were observed which is attributed to the pre-break-up of the MWCNT bundles by the PFT (45). Mostly, epoxy nanocomposites have been prepared by using *in situ* polymerization method. Polyamide 6 (PA6)/CNT composites have been prepared using pristine and carboxylated MWNT via *in situ* polymerization. In this study, the tensile strength and storage modulus of PA6/CNT composites were found to be improved and also the addition of small amount of CNTs had an effect on the crystallization and glass transition properties (46). The MWNT-reinforced PA1010 composites prepared by Zeng et al. by *in situ* polymerization method were found to possess excellent mechanical properties. Young's modulus increases with the addition of MWNTs, for example, 30 wt% MWNT improved Young's modulus of PA1010 by 87.3% (Table 1.1) (39).

In recent times, the unique properties of CNTs and Nylon-6 were combined to obtain nanocomposites through *in situ* polymerization. In this study, the plasma-polymerized CNTs and raw CNTs were added to *in situ* polymerization of ε-caprolactam catalyzed by 6-aminocaproic acid to obtain nanocomposites with 2 and 4 wt% of nanotubes. The electrical conductivity of nanocomposites with 4 wt% reached a maximum of

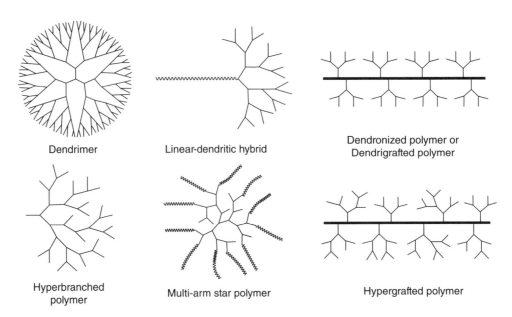

Figure 1.4 Schematic description of dendritic polymers. (*Source*: Reproduced with permission from Reference 51).

10^{-4} S/cm, suggesting a semiconductive material. The method used in this work was effective for the dispersion of CNTs which can be used to form nanocomposites with improved properties compared with pure Nylon-6 (47). At present, *in situ* method finds its use in many processing techniques including the production of strong nanocomposites and fibers. Hu et al. successfully prepared poly(*p*-phenylene benzobisoxazole) (PBO)–CNT fibers that are still stronger by *in situ* polymerization method. Solubility, dispersivity, reactivity, and interfacial adhesion of CNTs in polymer matrix have been greatly enhanced by this method. Dry-jet wet-spinning technique was used to fabricate continuous CNT–PBO copolymer fibers and the observations concluded that (PBO–CNT) copolymers had the high tensile modulus (89.4 ± 6.6 GPa), tensile strength (2.17 ± 0.19 GPa), and thermal stability (11% higher) compared to the PBO fibers (48). Furthermore, covalently functionalized unzipped CNTs (uCNTs) with PMMA (uCNTs-P) via *in situ* free radical polymerization has been studied recently by Wang et al., in which the mechanical reinforcement of uCNTs-P was higher than uCNTs and functionalized MWNTs (MWNTs-P) (49). Also, a combination of *in situ* polymerization and electrospinning processes has been tried to prepare CNT/polyaniline (PANI) thermoelectric composite nanofibers. Highly ordered structure of PANI backbone chains also increases the effective degree of electron delocalization in addition to reducing the π–π conjugated defects in the polymer backbone that helps in improving the thermoelectric properties (50).

1.1.3 Dendritic Polymer-Grafted CNTs

Dendrimers have prospective applications in a broad range of fields from drug delivery to material/metal coatings due to their high soluble nature, low melting viscosity, and profusion of functional groups. In general, dendrons and dendrimers possess high regularity and controlled molecular weight, and their synthesis involves step-by-step convergent and divergent approaches. Appropriately, a dendronized polymer is referred to as a linear polymer connected with side dendrons. Dendronized polymer can be obtained by direct polymerization of dendritic monomers or by linking linear polymeric core and dendrons. A review by Gao et al. reported the synthesis and applications of various polymers including the dendrimers under hyperbranched polymers in which the formation mechanism has been extensively explained (Fig. 1.4) (51).

Fu et al. introduced the dendritic polyethylene glycol with its hydroxyl group bound with the functionalized CNTs by "grafting to" method. The dendra-functionalized CNTs were hydrolyzed by trifluoroacetic acid (TFA) to defunctionalize them to get soluble CNTs. The defunctionalization of dendron-f-CNTs showed a decrease in UV/vis absorption, which leads to estimate the average absorptivities of the CNT species in solution. The observed average absorptivity for SWNTs at 500 nm is 97 ± 31 (mg/ml)/cm, a higher value obtained than that for fullerene (C60) (29). Various dendrons (52) and dendrimers (53) were synthesized and grafted onto CNTs by different methods. A different type of star-shaped nanostructures was observed resulting from the reaction of the outer surface of a multifunctional dendrimer with nanotubes. Lipophilic and hydrophilic dendron species were used to functionalize CNTs under amidation and esterification reaction conditions by Sun et al. (52). The dendrons were terminated with long alkyl chains and oligomeric PEG moieties before grafting on to CNTs. Defunctionalization of the grafted materials proved an evidence for the presence of ester linkages between the dendron functionalities and CNTs. The lipophilic dendron-functionalized materials were soluble in weakly polar solvents and oligomeric PEG functionalized CNTs were soluble in organic solvents and water. These results are helpful for exploring the solubility nature of nanotubes that in turn is considered to be important to study their optical properties.

Multihydroxyl dendritic polymers were grown on the convex surfaces of MWCNTs by *in situ* ring-opening polymerization (ROP) method by "grafting from" approach to produce one-dimensional nanocomposites by our group (54). High functionalization and possibility of

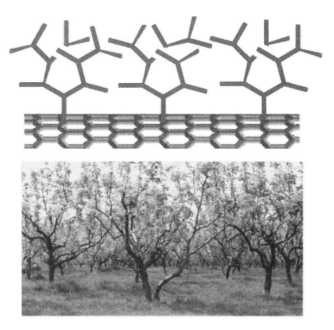

Figure 1.5 HP-grafted CNTs – branched-like trees. Graphical representation of a section of MWNTHP nanohybrids presented in this chapter (top) and a photograph of the trees grown on a hillside (bottom). (*Source*: Reproduced with permission from Reference 54).

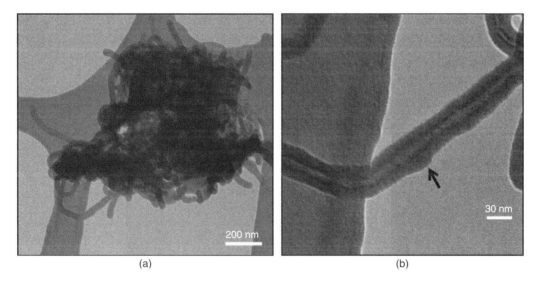

(a)   (b)

Figure 1.6 TEM images of MWNT-HP5 at (a) low magnification and (b) high magnification. (*Source*: Reproduced with permission from Reference 54).

extending functionalization at the ends of the tubes to build more complex nanohybrids with polymers (Fig. 1.5) were considered to provide interesting applications in designing nanomaterials and nanodevices. The dispersibility of the functionalized CNTs was good in polar solvents and the increase in the content of the grafted polymer relatively increased the dispersibility. Also, M_n of the grafted hyperbranched polymer (HP) was found be greater than that of free HPs. The grafting seemed to have been well enough bonded with the MWNTs as there considered to be no chance of termination than that would be in polymer solution alone. The reason for the high molecular weight remains as it could suggest another factor that MWNT-OH groups were activated initially by the catalyst before the monomer addition. Moreover, the topology of hyperbranched polymers depends on molecular weight (MW), degree of branching (DB), and unit density. Core-shell structures of HP-grafted MWNTs were observed from TEM images as shown in the figure and the thickness of the shell range from ~12 to 18, ~6 to 8, and ~7 to 10 nm, respectively. Polymers grown on the nanotube surface can be seen as in the TEM images (Fig. 1.6).

In order to avoid the rigid structure of the hyperbranched polymer, to make it flexible, dendrons of AB_2 type was prepared prior to the grafting to CNTs by Feng et al. The AB_2 monomer was initially synthesized from trimelliticanhydride and 1,6-diaminohexane and the resulted

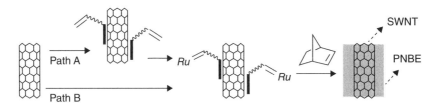

Figure 1.7 Complete functionalization strategy of ROP. (*Source*: Reproduced with permission from Reference 60).

products contained phenyl rings and aliphatic chains. Then, thermal polymerization of the AB$_2$ monomer yielded the PAI hyperbranched polymer. Following this, the HP polymer was grafted to MWCNT-COCl. To investigate the grafting nature and the interaction between the CNTs and polymer, the polymer-grafted functionalized CNTs were absorbed on NH$_2$–Si substrate. It was observed that there is a strong absorption of CNTs over the Si substrate, which confirmed the strong grafting of polymer to CNTs (55). Pan et al. reported the growth of multiamine terminated poly(amidoamine) dendrimers on the surface of CNTs. Covalent sidewall derivatization of nanotubes was confirmed by thermal degradation studies and the weight of the dendrimer increased with M_n, linearly from 2.0 to 5.0. Finally, from the observations, the amount of polymer that was grafted to CNTs could be controlled to about 10–50 wt% (56).

A different methodology was adopted for the SWCNTs grafting with poly(amidoamine) dendrimers by Campidelli et al. (57). Grafting involved two steps: in the first step, SWNTs were reacted with amino acid and paraformaldehyde, through a typical 1,3-dipolar cycloaddition reaction in DMF, to yield pyrrolidine ring functionalized with *N*-tert-butoxycarbonyl (Boc)-protected amine group, which was grafted to nanotubes in the second step. From thermogravimetric analysis (TGA) results, a weight loss of 62% was observed for SWNT nanoconjugates which revealed that unreacted amine groups must have been involved to suggest an approximate average of two porphyrins on each dendrimer. By this method more functionality can be loaded over CNTs without any deterioration of electronic properties of CNTs. MWCNT–PAMAM–PEG–PAMAM linear-dendritic copolymer composite was reported by Adeli et al. (58). Most importantly, the interaction between CNT and water has been considered to obtain better grafted CNTs with polymer molecules. In this regard, recent work on dispersing the CNTs in water by a two-step process was employed by Sohn et al. Here, poly(ether-ketone) (HPEK), a dendritic hyperbranched polymer, was grafted onto the surface of the MWCNT to avoid oxidative damages of CNTs by introducing copious reactive sites. On sulfonation of the resulting HPEK-grafted MWCNT, a zeta potential value of −57.8 mV was observed and the effect of the presence of sulfonic acids on the grafted CNTs makes them hydrophilic to get the oxygen being absorbed frequently (59).

1.1.3.1 In Situ *Ring-Opening Approach*

Among various approaches in grafting polymer to CNTs, *in situ* ring-opening approach is another strategy that has been used to grow specific polymers from CNTs. Noncovalent functionalization of SWNTs with polymer by ROP has been carried out by Gomez et al., who proposed a selective coating of CNTs with organic polymer. Such a method was believed to be a new opening to modify the CNT structure without any destruction (Fig. 1.7) (60).

The method involves the functionalization of SWNTs with ruthenium alkylidenes through cross-metathesis reactions over the walls of the CNTs and treated with norbornene (NBE). This technique was followed by two routes: (i) initially, organic precursors being adsorbed on to the CNTs and in turn with a ruthenium alkylidene with cross-metathesis, and (ii) pyrene-substituted ruthenium alkylidene absorbed over CNTs. AFM results showed that an ideal nonfunctionalized nanotube has the diameter of 1.4 nm and the diameter of polynorbornene (PNB)-grafted SWNT was found to be 9.1 nm, and in addition, a high selectivity of functionalization was achieved by this method. Moreover, the thickness of the PNB layer on nanotube was found to increase with the amount of monomer used and the time experienced no remarkable change during the polymerization. Also, the solubility of the polymer was good as the polymer growth seems to be longer.

Although the aforementioned method finds a novel pathway, it needs some modifications to avoid the polymer desorbed from the nanotubes. This case in general is considered to be more concerned when the polymers with high molecular weight are used for coating on CNTs. Following this, Liu et al. reported the grafting of high-molecular-weight polymers to the CNT surface by *in situ* ROP. Their experiments involve the covalent attachment of ring-opening metathesis polymerization (ROMP) catalysts to the CNTs and controlled polymerization of norbornene. Raman spectra, TEM, and AFM studies revealed the consistent properties and characteristics associated with grafted CNTs. In order to predict the effect of time over molecular weight during polymerization, polymer-grafted CNTs were treated with KOH/18 crown 6 in THF, resulting in the cleavage of the ester linkages, leaving the polymer in the solution. A linear increase in molecular weight with time was observed for 5–180 min. Also, glass transition temperature was found to be 27 °C obtained after cleavage, which was obvious with the literature values for polynorbornene (61).

In situ cationic ROP was used to grow multihydroxyl hyperbranched polymers from CNTs by using 3-ethyl-3-(hydroxymethyl) oxetane (EHOX) by Gao et al. (Fig. 1.8). Highly functionalized CNTs could be obtained by hyperbranched polymers on grafting CNTs with retaining the end-group functionalization, which can be surface modified further. Good compatibility of CNT with hyperbranched polymer matrix helps to extend the idea to achieve additional functionalization in the form of producing molecular nanoforests (discussed previously in dendritic polymer-grafted CNTs) (54).

Subsequently, poly(*e*-caprolactone) (PCL) was employed by Gao's group by "grafting from" approach based on *in situ* ROP (Fig. 1.9) (62). CNTs as the "hard" core and the hairy polymer layer as the "soft" shell constitute the core–shell structures in the presence of grafted polymer chains in high density. CNT-grafted-PCL was further investigated for its biodegradability, which often required for a modified nanomaterial.

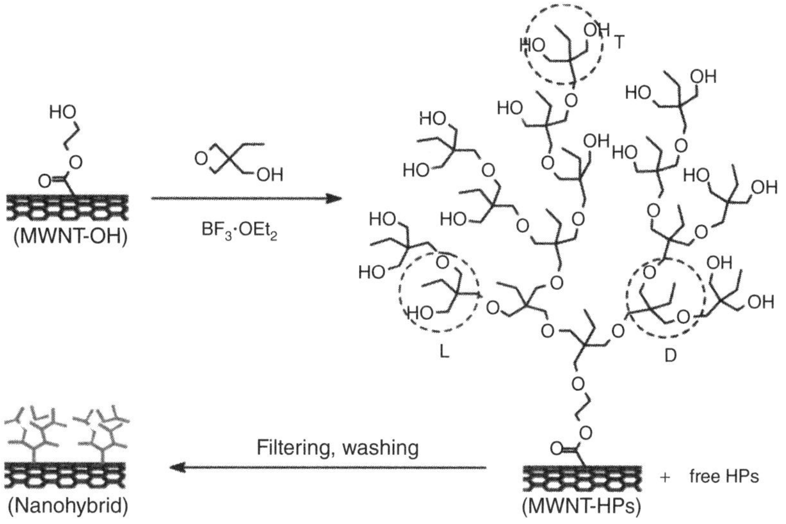

Figure 1.8 Synthesis of dendritic HP-MWNT nanohybrid through *in situ* ROP. (*Source*: Reproduced with permission from Reference 54).

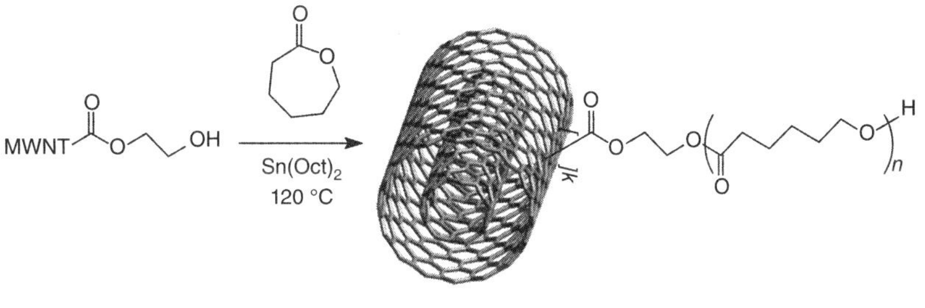

Figure 1.9 Grafting of PCL onto MWNTs. (*Source*: Reproduced with permission from Reference 62).

The results from the Raman spectra were interesting and showed the presence of two peaks of G band for PCL-functionalized MWNTs, implying that a large amount of polymer is grafted on to the CNTs and the additional peak corresponds to the degree of disorder in CNTs. Biodegradation nature of MWNT-*g*-PCL materials was tested by a bioactive enzyme catalyst, pseudomonas (PS) lipase. In the experiment, the samples were collected at regular intervals and analyzed by SEM and TEM. The results proved that there was no significant degradation for MWNT-*g*-PCL in the absence of the enzyme. At this stage, the CNTs were found to be thick with diameters of ~90.3 and 85.5 nm, respectively. Degradation occurred in the presence of PS lipase and resulted in thinner tubes than before, ~83.3 and 75.7 nm, respectively). For 48 h, the tubes were still thinner and after 96 h of degradation, residual polymer lies on the surfaces of MWNTs while the nanotubes were found with diameters of ~20–30 nm, which is more or less equal to that of the purified tubes. Covalently grafted biodegradable PCL onto CNT surfaces retains the biodegradability of conventional PCL, which suggests that it could be completely biodegraded by PS lipase within 4 days. It is evident that grafted CNTs could support cell growth without any damage of the enzyme.

Other works involved the use of ROP to build uniform and efficient grafting to CNTs and find specific applications depending upon the properties of polymer and CNTs. Grafting of poly(L-lactide) (PLLA) on CNTs by surface-initiated polymerization was carried out by Chen et al. (63). Purified MWNTs were acid functionalized and then hydroxyl functionalized with SOCl$_2$ to yield MWNT-OH. PLLA-grafted CNTs were obtained by the addition of L-lactide to MWNTs-OH with tin(II) 2-ethylhexanoate at various temperatures under nitrogen atmosphere. The materials were found to have good solubility in organic solvents such as chloroform and DMF due to the higher PLLA content. The measured weight of the PLLA-grafted CNTs increased after polymerization, which confirms the effective grafting. Raman spectroscopy revealed that I_D/I_G ratio for the grafted CNTs was 0.69, higher than that of the pristine MWNT (0.61), which is attributed to the grafting of PLLA. The work suggests that the Raman studies still needed further evidences to understand the interaction between CNTs and the polymer onto it. In another method adopted by Feng et al., PLLA brushes were grafted on MWNTs and the grafted polymers were found to possess superparamagnetic properties. A uniform layered grafting was achieved by this method and the layer depends on the feed ratio of the initial monomer to CNTs (64).

Gao et al. reported the functionalization of CNTs with hyperbranched polyglycerol (HPG) by anionic ROP (Fig. 1.10). Water-soluble and biocompatible HPG as a polymer with multihydroxyl groups was assumed to be more suitably prepared by anionic ROP technique. More particularly, a stable macroinitiator was prepared in the presence of anions by one-step [2+1] cycloaddition of nitrenes to the MWNTs.

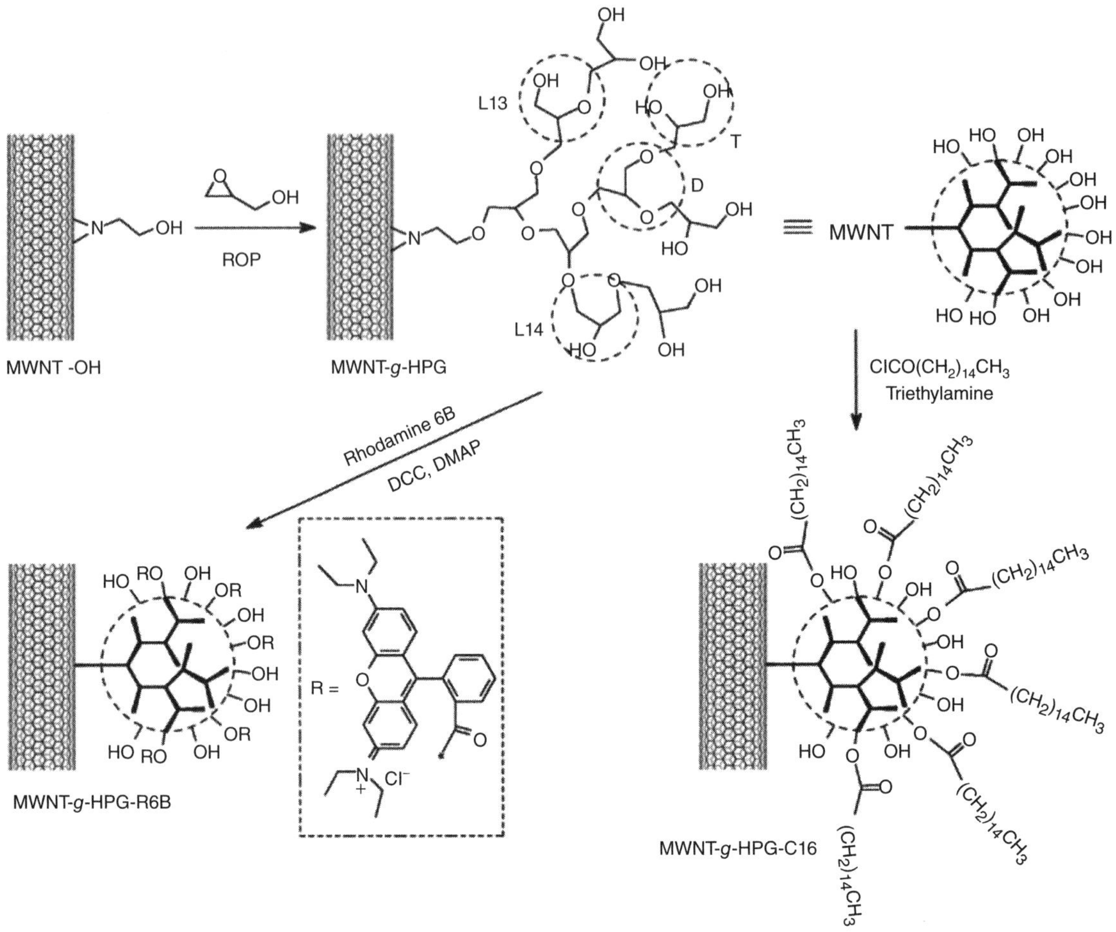

Figure 1.10 Functionalization of multiwalled carbon nanotubes (MWCNTs) with hyperbranched polyglycerol (HPG) by anionic ring-opening polymerization (ROP) and modification of the grafted HPG. (*Source*: Reproduced with permission from Reference 65).

The presence of hydroxyl groups in the MWNT-grafted HPGs contributes to the better formation of multifunctional nanohybrids. This was better understood by synthesizing the amphiphilic hyperbranched polymer-grafted MWNTs, on the addition of palmitoyl chloride with MWNT-g-HPGs in triethylamine. The selected results of HPG-grafted MWNTs (MWNT-g-HPG4) are shown in Table 1.2. It was observed that the grafted polymer content increases from ~22.8 to 90.8 wt% for the weight feed ratio (R_{wt}) rising from ~5.75:1 to 103.5:1. This is attributed to the fact that the weight fraction of grafted polymer (f_{wt}) could be well adjusted in a wide range by R_{wt} with the grafting efficiency up to 90.8 wt%. The efficiency was high when compared to that observed for other methods to obtain dendritic polymer-grafted CNTs. Also, the solubility was found to be good in weakly polar or nonpolar solvents such as THF, chloroform, and dichloromethane than that observed in polar solvents. Such MWNT-g-HPG hybrids synthesized by ROP method with tunable properties proposed a resourceful platform for bionanotechnology (65).

Moreover, MWNT-g-PCL was synthesized by ROP and compared for the demonstration of the reactivity and functions of the immobilized organic moieties with various CNT–polymer substrates (66).

Functional groups have been covalently anchored to the MWNTs. However, it is not clear in some aspects like, since the surface functionalities could not be distinguished even by HR-TEM, to predict whether the functional groups are anchored on the MWNTs uniformly or uneven or around the tips. Second, whether the chemical reactivity of functional groups would be retained and finally is that possible to graft polymer directly from the tube surfaces to propose a simple and cost-effective method. In order to clarify these doubts, various chemical reactions were performed on the f-MWNTs, as shown in Figure 1.11.

In situ ROP of ε-caprolactone has been reviewed by Han and Gao and the review presents the highlights of functionalization of CNTs by azide chemistry (2). An *in situ* ring-opening copolymerization of L-lactide (LA) and *e*-caprolactone (CL) was reported by Chakoli et al. using stannous octanoate and hydroxylated MWCNTs (MWCNT-OHs) as the initiators (67). Although there is an increase in the mechanical strength of the polymers due to the presence of MWCNTs, the elongation properties of the resulting material were found to decrease. Apart from compatibility between CNTs and polymer matrix, the fabrication of CNTs with appropriate functional groups associated with controlled thickness is extremely a subtle task. Recently, Wua et al. reported the feasibility of ring-opening approach for grafting

TABLE 1.2   Reaction Conditions and Selected Results for Grafting HPG from MWNT Surface

Sample	$R_{wt}{}^a$	$R_{mole}{}^a$	$f_{wt}{}^c$	$M_{n.GPC}{}^d$ g/mol	PDI^d	$M_{n.TGA}{}^e$/ g/mol	$m_{nongra}/m_{gra}{}^f$
MWNT-g-HPG1	5.75/1	56/1	22.8	1,300	1.37	1,230	11.6
MWNT-g-HPG2	13.8/1	134/1	39.2	8,300	1.79	4,640	7.2
MWNT-g-HPG3	34.5/1	335/1	49.6	21,100	1.74	7,080	4.8
MWNT-g-HPG4	63.25/1	614/1	66.1	57,600	1.54	14,200	5.1
MWNT-g-HPG5	74.75/1	726/1	74.5	81,300	1.69	21,020	4.5
MWNT-g-HPG6	103.5/1	1005/1	90.8	100,600	1.58	71,000	4.2

Source: Reproduced with permission from Reference 65.

[a] Weight feed ratio of glycidol to MWNT-OH.

[b] Mole feed ratio of glycidol to hydroxyl groups of the MWNT-OH.

[c] Weight fraction of grafted polymer in the product of MWNT-g-HPG calculated from corresponding TGA data between 200 and 500 °C.

[d] Number-average molecular weight (M_n) and polydispersity index (PDI) of the free polymer collected from the centrifugated solution, measured by GPC with DMF as the eluent and PS as the calibration.

[e] Average molecular weight of the grafted HPG calculated from TGA data is $M_{n, TGA} = m_{gra}/n_{gra} = f_{wt}/[(1 - f_{wt}) . 1.39 \times 10^{-3}]$, where 1.39 represents the concentration of initiating sites per gram of MWNTs (mmol g^{-3}), m_{gra} and n_{gra} represent the mass and molar amount of the grafted HPG, respectively. M_n;TGA for MWNT-g-HPG1 is the molar mass of neat HPG units excluding the weight fraction of organic moieties attached onto MWNT-OH, because of the great influence of such weight fraction on the calculation for such a low R_{mole}.

[f] Mass ratio of nongrafted polyglycerol to grafted amount.

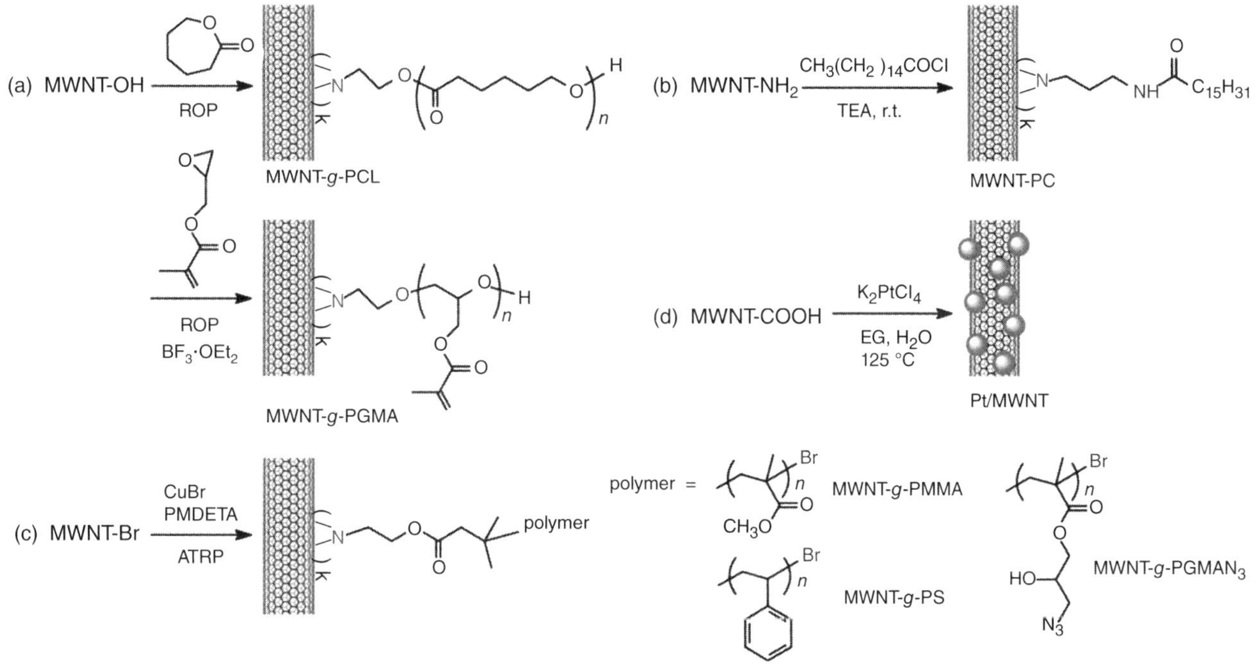

Figure 1.11   Chemical reactions on f-MWNTs. (*Source*: Reproduced with permission from Reference 66).

polybutylene terephthalate on MWNTs by using cyclic butylene terephthalate oligomers (68). Adeli et al. successfully modified MWCNTs using highly branched polyglycerol (PG) through *in situ* ROP of glycidol onto their surface (69). High biocompatibility and water solubility of MWCNT-g-PG hybrid materials was used in biological systems to test the level of toxicity of CNTs. *In vitro* cytotoxicity tests and hemolysis assay results showed no adverse effects on the HT1080 cell and red blood cells up to 1 mg/ml concentration. Polyglycerol functionalized CNTs found to decrease *in vitro* cytotoxicity of CNT.

1.1.3.2 Self-Condensing Vinyl Polymerization Approach   Self-condensing vinyl polymerization (SCVP) has found to be very useful process since its discovery by Frechet et al. (70). The process involves the polymerization of double bonds initiated by the active center of the functional group B* and A being the double bond, where AB* is an inimer. Later, the kinetic and mechanism for the SCVP was proposed by Muller et al. (71). A clear review on SCVP method involving its origin, its mechanism, and its uses has been reported by Gao and

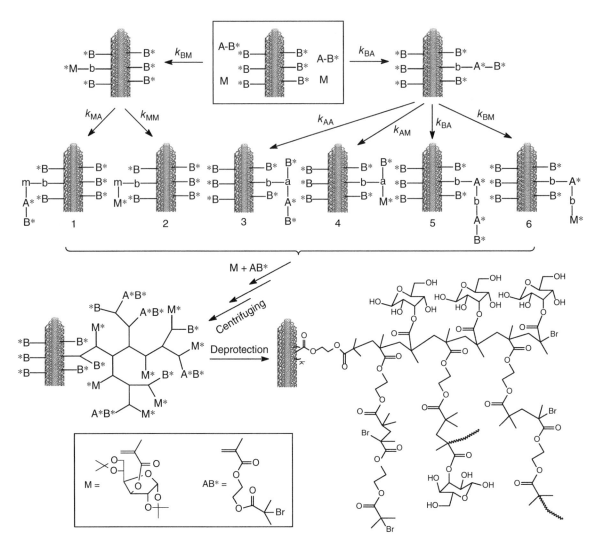

Figure 1.12   Synthetic strategy for grafting hyperbranched glycopolymer from surfaces of MWNTs by self-condensing vinyl copolymerization (SCVCP) of inimer (AB*) and monomer (M) via ATRP. (*Source*: Reproduced with permission from Reference 23).

Yan (51), including the first attempt to synthesize hybrid nanoparticles with hyperbranched polymer shells with the use of SCVP by Muller et al. (71).

Earlier work by Yan et al. initiated the use of SCVP via ROP to graft hyperbranched polymer on MWNTs. It was suggested that the ATRP would be helpful for the syntheses of hyperbranched polymers by SCVP (54). The grafting of HPs to CNTs finds some difficulties when inimer is employed in the polymerization as free polymer. An attempt to grow polymers on the CNTs without removing the monomer from the solution was carried out by Hong et al. (72). This process involves the preparation of bromoisobutyryl fragment-immobilized MWNTs to get MWNT-Br (ATRP surface initiator) and was further used to attach a hyperbranched polymer shell to obtain a polyfunctional initiator (B*f) in the SCVP to generate hyperbranched macromolecule functionalized MWNTs. By slow addition of AB* inimer, immobilization of B* on the surface of MWNTs helps in initiating SCVP. Also, hyperbranched polymer-grafted MWNTs were found to be dispersed in solvents even without observing sedimentation for a long period of time.

Gao et al. reported the grafting of linear and hyperbranched glycopolymers from the surface of MWCNTs by self-condensing vinyl copolymerization of 3-*O*-methacryloyl-1,2:5,6-di-*O*-isopropylidene-D-glucofuranose (MAIG) and AB* inimer, 2-(2-bromoisobutyryloxy)ethyl methacrylate (BIEM) (Fig. 1.12) (23). The kinetic study of SCVP showed that the molecular weight and polydispersity index (PDI) increased with conversion for lower γ (0.5 and 1) and for higher γ (2.5 and 5). After maximum conversion (~90%), the molecular weight reached a stand point and then multihydroxy hyperbranched glycopolymer-grafted MWNTs were obtained by deprotection. PDI was observed to be broader as after 45–50% conversion, which might be due to the polymer coupling in the reaction system.

The grafted polymer content and other results are shown in Table 1.3. DB of the hyperbranched glycopolymers is an important factor and was determined for the polymer-grafted CNTs by ^{1}H NMR. It can be seen that the DB decreases from 0.49 to 0.21 with increase in γ from 0.5 to 5, which is well in agreement with the theory. It was concluded that the kinetics observed for sacrificial initiator (first-order time

TABLE 1.3   Reaction Conditions and Results for the Self-Condensing Vinyl Copolymerization (SCVCP) of MAIG and BIEM in the Presence of MWNT-Br

γ^a	Time, h[b]	Convn, %[c]	$M_{n,app}{}^d$	PDI[d]	f_{wt}, %[e]	$MW_{TGA}{}^f$	DB_{NMR}	DB_{theo}
0	4.5	~95.0	4000	1.88	0.38	1165		0.465
0.5	21.5	73.8	3660	1.77	0.40	1270	0.49	0.50
1	4.5	~95.7	5630	2.03	0.42	1380	0.43	0.49
2.5	22.0	~93.0	4200	1.86	0.46	1680	0.34	0.40
5	29.5	~90.0	4370	1.81	0.53	2140	0.21	0.24

Source: Reproduced with permission from Reference 23.

[a]The feed mole ratio of MAIG to BIEM.

[b]A sample is taken from the reaction system after a given time to determine the conversion by NMR. The final reaction time set is dependent on the conversion of vinyl groups and the viscosity of the reaction system. In the case of either quite high viscosity (so it is difficult to take a sample by the syringe from the reaction system) or the small conversion difference after a relatively long reaction time, the reaction would be stopped. The initial feed ratio of BIEM to ethyl acetate is approximately 1/2 ($\gamma = 0$), 1/5 ($\gamma = 0.5$), 1/2 ($\gamma = 1$), 1/2.5 ($\gamma = 2.5$), and 1/5 ($\gamma = 5$) g/ml.

[c]The conversion of vinyl groups determined by 1H NMR.

[d]The apparent number-average molecular weight (M_n) and polydispersity index (PDI) of soluble polymer measured by GPC.

[e]The polymer content grafted on MWNTs, determined by TGA.

[f]The average molecular weight evaluated from TGA.

conversion plot) is comparable with the ATRP of MAIG initiated by MWNT-Br. These observations on polymer–CNT nanohybrids could be well enough expected to open new ventures in bionanotechnology.

1.2   FUNCTIONALIZATION OF CARBON NANO-ONIONS

Among the fullerene family, onion-like carbons consisting of multiple concentric graphitic shells to form encapsulated structures called as carbon nano-onions (CNOs) have been found by Ugarte in 1992 (73). Also, multishelled fullerenes or CNOs were discovered by Iijima and Ichihashi in 1993, at the same time they invented CNTs (74). Like CNTs, CNOs possess interesting properties such as high surface area, density, and multilayer morphology (75), and their rich attributes led to applications in different fields including catalysis, fuel cells, optical limiting, electromagnetic shielding, energy storage, tribology, tumor therapy, and supercapacitors (76–90). Till now, CNOs have not been explored as only few reports exist. It is considered that CNOs find importance for tribological applications due to their spherical shape (91). In particular, NASA has shown much interest on CNOs for their specific aerospace applications (84).

Ozawa et al. suggested the formation of spiroid-to-onion transformation, but the formation mechanism of CNOs has been left unproved till now (92). The transformation takes place through interlayer valence isomerization in which the delocalized radical centers propagate radially in a zipper-like fashion. The structure resembles a sphere with regular shells, which are the primary particles of carbon black. In recent times, CNOs has been found to exhibit dislocation dynamics in a different manner apart from bulk crystalline phase (93). The observed spherical topology results in an asymmetry that would have been attributed to an unexpected attraction of dislocations toward the core. The dislocation is mainly considered to be important for mechanical properties. The isolation and purification of CNOs are difficult and thus the clear mechanism has not been proved yet. The production of extremely pure CNOs in high yields by annealing carbon nanodiamond particles at temperatures above 1200 °C was reported by Kuznetsov et al. (94). The production of nano-onions to high yield have also been employed by Sano et al. in water by arcing between two graphite electrodes which involves without vacuum (95). The nonvacuum method involves the use of carbon arc sustained in deionized water by initiating arc discharge by contacting a pure graphite anode (tip diameter, 5 mm) with the carbon cathode (tip diameter, 12 mm) and the discharge voltage and current were 16–17 V and 30 A, respectively (Fig. 1.13). The nano-onions were collected on the water surface with high purity. In this study, they were able to produce 3 mg/min with average diameter of 25–30 nm (range 5–40 nm), a useful size range for many lubrication applications. Similarly to CNTs, CNOs are surface functionalized to make it more appropriate to be available for various applications.

Functionalization of carbon materials by diazonium-based compounds has been used for CNTs before and the same method was followed to functionalize CNOs by Flavin et al. (96). Pristine CNOs in DMF were dispersed and aniline derivative and isoamyl nitrite were added under inert nitrogen atmosphere and stirred at 60 °C overnight. The functionalized CNOs were obtained by centrifuging finally. Degree of derivatization of the f-CNOs was predicted by TGA and the observations were useful in differentiating the graphitic structure of CNOs. In addition, Raman spectra presented the D-band at 1354 cm^{-1}, which relates to the conversion of sp^2 to sp^3 carbon.

Ontoria et al. reported the covalent functionalization of CNOs with hexadecyl chains (Fig. 1.14) (97). The functionalization was achieved by reducing CNOs with Na/K alloy in a solvent under vacuum followed by nucleophilic substitution by using 1-bromohexadecane. The covalent functionalization was evidenced primarily by TGA and Raman spectroscopy. It was observed from the Raman studies that the pristine CNOs presents a varying ratio of $I(D)/I(G)$ with respect to the particle size. $I(D)/I(G)$ of 0.8 and 1.4 were observed for large CNOs prepared by arc discharge of graphite under water and that obtained by annealing method of nanodiamond particles to produce small CNOs.

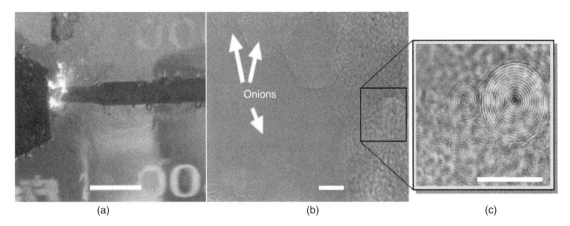

(a)   (b)   (c)

Figure 1.13 Carbon nano-"onions" created by arc discharge in water. (a) Image of a carbon arc discharge in water. Scale bar, 12 mm. (b, c) Low- and high-magnification electron micrographs of carbon nano-onions floating on the water surface after their production. Scale bars, 10 nm. (*Source*: Reproduced with permission from Reference 95).

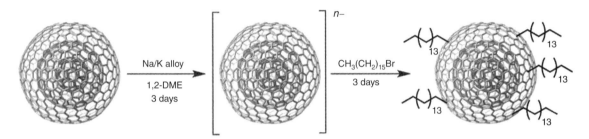

Figure 1.14 Reductive treatment of CNOs by a Na–K alloy in 1,2-DME and subsequent alkylation using 1-bromohexadecane. (*Source*: Reproduced with permission from Reference 97).

Functionalization of the CNOs with 1-bromohexadecane helps homogeneous dispersion and also the structure obtained with more consistent porous morphology and definite channels.

A different functionalization of CNOs was attempted by modifying CNOs with varying concentrations of RuO_2 by a controlled chemical method by Borgohain et al. (98). The f-CNOs was used as electrodes and found to enhance the specific capacitance, high power, and high energy density. The method proposed the synthesis of CNOs with 5–7 nm diameter in a graphitization furnace by annealing nanodiamond (Dynalene NB50) at 1650 °C for 1 h under helium. Then the CNOs were functionalized with RuO_2 in the presence of glycolic acid. CNOs and chemically modified CNOs have good electrical conductivity, ideal mesoporosity for ion transport, and high electrochemical and thermal stability. It was suggested that sp^2 carbon in CNOs could enhance the charge-transfer processes in metal oxide redox based supercapacitors. This method displays a relatively high specific capacitance of 334 F/g for 67.5 wt% RuO_2 loading which in turn can deliver an energy density of 11.6 W h/kg. In another work, porosity on CNOs was introduced by KOH activation, thereby improving remarkable changes in structure, surface area, and pore size (99). In this process CNOs were synthesized by combustion in air on laser resonant excitation of ethylene. Highly concentric pristine CNOs obtained at a laser wavelength of 10.532l m were used for further activation and tested for electrochemical performance. The activated CNOs demonstrate high charge/discharge rates at scan rates up to 5000 mV/s and the high capacitance retention ratio of about 71% was observed for an increase in the current density from 0.75 to 25 A/g. The considerable improvement of capacitance is ascribed to the high surface area, porosity, and hydrophilic surface upon KOH activation. It was concluded that the activated CNOs may find its objective in capacitor field which have a high knee frequency (825 Hz) and a smaller relaxation time constant (82.5 ms). Research on CNOs is still to be explored and the mechanism is far from the accomplishment to explain from the origin to different intermediates. However, there is an urge to this type of nanostructures in applying them in different fields, more particularly, as emerging lubricants up-to-date.

1.3 CONCLUSIONS AND FUTURE SCOPE

CNTs have been considered to be one of the ideal forms of carbon materials for various applications due to its extremely high electrical conductivity, and strong mechanical strength. Grafting of polymer on the surface of CNTs has been in practice to enhance mechanical and electrical properties; however, massive efforts are still needed to implement effective functionalization methodologies. The most essential problem lies in how to prepare homogeneous composition of CNT-grafted polymers and maintain control over molecular weight, polymerization time, grafting density, and molecular weight distribution. High molecular weight and the high grafting density of the CNT-grafted polymers make a subtle task in controlling them to achieve grafted materials with average molecular weight. Functionalization of CNTs and related nanostructures like CNOs have already emerged as an important aspect in applying hybrid materials in various applications from energy field to medicine. Future directions of research on modifying and developing novel functionalization strategies would pave the way for such nanomaterials to be used in various fields.

REFERENCES

1. Iijima S. Helical microtubules of graphitic carbon. Nature 1991;354(6348):56–58.

2. Han J, Gao C. Functionalization of carbon nanotubes and other nanocarbons by azide chemistry. Nano-Micro Lett 2010;2(3):213–226.

3. Baughman RH, Zakhidov AA, de Heer WA. Carbon nanotubes--the route toward applications. Science 2002;297(5582):787–792.

4. Huxtable ST, Cahill DG, Shenogin S, Xue L, Ozisik R, Barone P, Keblinski P. Interfacial heat flow in carbon nanotube suspensions. Nat Mater 2003;2(11):731–734.

5. Calvert P. Nanotube composites: a recipe for strength. Nature 1999;399(6733):210–211.

6. Down WB, Baker RTK. Modification of the surface properties of carbon fibers via the catalytic growth of carbon nanofibers. J Mater Res 1995;10(03):625–633.

7. Thostenson ET, Li WZ, Wang DZ, Ren ZF, Chou TW. Carbon nanotube/carbon fiber hybrid multiscale composites. J Appl Phys 2002;91(9):6034–6037.

8. Spitalsky Z, Tasis D, Papagelis K, Galiotis C. Carbon nanotube–polymer composites: chemistry, processing, mechanical and electrical properties. Prog Polym Sci 2010;35(3):357–401.

9. Pasquali M. Polymer composites: Swell properties and swift processing. Nat Mater 2004;3(8):509–510.

10. Ruoff RS, Tersoff J, Lorents DC, Subramoney S, Chan B. Radial deformation of carbon nanotubes by van der Waals forces. Nature 1993;364(6437):514–516.

11. Walther JH, Jaffe R, Halicioglu T, Koumoutsakos P. Carbon nanotubes in water: structural characteristics and energetics. J Phys Chem B 2001;105(41):9980–9987.

12. Jin Z, Sun X, Xu G, Goh SH, Ji W. Nonlinear optical properties of some polymer/multi-walled carbon nanotube composites. Chem Phys Lett 2000;318(6):505–510.

13. Curran SA, Ajayan PM, Blau WJ, Carroll DL, Coleman JN, Dalton AB, Davey AP, Drury A, McCarthy B, Maier S, Strevens A. A composite from poly (m-phenylenevinylene-co-2, 5-dioctoxy-p-phenylenevinylene) and carbon nanotubes: A novel material for molecular optoelectronics. Adv Mater 1998;10(14):1091–1093.

14. Coleman JN, Curran S, Dalton AB, Davey AP, McCarthy B, Blau W, Barklie RC. Percolation-dominated conductivity in a conjugated-polymer-carbon-nanotube composite. Phys Rev B 1998;58(12):R7492.

15. Chamberlain TW, Meyer JC, Biskupek J, Leschner J, Santana A, Besley NA, Bichoutskaia E, Kaiser U, Khlobystov AN. Reactions of the inner surface of carbon nanotubes and nanoprotrusion processes imaged at the atomic scale. Nat Chem 2011;3(9):732–737.

16. Dai H. Carbon nanotubes: synthesis, integration, and properties. Acc Chem Res 2002;35(12):1035–1044.

17. De Volder MF, Tawfick SH, Baughman RH, Hart AJ. Carbon nanotubes: present and future commercial applications. Science 2013;339(6119):535–539.

18. Riggs JE, Guo Z, Carroll DL, Sun YP. Strong luminescence of solubilized carbon nanotubes. J Am Chem Soc 2000;122(24):5879–5880.

19. O'Connell MJ, Boul P, Ericson LM, Huffman C, Wang Y, Haroz E, Kuper C, Tour J, Ausman KD, Smalley RE. Reversible water-solubilization of single-walled carbon nanotubes by polymer wrapping. Chem Phys Lett 2001;342(3):265–271.

20. Koshio A, Yudasaka M, Zhang M, Iijima S. A simple way to chemically react single-wall carbon nanotubes with organic materials using ultrasonication. Nano Lett 2001;1(7):361–363.

21. Lin Y, Zhou B, Shiral Fernando KA, Liu P, Allard LF, Sun YP. Polymeric carbon nanocomposites from carbon nanotubes functionalized with matrix polymer. Macromolecules 2003;36(19):7199–7204.

22. Gao C, Jin YZ, Kong H, Whitby RL, Acquah SF, Chen GY, Qian H, Hartschuh A, Silva SRP, Henley S, Fearon P, Kroto HW, Walton DR. Polyurea-functionalized multiwalled carbon nanotubes: synthesis, morphology, and Raman spectroscopy. J Phys Chem B 2005;109(24):11925–11932.

23. Gao C, Muthukrishnan S, Li W, Yuan J, Xu Y, Müller AH. Linear and hyperbranched glycopolymer-functionalized carbon nanotubes: synthesis, kinetics, and characterization. Macromolecules 2007;40(6):1803–1815.

24. Wu HX, Tong R, Qiu XQ, Yang HF, Lin YH, Cai RF, Qian SX. Functionalization of multiwalled carbon nanotubes with polystyrene under atom transfer radical polymerization conditions. Carbon 2007;45(1):152–159.

25. Adeli M, Hakimpoor F, Ashiri M, Kabiri R, Bavadi M. Anticancer drug delivery systems based on noncovalent interactions between carbon nanotubes and linear–dendritic copolymers. Soft Matter 2011;7(8):4062–4070.

26. Hijazi I, Jousselme B, Jégou P, Filoramo A, Campidelli S. Formation of linear and hyperbranched porphyrin polymers on carbon nanotubes via a CuAAC "grafting from" approach. J Mater Chem 2012;22(39):20936–20942.

27. Qin S, Qin D, Ford WT, Herrera JE, Resasco DE, Bachilo SM, Weisman RB. Solubilization and purification of single-wall carbon nanotubes in water by in situ radical polymerization of sodium 4-styrenesulfonate. Macromolecules 2004;37(11):3965–3967.

28. Liu Y, Yao Z, Adronov A. Functionalization of single-walled carbon nanotubes with well-defined polymers by radical coupling. Macromolecules 2005;38(4):1172–1179.

29. Fu K, Huang W, Lin Y, Riddle LA, Carroll DL, Sun YP. Defunctionalization of functionalized carbon nanotubes. Nano Lett 2001;1(8):439–441.

30. Qin S, Qin D, Ford WT, Resasco DE, Herrera JE. Functionalization of single-walled carbon nanotubes with polystyrene via grafting to and grafting from methods. Macromolecules 2004;37(3):752–757.

31. (a) Lou X, Detrembleur C, Sciannamea V, Pagnoulle C, Jérôme R. Grafting of alkoxyamine end-capped (co) polymers onto multi-walled carbon nanotubes. Polymer 2004;45(18):6097–6102. (b) Lou XD, Detrembleur C, Pagnoulle C, Jérôme R, Bocharova V, Kiriy A. Surface modification of multiwalled carbon nanotubes by poly (2-vinylpyridine): Dispersion, selective deposition, and decoration of the nanotubes. Adv Mater 2004;16(23–24):2123–2127.

32. Coleman JN, Dalton AB, Curran S, Rubio A, Davey AP, Drury A, McCarthy B, Lahr B, Ajayan PM, Roth S, Barklie RC, Blau WJ. Phase separation of carbon nanotubes and turbostratic graphite using a functional organic polymer. Adv Mater 2000;12(3):213–216.

33. Jia Z, Wang Z, Xu C, Liang J, Wei B, Wu D, Zhu S. Study on poly (methyl methacrylate)/carbon nanotube composites. Mater Sci Eng, A 1999;271(1):395–400.

34. Viswanathan G, Chakrapani N, Yang H, Wei B, Chung H, Cho K, Ryu Y, Ajayan PM. Single-step in situ synthesis of polymer-grafted single-wall nanotube composites. J Am Chem Soc 2003;125(31):9258–9259.

35. Guldi DM, Rahman GNA, Ramey J, Marcaccio M, Paolucci D, Paolucci F, Qin S, Ford WT, Balbinot D, Jux N, Tagmatarchis N, Prato M. Donor–acceptor nanoensembles of soluble carbon nanotubes. Chem Commun 2004;18:2034–2035.

36. Guldi DM, Rahman GMA, Prato M, Jux N, Qin S, Ford W. Single-Wall Carbon Nanotubes as Integrative Building Blocks for Solar-Energy Conversion. Angew Chem Int Ed 2005;44:2015–2018.

37. Coleman JN, Cadek M, Blake R, Nicolosi V, Ryan KP, Belton C, Fonseca A, Nagy JB, Gun'ko YK, Blau WJ. High performance nanotube-reinforced plastics: understanding the mechanism of strength increase. Adv Funct Mater 2004;14(8):791–798.

38. Coleman JN, Cadek M, Ryan KP, Fonseca A, Nagy JB, Blau WJ, Ferreira MS. Reinforcement of polymers with carbon nanotubes. The role of an ordered polymer interfacial region. Experiment and modeling. Polymer 2006;47(26):8556–8561.

39. Zeng H, Gao C, Wang Y, Watts PC, Kong H, Cui X, Yan D. In situ polymerization approach to multiwalled carbon nanotubes-reinforced nylon 1010 composites: mechanical properties and crystallization behavior. Polymer 2006;47(1):113–122.

40. Park SJ, Cho MS, Lim ST, Choi HJ, Jhon MS. Synthesis and Dispersion Characteristics of Multi-Walled Carbon Nanotube Composites with Poly (methyl methacrylate) Prepared by In-Situ Bulk Polymerization. Macromol Rapid Commun 2003;24(18):1070–1073.

41. Fan J, Wan M, Zhu D, Chang B, Pan Z, Xie S. Synthesis, characterizations, and physical properties of carbon nanotubes coated by conducting polypyrrole. J Appl Polym Sci 1999;74(11):2605–2610.

42. Tang BZ, Xu H. Preparation, alignment, and optical properties of soluble poly (phenylacetylene)-wrapped carbon nanotubes. Macromolecules 1999;32(8):2569–2576.

43. Velasco-Santos C, Martinez-Hernandez AL, Lozada-Cassou M, Alvarez-Castillo A, Castano VM. Chemical functionalization of carbon nanotubes through an organosilane. Nanotechnology 2002;13(4):495–498.

44. Kong H, Gao C, Yan D. Controlled functionalization of multiwalled carbon nanotubes by in situ atom transfer radical polymerization. J Am Chem Soc 2004;126(2):412–413.

45. Bonduel D, Mainil M, Alexandre M, Monteverde F, Dubois P. Supported coordination polymerization: a unique way to potent polyolefin carbon nanotube nanocomposites. Chem Commun 2005;6:781–783.

46. Zhao C, Hu G, Justice R, Schaefer DW, Zhang S, Yang M, Han CC. Synthesis and characterization of multi-walled carbon nanotubes reinforced polyamide 6 via in situ polymerization. Polymer 2005;46(14):5125–5132.

47. Cruz-Delgado VJ, España-Sánchez BL, Avila-Orta CA, Medellín-Rodríguez FJ. Nanocomposites based on plasma-polymerized carbon nanotubes and Nylon-6. Polym J 2012;44(9):952–958.

48. Hu Z, Li J, Tang P, Li D, Song Y, Li Y, Zhao L, Li C, Huang Y. One-pot preparation and continuous spinning of carbon nanotube/poly (p-phenylene benzobisoxazole) copolymer fibers. J Mater Chem 2012;22(37):19863–19871.

49. Wang J, Shi Z, Ge Y, Wang Y, Fan J, Yin J. Functionalization of unzipped carbon nanotube via in situ polymerization for mechanical reinforcement of polymer. J Mater Chem 2012;22(34):17663–17670.

50. Wang Q, Yao Q, Chang J, Chen L. Enhanced thermoelectric properties of CNT/PANI composite nanofibers by highly orienting the arrangement of polymer chains. J Mater Chem 2012;22(34):17612–17618.

51. Gao C, Yan D. Hyperbranched polymers: from synthesis to applications. Prog Polym Sci 2004;29(3):183–275.

52. Sun YP, Huang W, Lin Y, Fu K, Kitaygorodskiy A, Riddle LA, Yu YJ, Carroll DL. Soluble Dendron-Functionalized Carbon Nanotubes: Preparation, Characterization, and Properties. Chem Mater 2001;13(9):2864–2869.

53. Sano M, Kamino A, Shinkai S. Construction of carbon nanotube "stars" with dendrimers. Angew Chem 2001;113(24):4797–4799.

54. Xu Y, Gao C, Kong H, Yan D, Jin YZ, Watts PC. Growing multihydroxyl hyperbranched polymers on the surfaces of carbon nanotubes by in situ ring-opening polymerization. Macromolecules 2004;37(24):8846–8853.

55. Feng QP, Xie XM, Liu YT, Zhao W, Gao YF. Synthesis of hyperbranched aromatic polyamide–imide and its grafting onto multiwalled carbon nanotubes. J Appl Polym Sci 2007;106(4):2413–2421.

56. Pan B, Cui D, Gao F, He R. Growth of multi-amine terminated poly (amidoamine) dendrimers on the surface of carbon nanotubes. Nanotechnology 2006;17(10):2483.

57. Campidelli S, Sooambar C, Lozano Diz E, Ehli C, Guldi DM, Prato M. Dendrimer-functionalized single-wall carbon nanotubes: synthesis, characterization, and photoinduced electron transfer. J Am Chem Soc 2006;128(38):12544–12552.

58. Adeli M, Beyranvand S, Kabiri R. Preparation of hybrid nanomaterials by supramolecular interactions between dendritic polymers and carbon nanotubes. Polym Chem 2013;4(3):669–674.

59. Sohn GJ, Choi HJ, Jeon IY, Chang DW, Dai L, Baek JB. Water-dispersible, sulfonated hyperbranched poly (ether-ketone) grafted multiwalled carbon nanotubes as oxygen reduction catalysts. ACS Nano 2012;6(7):6345–6355.

60. Gómez FJ, Chen RJ, Wang D, Waymouth RM, Dai H. Ring opening metathesis polymerization on non-covalently functionalized single-walled carbon nanotubes. Chem Commun 2003;(2):190–191.

61. Liu Y, Adronov A. Preparation and utilization of catalyst-functionalized single-walled carbon nanotubes for ring-opening metathesis polymerization. Macromolecules 2004;37(13):4755–4760.

62. Zeng HL, Gao C, Yan DY. Poly (ε-caprolactone)-Functionalized Carbon Nanotubes and Their Biodegradation Properties. Adv Funct Mater 2006;16(6):812–818.

63. Chen GX, Kim HS, Park BH, Yoon JS. Synthesis of Poly (L-lactide)-Functionalized Multiwalled Carbon Nanotubes by Ring-Opening Polymerization. Macromol Chem Phys 2007;208(4):389–398.

64. Feng J, Cai W, Sui J, Li Z, Wan J, Chakoli AN. Poly (L-lactide) brushes on magnetic multiwalled carbon nanotubes by in-situ ring-opening polymerization. Polymer 2008;49(23):4989–4994.

65. Zhou L, Gao C, Xu W. Efficient Grafting of Hyperbranched Polyglycerol from Hydroxyl-Functionalized Multiwalled Carbon Nanotubes by Surface-Initiated Anionic Ring-Opening Polymerization. Macromol Chem Phys 2009;210(12):1011–1018.

66. Gao C, He H, Zhou L, Zheng X, Zhang Y. Scalable functional group engineering of carbon nanotubes by improved one-step nitrene chemistry. Chem Mater 2008;21(2):360–370.

67. Chakoli AN, Wan J, Feng JT, Amirian M, Sui JH, Cai W. Functionalization of multiwalled carbon nanotubes for reinforcing poly (l-lactide-co-ε-caprolactone) biodegradable copolymers. Appl Surf Sci 2009;256(1):170–177.

68. Wu F, Yang G. Poly (butylene terephthalate)-functionalized MWNTs by in situ ring-opening polymerization of cyclic butylene terephthalate oligomers. Polym Adv Technol 2011;22(10):1466–1470.

69. Adeli M, Mirab N, Alavidjeh MS, Sobhani Z, Atyabi F. Carbon nanotubes-graft-polyglycerol: Biocompatible hybrid materials for nanomedicine. Polymer 2009;50(15):3528–3536.

70. Frechet JM, Henmi M, Gitsov I, Aoshima S, Leduc MR, Grubbs RB. Self-condensing vinyl polymerization: an approach to dendritic materials. Science 1995:1080.

71. Mori H, Seng DC, Zhang M, Müller AH. Hybrid nanoparticles with hyperbranched polymer shells via self-condensing atom transfer radical polymerization from silica surfaces. Langmuir 2002;18(9):3682–3693.

72. Hong CY, You YZ, Wu D, Liu Y, Pan CY. Multiwalled carbon nanotubes grafted with hyperbranched polymer shell via SCVP. Macromolecules 2005;38(7):2606–2611.

73. Ugarte D. Curling and closure of graphitic networks under electron-beam irradiation. Nature 1992;359(6397):707–709.

74. Iijima S, Ichihashi T. Single-shell carbon nanotubes of 1-nm diameter. Nature 1993;363:603–605.

75. Cioffi CT, Palkar A, Melin F, Kumbhar A, Echegoyen L, Melle-Franco M, Zerbetto F, Rahman GMA, Ehli C, Sgobba V, Guldi DM, Prato M. A Carbon Nano-Onion–Ferrocene Donor–Acceptor System: Synthesis, Characterization and Properties. Chem A Eur J 2009;15(17):4419–4427

76. Keller N, Maksimova NI, Roddatis VV, Schur M, Mestl G, Butenko YV, Kuznetsov VL, Schlögl R. The Catalytic Use of Onion-Like Carbon Materials for Styrene Synthesis by Oxidative Dehydrogenation of Ethylbenzene. Angew Chem Int Ed 2002;41(11):1885–1888.

77. Langlet R, Lambin P, Mayer A, Kuzhir PP, Maksimenko SA. Dipole polarizability of onion-like carbons and electromagnetic properties of their composites. Nanotechnology 2008;19(11):115706.

78. Sano N, Wang H, Alexandrou I, Chhowalla M, Teo KBK, Amaratunga GAJ, Iimura K. Properties of carbon onions produced by an arc discharge in water. J Appl Phys 2002;92(5):2783–2788.

79. Hirata Y, Mataga N. Photoionization of aromatic diamines in electron-accepting solvents: formation of short-lived ion pairs. J Phys Chem 1984;88(14):3091–3095.

80. Joly-Pottuz L, Vacher B, Ohmae N, Martin JM, Epicier T. Anti-wear and friction reducing mechanisms of carbon nano-onions as lubricant additives. Tribol Lett 2008;30(1):69–80.

81. Joly-Pottuz L, Bucholz EW, Matsumoto N, Phillpot SR, Sinnott SB, Ohmae N, Martin JM. Friction properties of carbon nano-onions from experiment and computer simulations. Tribol Lett 2010;37(1):75–81.

82. Ding L, Stilwell J, Zhang T, Elboudwarej O, Jiang H, Selegue JP, Cooke PA, Gray JW, Chen FF. Molecular characterization of the cytotoxic mechanism of multiwall carbon nanotubes and nano-onions on human skin fibroblast. Nano Lett 2005;5(12):2448–2464.

83. Pech D, Brunet M, Durou H, Huang P, Mochalin V, Gogotsi Y, Taberna PL, Simon P. Ultrahigh-power micrometre-sized supercapacitors based on onion-like carbon. Nat Nanotechnol 2010;5(9):651–654.

84. Street KW, Marchetti M, Vander Wal RL, Tomasek AJ. Evaluation of the tribological behavior of nano-onions in Krytox 143AB. Tribol Lett 2004;16(1–2):143–149.

85. Zhang C, Li J, Liu E, He C, Shi C, Du X, Hauge RH, Zhao N. Synthesis of hollow carbon nano-onions and their use for electrochemical hydrogen storage. Carbon 2012;50(10):3513–3521.

86. Rettenbacher AS, Elliott B, Hudson JS, Amirkhanian A, Echegoyen L. Preparation and Functionalization of Multilayer Fullerenes (Carbon Nano-Onions). Chem A Eur J 2006;12(2):376–387.

87. Rettenbacher AS, Perpall MW, Echegoyen L, Hudson J, Smith DW. Radical addition of a conjugated polymer to multilayer fullerenes (carbon nano-onions). Chem Mater 2007;19(6):1411–1417.

88. Matsumoto N, Joly-Pottuz L, Kinoshita H, Ohmae N. Application of onion-like carbon to micro and nanotribology. Diamond Relat Mater 2007;16(4):1227–1230.

89. Joly-Pottuz L, Matsumoto N, Kinoshita H, Vacher B, Belin M, Montagnac G, Martin JM, Ohmae N. Diamond-derived carbon onions as lubricant additives. Tribol Int 2008;41(2):69–78.

90. Joly-Pottuz L, Ohmae N. Carbon-based nanolubricants. In: Martin JM, Ohmae N, editors. *Nanolubricants*. Chichester, UK: John Wiley & Sons; 2008. p 93–147.

91. Hirata A, Igarashi M, Kaito T. Study on solid lubricant properties of carbon onions produced by heat treatment of diamond clusters or particles. Tribol Int 2004;37(11):899–905.

92. Ozawa M, Goto H, Kusunoki M, Osawa E. Continuously growing spiral carbon nanoparticles as the intermediates in the formation of fullerenes and nanoonions. J Phys Chem B 2002;106(29):7135–7138.

93. Akatyeva E, Huang JY, Dumitrică T. Edge-mediated dislocation processes in multishell carbon nano-onions? Phys Rev Lett 2010;105(10):106102.

94. Kuznetsov VL, Chuvilin AL, Butenko YV, Mal'kov IY, Titov VM. Onion-like carbon from ultra-disperse diamond. Chem Phys Lett 1994;222(4):343–348.

95. Sano N, Wang H, Chhowalla M, Alexandrou I, Amaratunga GAJ. Nanotechnology: Synthesis of carbon 'onions' in water. Nature 2001;414(6863):506–507.

96. Flavin K, Chaur MN, Echegoyen L, Giordani S. Functionalization of Multilayer Fullerenes (Carbon Nano-Onions) using Diazonium Compounds and "Click" Chemistry. Org Lett 2010;12(4):840–843.

97. Molina-Ontoria A, Chaur MN, Plonska-Brzezinska ME, Echegoyen L. Preparation and characterization of soluble carbon nano-onions by covalent functionalization, employing a Na–K alloy. Chem Commun 2013;49(24):2406–2408.

98. Borgohain R, Li J, Selegue JP, Cheng YT. Electrochemical study of functionalized carbon nano-onions for high-performance supercapacitor electrodes. J Phys Chem C 2012;116(28):15068–15075.

99. Gao Y, Zhou YS, Qian M, He XN, Redepenning J, Goodman P, Li HM, Jiang L, Lu YF. Chemical activation of carbon nano-onions for high-rate supercapacitor electrodes. Carbon 2013;51:52–58.

2

THE FUNCTIONALIZATION OF GRAPHENE AND ITS ASSEMBLED MACROSTRUCTURES

Haiyan Sun, Zhen Xu, and Chao Gao

MOE Key Laboratory of Macromolecular Synthesis and Functionalization, Department of Polymer Science and Engineering, Zhejiang University, Hangzhou, People's Republic of China

2.1 INTRODUCTION

Graphene is a one-atom-thick planar sheet of sp^2-bonded carbon atoms densely packed in a honeycomb crystal lattice (1). It possesses a wealth of fascinating properties (2, 3), such as excellent electrical conductivity, giant carrier mobility, extremely high thermal conductivity, high surface area, good chemical and environmental stability as well as extraordinary elasticity and stiffness. Increasing interest has been ignited among researchers to prepare graphene with desirable properties in large scale. Since graphene was firstly mechanically exfoliated in 2004 (4), many protocols have been established to get this two-dimensional (2D) carbonaceous nanomaterial (5–7), mainly including epitaxial growth on electrically insulating surfaces such as silicon carbide (SiC) (8), chemical vapor deposition (CVD) (9, 10), bottom-up synthesis for pure and heteroatom (boron, phosphorus, or nitrogen)-substituted 2D carbon scaffolds (11), unzipping of carbon nanotubes (CNTs) (12), and solution-based reduction of graphene oxide (GO) (13). The strong interlayer van der Waals interaction makes graphene poorly dispersible in solvents and intangible in graphite crystals to react with other molecules, hindering their scalable applications in macroscopic materials. Therefore, the functionalization is becoming a greatly important alternative for processable graphene in large scale. GO, prepared by chemical exfoliation of graphite with acids and strong oxidants, has emerged as an important precursor of graphene. From the point of view in chemistry, reactive structures in GO can be categorized into two parts, that is, oxygen-containing functional groups and conjugated domains. On the one hand, the oxygen-containing functional groups not only determine the good dispersibility of graphene derivatives in single-layer state but also facilitate further the chemical functionalization on both the planes of GO, through covalent bonds, hydrogen bonds, or electrostatic interaction. On the other hand, the residual conjugated domain in GO plane sets the platform for surface modification by hydrophobic interaction and π–π interaction. To date, a myriad of fruits have been picked in the field of processable graphenes from the GO precursor and their latent realistic applications (14–16).

2.2 NONCOVALENT FUNCTIONALIZATION

Noncovalent functionalization has been widely used for the surface modification of the sp^2 network of carbonaceous nanomaterials, such as CNTs, showing its advantage in causing minimal damage to the pristine chemical structure of nanomaterials, so as to their electronic properties. In the case of graphene, π–π attraction, hydrogen bond, and hydrophobic interaction between the absorbed molecules and the graphene surface can be utilized to realize the noncovalent functionalization, which can be generally achieved by polymer wrapping, adsorption of surfactants or small aromatic molecules, and π–π interaction with porphyrins or biomolecules such as deoxyribonucleic acid (DNA) and peptides. Table 2.1 shows different noncovalent modifications of graphene using various modifying reagents and the resultant dispersibility in various solvents.

2.2.1 π–π Stacking

To form graphene sheets by chemical oxidation–reduction method, mineral or synthetic graphite was treated into GO, followed by the chemical reduction or annealing at high temperature. The reduction chemical reagents largely remove the oxygen-containing functional groups and partly restore the conjugate structure, which should increase π stacking between monolayer reduced graphene sheets. As a result, GO possesses good solubility, but its reduction counterpart is prone to irreversible aggregate. To circumvent the interlayer interaction and the irreversible

Nanomaterials, Polymers and Devices: Materials Functionalization and Device Fabrication, First Edition. Edited by Eric S. W. Kong.

TABLE 2.1   Different Noncovalent Modifications of Graphene Using Various Modifying Reagents, Interactions, and Their Dispersion Stabilities in Various Solvents and the Performed Dispersibility

Modifying Agent	Interaction	Dispersing Medium	Dispersibility (mg/ml)	Refs.
PSS	π–π stacking	Water	1	17
1-pyrenebutyrate	π–π stacking	Water	0.1	18
PNIPAAm	π–π stacking	Water	–	19
PmPV	π–π stacking	DCE	∼ 0.1	20
DXR	Hydrogen bonding	Water	0.145	21
DNA	Hydrophobic interaction, weak electrostatic/hydrogen bonding interactions	Water	2.5	22
PVA	Hydrogen bonding	Water	5	23
TMPyP	Electrostatic, π–π stacking	Water	0.25	24
Quaternary ammonium salts	Ionic interactions	Chloroform	0.05	25
PEO-b-PPO-b-PEO	Hydrophobic interaction	Water	1	26
DSPE-mPEG	Hydrophobic interaction	DMF	3	27

aggregation, the introduction of modifiers with large conjugated moieties to adsorb on the graphene surface by π–π stacking can bring graphene the good dispersibility in water and organic solvents of modifiers, depending on the attribute of the modifying reagents. Actually, the surface modification based on π–π interactions has been widely demonstrated as a convenient technique in CNT surface modification (28–30). After introduction of CNTs, considerable efforts have been addressed for the modification of graphene by π–π interactions. In general, the organic molecules or polymers containing conjugated moieties (fused aromatic ring compounds such as naphthalene, pyrene, perylene, etc.; porphyrin; ionic liquids; polyelectrolytes) were exploited as modifier molecules to improve the dispersibility of graphene by the π–π interactions with the large π electron system of the graphene surface. Stankovich et al. prepared stable aqueous dispersions of graphitic nanoplatelets via an exfoliation/*in-situ* reduction of GO in the presence of amphiphilic polymer poly(sodium 4-styrenesulfonate) (PSS) (17). The absorption of PSS by strong π–π interactions prevented the aggregation of graphene and resulted in good dispersibility of graphene in aqueous system with a high concentration of up to 1 mg/ml.

The conjugated pyrene and its derivatives are more attractive for noncovalent functionalization of graphene due to the strong π–π interaction of the large conjugated moieties. Xu et al. exploited the pyrenebutyrate as the modifying reagent to prepare stable aqueous dispersions of functionalized graphene sheets. After reduction by hydrazine monohydrate, pyrenebutyrate-functionalized graphene with good processibility can be assembled into films with seven orders of magnitude increase in conductivity than that of GO precursor (18). Liu et al. prepared thermosensitive graphene colloids by attaching poly(*N*-isopropylacrylamide) (PNIPAAm) with pyrene terminal moieties onto the basal plane of chemically reduced graphene (CRG) sheets via π–π stacking. As shown in Figure 2.1, the pyrene-terminated PNIPAAm was dissolved in water and then mixed with the aqueous suspension of CRG. The resultant PNIPAAm-warped graphene colloids showed a reversible lower critical solution temperature (LCST)-induced dispersibility at 24 °C (19).

The π–π interaction is also workable for the exfoliated graphite functionalization. The poly(*m*-phenylenevinylene-*co*-poly(*m*-phenylenevinylene-*co*-2,5-dioctoxy-*p*-phenylene-vinylene) (PmPV) was applied to disperse the exfoliated graphite in a 1,2-dichloroethane (DCE) solution, forming functionalized graphene nanoribbons (GNRs) (Fig. 2.2). They were solution phase-derived and stably dispersed in solvents and exhibited ultrasmooth edges with possibly well-defined zigzag or armchair-edge structures (20).

2.2.2   Hydrogen Bonding

A large number of polar groups such as –OH and –COOH groups on the surface of GO behave as the active sites for hydrogen bonding with the modifying molecules. After chemical/thermal reduction of GO, the absorbed molecules impart good dispersibility to CRGs, which were ought to possess irreversible aggregates.

For instance, polyvinylalcohol (PVA) with rich hydroxyl groups in long chain is able to form hydrogen bonding with GO sheets. Bai et al. prepared graphene composite hydrogels with PVA as weight ratios of PVA to GO in the range of 1:10 to 1:2. As shown in Figure 2.3, the GO/PVA hydrogels exhibited pH-induced gel–sol transition and could be used for loading and selectively releasing drugs in an environment at the physiological pH (23).

Yang et al. successfully controlled the loading and release of doxorubicin hydrochloride (DXR) on GO via a pH condition. The amino and hydroxyl groups on DXR formed strong hydrogen-bonding interaction with GO. As shown in Table 2.2, hydrogen bonding changed in different pH surroundings (21). At the same time, the fluorescence spectrum and electrochemical characterization results showed that strong π–π stacking interactions also existed between GO and DXR. Patil et al. stabilized single-layer graphene sheets with single-stranded

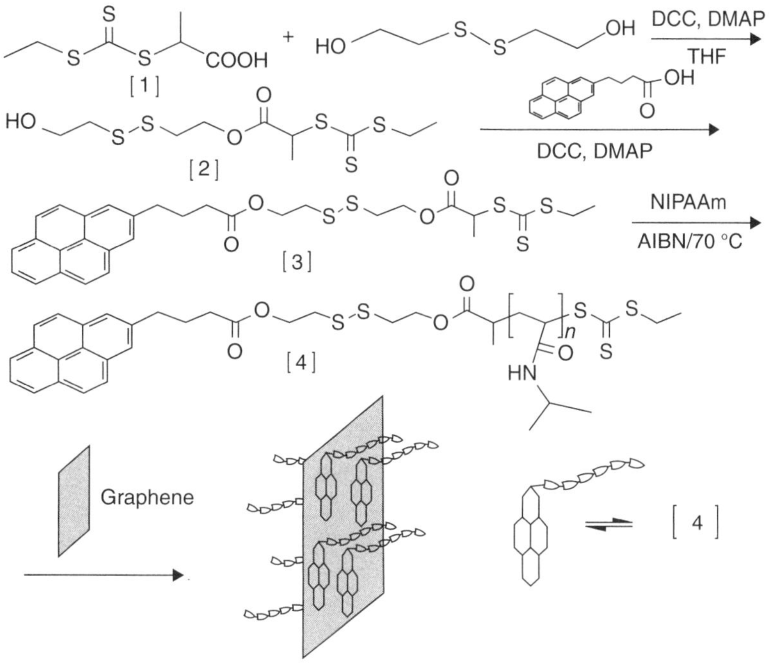

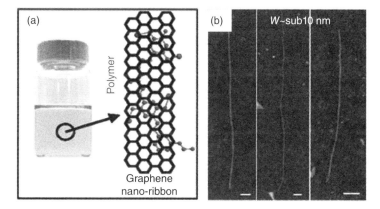

Figure 2.1 A schematic depicting the synthesis of pyrene-terminated PNIPAAm using a pyrene-functional RAFT agent and the subsequent attachment of the polymer to graphene. (*Source*: Reproduced with permission from (19)).

Figure 2.2 (a) (Left) Photograph of a polymer PmPV/DCE solution with GNRs stably suspended in the solution. (Right) Schematic drawing of a graphene nanoribbon with two units of a PmPV polymer chain adsorbed on top of the graphene via π-stacking. (b) AFM images of selected GNRs with widths in the sub-10-nm regions.

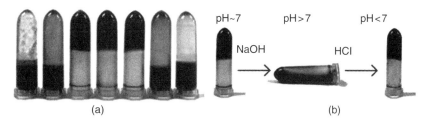

Figure 2.3 (a) Photographs of GO/PVA mixtures with varied content ratio. (b) Photographs of the pH-induced gel–sol transition. (*Source*: Reproduced with permission from (23)).

TABLE 2.2   Groups That Can Form Hydrogen Bonds in GO and DXR at Different pH Values

pH Value	GO	DXR
2	–OH, –COOH	–OH
7	–OH, –COOH	–OH, –NH$_2$
10	–OH	–OH, –NH$_2$

Source: Reproduced with permission from (21).

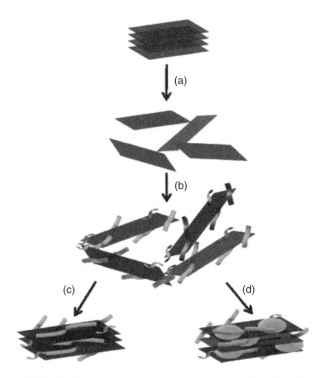

Figure 2.4   Schematic showing synthesis of DNA-stabilized graphene aqueous suspensions and fabrication of lamellar multifunctional nanocomposites. (a) Oxidative treatment of graphite yield delaminated nanometer-thick sheets of GO. (b) Chemical reduction of GO sols with hydrazine in the presence of freshly prepared single-stranded DNA (ssDNA) produced a stable aqueous suspension of ssDNA-functionalised graphene sheets (ssDNA-G). (c,d) Processing of ssDNA-G dispersions to produce ordered layered nanocomposites; (c) evaporation induced deposition and self-assembly on flat substrates results in lamellar nanocomposite films with intercalated ssDNA molecules, and (d) co-assembly of negatively charged ssDNA-G sheets and positively charged cytochrome *c* produces co-intercalated multifunctional layered nanocomposites.

DNA in water with a high stable concentration of up to 2.5 mg/ml. The high stability was realized principally through hydrophobic as well as weak electrostatic/hydrogen bonding between primary amines of the nitrogen bases and the carboxylic and hydroxyl groups of GO. The as-prepared negatively charged biofunctionalized graphene sheets were assembled into layered hybrid nanocomposites containing intercalated DNA molecules or cointercalated mixtures of DNA and the redox protein cytochrome c (Fig. 2.4) 22.

2.2.3   Electrostatic Interaction

The dispersive stability of GO colloids should be attributed to electrostatic repulsion, rather than their hydrophilicity only, as demonstrated by Li et al. The carboxylic acid groups are unlikely to be completely reduced, and the surface of the reduced graphene sheets in aqueous solution is still negatively charged. The electrostatic repulsion mechanism that works in stable aqueous dispersions of GO could also enable the formation of well-dispersed reduced graphene colloids, but just at limited concentrations (31).

To make dispersion of accessible CRGs stable in high concentrations, further modifications are needed. The electrostatic interaction between the carboxyl group on graphene panel and the cationic ion of the modifier molecules could also be used to achieve the noncovalent functionalization of graphene sheets (32, 33). Matsuo et al. allowed various quaternary alkylammonium or alkylpyridinium ions to intercalate into GO, reaching a maximum saturated amount of intercalated surfactant ions (0.56 mol/100 g GO). The proposed structure models of the composite suggest that it is possible to control the space between the surfactants in GO layers to a wide range by changing the kind and amount of surfactants (Fig. 2.5).

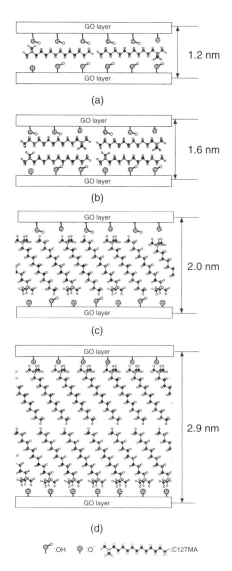

Figure 2.5 Structure models of lauryltrimethylammonium chloride-intercalated GOs with interlayer spacings of 1.2 nm (a), 1.6 nm (b), Ic = 1.9 nm (c), 2.9 nm (d).

Xu et al. prepared a complexation of cationic 5,10,15,20-tetrakis(1-methyl-4-pyridinio)porphyrin (TMPyP) and negatively charged CRG, attributing to the electrostatic and π–π stacking cooperative interactions between the flat TMPyP molecules and CRG. The complex of TMPyP and CRG could be an optical probe for rapid and selective detection of Cd^{2+} ions in aqueous media (24).

Specially, Liang et al. dispersed large-scale processable graphene sheets in organic solvents based on a transfer process assisted by ionic interactions. GO or CRG with negative charges (COO^-) started to mix with amphiphilic molecules carrying positive charge (NR_4^+), and chloroform was then added to the aforementioned solution followed by shaking, transferring the GO or reduced graphene oxide (RGO) from the aqueous phase into the organic phase. Then, a thin graphene film with the conductivity of 350 S/m was fabricated from the organic solution (25).

2.2.4 Hydrophobic Interaction

Together with the hydrophilic oxygen groups such as COO^- groups, the hydrophobic basal aromatic domains in GO or CRG sheets determine the amphiphilic attribute in liquid phases. Although inducing the aggregation tendency of graphene in aqueous system, the hydrophobic character could be used to modify graphene sheets through the hydrophobic interaction between the graphene and hydrophobic groups in the amphiphilic molecules. The hydrophobic groups combine the graphene sheets through hydrophobic interactions, and the hydrophilic groups form solvation with water molecules, improving the dispersibility of graphene.

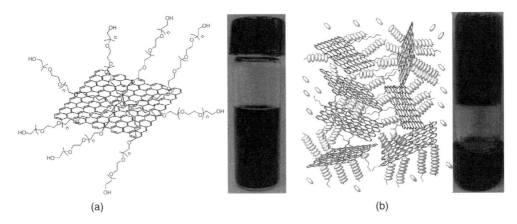

Figure 2.6 Proposed structure of the copolymer coated graphene (a) and supramolecular well-dispersed graphene sheet containing hybrid hydrogel (b).

Zu et al. employed triblock copolymers (PEO-*b*-PPO-*b*-PEO) to prepare well-dispersed CRG in water through *in situ* reduction. The obtained stable aqueous dispersion of copolymer-coated CRG was macroscopically homogeneous, but microscopically heterogeneous, in which the hydrophobic PPO segments were bound to the hydrophobic surface of graphene via hydrophobic effect, whereas the hydrophilic PEO chains extended into water to form supramolecular hydrogel with α-cyclodextrin through the penetration of PEO chains into the cyclodextrin cavities (Fig. 2.6). The supramolecular hybrid hydrogel was of the shear-thinning and temperature response properties.

Hydrophobic interaction was used by Li et al. to make high-quality graphene sheets (GS). As shown in Figure 2.7, commercial expandable graphite was exfoliated by brief heating to 1000 °C and inserted the exfoliated graphite with oleum and tetrabutylammonium hydroxide, followed by sonication with 1,2-distearoyl-sn-glycero-3-phosphoethanolamine-*N*-[methoxy(polyethyleneglycol)-5000] (DSPE-mPEG) for

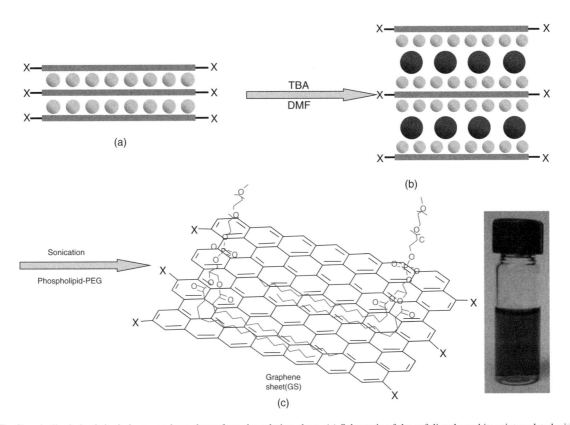

Figure 2.7 Chemically derived single-layer graphene sheets from the solution phase. (a) Schematic of the exfoliated graphite reintercalated with sulphuric acid molecules (small spheres) between the layers. (b) Schematic of tetrabutylammonium hydroxide (TBA, big spheres) insertion into the intercalated graphite. (c) Schematic of GS coated with DSPE-mPEG molecules and a photograph of a DSPE-mPEG/DMF solution of GS.

60 min. This method resulted in the suspension of large amount of GS in *N,N*-dimethylformamide (DMF), and they could be transferred to other solvents including water and organic solvents (27).

2.3 COVALENT FUNCTIONALIZATION

To circumvent the instability of noncovalent functionalization of graphene, covalent functionalization has been promoted and widely studied to establish the rich chemistry of graphene. Ideal graphene is of perfect lattice structure, and it is difficult to realize the covalent modification because of the high energy barrier of disrupting the conjugation of the graphene sheets. GO, an important derivative of graphene with rich oxygen-containing pedant groups and interrupted conjugated domains, is alternatively available for further modification to form permanent bonding between the graphene and the modifier (34). Covalent functionalization can not only improve the solubility of CRG, but also introduce new functional groups to prepare special materials with tunable functionalities. Detailed modification via covalent bonds of graphene will be discussed below.

2.3.1 "Attach to" Approach

"Attach to" approach has been regarded as one common method to tailor the function of graphene, which involves three steps: (i) modifying graphene to obtain appropriate functional group most examples started from GO with significant quantity of functionalities available for covalent reactions; (ii) preparing small molecules or polymer chains with matched functional groups; and (iii) connecting the above two building blocks together through designed reactions.

Chen's group modified GO with amine-terminated porphyrin. As shown in Figure 2.8, GO was acylated with thionyl chloride to introduce acyl chloride groups on the surface. And then, 5-4 (aminophenyl)-10, 15, 20-triphenyl porphyrin (TPP) molecules were bonded onto GO via amide bond using the amine-functionalized prophyrin (TPP-NH$_2$) in DMF. Attachment of TPP-NH$_2$ significantly improved the dispersion stability of graphene sheets in organic solvents. The donor–acceptor nanohybrid material additionally showed a superior optical limiting effect (35).

Pan's group fabricated stable aqueous dispersions of CRGs with the assistance of hydroxypropyl cellulose or chitosan covalently grafted on graphenes (36). Dai's group also covalently modified GO with six-armed polyethylene glycol (PEG) star molecules via carbodiimide catalyzed amide reaction. The resulted materials were attempted to deliver hydrophobic aromatic cancer drugs into water, loaded the drug by noncovalent interaction via π–π stacking (37). It is easy to obtain different polymer/graphene composites through the introduction of various polymers. PVA was attached to graphene through the carbodiimide-activated esterification reaction between the carboxylic acid moieties on the nanosheets and hydroxyl groups on polymer chains. The resulting functionalized sample became readily soluble in aqueous and good organic solvents for PVA, allowing solution-phase processing for various purposes such as the fabrication of polymer-carbon nanosheets composites (38).

Besides the approach of transferring –COOH into ester linkage or amide bonding, heterocyclization reaction was developed as an important methodology to realize the covalent functionalization of graphene (Fig. 2.9). Huang's group synthesized benzoxazole- and benzimidazole-grafted graphene via heterocyclization reaction of carboxylic groups on GO with the hydroxyl and amino groups on *o*-aminophenol and *o*-phenylenediamine, followed by reduction with hydrazine. The functionalized graphene materials showed corrugation

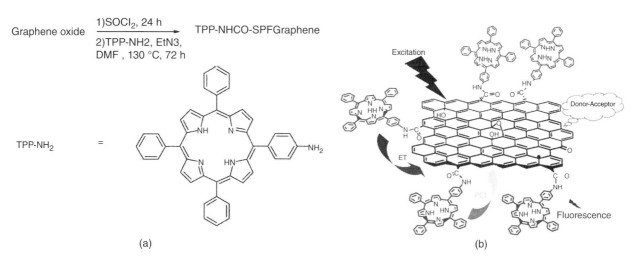

(a)   (b)

Figure 2.8 (a) Synthesis scheme of TPP-NHCO-SPFGraphene. (b) Schematic representation of part of the structure of the covalent TPP-NHCO-SPFGraphene. (*Source*: Reproduced with permission from (35)).

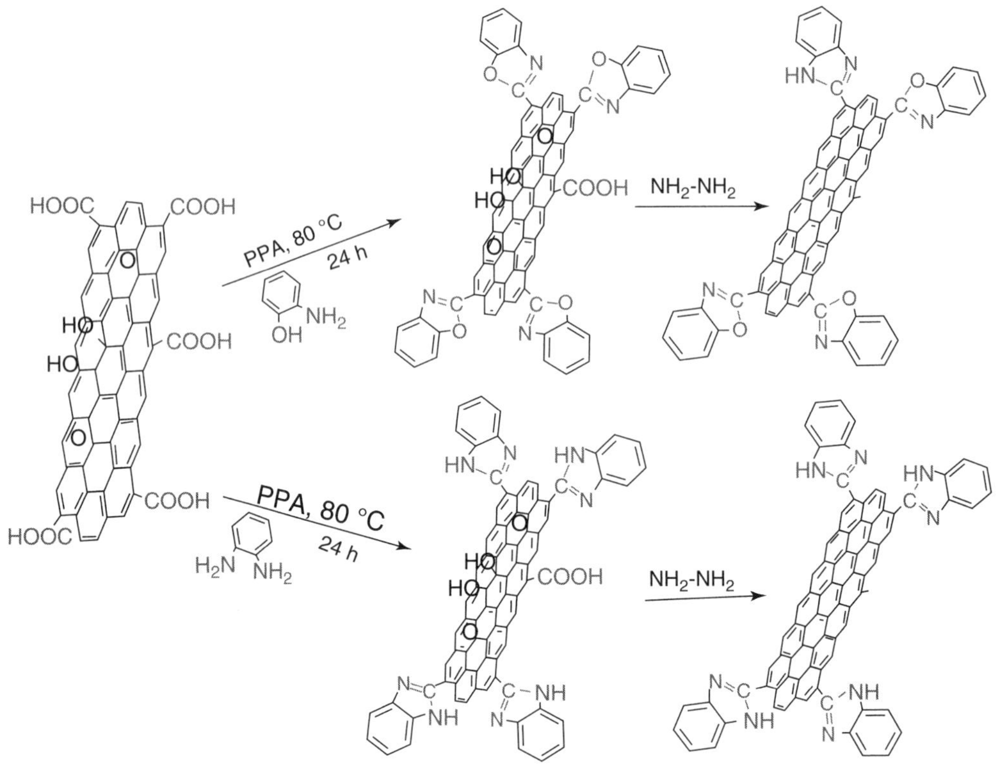

Figure 2.9 Synthesis procedure for benzoxazole- and benzimidazole-grafted graphene via heterocyclization reaction. (*Source*: Reproduced with permission from (39)).

and scrolling morphologies, and less restacking/aggregation. When applied as supercapacitor electrodes, they exhibited good electrochemical performance in terms of high specific capacitance (730 and 781 F/g for benzoxazole- and benzimidazole-grafted graphene, respectively, at a current density of 0.1 A/g), implying their potential for energy storage applications (39).

By utilizing the epoxy groups as reactive sites to process covalent functionalization, many efforts have been made to obtain graphene sheets with good dispersibility. Zhou et al. synthesized cysteamine-functionalized GO based on nucleophilic ring-opening reaction between the epoxy of GO and the amino group of the cysteamine carried out in potassium hydroxide (KOH) solution (Fig. 2.10). After deposition onto the Au electrode surface through the formation of Au–S bonds, the resultant graphene sheets showed selectively and sensitively the quantitative detection of Hg^{2+} (40). Yang et al. functionalized graphene nanosheets via the ring-opening reaction between the epoxy of GO and the amino group of 3-aminopropyltriethoxysilane (APTS). The obtained graphene could be dispersed into water, ethanol, DMF, dimethyl sulfoxide (DMSO), and APTS to form stable and homogeneous dispersions after ultrasonic treatment (41). According to the same rationale, the amine-terminated ionic liquid was used to functionalize graphene, catalyzed by the KOH. The introduction of charge and the widely soluble ionic liquid units resulted in CRGs with good stabilities in water, DMF, and DMSO (42). Ma's group also covalently functionalized graphene with amine-bearing polymer via a facile ring-opening reaction (43).

Besides the carboxyl group and epoxy group, the hydroxyl group of GO also could be applied for the covalent modification. The –OH groups are active in the reaction with the silylating reagent to form Si–O bonding (44, 45). Matsuo's group prepared various silylated GOs by the reaction between GO and alkylchlorosilanes with different chain lengths in the presence of butylamine and toluene. The interlayer spacing and silicon contents of silylated GOs varied in wide ranges. In the process, the butylamine played important roles, not only exfoliating GO layer but also scavenging HCl molecule which caused the decomposition of silylated GO (46).

The isocyanate treatment could synchronously functionalize the carboxyl and hydroxyl groups of GO via the formation of amide and carbamate esters (47, 48). Zhang et al. synthesized toluene-2,4-diisocyanate (TDI)-functionalized GO and then grafted the presynthesized S-1-dodecyl-S'-(α,α'-dimethyl-α''-acetic acid) trithiocarbonate-poly(*N*-vinylcarbazole) (DDAT-PVK) polymer onto GO surface with TDI as a coupling reagent (Fig. 2.11). The resultant modified GO exhibited an enhanced solubility of up to 10 mg/ml in organic solvents and a significant energy bandgap of ~2.0 eV, giving a donor–acceptor-type GO-based optoelectronic material (49).

2.3.2 "Grafting from" Approach

Compared with the "attaching to" approach, which involves the preparation of modifier molecules with matched functional groups prior to grafting, "grafting from" approach shows to be an effective method to achieve higher degree of functionality and higher density of grafted chains. "Grafting from" approach commonly includes two steps: surface activation and graft modification.

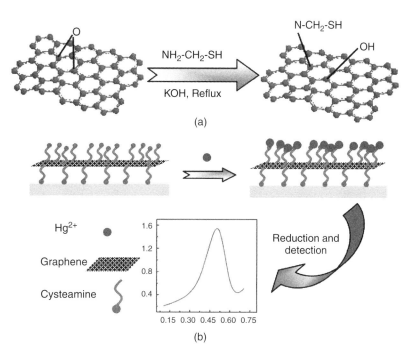

Figure 2.10 Schematic illustration of the reaction employed for (a) functionalization of graphene oxide with cysteamine and (b) procedures for Hg^{2+} determination. (*Source*: Reproduced with permission from (40)).

Atom transfer radical polymerization (ATRP) has been promoted as an effective method to achieve the excellent control over the polymer's chain length and architecture. In the field of graphene covalent functionalization, GO is often used to couple suitable initiators for surface-initiated ATRP of different vinyl monomers. According to Marques group's report, the free carboxylic acid groups of GO were converted to acyl chloride groups, which yielded rich hydroxyl groups upon treatment with ethylene glycol. Then the ATRP initiator was grafted onto GO sheets and finally polymethyl methacrylate (PMMA) chains were grown via ATRP (Fig. 2.12). The resultant GO sheets were readily dispersed in organic solvents and used as reinforcement fillers in the preparation of PMMA composite (51). Similarly, Lee et al. functionalized the hydroxyl groups of GO with ATRP initiators, and then polymers of styrene, butyl acrylate, and methyl methacrylate were grown directly via ATRP. The controlled polymerization process was versatile and facile to the GO derivatives that exhibited improved dissolution properties and chemical compatibility (50). Alternatively, the spontaneous grafting of diazonium salts onto CRG sheets was used by Fang et al. to attach initiators for ATRP. As shown in Figure 2.13, the initiator molecules were covalently bonded to the CRG surface via diazonium addition and the succeeding ATRP linked polystyrene chains to the CRG nanosheets. The functionalization resulted in 82 wt% polymer grafting efficiency and a 15 °C increase in the glass transition temperature (T_g) compared with pure polymer (52). Controlled diazonium addition was also examined to be effective for tuning the grafting density of PS on graphene nanosheets (53).

Alternatively, Ou et al. introduced ATRP initiators to CRG sheets, started from 1,3-dipolar cycloaddition with phenol groups. These CRG-based macroinitiators were exposed to MMA to achieve PMMA-functionalized graphene (Fig. 2.14). Specially, in this approach, two hydroxyl groups can be covalently attached to the CRG in a single step without degrading its electronic properties. So it was easy to achieve the functionalized graphene with high grafting densities and wide applications especially in conductive nanocomposites (54).

Deng et al. attached tris(hydroxymethyl) aminomethane (TRIS) to GO surface to increase the amount of hydroxyls on GO via the ring-opening reaction at room temperature. Then these hydroxyls were converted to Br-containing initiating groups through a mild esterification for initiating single-electron transfer-living radical polymerization (SET-LRP) of poly(ethylene glycol) ethyl ether methacrylate (PEGEEMA) macromonomer with $CuBr/Me_6TREN$ as the catalytic system (Fig. 2.15). This TRIS-GO-PPEGEEMA hybrid material showed improved dispersibility in various solvents and the temperature responsive attributes. It reversibly transferred between the self-assemble and de-assemble states in water by switching temperature at about 34 °C (55).

2.3.3 *In Situ* Polymerization Approach

Both "attaching to" and "grafting from" approaches are primarily based on the pendant sites of GO, such as hydroxyl, carboxyl, epoxy, and ketone. Both the methods need extra steps before introducing polymers, making the procedure relatively complex. More seriously, the multistep reactions raise the aggregation risk of graphene, which is adverse for the target of dispersibility as individual sheets. In this case, *in situ* polymerization is possibly the best choice for the synthesis of polymer-grafted graphene, for its simple procedure.

The Lerf model (56–58) represents that the conjugate region isolated by sp^3 network shows a certain chemical activity, and some isolated double bonds with high chemical activity could also be found. Thus, GO possesses the chemistry of double bonds (59–64). The isolated

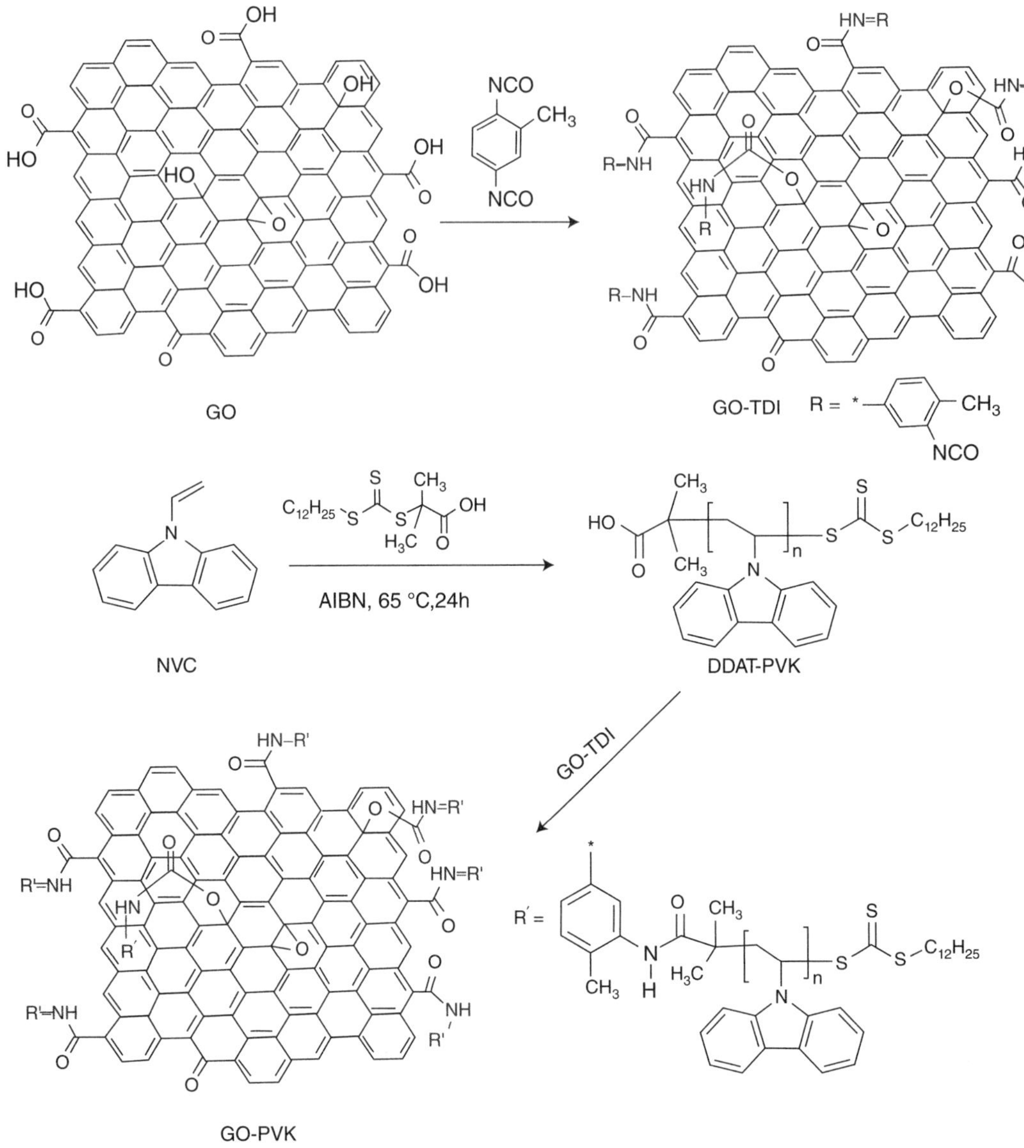

Figure 2.11   Synthesis of GO-TDI, DDAT-PVK, and GO-PVK. (*Source*: Reproduced with permission from (49)).

double bonds on the GO panel could be utilized for the *in situ* vinyl polymerization. Ye and coworkers prepared amphiphilic polymer-grafted graphene from polystyrene and polyacrylamide macroinitiators via radical coupling onto the vinyl bonds of GO (65). Gao's group applied conventional free radical polymerization (FRP) to obtain 2D molecular brushes with graphene as flat macromolecular backbones (Fig. 2.16). The approach was workable for almost all of the vinyl monomers, producing various 2D molecular brushes with multifunctional arms covering from polar to apolar, water-soluble to oil-soluble, and acidic to basic. The growing process of 2D molecular brushes was clearly visualized by Atomic Force Microscopy (AFM) and thermal gravimetric analysis (TGA) characterizations. The resultant giant 2D brushes showed a high arm density of up to 1.59×10^4 arms per μm^2 on single side of graphene and, as a result, possessed high solubility ($\sim$15 mg/ml), low intrinsic viscosity ($\sim$100 ml/g), and fine electrical conductivity ($\sim$8.4 $\times$ 10^{-3} S/cm) (66).

Ce^{4+}-alcohol redox pair has been demonstrated as an effective initiator for radical polymerization of a variety of water-soluble vinyl monomers. Replacing the alcohol with the hydroxyl groups on GO, this polymerization system could be established for the covalent functionalization of GO with acrylic acid (AA) and *N*-isopropylacrylamide (NIPAM) in aqueous solution at mild temperature. The graft ratio of PAA and PNIPAM could be controlled by varying the feeding amount of monomers. The solutions of obtained PAA-grafted GO (GO-PAA) and PNIPAM-grafted GO (GO-PNIPAM) were obviously pH and thermal-responsive in water, respectively (Fig. 2.17) (67).

Another source of *in situ* covalent functionalizing graphene is polycondensation. Wang et al. demonstrated the preparation of a graphene/polyaniline composite paper (GPCP) by *in situ* anodic electropolymerization (AEP) of aniline monomers to form PANI layers on CRG papers. This composite paper combined flexibility, conductivity, and electrochemical activity and exhibited excellent gravimetric

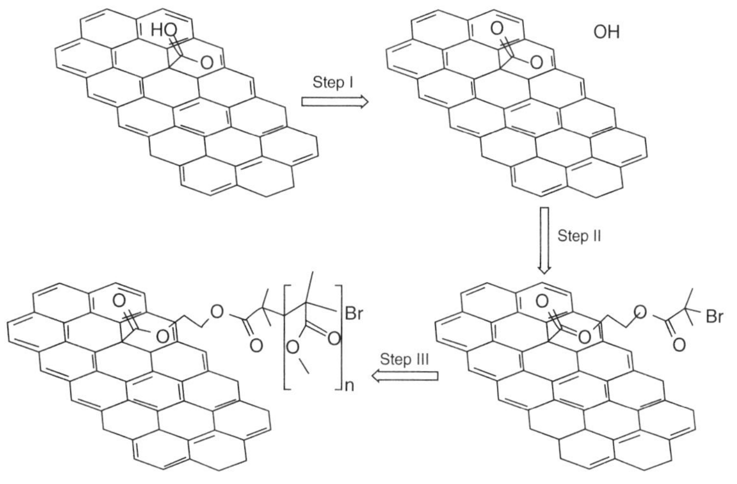

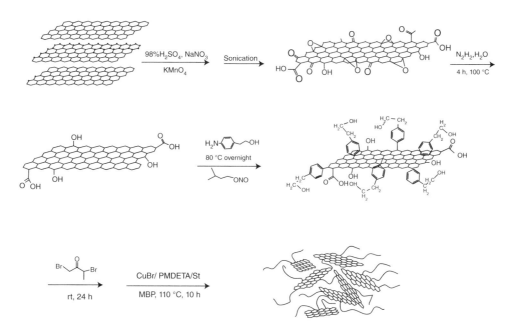

Figure 2.12 Reaction pathway involved in the functionalization of GO nanosheets with polymer chains. (*Source*: Reproduced with permission from (51)).

Figure 2.13 Synthesis route of polystyrene-functionalized graphene nanosheets. (*Source*: Reproduced with permission from (52)).

capacitance of 233 F/g and volumetric capacitance of 135 F/cm^3, outperforming many other currently available carbon-based freestanding electrodes (68). Yan et al. reported that graphene played two roles in the graphene/polyaniline composites, providing more active sites for nucleation of PANI as well as offering excellent electron transfer path (69).

Gao's group reported an effective protocol to prepare nylon-6- (PA6-) graphene (NG) composites on the basis of carboxyl groups of GO by *in situ* polymerization with simultaneous thermal reduction of GO (Fig. 2.18). The grafting ratio of PA6 arms on graphene sheets was up to 78 wt%. The modified graphene sheets were homogeneously dispersed in PA6 matrix with a depression effect on the crystallization of PA6 chains. The melt-spun composite fibers showed a 1.1-fold increase in tensile strength and a 1.4-fold increase in Young's modulus at the graphene loading of 0.1 wt% only, revealing an excellent reinforcement to composites of graphene (70).

Additionally, ring-opening polymerization has also been widely investigated for the graphene *in situ* modification. As shown in Figure 2.19, Pham et al. functionalized GO sheets by *in situ* ring-opening polymerization of glycidol, and the as-prepared materials acted as templates for further loading boronic acid functionalized Fe-core/Au-shell nanoparticles (B-*f*-MNPs) through boroester bonds. The synthesized novel hybrid nanostructures could be stably dispersed in water for at least 3 months. The combination of magnetic nanoparticles and biocompatible polymers, polyglycerol onto the GO sheets surface made the hybrid materials potential for a wide variety of applications (71).

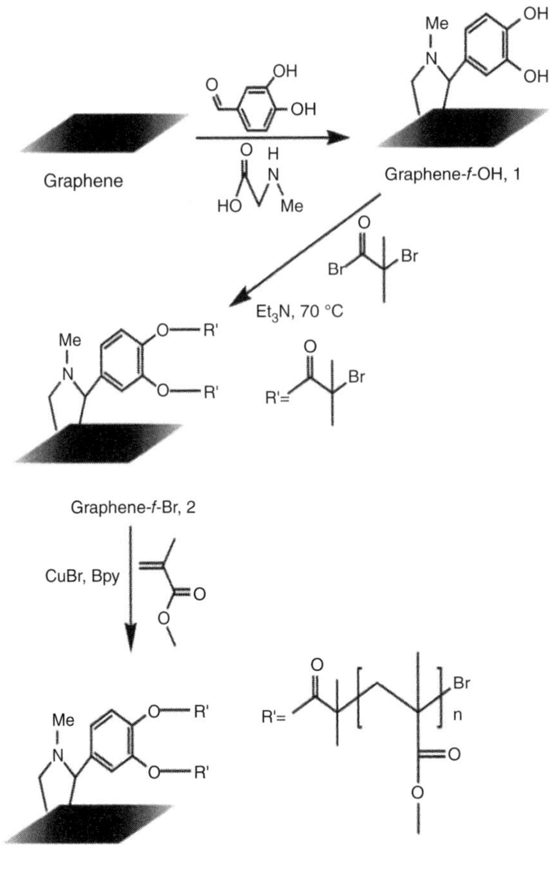

Figure 2.14 Schematic illustration of the three-step process for functionalization of graphene with PMMA by ATRP at room temperature. (*Source*: Reproduced with permission from (54)).

2.4 MACROSCOPIC ASSEMBLED GRAPHENE FIBERS, FILMS, AND FOAMS

Integration of individual 2D graphene sheets into macroscopic structures is essential for the application of graphene. Both noncovalent and covalent functionalization offer graphene expected dispersibility, which is the foundation of macroscopic materials by efficient fluid assembly method. Furthermore, the functionalization could provide graphene and the macroscopic materials thereof new designed functional properties. To date, a series of graphene-based composites and macroscopic structures have been fabricated from solvated graphene sheets, mainly including GO and its reduced counterpart.

2.4.1 Graphene Fibers

The extremely high mechanical strength of graphene promises this 2D carbonaceous nanomaterial as future candidate for the preparation of high-performance fibers. By analogy with the fabrication of high-performance polymeric fibers, graphene fiber can be spun from graphene liquid crystals (LCs), which have the ordered structure and the fluid attribute simultaneously. Recently, a couple of research groups reported graphene liquid crystals and the wet-spinning of graphene fibers, making the dream of turning graphite to continuous fibers come true.

Xu et al. found that GO sheets had rich liquid crystalline behaviors in their dispersions, including conventional nematic mesophase and a new chiral mesophase. Starting from GO LCs with ordered structure and fluid attribute, continuous neat graphene fibers were fabricated by industrially viable wet-spinning and the following chemical reduction. In graphene fibers, uniform alignment of graphene sheet, inherited from the intrinsic order of LCs, provided strong interactions between contacted sheets that are responsible for the strong mechanical strength ($\sim$140 MPa) together with high conductivity ($\sim$2.5 $\times$ 10^4 S/m), and the locally crumbled structures made the graphene fibers flexible (the ultimate elongation of 5.8 %). More remarkably, the graphene fibers were able to be fastened into tight knots without breakage and can be woven into designed patterns and complex textiles (Fig. 2.20) (72). Subsequently, considerable efforts in this group have been devoted to improve the mechanical strength of graphene fibers, such as the choice of giant sizes of graphene sheets, the introduction of divalent

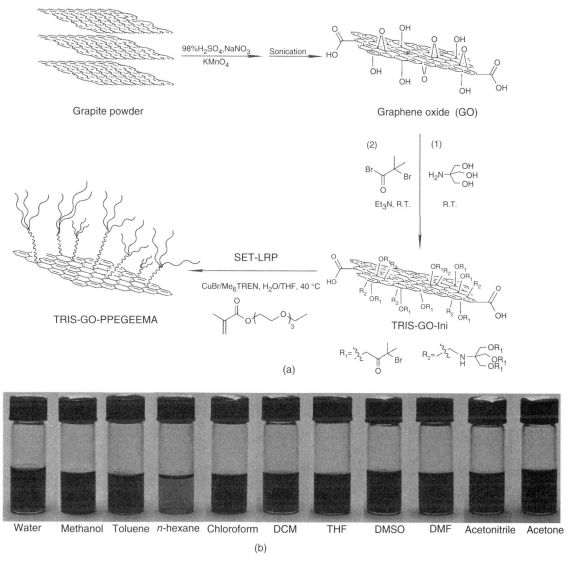

Figure 2.15   (a) In situ growing of PPEGEEMA polymer chains via SET-LRP from the surface of "TRIS" modified graphene oxide sheets. (b) Dispersion of TRIS-GO-PPEGEEMA in various solvents with a concentration of about 5.0 mg/ml. (*Source*: Reproduced with permission from (55)).

ions cross-linking, and the optimization of the spinning process (73–75). Since neat graphene fibers, the first continuous, conductive, flexible artificial nacre fibers with the length up to tens of meters and perfect brick-mortar structures were fabricated by wet-spinning methodology, in which graphene sheets acted as rigid platelets and hyperbranched polyglycerol (HPG) binders as elastic glue to provide hydrogen-bonding arrays between graphene interlayers (Fig. 2.21). The supramolecular fibers exhibited eminent properties such as high tensile strength comparable to nacre and bone, fine conductivity, and excellent resistance to corrosion (76). Aside from the hyperbranched architecture binder, the linear oxygen-containing polymers were additionally utilized as fine binders, reported by Kou et al. Hundreds of meters-long graphene composite fibers with "brick and mortar" layered structure were continuously wet-spun with PVA-coated graphene as building blocks. The effect of feed ratios of PVA to graphene and the molecular weight (MW) of PVA on the final fibers were systematically investigated (Table 2.3). The mechanical strength and electrical conductivity of the nacre-mimicking fibers were related to the coated PVA content. Distinguished from typical polymers and composites, MW of PVA had no effect on the mechanical performance of the fibers because the motion of PVA chains was confined to the nanochannels between the adjacent CRG sheets (77).

Wang et al. prepared the composite fibers containing CNTs, GO, or graphene by PVA-based coagulation spinning method. The resultant fibers had both high strength and conductivity, and showed excellent actuation performance. Upon the removal of carboxyl and hydroxyl groups, graphene fibers exhibited high toughness and the toughness could be tuned by mixing single-walled carbon nanotubes (SWCNTs) with graphene in different ratios (78).

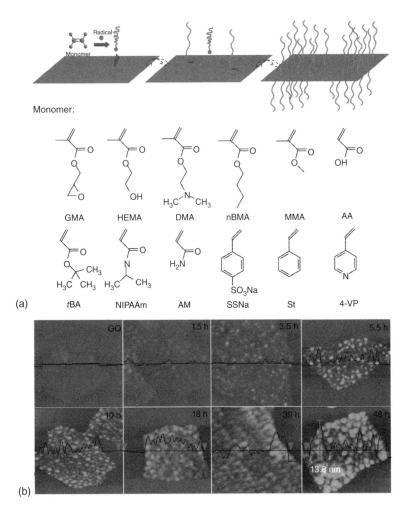

Figure 2.16 (a) Scheme of synthesizing 2D macromolecular brushes by free radical polymerization of various monomers with the backbone of GO sheets. The sheet represents GO, and the dots and double bonds represent radicals and active grafting points, respectively. (b) AFM height images of GO and GO-*g*-PGMA 2D brushes at different reaction time (size for all images: 1 μm × 1 μm). (*Source*: Reproduced with permission from (66)).

TABLE 2.3   Preparation Conditions and Selected Results for CRG@PVA Fibers

Fiber Code	M_w(g/mol)	R	f_{PVA} (wt%)	d(nm)	σ(MPa)	E(GPa)	Conductivity(S/m)
CRG@PVA 1	100 K	1.75	53.1	2.01	81	8.9	0.860
							350[a]
CRG@PVA 2	100 K	2.125	56.5	2.06	88	8.0	0.790
CRG@PVA 3	100 K	2.5	60.1	2.45	102	9.9	0.504
CRG@PVA 4	100 K	5	65.6	2.89	122	10.4	0.107
CRG@PVA 5	100 K	10	65.7	3.04	138	7.9	0.103
CRG@PVA 6	100 K	20	65.8	3.31	161	9.9	0.013
					199[b]	17.1[b]	
					140[c]	5.4[c]	
CRG@PVA 7	50 K	20	59.1	2.51	162	10.0	0.022
CRG@PVA 8	16 K	20	58.7	2.42	158	10.8	0.036
CRG@PVA 9	2 K	20	56.9	2.17	162	10.4	0.103

Source: Reproduced with permission from (77).
[a]The fiber was reduced by hydroiodic acid.
[b]The fiber was further immersed in 5 wt% aqueous PVA.
[c]The resulting fiber was immersed in water.

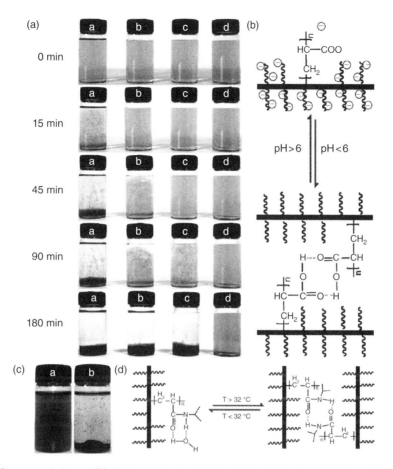

Figure 2.17 (a) Photographs of aqueous solutions of GO-PAA upon standing for 3 h at different pH values. (b) Schematic illustration of pH responsibility of GO-PAA. (c) Photographs of aqueous solution of GO-PNIPAM at (a) 25 °C for 3 h and (b) 35 °C for 10 min. (d) Schematic illustration of thermal responsibility of GO-PNIPAM. (*Source*: Reproduced with permission from (67)).

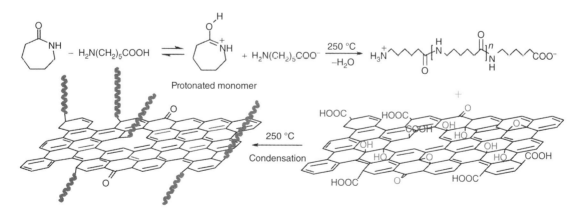

Figure 2.18 Scheme of synthesizing NG composites by in situ ring-opening polymerization of caprolactam in the presence of GO. The grey bonds in the nylon 6-grafted graphene represent restored conjugated bonds through high-temperature reduction; the curves represent the grafted PA6 macromolecular chains. (*Source*: Reproduced with permission from (70)).

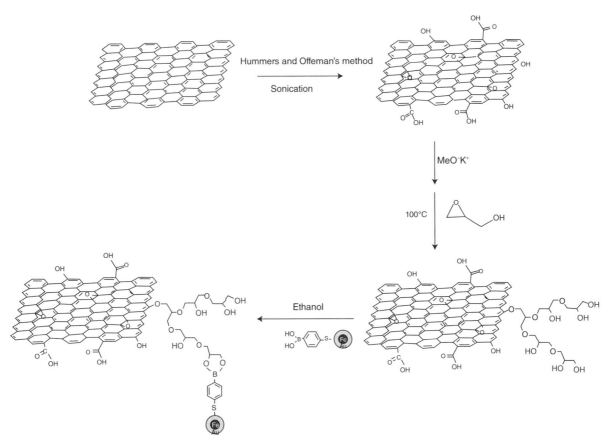

Figure 2.19   Synthetic route of water-dispersible PG-*g*-GO nanostructure and its immobilization with B-*f*-MNPs. (*Source*: Reproduced with permission from (71)).

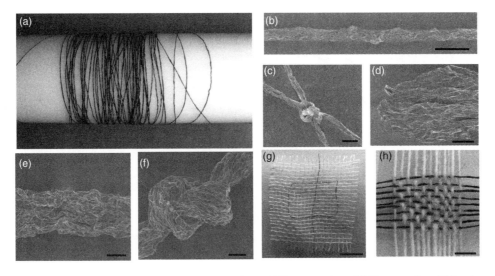

Figure 2.20   Macroscopic neat GO fibers and chemically reduced graphene fibers. (a) Four-meter-long GO fiber wound on a Teflon drum (diameter, 2 cm). (b) SEM image of the fiber. (c) The typical tighten knots of GO fibers. (d) The fracture morphology of GO fiber after tensile tests. The surface wrinkled morphology (e) and the tighten knot (f) of chemically reduced graphene fiber. (g) A Chinese character ('中', Zhong) pattern knitted in the cotton network (white) using two graphene fibers (black). (h) A mat of graphene fibers (horizontal) woven together with cotton threads (vertical). Scale bars, 50 μm (b–f) and 2 mm (g, h). (*Source*: Reproduced with permission from (72)).

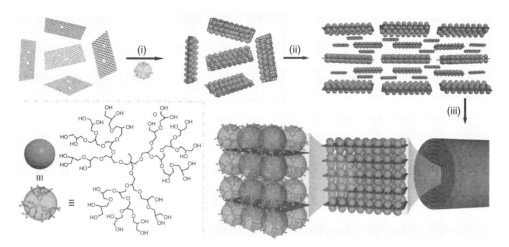

Figure 2.21 A schematic depicting the synthesis of HPG/graphene hybrid fiber with brick-mortar structures by the wet spinning. (i) Synthesis of HPG-enveloped graphene sheets sandwich building blocks via reaction of GO and HPG at 160 °C for 18 h. (ii) Pre-alignment of HPG-enveloped graphene sheets in highly concentrated liquid crystalline spinning dope. (iii) Formation of hierarchically assembled continuous artificial nacre fibres via wet-spinning. (*Source*: Reproduced with permission from (76)).

Another one-step dimensionally-confined hydrothermal strategy for graphene fibers was proposed by Qu's group. The aqueous GO suspension was injected into a glass pipeline, sealing up the two ends of the pipeline, baking it at 230 °C for 2 h, and then giving a graphene fiber matching the pipe geometry. To endow multifunctions to the fiber, they integrated the graphene fibers with Fe_3O_4 and TiO_2 and obtained magnetic and photoelectric response of functional graphene fibers (79).

The reduced graphene oxide nanoribbon (RGONR) can also be used to fabricate graphene fibers by an electrophoretic assembly method. The RGONR solution was injected into a 5 mm diameter cylindrical Teflon vessel with a copper plate as the negative electrode and a sharpened graphitic tip as the positive electrode. A gel-phase RGONR fiber was obtained when the sharpened graphitic tip was withdrawn from the RGONR solution at a velocity of 0.1 mm/min with a constant voltage of 1–2 V between the two electrodes. The thermally annealed fiber showed superior field emission performance with a low potential for field emission (0.7 V/μm) and a giant field emission current density (400 A/cm^2). Moreover, the fiber maintained a high current level of 300 A/cm^2 corresponding to 1 mA during long-term operation (80).

2.4.2 Graphene Films

As a typical 2D nanosheet, graphene is quite reasonable to be expected to construct 2D film or paper like material with the same topology. Actually, the graphene film or paper was the first macroscopically assembled grapheme-based material, even before the discovery of graphene (81). To date, many techniques have been established to fabricate the graphene films, such as spin-coating, interfacial self-assembly, electrophoretic deposition, vacuum filtration, layer-by-layer (LbL) spray-coating, Langmuir–Blodgett assembly, controlled pyrolysis, flame synthesis, drop-coating, dip-coating, and CVD. The basic properties of graphene films are summarized in Table 2.4.

Both GO and CRG are negatively charged nanosheets, and many efforts have been exerted on the fabrication of graphene-based films by the LbL assembly through the electrostatic interaction (103). Ye's group introduced negative and positive charge on the surface of CRG by covalent grafting of poly(acrylic acid) and poly(acryl amide), respectively, and then alternatively assembled the two components together to form a multilayer structure through electrostatic interaction (104). Lee et al. constructed a thin graphene film with the negatively charged GO and the positively charged ethylenediamine-grafted GO. The resultant films were of controllable thickness, transmittance, and sheet resistance and they could be applied in organic light-emitting diode (OLED) devices (98). Ji et al. reduced GO in the presence of ionic liquids at room temperature and then assembled LbL into graphene/ionic liquid films with poly(sodium styrenesulfonate) (PSS) on appropriate solid supports. The increased graphene layer spacing brought the clear enhancement of aromatic gas adsorption and the highly selective sensing performance (99).

Besides the electrically charged particles, GO and CRG also act as planar amphiphilic molecules. Driven by the minimization of interfacial free energy, they were preferentially adsorbed at the two immiscible liquid interfaces to obtain a monolayer of ultrathin film. Biswas et al. prepared graphene films with high electrical conductivity above 1000 S/cm and an optical transmission exceeding 70% at the wavelength of 550 nm (88). Self-assembly of solvated graphene could occur on a liquid/air interface to form CRG films. Cheng's group assembled GO sheets at the air–water interface, followed by the reduction with HI acid. The sheet resistance of CRG films decreased with increasing sheet area of the CRG at the same transmittance because of the decrease in the number of intersheet tunneling barriers. Typically, the CRG film made from GO sheets with an average area of 7000 μm^2 showed a sheet resistance of 840 Ω/sq at 78% transmittance, which was much lower than that of the CRG film made from small-area GO sheets of about 100–300 μm^2 (19.1 kΩ/sq at 79% transmittance), and comparable to that of graphene films grown on Ni by CVD (89). Air-DMF interfaces were utilized for the assembly of electrochemical-method-made graphene sheets with strong surface hydrophobicity, and the graphene films exhibited ultratransparency (~96% transmittance) and sheet resistance (1000 Ω/□) after a simple HNO$_3$ treatment, superior to those based on CRG or graphene sheets by other exfoliation methods (90).

TABLE 2.4   The Basic Properties of Graphene Films Prepared by Various Methods

Fabrication Method	Thickness	Conductance/ Resistance	Transmittance (wavelength)	Application	Refs.
CVD	Several layers	280 Ω/□	80% (550 nm)	Stretchable transparent electrodes	82
	6–8 layers	4 Ω/□	–	Gas sensor	83
	4 layers	1150-200 Ω/□	90.5% (350–2200 nm)	Front electrode of CdTe solar cells	84
Spin-coating	~3 nm	100–1000 Ω/□	80% (550 nm)	Transparent conductors	85
	~6 nm	5.4×10^{-6} S/m	84% (550 nm)	Thin film field-effect phototransistors	86
	>4 nm	–	–	Nanomechanical devices	87
Interfacial self-assembly	3~4 nm	10^5 S/m	70% (550 nm)	Optoelectronics applications	88
	–	840 Ω/□	78%	–	89
Electrophoretic deposition	–	~210 Ω/□	96%	–	90
	6 μm	15,000 S/m	–	–	91
	~4 μm	14,300 S/m	–	–	92
Vacuum filtration	10 μm	2.0×10^7 Ω/□	96%	–	31
	10 μm	35,100 S/m	–	–	93
	65 μm	58 S/m	–	Electroconductive hydrogel films	94
	22–53 nm	–	–	Nanofiltration membrane	95
	14 nm	2000 Ω/□	80% (550 nm)	–	96
	–	5000 Ω/□	80% (550 nm)	–	97
Layer-by-layer	5–15 nm	8600 Ω/□	86% (550 nm)	OLED device	98
		32,000 Ω/□	91% (550 nm)		
	2–4 layers	178 Ω/□	–	Gas sensor	99
Spray-coating	5 mm	8500 Ω/□	–	–	100
	–	2200 Ω/□	84% (550 nm)	–	101
Langmuir–Blodgett assembly	1 layer	150,000 Ω	93% (1000 nm)	–	27
	2 layers	20,000 Ω	88% (1000 nm)		
	3 layers	8000 Ω	83% (1000 nm)		
	18.5 nm	459 Ω/□	90%	–	102

For their partly hydrophobic attribute, CRG sheets in organic solvents were taken to make large-area transparent conducting films by Langmuir–Blodgett assembly process in countable cycles (27). A monolayer is adsorbed homogeneously with each immersion or emersion step, and thus films with very accurate thickness can be formed. Zheng et al. produced transparent conductive films from ultralarge GO (UL-GO) sheets. The density and degree of wrinkling of the UL-GO monolayers were turned from dilute, close-packed flat UL-GO to GO wrinkles and concentrated GO wrinkles by controlling the processing conditions. A remarkable sheet resistance of ~500Ω/□ at 90% transparency was obtained (102).

The electrophoretic deposition method has been demonstrated as an economical and versatile processing technique in the preparation of thin films from charged colloidal suspensions, which should be useful to be applied for graphene films from charged GO and CRG colloids. Additionally, this electrophoretic deposition method combines a number of advantages such as high deposition rate, excellent thickness controllability, uniformity, and simplicity of scaling up. Chen et al. prepared positively charged CRG with p-phenylene diamine as reducing reagent, which was well stable in ethanol and then was fabricated into a film on indium tin oxide (ITO)-coated conductive glass by electrophoretic deposition, exhibiting high conductivity of about 150 S/cm (91). By selecting the appropriate suspension pH and deposition voltage, the negatively charged GO sheets were deposited into either a smooth "rug" microstructure on the anode or a porous "brick" microstructure on the cathode. The brick films (79°) were more hydrophobic than the rug films (41°), primarily attributing to the distinctive microstructures (105). Lee et al. compared the electrophoretic deposition from GO and CRG components and demonstrated that the reduction with hydrazine for CRG yields smoother graphene films than electrophoretically deposited GO films followed by reduction (106). Furthermore, the reduction of GO frequently coexisted with the electrophoretic deposition process. The electrophoretically deposited GO film showed improved electrical conductivity (1.43×10^4 S/m) over GO papers (0.53×10^{-3} S/m) (92).

Owing to the excellent dispersibility of GO, it could be expediently coated on the appropriate substrates. The thickness of deposited layers can be tuned by varying the concentration of the GO dispersion. Chen's group spin-coated GO thin films on quartz, with dilute dispersions (2 mg/ml) giving films ~3 nm thick and higher concentration dispersions (12–15 mg/ml) yielding films ~20 nm thick. After a thermal graphitization procedure, the films showed low sheet resistances of 10^2–10^3 $\Omega/\square$ with 80% transmittance at 550 nm (85). Chang et al. constructed thin films by ambient spin-coating and controlled their bandgaps through the simple low-temperature annealing, ranging from 2.2 to 0.5 eV. The bandgap energy was strongly related to the electronic properties of reduced GO film, and the conductivity increased by 5.7×10^6 times after 260 min thermal annealing (~5.4×10^{-6} S/cm), compared with as-prepared GO films (~9.4×10^{-11} S/cm) (86). Additionally, accelerating solution evaporation by blowing dry nitrogen could be added into the spin-casting process to result in continuous films. Through thermal annealing the frequency response of suspended membranes can be systematically tuned via tension release, allowing for the identification of Young's modulus for RGO thin films at 185 GPa (87). Compared with the spin-coating method, the spraying coating and drop-coating methods are more suitable for fabricating films with continuous size. Zhou et al. coupled the spray coating technique with electrochemical method to produce electrochemically reduced GO films. The thicknesses of films ranged from a single monolayer to several microns, achieved on various conductive and insulating substrates (100). Pham et al. mixed GO dispersion with excess amount of hydrazine monohydrate, followed by spray deposition on preheated substrate to obtain a large CRG film, carrying out the evaporation of solvents and reduction of GO simultaneously. The prepared CRG films had a low sheet resistance of 2.2×10^3 $\Omega/\square$ and a high transmittance of 84% at a wavelength of 550 nm (101). Rafiee et al. deposited graphene on a solid surface by drop-coating approach. Depending on that whether water, acetone, or a combination of water and acetone was used as solvents, the contact angle of the surface can be tailored over a wide range from superhydrophobic to superhydrophilic performances (107).

Vacuum filtration is another widely used methodology for the fabrication of GO or graphene films (96, 97, 108), and the thickness can be precisely controlled by the feed amount of dispersion. Ruoff and coworkers duplicated this technique to give GO paper with high tensile modulus (32 GPa, and the highest being 42 ± 2 GPa) and fracture strength, resulting from a unique interlocking-tile arrangement of the nanoscale GO sheets (109). Li's group prepared uniform graphene films using a similar filtration strategy of CRG dispersions. The resultant films or papers were bendable and exhibited a shiny metallic luster and conductivity of ~7200 S/m at room temperature (31). After moderate thermal annealing, the mechanical stiffness and strength as well as electrical conductivity of the films could be largely enhanced. The conductivity of the papers treated at 220°C and 500°C exhibited the conductivity of 118 and 351 S/cm, respectively. Specially, cell culture experiments indicated that the CRG papers were biocompatible and therefore could be suitable for biomedical applications (93). They also demonstrated that CRG can self-gel at the solution–filter membrane interface in an ordered manner during filtration. This unusual gelation behavior provided an amazingly simple strategy to create a new class of mechanically strong, highly conductive, and anisotropic hydrogel films (94). Subsequently, they investigated the potential application of a wet CRG membrane in nanofiltration for nanoparticles and dyes. The extent of corrugation of CRG can be readily controlled in the nanometer scale by the hydrothermal treatment to turn the water permeation rate and rejection yield of the CRG membrane. Size exclusion tests indicated the presence of channels larger than 13 nm in the 150°C treated membrane; while the channels of 120°C and 100°C treated membranes were found to be between 3 and 13 nm, the channels of 90°C treated membrane were less than 3 nm (110). On the contrary, Geim and coworkers reported that micrometer-thick GO membranes were completely impermeable to liquids, vapors, and gases (even helium), but solely allowed unimpeded evaporation of H_2O (H_2O permeates through the membranes at least 1010 times faster than He); after thermal reduction, the dried graphene membrane became impermeable to any substance including H_2O vapor molecules. They attributed these seemingly incompatible observations to a low-friction flow of a monolayer of water through 2D capillaries formed by closely packed graphene sheets. Diffusion of other molecules was blocked by reversible narrowing of the capillaries in low humidity and/or by their clogging with water (111). Recently, Han et al. designed ultrathin (22–53 nm thick) graphene membranes with 2D nanochannels by filtering the extremely dilute base-refluxing reduced GO (bRGO) dispersion on commercialized microfiltration membranes as supporting substrates. And the resulting ultrathin graphene membranes were successfully applied as nanofiltration membranes for water purification with the relative high pure water flux (as high as 21.8 L/m² h bar), high retention rates for organic dyes, and moderate retention rates for salts. Figure 2.22 showed the proposed rejection mechanism of separation process for 2D nanochannels in thin bRGO films (95).

Besides the aforementioned liquid assembly methods, CVD also plays an important role in the preparation of graphene films (82–84 112, 113). Although it requires special equipment and complex procedures, CVD technique is possibly the most promising scalable method to produce graphene films with extremely low resistance and high optical transparency.

2.4.3 Graphene Foams

Graphene-based 3D porous macrostructures, coined as foams, aerogels, frameworks, or sponges, are believed to be of great importance in various applications, such as sensors, actuators, energy storage devices, and high-performance nanocomposites. Until now, researchers have successfully constructed 3D graphene foams by different approaches. The basic properties of graphene aerogels are summarized in Table 2.5.

For the GO dispersion, Wang et al. first attempted to make graphene aerogels by just using simple freeze-drying or critical point drying techniques. This template-free assemble strategy is desirable for scalable manufactures. The mechanical and electrical properties of the resultant aerogels were improved by adding water-soluble polymers (i.e., PVA) to the dispersion precursor and by controlling thermal treatment, respectively. When acted as a supercapacitor, it showed a specific capacitance of 70–90 F/g (118). Liu's group also applied freeze-drying method to prepare 3D GO foams, followed by thermal reduction (119). The reduced graphene foam was applied as an anode

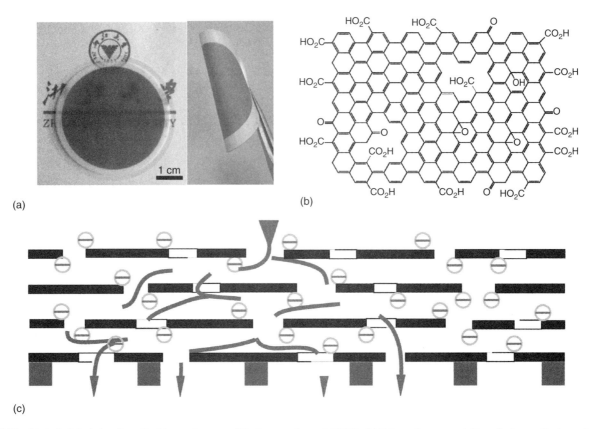

(a)

(b)

(c)

Figure 2.22 (a) A digital photo of an ultrathin graphene nanofiltration membrane (uGNM). (b) Schematic representation of a base-refluxing reduced GO (bRGO). (c) Schematic view for possible permeation route: water molecules go through the nanochannels of the uGNMs and the holes on the graphene sheets and at last reach the pores of the supporting membranes. The blank squares present the holes on the graphene sheets (black line). The edges of the bRGO and the periphery of the holes are negatively charged. (*Source*: Reproduced with permission from (95)).

TABLE 2.5 The Basic Properties of Graphene Aerogels Prepared by Various Methods

Fabrication Method	ρ (mg/cm^3)	Specific Surface Area	Conductance (S/m)	Additive	Application	Refs.
Template-directed CVD	0.18	–	0.2	–	–	114
	0.5	850 m^2/g	1000	PDMS	Flexible conductor	115
	0.5	850 m^2/g	1000	–	Gas sensor	116
	9.6	–	–	–	Thermal conductivity solid	117
Freeze-drying	0.5	–	–	–	Supercapacitor	118
	100	–	–	–	Anode for Li-ion batteries	119
	2	–	–	–	Catalyst of SO$_2$ oxidation	120
	0.5	–	12	–	–	121
	3	–	–	Ethylenediamine	–	122
	0.16	272	0.6	CNT	Oil absorption and phase-change energy storage materials	123
Sol–gel	2.1 ± 0.3	280	1200 ± 200	Pyrrole	Supercapacitor and oil absorption	124

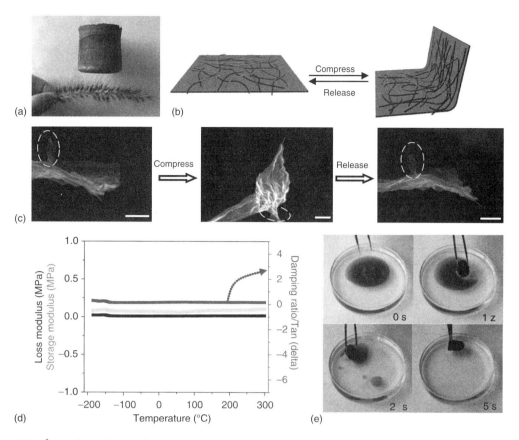

Figure 2.23 (a) A 100 cm³ ultra-flyweight superelastic carbonaceous aerogel (UFA) cylinder standing on a flowerlike dog's tail (*Setaira viridis* (L.) *Beauv*). (b, c) The schematic model and *in situ* SEM observations of a single graphene@CNTs cell wall in the process of compressing and releasing shows the elastic mechanism of the UFA: the deformation of cell walls rather than the sliding between them. (d) Temperature dependence of the storage modulus (light grey), loss modulus (black), and damping ratio (dark grey) of the UFA ($\rho = 7.6$ mg/cm³). (e) Absorption process of toluene (stained with Sudan Black B) on water by the UFA within 5 s. (*Source*: Reproduced with permission from (123)).

for Li-ion batteries, showing a good rate capability and cycle stability. The unique morphology and structure of graphene foam brought forth the better charge–discharge performance than previously reported powder-like graphene for anode-active materials (125). Long et al. unidirectionally froze GO dispersions from the bottom to the top by introducing a uniaxial thermal gradient, and obtained a perfectly exquisite 3D structure. The 3D material was used to explore the adsorption of SO_2 into the GO foam and the reaction between GO and SO_2 (120). Most recently, the combination of low density and elasticity is a new research target in the field of 3D aerogels. Li and coworkers designed regular cellular structures to realize the superelasticity of ultralight graphene-based aerogels, by careful control over the reduction process before freezing-drying. The obtained aerogels had biomimetic cellular structures similar to those in natural cork (121). Gao's group used 2D graphene sheets with giant size and 1D multiwalled carbon nanotubes (MWCNTs) as the two synergistic building blocks to scalably construct ultra-flyweight superelastic carbonaceous aerogel (Fig. 2.23) with the recorded lowest density down to 0.16 mg/cm³, with super reversible compressibility in the wide temperature range between the extremely low −196 °C in liquid nitrogen and high testable 300 °C and extremely high absorption capacities for organic liquids (123). More interestingly, the recovery under cycled compression of single MWCNT–graphene hybrid sheet was directly observed under scanning electron microscopy (SEM), which indicated that the origin of the superelasticity in this material was based on the elasticity of a single hybrid sheet rather than the sliding between sheets.

Another kind of template-free assemble strategy access to the aerogel is sol–gel method followed by freeze-drying or critical point drying (124, 126–130). Hu et al. achieved the similar cork-like structure by hydrothermal annealing and freeze-drying of GO dispersions in the presence of ethylenediamine (122). Sun's group made graphene aerogels by reducing a suspension of GO platelets followed by molding at 180 °C for 24 h. This shape-moldable and nanoporous material was of a high specific surface area and could be used as a versatile and recyclable sorbent material (131).

Template-directed CVD is also an important method to obtain graphene aerogels with high quality of constituent graphene sheets, bringing the merits of ordered and hierarchical structures (114–117, 132–134). The resultant graphene foam was of low density, with outstanding electrical and mechanical properties, and could be applied in flexible electronics and chemical sensors.

2.5 CONCLUSIONS AND FUTURE SCOPE

Since graphene was discovered in 2004, many researchers have focused on the large-scale production of processable graphene, and various functionalization methods have been established. Graphene and its derivatives performed rich chemistry and catalyzed the considerable advances in the functionalization of graphene, including covalent and noncovalent methodologies. For the limited stability of noncovalent functionalization and the deterioration of the graphene electronic properties in covalent functionalization, further researches are still required to establish new facile synthetic routines to keep balance between acceptable dispersibility and utmostly preserved electronic properties. The processable high-quality solvated graphene should further upgrade the combination properties of macroscopic graphene materials, such as fiber, film, paper, and aerogel, enabling their wider potential applications in daily production and life, including high-performance fiber, bulletproof vest, sewage disposal system, energy storage, catalysis, sensing, and composites.

REFERENCES

1. Novoselov KS, Jiang D, Schedin F, Booth TJ, Khotkevich VV, Morozov SV, Geim AK. Proc Natl Acad Sci U S A 2005;102:10451–10453.
2. Geim AK. Science 2009;324:1530–1534.
3. Wei W, Qu X. Small 2012;8:2138–2151.
4. Novoselov KS, Geim AK, Morozov SV, Jiang D, Zhang Y, Dubonos SV, Grigorieva IV, Firsov AA. Science 2004;306:666–669.
5. Singh V, Joung D, Zhai L, Das S, Khondaker SI, Seal S. Prog Mater Sci 2011;56:1178–1271.
6. Park S, Ruoff RS. Nat Nanotechnol 2009;4:217–224.
7. Zhu Y, James DK, Tour JM. Adv Mater 2012;24:4924–4955.
8. Forbeaux I, Themlin JM, Debever JM. Phys Rev B 1998;58:16396–16406.
9. Land TA, Michely T, Behm RJ, Hemminger JC, Comsa G. Surf Sci 1992;264:261–270.
10. Nagashima A, Nuka K, Itoh H, Ichinokawa T, Oshima C, Otani S. Surf Sci 1993;291:93–98.
11. Wu J, Pisula W, Mullen K. Chem Rev 2007;107:718–747.
12. Cai J, Ruffieux P, Jaafar R, Bieri M, Braun T, Blankenburg S. Nature 2010;466:470–473.
13. Hummers WS Jr,, Offeman RE. J Am Chem Soc 1958;80:1339.
14. Kuila T, Bose S, Mishra AK, Khanra P, Kim NH, Lee JH. Prog Mater Sci 2012;57:1061–1105.
15. Gorjizadeh N, Kawazoe Y. J Nanomater 2010;2010:513501.
16. Boukhvalov DW, Katsnelson MI. J Phys Condens Matter 2009;21:344205.
17. Stankovich S, Piner RD, Chen X, Wu N, Nguyen ST, Ruoff RS. J Mater Chem 2006;16:155–158.
18. Xu YX, Bai H, Lu GW, Li C, Shi GQ. J Am Chem Soc 2008;130:5856–5857.
19. Liu J, Yang W, Tao L, Li D, Boyer C, Davis TP. J Polym Sci Part A Polym Chem 2010;48:425–433.
20. Li XL, Wang XR, Zhang L, Lee S, Dai H. J Sci 2008;319:1229–1231.
21. Yang XY, Zhang ZY, Liu ZF, Ma YF, Huang Y, Chen YS. J Phys Chem C 2008;112:17554–17558.
22. Patil AJ, Vickery JL, Scott TB, Mann S. Adv Mater 2009;21:3159–3164.
23. Bai H, Li C, Wang X, Shi GQ. Chem Commun 2010;46:2376–2378.
24. Xu YX, Zhao L, Bai H, Hong WJ, Li C, Shi GQ. J Am Chem Soc 2009;131:13490–13497.
25. Liang YY, Wu DQ, Feng XL, Müllen K. Adv Mater 2009;21:1679–1683.
26. Zu SZ, Han BH. J Phys Chem C 2009;119:13651–13657.
27. Li XL, Zhang GY, Bai XD, Sun XM, Wang XR, Wang E, Dai HJ. Nat Nanotechnol 2008;3:538–542.
28. Chen RJ, Zhan YG, Wang DW, Dai HJ. J Am Chem Soc 2001;123:3838–3839.
29. Nakashima N, Tomonari Y, Murakami H. Chem Lett 2002;31:638–639.
30. Nakayama-Ratchford N, Bangsaruntip S, Sun X, Welsher K, Dai HJ. J Am Chem Soc 2007;129:2448–2449.
31. Li D, Müller MB, Gilje S, Kaner RB, Wallace GG. Nat Nanotechnol 2008;3:101–105.
32. Matsuo Y, Niwa T, Sugie Y. Carbon 1999;37:897–901.
33. Matsuo Y, Hatase K, Sugie Y. Chem Commun 1999;1:43–44.
34. Dreyer DR, Park S, Bielawski CW, Ruoff RS. Chem Soc Rev 2010;39:228–240.
35. Xu YF, Liu ZB, Zhang XL, Wang Y, Tian JG, Huang Y, Ma YF, Zhang XY, Chen YS. Adv Mater 2009;21:1275–1279.
36. Yang Q, Pan XJ, Clarke K, Li KC. Ind Eng Chem Res 2012;51:310–317.

37. Liu Z, Robinson JT, Sun XM, Dai HJ. J Am Chem Soc 2008;130:10876–10877.

38. Veca LM, Lu FS, Meziani MJ, Cao L, Zhang PY, Qi G, Qu LW, Shrestha M, Sun YP. Chem Commun 2009;2565–2567.

39. Ai W, Zhou WW, Du ZZ, Du YP, Zhang H, Jia XT, Xie LH, Yi MD, Yu T, Huang W J. Mater Chem 2012;22:23439–23446.

40. Zhou H, Wang X, Yu P, Chen XM, Mao LQ. Analyst 2012;137:305–308.

41. Yang HF, Li FH, Shan CS, Han DX, Zhang QX, Niu L, Ivaska A. J Mater Chem 2009;19:4632–4638.

42. Yang HF, Shan CS, Li FH, Han DX, Zhang QX, Niu L. Chem Commun 2009:3880–3882.

43. Hsiao MC, Liao SH, Yen MY, Liu PI, Pu NW, Wang CA, Ma CCM. ACS Appl Mater Interfaces 2010;2:3092–3099.

44. Matsuo Y, Fukunaga T, Fukutsuka T, Sugie Y. Carbon 2004;42:2117–2119.

45. Melucci M, Treossi E, Ortolani L, Giambastiani G, Morandi V, Klar P, Casiraghi C, Samorì P, Palermo V. J Mater Chem 2010;20:9052–9060.

46. Matsuo Y, Tabata T, Fukunaga T, Fukutsuka T, Sugie Y. Carbon 2005;43:2875–2882.

47. Stankovich S, Piner RD, Nguyen ST, Ruoff RS. Carbon 2006;44:3342–3347.

48. Xu C, Wu X, Zhu J, Wang X. Carbon 2008;46:365–389.

49. Zhang B, Chen Y, Zhuang XD, Liu G, Yu B, Kang ET, Zhu JH, Li Y. J Polym Sci Polym Chem 2010;48:2642–2649.

50. Lee SH, Dreyer DR, An J, Velamakanni A, Piner RD, Park S, Zhu Y, Kim SO, Bielawski CW, Ruoff RS. Macromol Rapid Commun 2010;31:281–288.

51. Gonçalves G, Marques PAAP, Barros-Timmons A, Bdkin I, Singh MK, Emami N, Grácio J. J Mater Chem 2010;20:9927–9934.

52. Fang M, Wang K, Lu H, Yang Y, Nutt S. J Mater Chem 2009;19:7098–7105.

53. Fang M, Wang KG, Lu HB, Yang YL, Nutt S. J Mater Chem 2010;20:1982–1992.

54. Ou BL, Zhou ZH, Liu QQ, Liao B, Yi SJ, Ou YJ, Zhang X, Li DX. Polym Chem 2012;3:2768–2775.

55. Deng Y, Li YJ, Dai J, Lang MD, Huang XY. J Polym Sci Polym Chem 2011;49:4747–4755.

56. Loh KP, Bao QL, Eda G, Chhowalla M. Nat Chem 2010;2:1015–1024.

57. Lerf A, He HY, Forster M, Klinowski J. Structure of graphite oxide revisited II. J Phys Chem 1998;102:4477–4482.

58. Buchsteiner A, Lerf A, Pieper J. J Phys Chem B 2006;110:22328–22338.

59. Chua K, Pumera M. Chem Soc Rev 2013;42:3222–3233.

60. Sarkar S, Bekyarova E, Haddon RC. Acc Chem Res 2013;45:673–682.

61. Huang P, Jing L, Zhu HR, Gao XY. Acc Chem Res 2013;46:43–52.

62. Park J, Yang MD. Acc Chem Res 2013;46:181–189.

63. Lomeda JR, Doyle CD, Kosynkin DV, Hwang WF, Tour JM. J Am Chem Soc 2008;130:16201–16206.

64. Zhong X, Lin J, Li SW, Niu ZY, Hu WQ, Li R, Ma JT. Chem Commun 2010;46:7340–7342.

65. Shen JF, Hu YZ, Li C, Qin C, Ye MX. Small 2009;1:82–85.

66. Kan LY, Xu Z, Gao C. Macromolecules 2011;44:444–452.

67. Wang BD, Yang D, Zhang JZ, Xi CB, Hu JH. J Phys Chem C 2011;115:24636–24641.

68. Wang DW, Li F, Jinping Zhao JP, Ren WC, Chen ZG, Tan J, Wu ZS, Ian Gentle I, Lu GQ, Cheng HM. ACS Nano 2009;3:1745–1752.

69. Yan J, Wei T, Shao B, Fan ZJ, Qian WZ, Zhang ML, Wei F. Carbon 2010;48:487–493.

70. Xu Z, Gao C. Macromolecules 2010;43:6716–6723.

71. Pham TA, Kumar NA, Jeong YT. Synth Met 2010;160:2028–2036.

72. Xu Z, Gao C. Nat Commun 2011;2:571.

73. Xu Z, Sun HY, Zhao XL, Gao C. Adv Mater 2013;25:188–193.

74. Xu Z, Zhang Y, Li PG, Gao C. ACS Nano 2012;6:7103–7113.

75. Cong HP, Ren XC, Wang P, Yu SH. Sci Rep 2012;2:613.

76. Hu XZ, Xu Z, Gao C. Sci Rep 2012;2:767.

77. Kou L, Gao C. Nanoscale 2013;5:4370–4378.

78. Wang RR, Sun J, Gao L, Xu CH, Zhang J. Chem Commun 2011;47:8650–8652.

79. Dong ZL, Jiang CC, Cheng HH, Zhao Y, Shi GQ, Jiang L, Qu LT. Adv Mater 2012;24:1856–1861.

80. Jang EY, Carretero-González J, Choi A, Kim WJ, Kozlov ME, Kim T, Kang TJ, Baek SJ, Kim DW, Park YW, Baughman RH, Kim YH. Nanotechnology 2012;23:235601.

81. Kovtyukhova NI, Ollivier PJ, Martin BR, Mallouk TE, Chizhik SA, Buzaneva EV, Gorchinskiy AD. Chem Mater 1999;11:771–778.

82. Kim KS, Zhao Y, Jang H, Lee SY, Kim JM, Kim KS, Ahn J-H, Kim P, Choi J-Y, Hong BH. Nature 2009;457:706–710.

83. Bae S, Kim H, Lee Y, Xu XF, Park JS, Zheng Y, Balakrishnan J, Lei T, Kim HR, Song YI, Kim YJ, Kim KS, Ozyilmaz B, Ahn JH, Hong BH, Iijima S. Nat Nanotechnol 2010;5:574–578.

84. Bi H, Fuqiang Huang FQ, Liang J, Xie XM, Jiang MH. Adv Mater 2011;23:3202–3206.

85. Becerril HA, Mao J, Liu Z, Stoltenberg RM, Bao Z, Chen YS. ACS Nano 2008;2:463–470.

86. Chang HX, Sun ZH, Yuan QH, Ding F, Tao XM, Yan F, Zheng ZJ. Adv Mater 2010;22:4872–4876.

87. Robinson JT, Zalalutdinov M, Baldwin JW, Snow ES, Wei ZQ, Sheehan P, Houston BH. Nano Lett 2008;8:3441–3445.

88. Biswas S, Drzal LT. Nano Lett 2009;9:167–172.

89. Zhao JP, Pei SF, Ren WC, Gao LB, Cheng HM. ACS Nano 2010;4:5245–5252.

90. Su CY, Lu AY, Xu YP, Chen FR, Khlobystov AN, Li LJ. ACS Nano 2011;5:2332–2339.

91. Chen Y, Zhang X, Yu P, Ma YW. Chem Commun 2009;4527–4529.

92. An SJ, Zhu Y, Lee SH, Stoller MD, Emilsson T, Park S, Velamakanni A, An J, Ruoff RS. J Phys Chem Lett 2010;1:1259–1263.

93. Chen H, Muller MB, Gilmore KJ, Wallace GG, Li D. Adv Mater 2008;20:3557–3561.

94. Yang X, Qiu L, Cheng C, Wu Y, Ma ZF, Li D. Angew Chem Int Ed 2011;50:7325–7328.

95. Han Y, Xu Z, Gao C. Adv Funct Mater 2013;23:3693–3700.

96. Wang SJ, Geng Y, Zheng QB, Kim J-K. Carbon 2010;48:1815–1823.

97. Geng J, Jung HJ. Phys Chem C 2010;114:8227–8234.

98. Lee DW, Hong TK, Kang D, Lee J, Heo M, Kim JY, Kim BS, Shin HS. J Mater Chem 2011;21:3438–3442.

99. Ji Q, Honma I, Paek SM, Akada M, Hill JP, Vinu A, Ariga K. Angew Chem Int Ed 2010;49:9737–9739.

100. Zhou M, Wang Y, Zhai Y, Zhai J, Ren W, Wang F, Dong S. Chem-Eur J 2009;15:6116–6120.

101. Pham VH, Cuong TV, Hur SH, Shin EW, Kim JS, Chung JS, Kim EJ. Carbon 2010;48:1945–1951.

102. Zheng Q, Ip WH, Lin X, Yousefi N, Yeung KK, Li Z, Kim J-K. ACS Nano 2011;5:6039–6051.

103. Ariga K, Hill JP, Ji Q. Phys Chem Chem Phys 2007;9:2319–2340.

104. Shen JF, Hu YZ, Li C, Qin C, Shi M, Ye MX. Langmuir 2009;25:6122–6128.

105. Hasan SA, Rigueur JL, Harl RR, Krejci AJ, Gonzalo-Juan I, Rogers BR, Dickerson JH. ACS Nano 2010;4:7367–7372.

106. Lee V, Whittaker L, Jaye C, Baroudi KM, Fischer DA, Banerjee S. Chem Mater 2009;21:3905–3916.

107. Rafiee J, Rafiee MA, Yu ZZ, Koratkar N. Adv Mater 2010;22:2151–2154.

108. Akhavan O. Carbon 2010;48:509–519.

109. Dikin DA, Stankovich S, Zimney EJ, Piner RD, Dommett GHB, Evmenenko G, Nguyen ST, Ruoff RS. Nature 2007;448:457–460.

110. Qiu L, Zhang XH, Yang WR, Wang YF, Simon GP, Li D. Chem Commun 2011;47:5810–5812.

111. Nair RR, Wu HA, Jayaram PN, Grigorieva IV, Geim AK. Science 2012;335:442–444.

112. Li X, Cai WW, An J, Kim S, Nah J, Yang DX, Piner R, Velamakanni A, Jung I, Tutuc E, Banerjee SK, Colombo L, Ruoff RS. Science 2009;324:1312–1314.

113. Joshi RK, Gomez H, Alvi F, Kumar AJ. Phys Chem C 2010;114:6610–6613.

114. Mechlenburg M, Schuchardt A, Mishra YK, Kaps S, Adelung R, Lotnyk A, Kienle L, Schulte K. Adv Mater 2012;24:3486–3490.

115. Chen ZP, Ren WC, Gao LB, Liu BL, Pei SF, Cheng HM. Nat Mater 2011;10:424–428.

116. Yavari F, Chen ZP, Thomas AV, Ren WC, Cheng HM, Koratkar N. Sci Rep 2011:1.

117. Pettes MT, Ji HX, Ruoff RS, Shi L. Nano Lett 2012;12:2959–2964.

118. Wang J, Ellsworth MW. ECS Trans 2009;19:241–247.

119. Zhou XF, Liu ZP. IOP Conf Ser: Mater Sci Eng 2011;18:062006.

120. Long Y, Zhang CC, Wang XX, Gao JP, Wang W, Liu Y. J Mater Chem 2011;21:13934–13941.

121. Qiu L, Liu JZ, Chang SLY, Wu YZ, Li D. Nat Commun 2013;3:1241.

122. Hu H, Zhao ZB, Wan WB, Gogotsi Y, Qiu JS. Adv Mater 2013;25:2219–2223.

123. Sun HY, Xu Z, Gao C. Adv Mater 2013;25:2554–2260.

124. Zhao Y, Hu CG, Hu Y, Cheng HH, Shi GQ, Qu LT. Angew Chem Int Ed 2012;51:11371–11375.

125. Yoo E, Kim J, Hosono E, Zhou H, Kudo T, Honma I. Nano Lett 2008;8:2277–2282.

126. Wu ZS, Yang SB, Sun Y, Parvez K, Feng XL, Müllen K. J Am Chem Soc 2012;134:9082–9085.

127. Zhang XT, Sui ZY, Xu B, Yue SF, Luo YJ, Zhan WC, Liu B. J Mater Chem 2011;21:6494–6497.

128. Zui ZY, Meng QH, Zhang XT, Ma R, Cao B. J Mater Chem 2012;22:8767–8771.

129. Jiang X, Ma Y, Li J, Fan Q, Huang WJ. Phys Chem C 2010;114:22462–22465.

130. Xie X, Zhou Y, Bi H, Yin K, Wan S, Sun L. Sci Rep 2013;3:2117.

131. Bi HC, Xie X, Yin KB, Zhou YL, Wan S, He LB, Xu F, Banhart F, Sun LT, Ruoff RS. Adv Funct Mater 2012;22:4421–4425.

132. Lee JS, Ahn HJ, Yoon JC, Jang JH. Phys Chem Chem Phys 2012;14:7938–7943.

133. Hu C, Zhao Y, Cheng H, Hu Y, Shi G, Dai L, Qu L. Chem Commun 2012;48:11865–11867.

134. Huang H, Chen P, Zhang X, Lu Y, Zhan W. Small 2013;9:1397–1404.

3

DEVICES BASED ON GRAPHENE AND GRAPHANE

XIAO-DONG WEN[1,2], TAO YANG[3], AND ERIC S. W. KONG[4]

[1]State Key Laboratory of Coal Conversion, Institute of Coal Chemistry, Chinese Academy of Sciences, Taiyuan, People's Republic of China
[2]Synfuels China Co Ltd, Huairou, Beijing People's Republic of China
[3]Baker Laboratory, Department of Chemistry and Chemical Biology, Cornell University, Ithaca, NY, USA
[4]Research Institute of Micro/Nanometer Science & Technology, Shanghai Jiao Tong University, Shanghai, People's Republic of China

3.1 GRAPHENE, A GROUNDBREAKING MATERIAL

Graphene is an allotrope of carbon, which is a flat monolayer of carbon atoms tightly packed into a two-dimensional (2D) honeycomb lattice (see Fig. 3.1), and is a basic building block for graphitic materials of all other dimensionalities. The Nobel Prize in Physics for 2010 was awarded to Andre Geim and Konstantin Novoselov at the University of Manchester "for groundbreaking experiments regarding the two-dimensional material graphene." As a material, it is completely – not only the thinnest ever but also the strongest. As a conductor of electricity, its performance is similar to copper. As a conductor of heat, it outperforms all other known materials. It is almost completely transparent, yet so dense that not even helium, the smallest gas atom, can pass through it. Carbon, the basis of all known life on earth, has surprised us once again.

Theoretically, graphene has been studied for more than 60 years (1); experimentally, it was synthesized in the year 2004 (2). After 2004, graphene has been studied extensively both in theory and experiment. Several papers are published every day, and many reviews (3–5), and book chapters (6, 7), have appeared in the last 8 years. Until 2012, more than 20,000 research papers were published (see Fig. 3.2 from http://www.silicene.com/).

These studies range from electronic to optical to excitonic to thermal to mechanical to spin transport to quantum hall effect to functionalization of graphene, and so on. Several potential applications for graphene are under development, and many more have been proposed. These include lightweight, thin, flexible, yet durable display screens, electric circuits, and solar cells, as well as various medical, chemical, and industrial processes enhanced or enabled by the use of new graphene materials. Here, we would like to point out that more 2D planar carbon structures, the so-called graphyne (8–11), have also been predicted to exist, but until now only molecular fragments have been synthesized (9). Unlike graphene, graphyne contains double and triple bonds and its atoms do not always have a hexagonal arrangement (see Fig. 3.3). Indeed, there may be a vast number of possible graphynes, each with the double and triple bonds in slightly different arrangements. Theorists have been studying graphynes since 1980s. Here, we would not like to focus on the topic of graphyne.

Graphene makes experiments possible that give new twists to the phenomena in quantum physics. Also, a vast variety of practical applications now appear possible, including the creation of new materials and the manufacture of innovative electronics. Graphene transistors are predicted to be substantially faster than today's silicon transistors and result in more efficient computers. As it is practically transparent and a good conductor, graphene is suitable for producing transparent touchscreens, light panels, and maybe even solar cells. When mixed with plastics, graphene can turn them into conductors of electricity while making them more heat resistant and mechanically robust. This resilience can be utilized in new super strong materials, which are also thin, elastic, and lightweight. In future, satellites, airplanes, and cars could be manufactured out of the new composite materials (12).

In this chapter, we will mainly focus on the devices based on graphene, an important application of graphene in our world. We will start with the synthesis and the electronic properties of graphene as these are the basis of application.

3.1.1 The Discovery and Synthesis of Graphene

Graphene had already been studied theoretically in 1947 by P.R. Wallace (1) as a text book example for calculations in solid-state physics. He predicted the electronic structure and noted the linear dispersion relation. In 1956, the wave equation for excitations was written down by

Nanomaterials, Polymers and Devices: Materials Functionalization and Device Fabrication, First Edition. Edited by Eric S. W. Kong.
© 2015 John Wiley & Sons, Inc. Published 2015 by John Wiley & Sons, Inc.

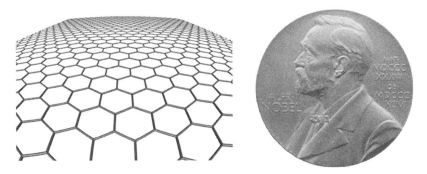

Figure 3.1   Two-dimensional graphene honeycomb lattice.

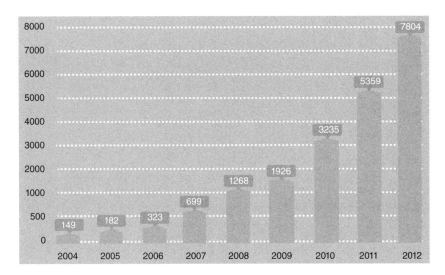

Figure 3.2   Published papers on graphene.

J.W. McClure (14), and in 1984, the similarity to the Dirac equation was discussed by G.W. Semenoff (15); also by DiVincenzo and Mele (16). Though graphene-like structures were already predicted over 60 years, there were experimental difficulties in isolating single layers in such a way that electrical measurements could be performed on them, and there were doubts that this was practically possible.

It came as a surprise to the physics community when Andre Geim, Konstantin Novoselov, and their collaborators from the University of Manchester (UK) and the Institute for Microelectronics Technology in Chernogolovka (Russia) presented their results on graphene structures. They published their results in October 2004 in Science (2). In their study, they described the fabrication, identification, and Atomic Force Microscopy (AFM) characterization of graphene. They used a simple but effective mechanical exfoliation method for extracting thin layers of graphite from a graphite crystal with Scotch tape and then transferred these layers to a silicon substrate. This method was first suggested and tried by R. Ruoff's group (17), who were, however, not able to identify any monolayers. The Manchester group succeeded by using an optical method with which they were able to identify fragments made up of only a few layers. An AFM picture of one such sample is shown in Figure 3.4 c. In some cases, these flakes were made up of a single layer, that is, graphene was identified. Furthermore, they managed to pattern samples containing only a few layers of graphene into a Hall bar and connect electrodes to it.

So far, Graphene has been synthesized in various ways and on different substrates, as many groups have reviewed (18, 19). The synthesis approach of graphene includes the exfoliation and cleavage technique, thermal chemical vapor deposition (CVD) techniques, plasma-enhanced chemical vapor deposition techniques, chemical methods including carbon dioxide reduction and graphene oxide reduction methods, thermal decomposition of silicon carbide (SiC), thermal decomposition on other substrates (such as metals), unzipping CNTs, and so on. In this chapter, we would like to introduce some of these techniques followed by a step-by-step procedure (20).

1. *Exfoliated Graphene*. Graphene was first exfoliated mechanically from graphite in 2004, as discussed earlier. This simple, low-budget technique has been widely credited for making graphene. One can follow the step-by-step guide to make graphene layers:

   □ Take a piece of scotch tape and some highly ordered graphite.
   □ Put the piece of scotch tape on the graphite and then rip it off.

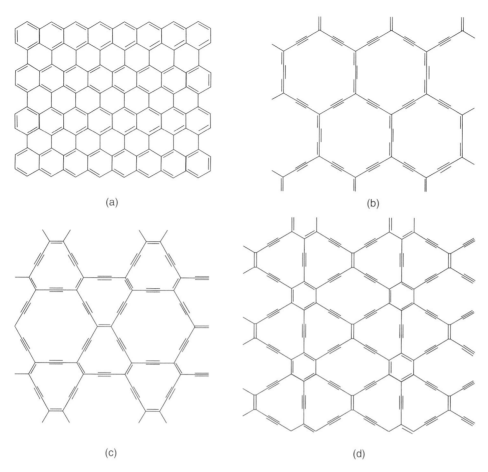

Figure 3.3 Structures of graphene and graphynes. (a) Graphene. (b) α-graphyne. (c) β-graphyne. (d) 6,6,12-graphyne. In all these cases, only one resonance structure, that is, one of several equivalent Lewis structures, is shown.

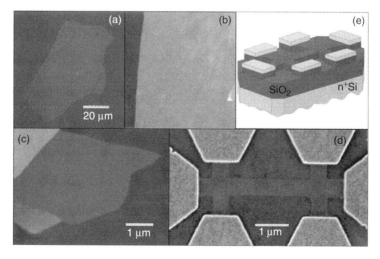

Figure 3.4 Graphene films. (a) Photograph of a relatively large multilayer graphene flake with thickness ~3 nm on top of an oxidized Si wafer. (b) Atomic force microscope (AFM) image of 2 μm by 2 μm area of this flake near its edge. gray for graphene and dark gray for SiO_2 surface. (c) AFM image of single-layer graphene. For details of AFM imaging of single-layer graphene, please check the original figure in color. (d) Scanning electron microscope image of one of our experimental devices prepared from FLG. (e) Schematic view of the device in (d).

- □ Repeat a lot of times.
- □ Deposit on a SiO_2 substrate.
- □ Look for graphene layers.

The graphene obtained by this approach produced very uneven films, meaning that it is very time-consuming to find where the graphene (as opposed to graphite) is. Also, it is labor-intensive, which means that it is used for education or research purpose but not for industry.

2. *CVD Graphene.* Graphene and few-layer graphene have been grown by CVD from C-containing gases on catalytic metal surfaces and/or by surface segregation of C dissolved in the bulk of such metals. Depending on the solubility of C in the metal, the former or the latter can be the dominant growth process, or they can coexist. The method is good for making large amounts of film. The step-by-step guide would be:

- □ Pump in hydrocarbon gas (usually CH_4), sometimes under vacuum.
- □ Watch as the carbon arranges into graphene on the surface, such as Ni, Fe, Pt, Pd, Cu, or Co.
  These graphenes often produced unpredictably arranged multilayers, with defects being linked to the substrate used. In addition, the metal surfaces might affect the graphene layers.

3. *Epitaxial Graphene.* The graphitization of hexagonal SiC crystals during annealing at high temperatures in vacuo was reported (21). Under such annealing conditions, the top layers of SiC crystals undergo thermal decomposition, Si atoms desorb, and the carbon atoms remaining on the surface rearrange and re-bond to form epitaxial graphene layers. This technique can produce the most even films among these methods and is a simple process. One can obtain these even films following this guide:

- □ Take a SiC wafer.
- □ Heat it to 1100 °C.

The aforementioned approach produces few-layered graphene, usually 5–10 layers, and produces epitaxial graphene with various dimensions depending on the size of the SiC substrate (wafer). The face of the SiC used for graphene formation, that is, silicon- or carbon-terminated, highly influences the thickness, mobility, and carrier density of the graphene.

4. *Graphene Oxide Approach.* The general idea in this process is to dissolve carbon atoms inside a transition metal melt at a certain temperature and then allow the dissolved carbon to precipitate out at lower temperatures as single-layered graphene. This method is more versatile than epitaxial methods, less time-consuming, and easier to scale up than exfoliation methods. One can try it, as shown below:

- □ Oxidize highly ordered graphite with HNO_3 and H_2SO_4
- □ Sonicate it, and then purify via centrifuging
- □ Reduce to graphene material, and then put on substrate, or
- □ Put on substrate, and then reduce to graphene material.

However, it is difficult to keep solution from re-aggregating into graphite; after reduction, graphene layers are still partially oxidized, potentially changing electronic, optical, and mechanical properties.

3.1.2   Properties of Graphene

The carbon–carbon bond length in graphene is about 0.142 nm. Graphene sheets stack to form graphite with an interplanar spacing of 0.335 nm. Graphene is the basic structural element of some carbon allotropes including graphite, charcoal, carbon nanotubes, and fullerenes. It can also be considered as an indefinitely large aromatic molecule, the limiting case of the family of flat polycyclic aromatic hydrocarbons.

3.1.2.1   Electronic Properties   Graphene differs from most conventional three-dimensional (3D) materials. Intrinsic graphene is a semimetal or a zero-gap semiconductor. Understanding the electronic structure of graphene is the starting point for finding the band structure of graphite. It was realized as early as 1947 by P. R. Wallace that the E–k relation is linear for low energies near the six corners of the 2D hexagonal Brillouin zone, leading to zero effective mass for electrons and holes.

Its Fermi surface is characterized by six double cones, as shown in Figure 3.5 (left) (22); a more readable density of state and band structures are shown on the right. In intrinsic (undoped) graphene, the Fermi level is situated at the connection points of these cones. The graphene is the so-called semimetal instead of semiconductor, insulator, or metal. As the density of states of the material is zero at that point, the electrical conductivity of intrinsic graphene is quite low and is of the order of the conductance quantum $\sigma \sim e^2/h$; the exact prefactor is still debated. The Fermi level can however be changed by an electric field so that the material becomes either n-doped (with electrons) or p-doped (with holes) depending on the polarity of the applied field. Graphene can also be doped by adsorbing, for example, water or ammonia on its surface. The electrical conductivity for doped graphene is potentially quite high, at room temperature it may even be higher than that of copper. Close to the Fermi level, the dispersion relation for electrons and holes is linear. Since the effective masses are given by the curvature of the energy bands, this corresponds to zero effective mass. In contrast to low-temperature 2D systems based on semiconductors, graphene maintains its 2D properties at room temperature.

In 2009, Castro Neto et al. (5) reviewed the basic theoretical aspects of graphene with unusual 2D Dirac-like electronic excitations. The Dirac electrons can be controlled by application of external electric and magnetic fields, or by altering sample geometry and/or topology. The

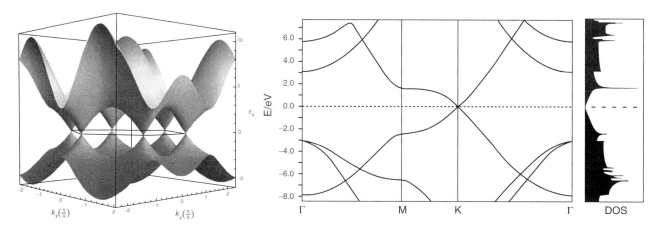

Figure 3.5   (a) The energy, E, for the excitations in graphene as a function of the wave numbers, k_x and k_y, in the x- and y directions. The black line represents the Fermi energy for an undoped graphene crystal. Close to this Fermi level, the energy spectrum is characterized by six double cones where the dispersion relation (energy vs momentum, $\hbar k$) is linear. (b) The computed band structure and density of states of graphene using DFT.

Dirac electrons behave in unusual ways in tunneling, confinement, and the integer quantum Hall effect. In their review, the electronic properties of graphene stacks were discussed and vary with stacking order and number of layers. And they also summarized that edge surface states in graphene depend on the edge termination zigzag or armchair and affect the physical properties of nanoribbons. Different types of disorder modify the Dirac equation leading to unusual spectroscopic and transport properties. The effects of electron–electron and electron–phonon interactions in single-layer and multilayer graphenes were reviewed.

Graphene is a unique system in many ways. It is truly 2D, has unusual electronic excitations described in terms of Dirac fermions that move in a curved space, is an interesting mix of a semiconductor zero density of states and a metal gaplessness, and has properties of soft matter. The electrons in graphene seem to be almost insensitive to disorder and electron–electron interactions and have very long mean free paths. Hence, graphene's properties are different from what is found in usual metals and semiconductors. Graphene has also a robust but flexible structure with unusual phonon modes that do not exist in ordinary 3D solids. In some sense, graphene brings together issues in quantum gravity and particle physics, and also from soft and hard condensed matter. Interestingly enough, these properties can be easily modified with the application of electric and magnetic fields, addition of layers, control of its geometry, and chemical doping. Moreover, graphene can be directly and relatively easily probed by various scanning probe techniques from mesoscopic down to atomic scales, because it is not buried inside a 3D structure. This makes graphene be one of the most versatile systems in condensed-matter research.

3.1.2.2   Density of Graphene   The unit hexagonal cell of graphene contains two carbon atoms and has an area of $0.052\ nm^2$. We can thus calculate its density as being $0.77\ mg/m^2$. A hypothetical hammock measuring $1\ m^2$ made from graphene would thus weigh 0.77 mg.

3.1.2.3   Optical Transparency of Graphene   Graphene is almost transparent, absorbs only 2.3% of the light intensity, and is independent of the wavelength in the optical domain. This number is given by $\pi\alpha$, where α is the fine structure constant. Thus suspended graphene does not have any color. This is "a consequence of the unusual low-energy electronic structure of monolayer graphene that features electron and hole conical bands meeting each other at the Dirac point … is qualitatively different from more common quadratic massive bands" (23).

3.1.2.4   Strength of Graphene   Graphene has a breaking strength of $42\ N/m^2$. Steel has a breaking strength in the range of 250–1200 MPa $= 0.25$–$1.2 \times 10^9\ N/m^2$. For a hypothetical steel film of the same thickness as graphene (which can be taken to be $3.35\ \text{Å} = 3.35 \times 10^{-10}\ m$, i.e., the layer thickness in graphite), this would give a 2D breaking strength of 0.084–$0.40\ N/m^2$. Thus, graphene is 100 times more stronger than the strongest steel.

In our $1\ m^2$ hammock tied between two trees you could place a weight of ~ 4 kg before it would break. It should thus be possible to make an almost invisible hammock out of graphene that could hold a cat without breaking. The hammock would weigh less than 1 mg, corresponding to the weight of one of the cat's whiskers.

3.1.2.5   Electrical Conductivity of Graphene   The sheet conductivity of a 2D material is given by $\sigma = en\mu$. The mobility is theoretically limited to $\mu = 200{,}000\ cm^2/Vs$ by acoustic phonons at a carrier density of $n = 1012\ cm^{-2}$. The 2D sheet resistivity, also called the resistance per square, is then 31 Ω. Our fictional hammock measuring $1\ m^2$ would thus have a resistance of 31 Ω. Using the layer thickness, we get a bulk conductivity of $0.96 \times 106\ \Omega^{-1}cm^{-1}$ for graphene. This is somewhat higher than the conductivity of copper, which is $0.60 \times 106\ \Omega^{-1}cm^{-1}$. More details can be obtained from the review paper by Das Sarma et al. (24).

Despite the zero carrier density near the Dirac points, graphene exhibits a minimum conductivity on the order of $4e^2/h$. The origin of this minimum conductivity is still unclear. Several theories suggest that the minimum conductivity should be $4e^2/\pi h$; however, most measurements are of order $4e^2/h$ or greater (3) and depend on impurity concentration (25).

3.1.2.6   Thermal Conductivity   The thermal conductivity of graphene is dominated by phonons and has been measured to be ~5000 W/mK. Copper at room temperature has a thermal conductivity of 401 W/mK. Thus, graphene conducts heat 10 times better than copper.

Besides these unique properties, graphene also exhibits the interesting saturable absorption, nonlinear Kerr effect, excitonic, spin transport, strong magnetic fields, and so on. The outstanding properties of graphene make it attractive for many applications in our life. Next, we will learn more about the applications of graphene.

3.1.3   Applications of Graphene: Devices Based on Graphene

Since graphene's discovery 9 years ago, several potential applications of graphene are under development and an enormous number of potential applications have been identified. These include lightweight, thin, flexible, yet durable display screens, electric circuits, and solar cells, as well as various medical, chemical, and industrial processes enhanced or enabled by the use of new graphene materials.

3.1.3.1   Graphene Transistors   Due to its high electronic quality, graphene has also attracted the interest of technologists who see them as a way of constructing ballistic transistors. Graphene exhibits a pronounced response to perpendicular external electric fields, allowing one to build field-effect transistors (FETs). In 2004, Novoselov and his coworkers (2) demonstrated FETs with a "rather modest" on–off ratio of ~30 at room temperature. In 2006, Georgia Tech researchers, led by Walter de Heer, announced that they had successfully built an all-graphene planar FET with side gates (26). Their devices showed changes of 2% at cryogenic temperatures. The first top-gated FET (on–off ratio of <2) was demonstrated by researchers of AMICA and RWTH Aachen University in 2007 (see Fig. 3.6) (27), which exhibits extraordinary carrier mobility that exceeds the universal mobility of silicon-based materials.

Intrinsic graphene is a semimetal, implying a very small current on–off ratio of graphene transistors. If graphene devices were to replace conventional Si devices, new approaches in bandgap engineering or circuit design would be needed, with the most attractive possibility being to implement the same functionality with fewer transistors. Facing the fact that current graphene transistors show a very poor on–off ratio, researchers are trying to find ways for improvement. In 2008, researchers of AMICA and University of Manchester demonstrated a new switching effect in graphene field-effect devices. This switching effect is based on a reversible chemical modification of the graphene layer and gives an on–off ratio of greater than six orders of magnitude. These reversible switches could potentially be applied to nonvolatile memories (28). The device structure shown in Figure 3.7 is identical to conventional silicon-on-insulator and graphene MOSFETs and may potentially be seen as candidates for future nonvolatile memory applications. Their experiments show good cyclability and reset times of 80 μs. However, to better understand the involved processes, it is necessary to improve the switching times and reliability, including finding a reliable source for the species involved in the switching mechanism.

In 2009, researchers at the Politecnico di Milano demonstrated four different types of logic gates, each composed of a single graphene transistor (Fig. 3.8 left) (29). Single-transistor operation is obtained in a circuit designed to exploit the charge neutrality point of graphene to perform Boolean logic. Although further improvements are required to approach the performance of conventional Si CMOS logic gates, the fabricated logic gates offer an attractive alternative to conventional gates due to their minimal transistor count. In the same year, the Massachusetts Institute of Technology researchers (30) demonstrated for the first time usage of the ambipolar transport properties of graphene in the fabrication of nonlinear electronics for full-wave rectification and frequency doubling of electrical signals (Fig. 3.8 right). Although the measurement setup limited the input frequency to 10 kHz, these devices are expected to work at much higher frequencies due to the outstanding transport properties of graphene and the simplicity of the proposed circuit, which significantly reduce parasitic capacitances and resistances. They also pointed out that the ambipolar FETs might lead the next revolution in communications technology.

Figure 3.6   SEM image of a graphene transistor.

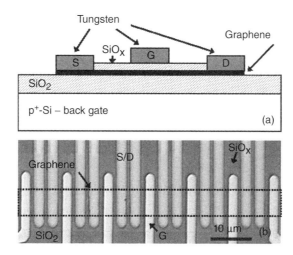

Figure 3.7 (a) Schematic of a double-gated graphene FED used in the experiments. (b) Optical micrograph of several FEDs fabricated from one graphene flake.

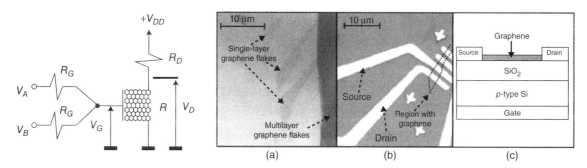

Figure 3.8 Left: two-input (*A* and *B*) logic gate incorporating one monolayer graphene transistor. *R* is the output resistance of the graphene transistor, which depends on the gate voltage V_G. Right: Structure of the fabricated G-FET. (a) Optical micrograph showing single-layer graphene flakes. (b) Optical micrograph of the structure of the final device. The device is back gated through the p-type Si wafer. (c) Schematic of the vertical structure of the device.

Although these graphene chips open up a range of new applications, their practical use is limited by a very small voltage gain (typically, the amplitude of the output signal is about 40 times less than that of the input signal). Moreover, none of these circuits was demonstrated to operate at frequencies higher than 25 kHz. In the same year, tight-binding numerical simulations obtained by means of the open-source software NanoTCADViDES had demonstrated that the bandgap induced in graphene bilayer field-effect transistors is not sufficiently large for high-performance transistors for digital applications, but it can be sufficient for ultralow-voltage applications, when exploiting a tunnel-FET architecture (Fig. 3.9) (31).

In February 2010, researchers at IBM reported that they have been able to create graphene transistors (Fig. 3.10) with an on and off rate of 100 gigahertz, far exceeding the rates of previous attempts, and exceeding the speed of silicon transistors with an equal gate length. Uniform and high-quality graphene wafers were synthesized by thermal decomposition of a SiC substrate. The graphene transistor itself utilized a metal top-gate architecture and a novel-gate insulator stack involving a polymer and a high dielectric constant oxide. The 240 nm graphene transistors made at IBM were made using extant silicon-manufacturing equipment, meaning that for the first time graphene transistors are a conceivable-though still fanciful-replacement for silicon (32, 33). This accomplishment is a key milestone for the Carbon Electronics for RF Applications (CERA) program funded by DARPA, in an effort to develop next-generation communication devices.

In May 2012, Samsung Advanced Institute of Technology developed a new transistor structure utilizing graphene (see Fig. 3.11). Large modulation on the device current (on–off ratio of 10^5) was achieved by adjusting the gate voltage to control the graphene–silicon Schottky barrier without degrading its mobility. This device was named "barristor," after its barrier-controllable feature. By using barristors, the most basic logic gate (inverter) and logic circuits (half-adder) were fabricated, and basic operation (adding) was demonstrated (34, 35). Furthermore, the on–off ratio and the current density can be improved with well-developed semiconductor processes because there is no fundamental (or structural) limit.

In July 2012, researchers from the UCLA reported a scalable method for fabricating self-aligned graphene transistors with transferred gate stacks (see Fig. 3.12). By performing the conventional lithography, deposition and etching steps on a sacrificial substrate before integrating with large-area graphene through a physical transferring process, the new approach addresses and overcomes the challenges of conventional

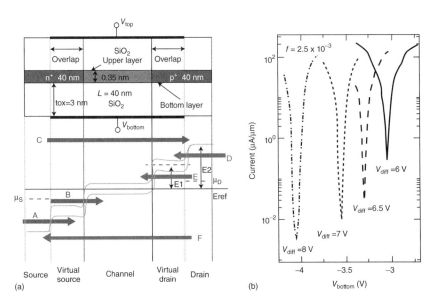

Figure 3.9   (a) Sketch of the double-gate bilayer graphene TFET: the channel length is 40 nm, and n^+ and p^+ reservoirs are 40 nm long with molar fraction f. The device is embedded in 3 nm-thick SiO_2 dielectric. V_{top} and V_{bottom} are the voltages applied the top and bottom gate, respectively. Gate overlap has also been considered. Below, band edge profile of the device in the OFF state; (b) Transfer characteristics of the double-gate BG-TFET for different V_{diff}. f is equal to 2.5×10^{-3} and VDS = 0.1 V.

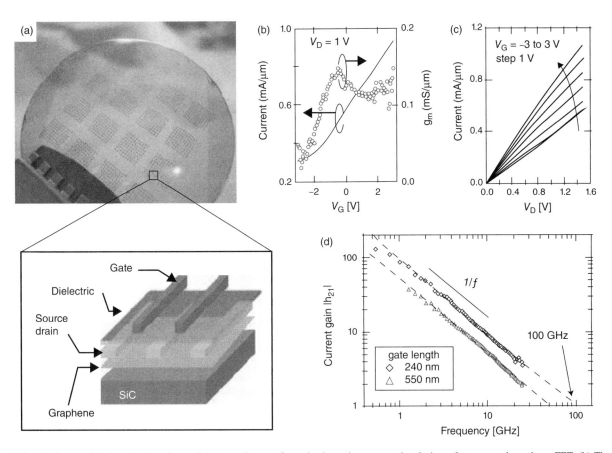

Figure 3.10   (a) Image of devices fabricated on a 2-inch graphene wafer and schematic cross-sectional view of a top-gated graphene FET. (b) The drain current, ID, of a graphene FET (gate length LG = 240 nm) as a function of gate voltage at drain bias of 1 V with the source electrode grounded. The device transconductance, gm, is shown on the right axis. (c) The drain current as a function of VD of a graphene FET (LG = 240 nm) for various gate voltages. (d) Measured small-signal current gain |h21| as a function of frequency f for a 240-nm-gate ($\diamond$) and a 550-nm-gate ($\triangle$) graphene FET at VD = 2.5 V. Cutoff frequencies, fT, were 53 and 100 GHz for the 550-nm and 240-nm devices, respectively.

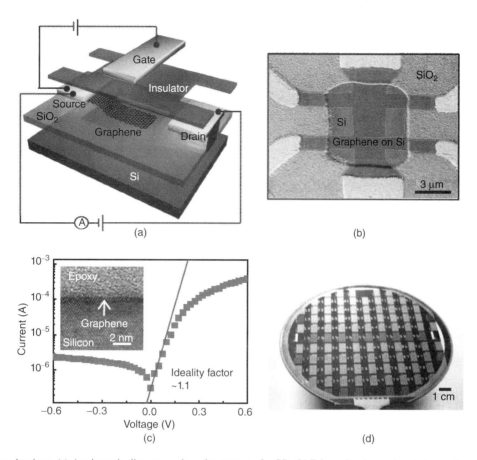

Figure 3.11 Graphene barristor. (a) A schematic diagram to show the concept of a GB. (b) False-colored scanning electron microscopy image of the GB before the top-gate fabrication process. (c) Current versus bias voltage characteristic of a GB at a fixed gate voltage $V_{gate} = 0$ V, showing a Schottky diode characteristic. The inset shows a TEM image of graphene/silicon junction. No native oxide or defect is seen in the image. (d) A photograph of ~2000 GB arrays implemented on a 6-inch wafer.

fabrication. With a damage-free transfer process and a self-aligned device structure, this method has enabled self-aligned graphene transistors with the highest cutoff frequency to date – greater than 400 GHz (36, 37).

In 2013, researchers demonstrated that the graphene transistor under a magnetic field is capable of detecting THz and IR waves in a very wide band of frequencies (0.76–33 THz) and that the detection frequency is tuned by changing the magnetic field (38). Researchers at University of Manchester created a terahertz-speed transistor with bistable characteristics, which means that the device can spontaneously switch between two electronic states. The device consists of two layers of graphene separated by an insulating layer of boron nitride just a few atomic layers thick (see Fig. 3.13). Electrons move through this barrier by quantum tunneling. These new transistors exhibit "negative differential conductance," whereby the same electrical current flows at two different applied voltages (39).

Electronic engineers at Japan's GNC and AIST research centers have successfully created CMOS-compatible, 30 nm programmable graphene transistor (see Fig. 3.14) that are constructed and operated in a way that redefines 50 years of transistor development. These graphene transistors can be built using conventional CMOS processes, and could potentially be many times smaller, hundreds of times faster, and consume much less power than silicon transistors (40). In the developed transistor, two electrodes and two top gates are placed on graphene, and graphene between the top gates is irradiated with a helium ion beam to introduce crystalline defects. Gate biases are applied to the two top gates independently, allowing carrier densities in the top-gated graphene regions to be effectively controlled. An electric current on–off ratio of approximately four orders of magnitude was demonstrated at 200 K (approximately −73 °C). In addition, its transistor polarity can be electrically controlled and inverted, which to date has not been possible for transistors. This technology can be used in the conventional production technology of integrated circuits based on silicon and is expected to contribute to the realization of ultralow-power-consumption electronics by reducing operation voltage in the future.

There are many reviews on graphene transistor (41, 42). To conclude the section, graphene has changed from being the exclusive domain of condensed-matter physicists to being explored by those in the electron-device community. In particular, graphene-based transistors have developed rapidly and are now considered an option for post-silicon electronics. However, many details about the potential performance of graphene transistors in real applications remain unclear. The excellent mobility of graphene may not, as is often assumed, be its most compelling feature from a device perspective. Rather, it may be the possibility of making devices with channels that are extremely thin

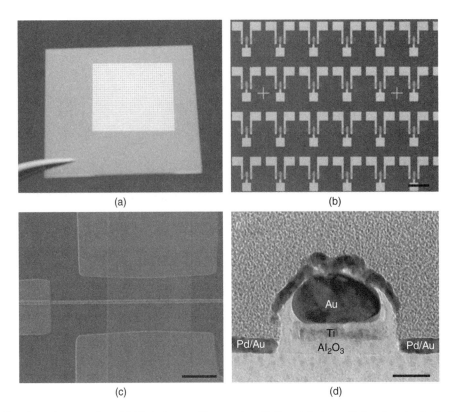

Figure 3.12   The self-aligned graphene transistor. (a) Photo image of large-scale self-aligned devices with transferred gate stacks on glass substrate. (B) Optical image of self-aligned graphene transistors on 300-nm SiO_2/Si substrate. Scale bar, 100 μm. (c) SEM image of a graphene transistor with transferred gate stack. Scale bar, 2 μm. (d) Cross-sectional TEM image of the overall device layout. Scale bar, 30 nm.

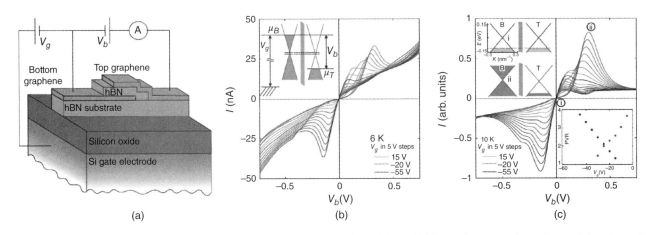

Figure 3.13   Graphene-BN resonant tunneling transistor. (a) Schematic diagram of the devices. (b) Measured current–voltage characteristics of one of the devices (device A) at 6 K. (c) Theoretical simulation of device A obtained by using the Bardeen model and including the effect of doping in both graphene electrodes.

that will allow graphene field-effect transistors to be scaled to shorter channel lengths and higher speeds without encountering the adverse short-channel effects that restrict the performance of existing devices. Outstanding challenges for graphene transistors include opening a sizeable and well-defined bandgap in graphene, making large-area graphene transistors that operate in the current saturation regime, and fabricating graphene nanoribbons with well-defined widths and clean edges.

3.1.3.2   *Integrated Circuits Based on Graphene*   Graphene has the ideal properties to be an excellent component of integrated circuits. Graphene has a high carrier mobility, as well as low noise, allowing it to be used as the channel in a field-effect transistor. The issue is that single sheets of graphene are hard to produce, and even harder to make on top of an appropriate substrate. Researchers are looking into

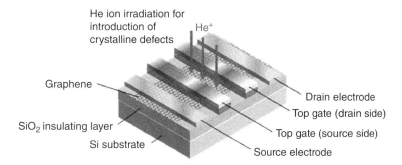

Figure 3.14   CMOS-compatible, 30 nm programmable graphene transistor.

methods of transferring single graphene sheets from their source of origin (mechanical exfoliation on SiO_2/Si or thermal graphitization of a SiC surface) onto a target substrate of interest (43).

In 2008, the smallest transistor so far, that is, one atom thick and 10 atoms wide, was made of graphene (44). IBM announced on December 2008 that they fabricated and characterized graphene transistors operating at GHz frequencies (see Fig. 3.15) (45, 46). The short-circuit current gain showed the ideal 1/*f* frequency dependence, confirming the measurement quality and the FET-like behavior for graphene devices. A peak cutoff frequency fT as high as 26 GHz was measured for a 150 nm-gate graphene transistor, establishing the state of the art for graphene transistors. These results also indicate that if the high mobility of graphene can be preserved during the device fabrication process, a cutoff frequency approaching THz may be achieved for graphene FET with a gate length of just 50 nm and a carrier mobility of 2000 cm²/Vs.

In May 2009, an n-type transistor was announced meaning that both n- and p-type transistors have now been created with graphene (see Fig. 3.16) (47). A functional graphene-integrated circuit was also demonstrated – a complementary inverter consisting of one p- and one n-type graphene transistor (48). The voltage transfer characteristics of the fabricated inverter exhibit clear voltage inversion. Dynamic pulse measurements display characteristic NOT functionality when the inverter is operated with a CMOS input voltage swing and supply voltage. Although application of the present inverter is limited by power consumption and inability for direct cascading, its realization demonstrates feasibility of using graphene as a substrate on which complete electronic circuits can be integrated. However, this inverter also suffered from a very low voltage gain.

According to a January 2010 report (49), graphene was epitaxially grown on SiC in a quantity and quality suitable for mass production of integrated circuits (see Fig. 3.17). They reported quantum Hall resistance quantization accurate to a few parts in a billion at 300 mK in a large-area epitaxial graphene sample. Several more devices have been studied at 4.2 K, demonstrating quantization within accuracy of some tens in 10^9, confirming the robustness of the QHE in graphene synthesized on the silicon-terminated face of SiC. This remarkable precision constitutes an improvement of four orders of magnitude over the best previous results obtained in exfoliated graphene, and is similar to the accuracy achieved in the established semiconductor resistance standards.

In June 2011, IBM researchers announced that they had succeeded in creating the first graphene-based integrated circuit (see Fig. 3.18), a broadband radio mixer (50) – another step toward overcoming the limitations of silicon and a potential path to flexible electronics. The circuit, built on a wafer of silicon carbide, consists of FETs made of graphene, a highly conductive chicken-wire-like arrangement of carbon that is a single atomic layer thick. The IC also includes metallic structures, such as on-chip inductors and the transistors' sources and drains. These graphene circuits exhibit outstanding thermal stability with little reduction in performance (<1 dB) between 300 and 400 K. These results open up possibilities of achieving practical graphene technology with more complex functionality and performance. Researchers proposed that graphene, which has the potential to make transistors that operate at terahertz speeds, could one day supplant silicon as the basis for computer chips.

Considering the still early stage of graphene electronics, this is already a rather impressive progress. However, although the speed of single graphene transistors is competitive to the current state-of-the-art, the performance integrated circuits based on graphene transistors lack far behind as they show functionality only at a couple of GHz. Additionally, the complexity of graphene-based integrated circuits is still very limited, involving only one or two graphene transistors.

In 2012, researches in University of Michigan reported a fully bendable all-graphene modulator circuit with the capability to encode a carrier signal with quaternary digital information for the first time (see Fig. 3.19) (51). By exploiting the ambipolarity and the nonlinearity in a graphene transistor, they demonstrated two types of quaternary modulation schemes: quaternary amplitude-shift keying and quadrature phase-shift keying. Remarkably, both modulation schemes can be realized with just 1 and 2 all-graphene transistors, respectively, representing a drastic reduction in circuit complexity when compared with conventional modulators. In addition, the circuit is not only flexible but also highly transparent (~95% transmittance) owing to their all-graphene design with every component (channel, interconnects, load resistor, and source/drain/gate electrodes) fabricated from graphene films.

In 2013, researchers in the United States and Italy succeeded in making the first integrated graphene digital circuits that function at gigahertz (GHz) frequencies (see Fig. 3.20) (52). The circuits in question are ring oscillators and the work is an important step toward realizing all-graphene microwave circuits. The highest oscillation frequency was 1.28 GHz, while the largest output voltage swing was 0.57 V. Both values remain limited by parasitic capacitances in the circuit rather than intrinsic properties of the graphene transistor components, suggesting

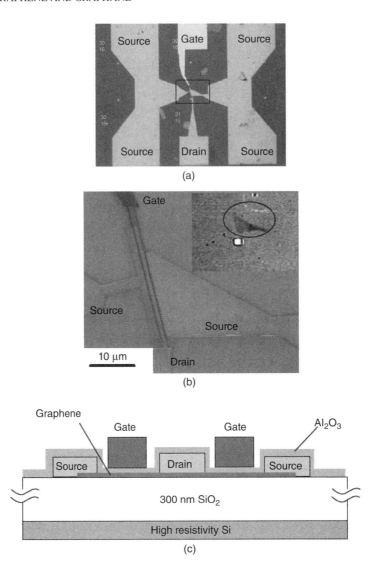

Figure 3.15 (a) Optical image of the device layout with ground-signal accesses for the drain and the gate. (b) (False color) SEM image of the graphene channel and contacts. The inset shows the optical image of the as-deposited graphene flake (circled area) prior to the formation of electrodes. (c) Schematic cross-section of the graphene transistor. Note that the device consists of two parallel channels controlled by a single gate in order to increase the drive current and device transconductance.

further improvements are possible. As a demonstration, we also realized the first stand-alone graphene mixers that do not require external oscillators for frequency conversion. The first gigahertz multitransistor graphene integrated circuits demonstrated here pave the way for application of graphene in high-speed digital and analog circuits in which high operating speed could be traded off against power consumption.

Currently, graphene transistors look extremely promising to revolutionize both high-frequency electronics and plastic electronics, especially as the expected performance has been proven recently at least in single devices. However, the major crux for becoming a real player in the semiconductor universe will be the move from the single device level toward sophisticated integrated circuits, which will require extensive efforts not only from academia but also from industry.

3.1.3.3 Graphene Electrodes Based on Graphene Graphene's high electrical conductivity and high optical transparency make it a candidate for transparent conducting electrodes and optical electrodes, required for such applications as touchscreens, liquid crystal displays, organic photovoltaic cells, and organic light-emitting diodes. In particular, graphene's mechanical strength and flexibility are advantageous compared to indium tin oxide (ITO), which is brittle, and graphene films may be deposited from solution over large areas (53).

Transparent, conductive, and ultrathin graphene films, as an alternative to the ubiquitously employed metal oxides window electrodes for solid-state dye-sensitized solar cells, are demonstrated. These graphene films are fabricated from exfoliated graphite oxide, followed by thermal reduction. The obtained films exhibit a high conductivity of 550 S/cm and a transparency of more than 70% over 1000–3000 nm. Furthermore, they showed high chemical and thermal stabilities as well as an ultrasmooth surface with tunable wettability (see Fig. 3.21)

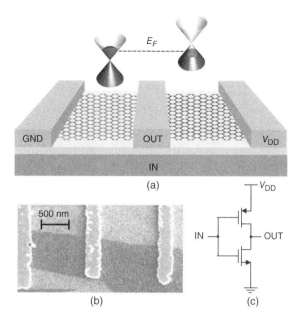

Figure 3.16 Integrated complementary graphene inverter. (a) A schematic of the fabricated inverter. (b) Scanning electron microscopy image of the fabricated inverter. (c) The circuit layout (power supply V_{DD} = 3.3 V).

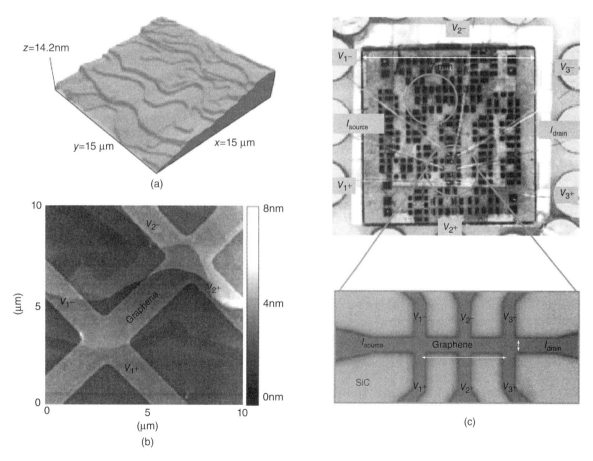

Figure 3.17 Sample morphology and layout. (a) AFM images of the sample: large flat terraces on the surface of the Si-face of a 4H-SiC(0001) substrate with graphene after high-temperature annealing in an argon atmosphere. (b) Graphene patterned in the nominally 2-mm-wide Hall bar configuration on top of the terraced substrate. (c) Layout of a 7 mm² wafer with 20 patterned devices. Encircled are two devices with dimensions L ¼ 11.6 mm and W ¼ 2 mm (wire bonded) and L ¼ 160 mm and W ¼ 35 mm. The contact configuration for the smaller device is shown in the enlarged image. To visualize the Hall bar this optical micrograph was taken after oxygen plasma treatment, which formed the graphene pattern, but before the removal of resist.

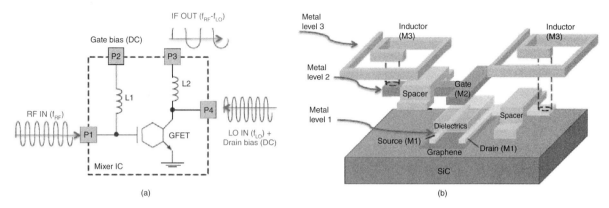

(a)   (b)

Figure 3.18 (a) Circuit diagram of a four-port graphene RF frequency mixer. The scope of the graphene IC is confined by the dashed box. The hexagonal shape represents a graphene FET. (b) Schematic exploded illustration of a graphene mixer circuit.

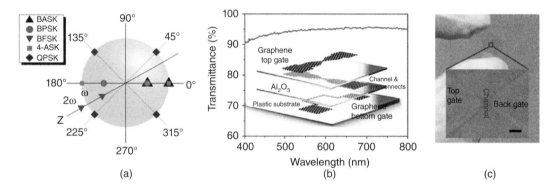

(a)   (b)   (c)

Figure 3.19 Modulation mechanism and device characteristics of flexible and transparent all-graphene transistors. (a) A constellation diagram depicting five different digital modulation techniques demonstrated in this work. The z-axis, representing the frequency, is included to show the frequency modulated signals. (b) A plot of the transmittance as a function of the wavelength and an illustration of the all-graphene transistor structure (inset). (c) a photograph of graphene circuit on a transparent and bendable plastic substrate, and a microscopic image of an all-graphene transistor (inset). The scale bar is 10 μm.

(54). Large-area, continuous, transparent, and highly conducting few-layered graphene films were produced by chemical vapor deposition and used as anodes for application in photovoltaic devices. A power conversion efficiency (PCE) up to 1.71% was demonstrated, which is 55.2% of the PCE of a control device based on indium tin oxide (55). Transparent, flexible electrode made from graphene could see a one-atom thick honeycomb of carbon, and replace other high-tech materials used in displays (see Fig. 3.22). It could even be used instead of silicon in electronics. Researchers from Sungkyunkwan University in Suwon, Korea, transferred a wafer-thin layer of graphene, etched into the shape needed to make an electrode, onto pieces of polymer. The polymers they used are transparent, and one – polyethylene terephthalate (PET) – can be bent, whereas the other – polydimethylsiloxane (PDMS) – is stretchable. The resulting films conduct electricity better than any other sample of graphene produced in the past (56, 57).

Organic light-emitting diodes (OLEDs) with graphene anodes have also been demonstrated. The electronic performance and optical performance of devices based on graphene are shown to be similar to devices made with indium tin oxide (58). An all carbon-based device called a light-emitting electrochemical cell (LEC) was demonstrated with chemically derived graphene as the cathode and the conductive polymer PEDOT as the anode by Matyba et al. (59). Unlike its predecessors, this device contains no metal, but only carbon-based electrodes. The use of graphene as the anode in LECs was also verified in the same publication. One layer of graphene absorbs 2.3% of white light. This property was used to define the Conductivity of Transparency that combines the sheet resistance and the transparency. This parameter was used to compare different materials without the use of two independent parameters (60).

In a recent article published in journal *Advanced Functional Materials* (61), researchers described a new graphene-coated transparent electrode made of silver nanowires (see Fig. 3.23). Because of its ability to bend without breaking, the new invention can be used to create flexible solar cells, computer and consumer electronics displays, and future "optoelectronic" circuits for sensors and information processing. The hybrid material shows promise as a possible replacement for indium tin oxide, or ITO, used in transparent electrodes for touchscreen monitors, cell-phone displays, and flat-screen televisions. Industry is seeking alternatives to ITO because of drawbacks: It is relatively expensive due to limited abundance of indium, and it is inflexible and degrades over time, becoming brittle and hindering performance.

Stable Li-ion cycling has recently been demonstrated in bi- and few-layered graphene films grown on nickel substrates (62), while single-layered graphene films have been demonstrated as a protective layer against corrosion in battery components such as the steel casing.

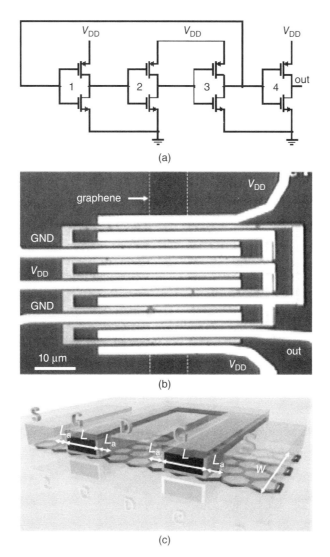

(a)

(b)

(c)

Figure 3.20   Integrated monolayer graphene ring oscillator (RO). (a) Circuit diagram of a three-stage RO. (b) Optical microscope image of a small RO. (c) Schematic of a complementary graphene inverter composed of two FETs.

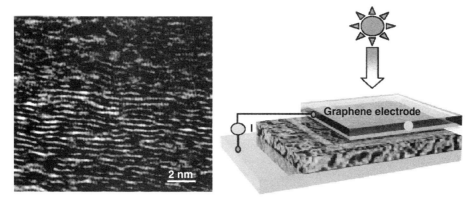

Figure 3.21   Graphene transparent electrodes for dye-sensitized solar cells.

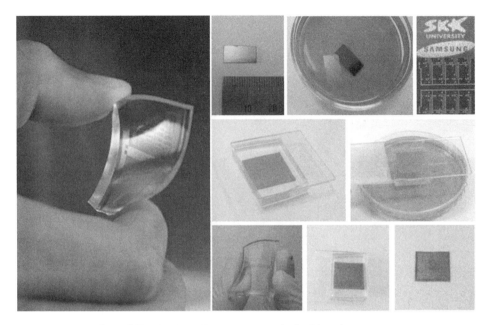

Figure 3.22   Graphene electrodes can now be flexible and transparent.

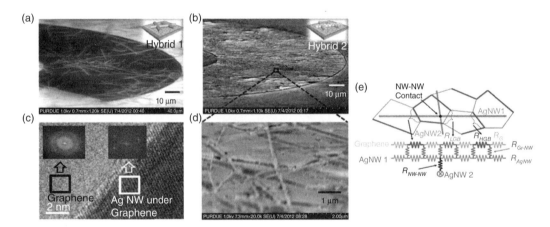

Figure 3.23   Electron microscope images show a new material for transparent electrodes that might find uses in solar cells, flexible displays for computers and consumer electronics, and future "optoelectronic" circuits for sensors and information processing. The electrodes are made of silver nanowires covered with a material called graphene. At bottom is a model depicting the "co-percolating" network of graphene and silver nanowires. (a,b) FESEM images and corresponding schematics (insets) of Hybrid 1 and Hybrid 2 films, within circular transfer length measurement structures; in each case, concentric metal rings and gap where hybrid film is exposed are visible. (c) HRTEM image of Hybrid 2 film, showing edge of Ag (beneath SLG) and nearby SLG region. The silver lattice planes can be clearly seen underneath the transparent SLG. (d) Magnified view of FESEM image for Hybrid 2 showing the wrapping of underlying AgNWs by SLG. (e) A resistor network model for graphene-AgNWs co-percolating system.

This opens up the possibilities for flexible electrodes for microscale Li-ion batteries where the anode acts as the active material as well as the current collector. Over the last couple of years, research to improve lithium-ion (Li-ion) batteries have been turning to graphene, particularly after researchers (63) at Northwestern University successfully sandwiched a layer of silicon between graphene sheets in the anodes of Li-ion batteries. When the researchers used the graphene films to fashion the negative electrode of a Li-ion battery, they discovered that graphene formed from the copper did not cycle lithium ions and had negligible capacity. However, the graphene electrode created from nickel had far superior performance to the copper version.

Up to now, one can find more reviews on the graphene electrodes (64, 65). Graphene is a promising next-generation conducting material with the potential to replace traditional electrode materials such as indium tin oxide in electrical and optical devices. It combines several advantageous characteristics including low sheet resistance, high optical transparency, and excellent mechanical properties. Recent research has coincided with increased interest in the application of graphene as an electrode material in transistors, light-emitting diodes, solar cells, and flexible devices. However, for more practical applications, the performance of devices should be further improved by the engineering of graphene films, such as through their synthesis, transfer, and doping.

3.1.3.4 Solar Cells Based on Graphene

Graphene has a unique combination of high electrical conductivity and optical transparency, which make it a good candidate for using in solar cells. A single sheet of graphene is a zero-bandgap semiconductor whose charge carriers are delocalized over large areas, implying that carrier scattering does not occur. Because this material only absorbs 2.3% of visible light, it is a candidate for applications as a transparent conductor (66). Graphene can be assembled into a film electrode with low roughness. However, in practice, graphene films produced via solution processing contain lattice defects and grain boundaries that act as recombination centers and decrease the electrical conductivity of the material. Thus, these films must be made thicker than one atomic layer in order to obtain sheet resistances that are sensible. This added resistance can be combatted by incorporating conductive filler materials, such as a silica matrix.

In order for graphene to be put to usage in solar cells commercially, large-scale production of the material would need to be achieved. However, the peeling of pyrolytic graphene does not seem to be a simple process to scale up. An alternative method with potential for scalable production of graphene that has been suggested is the thermal decomposition of silicon carbide.

Other than graphene's use as a transparent conducting oxide (TCO), it has also exhibited high charge mobilities that lead one to conclude that it could be put to use as a charge collector and transporter in PVs. The use of graphene in OPVs as a photoactive material requires the bandgap to be tuned to within the range of 1.4–1.9 eV. In 2010, Yong & Tour reported single-cell efficiencies of nanostructured graphene-based PVs of over 12%. According to P. Mukhopadhyay and R. K. Gupta, the future of graphene in OPVs could be "devices in which semiconducting graphene is used as the photoactive material and metallic graphene is used as the conductive electrodes" (67).

The USC Viterbi School of Engineering lab reported the large-scale production of highly transparent graphene films by chemical vapor deposition in 2008. In this process, researchers create ultrathin graphene sheets by first depositing carbon atoms in the form of graphene films on a nickel plate from methane gas. Then they lay down a protective layer of thermoplastic over the graphene layer and dissolve the nickel underneath in an acid bath. In the final step, they attach the plastic-protected graphene to a very flexible polymer sheet, which can then be incorporated into an OPV cell (graphene photovoltaics). Graphene/polymer sheets have been produced that range in size up to 150 cm^2 and can be used to create dense arrays of flexible OPV cells. It may eventually be possible to run printing presses laying extensive areas covered with inexpensive solar cells, much like newspaper presses print newspapers (roll-to-roll) (68, 69).

In 2013, researchers (70, 71) from the University of Manchester built a superefficient solar cell made from the atom-thick carbon material known as graphene (see Fig. 3.24). The panel is composed of photovoltaic material created by vertically stacking graphene with atom-thick transition metal dichalcogenides (TMDCs). The resulting paper-thin material can be used as an electricity-producing coating for buildings, mobile phones, and other devices.

The stack is composed of two outer layers of graphene sandwiching the TMDC layers. The graphene functions as an extremely efficient conductive layer while the TMDC acts as a very sensitive light absorber. The researchers have found that peppering the graphene with particles of gold increases light absorption. The device has a quantum efficiency of 30%.

The scientists believe that entire buildings could be powered by coating their exposed surfaces with the panels. Furthermore, the energy produced by the panels could then be used to alter the transparency and reflectivity of windows and fixtures. Ultrathin, transparent, and flexible solar-powered mobile phones will also be made possible.

Figure 3.25 shows EROI values (which points out the potential usage of various sources of energy) for different sources of energy including graphene solar cells which new research points to a potential EROI of 20.4, which for the first time in history puts solar power on the same footing as crude oil imports' EROI (12.5). Incidentally, oil has been a dominating source of energy since the late 1800s because of its EROI (72).

According to recent research, graphene solar cells are in the near future. Recent research published in February 2013 in Nature Physics (73) shows that graphene is quite efficient at turning photons into electricity rather than heat and a little bit of electricity. In April 2012, a group of scientists from Portugal filed patent for the manufacturing of graphene solar cells with the World Intellectual Property Office. These are only two examples of groups pushing for graphene solar cells. In fact, there are more such studies around.

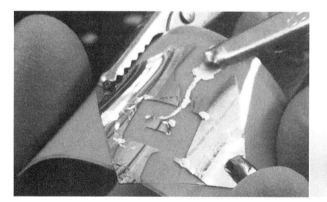

Figure 3.24 Graphene solar cell diagram.

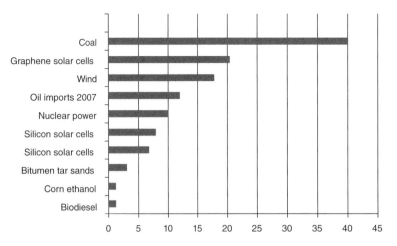

Figure 3.25   Energy returned on energy invested.

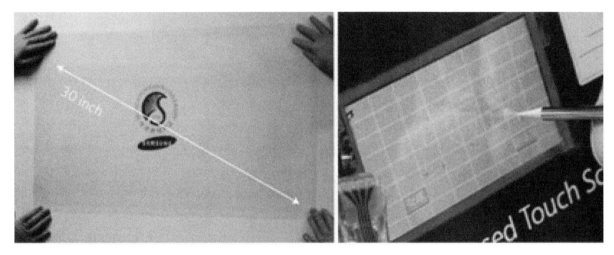

Figure 3.26   Left: A transparent graphene film transferred on a 35-inch PET sheet. Right: A graphene-based touchscreen panel connected to a computer.

While silicon has long been the standard for commercial solar cells, new research from the Institute of Photonic Sciences (ICFO) in Spain has shown that graphene could prove far more efficient when it comes to transforming light into energy. The study found that unlike silicon, which generates only one current-driving electron for each photon it absorbs, graphene can produce multiple electrons. Solar cells made with graphene could offer 60% solar cell efficiency – double the widely regarded maximum efficiency of silicon cells (74).

3.1.3.5   Optical Electronics Based on Graphene   One particular area in which we will soon begin to see graphene used on a commercial scale is in optoelectronics, specifically touchscreens, liquid crystal displays (LCD), and OLEDs. For a material to be able to be used in optoelectronic applications, it must be able to transmit more than 90% of light and also offer electrical conductive properties exceeding $1 \times 10^6 \, \Omega 1 m1$ and therefore low electrical resistance. Graphene is an almost completely transparent material and is able to optically transmit up to 97.7% of light. It is also highly conductive, as we have previously mentioned, and so it would work very well in optoelectronic applications such as LCD touchscreens for smartphones, tablet and desktop computers, and televisions.

In 2010, researchers in Korea and Japan fabricated films of graphene – planar sheets of carbon one atom thick – measuring tens of centimeters (see Fig. 3.26). The researchers engineered these large graphene films into transparent electrodes, which were incorporated into touchscreen panel devices. The new work represents another milestone in the astonishing technological advance of graphene from its initial isolation only a few years ago. Experts predict that graphene will be found in consumer products within a couple of years (75).

The researchers used this technique to create a rectangular graphene film measuring 30 inches (76 cm) in the diagonal. The graphene was doped by treating with nitric acid and in this form the graphene sheet can act as a large, transparent electrode and was demonstrated to work in a touchscreen device. Typically, transparent electrodes used in such applications are made from indium tin oxides).

Samsung advances its dominance of the graphene touchscreen sector with a patent to protect the use of graphene networks in capacitive touchscreens. US patent number 8390589 describes the use of a network of nanostructures as the conductive layer of a touchscreen. The nanostructures could include carbon nanotubes, nanowires, nanoparticles, and/or graphene flakes. Graphene and other nanostructures are being studied as possible replacements for the costly indium tin oxide in touchscreens, which also has a disadvantage of being brittle, making it unsuitable for flexible touchscreens. A network of graphene nanostructures provides an inexpensive alternative, which is also flexible and stretchable.

Currently, the most widely used material is indium tin oxide, and the development of manufacturing this ITO over the last few decades has resulted in a material that is able to perform very well in this application. However, recent tests have shown that graphene is potentially able to match the properties of ITO, even in current (relatively under-developed) states. Also, it has recently been shown that the optical absorption of graphene can be changed by adjusting the Fermi level. While this does not sound like much of an improvement over ITO, graphene displays additional properties which can enable very clever technology to be developed in optoelectronics by replacing the ITO with graphene. The fact that high-quality graphene has a very high tensile strength, and is flexible (with a bending radius of less than the required 5–10 mm for rollable e-paper), makes it almost inevitable that it will soon become utilized in these aforementioned applications.

In terms of potential real-world electronic applications, we can eventually expect to see such devices as graphene based e-paper with the ability to display interactive and updatable information and flexible electronic devices including portable computers and televisions.

3.1.3.6 *Biodevices Based on Graphene*

Bioengineering will certainly be a field in which graphene will become a vital part in the future, though some obstacles need to be overcome before it can be used. Current estimations suggest that it will not be until 2030 when we will begin to see graphene widely used in biological applications, as we still need to understand its biocompatibility (and it must undergo numerous safety, clinical, and regulatory trials, which, simply put, will take a very long time). However, the properties that it displays suggest that it could revolutionize this area in a number of ways. With graphene offering a large surface area, high electrical conductivity, thinness, and strength, it would make a good candidate for the development of fast and efficient bioelectric sensory devices, with the ability to monitor such things as glucose levels, hemoglobin levels, cholesterol, and even DNA sequencing. Eventually, we may even see engineered "toxic" graphene that is able to be used as an antibiotic or even anticancer treatment. Also, due to its molecular makeup and potential biocompatibility, it could be utilized in the process of tissue regeneration.

Graphene's modifiable chemistry, large surface area, atomic thickness, and molecularly gatable structure make antibody-functionalized graphene sheets excellent candidates for mammalian and microbial detection and diagnosis devices (76). The most ambitious biological application of graphene is for rapid, inexpensive electronic DNA sequencing. Integration of graphene (thickness of 0.34 nm) layers as nanoelectrodes into a nanopore can solve one of the bottleneck issues of nanopore-based single-molecule DNA sequencing (77).

3.1.3.7 *Other Devices Based on Graphene*

Here, we mainly reviewed several important applications of graphene. In fact, there are more applications, such as electrochromic devices (78), single-molecule gas detection (79), quantum dots (80), frequency multiplier (30), optical modulators (81), ultracapacitors (82), and engineered piezoelectricity (83).

3.2 BEYOND GRAPHENE: GRAPHANE, A TWO-DIMENSIONAL HYDROCARBON

Graphene, a single layer of hexagonally arranged sp^2-hybridized carbon atoms, has been well studied in recent years because of its interesting electronic properties. Now, researchers are not only focusing on graphene studies but also perusing the chemical modification of graphene to create derivatives with different structures and properties.

One among those chemical modifications is the addition of hydrogen atoms to graphene – creating "graphane" – altering the sp^2 carbon atoms to sp^3 and thus changing the structure and electronic properties. Graphane is a fully saturated molecular CH sheet, four-coordinate at C. Graphane's carbon bonds are in sp^3 configuration, as opposed to graphene's sp^2 bond configuration.

In fact, graphane can be regarded as one type of functionalization of graphene, which can be performed by covalent and noncovalent modification techniques (84, 85). For instance, the covalent functionalization can be thought of as the binding of organic functionalities like free radicals and dienophiles on pristine graphene, and attachment through the chemistry of oxygen groups, hydrogen and halogens of graphene. The noncovalent functionalization by π-interactions (as in the case of graphene/carbon nanotubes) is an attractive system, because it offers the possibility of attaching functional groups to graphene without disturbing the electronic network.

It has been found that both the covalent and noncovalent modification techniques are very effective in the preparation of processable graphene. However, the electrical conductivity of the functionalized graphene has been observed to decrease significantly compared to pure graphene. Moreover, the surface area of the functionalized graphene prepared by covalent and noncovalent techniques decreases significantly due to the destructive chemical oxidation of flake graphite followed by sonication, functionalization, and chemical reduction.

In this chapter, we would like to focus on the hydrogenated graphene, the so-called graphane.

3.2.1 The Discovery and Synthesis of Graphane

In 2003, the stability of a new extended 2D hydrocarbon on the basis of first-principles total energy calculations was predicted (86). The compound was called graphane and it was a fully saturated hydrocarbon derived from a single graphene sheet with formula CH. All of the carbon atoms are in sp^3 hybridization forming a hexagonal network and the hydrogen atoms are bonded to carbon on both sides of the plane in an alternating manner (see Fig. 3.27). Graphane is predicted to be stable with a binding energy comparable to other hydrocarbons such as benzene, cyclohexane, and polyethylene (87).

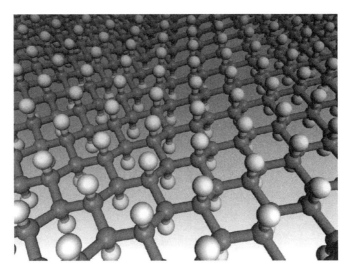

Figure 3.27   Structure of graphane in the chair conformation. The carbon atoms are shown in gray and the hydrogen atoms in white. The figure shows the hexagonal network with carbon in the sp^3 hybridization.

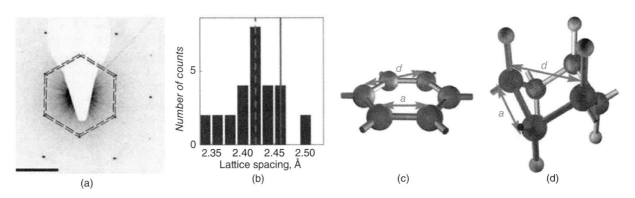

Figure 3.28   Structural studies of graphane via TEM. (a) Changes in the electron diffraction after ~4 h exposure of graphene membranes to atomic hydrogen. (c) Distribution of the lattice spacing d found in hydrogenated membranes. The green dashed line marks the average value, whereas the red solid line shows (d) always observed for graphene (both before hydrogenation and after annealing). (c and d) Schematic representation of the crystal structure of graphene and theoretically predicted graphane.

However, making graphane had proven to be difficult. The problem is that the hydrogen molecules must first be broken into atoms and this process usually requires high temperatures that could alter or damage the crystallographic structure of the graphene. In 2009, a team led by Geim and Novoselov has worked out a way to make graphane by passing hydrogen gas through an electrical discharge (Fig. 3.28) (88). This creates hydrogen atoms, which then drift toward a sample of graphene and bond with its carbon atoms. The team studied both the electrical and structural properties of graphane and concluded that each carbon atom is bonded with one hydrogen atom. It appears that alternating carbon atoms in the normally flat sheet are pulled up and down – creating a thicker structure that is reminiscent of how carbon is arranged in a diamond crystal. And, like diamond, the team found that graphane is an insulator – a property that could be very useful for creating carbon-based electronic devices.

In fact, in 2008, Ryu et al. (89) reported the chemical reaction of single-layered graphene with hydrogen atoms, generated *in situ* by electron-induced dissociation of hydrogen silsesquioxane (HSQ) (see Fig. 3.29). They suggested that 1 L (single-layered) graphene can be more easily hydrogenated than 2 L (double-layered) graphene near room temperature. This enhanced reactivity is attributed to the lack of π-stacking and/or out-of-plane deformation needed to stabilize the transition state of the hydrogenation reaction. The hydrogenated graphene can be restored by thermally desorbing bound hydrogen atoms. They pointed out that this functionalization of graphene can be exploited to manipulate electronic and charge transport properties of graphene devices.

Pumera et al. (90) reviewed the synthesis of hydrogenated graphene and summarized that there are two conceptually different approaches: top-down or bottom-up (see Fig. 3.30). Top-down approaches include solution-based reduction of graphite as well as hydrogenation of sp^2 carbon materials (multiwalled carbon nanotubes, graphene or graphite oxide) in a hydrogen gas/plasma atmosphere. The bottom-up approach involves CVD fabrication of hydrogenated graphene. In detail, these synthesis methods include gas-phase hydrogenation (91–94), liquid-based hydrogenation (95, 96), plasma deposition (97), and patterning of hydrogenated graphene strips in a graphene lattice (98, 99).

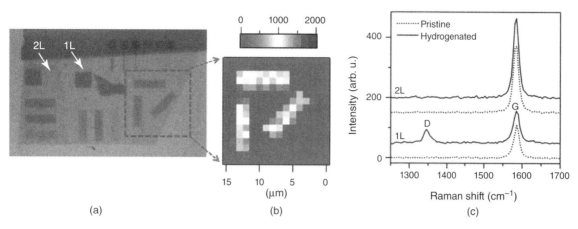

Figure 3.29 (a) Optical micrograph of e-beam patterned sample, which contains 1 L, 2 L, and thick sheets of graphene. The squares and rectangles are cross-linked HSQ etch masks. Non-cross-linked HSQ has been removed by the developer. The 1 L area in the dashed square is 15×15 μm^2 in size. (b) The D band intensity of Raman map for the dashed square in (a). (c) Raman spectra taken at the center of 1 L and 2 L graphene squares (area: 4×4 μm^2) shown in panel (a) before (dotted) and after (solid, displaced for clarity) hydrogenation. Data in panels (b) and (c) were obtained in ambient conditions with λexc = 514.5 nm. The employed laser power was 3 mW and focused onto a spot of ~1 μm in diameter. The integration time for each pixel was 20 s.

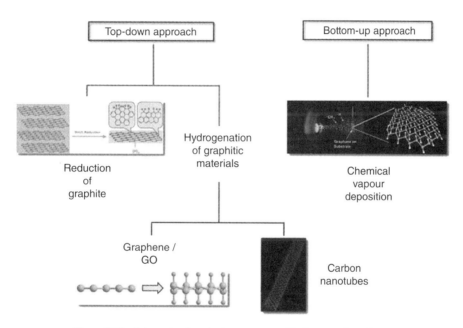

Figure 3.30 Schematic of synthetic routes toward hydrogenated graphene.

So far, the pure graphane has not yet been made experimentally, even after numerous attempts. Gas-phase methods for hydrogenating sp^2 carbon seem to be less successful in total hydrogen coverage than methods using wet chemistry, such as Birch reduction. Reduction of graphene in Na/NH$_3$ with consequent reaction with alcohol can produce highly hydrogenated graphene with a formula of (C1.3H1)n. Several methods based on e-beam patterning/lithography of graphene have been presented which lead to the creation of small strips of hydrogenated graphene within a graphene lattice. One might expect that further hydrogenation methods will be developed and pure graphane eventually fabricated. Whether it will possess the predicted properties or not, only time will tell.

Here, we would like to emphasize two approaches for synthesizing graphane. One is by Wang et al. (97) who came up with an original, direct synthetic route for hydrogenated graphene via plasma deposition (see Fig. 3.31). Ti/Cu layers were first deposited on a Si/SiO$_2$ substrate. CH$_4$ (5%) in H$_2$ was introduced into an ultrahigh vacuum chamber under high-frequency (13.5 MHz) plasma. The deposited film of hydrogenated graphene was then characterized by TEM, scanning tunneling microscope (STM), and near edge X-ray absorption fine structure (NEXAFS). The advantage of such hydrogenated graphene preparation is that it yields large sheets of hydrogenated graphene; this technology is thus compatible with wafer-scale production. It was found that heating the hydrogenated graphene to 600 °C for 20 min caused dehydrogenation and the subsequent complete recovery of a Moiré superlattice characteristic of graphene.

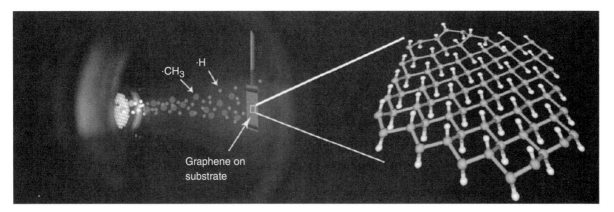

Figure 3.31 The growth of large-area graphane-like film by RF plasma beam deposition in high vacuum conditions. Reactive neutral beams of methyl radicals and atomic hydrogen effused from the discharged zone and impinged on the Cu/Ti-coated SiO_2/Si samples placed remotely. A substrate heating temperature of 650 °C was applied.

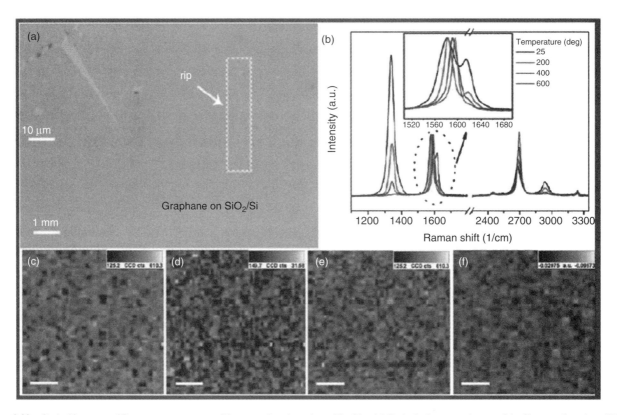

Figure 3.32 Optical images and Raman measurements of the as-produced graphane-like film. (a) Optical microscope image of the film transferred to a SiO_2/Si substrate. The inset shows ripping along crystallographic directions. (b) The evolution of Raman spectra (532 nm laser wavelength) with increasing annealing temperature for as-produced graphane sample. (c–e) Raman maps of the D(1300–1400), G+D' = (1550_1650), and 2D (2650–2750). Scale bar, 4 μm. (f) Optical contrast image shows homogeneous monolayer graphane film. Scale bar, 4 μm.

They also reported that the obtained graphane was crystalline with the hexagonal lattice, but its period was shorter than that of graphene. The reaction with hydrogen was found reversible, so that the original metallic state, the lattice spacing, and even the quantum Hall effect can be restored by annealing (100). The degree of hydrogenation of graphene layers was analyzed by changing Moiré pattern and the interconversion of graphane to graphene is presented in Fig. 3.32.

Another approach developed by Zheng et al. (101) is the so-called catalytic hydrogenation of graphene films which is not mentioned in the review by Pumera et al. They developed a facile method to hydrogenate graphene by using conditioning catalyst at the upstream of the

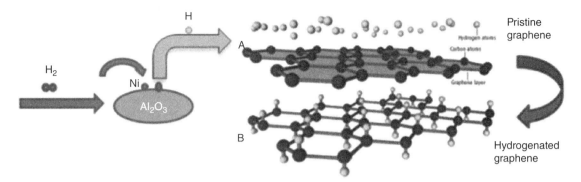

Figure 3.33 Schematic illustration for catalytic hydrogenation of graphene.

graphene sample to generate atomic hydrogen. The efficient hydrogenation of graphene can also be possible by catalytic conversion where catalyst provides a new reduction pathway of graphene through dissociation of graphene molecule. The intercalated Ni nanoparticles inside the alumina matrix with the flow of H_2 at 820 °C demonstrated the catalytic hydrogenation of graphene films (Fig. 3.33).

It should be said right away that graphane, CH, is very much related to CF, a material with an older experimental and theoretical history (102–105). A number of structural proposals for graphane have antecedents in the fluorinated graphite literature. Certainly, the graphane literature exploded after Sofo et al. and Elias et al. papers. In this chapter, we mention only some of the dozens of experimental and theoretical investigation of graphane that had been followed (106–108).

In experiment, the only chair graphane has been observed. Can other isomers be made? So far, we can answer the question only from the theoretical point of view. Two of the authors of the book chapter (109) theoretically predicted eight isomeric 2D graphane (CH) sheets. By the way, the story of isomeric graphanes can be learnt from our previous study on benzene under high pressure, entitled "Benzene under High Pressure: a Story of Molecular Crystals Transforming to Saturated Networks, with a Possible Intermediate Metallic Phase" (110).

We found four of these nets – two built on chair cyclohexanes and two on boat – are more stable thermodynamically than the isomeric benzene, or polyacetylene. The four isomeric 2D sheets of stoichiometry CH, labeled A ("chair1"), B ("chair2"), C ("boat1"), and D ("boat2"), are shown in Figure 3.34. Interestingly, these isomers are more stable thermodynamically than even the archetypical aromatic benzene (111). Dr. Roald Hoffman, winner of the 1981 Nobel Prize in Chemistry, wrote an article entitled "One Shocked Chemist" and described the story on graphane sheets.

At P = 1 atm the chair–boat differential, long known and understood in organic chemistry, governs the relative energy of the various graphane sheets. At higher pressures, for example, 300 GPa, lattices built from two sheet types that are not so stable at atmospheric pressure,

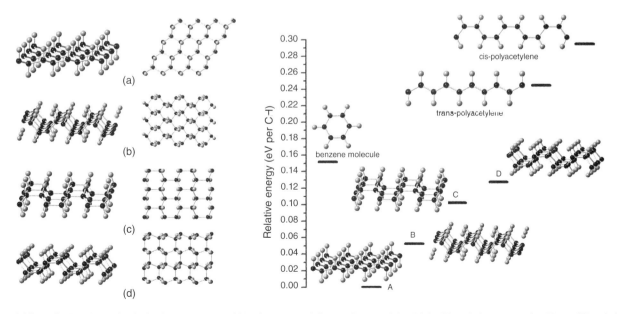

Figure 3.34 Left: Four isomeric single-sheet graphanes. Side views are at left, top views at right. Right: The relative energy (in eV per CH; relative to single-sheet graphane A, 0 K) of some CH structures.

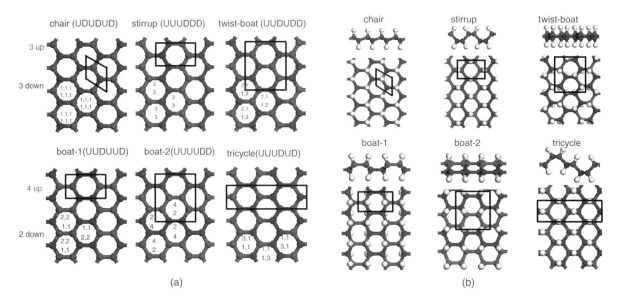

Figure 3.35 (a) Schematic diagram of six possible configurations of hydrogenated graphene with equivalent hexagonal hydrocarbon rings. (b) Crystal structures (side and top views) of graphane with chair, stirrup, twist-boat, boat-1, boat-2, and tricycle configurations, respectively. In figures, the gray balls correspond to carbon atoms with up and down hydrogenation, respectively, and the white balls are hydrogen atoms.

chair2 and boat1, become enthalpically favored in our calculations. To put it another way, the high pressure might offer a novel way to synthesize other graphane isomers.

After the study, He et al. (112) proposed a new allotrope of graphane, named tricycle graphane (see Fig. 3.35), which is a stable phase in between the previously proposed chair1 and boat1 allotropes. In fact, the tricycle graphane can be thought of as the mixture of chair1 and boat1.

Using ab initio and reactive molecular dynamics simulations, Flores et al. (106) have investigated the process from graphene to graphane, and the role of H frustration (breaking the H atoms' up and down alternating pattern) in graphane-like structures. The results indicate that a significant percentage of uncorrelated H-frustrated domains are formed in the early stages of the hydrogenation process leading to membrane shrinkage and extensive membrane corrugations. They suggested that large domains of perfect graphane-like structures are unlikely to be formed (see Fig. 3.36), as H-frustrated domains are always present.

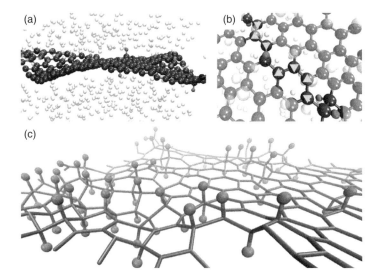

Figure 3.36 (a) Representative snapshot of the early hydrogenation stages from ReaxFF molecular simulations at 500 K. Nonbonded atomic H atoms are indicated in white and C-bonded ones in green. (b) Zoomed region indicating H frustrated domains formed. The triangle path shows that a sequence of up and down H atoms is no longer possible. (c) Representative snapshot of the final hydrogenation states. Extensive hydrogenation and multiple formed H domains are clearly visible.

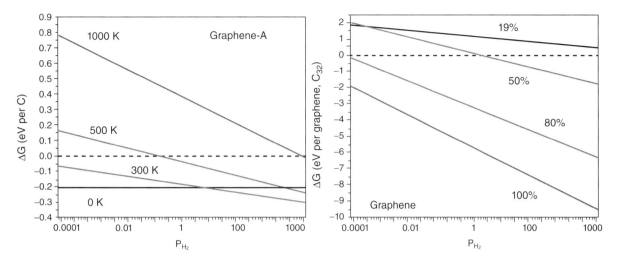

Figure 3.37   (a): Free energies (ΔG) per carbon for graphane-A (chair1 type) as a function of the P_{H_2} at various temperature; (b): Free reaction energies (ΔG) per graphene for partially hydrogenated graphene configurations as a function of the P_{H_2} at 300 K.

However, a recent theoretical study by Wen et al. (113) (one of the authors) investigated the reaction (graphene + H$_2$ = graphane), and concluded that graphane might be synthesized at low temperature and high partial pressure of H$_2$ gas (see Fig. 3.37). For the partially hydrogenated graphanes, the more the hydrogenation is (of graphene or the all-C nanotube), the more stable is the resulting structure. The interest in these partially hydrogenated structures lies in the opportunity they provide of engineering bandgaps.

In the study by Wen et al., the authors theoretically predicted the one-dimensional graphane nanotubes (GN, stoichiometry CH, see Fig. 3.38) built from 2D single-sheet graphanes. GN(10,10)-A, the armchair type with chair1 configuration, is found to be the most stable configuration among the GN structures considered. Similar to graphane, the graphane nanotubes are predicted to be wide bandgap insulators. In 2001, Stojkovic et al. (114) explored hydrogenated carbon nanotubes using tight-binding model, and they suggested that the resulting structures possess unique properties inaccessible in their sp^2, a very large insulating bandgap, high stability at an extremely small diameter, and Young's moduli exceeding 1.5 TPa.

3.2.2   Properties of Graphane

The carbon–carbon bond length in graphane (chair1 type) is about 1.54 Å. The shortest H–H distance is around 2.54 Å. More detailed information can be found in the study by Wen et al. (113). The unit hexagonal cell of graphane contains two carbon atoms and two hydrogen atoms and has an area of 0.092 nm^2. We can thus calculate its density as 0.59 mg/m^2, which is lower than graphene, that is, 0.77 mg/m^2. Due to the limitation of sample quality and synthesis approach for graphane, fully experimental characterizations are still lacking.

3.2.2.1   Electronic Properties   The experiments by Elias et al. (88) show that graphane is an insulator, but does not possess any value for the bandgap. Theoretically, using generalized gradient approximation (GGA), Sofo et al. (87) predicted that graphane should exhibit semiconducting properties, with a bandgap of 3.5 eV for the chair conformer and 3.7 eV for the boat conformer. This bandgap also depends on the graphane sheet (or ribbon) dimensions, with bandgaps decreasing with increase in the width of graphane nanoribbon (115). The bandgap can also be tuned by the extent of hydrogenation (113). Fokin et al. (116) studied multilayered graphane and suggested that decreasing the bandgap can be done by increasing the size of graphane sheet as well as through van der Waals interactions between graphane sheets.

We should point out that normal DFT calculations, such as PBE or LDA, systematically underestimate bandgaps (117). So far, the screened hybrid functional calculations, the so-called Heyd–Scuseria–Ernzerhof (HSE) hybrid function, generally produce a more realistic bandgap (118–120). Wen et al. (113) calculated HSE bandgaps for several graphane sheets; the bandgap for the graphane sheets given is around 4.0 eV which is higher than the PBE gap at 3.5 eV. There are no configuration-dependent bandgaps for these sheets, which is unlike the results by Sofo et al. However, a GW calculation reported a bandgap larger than 5 eV (121).

For stacking graphane crystals, Wen et al. (109) found that the bandgaps for the four graphanes increase at first with elevating pressure and reach a maximum at ~20 GPa for I and II and ~50 GPa for III and IV. Something similar also happens for diamond, where the bandgap also initially rises with pressure and even the computed pressure derivative for the bandgap is quantitatively similar, ~0.55 meVGPa (122). At higher pressures, the bandgaps decrease (see Fig. 3.39).

3.2.2.2   Mechanical Properties of Graphane   The theoretical Young's modulus of graphane is 0.61 and 0.59 TPa for the armchair and the zigzag direction, respectively (as compared to diamond values of ~1.20 TPa and graphene of 0.86 TPa) (123). Topsakal et al. (124) investigated the elastic constants of graphane, and suggested that that it has quite high in-plane stiffness and very low, perhaps the lowest, Poisson's ratio

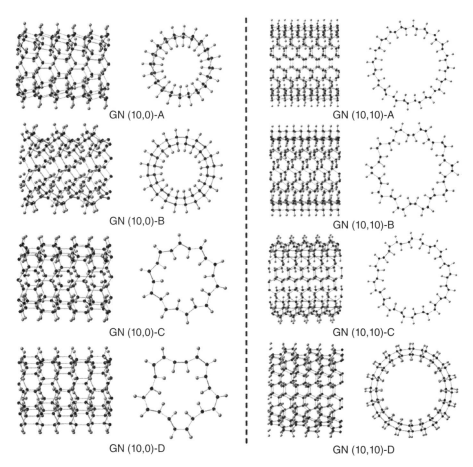

Figure 3.38 Left: Four isomeric graphane nanotube (10,0) structures. Right: Four isomeric graphane nanotube (10,10) structures. For each type GN, front views are at left, side views at right.

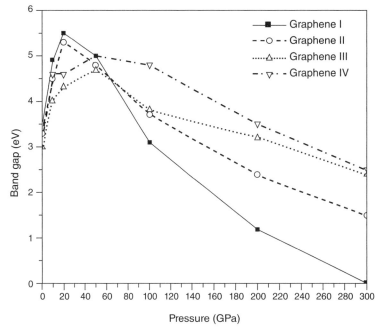

Figure 3.39 Bandgap of four stacking graphanes (3D) as a function of pressure.

among known monolayer honeycomb structures. The bandgap of graphane can be modified significantly by applying strain in the elastic range. Peng et al. (125) studied the mechanical response of graphane under various strains using first-principles calculations based on the density functional theory with van der Waals interactions and found that the van der Waals interactions have little effect on the geometry, ultimate strengths, ultimate strains, in-plane stiffness, and Poisson's ratio.

3.2.2.3 Lattice Thermal Properties of Graphane Neek-Amal and Peeters (126) studied the roughness and the thermal properties of a suspended graphane sheet using atomistic simulations. As compared to graphene, they found that (i) hydrogenated graphene has a larger thermal contraction, (ii) the roughness exponent at room temperature is smaller, that is, $\sim$1.0 versus $\sim$1.2 for graphene, (iii) the wavelengths of the induced ripples in graphane cover a wide range corresponding to length scales in the range 30–125 Å at room temperature, and (iv) the heat capacity of graphane is estimated to be 29.32 $\pm$ 0.23 J/mol K, which is 14.8% larger than that for graphene, that is, 24.98 $\pm$ 014 J/mol K.

3.2.3 Applications of Graphane: Devices Based on Graphane

Graphane has the same honeycomb structure as graphene, except that it is "spray-painted" with hydrogen atoms that attach themselves to the carbon. The resulting bonds between the hydrogen and carbon atoms effectively tie down the electrons that make graphene so conducting. Yet graphane retains the thinness, superstrength, flexibility, and density of its older chemical cousin (127).

One advantage of graphane is that it could actually become easier to make the tiny strips of graphene needed for electronic circuits. Such structures are currently made rather crudely by taking a sheet of the material and effectively burning away everything except the bit you need. But now, such strips could be made by simply coating the whole of a graphene sheet – except for the strip itself – with hydrogen. The narrow bit left free of hydrogen is your conducting graphene strip, surrounded by a much bigger graphane area that electrons cannot go down.

Perhaps most important of all is the discovery of graphane that opens the flood gates to further chemical modifications of graphene. With metallic graphene at one end and insulating graphane at the other, can we fill in the divide between them with, say, graphene-based semiconductors or by, say, substituting hydrogen for fluorine?

Now, let us give some brief discussions on the potential applications of graphane. We have to say that these potential applications of graphane are still staying at the theoretical level.

3.2.3.1 Hydrogen Storage Raman studies reveal that the hydrogenation interrupts the π-bonding system of graphene through the formation of sp^3 carbon–hydrogen bonds. Transmission electron microscopy studies indicate that the original hexagonal bonding arrangement is retained, but has a much smaller lattice constant. The hydrogenation is reversible through annealing, thereby restoring the conductivity and structure of graphene. This reversibility also creates the possibility of using such materials for hydrogen storage (128).

Krishna et al. (129) in their book chapter (entitled "Hydrogen Storage for Energy Application") summarized the methods of hydrogen storage, and point out the potential application of graphane on hydrogen storage.

Based on the first-principle density functional calculations, Hussain et al. (130) predicted that Li-doped graphane (prehydrogenated graphene) can be a potential candidate for hydrogen storage (Fig. 3.40 left). The calculated Li-binding energy on graphane is significantly higher than the Li bulk's cohesive energy ruling out any possibility of cluster formations in the Li-doped graphane. The study shows that even with very low concentration (5.56%) of Li doping, the Li-graphane sheet can achieve a reasonable hydrogen storage capacity of 3.23 wt%. The van der Waals corrected H$_2$ binding energies fall within the range of 0.12–0.29 eV, suitable for practical H$_2$ storage applications. The same authors studied the stability, electronic structure, and hydrogen storage capacity of a monolayer calcium-doped graphane (CHCa) (131) (Fig. 3.40 right). The binding energy of Ca on graphane sheet was found to be higher than its bulk cohesive energy, indicating the stability of CHCa. They found that with a doping concentration of 11.11% of Ca on graphane sheet, a reasonably good H$_2$ storage

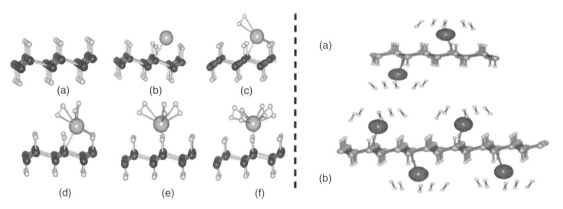

Figure 3.40 Left: The relaxed structures (van der Waals corrected) of (a) CH, (b) LiCH, (c) LiCH+H$_2$, (d) LiCH+2H$_2$, (e) LiCH+3H$_2$, and (f) LiCH+4H$_2$; Right: (a) and (b) are the side and top view of the optimized structures of pure CH, respectively.

Figure 3.41   Schematic graphane quantum dots.

capacity of 6 wt% could be attained. The adsorption energies of H_2 were found to be 0.1 eV, within the range of practical H_2 storage applications.

3.2.3.2   Graphane: A Route to Quantum Dots   Singh et al. (132, 133) have discovered that the strategic extraction of hydrogen atoms from a 2D sheet of graphane naturally opens up spaces of pure graphene that look – and act – like quantum dots (see Fig. 3.41). They found that the shape and size of such dots depend crucially on the graphene–graphane interface energy and the degree of aromaticity. Furthermore, these graphene dots show a pronounced quantum confinement effect of Dirac fermions, with energy gaps lying in the optically relevant range. That opens up a new world of possibilities for an ever-shrinking class of nanoelectronics that depend on the highly controllable semiconducting properties of quantum dots, particularly in the realm of advanced optics. Along with optical applications, the dots may be useful in single-molecule sensing and could lead to very tiny transistors or semiconductor lasers.

3.2.3.3   Hydrogenated Graphene Makes Good Transistors   Graphene is extremely promising for use in fast and ultralow-power devices thanks to the fact that electrons move through the material at extremely high speeds, as discussed in the section on graphene. Indeed, it is often touted to replace silicon as the material of choice in future electronics. However, the material does suffer from a major drawback in that it is a "zero-gap" semiconductor (it lacks an energy gap) and so cannot be used for digital applications. To open a gap of just 1 eV in the material would mean making graphene ribbons smaller than 2 nm across with single atom precision, something that is difficult using current fabrication technology.

Luckily, graphene can easily be hydrogenated by exposing it to a stream of hydrogen atoms to produce graphane. This process opens up a large energy gap in the material of a few electron-volts. Hydrogenated or semihydrogenated graphene (called graphane and graphone, respectively) could be better suited to making electronic devices than graphene itself. A team of researchers (134, 135) in Europe calculated that graphane and graphone could be used to make transistors with excellent properties. These come thanks to the direct bandgap of 5.4 eV for graphane and an indirect bandgap of 3.2 eV for graphone from GW calculations. The gaps strongly reduce band-to-band tunneling in the material, which normally degrades the electronic properties of monolayer and bilayer graphene transistors.

The simulations show that graphane/graphone-based transistors could comply with ITRS requirements for 2015–2016 technology because the transistors possess a large on–off current ratio – a parameter that shows how good the device is when behaving as a switch. These could open the way for making graphene nanoelectronics a real alternative to silicon CMOS (see Fig. 3.42).

3.2.3.4   Multifunctional Electrical Circuits and Chemical Sensors   Sun et al. (98) demonstrated the controlled patterning of graphane/graphene superlattices within a single sheet of graphene (see Fig. 3.43). By exchanging the sp^3 C–H bonds in graphane with sp^3 C–C bonds through functionalization, sophisticated multifunctional superlattices can be fabricated on both macroscopic and microscopic scales, which could be visualized using fluorescence quenching microscopy techniques and confirmed using Raman spectroscopy. By tuning

Figure 3.42 Sketch of the simulated device. The channel here shown is graphane, but the very same structure has been considered for graphone-based FET. In the inset, the device transversal cross-section is shown.

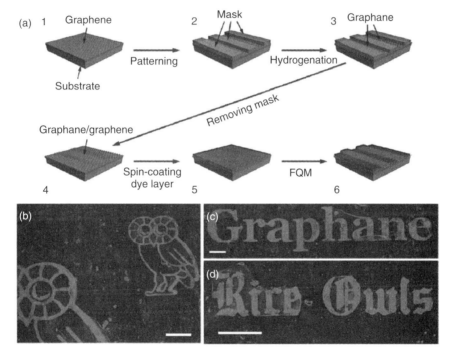

Figure 3.43 Graphane/graphene superlattices fabrication and imaging. (a) Schematic illustration of fabrication of the graphane/graphene superlattices and subsequent fluorescence quenching microscopy (FQM) imaging. (b–d) FQM imaging of the graphene with different graphane/graphene patterns. The scale bars in b–d are 200 μm.

the extent of hydrogenation, the density of the sp^3 C functional groups on graphene's basal plane can be controlled from 0.4% to 3.5% with this two-step method. Using such a technique, which allows for both spatial and density control of the functional groups, a route to multifunctional electrical circuits and chemical sensors with specifically patterned recognition sites might be realized across a single graphene sheet, facilitating the development of graphene-based devices.

3.2.3.5   p-Doped Graphane Could Be High-Temperature Superconductor   Savini et al. (107) predicted that p-doped graphane is postulated to be a high-temperature BCS theory superconductor with a *Tc* above 90 K using first-principles calculations. To put it another way, p-doped graphane is an electron–phonon superconductor with a critical temperature above the boiling point of liquid nitrogen. In theory, the unique strength of the chemical bonds between carbon atoms and the large density of electronic states at the Fermi energy arising from the reduced dimensionality give rise to a giant Kohn anomaly in the optical phonon dispersions and push the superconducting critical temperature above 90 K. They also found that Tc is rather insensitive to doping (see Fig. 3.44). This is important for the practical realization of superconducting graphane.

Loktev and Turkowski (136) predicted a possible superconductivity in the hole-doped system of layered hydrogenized graphene by taking into account thermal fluctuations of the order parameter. They demonstrated that in the one-layered case, the values of the high mean-field (MF) critical temperature are ~80–90 K, which is the same as predicted by Savini et al. In the case of multilayered system, when the coupling between the layers stabilizes the superconducting phase in the form of fluxon superconductivity, the critical temperature Tc can increase dramatically to the values of ~150 K, higher than the corresponding values in cuprates under ambient pressure.

3.3   GRAPHENE AND GRAPHANE ANALOGUES

3.3.1   Graphene Analogues

Due to the fascinating properties and extensive applications of graphene, people are perusing and discovering graphene analogues or the so-called graphene-like two dimensional materials (137, 138). A chemical perspective is that one moves down Group 14. Silicene and germanene, the silicon- or germanium-based counterparts of graphene, have been predicted in theory and synthesized in experiment. They are found to exhibit electronic characteristics similar to graphene.

One of our authors (Wen) theoretically explored group 14 structures across from 1D to 2D to 3D (139). They suggested that the graphene-like sheets of Si, Ge, Sn, and Pb are most unlikely to have an independent existence. If they are to be made, they will have to be intercalated by other atoms or molecules, or substrates, or otherwise protected from reacting with each other. Cahangirov et al. (140) predict that silicon and germanium can have stable, 2D, *low-buckled*, honeycomb structures, which are more stable than their corresponding planar-layer-type structures.

Though theorists had predicted the existence of silicene, experimentalists first synthesized silicene in 2010 (141, 142). Using the STM, they studied self-assembled silicene nanoribbons and silicone sheets deposited onto a Ag surface, Ag(110) (see Fig. 3.45 left) and Ag(111) (see Fig. 3.45 right), with atomic resolution. Recently, silicene has been reported to grow on a ZrB_2 substrate (143) and Ir (111) (144). Though the electronic properties and charge carriers of silicene are similar to graphene, the applications of silicene are still under study.

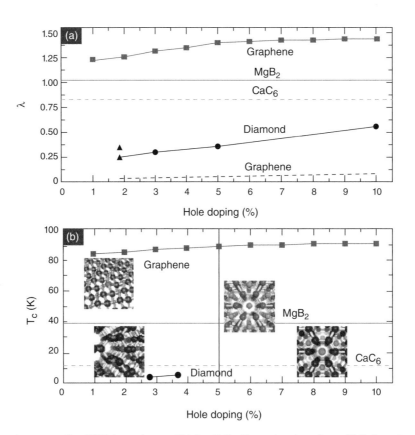

Figure 3.44   (a) Total electron phonon coupling (EPC) in graphane and MgB_2, CaC_6, diamond, and graphene; (b) Tc from the modified McMillan formula.

3.3.2 Graphane Analogues

Silicane is hydrogenated silicene, an analogue of graphane. Theoretically, silicane and germanane are predicted to be wide bandgap semiconductors; the type of gap in silicane (direct or indirect) depends on its atomic configuration, while germanane is predicted to be a direct-gap material, independent of its atomic configuration, with an average energy gap of about 3.2 eV (145). However, there are still no experimental evidences for the existence of silicane.

Interestingly, germanane (GeH, see Fig. 3.46), a graphane analogue, was synthesized for the first time by Bianco et al. (146). The germanane is thermally stable up to 75 °C; however, above this temperature amorphization and dehydrogenation begin to occur. These sheets can be mechanically exfoliated as single and few layers onto SiO_2/Si surfaces. This material represents a new class of covalently terminated graphane analogues and has great potential for a wide range of optoelectronic and sensing applications, especially as theory predicts a direct bandgap of 1.53 eV and an electron mobility about five times higher than that of bulk Ge. The applications of germanane are still under process.

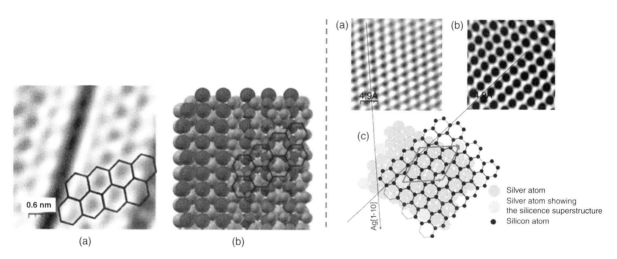

Figure 3.45 Left: Silicene on Ag (110), (a) high resolution filled state STM image revealing honeycomb arrangement (few honeycombs are drawn on the image); (b) ball model of the corresponding calculated atomic structure (dark grey balls for Ag atoms in the first layer; small grey balls for the top most Si atoms); Right: silicene on Ag (111) (a) Filled-state atomically resolved STM image of the clean Ag(111) surface. (b) Filled-state atomically resolved STM image of the same sample (without any rotation) after deposition of one silicon monolayer. (c) Proposed ball model of silicene on Ag(111) derived from both STM images (a) and (b) and from the observed $(2\sqrt{3} \times 2\sqrt{3})R30°$ LEED pattern.

Figure 3.46 Synthesized germanane in experiment.

3.4 CONCLUSION REMARKS

We are excited about the new physics, new opportunities, and new applications of graphene and its derivatives, such as graphane. As we create more and more complex heterostructures, the functionalities of the devices will become richer, entering the realm of multifunctional devices. However, many questions remain before the commercialization of graphene and graphane can be realized: What are the timelines and roadmaps for these applications? Will material capability translate into device performance? How much will be the material cost and in what qualities and quantities will it be available? Is production scalable? Let us try to answer these questions either in theory or by experiment!

ACKNOWLEDGMENT

This work was supported by the 100 Talents Program of the Chinese Academy of Sciences, the 100 Talents Program of Shanxi Province, the Institute of Coal Chemistry, Chinese Academy of Sciences (Taiyuan, Shanxi, P. R. China) and Synfuels China Co. Ltd. Huairou, Beijing, China. We are acknowledged for the National Natural Science Foundation of China (no. 21473229) and the innovation foundation of Institute of Coal Chemistry, Chinese Academy of Sciences (No. Y4SC821981).

REFERENCES

1. Wallace PR. Phys Rev 1947;71:622–634.
2. Novoselov KS, Geim AK, Morozov SV, Jiang D, Zhang Y, Dubonos SV, Grigorieva IV, Firsov AA. Science 2004;306:666–669.
3. Geim AK, Novoselov KS. Nat Mater 2007;6:183–191.
4. Geim AK. Science 2009;324:1530–1534.
5. Castro Neto AH, Guinea F, Peres NMR, Novoselov KS, Geim AK. Rev Mod Phys 2009;81:109–162.
6. Murali R, editor. *Graphene Nanoelectronics*. Springer; 2012.
7. Choi W. In: Lee J, editor. *Graphene: Synthesis and Applications*. CRC Press; 2011.
8. Baughman RH, Eckhardt H, Kertesz MJ. Chem Phys 1987;87:6687.
9. Li G, Li Y, Liu H, Guo Y, Li Y, Zhu D. Architecture of graphdiyne nanoscale films. Chemical Communications 2010;46(19):3256–3258.
10. Baughman RH, Galvao DS, Cui C, Tomanek D. Chem Phys Lett 1993;204:8.
11. Malko D, Neiss C, Vines F, Gorling A. Phys Rev Lett 2012;108:086804.
12. Nobel Media. Available at http://www.nobelprize.org/nobel_prizes/physics/laureates/2010/press.html. Accessed 2014 Nov 14.
13. Boehm HP, Clauss A, Hofmann U, Fischer GO. Z Naturforsch, B 1962;17:150.
14. McClure JW. Phys Rev 1956;104:666.
15. Semenoff GW. Phys Rev Lett 1984;53:2449.
16. DiVincenzo DP, Mele EJ. Phys Rev B 1984;29:1685.
17. Lu XK, Yu MF, Huang H, Ruoff RS. Nanotechnology 1999;10:269.
18. Choi W, Lahiri I, Seelaboyina R, Kang YS. Solid State Mater Sci 2010;35:52–71.
19. Avouris P, Dimitrakopoulos C. Mater Today 2012;15:86–97.
20. Graphenelitreviews 0000. Available at http://graphenelitreviews.blogspot.com/2008/03/overview-of-graphene-synthesis.html. Accessed 2014 Nov 14.
21. Badami DV. Nature 1962;193:569.
22. Nobel Media 0000. Scientific background on the Nobel Prize in Physics 2010. Available at http://www.nobelprize.org/nobel_prizes/physics/laureates/2010/press.html. Accessed 2014 Nov 14.
23. Nair RR et al. Science 2008;320:1308.
24. Das Sarma S, Adam S, Hwang EH, Rossi E. Rev Mod Phys 2011;83:407–470.
25. Chen JH et al. Nat Phys 2008;4:377–381.
26. Toon, J. Carbon-based electronics: researchers develop foundation for circuitry and devices based on graphite. GeorgiaTech Research News (2006).
27. Lemme MC et al. IEEE Electron Device Lett 2007;28:282–284.
28. Echtermeyer TJ et al. IEEE Electron Device Lett 2008;29:952–954.
29. Sordan R, Traversi F, Russo V. Appl Phys Lett 2009;94:073305.
30. Wang H, Nezich D, Kong J, Palacios T. IEEE Electron Device Lett 2009;30:547–549.
31. Fiori G, Iannaccone G. Ultra-low-voltage bilayer graphene tunnel FET 2009 ; arXiv:0906.1254.
32. Lin Y-M et al. Science 2010;327:662.
33. IBM. ibm scientists demonstrate world's fastest graphene transistor. IBM News room (2010).

34. Yang H et al. Graphene barristor, a triode device with a gate-controlled Schottky barrier. 2012;336:1140–1143.

35. SAMSUNG. SAMSUNG Electronics presents a new graphene device structure. U. S. News Center (2012).

36. Cheng R et al. Proc Natl Acad Sci U S A 2012;109:11588–11592.

37. Marcus J. UCLA Researchers Devise Scalable Method for Fabricating High-Quality Graphene Transistors. UCLA Newsroom (2012).

38. Yukio K. Nanotechnology 2013;24:214004.

39. Britnell L et al. Nat Commun 2013;4:1794.

40. Phys.Org. Available at http://phys.org/news/2013-02-graphene-transistor-principle.html#jCp. Accessed 2014 Nov 14.

41. Schwierz F. Nat Nanotechnol 2010;5:487.

42. Chahardeh JB. Int J Adv Res Comput Commun Eng 2012;1:193.

43. Chen J-H et al. Adv Mater 2007;19:3623–3627.

44. Ponomarenko LA et al. Science 2008;320:356–358.

45. Lin Y-M et al. Nano Lett 2009;9:422–426.

46. IBM. IBM scientists develop world's fastest graphene transistor. IBM News room (2008).

47. Wang X et al. Science 2009;324:768–771.

48. Traversi F, Russo V, Sordan R. Appl Phys Lett 2009;94:223312.

49. NPL 2010-01-19. European collaboration breakthrough in developing graphene. Available at http://www.npl.co.uk/news/european-collaboration-breakthrough-in-developing-graphene. Accessed 2014 Nov 14.

50. IEEE Spectrum 2011-06-09. First graphene integrated circuit. Available at http://spectrum.ieee.org/semiconductors/devices/first-graphene-integrated-circuit. Accessed 2014 Nov 14.

51. Lee S, Lee K, Liu C-H, Girish S, Kulkarni GS, Zhong Z. Nat Commun 2012;3:1018.

52. Guerriero E, Polloni L, Bianchi M, Behnam A, Carrion E, Rizzi LG, Pop E, Sordan R. Gigahertz integrated graphene ring oscillators. ACS Nano 2013;7:5588–5594

53. Eda G, Fanchini G, Chhowalla M. Nat Nanotechnol 2008;3:270–274.

54. Wang X et al. Nano Lett 2008;8:323–327.

55. Wang Y et al. Appl Phys Lett 2009;95:063302.

56. Kim KS et al. Nature 2009;457:706–710.

57. NPG. Available at http://www.nature.com/news/2009/090114/full/news.2009.28.html. Accessed 2014 Nov 14.

58. Wu J et al. ACS Nano 2010;4:43–48.

59. Matyba P et al. ACS Nano 2010;4:637–642.

60. Eigler S. Carbon 2009;171:582–587.

61. Chen R, Das S, Jeong C, Khan M, Janes D, Alam M. Adv Funct Mater 2013. DOI: 10.1002/adfm.201300124.

62. Radhakrishnanz G, Cardema JD, Adams PM, Hyun I, Kim HI, Foran BJ. Electrochem Soc 2012;159:A752–A761.

63. David L, Bhandavat R, Kulkarni G, Pahwa S, Zhong Z, Singh G. ACS Appl Mater Interfaces 2013;5:546–552.

64. Jo G, Choe M, Lee S, Park W, Kahng YH, Lee T. Nanotechnology 2012;23:112001.

65. Huang X, Zeng Z, Fan Z, Liu J, Zhang H. Adv Mater 2012;24:5979–6004.

66. Mukhopadhyay P. *Graphite, Graphene and their Polymer Nanocomposites*. Boca Raton, FL: Taylor & Francis Group; 2013. p 202–213.

67. Mukhopadhyay P. *Graphite, Graphene and their Polymer Nanocomposites*. Boca Raton, FL: Taylor & Francis Group; 2013. p 211. ISBN: 978-1-4398-2779-6.

68. Science Daily. Graphene organic photovoltaics: flexible material only a few atoms thick may offer cheap solar power, 2010.

69. Walker, S. 2010-08-04. Use of graphene photovoltaics as alternate source of energy. Computer Talks at http://www.comptalks.com/use-of-graphene-photovoltaics-as-alternate-source-of-energy/. Accessed 2014 Nov 14.

70. Britnell L et al. Science 2013;340:1311–1314.

71. FutureLeap. Available at http://www.futureleap.com/news/nanoscale-graphene-solar-cell-material-could-paint-homes revolutionizesmartphones/. Accessed 2014 Nov 14.

72. InvestorIntel. Available at http://investorintel.com/graphite-graphene-intel/graphene-solar-cells-to-change-global-energy-balance-sourcing/. Accessed 2014 Nov 14.

73. Tielrooij KJ et al. Nat Phys 2013;9:248–252.

74. inhabitat.com cooperating with ICFO (Institute of Photonic Sciences)(2013-04-03.

75. RSC. Available at http://www.rsc.org/chemistryworld/news/2010/june/20061001.asp. Accessed 2014 Nov 14.

76. Mohanty N, Berry V. Nano Lett 2008;8:4469–4476.

77. Xu M, Xu S, Fujita D, Hanagata N. Small 2009;5:2638–2649.

78. Ekiz OO et al. ACS Nano 2011;5:2475–2482.

79. Schedin F et al. Nat Mater 2007;6:652–655.

80. Mohanty N, Moore D, Zh X, Sreeprasad TS, Nagaraja A, Rodriguez AA, Berry V. Nat Commun 2012;3:844.

81. Liu M, Yin X, Erick U-A, Geng B, Thomas Z, Long J, Feng W, Zhang X. Nature 2011;474:64–67.

82. Stoller MD, Park S, Zhu Y, An J, Ruoff RS. Nano Lett 2008;8:3498–3502.

83. Ong M, Reed EJ. ACS Nano 2012;6:1387–1394.

84. Georgakilas V et al. Chem Rev 2012;112:6156–6214.

85. Kuila T et al. Prog Mater Sci 2012;57:1061–1105.

86. Sluiter MHF, Kawazoe Y. Phys Rev B 2004;68:085410.

87. Sofo JO, Chaudhari AS, Barber GD. Phys Rev B 2007;75:153401.

88. Elias DC et al. Science 2009;323:610–613.

89. Ryu S et al. Nano Lett 2008;8:4597–4602.

90. Pumera M, Hong C, Wong A. Chem Soc Rev. 2013;21:5987–5995.

91. Burgess JS, Matis BR, Robinson JT, Bulat FA, Perkins FK, Houston BH, Baldwin JW. Carbon 2011;49:4420–4426.

92. Talyzin AV et al. ACS Nano 2011;5:5132–5140.

93. Luo Z, Yu T, Kim K, Ni Z, You Y, Lim S, Shen Z, Wang S, Lin J. ACS Nano 2009;3:1781–1788.

94. Poh HL, Sanek F, Sofer Z, Pumera M. Nanoscale 2012;4:7006–7011.

95. Birch AJ. Reduction by dissolving metals. Part I. J. Chem Soc 1944:430–436.

96. Yang Z, Sun Y, Alemany LB, Narayanan TN, Billups WE. J Am Chem Soc 2012;134:18689–18694.

97. Wang Y, Xu X, Lu J, Lin M, Bao Q, Ozyilmaz B, Loh KP. ACS Nano 2010;4:6146–6152.

98. Sun Z, Pint CL, Marcano DC, Zhang C, Yao J, Ruan G, Yan Z, Zhu Y, Hauge RH, Tour JM. Nat Commun 2011;2:559.

99. Jones JD, Mahajan KK, Williams WH, Ecton PA, Mo Y, Perez JM. Carbon 2010;38:2335–2340.

100. Wang Y, Xu X, Lu J, Lin M, Bao Q, Özyilmaz B, Loh KP. ACS Nano 2010;26:6146–6152.

101. Zheng L, Li Z, Bourdo S, Watanabe F, Ryerson CC, Biris AS. Chem Commun 2011;47:1213–1215.

102. Watanabe N, Nakajima T, Touhara H. *Graphite Fluorides*. Amsterdam, New York: Elsevier; 1988.

103. Rüdorff W, Rüddorf G. Anorg Allg Chem 1947;253:281–296.

104. Ebert LB, Brauman JI, Huggins RA. J Am Chem Soc 1974;96:7841–7842.

105. Charlier JC, Gonze X, Michenaud JP. Phys Rev B 1993;47:16162–16168.

106. Flores MZS, Autreto PAS, Legoas SB, Galvao DS. Nanotechnology 2009;20:465704.

107. Savini G, Ferrari AC, Giustino F. Phys Rev Lett 2010;105:037002.

108. Cadelano E, Palla PL, Giordano S, Colombo L. Phys Rev B 2010;82:235414.

109. Wen X-D, Hand L, Labet V, Yang T, Hoffmann R, Ashcroft NW, Artem RO, Andriy OL. Proc Natl Acad Sci U S A 2011;108:6833–6837.

110. Wen X-D, Ashcroft NW, Hoffmann R. J Am Chem Soc 2011;133:9023–9025.

111. Hoffmann R. Am Sci 2011;99:116–119.

112. He C, Zhang CX, Sun LZ, Jiao N, Zhang KW, Zhong J. Phys Status Solidi 2012;6:427–429.

113. Wen X-D, Yang T, Hoffmann R, Ashcroft NW, Martin RL, Rudin SP, Zhu J-X. ACS Nano 2012;6:7142–7150.

114. Stojkovic D, Zhang P, Crespi VH. Phys Rev Lett 2001;87:125502.

115. Samarakoon DK, Wang X-Q. ACS Nano 2009;3:4017–4022 and references therein.

116. Fokin AA, Gerbig D, Schreiner PR. J Am Chem Soc 2011;133:20036–20039.

117. Hafner J. J Comput Chem 2008;29:2044–2078.

118. Heyd J, Scuseria GE, Ernzerhof M. J Chem Phys 2003;118:8207.

119. Heyd J, Scuseria GE, Ernzerhof M. Erratum: hybrid functions based on a screened coulomb potential. J Chem Phys 2006;124:219906.

120. Heyd J, Peralta JE, Scuseria GE, Martin RL. J Chem Phys 2005;123:174101.

121. Lebegue S et al. Phys Rev B 2009;79:245117.

122. Fahy S, Chang KJ, Louie SG, Cohen ML. Phys Rev B 1987;35:5856–5859.

123. Pei QX, Zhang YW, Shenoy VB. Carbon 2010;48:898–904.

124. Topsakal M, Cahangirov S, Ciraci S. Appl Phys Lett 2010;96:091912.

125. Peng Q, Chen Z-F, De S. A density functional theory study of the mechanical properties of graphane with van der Waals corrections. Mech. Adv. Mater. and Strut. 2014; DOI: 10.1080/15376494.2013.839067.

126. Neek-Amal M, Peeters FM. Phys Rev B 2011;83:235437.

127. ScienceDaily. Available at http://www.sciencedaily.com/releases/2009/07/090731090011.htm. Accessed 2014 Nov 14.

128. Armstrong G. Graphene: Here comes graphane? Nat Chem 2009. DOI: 10.1038/nchem.116.

129. Krishna R, et al. In: Liu J, editor. *Hydrogen Storage for Energy Application "Hydrogen Storage"*. 2012. by InTech. ISBN 978-953-51-0731-6.

130. Hussain T, Pathak B, Maark TA, Araujo CM, Scheicher RH, Ahuja R. Europhys Lett 2011;96:27013.

131. Hussain T, Pathak B, Ramzan M, Maark TA, Ahuja R. Appl Phys Lett 2012;100:183902.

132. Singh AK, Penev ES, Yakobson BI. ACS Nano 2010;4:3510–3514.

133. ScienceDaily. Available at http://www.sciencedaily.com/releases/2010/05/100525133219.htm. Accessed 2014 Nov 14.

134. Fiori G, Lebègue S, Betti A, Michetti P, Klintenberg M, Eriksson O, Iannaccone G. Phys Rev B 2010;82:153404.

135. Nanotechweb. Available at http://nanotechweb.org/cws/article/tech/44090. Accessed 2014 Nov 14.

136. Loktev VM, Turkowski VJ. Low Temp Phys 2011;164:264–271.

137. Xu M, Liang T, Shi M, Chen H. Graphene-like two-dimensional materials. Chem Rev 2013;113:3766–3798.

138. Butler SZ et al. Progress, challenges, and opportunities in two-dimensional materials beyond graphene. ACS Nano 2013;7:2898–2926.

139. Wen X-D, Cahill T-J, Hoffmann R. Exploring group 14 structures: 1D to 2D to 3D. Chem Eur J 2010;16:6555–6566.

140. Cahangirov S, Topsakal M, Akturk E, Sahin H, Ciraci S. Two- and one-dimensional honeycomb structures of silicon and germanium. Phys Rev Lett 2009;102:236804.

141. Aufray B, Kara A, Vizzini S, Oughaddou H, Léandri C, Ealet B, Le Lay G. Graphene-like silicon nanoribbons on Ag(110): a possible formation of silicene. Appl Phys Lett 2010;96:183102.

142. Lalmi B, Oughaddou H, Enriquez H, Kara A, Vizzini S, Ealet B, Aufray B. Epitaxial growth of a silicene sheet. Appl Phys Lett 2010;97:223109.

143. Fleurence A, Friedlein R, Ozaki T, Kawai H, Wang Y, Yamada-Takamura Y. Experimental evidence for epitaxial silicene on diboride thin films. Phys Rev Lett 2012;108:245501.

144. Meng L et al. Buckled silicene formation on Ir(111). Nano Lett 2013;13:685–690.

145. Houssa M, Scalise E, Sankaran K, Pourtois G, Afanas'ev VV, Stesmans A. Electronic properties of hydrogenated silicene and germanene. Appl Phys Lett 2011;98:223107.

146. Bianco E et al. Stability and exfoliation of germanane: a germanium graphane analogue. ACS Nano 2013;7:4414–4421.

4

LARGE-AREA GRAPHENE AND CARBON NANOSHEETS FOR ORGANIC ELECTRONICS: SYNTHESIS AND GROWTH MECHANISM

HAN-IK JOH[1], SUKANG BAE[2], SUNGHO LEE[1], AND ERIC S.W. KONG[3]

[1]*Carbon Convergence Materials Research Center*
[2]*Soft Innovative Materials Research Center, Institute of Advanced Composite Materials, Korea Institute of Science and Technology, Eunha-ri, Bongdong-eup, Wanju-gun, Jeollabuk-do, South Korea*
[3]*Research Institute of Micro/Nanometer Science & Technology, Shanghai Jiao Tong University, Shanghai, China*

4.1 INTRODUCTION

Carbon materials with various dimensionalities have been widely used in real life for a long time. From bulk graphite to fullerene nanomaterials, except for two-dimensional (2D) carbon materials, these materials were discovered and subsequently synthesized for commercial applications by material scientists and engineers. For more than 80 years, numerous physicists have argued that based on the theoretical and experimental evidences, 2D crystals were thermodynamically unstable due to the thermal fluctuations in their crystal lattices; therefore, these structures could not exist unless they formed three-dimensional structures containing dozens of 2D material (atomic) layers (1–6). However, Novoselov et al. have experimentally discovered 2D carbon allotropes, specifically graphene, in 2004 (7). The mechanically exfoliated graphene exhibited extraordinary physical properties, such as ultrahigh intrinsic carrier (electron and hole) mobility, gate-tunable ambipolar carrier density, high current density capacity, high thermal conductivity, and a high elastic modulus (8). As expected, the mechanical exfoliation by scotch tape is limited to production of graphene in large sizes and quantities. Therefore, the breakthrough technologies suitable for wide usage or further commercialization, similar to the other carbon allotropes have been required for synthetic approaches of the graphene. There are two representative and alternative methodologies to synthesize the graphene for this purpose so far: chemical exfoliation and chemical vapor deposition (CVD). Chemical exfoliation can be utilized to produce large quantities of graphene through the sequential process of oxidation, exfoliation, and reduction (9–12). However, the graphene has sheet sizes with dimensions in microns and numerous irreversible defects due to the severe oxidation conditions, despite performing high temperature or toxic chemical reductions, leading to a low-quality graphene product. Graphene generated by CVD has been synthesized by introducing gasified hydrocarbons to metal catalysts at high temperatures. Because it facilitates the mass production of high-quality graphene on a large scale, CVD is a most promising method that satisfies the requirements; however, complicated processes such as the catalyst elimination and the transfer of the graphene into a new substrate, inevitably degrading the products with artifacts, must be improved.

Because the graphene grown on metal catalysts via CVD must undergo the processes as mentioned earlier, numerous researchers have focused on improving quality of the graphene by optimizing the process. However, some researchers have developed modified or entirely new routes for synthesizing graphene to circumvent the drawbacks of the conventional CVD process, specifically catalyst- and/or transfer-free growth, while maintaining graphene quality. In this chapter, we will introduce various methods and techniques, such as CVD-based or direct pyrolysis of carbon sources, developed for synthesizing graphene with a large scale. In addition, their growth mechanisms, which are related to the material's structural and physical properties, and their potential applications in electronic devices will be discussed.

Nanomaterials, Polymers and Devices: Materials Functionalization and Device Fabrication, First Edition. Edited by Eric S. W. Kong.
© 2015 John Wiley & Sons, Inc. Published 2015 by John Wiley & Sons, Inc.

4.2   GRAPHENE GROWN ON METAL CATALYSTS

4.2.1   Experimental Approaches and Growth Mechanism of Graphene

Studies describing the formation of carbon materials on metal catalysts have been reported since the mid-1960s. The materials were prepared by thermally decomposing hydrocarbons on a single crystalline metal, such as Pt and Ni, under ultrahigh vacuum (UHV) conditions (13–15). In particular, Blakely et al. observed a phase transition from a low coverage of carbon film to a graphite precipitate by varying the temperature to form a monolayer of carbon on the surface of a Ni single crystal (15). This exceptional finding might have become the foundation of the metal-catalyzed graphene synthesis used today, even though the researchers were focused on the structural and crystallographic properties of the materials.

The intrinsic properties of the metal catalysts affect the growth mechanism of the graphene regardless of whether the carbon source is gaseous or solid, leading to the formation of graphene with different structural, electronic, and morphological characteristics. Ni and Cu films have seen general use as catalysts for graphene growth. The binary phase diagram of carbonmetal revealed a significant difference of the carbon solubility in Ni from in Cu catalysts. Based on this difference and various experimental results, numerous researchers proposed the growth mechanisms of graphene over Ni (precipitation mechanism) and Cu (self-limiting mechanism) (16–18). Therefore, in this section, the properties of the resulting graphene and the proposed growth mechanism relative to the metal catalysts will be discussed in detail.

4.2.1.1   Precipitation of Graphene on Nickel   During the CVD process, gaseous species from a hydrocarbon source are injected into the reactor with a hot zone (~1000 °C), where the hydrocarbon precursors decompose to form various carbon radicals. When these species arrive at the surface of the Ni film they form single- and multilayered graphene during the cooling process (17) because the carbon is highly soluble in the Ni (19). Due to the high solubility of the carbon, decomposed carbon species diffused into the metal lattices at high temperatures (dissolution); graphene films are segregated on the surface of the Ni film (segregation) during the cooling process because the carbon solubility at lower temperatures is somewhat lower (<800 °C) due to a precipitation mechanism, as previously reported (17). The formation of graphite on metal surfaces has already been extensively studied over the years (20, 21); however, more recently, these methods have been developed to control the graphene layers on Ni thin films (22). The graphene grown on Ni thin films will be discussed below.

Figure 4.1a provides a schematic diagram of graphene formation on a Ni substrate (22). Several parameters are effective for controlling the quality of the graphene films grown on Ni surfaces, such as the ratio of mixture gases (23), the thickness of the Ni film (22), the growth time (22), and the cooling rate (17). In one report, graphene was grown over varying thicknesses of the Ni film and growth times; when the Ni layer was more than 300 nm thick, a large amount of the carbon source material was absorbed by the Ni substrate, forming a graphitic layer. However, for the Ni films less than 300 nm thick, the graphene monolayer was grown after a shorter growth time (22). Therefore, the quantity of carbon absorbed into Ni could be limited by controlling the thickness of the Ni layer and the growth time. Figure 4.1b exhibits the optical images of graphene films transferred on 300 nm SiO_2/Si substrates; these films were grown using different thicknesses of Ni and growth times. The graphene films grown over 100-nm-thick Ni films for 30 s are usually continuous with monolayer and a few multilayer regions. Figure 4.1c presents the scanning electron microscope (SEM) images of graphene films grown on a 300 nm Ni film and a thick Ni foil (inset). After transferring the product to a 300 nm SiO_2 substrate, the distribution of the graphene layers was observed by optical microscope (Fig. 4.1d), revealing that the ripple structures arose due to the different thermal expansion coefficients between the nickel and graphene layers (22). Figure 4.1e displays the Raman spectra of the graphene films grown on Ni substrates that were measured at the marked locations in Figure 4.1d. The lower intensity of D-band peaks in Figure 4.1e indicates that the quality of the graphene films grown on the Ni substrate is excellent. When the graphene grown on a SiO_2/Si substrate with deposited Ni was transferred to a SiO_2/Si substrate, not only did it exhibit mobility values approaching ~3650 cm²/(V s), but it also initially exhibited a half-integer quantum Hall effect when using the CVD synthesis method (22).

However, most of the multilayer graphene regions exist at the Ni grain boundaries; these areas are defects in the polycrystalline Ni substrates (19), meaning that annealing the Ni substrate under a hydrogen atmosphere increases the singlecrystalline Ni grain size and removes impurities from the Ni to improve the graphene quality (17). To improve the uniformity of the monolayer graphene region, a single crystalline Ni (111) substrate with smooth surface morphology is employed by Zhang et al. for the synthesis of graphene (18). The graphene grown on Ni (111) films are usually monolayer graphene films (Fig. 4.2b and c), while multilayer graphene areas are more common on polycrystalline Ni films (Fig. 4.2e and f) because a high density of carbon atoms is diffused through the grain boundaries in the Ni film, as displayed in Figure 4.2a and d. In the monolayer and bilayer graphene grown on Ni (111), the substrate is generated on approximately 91.4% of films; this value is much higher than in previous reports (22–24) using the polycrystalline Ni substrate, as indicated by Figure 4.2.

Modified CVD methods have been reported that use solid carbon sources that are evaporated or decomposed to deposit on the metal catalysts before a growth reaction takes place in a high temperature chamber (25–27). Notably, the original structure of the evaporated or decomposed molecules on the substrates from the solid sources influenced the quality of the graphene produced by this method. Kalita et al. reported a method to synthesize graphene using a green and renewable carbon source, namely, camphor ($C_{10}H_{16}O$); camphor contains hexagonal and pentagonal rings that may be converted into a graphitic structure easily using pyrolysis (26, 27). Similar to the CVD growth mechanism for the gaseous carbon sources, the graphene growth during this process strongly depends on its solubility in the metal catalysts. Therefore, monolayer and bilayer graphene might be synthesized via the simple pyrolysis of natural solid carbon sources.

The specific CVD process conditions that were employed to control the graphene layers are listed in Table 4.1 (17, 18, 22, 24, 29–32). The fundamental limitations of using Ni as the catalyst layer due to high solubility of carbon hinders uniform graphene film formation (17, 22, 24). It is well known that monolayer dominated graphene films are easily obtained on Cu substrates because of carbon's extremely

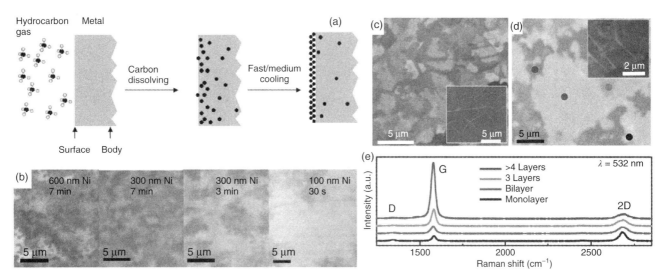

Figure 4.1 (a) Illustration of carbon segregation at metal surface. (b) Optical microscope images of graphene films transferred on 300 nm SiO$_2$/Si substrates. Images of the graphene films grown on different thicknesses of Ni layers and growth time. The sample grown on 100 nm thick Ni for 30 s shows monolayer domains as large as 20×20 m^2. Various spectroscopic analyses of the large-scale graphene films grown by CVD. (c) SEM images of as-grown graphene films on thin (300 nm) nickel layers and thick (1 mm) Ni foils (inset). (d) An optical microscope image of the graphene film transferred to a 300-nm-thick silicon dioxide layer. The inset AFM image shows typical rippled structures. (e) Raman spectra (532-nm laser wavelength) obtained from the corresponding colored spots in (d).

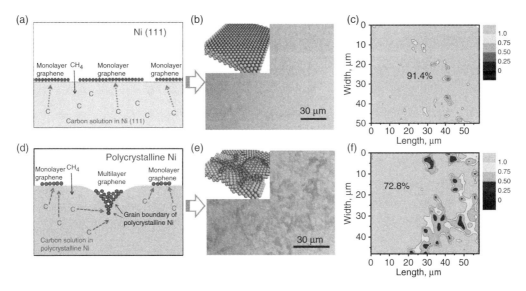

Figure 4.2 Schematic diagrams of graphene growth mechanism on Ni (111) (a) and polycrystalline Ni surface (d). (b) Optical image of a graphene/Ni (111) surface after the CVD process. The inset is a three-dimensional schematic diagram of a single graphene layer on a Ni (111) surface. (e) Optical image of a graphene/polycrystalline Ni surface after the CVD process. The inset is a three-dimensional schematic diagram of graphene layers on polycrystalline Ni surface. Multiple layers formed from the grain boundaries. (c) Maps of $I_{G'}/I_G$ of 780 spectra collected on a 60×50 µm^2 area on the Ni (111) surface and (f) 750 spectra collected on a 60×50 µm^2 area on the polycrystalline Ni surface.

low solubility (nearly 0% at 1000 °C) (28) and different growth mechanisms (16). Therefore, many researchers have attempted to make uniform graphene products on Cu substrates relative to the graphene growth on the Ni film. Detailed studies of graphene growth on Cu films are discussed in the following.

4.2.1.2 *Self-Limited Growth of Graphene on Copper*
In contrast to Ni, Cu exhibits a different carbon solubility and growth mechanism; carbon atoms are precipitated from Ni, but the graphene grows by surface reactions on a Cu surface. Predominantly monolayer graphene films

TABLE 4.1 Specific CVD Process Conditions

Date	Ref.	Ni Film	Annealing	Growth	Cooling	Layer
2008	(17)	Polycrystalline Ni foil (0.5 mm)	1000 °C, 1 h in H_2	CH_4 = 15 sccm, H_2 = 100 sccm, Ar = 200 sccm 1 atm, 20 min, 1000 °C	Fast: 20 °C/s Medium: 10 °C/s Slow: 0.1 °C/s	3–4 layer
	(24)	Evaporated Ni film (500 nm)	900–1000 °C, 10–20 min Ar = 600 sccm, H_2 = 500 sccm	CH_4 = 5–25 sccm, H_2 = 1500 sccm, 1 atm, 5–10 min, 900–1000 °C		1–12 layer Single/bilayer region ~20 µm
2009	(29)	Evaporated Ni film (100 nm)	800 °C, Ar/H_2 = 10/1	CH_4 = 100 sccm, H_2 = 600 sccm, 1 atm, 8 min, 800 °C	0.15 °C/min in Ar	2–3 layer
	(22)	Evaporated Ni film (300 nm)	1000 °C in Ar	CH_4 = 550 sccm, H_2 = 65 sccm Ar = 200 s 7 min, 1000 °C	10 °C/s in Ar	Predominantly Mono and bilayer
	(18)	Single crystal Ni and polycrystalline Ni (500 nm)	Fast: 175 °C/min, 5 min Medium: 58 °C/min, 20 min Slow: 27 °C/min, 20 min annealing	CH_4 = 80 sccm, H_2 = 600 sccm Atmospheric pressure, 10 min, 900 °C	16 °C/min	Mono and bilayer

Year	Ref.	Catalyst/Substrate	Conditions	Growth	Cooling	Result
2010	(30)	Sputtered Ni film (500 nm) Deposited temperature: 50–450 °C Ar pressure: 3–15 mTorr	900–1100 °C, H$_2$ = 200 sccm, Ar = 800 sccm 1–40 min	CH$_4$ = 4 sccm, H$_2$ = 1300 sccm 5 min, 800 °C	Cooling in H$_2$, Ar	
	(31)	Carbon implantation on 200 nm Ni film		900 °C, 30 min	0.5 °C/s to 725 °C Quenched to room temperature	~4 layer
2011	(32)	Polycrystalline Ni film (200 mm) Evaporated on SiO$_2$/Si with ~2.6 at.% C in bulk	1100 °C in vacuum	1100 °C, 0–100 min 0.4–0.004 Pa	2–50 °C/min	Mono or bilayer Up to 95%

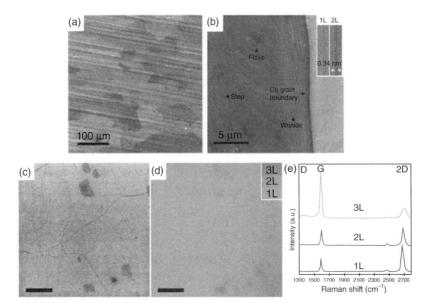

Figure 4.3   (a) SEM image of graphene on a copper foil with a growth time of 30 min. (b) High-resolution SEM image showing a Cu grain boundary and steps, two- and three-layer graphene flakes, and graphene wrinkles. Inset in (b) shows TEM images of folded graphene edges. 1L, one layer; 2L, two layers. (c) SEM image of graphene transferred on SiO$_2$/Si (285-nm-thick oxide layer) showing wrinkles, as well as two- and three-layer regions. (d) Optical microscope image of the same regions as in (c). (e) Raman spectra from the marked spots with corresponding colored circles or arrows showing the presence of one, two, and three layers of graphene.

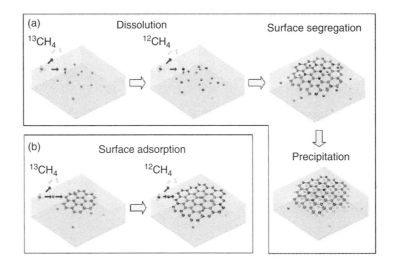

Figure 4.4   Schematic diagrams of the possible distribution of C isotopes in graphene films based on different growth mechanisms for sequential input of C isotopes. (a) Graphene with randomly mixed isotopes such as might occur from surface segregation and/or precipitation. (b) Graphene with separated isotopes such as might occur by surface adsorption.

are easily synthesized on Cu substrates due to the extremely low carbon solubility (28). Li et al. first reported that graphene was grown on Cu using a surface-catalyzed process, and the monolayer accounted for >95% of the graphene films. A SEM image of graphene grown on copper foil is displayed in Figure 4.3a and b; the copper grain boundaries, step edges of copper and wrinkles of graphene are clearly visible due to the different thermal expansion coefficients of copper and graphene (28). The inset in Figure 4.3b also displays the transmission electron microscopy (TEM) images of graphene grown on copper foil. To investigate the graphene grown on copper further, Raman spectroscopy is used to evaluate quality of the graphene after transferring it to the SiO$_2$/Si substrate. Figure 4.3c and d displays the SEM and optical images of the graphene transferred on a SiO$_2$/Si substrate. The Raman spectra in Figure 4.3e indicate that the graphene films grown on copper foil are high in quality and uniform, while the small dark portions are multilayer graphene.

Figure 4.4 presents a schematic diagram of different graphene growth mechanisms accessed using the CVD method (16). To investigate the different mechanisms and kinetics of graphene growth on Ni and Cu substrates, Li et al. isotopically labeled the carbon source with ^{12}C

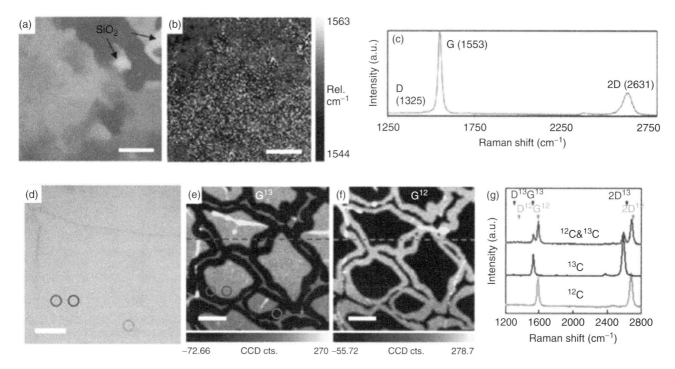

Figure 4.5 Optical micrograph and distribution of C isotopes in a FLG film grown on Ni. (a) An optical micrograph of a FLG film transferred onto a SiO$_2$/Si wafer. (b) The corresponding Raman map of location of the G ba and (c) a typical Raman spectrum from this film, showing the film consists of randomly mixed isotopes (with an overall composition of 45% ^{13}C and 55%. Scale bars are 5 μm. Micro-Raman characterization of the isotope-labeled graphene grown on Cu foil and transferred onto a SiO$_2$/Si wafer. (d) An optical micrograph of the identical region analyzed with micro-Raman spectroscopy. Integrated intensity Raman maps of (e) G^{13}(1500−1560 cm^{-1}), (f) G^{12}(1560−1620 cm^{-1}) of the area shown in (d). Scale bars are 5 μm.

and ^{13}C introduced in sequence (16) under the same growth conditions (Fig. 4.4). Figure 4.5 reveals the Raman mapping of the graphene grown on Cu and Ni substrates after transferring the graphene films to a SiO$_2$/Si substrate. For the graphene grown on a Ni substrate, there is no distinguishable separation of the different isotopes, suggesting a precipitation mechanism occurs during the cooling process (Fig. 4.5a and b). The Raman spectra indicate that the graphene is not uniform, but it formed multiple layers (Fig. 4.5c). However, optical microscopy images of the graphene grown on Cu foils reveal a uniform and dominant graphene monolayer (Fig. 4.5d), but the Raman spectra (Fig. 4.5e–g) indicate that both ^{12}C and ^{13}C regions are clearly distinguished in the optical images. This result supports the hypothesis that graphene grows progressively outwards from the initial nucleation site on Cu through a surface adsorption and reaction mechanism (33). More interestingly, the graphene growth seemed to be self-limited on the Cu surface when grown under vacuum; a single layer of graphene with very few bi- or trilayer islands are produced uniformly (34).

However, the origin of the adlayer graphene formation mechanism is not fully understood and remains under discussion (35, 36). Recently, Li and their collaborators studied and reported the origin of the bi- and multilayer graphene grown on copper substrates with the previously described carbon isotopic labeling method using micro-Raman mapping and time-of-flight secondary ion mass spectrometry (TOF-SIMS) (37). They surmised that the adlayer graphene can be grown on the underlying graphene surface on the copper foil by diffusing the decomposed carbon precursors at high temperatures (Fig. 4.6a). To identify the proposed mechanism of ad-layer graphene formation, they employed a TOF-SIMS depth profile analysis on decomposed carbon isotope images to determine the order of layer growth. Each atomic carbon layer could be removed with fine control to achieve high sensitivity (down to parts-per-billion) and mass resolution (m/δm > 7000), allowing the detection of all components in a sample (37). Figure 4.6b–f presents the TOF-SIMS mapping image of the ^{12}C and ^{13}C secondary ion intensities. As the Cs$^+$ sputtering (1 kV energy) time increases, the upper layer of the graphene is removed progressively; the smallest region therefore contains the most recently grown graphene flake, as displayed in Figure 4.6b–f. The total intensity image for the multilayer graphene is presented in Figure 4.6g; the cross-sectional images with the marked lines in Figure 4.6h and i indicate that the ad-layers grow below the upper graphene layers. According to the given results, they suggest that the graphene adlayers appeared as interface layers between graphene and Cu. In addition, this evidence has also been reported that gas molecules can diffuse into the interface between the graphene and the substrate (38, 39). The molecules in the interface layers are explained by the weak interactions between graphene and the substrate (39). Using this method, they also generated submillimeter bilayer graphene films using a slower growth rate condition than other research groups (40–42).

As the interest in graphene grown on metal substrates by CVD increases due to the high performance, unique properties and unlimited scalability of monolayer graphene film production, many research groups have focused on tuning the growth conditions to decrease the

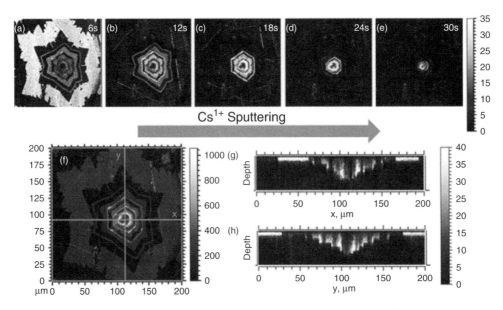

Figure 4.6   (a) Schematic of growth mechanism of the adlayer graphene. TOF-SIMS mapping ($200 \times 200 \ \mu m^2$) of isotopically labeled multilayer graphene on Cu foil. (b–f) The ^{12}C isotope distribution images of graphene by TOF-SIMS after 6 s (b), 12 s (c), 18 s (d), 24 s (e), and 30 s (f) 1 kV Cs^{1+} ion beam sputter. (g) The overall sum image of 36 total images. (h, i) The cross-sectional views of ^{12}C graphene from the marked x (h) and y (i) lines in (g). The color scale represents secondary ion intensity.

defects caused by growth mechanisms over various metal substrates. Researchers have studied growth mechanism of the graphene to produce graphene films on a large scale with few defects. However, theoretical studies have predicted that utilizing hydrocarbons on the Cu substrate is unfavorable (43).

Bhaviripudi et al. reported the kinetic models of graphene growth on Cu with different cooling rates and pressure conditions; ambient-pressure CVD, low-pressure CVD, and UHV CVD are discussed to understand the pressure effects (34). When studying the graphene growth under ambient pressure, thick graphene with nonuniform surfaces appeared, especially at high methane flow rates (lowering the flow rate somewhat increased the surface uniformity and decreased the layer thickness). However, low pressure and UHV conditions facilitate the typical growth of uniform single-layer graphene, as previously reported (28). Therefore, the previously reported self-limiting characteristic of graphene growth on Cu was restricted to low pressure growth conditions (33).

Concurrently, Li and collaborators also reported their optimized growth conditions using a two-step CVD process to produce high-quality graphene films. They defined parameters, temperature (T), methane flow rate (J_{Me}), and methane partial pressure (P_{Me}) and excluded Cu grain orientation, which affects the domain size of the graphene grown on Cu foil (44). The SEM images of partially grown graphene flakes with controlled variables, such as T, J_{Me}, P_{Me}, are shown in Figure 4.7. When the temperature increased or J_{Me} and P_{Me} decreased, the number of graphene nuclei decreased, as displayed in Figure 4.7a–d. These results suggest that domain size of graphene increased due to the low density of graphene nuclei sites under relatively high T, as well as low J_{Me} and P_{Me} (44). They also attempted to grow ultralarge single crystalline graphene domains up to the submillimeter scale (~few hundreds of micro meters) with the enclosed structure of copper foil due to the much lower partial pressure of carbon species; the lower partial pressure allows these growth conditions to produce high-quality graphene films on a large scale (45).

Vlassiouk et al. reported critical role of hydrogen gas flow under the CVD growth process (46). Before this paper, most researchers mentioned that the transition metal substrates required treatment with H_2 gas, reducing and annealing the metal surface during growth process due to the oxidative surface of the as-prepared metal substrate. According to the report, hydrogen gas strongly contributes to graphene growth because it activates the surface reaction and controls the size and morphology of the graphene domains (46). In addition, the morphology and size of the graphene domains could be varied using different hydrogen partial pressures. At first, graphene could not be grown at very low hydrogen pressures (just enough to anneal the metal surface effectively), no matter how high methane pressure was. However, once the hydrogen pressure exceeded 2 Torr, graphene growth was observed, even with a low methane pressure. Moreover, hydrogen gas also acts as an etching reagent to modify the size and shape of the graphene grains. The researchers observed highly irregular domain shapes under moderate hydrogen pressure (P_{H2} of 4 and 6 Torr), while perfect hexagons (Fig. 4.8) appeared under high hydrogen pressure (P_{H2} of 19 Torr). High hydrogen pressures also help form the largest domain size and uniform zigzag edge characteristics (under moderate hydrogen pressure, the edges were mixed zigzags and armchairs). Therefore, hydrogen pressure during graphene growth is critical for controlling crystalline shape and edges of the graphene grains (46).

Yu et al. (47) have successfully formed a uniform hexagonal graphene domain using a nucleation/seeded growth method. The group believes that, through further study, generating single crystalline and boundaryless graphene sheets with fewer defects may be possible.

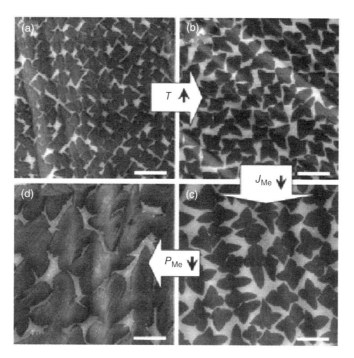

Figure 4.7   SEM images of partially grown graphene under different growth conditions: T (°C)/J_{Md} (sccm)/P_{Me} (mTorr): (a) 985/35/460, (b) 1035/35/460, (c) 1035/7/460, (d) 1035/7/160. Scale bars are 10 μm.

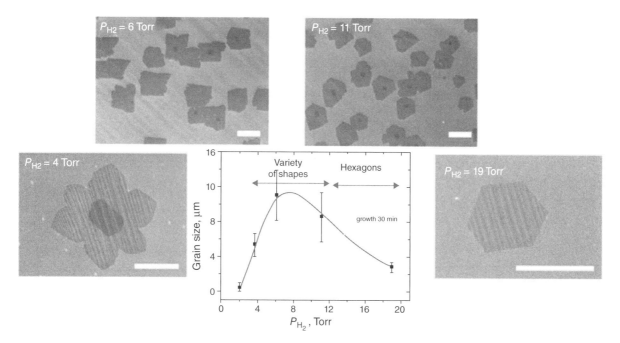

Figure 4.8   The average size of graphene grains grown for 30 min at 1000 °C on Cu foil using 30 ppm methane in Ar mixture at 1 atm, as a function of partial pressure of hydrogen. The inserts illustrate SEM images of the typical shapes under these different conditions. Note that perfect hexagons are observed only at higher hydrogen pressures. Irregularly shaped grains grown at low hydrogen pressure have smaller size second layers (and even third layers on some) in the centers of grains. Scales bars are 10 μm (top two images) and 3 μm (bottom two images).

Since then, Geng et al. has suggested another method to synthesize graphene on a liquid copper substrate, allowing more effective control over nucleation density of the graphene by the CVD process for self-aligned graphene with a large single domain (48). Liquid Cu surfaces

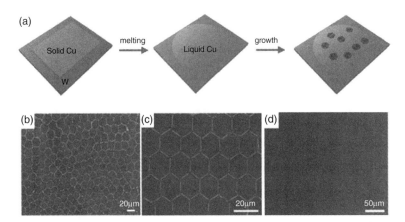

Figure 4.9   The growth of HGFs on flat liquid Cu surfaces on W substrates. (a) Scheme showing CVD process for the synthesis of HGFs on liquid Cu surface. (b) SEM image of HGFs showing a compact assembly of HGFs in which the dark and bright parts represent HGFs and the Cu surface, respectively. (c) SEM image of a near-perfect 2D lattice composed of similar-sized HGFs. (d) SEM image of the sample for 2 h growth showing the continuous graphene film with uniform contrast.

allow graphene domains that are more homogeneous and uniform because there is no grain boundary, unlike the solid copper substrate. The schematic formation of graphene on a liquid Cu substrate is illustrated in Figure 4.9. As the number of hexagonal graphene flakes (HGFs) on the Cu surface increased, the selfaligned HGFs form an ordered structure with the most compact packing arrangement (Fig. 4.9b). The low-angle grain boundaries were formed through an edge-to-edge alignment of the HGFs. Surprisingly, when the HGFs have a similar size, perfectly ordered 2D lattice structured HGFs were obtained (Fig. 4.9c). This finding indicates that the translation or rotation of HGFs on a liquid Cu surface occurs during the self-assembly of their ordered structures, and the minimized total HGF surface/edge energy on liquid Cu is the driving force for the alignment. A typical SEM image depicting a large area of continuous graphene film grown on Cu films for 2 h is displayed in Figure 4.9d and exhibits very similar images to those grown at low-pressure CVD; however, the original shape and edges of the HGFs disappear. This improved fabrication process allows the production of self-aligned, large-sized, and uniform single crystal monolayer graphene films (48).

High-quality monolayer or bilayer graphene covering a large area has been obtained using the CVD method with CH_4 or C_2H_2 gases on Cu or other metal substrates (17, 22, 28). However, the conventional CVD process is limited to gaseous carbon sources under vacuum, hindering the use various potential feedstocks for producing graphene. To overcome the limits and drawbacks of CVD, new methodology has been developed to grow graphene using different carbon sources, such as poly(methyl methacrylate) (PMMA), sucrose, benzene, methanol, and ethanol, that are flowed or deposited on a Cu layer (25, 49, 50). Sun et al. utilize solid carbon sources, such as polymer films or small molecules deposited on the Cu catalyst substrate, and pyrolyze them at temperatures as low as 800 °C to synthesize high-quality graphene with a large area and controllable thickness, as displayed in Figure 4.10a (49). The thickness of the poly(methyl methacrylate)-derived graphene was controlled by changing the Ar and H_2 gas flow rates. Monolayer graphene exhibited a 97.1% transmittance at 550 nm and a sheet resistance of 1200 Ω/sq. The Raman spectra revealed that the 2D peak can be fitted with a single sharp Lorentz peak (Fig. 4.10b), and the ratio of the G and 2D peaks is below 0.4 for more than 95% of the graphene, demonstrating that the high-quality graphene is comparable to conventional graphene produced using gaseous carbon sources. Note that the low concentration and solubility of the carbon source in Cu enable the formation of monolayer graphene. When using solid carbon sources, hydrogen is important for forming the high-quality graphene. Authors emphasized that hydrogen acts as both the reducing reagent and a carrier gas to remove the carbon atoms produced from the decomposing PMMA during growth. Therefore, a slower H_2 flow, as well as highly concentrated and uniformly dispersed carbon sources on substrates, favors multilayer graphene. Ji et al. also described the effects of hydrogen on graphene grown using a solid-state carbon feedstock (25). Graphene forms only in the presence of H_2, strongly suggesting that the gaseous hydrocarbons and/or their intermediates are important for synthesizing graphene on Cu through the reaction between H_2 and the amorphous carbon. However, Suziki et al. argued that graphene converted from polymers on metal substrate, such as polystyrene and polyaniline, can be grown using heat in an Ar atmosphere or under vacuum (51, 52). In the Ar atmosphere, precise control over the initial polymer film thickness (< 2nm) induced only a few layers of graphene growth; however, under vacuum, multilayer graphene films were obtained regardless of the initial thickness of polymer film because the excess carbon atoms are removed from the surface when the polymer is thermally decomposed.

Despite the efforts to improve the quality of graphene films, developing the transfer process to practically use the graphene is also important for the future industrial applications that are discussed in later sections.

4.2.1.3   *Graphene Growth on Other Metal Catalysts*   As mentioned in Sections 4.2.1.1 and 4.2.1.2, Ni and Cu are widely used to grow high-quality graphene and their growth mechanisms vary depending on the intrinsic properties of the metal catalysts and the solubility of the carbon sources. Group 8–10 transition metals can act as catalysts to grow carbon nanotubes (CNT) (53, 54). These transition metals have different solubilities relative to their growth temperature, which is a major variable for modulating the quality of carbon materials. Therefore,

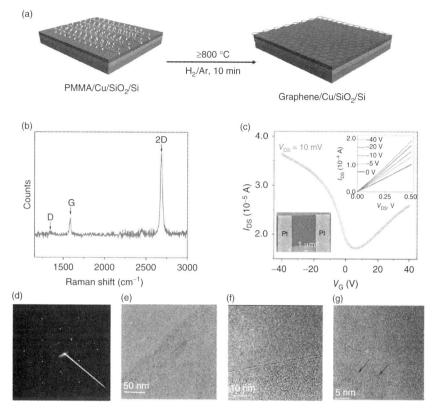

Figure 4.10 (a) Monolayer graphene is derived from solid PMMA films on Cu substrates by heating in an H_2/Ar atmosphere at 800 °C or higher (up to 1000 °C). (b) Raman spectrum (514 nm excitation) of monolayer PMMA-derived graphene obtained at 1000 °C. See text for details. (c) Room temperature I_{DS}–V_G curve from a PMMA-derived graphene-based back-gated FET device. Top inset, I_{DS}–V_{DS} characteristics as a function of V_G; V_G changes from 0 V (bottom) to −40 V (top). Bottom inset, SEM (JEOL-6500 microscope) image of this device where the PMMA-derived graphene is perpendicular to the Pt leads. I_{DS}, drain–source current; V_G, gate voltage; V_{DS}, drain–source voltage. (d) SAED pattern of PMMA-derived graphene. (e–g) HRTEM images of PMMA-derived graphene films at increasing magnification. In (g), black arrows indicate Cu atoms.

single-walled carbon nanotubes could be synthesized by finely controlling the experimental conditions based on properties of the metals and growth mechanisms. Numerous researchers have attempted to use transition metals, such as Fe (55, 56), Ru (57–59), Co (60–63,), Rh (64, 65), Ir (66, 67), Pd (68, 69), Pt (70, 71), and Au (72) and numerous alloys including Cu-Ni (32, 42, 73, 74), Au-Ni (75), and Ni-Mo (76) for growing graphene. In this subsection, transition metals, excluding those on which graphene is epitaxially grown (Ru, Rh, and Ir), will be discussed.

Among the noble metals, such as Pt, Au, and Pd, the growth mechanism of graphene on Pd metal catalyst stands out due to the relatively high solubility of carbon in Pd (up to 1.4% at 1023 K) compared to Pt (~0.06 atom%) and Au (~0.06 atom%). Kodambaka et al. demonstrated *in-situ* graphene growth on Pd using ethylene (C_2H_2) gas in an UHV multichamber (base pressure < 2×10^{-10} Torr) STM system (68). The graphene islands on the Pd surfaces form via a segregation mechanism with dissolved carbon atoms in the lattice because of the high solubility of carbon in Pd similar to graphene on Ni described in Section 4.2.1.1.

Other metals, such as Co and Fe, with similarly high carbon solubility (~1 atom% at 1000 °C) are used to grow graphene using the rapid thermal CVD method with a precipitation mechanism. Ramon et al. reported that the graphene on Co films could be grown under a low decomposition temperature using acetylene gas (C_2H_2) because the interfacial electronic coupling between the graphene π-state and the Co d-states is quite strong compared to the other transition metals (61). A number of graphene layers increased as the thickness of Co films increase, which was observed with graphene growth on Ni films. Therefore, they controlled the thickness of the Co film to synthesize predominantly monolayer graphene, covering over 80% of the areas on Co film with the 100 nm thickness. This approach is very useful and can be used in spintronic devices due to the ferromagnetic properties of Co.

As described, Fe also complies with the precipitation mechanism due to its high solubility of carbon. There have been numerous studies describing CNT growth on Fe catalysts (56). Recently, the selective growth of vertically aligned CNTs and graphene films on Fe films also have been reported by Yudasaka et al. (77) and Fujita et al. (78), but the details remain unclear (56). To undertake a further systematic study, Kondo et al. (56) demonstrated that selective growth could be achieved by changing the thickness of the Fe films at 620 °C; this temperature is lower than that in other reports describing a conventional CVD (22, 28, 88). Figure 4.11 displays the TEM analysis of the carbon allotropes on various Fe films ranging in thickness from 2.5 nm to 200 nm. Except for the image of the 2.5 nm Fe films presented in Figure 4.11a, carbon

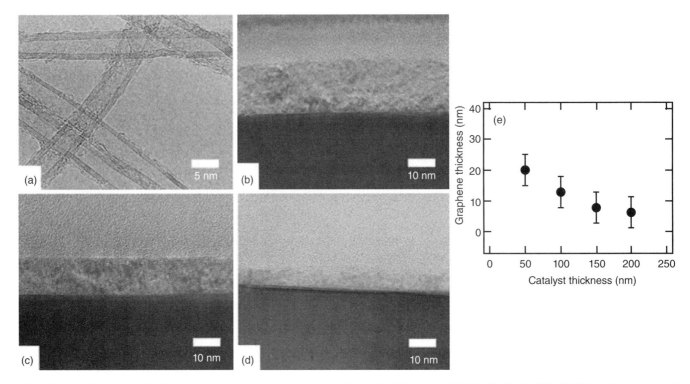

Figure 4.11 TEM images of carbon materials obtained from iron catalyst films with thicknesses of (a) 2.5, (b) 50, (c) 100, (d) 200 nm, (e) thickness of multilayer graphene as a function of catalyst thickness.

and graphene films were observed on every Fe film (50, 100, and 200 nm thick) (Fig. 4.11b–d). Additionally, only CNTs can be prepared on the Fe films less than 5 nm thick. Interestingly, as the thickness of the Fe films increased, the thickness of the graphene layers gradually decreased (Fig. 4.11e). It is speculated that the same amount of carbon is dissolved in an Fe film regardless of the film's thickness, leading to a lower fraction of carbon in thicker Fe films. In addition, the thickness of multilayer graphene resulted from the carbon fraction in the Fe film. Therefore, monolayer graphene could be obtained from Fe films up to 500 nm or more. These results indicate that the vertically aligned CNTs and multilayered graphene films could be obtained using the same substrate, suggesting a simple way to produce nanomaterials containing CNTs and graphene for composites.

Graphene films can also be synthesized on Pt catalyst layers using the conventional CVD method (70, 71). The carbon solubility in Pt is very similar to that of Cu and much smaller than that of a representative transition metal, such as Ni; therefore, the graphene was expected to display the self-limiting growth mechanism on Pt films, as reported by Gao et al. (70).

Oznuluer et al. reported that polycrystalline gold foil could be used to form graphene using the CVD method (72). Carbon solubility in Au (0.06 at%) is slightly higher than that of Cu (0.04 at%) but significantly lower than that of Ni (2.7 at%) (79). They proposed that the mechanism on Au is similar to that of Cu based on the solubility and various experimental results under given synthetic conditions. The Raman spectra revealed that the 2D peak was fitted with a symmetric Lorentzian shape and a FWHM of 37 cm^{-1}, indicating single layer graphene. However, the intensity of the D peak in the Raman spectra of the graphene is higher than that of the G peak. This observation is different from that for the graphene grown on Ni or Cu. The intensive D peak may result from the large lattice mismatch between graphene and Au. Although the graphene grown on Au exhibited defective graphene properties, the on/off ratio and the field-effect mobility of FET devices using this graphene were approximately 2 and 20 cm^2/(V s), respectively.

Generally, the carbon source decomposes on the catalyst, and a precipitation or self-limiting mechanism governs the graphene growth, as discussed for a popular mono metal catalyst system. Understanding the detailed mechanism is extremely important when optimizing experimental variables to generate high-quality graphene and inventive ideas for more cost-effective and simplified processes and materials. Researchers in the field of graphene synthesis have designed a new metallurgical model (viz., binary metal system) that maximized the advantageous effects of each metal even though the use of metal alloys is a well-known approach for controlling the activity and selectivity for multiple complex reactions in heterogeneous catalysis.

Cu–Ni alloy metals as a catalyst to grow the graphene have been used (32, 42, 73, 74). Liu et al. demonstrated a batch production of 4 inch wafer-scale graphene using segregation technique, which precipitated carbon atoms directly from Ni film containing 5.5 at% of carbon (32). In this process, key factors to grow high-quality graphene are carbon solubility, melting point, lattice mismatch between metal and graphene, and chemical stability of metals. Over 82% of whole area showed 1–3 layer graphene, leading to approximately 750 Ω/sq of sheet resistance and 94% of transmittance at 550 nm. In Cu/Ni/SiO$_2$/Si sandwiched structure, Cu and Ni layer play a role in segregation and thickness controller, respectively, due to the distinct difference of the solubility.

An interesting study to grow high-quality graphene using polycrystalline Ni film with Au at low temperature (~450 °C) was reported (75). The graphene on Au–Ni alloy catalyst showed monolayer coverage of approximately 74% and lateral domain in excess of 15 μm. The authors suggested the role of Au based on the in-situ and ex-situ experimental results that thermally evaporated Au modified lots of highly reactive sites of Ni surfaces, such as step edges, leading to decrease in the nucleation density for graphene growth compared to the elemental Ni catalyst. Thus, homogeneous nuclei on Au–Ni alloy metal catalyst promote the growth of monolayer graphene. These further studies on various transition metals and binary metal systems to understand the mechanism of graphene growth provide to obtain high-quality graphene films by CVD system for electrical device applications.

4.2.2 Transfer Methods for Electronic Device Applications

High-quality graphene grown on metal catalysts must be transferred from the metal to a desirable substrate to use it in practical devices. Atomically thin graphene cannot avoid some types of damage, such as folding, wrinkles, tearing, cracking, and rolling, which degrade the quality. For the wide use and commercialization of graphene, an innovative idea for a transfer method that minimizes damage is as important as graphene growth. In this section, a brief introduction to the transfer methods and the properties of transferred graphene to target substrates relative to the protective layer will be explained.

PMMA is a popular polymer used as a support material to transfer the graphene film onto an arbitrary substrate (16, 24, 28, 44, 45, 80, 81). The transfer of CNT has already been described in a previous report (81). Reina et al. first demonstrated in 2008 that CVD-graphene films can be easily transferred onto an arbitrary substrate, such a SiO_2/Si substrate with a PMMA support, using a wet-etching process on the metal catalyst layer (80). After transferring the PMMA/graphene layers onto the desired substrate, the PMMA polymer support can be easily removed in an organic solvent, such as acetone. Consequently, many research groups have employed the PMMA polymer-based transfer method for large-scale CVD-graphene film and improved this transfer method. Li et al. found that CVD graphene films can crack during the transfer process due to the poor mechanical properties of a monolayer graphene film on a Cu substrate (82). After transferring the graphene films to a SiO_2/Si substrate, some small air gaps remain between the graphene and the substrate surface and the graphene does not make full contact with the SiO_2/Si substrate. In this case, the unattached regions of the graphene films may be easily broken and form cracks when the PMMA polymer support is removed with acetone. To prevent this occurrence, they suggest that an additional curing process with PMMA solution be applied to mechanically relax the underlying graphene films, allowing better contact between graphene and the substrate. Figure 4.12 displays the transfer process for the graphene films using PMMA. This additional process improved transfer method, showing better quality than the old method (82).

Afterward, Wang et al. suggests another method: large-area delamination of graphene films with PMMA from Cu foils using an electrochemical process, as presented in Figure 4.13 (83). The H_2 bubbles from the electrolysis of water permeate between the PMMA/graphene and copper foil from the edges of the copper foil under an electrochemical process with an electrolyte such as $K_2S_2O_8$ (Fig. 4.13b and c). In the peeling process, the PMMA plays an important role in supporting the graphene films to prevent rolling or tearing of them, suggesting that this etching and transfer process afforded many numerous advantages, such as efficiency, low cost, and reusability of the copper substrate for high-quality graphene films produced on a large scale (Fig. 4.13d).

Another alternative polymer support is polydimethylsiloxane (PDMS) for CVD-graphene transfer. PDMS has many advantageous properties, such as its durability, moldability, and stretchability suitable for various applications. This material can also be employed for soft lithography using the stamping method due to its low adhesion force on interfacial surfaces (84). Kim et al. first demonstrated the graphene transfer process with PDMS polymer support using the stamping process (22). Figure 4.14a demonstrates the fabrication process of large-scale and patterned graphene films. At the first stage, the PDMS support is attached to graphene grown on a Ni-deposited SiO_2/Si substrate (Fig. 4.14b) and then the Ni catalyst layer is etched away in an $FeCl_3$ solution (Fig. 4.14c). Afterward, the PDMS/graphene layers are separated from the SiO_2/Si substrate (Fig. 4.14d) and attached to the desire substrate (in here; SiO_2/Si) for the soft lithography process (Fig. 4.14e). Graphene films can be released from PDMS to the silica substrate due to the relatively lower surface energy of the SiO_2/Si substrate (Fig. 4.14f).

However, this process etches the wafer-scale graphene films very slowly, and Lee et al. suggest a modified method to overcome the long process time (85). One of the key differences is that the PDMS/graphene/Ni/SiO_2/Si substrate is soaked in water with mild sonication, as shown in Figure 4.15a. The PDMS/graphene layers can be easily detached from the SiO_2/Si substrate due to the different wettability between the metal and the SiO_2 substrate. This approach requires only a few minutes to remove the metal catalyst layers completely, compared to the previous methods (>few hours) (22, 24). The separated graphene films on the PDMS can be patterned using conventional photolithography, and prepatterned graphene films can be transferred onto a target substrate using this method, as illustrated by Figure 4.15b and c. This transfer method provides a high-throughput route for the wafer-scale production of graphene films for practical applications. Even with the enhanced transfer methods, the scalability problem for graphene films remains in industrial production.

Thermal release tape can alter the polymers mentioned earlier and is the one promising material used as a support during the industrial graphene transfer process. For example, as demonstrated by the works of Caldwell et al., epitaxially grown graphene on SiC can be transferred using thermal release tape and a steel pressure plate (86). The adhesive force pulls the graphene from the SiC substrate, while the graphene can be placed by a simple release process (87). Bae et al. first developed a graphene transfer method using a roll-to-roll process with thermal release tape (Jinsung Chemical Co. and Nitto Denko Co.) (88). Due to the flexibility of graphene and copper foils, efficient etching and transfer processes that are cost- and time effective, such as roll-to-roll production, are possible. Transfer processes are simple, as shown in Figure 4.16a. Graphene grown on copper foil is pressed with thermal release tape (TRT) to adhere the TRT to the graphene on the copper foil. After the copper layers are etched away, the graphene films on the TRT support can be easily transferred onto target substrates, such

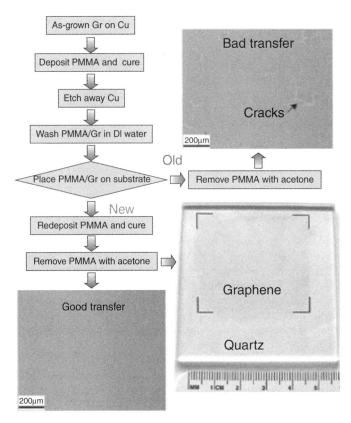

Figure 4.12   Processes for transfer of graphene films ("Gr" = graphene). The top-right and bottom-left insets are the optical micrographs of graphene transferred on SiO_2/Si wafers (285 nm thick SiO_2 layer) with "bad" and "good" transfer, respectively. The bottom-right is a photograph of a $4.5 \times 4.5 \, cm^2$ graphene on quartz substrate.

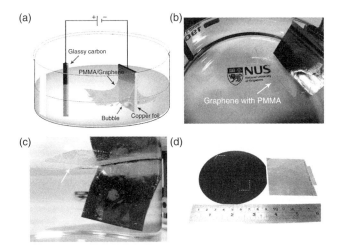

Figure 4.13   Electrochemical exfoliation of graphene from Cu foil. (a) Schematic diagram of electrochemical cell used for the electrochemical exfoliation. (b–c) Optical images showing the "whole film" peeling of PMMA-covered graphene from the copper foil. (d) Graphene film was transferred onto a 4 inch wafer by the wetting transfer process.

as polyethylene terephthalate (PET), during the pressing process with mild heating (~120 °C) (Fig. 4.16b). In addition, to avoid expanding and to minimize the damage to the PET substrate, they used the lowest released temperature of TRT because the glass transition temperature of the PET substrate (~70 °C) is lower than the release temperature (~120 °C). In this approach, up to a 30 inch scale graphene film can be transferred onto target substrates (Fig. 4.16c). Figure 4.16d reveals the first demonstrated practical application of roll-to-roll graphene transfer on PET for a flexible touch screen device (88). The resulting graphene-based touch panel application means it is possible to make a product

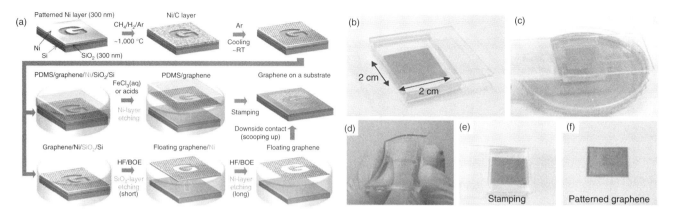

Figure 4.14 (a) Synthesis of patterned graphene films on thin nickel layers. Etching using FeCl$_3$ (or acids) and transfer of graphene films using a PDMS stamp. Etching using BOE or hydrogen fluoride (HF) solution and transfer of graphene films. RT, room temperature (25 °C). The dry transfer method based on a PDMS stamp is useful in transferring the patterned graphene films. After attaching the PDMS substrate to the graphene (b), the underlying nickel layer is etched and removed using FeCl$_3$ solution (c). (d) Graphene films on the PDMS substrates are transparent and flexible. The PDMS stamp makes conformal contact with a silicon dioxide substrate. Peeling back the stamp (e) leaves the film on a SiO$_2$ substrate (f).

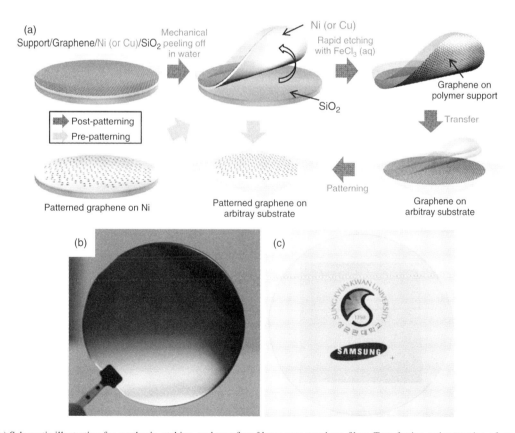

Figure 4.15 (a) Schematic illustration for synthesis, etching, and transfer of large-area graphene films. Transferring and patterning of graphene films grown on a metal/SiO$_2$/Si wafer. Graphene/metal layers supported by polymer films are mechanically separated from a SiO$_2$/Si wafer. After fast etching of metal, the graphene films can be transferred to arbitrary substrates and then patterned using conventional lithography. (b) A image of as-grown graphene film on 3 inch 300 nm thick Ni on a SiO$_2$/Si substrate. (c) A transferred wafer-scale graphene film on a PET substrate.

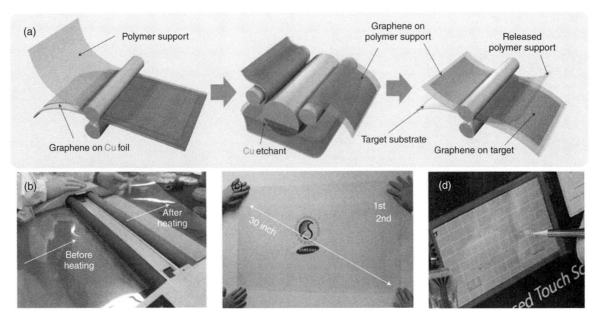

Figure 4.16 Schematic of the roll-based production of graphene films grown on a copper foil. (a) The process includes adhesion of polymer supports, copper etching (rinsing), and dry transfer printing on a target substrate. A wet-chemical doping can be carried out using a set-up similar to that used for etching. (b–d) Photographs of the roll-based production of graphene films. (b) Roll-to-roll transfer of graphene films from a thermal release tape to a PET film at 120 °C. (c) A transparent ultralarge-area graphene film transferred on a 35 inch PET sheet. (d) A graphene-based touchscreen panel connected to a computer with control software.

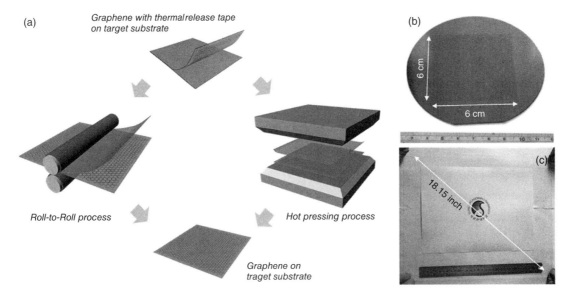

Figure 4.17 (a) Schematic illustration of graphene transfer by R2R and hot pressing; (b) photograph of a $6 \times 6\,cm^2$ graphene film transferred onto a SiO_2/Si wafer by hot pressing; and (c) photograph of an 18 inch graphene film transferred on a glass substrate by hot pressing.

with graphene films on a flexible substrate for industrial applications in the near future after these methods are further improved to reduce defects during fabrication.

When the rolling speed is too fast or when the transfer substrate is rigid, shear stress damages the graphene layers, creating cracks or holes during the roll-to-roll process. Additionally, when the TRT releases, it leaves its adhesives, similar to PMMA. To improve the dry transfer method using TRT to transfer graphene to rigid substrates, Kang et al. used a hot pressing process (89) at a controlled temperature instead of the roll-to-roll process (88). Figure 4.17a depicts a diagram of the hot pressing method and photographs of the graphene transferred to the

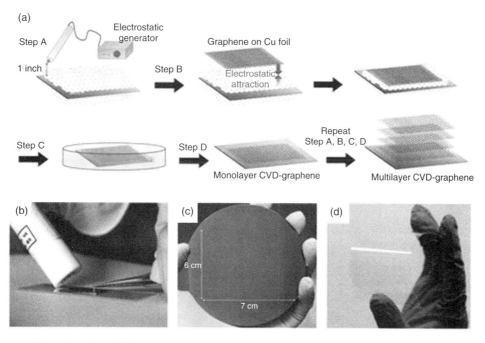

Figure 4.18 (a) Schematic illustrations of the CLT processes of as-grown graphene on Cu foil onto a substrate. Large-area graphene films transferred onto (b) Photograph of electrostatic discharging on the substrate using an electrostatic generator. (c) A SiO$_2$/Si wafer and (d) a flexible PET substrate using the CLT technique.

SiO$_2$/Si and PET substrates (Fig. 4.17b and c). They also reported that the mechanical stress applied to the graphene on the SiO$_2$/Si substrate when releasing the TRT is lower during the hot pressing process than the roll-to-roll process onto a SiO$_2$/Si substrate in a theoretical simulation (89). The roll-to-roll transfer and hot pressing techniques with continuous graphene synthesis will allow for the industrial-scale production of graphene in the near future (88).

However, these transfer methods inevitably contaminate the transferred graphene surface onto arbitrary substrate with organic polymer layers and might considerably degrade the properties of graphene films relative to the as-grown graphene films. Chen et al. developed a unique transfer technique for graphene films without using an organic support or adhesive; instead, electrostatic force is used, and it allows a graphene film to be easily attached to the target substrate (90). Figure 4.18 presents a photograph and scheme for the fabrication of graphene films using an electrostatic generator. At first, the graphene is grown on copper layers, and then graphene/Cu film is attached to a target substrate, such as PET, glass, or a SiO$_2$/Si substrate treated with induced electrostatic forces that accumulate charges on the surface. Subsequently, a pressing process is employed to ensure that the graphene is attached to the substrate. Finally, the Cu foils are etched away using FeNO$_3$, and the graphene films can be transferred onto the target substrate without the support (Fig. 4.18b–d). This new approach provides an efficient route for producing residue-free graphene films on both rigid and flexible substrates for high-quality graphene-based electronic applications.

As the technologies for graphene synthesis gradually develop, allowing the mass production of high-quality graphene films supported on desirable substrates. However, improving technology for the synthesis and transfer should be continuously studied for industrial graphene-based electronic applications.

4.3 DIRECT SYNTHESIS OF GRAPHENE ON AFFORDABLE SUBSTRATES

High-quality, large-area graphene films are usually synthesized via a CVD and/or a solid deposition method at elevated temperatures (~1000 °C) on polycrystalline metal surfaces. After growing the graphene, the catalyst layers below graphene are eliminated using toxic chemicals and the desired material is transferred to another substrate, as is inevitably required for various applications. During transfer processes, the graphene quality degrades because the incomplete elimination of catalytic metals and the formation of defects, wrinkles, and cracks cannot be avoided. In practice, it is not manageable to both completely remove the metal catalysts and prevent damage to the graphene during the metal elimination processes. In this section, to overcome those drawbacks and develop an affordable process, transfer- and catalyst-free processes for synthesizing graphene will be discussed.

4.3.1 Graphene from Transfer-Free Processes Assisted by Catalysts

As mentioned previously, transfer processes introduce numerous complex steps during which graphene can be damaged. The polymers and TRT used for the straightforward approach to detach graphene from the catalyst layer can leave residue and the same artifacts are observed

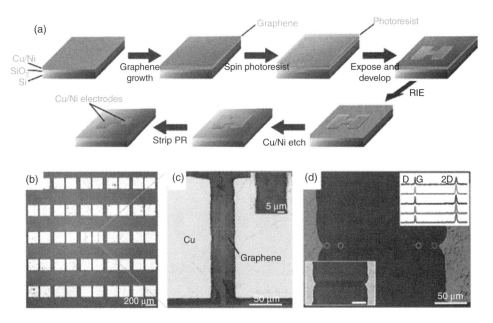

Figure 4.19 Fabrication of SLG device array. (a) Schematic representation of device fabrication procedure (see main text for details). (b) Brightfield optical image of a typical sample substrate after fabrication. (c) Close-up brightfield image of the same sample. Graphene connecting the copper pad is just visible (boxed). Inset: Image of the device channel (100×, NA = 0.9). (d) Differential interference contrast image of a longer device. Upper inset: Raman spectra across the length of the graphene strip are highly uniform. Lower inset: brightfield image of the sample. All brightfield images have been contrast enhanced.

in other physical transfer methods, such as ripping and tearing. Due to the disadvantages of transferring synthesized graphene, even after improving the existing transfer methods and developing new approaches, several researchers have developed transfer-free graphene growth techniques to circumvent these issues (91–95). To eliminate transfer steps, they have discovered how to grow graphene directly onto the desired substrate.

Levendorf et al. reported using CVD to grow graphene on 500 nm Cu (with a thin layer of Ni for adhesion) for device fabrication on SiO_2/Si substrates (Fig. 4.19) (91). Subsequently, instead of transferring the graphene to the new substrate, they directly pattern the graphene while removing a metal layer. Through this process, Cu layers thicker than 500 nm produced high-quality graphene. After the graphene is grown, a conventional photoresist process is applied to protect the areas to be used in device fabrication. Oxygen plasma was used and any exposed areas were removed. With the photoresist still intact, the patterned graphene on the Cu/Ni metal and SiO_2/Si are placed in a continuously refreshed etching chamber to remove any exposed metal. Therefore, any remaining metals reside beneath a thin strip of graphene and remain as part of the device. In the last step, the photoresist layer on the graphene strip is removed, and the final device is tested to demonstrate its performance with good electron mobility ($\sim$700 cm^2/(V s)). In addition, excellent quality of graphene was evident by Raman spectroscopy with universally low D peaks as shown in Figure 4.19d. Although the method still requires etching to remove the metal layers, the authors claim that eliminating the transfer step can reduce the risks of tearing holes in the device, enabling the large-scale production of devices arrays.

Ismach et al. suggested a Cu etching method in a high temperature oven after graphene synthesis (92). They grew graphene directly on Cu-deposited insulating substrates using CVD (Fig. 4.20a) and kept the samples from 15 min to 7 h at high temperatures, leading to the controlled dewetting and evaporation of the Cu layer. To remove the Cu layer entirely, more than 6 h was needed. To investigate structure of graphene after the 6 h treatment, Raman spectroscopy was performed, and the D peak of graphene was significantly larger than that of graphene kept in the reactor for fewer hours. The problem was that the graphene began to degrade after only 2 h under the given high temperature conditions. Although the method completely avoids transfer process, completely removing Cu requires long reaction times and degrades the graphene.

Another group demonstrated a different approach for growing graphene on SiO_2/Si substrates. Su et al. speculated that carbon atoms could travel through the Cu grain boundaries when graphene was synthesized on Cu layers via CVD as shown in Figure 4.20b (93). As discussed in Section 4.2.1.2, graphene grows by surface reaction (self-limiting mechanism) on Cu catalyst. However, they conjectured that graphene could grow underneath the deposited Cu if the carbon atoms could pass through grain boundary on the other side of the Cu layer. Therefore, this synthetic approach does not require transfer steps, but the Cu layer must be still peeled off with a tape. The applied mechanical force tears the graphene along random crack lines, partially damaging it; however, this damage enables control over the size and quality of the remaining graphene on the SiO_2/Si substrate. When a small area of graphene is required, this method is capable of providing it without a lengthy transfer step. However, to produce high-quality graphene over a large area for producing devices, the Cu layer is etched away using common reagents.

Byun et al. first demonstrated a cost- and time-effective transfer-free process for directly forming the graphene on the target substrate using the metal catalyst as a capping layer (metal/polymer/SiO_2/Si structure) as illustrated in Figure 4.21 (94). There are two functions for metal capping layer to prevent from evacuation of decomposed molecules from the polymer and catalyze graphene growth. To discover the optimal

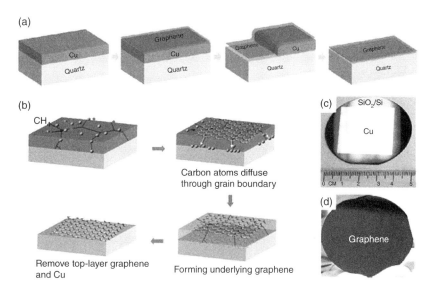

Figure 4.20 (a) Schematic representation of the process. Step 1: a thin layer of copper is evaporated on the dielectric surface, Step 2: During the CVD, Step 3: the metal dewets and evaporates, Step 4: leaving the graphene layer on the substrate. (b) The schematic illustration of the growth. Step 1: The CH_4 precursors dissociate into carbon species and migrate on Cu surfaces, where some of the carbon species diffuse downward through Cu grain boundary (GB). Step 2: The graphene layer formed on the Cu surface; meanwhile, the carbon atoms continue to diffuse through Cu GB and segregate at the Cu–insulator interface. Step 3: The graphitization of carbon atoms underlying Cu film leads to the formation of bottom layer graphene. Step 4: The large-area and continues graphene layers can be obtained directly on substrate after removing the top layer graphene, followed by wet-etching of Cu thin film. (c) Photo of 2 inch insulating substrates: Cu on SiO_2 (300 nm)/Si. (d) Photos of the as-grown bottom layer graphene on the corresponding substrate shown in (c).

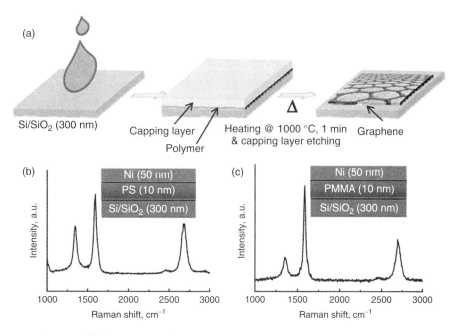

Figure 4.21 (a) Graphene growth process (b) Raman spectra after pyrolysis of PS polymer films with a 50 nm thick Ni capping layer at 1000 °C. (c) Raman spectra after pyrolysis of PMMA polymer films with a 50 nm thick Ni capping layer at 1000 °C.

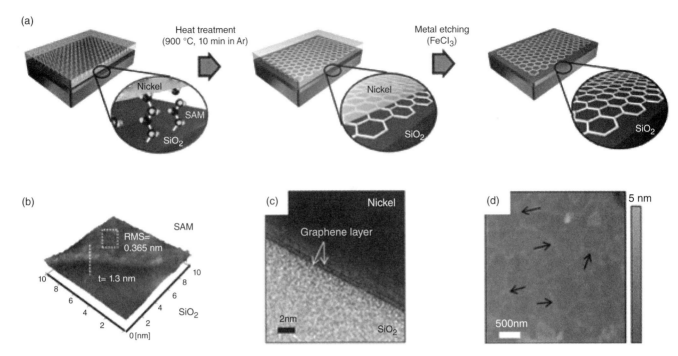

Figure 4.22   (a) Schematic illustration for transfer-free growth of graphene on a substrate. Carbon SAM materials are squeezed between the top metal layer and the substrate, where the top-most metal layer is etched after pyrolysis. (b) AFM image of octyl-SAM on a SiO$_2$/Si substrate. The tapping mode was used to measure the thickness of SAM. A step was formed by removing half of the SAM layer with UV/ozone cleaning. (c) Cross-sectional HRTEM image after pyrolysis of octyl-SAM, showing bilayer graphene. (d) AFM image of graphene originated by octyl-SAM on the substrate after metal etching.

conditions for producing high-quality graphene films, the authors grew the graphene films as the thickness of the Ni capping layers increased up to $\sim$100 nm. Few-layer graphene can be obtained under above 50 nm thick Ni layer, while 10 or 25 nm thick layers produce amorphous carbon film. 50 nm is the optimal thickness of Ni capping layer in this work. Figure 4.19b and c display the Raman spectra of the graphene film converted from polymers (PS: polystyrene and PMMA) by a 50 nm Ni capping layer. However, the Raman spectra reveal that the quality of the graphene films produced using this way is poor regardless of polymers compared to those produced using the conventional CVD system as discussed in previous section.

There are similar approaches toward synthesizing graphene underneath the metal layer reported by Shin et al. (96). They developed transfer-free process that graphene on a dielectric substrate grew by pyrolysis of self-assemble monolayer (SAM) placed between Ni and the substrate (Fig. 4.22a). Controlling layer thickness and stacking of SAM materials, such as octyl-, octadecyl-, and phenyl-SAM (aliphatic SAM versus aromatic SAM) provide high-quality graphene. The thickness and surface roughness of octyl-SAM, which can be theoretically obtained bilayer graphene, were approximately 1.3 nm and 0.36 nm, respectively (Fig. 4.22b). After heat treatment above 1000 °C, bilayer graphene between Ni and SiO$_2$ and grain boundary of the graphene surface can be observed as shown in Figure 4.22c and d, respectively. Interestingly, the authors argued that no wrinkles in the graphene were observed due to the transfer-free growth due to low differential of expansion coefficient between the graphene and SiO$_2$/Si substrate. The wrinkle-free graphene showed carrier mobility of $\sim$4400 cm^2/(V s) at -2×10^{12} cm^{-2}.

Considering that the CVD process occurs at high temperatures under an inert atmosphere, strong and solid insulating materials, such as SiO$_2$/Si wafer, quartz, and sapphire, may be used as substrates because they can endure the high temperatures, while flexible plastic substrates are unsuitable due to their low deformation and/or melting temperatures. Therefore, transferring the graphene grown on solid substrates onto flexible substrates remains necessary in order to use graphene for flexible, stretchable, and foldable devices (87, 88).

Kwak et al. describes a diffusion-assisted synthesis (DAS) method for graphene growth without transfer process at nearly room temperature (25 $\sim$ 160 °C) (97). Growth mechanism in the DAS process is basically similar to the prediscussed approach (93), where the deposited solid carbon sources on the Ni catalyst layer are diffused through the grain boundaries, making the graphene films at the Ni-substrate interface. One critical difference is relatively the low temperature ($\sim$160 °C), compared to previous reports (93) ($\sim$1000 °C), used to build the graphene films onto the desired substrate. In addition, controlling the grain sizes of Ni catalyst film could tune the optical- and electrical-properties of the graphene films. Forming different graphene grain sizes can also be controlled using wrinkle-free graphene films on SiO$_2$/Si, plastic and glass substrates that range from the nano- to micrometer scale; these materials depend on the target substrate due to the annealing process for the deposited poly-Ni films. Figure 4.23a depicts the schematic illustration of DAS process. The optical image and Raman spectra of the graphene attained using this method at T = 160 °C for 5 min are displayed in Figure 4.23b and c. In the optical image, the colored spots correspond to the mono- (red), bi- (blue), and multilayered (green) graphene shown in the Raman spectra (Fig. 4.23c) and mapping

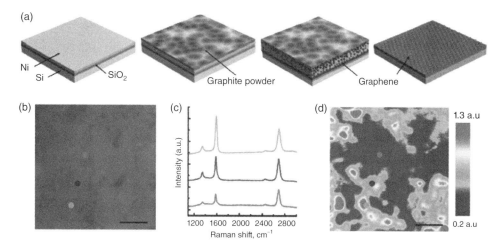

Figure 4.23 (a) Schematic drawing of the DAS process for directly depositing graphene films on nonconducting substrates. The diagrams represent (from left to right) the elementary steps in the DAS process, including deposition (and annealing) of Ni thin films on desired substrates (SiO_2/Si or PMMA, glass), preparation of diffusion couple of C–Ni/substrate, annealing in Ar or air (25–260 °C) to form C–Ni/graphene/substrate and formation of graphene on desired substrates by etching away C–Ni diffusion couple, respectively. Representative (b) optical microscopy (OM) image, (c) Raman spectra from red, blue, and green spots showing the presence of one, two, and three layers of graphene, respectively, (from bottom to top) and (d) Raman map image of the G/2D bands of graphene grown at temperature $T = 160$ °C for 5 min on SiO_2(300 nm)/Si substrate. Scale bars, 4 μm (b and d).

(Fig. 4.23d), respectively. The authors suggested that the multilayer graphene regions arise from the grain boundaries of the deposited Ni thin films because the morphology of mono- or bilayer graphene and ridges of multilayer graphene are similar to those of grain and grain boundary of Ni film, respectively. The graphene films exhibited wrinkle-free features, like graphene grown by transfer-free process.

Kim et al. developed a direct transfer method for graphene on polyimide (PI); the PI is a flexible substrate deposited and pre-patterned with Cu at low temperatures using inductively coupled plasma-CVD (ICP-CVD) (98). The Cu metal was etched away slowly after the ICP-CVD synthesis below 300 °C; the graphene and PI form π–π interactions and van der Waals forces, adhering them to each other. Even without the support, the graphene remained intact and in good condition. The fabrication procedures are shown in Figure 4.24a–d. Although this provides a transfer-free synthesis and reduces both artifacts and damage to the graphene during fabrication process, only relatively low-quality graphene was obtained due to the limits of the low process temperature. A high-intensity D peak in the Raman spectrum is observed (Fig. 4.24e), and the sheet resistance ranged from 400 to 8000 Ω/sq for the graphene grown over 30 to 180 s(76).

The quality of the graphene films produced using the aforementioned method would not suffice for making graphene-based electronic devices, but the simplified fabrication process facilitates low-cost and high-throughput mass production of graphene films for industrial applications after advancing and optimizing the graphene growth using transfer-free methods.

4.3.2 Catalyst-Free Synthesis

The catalyst-free processes generate graphene using a different growth mechanism than that for conventional CVD-based graphene. The substrate properties, including its affinity for carbon sources and degree of lattice mismatch between the substrate and the graphene, influence the structural perfection, electrical properties, and morphological characteristics of the graphene grown by catalyst-free synthesis. The representative substrates used for this processes include hexagonal boron nitride (h-BN), sapphire, SiC, silicon, etc.

h-BN is commonly referred to as white graphene because it has a similar hexagonal structure to graphene, albeit with a white color in the bulk state, while graphene is a black carbonaceous material. h-BN is an interesting dielectric material for graphene-based microelectronics and optoelectronics due to its wide-band-gap of 5.2–5.4 eV, a good lattice match with graphene (mismatch ~1.7%) and atomically smooth surface without dangling bonds and charge traps, which is capable of serving as a dielectric substrate for graphene-based electronics (99). Directly growing graphene on a single atomic h-BN buffer layer substrate is known: The h-BN buffer layer was prepared using a metal catalyzed CVD process on a Ru (0001) or Ni (111) film (100, 101). Interestingly, the significant charge donation from h-BN to graphene on Ru was not observed on Ni. The Raman spectrum of graphene/h-BN/Ru (0001) did not exhibit the characteristic peaks at ~1370 cm^{-1} and ~2700 cm^{-1} for h-BN and the graphene 2D peak, respectively (100). Therefore, direct graphene growth on an h-BN layer during CVD, which is different from the conventional CVD growth technique for graphene, reveals unexpected behavior when different metal layers are used beneath the h-BN atomic layer. Ding et al. demonstrated the CVD growth of graphene on h-BN single crystal flakes without any metal catalysts. The growth time and flow rate of the carbon source on the number of layers exerted a significant effect that was observed in the Raman spectra with the various I_{2D}/I_G values from 0.56 to 0.79 for the graphenes grown directly on top of h-BN; the disorder-induced D peak at 1350 cm^{-1} was absent. In addition, Raman measurements of the graphene on the large flakes reveal the conformal coating on the entire h-BN flake, while the 2D peak has a single-Lorentzian line shape with a full width at half maximum of 40–52 cm^{-1}, indicating the stacking was ordered for

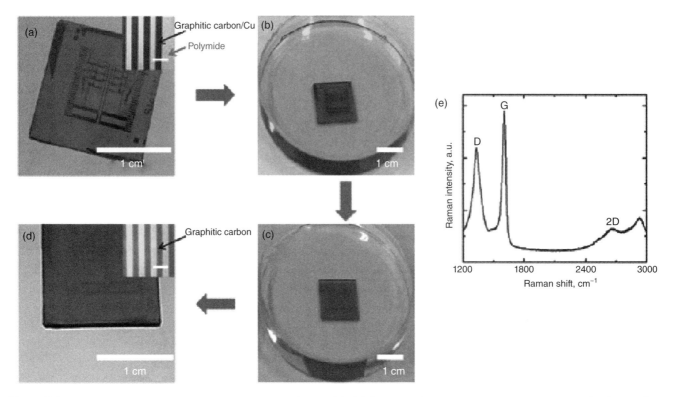

Figure 4.24 Photographs of the process used for etching and transferring of G–GC films onto plastic substrates. (a) As-synthesized G–GC films on Cu/PI substrates. The inset image displays a large array of patterned G–GC films formed on the PI substrates. (b) Wet-chemical etching of the underlying Cu layers by FeCl$_3$ solution. (c) Washing and cleaning with deionized water. The patterned films were not easily damaged or detached from the plastic substrates during the chemical etching and rinsing. (d) A transferred G–GC film on a PI substrate. The inset image shows that the transferred samples on the plastic substrates exhibit clear contrast between the G–GC and the substrates. (e) Raman spectrum of the G–GC films transferred to the SiO$_2$/Si substrates.

the multilayer graphene synthesized on the h-BN flakes (24). The authors speculated that the mechanism for graphene formation could be epitaxial growth because the lattice mismatch between the graphene and h-BN was only 1.7%.

Graphene with few layers on h-BN is derived from a solid carbon source, PMMA, and is synthesized using a hydrogen flame method (Fig. 4.25) (102). The ratio of I$_G$/I$_{2D}$ ($\approx$1.6) and the 2D peak at 2687 cm^{-1} in Raman spectra indicated that the number of the graphene layers is less than five (103). In addition, the D peak that originates from the defects or edges of the graphene was observed due to the small sheet size of the produced graphene (50–200 nm). TEM and selected area electron diffraction (SAED) analyses of the graphene reveal a $\sim$0.34 nm interspacing between the multilayer graphene sheets on h-BN, while the graphitic layers align with the h-BN ($\bar{1}$ 100) lattice planes and have an interspacing of $\sim$0.21 nm. The growth mechanism was proposed by anchoring of the graphene layers to the h-BN crystal, similar to the graphitic layers anchored to ZnS ribbons (104) or MgO (105) under the CVD method; the graphene growth is along the ($\bar{1}$ 100) plane of the h-BN substrate and is observed by TEM; it arises from the small lattice mismatch between the graphene and the h-BN (106), as well as thermodynamically stable and inert h-BN below 1200 °C (107, 108). Similar to the CVD growth using polymeric carbon sources on metal catalyst layers, the highly active edges, which were developed after the graphene domain had formed, received nascent carbon atoms from the decomposed PMMA without the assistance of the substrate or catalysts (25, 49, 109, 110). The imperfect and unstable carbon sites, as well as the residual active carbon and other species on the substrate, can be etched or removed using decomposed hydrogen and oxygen atoms under vacuum. The sheet resistance of graphene formed on h-BN varies from 27 to 126 Ω/sq, depending on the carbon content (70.4–21.9%, respectively); these are much lower than those of chemically or thermally reduced graphene oxide (>200 Ω/sq) (10, 12, 111). Son et al. demonstrated that the flat and regular pad shape of graphene was $\sim$0.5 nm thick on the h-BN substrate and had a surface coverage of >95%, while the yield of a single layer graphene pad exceeds 90% under the correct reaction conditions (112). Mechanically exfoliated h-BN 2–10 μm in diameter and 100–300 nm thick, which is prepared on a SiO$_2$/Si substrate with CVD using methane at 1000 °C, produced single layer graphene pads with regular round shapes and the pad size increases up to approximately 110 nm in diameter, beyond which the graphene layers grow (112). As usual, the process parameters, such as reaction temperature and precursor concentration significantly affected the population and size of the graphene pads, and three steps of growth during the time-dependent study are observed: the formation of a nucleate particle, a semi-bilayer graphene pad as an intermediate state, and the formation of single layer of graphene pads through peripheral growth on the h-BN surface.

The h-BN is atomically flat and has a good lattice, making it a promising substrate for growing high-quality graphene without catalysts because its morphological and structural properties contribute to those of graphene during CVD. However, the low lattice mismatch

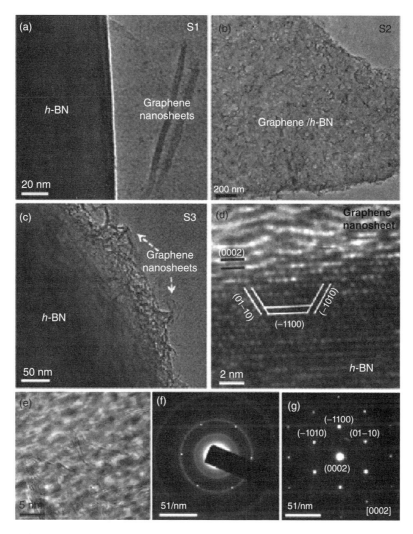

Figure 4.25 (a–c) TEM images of graphene nanosheet growth beyond the h-BN substrate, (d) HRTEM image of the interface between a graphene domain and h-BN substrate, (e) HRTEM image of a folded edge graphene layer enlarged from (c), and (f and g) SAED patterns of graphene/h-BN taken from (b) and (c), respectively.

(below 2%) between the graphene and h-BN prohibits lateral growth of the graphene on h-BN, generating small sizes of graphene ~150 nm in diameter. Considering this disadvantage, several researchers have studied the catalyst-free growth of graphene on a sapphire substrate with a higher degree of crystal mismatch with graphene (>50%) compared to h-BN. Fanton et al. reported that the CVD of methane at 1425–1600 °C on sapphire without a metal catalyst generates a high-quality monolayer graphene comparable to that of graphene formed on a single SiC semiconductor crystal wafer (113). A systematic understanding of the reactivity between carbon and sapphire was necessary because the graphene quality was highly dependent on the growth temperature (up to 1700 °C). Thermochemical modeling with Al, O, C, H, and Ar verified that no solid phase Al–C or O–C containing products were present in the system during the CVD of methane on sapphire, suggesting that no covalent bonding will occur between the graphene film and the substrate. Above 1200 °C, the gas concentrations for both CO and Al_2O are significant, and at temperatures above 1400 °C, methane could not be deposited on the substrate by conversion to CO and Al_2O, revealing two undesirable effects during the CVD of graphene on sapphire. CO forms faster than the decomposition rate of methane to carbon, and it could not generate enough atomic carbon to synthesize graphene on the substrate. In addition, the presence of Al_2O is an evident that the substrate etches, leading to not suitable surface rough for semiconductor device processing. Without flowing methane at 1550 °C under 10% H_2/Ar ambient conditions, the surface roughness of the substrates was 0.3–0.5 nm; however, the presence of 0.5% methane significantly increased the surface roughness to 2.9 nm at 1525 °C and to 6.3 nm at 1550 °C, indicating that adding a carbon source to the growth system helped etch the surface. The I_D/I_G ratio of the graphene grown at 1425–1450 and 1575 °C was 0.42 and 0.05 in the Raman spectra, respectively, and was greater than 1.5 of the I_{2D}/I_G ratio from all graphenes, suggesting the deposition of mono- or multilayer graphenes (Fig. 4.26).

Although the structural quality was improved at increasing temperatures, the graphene showed only partial coverage (<20%) or did not form above 1575 °C. As discussed above, this is responsible for the significant increase in the average surface roughness, indicating that the

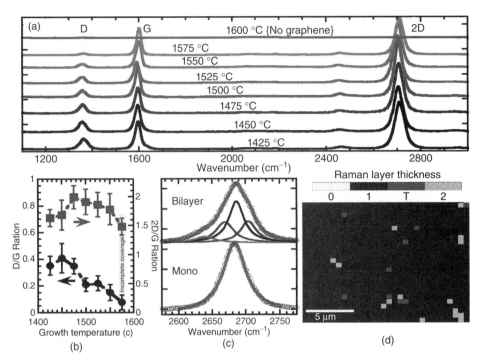

Figure 4.26 (a) Raman spectroscopy of graphene on sapphire indicates that structural quality improves as the growth temperature is increased from 1425 to 1575 °C. Additionally, the 2D/G ratio (b) remains equal to or greater than 1.5 with a significant fraction of the 2D Raman spectra being fit to one or four Lorentzian curves (c) suggesting the presence of monolayer and bilayer graphene. Finally, Raman mapping and subsequent peak fitting of the 2D peak for a film grown at 1525 °C indicates >90% monolayer coverage (d).

etching of the sapphire by the carbon atoms is faster than the deposition of graphene. The structural quality and carrier mobility are comparable to graphene on SiC (0001) but not to exfoliated graphene, despite the significant strain relief. The inferior behavior of graphene on sapphire in carrier mobility relative to the exfoliated graphene is related to the defects observed in the Raman spectra as a D peak and any remaining graphene/sapphire interaction as noted by the small presence of strain in the graphene, to a lesser extent.

Other studies regarding the growth of graphene on a sapphire substrate have been reported with various carbon sources and process temperatures in CVD (114–116). Hwang et al. have demonstrated the epitaxial growth of the graphitic carbon monolayer or limited multilayer graphene at temperatures from 1350 to 1650 °C using propane (114), while Miyasaka et al. have attempted to grow graphene on sapphire using alcohol CVD at relatively lower temperatures (800–1000 °C) (115). While they successfully formed graphene, more detailed investigations concerning its crystallinity, carrier mobility, the number of layers, size, and the growth mechanism remain to be studied. Song et al. synthesized a large-scale graphene grown directly on an α-Al$_2$O$_3$ (0001) wafer that could be up to 2 inches in diameter, and they applied it using a transfer-free process for fabricating top-gated FET devices that clearly exhibited the characteristic Dirac point (116). They prepared a single- or double-sided polished single crystalline α-Al$_2$O$_3$ (0001) substrate and precleaned it with acetone and isopropyl alcohol. The substrate was heat treated directly without a metal catalyst at 950 °C with 50 sccm of H$_2$ and 30 sccm of methane gases during the given reaction times. As displayed in Figure 4.27, small graphene dots appeared after 30 min of reaction, and the size of each dot gradually increased at longer reaction times, even though their sizes and shapes are quite irregular (Fig. 4.3a and b). After 90 min, the reaction produced relatively homogeneous graphene dots with an average diameter of 250 nm. The larger graphene structures arose from two merged graphene dots, which were not observed in the graphene grown on h-BN (112). There are two distinct merging patterns (Fig. 4.3c). First, the smooth lateral merging among several dots (dotted circles) generated seamless interfaces at which graphene grains could merge, as observed under the scanning TEM with graphene grown on Cu layer in CVD (117, 118). The second merging process is rough lateral merging, leading to solid wrinkles (solid circles) that resulted from overlays formed by collisions between graphene dots, while wrinkles found in the graphene grown on the metal catalyst were induced by thermal stress (119). A further reaction up to 120 min revealed the complete coverage of the substrate (>98%) by graphene containing several pinholes and an increased wrinkle population (Fig. 4.27d). The evolution from the graphene dots into a film enhanced the quality of the graphene, as verified by a gradual decrease in the D peak and an increase in the G' peak, indicating the reduction of amorphous carbons or defect sites (Fig. 4.27f). Finally, utilizing a 150 min reaction time generates a smoother surface with fewer wrinkles (Fig. 4.27e). Unfortunately, this morphological change degraded the quality of graphene and most likely increased the amorphous carbon content, as suggested by the larger D peak and the emergence of the D' peak (120) next to the G peak (bottom spectrum in Fig. 4.27f). The thickness and transmittance of the graphene are ~0.8 nm and 97.5% at the saturation level, respectively.

A large degree of mismatch (~50%) between the α-Al$_2$O$_3$ (0001) and the graphene rule out a lattice matched epitaxial growth, which is a well-known mechanism for the h-BN substrate. Instead, nucleation along the closely packed oxygen atom lines and further growth due to the

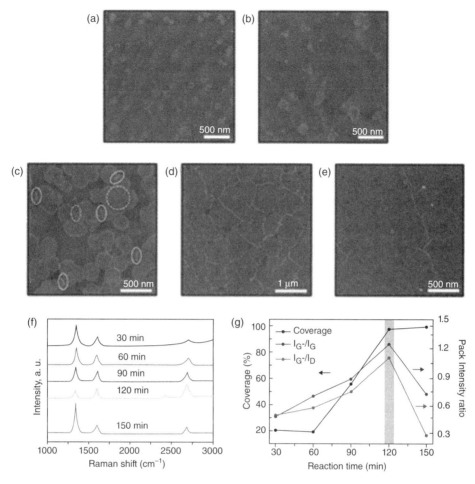

Figure 4.27 Time-dependent growths of graphene dots and films on α-Al2O3 substrates. (a–e) AFM images taken after the reactions for 30, 60, 90, 120, and 150 min, respectively. (f) Raman spectra measured from samples (a) through (e). All the spectra are normalized with G peak intensities. (g) Coverage, IG'/IG, and IG/ID as a function of reaction time.

pseudo-sp^2 hybridized electronic environment on the oxygen atoms might be a more suitable explanation. This environment was induced by the shortened interlayer distance between Al and O through the polishing process of the substrate, constraining Al to be coordinated to only three oxygen atoms (121), and similar pseudo-epitaxial growth of Cu (111) film on α-Al$_2$O$_3$ (0001) was observed (122).

Various attempts have also been made to investigate the catalyst-free growth of graphene on SiO$_2$/Si wafers in CVD for use in practical electronic devices. According to the phase diagram of C-Si, the solid solubility of carbon in silicon is very low (10^{-3}%) at 1400 °C; this value is too low to form the graphene on solid Si (123). Despite the low solubility, there is a van der Waals attractive interaction force when carbonaceous gas molecules are deposited on the surface of a solid substrate. Noticeably, strength of the attractive interaction varies over a wide range according to the origin of the interaction. Consequently, the gases adsorb on the surface as long as the sticking coefficient is not zero in CVD. Therefore, graphene can be grown on the surface of any material, even though a high temperature may be required to enhance the surface reactivity as well as diffusivity to increase the lateral growth rate and the extent of order. However, most methods for direct growth on the Si wafers, such as low-pressure CVD (124, 125), ambient-pressure CVD (126), PECVD (127), laser irradiation (123), and gas-source molecular beam epitaxy (128) produced low-quality graphene relative to the graphene on the other substrates or nanocrystalline graphene with submicron sizes.

Chen et al. reported a breakthrough method for directly synthesizing high-quality polycrystalline graphene with a large area on silicon wafer or quartz using oxygen aided CVD (129). In contrast to the typical procedure for graphene growth, the SiO$_2$/Si substrate was annealed at 800 °C under air atmosphere in order to remove any organic residues and to activate the growth sites before growth. The authors claimed that the presence of oxygen can enhance the adsorption of hydrocarbons on SiO$_2$, as indicated by density functional theory calculations (130, 131). CVD of methane for 3 h with a carrier gas mixture of H$_2$/Ar provided monolayer graphene islands on the SiO$_2$/Si substrate with a thickness of 0.659 nm (Fig. 4.28). The size and number of the graphene islands can easily be controlled by changing the carbon flow, deposition temperature, or growth time. Similar to the growth mechanism from other CVD processes, a longer deposition time (7 h) enlarges the size of the graphene sheets until they interconnect and form 2D graphene networks (Fig. 4.28e). After 8 h, graphene layers contain numerous

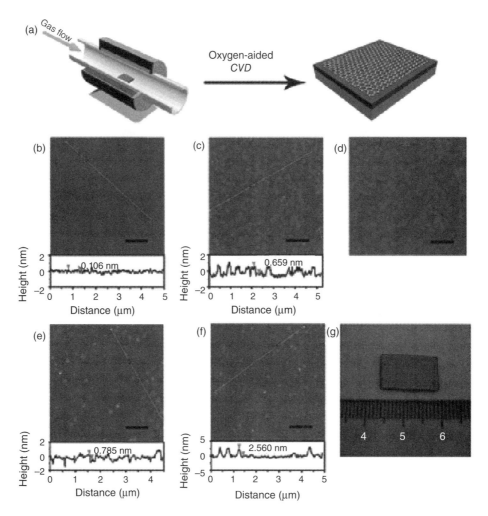

Figure 4.28 Synthesis process and morphological changes of graphene on SiO$_2$/Si substrates. (a) Schematic diagram of the oxygen-aided CVD growth of graphene on a SiO$_2$/Si substrate. (b) Initial surface of the SiO$_2$/Si substrate, characterized by a uniform flat surface. (c) AFM height image of graphene sheets with a thickness of 0.659 nm. (d) AFM phase image of graphene sheets. (e) AFM image of 2D interconnected graphene networks. (f) AFM image of continuous graphene films. (g) Photograph of a graphene film on SiO$_2$/Si substrate. The edge was removed using adhesive tape. Scale bar = 1 μm.

islands 2.560 nm thick, revealing multilayered graphene from a height profile measured by atomic force microscopy (Fig. 4.28f and g). For the quartz substrates under a similar process, the synthesized polycrystalline graphene exhibited ~91.2% transmittance and ~800 Ω/sq of sheet resistance. Considering the studies of SWCNT growth on nanosized SiO$_2$ particles (130), the graphene synthesis on SiO$_2$ substrates may occur through a vapor–solid–solid (VSS) growth mechanism. Chen et al. found that oxygen-aided CVD enhanced the number of graphene nucleation sites on the SiO$_2$ substrate, leading to the capture of more CH$_x$ ($x = 0$–4) fragments through C–O and H–O binding, as well as more opportunities for C–C coupling and graphene nucleation. Similar phenomena were observed during metal-catalyst-free SWCNT growth (130). To evaluate the characteristics of the synthesized graphene, field-effect transistor devices were fabricated using Au source and drain electrodes. Relatively high current "ON/OFF" ratios for pure graphene-film-based FET devices, such as ~2.51 in air and ~3.24 under N$_2$, were observed. In addition, the carrier mobilities that were calculated from the linear regime of the transfer characteristics are ~531 cm^2/(V s) in air and ~472 cm^2/(V s) in N$_2$ with a channel length (L) of ~65 μm and a channel width (W) of ~1200 μm. These results exceed those obtained using chemically reduced graphene oxide films (1 cm^2/(V s), $L = 21$ μm, $W = 400$ μm) (132) and also compare favorably with those of PMMA-derived graphene on Cu (410 cm^2/(V s), $L \approx 7.5$ μm, $W \approx 9$ μm) (49).

4.4 CARBON NANOSHEETS SIMILAR TO GRAPHENE

Large graphene sheets can be synthesized using the CVD of gasified carbon species onto transition metals, such as Ni or Cu catalyst films under a H$_2$/Ar or other inert atmospheres at temperatures as low as 800 °C, as described in the previous sections (17, 22, 24, 28). For graphene to be grown above or below the metal films, additional processes that eliminate the catalyst and transfer the graphene to another substrate are required for various applications. The graphene quality may degrade because the elimination of the metal films was incomplete and defects, wrinkles, and cracks formed unavoidably during these processes. Therefore, catalyst- and transfer-free processes using unaffordable sapphire (113, 116) and hexagonal boron nitride (112) substrates that are limited to electronic device applications have been developed. In this section, a novel and straightforward approach to the synthesis of atomically thin carbon nanosheets (CNS) similar to graphene on substrates such as silicon wafers and quartz, as well as the direct application of these CNS as electrodes for organic electronics will be discussed.

4.4.1 From Sources to Hexagonal Structure Formation

An ultrathin CNS has a typical 2D carbon nanostructure and characteristics including a large surface area, a smooth surface, flexibility, elasticity, high thermal conductivity, chemical stability, and low mass (132–134). To synthesize the CNS as an alternative to graphene, carbon sources are required to form the hexagonal structure by an additional process or that already possess the hexagonal moieties. Joh et al. have prepared CNS using polyacrylonitrile (PAN) as the carbon source to form CNS, which are widely known as a typical precursor for preparing carbon fiber (CF), which is composed mainly of carbon atoms and is approximately 7–13 μm in diameter (136). They synthesized an atomically thin CNS analogous to graphene with properties suitable for an organic thin film transistor (TFT). The PAN polymer had an open ring structure and was spin-coated in DMF onto a 100 nm SiO_2/Si substrate. The PAN film was stabilized at 250 °C in air, and the nitrile group ($-C\equiv N$) on the PAN was converted into a ladder structure through a cyclization reaction (Fig. 4.29). The cyclization reaction needed to stabilize the PAN polymer during CF production occurs at 230–325 °C (137). The cyclized thin films were successfully converted into CNS via carbonization at temperatures as low as 1200 °C under a H_2/Ar gas mixture. The synthesized CNS was used directly to produce electrodes for OTFTs, as illustrated in Figure 4.30. The morphological and electrical properties were easily controlled by changing the polymer concentration. The thinnest CNS exceeded 1600 S/cm electrical conductivity and 92% transmittance. These properties are superior to those of the chemically exfoliated and reduced graphene (12, 138, 139), although the graphene grown through CVD possesses the highest electrical conductivity due to its more ordered graphitic structure. Using a similar transfer process for graphene, the CNS on the Si wafer substrate can be transferred to the other substrate by eliminating the SiO_2 layer.

Based on the STM analysis, the CNS appears to be flat in the wide-scan area image, as shown in Figure 4.31. However, on the atomic scale, different types of defects were observed in Figure 4.31c and d, in addition to the well-ordered hexagonal pattern in Figure 4.31b. Both line defects (dislocation) and a superlattice pattern induced by the point defects are revealed in Figure 4.31c. The superlattice pattern may have been induced by vacancies or impurities (oxygen or nitrogen) in the layer (140) or intercalations between the layers (141). The size of the point defects may be less than a few Å in diameter; however, these defects enhance the charge density states near the Fermi energy level due to either a dangling bond or a deformed layer. This enhancement results in brighter and larger defect features in the STM images. A disordered area was also observed (Fig. 4.31d). The defective and disordered portions were usually observed near the CNS edges, while the hexagonal pattern was confined to the central area. The grain boundary that generally forms in CNS grown on a metal catalyst was not observed.

Petroleum-derived pitch is a complex mixture containing hundreds of aromatic hydrocarbons comprising structures with three- to eight-membered rings, alkyl side groups, and an average molecular weight of 300–400 (142). Na et al. reported that CNS was fabricated using pitch obtained by reforming naphtha cracking bottom oil as one of the cheapest precursors for CF (143). The pitch-based CNS was obtained by processes similar to those for the PAN-based CNS described above: spin-coating the pitch solution with small aromatic molecules, stabilization, and carbonization. CNS, having an imperfect sp^2-hybridized carbon network with some impurities or amorphous carbon, is easily obtained using inexpensive and solution-processable pitch. However, the pitch-derived CNS displayed comparable structural development to the chemically reduced graphene films displayed in Figure 4.32. The transmittance at 550 nm varied from ~89 to ~52%, depending on thickness of the CNS that was controlled by the pitch concentration in solution during spin-coating. The pitch-derived CNSs exhibited an overall average conductivity of ~300 S/cm, suggesting that they have a relatively higher conductivity compared to the chemically converted graphene materials (144, 145).

Botanical hydrocarbons were used as the carbon source for fabricating CNS. Kalita et al. reportedly prepared transparent graphene-based carbon films by pyrolysis of camphor ($C_{10}H_{16}O$), which is transparent (or white), waxy, and aromatic compound, obtained from a large evergreen tree (146). Camphor has a unique structure consisting of hexagonal and pentagonal rings, as well as a methyl carbon. Camphor evaporated at 200 °C was deposited on quartz substrates in a CVD reactor and subsequently pyrolyzed at 900 °C in argon to synthesize CNS that contained graphitic and highly disordered carbon, as observed by Raman spectroscopy and TEM. Although the CVD method was applied to these CNS, catalytic layers, such as Ni or Cu, were not deposited on the quartz, indicating that the pyrolysis of the camphor molecules generated a graphitic structure from the hexagonal or pentagonal rings in CNS, while the methyl carbon was detached (Fig. 4.33). The camphor-derived CNS was approximately 9 nm thick and exhibited a 90% transmittance at 550 nm, as well as a conductivity of approximately 350 S/cm.

Camphor and pitch have aromatic rings, while stabilized PAN also contains a cyclized network structure, even though the PAN polymer has linear chains; subsequently, pyrolysis generated a larger graphitic structure by eliminating all atoms except carbon. Therefore, generating a network similar to a graphitic structure that is as large as possible before pyrolysis could account for the development of graphene when solid sources were used. There was research aiming to synthesize CNS using large aromatic molecules as solid sources. Wang et al. synthesized extremely large polycyclic aromatic hydrocarbons (PAH), called nanographene molecules with alkyl side chains, as displayed in Figure 4.34a (hexadodecyl-substituted superphenalene C_{96}-C_{12}) (147). Those alkyl side chains on the periphery allowed PAH to dissolve, enabling the use of a simple solution process. When the solution was spin-coated onto the substrates, the resulting film contained a combination of "edge-on,"

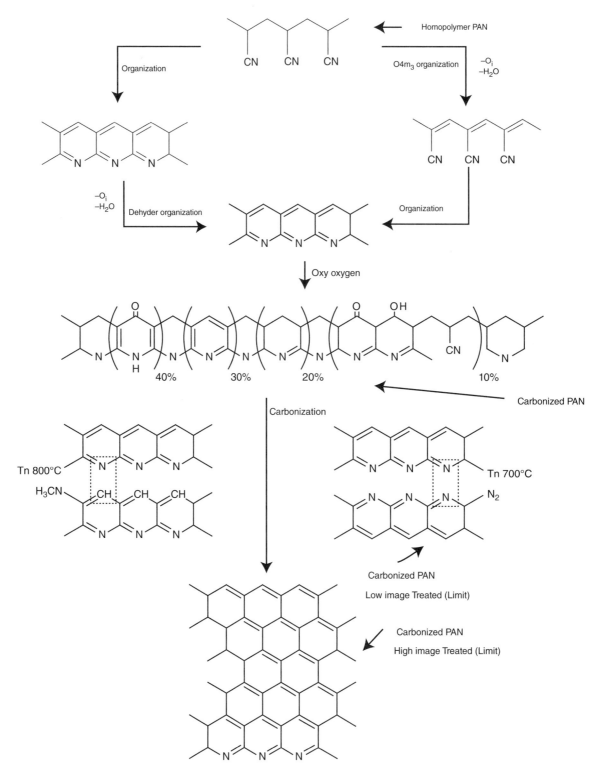

Figure 4.29   Proposed mechanism for growth of carbon fiber derived from polyacrylonitrile precursor.

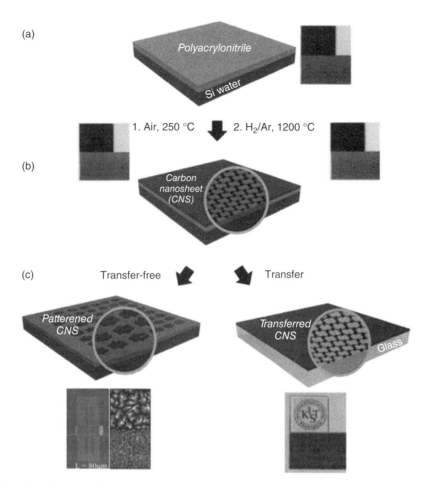

Figure 4.30 Schematic illustration of a catalyst-free process for graphene on a silicon wafer. (a) Polyacrylonitrile polymer solution is spin-coated on top of a silicon oxide substrate, forming the starting material. (b) The polymer film is stabilized under air atmosphere and then carbonized under a mixture of H_2/Ar gases, resulting in graphene film. (c) The as-grown graphene is directly applied as an electrode for a thin film transistor. And the graphene can be transferred onto another substrate through etching of the silicon wafer. The insets show photographs of the sample at each step in the process.

"face-on," and tilted alignment of nanographenes. This process formed packed graphene layers that were oriented toward different directions along the film and overlapped with one another to form a graphitic network. A controlled heating procedure, such as a slow ramp and annealing at 400 °C, was crucial for obtaining uniform and continuous CNS after carbonization because these conditions control the initiation of the alkyl chains' decomposition and subsequent cross-linking between the aromatic cores (151, 152). The transparency of the CNS on quartz could be tuned by controlling thickness of the film with the solution concentration; this control is one of advantages of using a solution process, as observed in previous CNS development. The 4 nm thick nanographene-derived CNS has a 90% transmittance and a conductivity below 200 S/cm.

One of the drawbacks of generating CNS from solid carbon sources was that the development of one layer must be controlled during pyrolysis. Therefore, SAM could be a promising material for generating single layers of CNS because the thickness and morphology of SAMs can be atomically controlled (153, 154). Turchanin et al. synthesized CNS using SAMs containing biphenylthiols or 1,1'-biphenyl-4-thiol on gold; these materials were cross-linked with electron irradiation (Fig. 4.35). Through the radiation-induced cross-linking, the aromatic SAMs could be converted into mechanically stable 2D CNS with a thickness of ∼1 nm. The CNS was released from its original substrate and prepared as a freestanding nanomembrane (153). In addition, the insulating CNS (sheet resistivity: ∼10^8 kΩ/sq) became conducting as it transformed into a nanocrystalline CNS (sheet resistivity: ∼10^2 kΩ/sq) during the vacuum annealing because the cross-linking reaction did not form the hexagonal structure (153, 155). For the cross-linked CNS, both HE-TEM images and SAED patterns revealed only amorphous materials. However, the annealed CNS exhibited curvy and nearly parallel fringes, indicating the presence of graphitic material. Therefore, the distinct structural changes that occur in the CNS after annealing improved the electronic properties. However, the quality of the CNS was not satisfactory as observed by Raman spectroscopy and TEM compared to the graphene grown by CVD.

Most of the syntheses of CNS converted carbon sources to graphitic structures by pyrolysis under an inert atmosphere, providing energy to initiate gas evolution from the graphitic network, depending on the carbon sources. CNS may also be synthesized using laser ablation. Laser ablation is one of the major techniques used to grow various carbon nanostructures, such as fullerenes, carbon nano-onions, and carbon

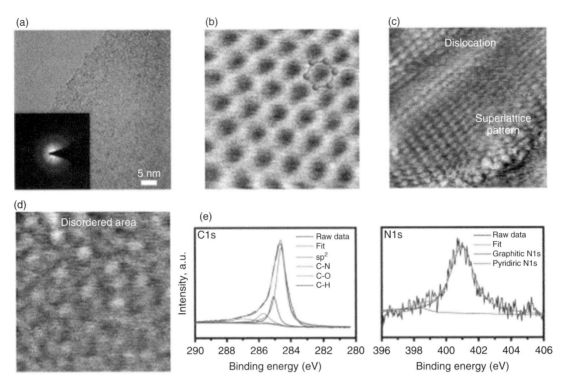

Figure 4.31 Structural analysis of the graphene prepared using a 0.5 wt% PAN polymer. (a) HRTEM image of the graphene. The inset shows the selected area electron diffraction (SAED) pattern of the graphene. (b–d) STM topographical images, which were obtained at a tunneling current of 10 pA-2 nA and a tip-bias voltage of 0.1–100 mV, of the graphene transferred to a HOPG. (b) Well-ordered hexagonal pattern (2×2 nm^2). (c) Dislocation and superlattice pattern induced by impurities (such as oxygen and nitrogen) and point defect, respectively (5×5 nm^2). (d) Disordered area (2×2 nm^2). e, XPS survey spectra (C1s and N1s) for the graphene.

nanotubes (156–158). Qian et al. prepared freestanding CNS from a preceramic polymer (poly(phenylcarbyne) (PPC)) using a pulsed Nd:YAG laser (159). The CNS was formed by a pulsed-laser method in three steps: ablation, carbonization, and landing. The pulsed laser interaction on the PPC achieved a polymer-to-carbon conversion with the formation of diamond-like and amorphous carbon species (160, 161). The CNS exhibited nanometer thickness and were microsized. The quality and size of the deposited carbon materials was controlled and classified into three categories relative to laser fluence: amorphous carbon (below 1.0 J/cm^2), ultrathin amorphous CNS (from 1.0 to 10.0 J/cm^2), and thick carbon films (above 10.0 J/cm^2). Laser fluence below 1.0 J/cm^2 was insufficient for producing sheet structures from polymer films. Figure 4.36a depicted a schematic of the thermal decomposition of the PPC polymer into smaller clusters, leading to the deposition of amorphous carbon. The high yield of the CNS using the laser fluence between 1.0 and 10.0 J/cm^2 was obtained because more macromolecules or large fragments of PPC were ablated from the PPC target; in the given process, ultrathin amorphous CNS was formed (Fig. 4.36b).

In general, CNS is a 2D carbon material deposited on the substrate even though its thickness, properties, and structure vary relative to its synthetic process and conditions. The synthetic approaches using PECVD to provide wrinkled or vertically oriented CNS have become promising candidates for mass producing CNS or graphene because it is a simple method that remains compatible with traditional semiconductor production processes (162, 163). The schematic of the radio-frequency (RF) PECVD system is displayed in Figure 4.37; the RF coupled into the deposition chamber is 13.56 MHz, and a sample holder acted as two plate electrodes. The PECVD process variables, such as RF power, temperature of deposition chamber, gaseous carbon sources, and catalysts, are important for determining the properties of CNS. Liu et al. reported that catalysts such as Ni, Ni/Co, Co/Ni, and Ni/Zn affected the morphologies and internal structures of CNS (162). Similar to a conventional CVD, the CNS was only grown on the catalyst layer, and the effect of the morphology and type of the catalysts on the morphology of CNS was significant. A Ni/Zn catalyst layer produced the CNS with densely distributed, catalyst-induced wrinkles, while the CNS produced on a Ni catalyst layer was flat. TEM micrographs of the wrinkles revealed a clear graphitic-layer structure with a thickness of ~9.

Zhu et al. also reported using PECVD to create wrinkled CNS using methane or acetylene gas as a feedstock (163). The acetylene-derived CNS had a self-assembled cellular-like structure and was more ordered, vertically orientated, and had a much more uniform sheet height distribution than the methane-derived CNS shown in Figure 4.38. Through HRTEM analysis, the acetylene-derived CNS had three to four parallel fringes, indicating 3–4 atomic layers that corresponded to an edge approximately 1 nm thick. The wrinkled or vertically oriented CNS generated by PECVD arose from the balance between deposition through the surface diffusion of carbon sources, etching by atomic hydrogen in the plasma to remove amorphous carbon, and cross-linking at the free edges of the growing CNS (Fig. 4.39) (164).

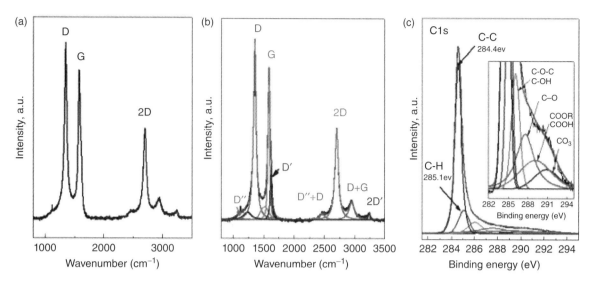

Figure 4.32   (a) Raman spectrum, (b) deconvoluted Raman spectrum, and (c) deconvoluted XPS C1s spectrum of the CNS synthesized from the 6 wt% pitch solution. The inset is an enlarged XPS spectrum.

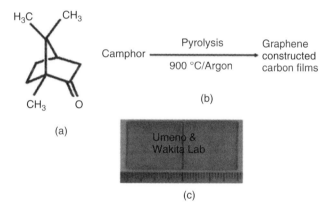

Figure 4.33   Representation of (a) the molecular structure of camphor (a botanical hydrocarbon), (b) pyrolysis of camphor in an argon atmosphere to form a graphene constructed carbon film, (c) a deposited film on the quartz substrate showing very good transparency.

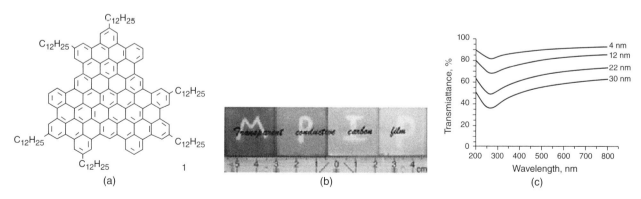

Figure 4.34   (a) Molecular structure of 1, the hexadodecyl-substituted superphenalene $C_{96}-C_{12}$; (b) 30, 22, 12, and 4 nm-thick TGFs on quartz (2.5×2.5 cm^2) with "M," "P," "I," and "P" letters inside, erased from the film before heat treatment; (c) transmission spectrum of the TGFs with different thicknesses.

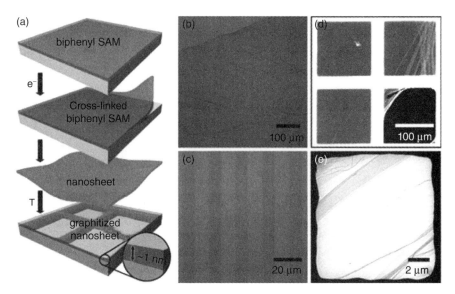

Figure 4.35   Fabrication scheme and microscopy images of supported and suspended carbon nanosheets. (a) A 1 nm thick SAM of biphenyl molecules is irradiated by electrons. This results in a mechanically stable cross-linked SAM (nanosheet) that can be removed from the substrate and transferred onto other solid surfaces. When transferred onto TEM grids, nanosheets suspend over holes. Upon fabrication scheme and microscopy images of supported and suspended carbon nanosheets. (b) Optical microscopy image of the section of a 5 cm^2 nanosheet that was transferred from a gold surface to an oxidized silicon wafer (300 nm SiO$_2$). Some folds in the large sheet are visible, and originate from wrinkling during the transfer process. (c) Optical microscopy image of a line pattern of 10 mm stripes of nanosheet. The pattern was fabricated by e-beam lithography and then transferred onto oxidized silicon. Note that the small lines are almost without folds. (d) SEM of four 130 mm × 130 mm holes in a TEM grid after a nanosheet (cross-linked biphenyl SAM) has been transferred onto the grid. Two left holes are covered by an almost unfolded nanosheet. The upper right hole shows some folds, whereas in the lower right hole the sheet has ruptured. (e) TEM image of a nanosheet transferred onto a TEM grid with 11 mm holes after pyrolysis at 1100 K. The hole is uniformly covered with an intact nanosheet. Some folds within the sheet are visible. Heating to T > 1000 K in vacuum (pyrolysis), nanosheets transform into a graphitic phase.

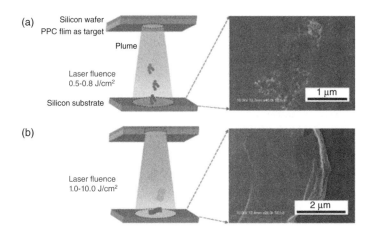

Figure 4.36   (a) Suggested model for the transition from PPC to amorphous carbon (b) Suggested model for the transition from PPC to ultrathin amorphous CNSs.

4.4.2   Potential Application Areas

CNS is a graphene alternative because it is transparent and conductive, although its optical and electrical properties are inferior to the graphene grown using CVD. The CNS derived from synthetic polymers, camphor, and polycyclic aromatic hydrocarbons, such as pitch and superphenalenes, may act as transparent electrodes for ITO-free organic solar cells (OSC). The electrical and optical properties of the CNSs and power conversion efficiencies (PCE) of the CNS-based OSC are listed in Table 4.2. Of the CNSs, the PAN-derived CNS on quartz exhibits the best PCE when employing a CNS-based OSC (165). An increase in the PAN concentration increased the fill factor (FF) of the OSC due to either a decrease in the series resistances (R_s), an increase in the shunt resistances (R_{sh}), or both as illustrated in Figure 4.40. A gradual

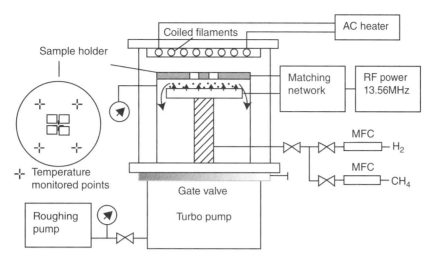

Figure 4.37 Apparatus used for the deposition of CNS.

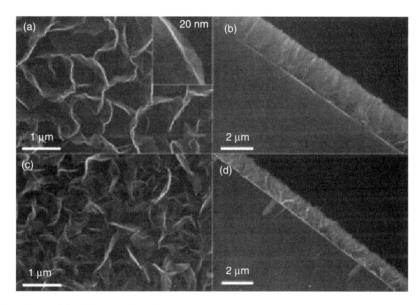

Figure 4.38 (a) Top view SEM image of CNSs deposited on a Si substrate using 80% C_2H_2 in H2, at 600 °C substrate temperature, 35 mTorr total pressure, and 1000 W RF power for 10 min. Inset: Enlarged SEM image shows the edge thickness of 1–2 nm. (b) Side view SEM of the C_2H_2 CNSs shown in (a). (c) Top view SEM of typical (40% CH_4 in H_2, 700 °C, 100 mTorr, and 900 W) CH4 CNSs deposited on a Si substrate for 20 min. (d) Side view SEM of the CH_4 CNSs shown in (c).

decease in the R_s values revealed in the inset of Figure 4.40 with increasing PAN concentration was observed, while the R_{sh} values did not exhibit a clear trend. The R_s and the FF of the OSCs depend mainly on the sheet resistance of the CNS electrodes. After accounting for the thickness of the PAN-derived CNS controlled by solution concentration, an inverse relationship exists between the FF and the current density or between the sheet resistance and transmittance, suggesting the thickness of the transparent electrode should be adjusted to balance the FF and current density, as well as to optimize the PCE. Approximately 1.7% PCE was obtained with ITO-free OSC using CNS derived from 1.5 wt% PAN. Although the cell-efficiency of the CNS-based OSC is lower than that of the ITO-based reference cell due to its higher sheet resistance and a lower transmittance, the CNS electrodes from PAN are comparable to the chemically derived graphene and the CVD-grown graphene electrodes; the PCE appeared from 0.1% to 1.27% and 1.23% to 2.6%, respectively (Table 4.3), (132, 144, 166, 167). The results demonstrate that transparent and conductive CNS have great potential as window electrodes in OSC cells relative to the graphene prepared by CVD and chemical exfoliation methods.

CNS has been used as electrodes on quartz substrates in electronic devices, particularly OSC. Joh et al. reported another electrode application for the CNS synthesized on a Si wafer: the direct use of CNS as the electrode for a pentacene-based TFT (136). The field-effect mobility of the pentacene-CNS TFT (0.25–0.35 cm²/(V s)) is superior to that of the pentacene-Au TFT (0.2–0.25 cm²/(V s)) at a given drain

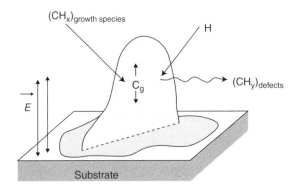

Figure 4.39 A schematic explanation of CNS growth model.: direction of the electric field near a substrate surface; CH_x: carbon-bearing growth species impinging from gas phase; C_g: growth species diffuse along CNS surface; H: atomic hydrogen impinging from gas phase; CH_y: defects removed from CNS by atomic hydrogen etching effects.

TABLE 4.2   Electrical and Optical Properties of the CNSs and Power Conversion Efficiency (PCE) of CNS-Based OSC

Carbon Sources	Thickness (nm)	Electrical Conductivity (S/cm)	Transmittance (%)	PCE (%)	Reference
PAN	8.0	~500	61.5	1.76	178
Pitch	21.0	~300	52.1	1.73	152
Superhenalene	12.0	~46	85	0.29	156
Champhor	17.0	357	81	1.21	155

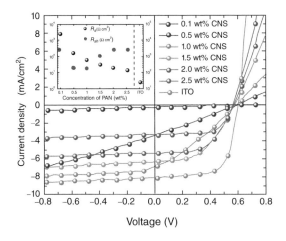

Figure 4.40 Current density–voltage (J–V) curves of ITO-free solar cells comprising different CNS anode films and the conventional ITO-based cell. The inset shows R_s and R_{sh} of various solar cells that varied only in their respective anodes.

(-40 V) and gate voltage (-40 V), even though the conductivity of the CNS ($\sim 1.6 \times 10^3$ S/cm) is two times lower than that of Au (4.1×10^5 S/cm). These results arose from the superior contact properties between the CNS and the pentacene than those between the Au and the pentacene in electrodes.

The 1 nm thick, vertically oriented CNS was synthesized using PECVD with methane as a carbon source under H_2, and its electron field emission properties were investigated (168). Other carbon materials, such as carbon nanotubes and carbon nanofibers, have been studied for the emission stability, which ranges from 12 to 72 h. The synthesized CNS-based back-gated triode emission device demonstrated reproducibility and stability over 200 h at a 1.3 mA emission current level, due to the high purity and uniform height distribution of the CNS. Over this time, no degradation was observed, the variability of the individual I–V curves was small among 7216 voltage cycles, and the standard deviation at the maximum current was less than 2.3%, indicating a great potential for applications in field emitter devices.

TABLE 4.3 Summary of graphene based electrodes and best PCEs employing them in OPV cells.

Graphene Material	Sheet Resistance	T (%)	Device structures[a]	PCE (%)	Ref.
rGO	17.9 kΩ/sq	69	G/PEDOT:PSS/P3HT:PCBM/LiF/Al	0.13	(37)
rGO	100–500 kΩ/sq	85–95	G/CuPc/C$_{60}$/BCP/Ag	0.4	(38)
rGO	1.6 kΩ/sq	55	GPEDOT:PSS/P3HT:PCBM/TiO$_2$/Al	0.78	(39)
rGO	40 kΩ/sq	64	G/PEDOT:PSS/P3HT:PCBM/Al	0.1	(40,41)
rGO	1 kΩ/sq	80	GPEDOT:PSS/P3HT:PCBM/Al	1.01	(42)
rGO-CNT	240 Ω/sq	86	G/PEDOT:PSS/P3HT:PCBM/Ca:Al	0.85	(43)
rGO-SWCNTs-CsCO$_3$	331 Ω/sq	65.8	G/P3HT:PCBM/V$_2$O$_5$/Al	1.27	(44)
rGO-Pys	916 S/cm	68	G/PEDOT:PSS/P3HT:PCBM/ZnO/Al	1.12	(45)
Bottom-up synthesis	18 kΩ/sq	85	G/P3HT:PCBM/Ag	0.29	(46)
CVD	210–1350 Ω/sq	72–91	G/PEDOT:PSS/P3HT:PCBM/LiF/Al	1.71	(50)
CVD	230 Ω/sq	72	G/PEDOT:PSS/CuPc/C$_{60}$/BCP/Al	1.27	(51)
CVD	606 Ω/sq	87	GPEDOT:PSS/P3HT:PCBM/TiO$_x$/Al	2.58	(52)
CVD	850 to 520 Ω/sq	90 to 85	G/WPF-6-oxy-F/P3HT:PCBM/PEDOT:PSS/Al	1.23	(53)
CVD	80 Ω/sq	90	G/MoO$_3$/PEDOT:PSS/P3HT:PCNM/LiF/Al	2.5	(54)
CVD-AuCl$_3$	500 to 300 kΩ/sq	97.1 to 91.2	G/PEDOT:PSS/CuPc/C$_{60}$/BCP/Ag	1.63	(35)

[a]"G" represents graphene based transparent electrode.

The applications of the CNSs are very circumscribed due to their defective structure and relatively poor electronic properties relative to the CVD-grown graphene materials, even though the CNSs have numerous advantages with regard to their synthesis and processability. Therefore, to use the CNS in various applications, the structural perfection and electrical conductivity of the CNS must be improved by optimizing the synthetic parameters and doping treatments, respectively.

4.5 SUMMARY

Ever since graphene was discovered in 2004, numerous researchers in physics, material science, chemical engineering, etc. have sought to use this material to improve human lives. Until now, for the mass production of large-sized and high-quality graphene, CVD was the most promising method to satisfy the requirements, even though complicated processes including catalyst elimination and transfer of graphene onto a new substrate, which inevitably degraded the products by producing artifacts, need improvement.

Numerous reports describe their attempts to improve quality of graphene by optimizing the synthetic and transfer processes, as well as their innovative ideas for transfer- and catalyst-free processes and CNS with properties similar to graphene. However, breakthrough technologies suitable for general use or further commercialization done for the other dimensional carbon allotropes are still needed for the synthetic approaches used in the mass production of high-quality graphene.

ACKNOWLEDGMENT

This work was supported by a grant from Korea Institute of Science and Technology institutional program, Republic of Korea

REFERENCES

1. Peierls RE. Ann I H Poincare 1935;5:177–222.

2. Landau LD. Phys Z Sowjetunion 1937;11:26–35.

3. Landau LD, Lifshitz EM. Statistical Physics, Part I. Oxford: Pergamon; 1980.

4. Mermin ND. Phys Rev 1968;176(1):250–254.

5. Venables JA, Spiller GDT, Hanbucken M. Rep Prog Phys 1984;47(4):399.

6. Evans JW, Thiel PA, Bartelt MC. Surf Sci Rep 2006;61(1–2):1–128.

7. Novoselov KS, Geim AK, Morozov SV, Jiang D, Zhang Y, Dubonos SV, Grigorieva IV, Firsov AA. Science 2004;306(5696):666–669.

8. First PN, Heer WA, Seyller T, Berger C, Stroscio JA, Moon JS. MRS Bull 2010;35:296.

9. Li D, Mueller MB, Gilje S, Kaner RB, Wallace GG. Nat Nanotechnol 2008;3:101–105.

10. Park S, Ruoff RS. Nat Nanotechnol 2009;4:217–224.

11. Stankovich S, Dikin DA, Dommett GHB, Kohlhaas KM, Zimney EJ, Stach EA, Piner RD, Nguyen ST, Ruoff RS. Nature 2006;442(7100):282–286.

12. Gao W, Alemany LB, Ci L, Ajayan PM. Nat Chem 2009;1(5):403–408.

13. May JW. Surf Sci 1969;17(1):267–270.

14. Hagstrom S, Lyon HB, Somorjai GA. Phys Rev Lett 1965;15(11):491–493.

15. Shelton JC, Patil HR, Blakely JM. Surf Sci 1974;43(2):493–520.

16. Li X, Cai W, Colombo L, Ruoff RS. Nano Lett 2009;9(12):4268–4272.

17. Yu Q, Lian J, Siriponglert S, Li H, Chen YP, Pei S-S. Appl Phys Lett 2008;93:113103.

18. Zhang Y, Gomez L, Ishikawa FN, Madaria A, Ryu K, Wang C, Badmaev A, Zhou C. J Phys Chem Lett 2010;1(20):3101–3107.

19. Zhang Y, Zhang L, Zhou C. Acc Chem Res 2013.

20. Banerjee BC, Hirt T, Walker P. Nature 1961;192:450–451.

21. Arsem W. Ind Eng Chem 1911;3(11):799–804.

22. Kim KS, Zhao Y, Jang H, Lee SY, Kim JM, Kim KS, Ahn J-H, Kim P, Choi J-Y, Hong BH. Nature 2009;457(7230):706–710.

23. Reina A, Thiele S, Jia X, Bhaviripudi S, Dresselhaus M, Schaefer J, Kong J. Nano Res 2009;2(6):509–516.

24. Reina A, Jia X, Ho J, Nezich D, Son H, Bulovic V, Dresselhaus MS, Kong J. Nano Lett 2008;9(1):30–35.

25. Li Z, Wu P, Wang C, Fan X, Zhang W, Zhai X, Zeng C, Li Z, Yang J, Hou J. ACS Nano 2011;5(4):3385–3390.

26. Kalita G, Masahiro M, Uchida H, Wakita K, Umeno M. Mater Lett 2010;64(20):2180–2183.

27. Kalita G, Wakita K, Umeno M. Physica E: Low-dimensional Systems and Nanostructures 2011;43(8):1490–1493.

28. Li X, Cai W, An J, Kim S, Nah J, Yang D, Piner R, Velamakanni A, Jung I, Tutuc E. Science 2009;324(5932):1312–1314.

29. De Arco LG, Yi Z, Kumar A, Chongwu Z. Nanotechnology, IEEE 2009;8(2):135–138.

30. Thiele S, Reina A, Healey P, Kedzierski J, Wyatt P, Hsu P-L, Keast C, Schaefer J, Kong J. Nanotechnology 2010;21(1):015601.

31. Baraton L, He Z, Lee CS, Maurice J-L, Cojocaru CS, Gourgues-Lorenzon A-F, Lee YH, Pribat D. Nanotechnology 2011;22(8):085601.

32. Liu N, Fu L, Dai B, Yan K, Liu X, Zhao R, Zhang Y, Liu Z. Nano Lett 2010;11(1):297–303.

33. Shin D, Bae S, Yan C, Kang J, Ryu J, Ahn J -H, Hong BH. Carbon Lett 2012;13:1.

34. Bhaviripudi S, Jia X, Dresselhaus MS, Kong J. Nano Lett 2010;10(10):4128–4133.

35. Shu N, Wei W, Shirui X, Qingkai Y, Jiming B, Shin-shem P, Kevin FM. New J Phys 2012;14(9):093028.

36. Kalbac M, Frank O, Kavan L. Carbon 2012;50(10):3682–3687.

37. Li Q, Chou H, Zhong J-H, Liu J-Y, Dolocan A, Zhang J, Zhou Y, Ruoff RS, Chen S, Cai W. Nano Lett 2013;13(2):486–490.

38. Starodub E, Bartelt NC, McCarty KF. J Phys Chem C 2010;114(11):5134–5140.

39. Mu R, Fu Q, Jin L, Yu L, Fang G, Tan D, Bao X. Angew Chem Int Ed 2012;51(20):4856–4859.

40. Yan K, Peng H, Zhou Y, Li H, Liu Z. Nano Lett 2011;11(3):1106–1110.

41. Lee S, Lee K, Zhong Z. Nano Lett 2010;10(11):4702–4707.

42. Chen S, Cai W, Piner RD, Suk JW, Wu Y, Ren Y, Kang J, Ruoff RS. Nano Lett 2011;11(9):3519–3525.

43. Zhang W, Wu P, Li Z, Yang J. J Phys Chem C 2011;115(36):17782–17787.

44. Li X, Magnuson CW, Venugopal A, An J, Suk JW, Han B, Borysiak M, Cai W, Velamakanni A, Zhu Y, Fu L, Vogel EM, Voelkl E, Colombo L, Ruoff RS. Nano Lett 2010;10(11):4328–4334.

45. Li X, Magnuson CW, Venugopal A, Tromp RM, Hannon JB, Vogel EM, Colombo L, Ruoff RS. J Am Chem Soc 2011;133(9):2816–2819.

46. Vlassiouk I, Regmi M, Fulvio P, Dai S, Datskos P, Eres G, Smirnov S. ACS Nano 2011;5(7):6069–6076.

47. Yu Q, Jauregui LA, Wu W, Colby R, Tian J, Su Z, Cao H, Liu Z, Pandey D, Wei D. Nature Mater 2011;10(6):443–449.

48. Geng D, Wu B, Guo Y, Huang L, Xue Y, Chen J, Yu G, Jiang L, Hu W, Liu Y. Proc Natl Acad Sci U S A 2012;109(21):7992–7996.

49. Sun Z, Yan Z, Yao J, Beitler E, Zhu Y, Tour JM. Nature 2010;468(7323):549–552.

50. Guermoune A, Chari T, Popescu F, Sabri SS, Guillemette J, Skulason HS, Szkopek T, Siaj M. Carbon 2011;49(13):4204–4210.

51. Suzuki S, Takei Y, Furukawa K, Hibino H. Appl Phys Express 2011;4(6):5102.

52. Suzuki S, Takei Y, Furukawa K, Webber G, Tanabe S, Hibino H. Jpn J Appl Phys 2012;51(6).

53. Saito Y, Nishikubo K, Kawabata K, Matsumoto T. J Appl Phys 1996;80(5):3062–3067.

54. Ichi-oka H-A, Higashi N-O, Yamada Y, Miyake T, Suzuki T. Diam Relat Mater 2007;16(4–7):1121–1125.

55. An H, Lee W-J, Jung J. Curr Appl Phys 2011;11(4, Supplement):S81–S85.

56. Kondo D, Yagi K, Sato M, Nihei M, Awano Y, Sato S, Yokoyama N. Chem Phys Lett 2011;514(4–6):294–300.

57. Sutter E, Albrecht P, Sutter P. Appl Phys Lett 2009;95(13):133109–133109-3.

58. Sutter P, Albrecht P, Sutter EA. Appl Phys Lett 2010;97(21):213101–213101-3.

59. Sutter E, Albrecht P, Camino FE, Sutter P. Carbon 2010;48(15):4414–4420.

60. Wang S, Pei Y, Wang X, Wang H, Meng Q, Tian H, Zheng X, Zheng W, Liu Y. J Phys D: Appl Phys 2010;43(45):455402.

61. Ramón ME, Gupta A, Corbet C, Ferrer DA, Movva HC, Carpenter G, Colombo L, Bourianoff G, Doczy M, Akinwande D. ACS Nano 2011;5(9):7198–7204.

62. Zhan N, Wang G, Liu J. Appl Phys A 2011;105(2):341–345.

63. Ago H, Ito Y, Mizuta N, Yoshida K, Hu B, Orofeo CM, Tsuji M, Ikeda K-i, Mizuno S. ACS Nano 2010;4(12):7407–7414.

64. Rut'kov E, Kuz'michev A. Phys Solid State 2011;53(5):1092–1098.

65. Roth S, Osterwalder J, Greber T. Surf Sci 2011;605(9):L17–L19.

66. Müller F, Grandthyll S, Zeitz C, Jacobs K, Hüfner S, Gsell S, Schreck M. Phys Rev B 2011;84(7):075472.

67. Vo-Van C, Kimouche A, Reserbat-Plantey A, Fruchart O, Bayle-Guillemaud P, Bendiab N, Coraux J. Appl Phys Lett 2011;98(18):181903–181903-3.

68. Kwon S-Y, Ciobanu CV, Petrova V, Shenoy VB, Bareno J, Gambin V, Petrov I, Kodambaka S. Nano Lett 2009;9(12):3985–3990.

69. Murata Y, Nie S, Ebnonnasir A, Starodub E, Kappes B, McCarty K, Ciobanu C, Kodambaka S. Phys Rev B 2012;85(20):205443.

70. Gao T, Xie S, Gao Y, Liu M, Chen Y, Zhang Y, Liu Z. ACS Nano 2011;5(11):9194–9201.

71. Kang BJ, Mun JH, Hwang CY, Cho BJ. J Appl Phys 2009;106(10):104309–104309-6.

72. Oznuluer T, Pince E, Polat EO, Balci O, Salihoglu O, Kocabas C. Appl Phys Lett 2011;98(18):183101–183101-3.

73. Liu X, Fu L, Liu N, Gao T, Zhang Y, Liao L, Liu Z. J Phys Chem C 2011;115(24):11976–11982.

74. Chen S, Brown L, Levendorf M, Cai W, Ju S-Y, Edgeworth J, Li X, Magnuson CW, Velamakanni A, Piner RD. ACS Nano 2011;5(2):1321–1327.

75. Weatherup RS, Bayer BC, Blume R, Ducati C, Baehtz C, Schlögl R, Hofmann S. Nano Lett 2011;11(10):4154–4160.

76. Dai B, Fu L, Zou Z, Wang M, Xu H, Wang S, Liu Z. Nature Commun 2011;2:522.

77. Yudasaka M, Kikuchi R, Matsui T, Ohki Y, Yoshimura S, Ota E. Appl Phys Lett 1995;67(17):2477–2479.

78. Fujita D, Ohgi T, Onishi K, Kumakura T, Harada M. Jpn J Appl Phys 2003;42:1391.

79. Okamoto H, Massalski T. J Phase Equilibria 1984;5(4):378–379.

80. Reina A, Son H, Jiao L, Fan B, Dresselhaus MS, Liu Z, Kong J. The J Phys Chem C 2008;112(46):17741–17744.

81. Jiao L, Fan B, Xian X, Wu Z, Zhang J, Liu Z. J Am Chem Soc 2008;130(38):12612–12613.

82. Li X, Zhu Y, Cai W, Borysiak M, Han B, Chen D, Piner RD, Colombo L, Ruoff RS. Nano Lett 2009;9(12):4359–4363.

83. Wang Y, Zheng Y, Xu X, Dubuisson E, Bao Q, Lu J, Loh KP. ACS Nano 2011;5(12):9927–9933.

84. Lee JN, Park C, Whitesides GM. Anal Chem 2003;75(23):6544–6554.

85. Lee Y, Bae S, Jang H, Jang S, Zhu S-E, Sim SH, Song YI, Hong BH, Ahn J-H. Nano Lett 2010;10(2):490–493.

86. Caldwell JD, Anderson TJ, Culbertson JC, Jernigan GG, Hobart KD, Kub FJ, Tadjer MJ, Tedesco JL, Hite JK, Mastro MA. ACS Nano 2010;4(2):1108–1114.

87. Kang J, Shin D, Bae S, Hong BH. Nanoscale 2012;4(18):5527–5537.

88. Bae S, Kim H, Lee Y, Xu X, Park J-S, Zheng Y, Balakrishnan J, Lei T, Kim HR, Song YI. Nature Nanotech 2010;5(8):574–578.

89. Kang J, Hwang S, Kim JH, Kim MH, Ryu J, Seo SJ, Hong BH, Kim MK, Choi J-B. ACS Nano 2012;6(6):5360–5365.

90. Wang D-Y, Huang IS, Ho P-H, Li S-S, Yeh Y-C, Wang D-W, Chen W-L, Lee Y-Y, Chang Y-M, Chen C-C, Liang C-T, Chen C-W. Adv Mater 2013;25(32):4521–4526.

91. Levendorf MP, Ruiz-Vargas CS, Garg S, Park J. Nano Lett 2009;9(12):4479–4483.

92. Ismach A, Druzgalski C, Penwell S, Schwartzberg A, Zheng M, Javey A, Bokor J, Zhang Y. Nano Lett 2010;10(5):1542–1548.

93. Su C-Y, Lu A-Y, Wu C-Y, Li Y-T, Liu K-K, Zhang W, Lin S-Y, Juang Z-Y, Zhong Y-L, Chen F-R. Nano Lett 2011;11(9):3612–3616.

94. Byun S-J, Lim H, Shin G Y, Han T-II, Oh SII, Ahn J-H, Choi HC, Lee T-W. J Phys Chem Lett 2011;2(5):493–497.

95. Lee, C. S.; Baraton, L.; He, Z.; Maurice, J.-L.; Chaigneau, M.; Pribat, D.; Cojocaru, C. S. SPIE Nanoscience and Engineering, International Society for Optics and Photonics: 2010; pp 77610P–77610P-7.

96. Shin HJ, Choi WM, Yoon SM, Han GH, Woo YS, Kim ES, Chae SJ, Li XS, Benayad A, Loc DD. Adv Mater 2011;23(38):4392–4397.

97. Kwak J, Chu JH, Choi J-K, Park S-D, Go H, Kim SY, Park K, Kim S-D, Kim Y-W, Yoon E. Nature Commun 2012;3:645.

98. Kim Y-J, Kim SJ, Jung MH, Choi KY, Bae S, Lee S-K, Lee Y, Shin D, Lee B, Shin H. Nanotechnology 2012;23(34):344016.

99. Ding X, Ding G, Xie X, Huang F, Jiang M. Carbon 2011;49(7):2522–2525.

100. Bjelkevig C, Mi Z, Xiao J, Dowben P, Wang L, Mei W-N, Kelber JA. J Phys Condens Matter 2010;22(30):302002.

101. Oshima C, Itoh A, Rokuta E, Tanaka T, Yamashita K, Sakurai T. Solid State Commun 2000;116(1):37–40.

102. Lin T, Wang Y, Bi H, Wan D, Huang F, Xie X, Jiang M. J Mater Chem 2012;22(7):2859–2862.

103. Ferrari A, Meyer J, Scardaci V, Casiraghi C, Lazzeri M, Mauri F, Piscanec S, Jiang D, Novoselov K, Roth S. Phys Rev Lett 2006;97(18):187401.

104. Wei D, Liu Y, Zhang H, Huang L, Wu B, Chen J, Yu G. J Am Chem Soc 2009;131(31):11147–11154.

105. Rummeli MH, Bachmatiuk A, Scott A, Borrnert F, Warner JH, Hoffman V, Lin J-H, Cuniberti G, Buchner B. ACS Nano 2010;4(7):4206–4210.

106. Giovannetti G, Khomyakov PA, Brocks G, Kelly PJ, van den Brink J. Phys Rev B 2007;76(7):073103.

107. Kubota Y, Watanabe K, Tsuda O, Taniguchi T. Science 2007;317(5840):932–934.

108. Zeng H, Kan Y-M, Zhang G-J. Mater Lett 2010;64(18):2000–2002.

109. Lin T, Huang F, Liang J, Wang Y. Energy & Environ Science 2011;4(3):862–865.

110. Lü X, Wu J, Lin T, Wan D, Huang F, Xie X, Jiang M. J Mater Chem 2011;21(29):10685–10689.

111. Gwon H, Kim H-S, Lee KU, Seo D-H, Park YC, Lee Y-S, Ahn BT, Kang K. Energy & Environ Sci 2011;4(4):1277–1283.

112. Son M, Lim H, Hong M, Choi HC. Nanoscale 2011;3(8):3089–3093.

113. Fanton MA, Robinson JA, Puls C, Liu Y, Hollander MJ, Weiland BE, LaBella M, Trumbull K, Kasarda R, Howsare C. ACS Nano 2011;5(10):8062–8069.

114. Hwang J, Shields VB, Thomas CI, Shivaraman S, Hao D, Kim M, Woll AR, Tompa GS, Spencer MG. J Cryst Growth 2010;312(21):3219–3224.

115. Miyasaka Y, Nakamura A, Temmyo J. Jap J Appl Phys 2011;50:04DH12.

116. Song HJ, Son M, Park C, Lim H, Levendorf MP, Tsen AW, Park J, Choi HC. Nanoscale 2012;4(10):3050–3054.

117. Huang PY, Ruiz-Vargas CS, van der Zande AM, Whitney WS, Levendorf MP, Kevek JW, Garg S, Alden JS, Hustedt CJ, Zhu Y. Nature 2011;469(7330):389–392.

118. Kim K, Lee Z, Regan W, Kisielowski C, Crommie M, Zettl A. ACS Nano 2011;5(3):2142–2146.

119. Chae SJ, Güneş F, Kim KK, Kim ES, Han GH, Kim SM, Shin HJ, Yoon SM, Choi JY, Park MH. Adv Mater 2009;21(22):2328–2333.

120. Casiraghi C, Hartschuh A, Qian H, Piscanec S, Georgi C, Fasoli A, Novoselov K, Basko D, Ferrari A. Nano Lett 2009;9(4):1433–1441.

121. Godin T, LaFemina JP. Phys Rev B 1994;49(11):7691.

122. Guo Q, Wang E. Sci Technol Adv Mater 2005;6(7):795–798.

123. Wei D, Xu X. Appl Phys Lett 2012;100(2):023110–023110-3.

124. Kim K-B, Lee C-M, Choi J. J Phys Chem C 2011;115(30):14488–14493.

125. Hong G, Wu Q-H, Ren J, Lee S-T. Appl Phys Lett 2012;100(23):231604–231604-5.

126. Bi H, Sun S, Huang F, Xie X, Jiang M. J Mater Chem 2012;22(2):411–416.

127. Zhang L, Shi Z, Wang Y, Yang R, Shi D, Zhang G. Nano Res 2011;4(3):315–321.

128. Maeda F, Hibino H. Jap J Appl Phys 2010;49(4):04DH13.

129. Chen J, Wen Y, Guo Y, Wu B, Huang L, Xue Y, Geng D, Wang D, Yu G, Liu Y. J Am Chem Soc 2011;133(44):17548–17551.

130. Liu B, Tang D-M, Sun C, Liu C, Ren W, Li F, Yu W-J, Yin L-C, Zhang L, Jiang C. J Am Chem Soc 2010;133(2):197–199.

131. Choi H. A Single Layer of Graphene Formation on Silicon Oxide Surface (001): Simulation and Analysis of a New Material. Massachusetts: ProQuest; 2008.

132. Eda G, Fanchini G, Chhowalla M. Nat Nanotechnol 2008;3(5):270–274.

133. Kuang Q, Xie S-Y, Jiang Z-Y, Zhang X-H, Xie Z-X, Huang R-B, Zheng L-S. Carbon 2004;42(8):1737–1741.

134. Manning TJ, Mitchell M, Stach J, Vickers T. Carbon 1999;37(7):1159–1164.

135. Chung D. J Mater Sci 2002;37(8):1475–1489.

136. Joh H-I, Lee S, Kim T-W, Hwang SY, Hahn JR. Carbon 2013;55:299–304.

137. Morgan P. Carbon fibers and their composites. Boca Raton, Florida: Taylor & Francis; 2005.

138. Park S, An J, Jung I, Piner RD, An SJ, Li X, Velamakanni A, Ruoff RS. Nano Lett 2009;9(4):1593–1597.

139. Moon IK, Lee J, Ruoff RS, Lee H. Nature Commun 2010;1:73.

140. Lee SM, Lee YH, Hwang YG, Hahn J, Kang H. Phys Rev Lett 1999;82(1):217–220.

141. Hahn J, Kang H. Phys Rev B 1999;60(8):6007–6017.

142. Matsumoto T. Pure Appl Chem 1985;57(11):1553–1562.

143. Na S-I, Lee J-S, Noh Y-J, Kim T-W, Kim S-S, Joh H-I, Lee S. Sol Energy Mater Sol Cells 2013;115:1–6.

144. Wan X, Long G, Huang L, Chen Y. Adv Mater 2011;23(45):5342–5358.

145. Yun JM, Yeo JS, Kim J, Jeong HG, Kim DY, Noh YJ, Kim SS, Ku BC, Na SI. Adv Mater 2011;23(42):4923–4928.

146. Kalita G, Matsushima M, Uchida H, Wakita K, Umeno M. J Mater Chem 2010;20(43):9713–9717.

147. Wang X, Zhi L, Tsao N, Tomović Ž, Li J, Müllen K. Angew Chem 2008;120(16):3032–3034.

148. Tomović Ž, Watson MD, Müllen K. Angew Chem Int Ed 2004;43(6):755–758.

149. Watson MD, Fechtenkötter A, Müllen K. Chem Rev 2001;101(5):1267–1300.

150. Schmidt-Mende L, Fechtenkötter A, Müllen K, Moons E, Friend R, MacKenzie J. Sci 2001;293(5532):1119–1122.

151. Zhi L, Wu J, Li J, Kolb U, Müllen K. Angew Chem Int Ed 2005;44(14):2120–2123.

152. Gherghel L, Kübel C, Lieser G, Räder H-J, Müllen K. J Am Chem Soc 2002;124(44):13130–13138.

153. Turchanin A, Beyer A, Nottbohm CT, Zhang X, Stosch R, Sologubenko A, Mayer J, Hinze P, Weimann T, Gölzhäuser A. Adv Mater 2009;21(12):1233–1237.

154. Nottbohm CT, Turchanin A, Beyer A, Stosch R, Gölzhäuser A. Small 2011;7(7):874–883.

155. Nottbohm CT, Turchanin A, Beyer A, Gölzhäuser A. J Vac Sci Technol, B 2009;27:3059–3062.

156. Ying Z, Hettich R, Compton R, Haufler R. J Phys B: At Mol Opt Phys 1996;29(21):4935–4942.

157. Park JB, Jeong SH, Jeong MS, Kim JY, Cho BK. Carbon 2008;46(11):1369–1377.

158. Radhakrishnan G, Adams PM, Bernstein LS. Appl Surf Sci 2007;253(19):7651–7655.

159. Qian M, Zhou YS, Gao Y, Park JB, Feng T, Huang SM, Sun Z, Jiang L, Lu YF. Carbon 2011;49(15):5117–5123.

160. Lu YF, Huang SM, Sun Z. J Appl Phys 2000;87(2):945–951.

161. Huang SM, Lu YF, Sun Z, Luo XF. Surf Coat Technol 2000;125(1):25–29.

162. Liu W, Dang T, Xiao Z, Li X, Zhu C, Wang X. Carbon 2011;49(3):884–889.

163. Zhu MY, Outlaw RA, Bagge-Hansen MJ, Chen HJ, Manos DM. Carbon 2011;49(7):2526–2531.

164. Zhu M, Wang J, Holloway BC, Outlaw RA, Zhao X, Hou K, Shutthanandan V, Manos DM. Carbon 2007;45(11):2229–2234.

165. Na S-I, Noh Y-J, Son S-Y, Kim T-W, Kim S-S, Lee S, Joh H-I. Appl Phys Lett 2013;102(4):043304–043304-5.

166. Lee S, Yeo J-S, Ji Y, Cho C, Kim D-Y, Na S-I, Lee BH, Lee T. Nanotechnology 2012;23(34):344013.

167. Geng JX, Liu LJ, Yang SB, Youn S-C, Kim DW, Lee J-S, Choi J-K, Jung H-T. J Phys Chem C 2010;114(34):14433–14440.

168. Wang S, Wang J, Miraldo P, Zhu M, Outlaw R, Hou K, Zhao X, Holloway BC, Manos D, Tyler T. Appl Phys Lett 2006;89(18):183103–183103-3.

5

FUNCTIONALIZATION OF SILICA NANOPARTICLES FOR CORROSION PREVENTION OF UNDERLYING METAL

Dylan J. Boday, Jason T. Wertz, and Joseph P. Kuczynski

International Business Machines Corporation, Tucson, AZ 85744, USA

5.1 INTORDUCTION

5.1.1 Corrosion in IT Environment

High-end servers must meet stringent reliability requirements in order to meet customer expectations. Typical unscheduled downtime for a server is in the range of 0.25-4.00 h/year (Fig. 5.1) (1).

As information technology (IT) equipment is increasingly used outside of traditional datacenters, coupled with a vast number of installations growing in the industrial regions of China and India where airborne corrosive gases are prevalent, recent attempts to understand the complex nature of gaseous corrosion in these environments have been undertaken. Although many environmental factors exist which may lead to premature hardware failure, the role of relative humidity, sulfidation (from various forms of sulfur-bearing gases), and synergistic reactants such as chlorine and NO_x have received considerable attention. Corrosion of metallurgy used for interconnections, solder joints, pin-through-hole components, and so on may often result in catastrophic failures. Consequently, there has been a tremendous focus on the mechanism of electrical contact corrosion of computer hardware.

5.1.2 Environmental Conditions Leading to Corrosion

Numerous studies have attempted to address which environmental conditions lead to corrosion. Although silver is stable in pure air and water, the oxidation of a copper surface to Cu_2O when exposed to humidified air has been studied by several authors (2–5). The oxidation rate was found to follow a logarithmic dependence on exposure time. For relative humidity conditions ranging from 40 to 80%, the thickness of the cuprous oxide layer was determined to be 0.2–9.6 nm, respectively, and was associated with an aqueous adlayer that grew from 0.3 nm at 40% RH to 1.1 nm at 80% RH (5). The kinetics of the growth of Cu_2O is enhanced by higher relative humidity that results in an increased amount of physisorbed water. However, the amount of this aqueous adlayer is approximately independent of exposure time. Atomic force microscopy studies revealed that Cu_2O deposits as small grains on the copper surface with an enhanced nucleation rate at higher relative humidity. This enhanced nucleation rate has been postulated for the increased oxidation rate at higher relative humidity. The oxidation rate is strongly correlated to the physisorbed water layer on the copper surface, which in turn is dependent on the relative humidity. Consequently, the effect of increased relative humidity can be summarized as follows:

- Adsorbed water generally forms clusters on clean metal surfaces (6–8), which act as nucleation sites for Cu_2O grains. The greater the amount of adsorbed water at elevated RH, the greater the number of nucleation sites.
- The physical properties of the aqueous adlayer change with thickness. The thicker the adlayer, the more bulk-like the physisorbed water becomes which results in higher gas solubility. Increased relative humidity results in greater bulk-like water properties, increased oxygen solubility, and an increased oxidation rate.
- Physisorbed water may interact with cuprous oxide (9) to form a defect-rich oxide layer. This defect-rich layer affects the transport properties of copper in the oxide and results in higher oxidation rates. Increased RH results in a greater amount of physisorbed water which enhances the oxidation rate.

Nanomaterials, Polymers and Devices: Materials Functionalization and Device Fabrication, First Edition. Edited by Eric S. W. Kong.

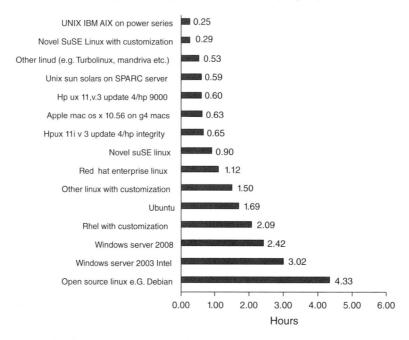

Figure 5.1   Unscheduled server downtime for various hardware/operating system configurations. (*Source*: Figure adapted from data from Reference 1.)

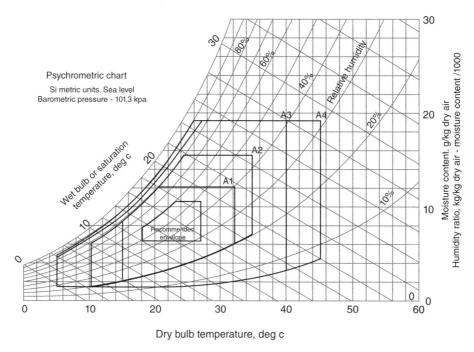

Figure 5.2   ASHRAE Environmental Classes for Datacenters. (*Source*: Reproduced with permission from Reference 10.)

The increased oxidation rate of copper, as the RH is increased, is of practical importance as the current trend in datacenters is to permit operation of IT equipment at higher temperature and relative humidity. Currently, most datacenters operate in American Society of Heating, Refrigerating and Air Conditioning Engineers (ASHRAE) Class A2 (see the psychrometric chart in Fig. 5.2) environments, temperature 10–35°C, humidity 8–80% RH (10). In order to enable the datacenter professional to operate in the most energy efficient mode while still achieving the required reliability demanded by their business, ASHRAE drafted new guidelines for IT equipment environments. Two new datacenter classes were created to achieve the most flexibility in the operation of the datacenter. ASHRAE Class A3 expands the temperature range to 5–40 °C while also expanding the moisture range from 8% RH and −12 °C dew point to 85% relative humidity. ASHRAE Class A4 expands the allowable temperature and moisture range even further than A3. The temperature range is expanded to 5–45 °C while the moisture range extends from 8% RH and −12 °C dew point to 90% RH. Since the oxidation rate of copper is dependent on both the temperature and relative humidity, migration to either Class A3 or Class A4 is expected to exacerbate copper corrosion.

Although the effect of relative humidity on metal corrosion is important regarding long-term IT equipment reliability, perhaps of even greater relevance is corrosion induced by atmospheric pollutants. Two common sulfur-bearing gases capable of corroding both copper and silver are hydrogen sulfide (H_2S) and carbonyl sulfide (OCS). In the presence of water, OCS decomposes to carbon dioxide and H_2S (11). The reaction of copper with H_2S has been shown by various authors to be dependent on relative humidity (12). At H_2S concentrations representative of atmospheric conditions, the sulfidation rate of copper was found to vary by an order of magnitude as the relative humidity was increased from 0 to 100%. Sharma (12) determined that a copper oxide film grown in air (several hundred Å thick) provided adequate protection against H_2S at low relative humidity, but little protection at high relative humidity. Although similar studies have not been conducted on copper sulfidation by OCS, decomposition of OCS in physisorbed water on copper surfaces is expected to result in the formation of H_2S. Corrosion of copper by OCS is therefore expected to be dependent on relative humidity, with the corrosion rate increasing with increasing RH.

Literature reports on the RH dependence of silver sulfidation by H_2S are conflicting. Backlund et al. (13) and Drott (14) both found a strong RH dependence on silver sulfidation using H_2S as the sole corrosive gas. These authors observed a marked increase in the corrosion rate above 70% RH. Lorenzen observed a similar result using mixed corrosive gases, including H_2S (15). However, more recent work by Rice et al. (16) using H_2S and other mixed corrosive gases observed no effect of RH on silver sulfidation rates. Graedel and coworkers conducted extensive experiments on the sulfidation of both copper and silver by H_2S and OCS over a wide range of relative humidities (17). The authors proposed the following corrosion reactions for copper and silver in the presence of H_2S (Scheme 5.1).

The reactions are most likely multistep processes with varying rate-limiting mechanisms operable at different times throughout the course of the reactions. A schematic representation of the sulfidation reactions is illustrated in Fig. 5.3). During the initial stage (1), sulfidation occurs on either silver or a Cu_2O surface (formed via oxidation in the aqueous adlayer of physisorbed water). Once a continuous sulfide film is formed (2), further growth of the film takes place on the sulfide layer. Over an extended time range, the sulfidation is limited by the rate at which H_2S is transported to the surface (Region G). This assertion is supported by the observation that short, high concentration exposures produce identical results at long, low concentration exposures (18, 19). For very long exposures, the rate becomes independent of the H_2S transport and is limited by ion diffusion through thick sulfide (Region D). As is the case for oxide growth on copper surfaces, an aqueous adlayer functions to dissolve the H_2S, rendering it more accessible to the metal. Since the physisorbed layer forms clusters, the quantity of absorbed H_2S will increase with increasing relative humidity.

A comprehensive study of electrical contact corrosion in mixed flowing gas environments was undertaken by Abbott (20). Both copper and porous gold (Au over Ni over Cu) coupons were subjected to various gas mixtures at various concentrations and the corrosion rate determined (Table 5.1). An important result of these studies is that only those environments which incorporate the sulfide–chloride interaction begin to approach conditions actually occurring in the field. In pure gaseous sulfide environments, even those at $[H_2S] > 1000$ ppb, Abbott determined

$$2Ag + H_2S \longrightarrow Ag_2S + H_2$$
$$2Cu + H_2S \longrightarrow Cu_2S + H_2$$

Scheme 5.1   Corrosion reaction for silver and copper in the presence of H_2S.

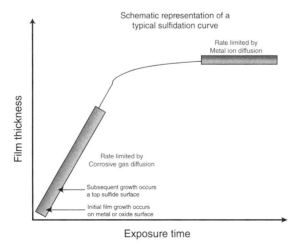

Figure 5.3   Representative sulfidation curve. (*Source*: Based on data from Reference 17.)

TABLE 5.1   Gas Effects on Corrosion

Corrosive Gas	Concentration (ppb)	Corrosion Product	
		Copper	Porous Gold
H_2S	<100	Cu_2O	ND^a
H_2S	>1000	Cu_2S	Cu_2S
SO_2	<100	Cu_2O	ND
SO_2	>1000	$Cu_xS_yO_z$	$Cu_xS_yO_z$
Cl_2	<10	$Cu_xCl_yO_z$	$Cu_xCl_yO_z$
Cl_2	>30	$Cu_xCl_yO_z$	$Cu_xCl_yO_z$
NO_2	<100	ND	ND
H_2S/SO_2	<100/<100	Cu_2O	Cu_2S
H_2S/SO_2	100–1000	Cu_2S	Cu_2S
H_2S/Cl_2	<10	$Cu_xCl_yO_z$	$Cu_xCl_yO_z$
H_2S/Cl_2	<100/<10	$Cu_2S + Cu_xCl_yO_z$	$Cu_2S + Cu_xCl_yO_z$
H_2S/NO_2	<100	Cu_2O	Cu_2S
H_2S/NO_2	100–1000	Cu_2S	Cu_2S
$H_2S/SO_2/NO_2$	<100	Cu_2O	Cu_2S
$H_2S/SO_2/NO_2$	100–1000	Cu_2S	Cu_2S
$H_2S/NO_2/Cl_2$	<10	$Cu_xCl_yO_z$	$Cu_xCl_yO_z$
$H_2S/SO_2/NO_2/Cl_2$	<100/<10/<10/<10	$Cu_2S + Cu_xCl_yO_z$	$Cu_2S + Cu_xCl_yO_z$

Source: Adapted from data referenced in Reference 20.
[a]ND, none detected.

that the corrosion was very low and not representative of field data. In sharp contrast, the addition of reactive chloride (e.g., Cl_2) greatly accelerates the corrosion reaction and results in corrosion products that are representative of those found in field failures. Unfortunately, a systematic evaluation of the effect of relative humidity on both the reaction kinetics and resulting corrosion products for the various gas mixtures was not undertaken.

It is obvious that the specific gas mixture, as well as the relative humidity, play an interactive role in both the reaction kinetics and the resulting corrosion products. Since the current trend is to permit operation of IT equipment in an ASHRAE Class 3 environment, mitigation of gaseous corrosion on metallic surfaces is of paramount importance.

5.1.3   Hardware Challenges in Corrosive Environments

The performance of electronic hardware continues to improve as a direct result of the miniaturization of circuitry and components. However, as package density increases, additional air flow is required to dissipate the thermal load. Increased packaging density may also prevent hermetic sealing of all of the components on a printed circuit board (PCB) resulting in the direct exposure of sensitive metallurgy to airborne contaminants (corrosive gases and dust). Moreover, as the line widths and spacing on circuit cards decrease, the potential increases for corrosion-induced shorts due to ion migration or creep. The purported increase of hardware failures in datacenters in geographies with an elevated concentration of atmospheric sulfur-bearing gases (India and China) led ASHRAE to issue a white paper on the subject recommending that in addition to temperature and humidity control, dust and airborne contamination be monitored and controlled as well (21). However, the literature cited in the ASHRAE paper (22–29) has been criticized for failure to provide definitive information of the cause and effect relationship between specific levels of gaseous contaminants and the damage they caused to electronic equipment within datacenters (30). Although the ASHRAE white paper failed to provide critical information and data relating the increased hardware failure rates in datacenters, gaseous corrosion-induced failures have been observed (31). This increase in gaseous corrosion has been attributed to the use of airside economizers (32). An airside economizer comprises numerous sensors, and dampers that control both the air volume and temperature supplied to the datacenter. Ambient air is used to directly cool the hardware but is generally maintained to a suitable environmental range given by published standards. Concurrent with the arrival of "free air cooling," datacenter corrosion inquiries reported by Purafil for 2010 (Table 5.2) numbered in excess of 1200 for countries around the globe.

Representative hardware failures attributed to both out-of-control humidity and airborne contamination were studied by Gouda et al. (33). The authors investigated hardware failures in datacenters located in Kuwait where malfunctions of the computer room air conditioning sensors resulted in an increase in the datacenter relative humidity from 50% to 75% RH. Humidity alone, however, could not be solely implicated as the environmental factor leading to copper corrosion on contact pins of a logic card. Analysis of the gaseous components present in the datacenter revealed that high levels of NO_x were present, but both H_2S and SO_2 were below the detection limit. Nevertheless, elemental analysis of the corrosion products identified the presence of sulfur-containing compounds along with copper oxide. Previous work on the corrosion of copper in indoor atmospheric environments found the major corrosion product to be Cu_2O. As this was a legacy installation, that is, prior to the use of "free air" cooling, these results highlight the need to mitigate corrosive gases in the datacenter.

TABLE 5.2   Datacenter Corrosion Inquiries (2010)

Country	Number of Cities	Locations	Monitoring Sites
USA	34	36	192
Mexico	1	1	3
England	1	4	10
Germany	1	1	2
Bulgaria	1	1	2
India	*15*	*47*	*237*
China	*12*	*28*	*621*
Singapore	1	1	2
Indonesia	1	1	4
Korea	1	1	5
Japan	2	4	7

Source: Adapted from Reference 31.

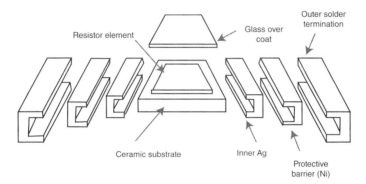

Figure 5.4   Diagram of standard thick film resistor construction.

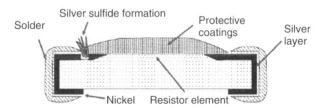

Figure 5.5   Cross-sectional schematic of a corrode resistor from within a sulfur-rich environment.

5.1.4   Effect of Sulfur on Resistor Corrosion

A common component affected by environmental conditions rich in sulfur is surface mount thick film resistors. A general construction of a thick film resistor is shown above in Figure 5.4. Within this construction, the resistive element is commonly made from silver, which is susceptible to corrosion.

When these resistors are exposed to environmental conditions rich in sulfur, a corrosive product is formed. In the above cross-sectional image (Fig. 5.5), silver sulfide corrosion products can be seen extending from the inner silver resistor layer. The cross section also shows a protective layer, which has been found to actually increase the sulfur concentration within standard silicone conformal coatings; this is described in detail in Section 5.1.7. To show an experimental corrosion failure of a thick film resistor in a sulfur-rich environment, an optical microscope image of a sulfur-corroded resistor is shown in Figure 5.6.

5.1.5   Industry Standards for Environmental Specifications

Gaseous composition environmental limits have been published by the International Society of Automation (ISA). ISA Standard 71.04, *Environmental Conditions for Process Measurement and Control Systems: Airborne Contaminants* (34) documents guidelines for specifying

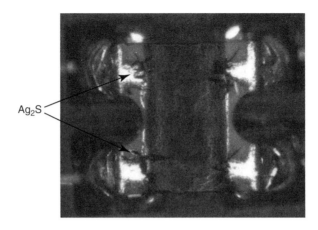

Figure 5.6   Optical image of resistor after being exposed to a sulfur-rich environment.

TABLE 5.3   Gaseous Contamination Corrosivity Levels

Severity Level	Copper Reactivity Level (Å/Month)
G1 mild	<300
G2 moderate	300–1000
G3 harsh	1000–2000
GX severe	>2000

Source: Adapted from ISA-71.04.

environmental cleanliness. Reactive environments are broken into one of the four classifications based upon the severity of the corrosion of copper coupons (Table 5.3).

As previously discussed, there is an industry trend to operate datacenters in the ASHRAE Class A3 environment. This environment specifies a recommended temperature/humidity envelope for IT equipment. In response to a purported increase in datacenter hardware failures arising from gaseous and/or particulate contamination, a white paper published by ASHRAE Technical Committee 9.9: *Mission Critical Facilities, Technology Spaces, and Electronic Equipment* recommended that gaseous contamination be controlled to a modified G1 severity level (35). The modified G1 severity level requires that both copper and silver corrosion rates must be less than 300 Å/month. The effect of the elevated temperature and humidity of a Class A3 environment on copper and silver corrosion has not yet been adequately studied. For copper corrosion testing, the ISA has defined the severity of the corrosion within datacenters based on the reactivity level of the copper coupon (Table 5.3) (34). Other real-time monitoring devices have been developed based on the silver thickness loss from the exposure to corrosive environments (36).

5.1.6   Solutions to Mitigate Corrosion in IT Hardware

Datacenter environments are typically continuously monitored and maintained in accordance to the ASHRAE guidelines pertaining to gaseous and particulate contaminations (21). As only guidelines though, it is not uncommon for datacenter operations to have environments which fall outside of the guidelines. Other factors such as datacenter location can also have an effect on the reliability of the IT hardware. These changes in environmental conditions can lead to failures unpredicted by IT hardware manufacturers and is one of the main reasons for a need to harden the product sets to prevent failures from occurring. In order to prevent these failures from occurring, there are many different solutions that exist today in attempts to mitigate the corrosion of IT hardware.

Particulate contamination is typically defined as a dust like material and further defined by the quantity of the dust present and the corrosiveness of that dust at different humidities (37). Typically, a sample of dust is collected from datacenter surfaces and filters and sent to a lab for analysis. The collected dust samples are placed onto an interdigitated comb coupon (Fig. 5.7) and placed in a temperature/humidity chamber for testing. In these experiments, the deliquescent relative humidity is measured to determine the corrosiveness of the dust. If the deliquescent relative humidity is low, then the dust can be considered corrosive due to its ability to wet whereas a high deliquescent relative humidity is considered to be benign.

In order to overcome particulate contamination based on the results of the deliquescent relative humidity testing, ASHRAE has recommended that datacenters be maintained according to ISO 14644-1 Class 8 (Table 5.4) (21, 38). To achieve this standard, particulate filters are installed to trap unwanted particulates from entering the datacenter environment. For room air, a minimum efficiency rating value (MERV) 8 filter is used while air that enters the datacenter can be filtered with MERV 11 or MERV 13 filters (Table 5.5) (39). Datacenters also sometimes use free air or air-economizers as mentioned previously, which also require the use of MERV 11 or MERV 13 filters depending on

Figure 5.7 Example of an interdigitated comb coupon for deliquescent relative humidity dust corrosiveness testing.

TABLE 5.4   Controlled Environments – Part 1: Classification of Air Cleanliness

Class	Concentration (p/m^3) for Each Particle Size					
	0.1 (μm)	0.2 (μm)	0.3 (μm)	0.5 (μm)	1.0 (μm)	5.0 (μm)
1	10	2				
2	100	24	10	4		
3	1,000	237	102	35	8	
4	10,000	2,370	1,020	352	83	
5	100,000	23,700	10,200	3,520	832	29
6	1,000,000	237,000	102,000	35,200	8,320	293
7				352,000	83,200	2,930
8				3,520,000	832,000	29,300
9				35,200,000	8,320,000	293,000

Source: Adapted from ISO 14644-1.

the specific conditions of the datacenter. While particulate filters are effective in removing unwanted dust, these filters also have disadvantages that can lead to costly maintenance and continuous monitoring.

One of the major disadvantages of using particulate filters to mitigate IT hardware failures is the cost of the filters. Typically, due to the high-efficiency filtration required, a single filter can cost hundreds of dollars to purchase and replace.

With the need to replace these filters over the life span of the datacenter, this solution can easily add up and become a burden on the customer for maintaining these strict guidelines. Another factor that may not usually be considered is the pressure drop that these high-efficiency filters create (40). As the MERV rating increases, the pressure drop becomes greater thus requiring the use of larger fans within the HVAC systems that can overcome these large pressure drops in order to maintain the same environmental conditions as those specified by the datacenter's design. While particulate contamination is a major concern within the datacenters, it is also important to focus on gaseous contamination.

Gaseous contamination within datacenters that lead to IT hardware failures, typically come from sulfur-bearing gases, such as sulfur dioxide and hydrogen sulfide (16). Monitoring for such corrosion is typically performed using corrosion coupon sets that are placed throughout the datacenter. These corrosion coupons are placed within the datacenter environment and allowed to react with the contaminated air source. After reaction, the coupons are analyzed through electrochemical methods to determine the rate of corrosion based on the growth of the corrosion layer on the coupons (41).

While monitoring of a datacenter's gaseous corrosive environment is effective in defining the problem, it lacks the solution to overcome the problem. One method for overcoming gaseous corrosion is the installation of gas-phase filters within the HVAC system of the datacenter. These filters help to bring the environmental conditions within or below the G1 severity level (21). In combination with the particulate filters,

TABLE 5.5   MERV Filter Parameters

MERV	Particle Size Removal Efficiency (%)			Typical Controlled Contaminant
	0.3–1 (μm)	1.0–3.0 (μm)	3.0–10.0 (μm)	
8			>70	Mold spores, dust mites body parts, hair spray, dusting aids
11		65–80	>85	Legionella, humidifier dust, lead dust, auto emission particles
13	<75	>90	>90	All bacteria, droplet nuclei, most smoke, insecticide dust, paint pigments

Source: Adapted from ASHRAE Standard 52.2.

these filters help to reduce the corrosive nature of the datacenter's environment but at an extreme cost to the datacenter operating cost. The cost of initial purchase and replacement filters over the life of the datacenter can become burdensome as these filters must be changed on a regular basis in order to maintain the environmental conditions.

Additionally, a common practice for preventing the IT hardware failure is the use of conformal coatings which is discussed in depth in Section 5.1.7.

5.1.7   Conformal Coatings Overview

Conformal coatings are thin protective layers which are commonly applied to PCBs or electronic arrays as the last major process step within manufacturing. Conformal coatings are applied to protect the various underlying metal structures or devices from subsequent processing, mechanical damages and environmental conditions. Although conformal coatings aid in protecting from subsequent processing and mechanical damages, today the environmental conditions offer the most significant challenges. Common environmental contaminates, which conformal coatings are used to separate the metals within a PCB or surface mount device from are dust, moisture, solvents, corrosive species such as ionic gaseous contaminates or other airborne corrosive species as well as other detrimental chemicals or materials that can be found within hardware environments. These corrosive and detrimental species can be exacerbated due to temperatures and humidity concentrations that can range significantly, thus increase temperature and humidity corrosion kinetics significantly (42). With the proliferation of datacenters across the globe, the environments which the hardware is exposed to vary significantly, therefore requiring versatile conformal coatings to provide protection in a plethora of environmental conditions.

As the conformal coating is applied during the last step in manufacturing, the yield and robustness must be very high as by this point the board or chip has reached a state of high value. There are several methods used to apply conformal coatings depending on the type of material, large area application, or if a more localized application is required. Briefly, many conformal coatings can be dipped, brushed, sprayed, needle dispensed, or nonatomized coated (43). The common thickness of the conformal coating ranges from microns to millimeters depending on the desired protection.

The materials used to prepare conformal coatings can vary depending on operation temperatures, curing conditions and what type of protection it is to provide. Commonly, the materials used to prepare conformal coatings consist of polyurethanes (44), polyimides (45), acrylates (46), Parylene (47), fluoropolymers (48), and silicone (49) conformal coatings. The advantages and disadvantages of each are summarized in Table 5.6. Of the previously mentioned materials, silicone conformal coatings are amongst the most widely used conformal coatings in the industry. As silicone conformal coatings are used extensively in our research, their further review is given in the following. If a more detailed review of various conformal coatings is desired, the authors direct the reader to general overview reviews (50).

5.1.7.1   Silicone Conformal Coatings   Silicone conformal coatings are often considered as being superior to other conformal coatings because of their superior application yields, application methodologies, speed of application, and cure cycles. Additionally, the mechanical properties of silicone conformal coatings offer outstanding toughness, low moisture uptake, electrical properties and purities. Since there are numerous electronic applications that operate at elevated temperatures, yet still require exceptional moisture resistance, only paraxylylene or silicone conformal coatings are considered acceptable in these instances. Due to the prohibitive cost of paraxylylene, silicones have become the conformal coating of choice for electronic applications. Due to these outstanding properties and ease of applications, silicone conformal coatings are the most commonly used conformal coatings today in the marketplace. It is estimated that silicone conformal coatings represent approximately 40% of the entire conformal coating market (51).

Silicone conformal coatings can be applied similarly to other common conformal coatings. They may be applied by any of a number of techniques including brush, dip, or spray coating to yield a void free, conformal coating of suitable thickness. Many of these applications are automated thus allowing for the silicone conformal coating to be applied in high volume applications.

Silicone conformal coatings are available in a multitude of formulations ranging from low viscosity (~300 cP) and single component RTVs to high viscosity (>50 K cP) versions. Figure 5.8 represents a common silicone conformal coating formulation.

Curing is accomplished via a condensation mechanism of silanol-terminated polymers. This typically occurs in a two-stage reaction sequence (52). In the first stage, a silanol-terminated polydimethylsiloxane (PDMS) is reacted with an excess of multifunctional silane which displaces the silanol. Although the organic group (–OR) of the silane may be selected from any number of functionalities (including acetoxy, amine, enoxy, etc.), the most common moiety is alkoxy (–OH), typically methoxy or ethoxy. This is due to the fact that upon moisture curing at

TABLE 5.6   Overview of the Advantages and Disadvantages of Conformal Coatings

Material Class	Advantages	Disadvantages
Acrylic	*Fast cure *High moisture resistance *Tough coating *Easily reworked	*Solvent cast *Moderate cost *Poor chemical resistance
Polyurethane	*Acceptable cure time *High chemical resistance *High moisture resistance *Low cost	*Difficult to rework *Poor high temperature stability
Epoxy	*Acceptable adhesion *Excellent solvent resistance *Acceptable moisture resistance *Excellent high temperature stability	*Difficult to rework *Low flexibility *May impart stress on components *Moderate cost
Paraxylylene	*Excellent uniformity *Excellent chemical resistance *Excellent moisture resistance	*Impossible to rework *Requires vacuum processing equipment *Very high cost
Silicone	*Excellent chemical resistance *Excellent moisture resistance *Excellent high temperature stability	*High cost *Rework may be difficult

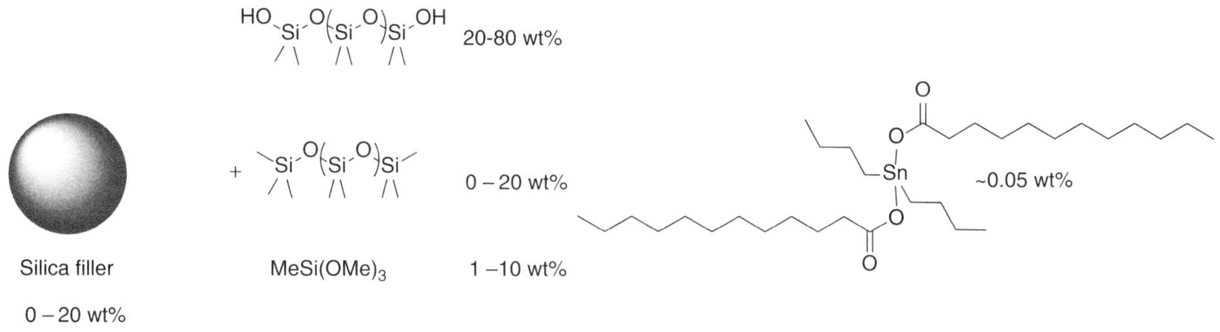

Figure 5.8   Common silicone conformal coatings contain silica fillers for rheology and various silicones and catalyst to allow cure of the conformal coating.

30–80% RH, an innocuous alcohol (methanol or ethanol) is liberated following rapid cross-linking of the silicone polymer. The cross-linking reaction is rapid and is catalyzed by titanates, frequently in combination with organotin compounds (53). The most common tin catalysts are dibutyltindilaurate and dibutyltinocanoate. The condensation reaction rate is accelerated by water and is dependent upon the type of catalyst and the functionality, structure, and concentration of the cross-linking silane. Cross-linking can be accomplished at room temperature (24 h for complete cure) in the previously specified relative humidity range, or at elevated temperature (60 °C for approximately 10 min) (54). The cured silicone conformal coating provides excellent moisture resistance and high temperature stability. A general cross-linking scheme for silicones can be found below (Scheme 5.2).

5.2   SILICA PARTICLES FOR CORROSION PREVENTION

5.2.1   Methods of Producing Silica Nanoparticles

There are many different methods of producing silica nanoparticles that can be used in conformal coatings as fillers. These methods commonly employ solution chemistry to generate the nanoparticles through aerosols, vapor-phase methods, and other various solution methods (55). To narrow down our focus as pertinent to the work presented here, we will concentrate on flame oxidation and sol–gel synthesis of silica nanoparticles.

The production of silica from flame oxidation is commonly referred to commercially as *fumed silica*. These nanoparticles are produced by the oxidation of halides in flame manufacturing methods. For silica, the flame is fueled by a mixture of H_2/O_2 and is the general processing fuel for Cab-o-Sil® and Aerosil®, which are the two most common fumed silica products on the market today. The reaction of the halide precursor with oxygen produces the silica giving off chlorine as a gas (Scheme 5.3).

In flame oxidation, the fuel is fed into the base of the burner through nozzles and lit in order to obtain a flame of around 1100 °C. Next, a solution precursor, for example tetrachlorosilane, is fed through a nozzle at the center of the flame and allowed to react with the oxygen being supplied to the flame (56). The reaction of the precursor then goes through a series of phases in order to obtain the final product (Fig. 5.9). As the precursor passes through the nozzle, the solvent evaporates and the precursor decomposes (57). A combustion reaction then occurs and the silica particles begin to nucleate. As the particles travel further along the flame, aggregation occurs and finally agglomeration of these smaller particles results in unaggregated, well-defined silica nanoparticles.

These well-defined silica nanoparticles are a direct result of the residence time the aggregates spend in the flame (55, 58). It has been shown that the longer the aggregates stay in the flame, the larger the spheres produced from this process. This is due to the coalescence of the aggregates over the short reaction time of milliseconds. Flame oxidation is an economical method for producing silica nanoparticles; however, the particles are generally aggregated (Fig. 5.10). Another common method is using a sol–gel technique to produce nearly monodisperse particles through solution chemistry (55).

To understand the sol–gel synthesis, it is best to first look at the mechanism from which monodisperse particles form. As can be seen in Figure 5.11 (narrow curve), a hydrous oxide becomes supersaturated and is increased continuously through temperature change or pH until the critical concentration, C_N, is reached (55, 59). At this stage, the rate of nucleation is rapid. As the reaction continues, the precipitation of

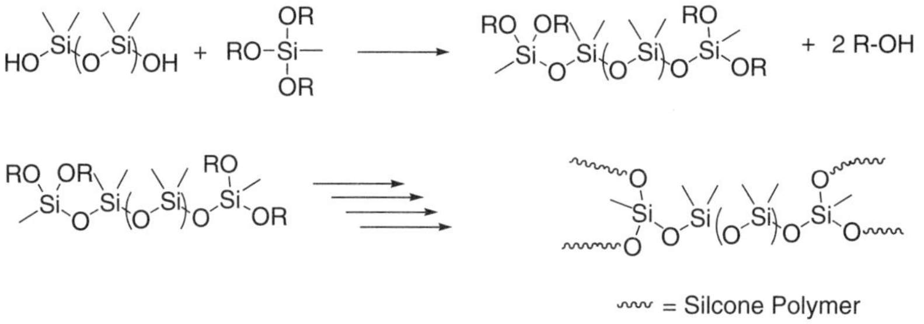

$$\text{www} = \text{Silcone Polymer}$$

Scheme 5.2   Example of silicone curing reaction commonly used in conformal coatings.

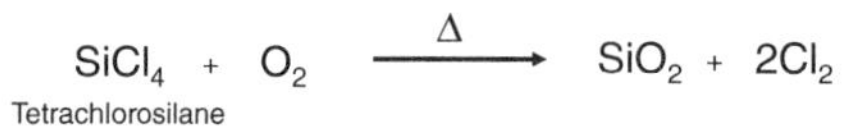

$$\text{SiCl}_4 \;+\; \text{O}_2 \;\xrightarrow{\;\Delta\;}\; \text{SiO}_2 \;+\; 2\text{Cl}_2$$
Tetrachlorosilane

Scheme 5.3   Reaction of a halide with oxygen to produce silica.

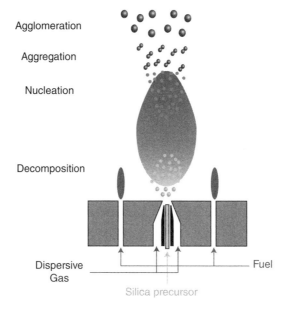

Figure 5.9   Flame oxidation process for the preparation of fumed silica.

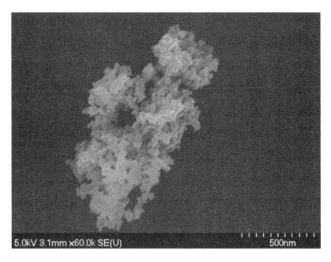

Figure 5.10   SEM image of aggregated fumed silica.

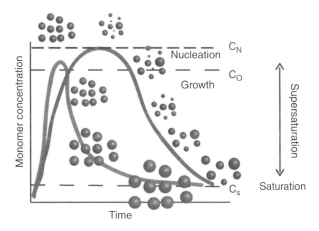

Figure 5.11   Nucleation and growth of particles. (*Source*: Adapted from References 55, 59, 60.)

$$Si(OC_2H_5)_4 + 4H_2O \longrightarrow Si(OH)_4 + 4C_2H_5OH \text{ (Hydrolysis RXN)}$$

Scheme 5.4   Hydrolysis reaction of alkoxysilane.

particles reduces the supersaturation to below C_0 and nucleation is no longer likely. Finally, as the reaction completes, growth of the nuclei continues until the concentration is reduced to that of the equilibrium solubility, C_s. As can be seen in Figure 5.11 (broad curve), there are two conflicting models resulting in a range of sizes, which is a result of new nuclei forming during the growth stage (60). While these two models exist, the process developed by Werner Stöber et al. (61) appears to be a combination of each and will be further discussed below.

Sol–gel synthesis of particles is typically formed under basic conditions with pH 7–10. Under these conditions, an alkoxysilane undergoes hydrolysis and condensation reactions (Schemes 5.4 and 5.5) (55) concurrently to form SiO_2. This was widely explored by Stöber, Fink, and Bohn (61) which is referred to as the Stöber process. In this reaction, Stöber et al. used a combination of an alkoxysilane (tetraethyl orthosilicate [TEOS]), water, alcohol, and a base to form nearly monodisperse particles in the micron size range (>1 μm). Typically, in the hydrolysis reaction, 4 moles of water to that of the alkoxysilane are used, or 2 moles if the condensation goes to completion (55). When preparing particles though, the ratio of water to alkoxysilane can be as large as 20:1 with a high pH in order to promote the condensation reaction.

The Stöber process has been thoroughly tested by varying various reaction components and conditions to determine a wide range of particles that can be formed from 20 nm to 2 μm (61–64). Modified Stöber processes have also been used (Scheme 5.6), where the amount of water is varied while alkoxysilane, ammonia, and solvent concentrations remain constant to produce nanometer-sized particles. In this process, an ammoniacal solution is mixed with a monomer solution and stirred at room temperature for ~24 h (Fig. 5.12) generating nearly monodisperse particles (Fig. 5.13). By changing the water concentration, nanometer-sized particles are formed from 20 to 500 nm and are indicated by a change in light scattering from translucent to opaque as the particle size becomes larger (Fig. 5.14).

$$Si(OH)_4 \longrightarrow SiO_2 + 2H_2O \quad \text{(Condensation RXN)}$$

Scheme 5.5   Condensation reaction of silicic acid.

$$Si(OEt)_4 \quad + \quad H_2O \quad \xrightarrow[\text{EtOH}]{\text{2 MNH}_3 \text{ in EtOH}} \quad SiO_2 \quad + \quad 4\ EtOH$$

tetraethoxysilane

Scheme 5.6   Stöber reaction using an alkoxysilane.

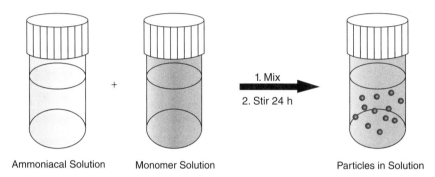

Ammoniacal Solution   Monomer Solution   Particles in Solution

Figure 5.12   Example preparation of SiO_2 by a modified Stöber process. By changing the concentration of water, various particle sizes in the nanometer region can be produced.

While many approaches exist for generating silica nanoparticles, our main focus for this work is based on the fumed silica due to the commercial interest and cost associated with their production.

5.2.2   Sulfur-Based Corrosion in IT Environment

As previously mentioned, IT equipment is increasingly deployed in datacenters that are in the industrial regions of China and India where airborne corrosive gases are prevalent. Although silicone conformal coatings have proven very successful in the past at protecting sensitive electronic hardware, the elevated levels of these airborne contaminants, in particular S_8, have underscored a deficiency in silicones that until recently has not posed a significant problem. As early as 1977, Berry and Susko (65) reported that sulfur vapor has an appreciable heat enthalpy of solution in various polymeric materials, most notably silicones. These authors demonstrated that sulfur can be sorbed into silicones and concentrated 100,000-fold over ambient atmospheric levels. Consequently, presumed innocuous levels of airborne sulfur can become highly corrosive once concentrated within a silicone conformal coating. Recent studies (66) have indicated that silicone-based conformal coatings actually accelerate silver metallurgy corrosion in resistor arrays. Based on these results, coupled with the fact that silicone conformal coatings still possess many desirable characteristics, it can be seen that a need exists to modify the silicone materials to mitigate sulfur-induced corrosion.

5.2.3   Methods to Remove Sulfur

The oil and gas industry produces more elemental sulfur (as a byproduct) than the market can absorb. In coal alone, sulfur content ranges from <1 wt% to >14 wt% (67). As coal ultimately undergoes combustion, the potential to release significant amounts of sulfur or sulfur-bearing gases into the environment is enormous. However, prior to combustion, various methods exist for the sequestration of sulfur from liquid and gaseous hydrocarbons. Dense petroleum fractions are often treated with hydrogen gas, sequestering sulfur as H_2S. For the case of natural gas, sulfur is typically captured together with carbon dioxide in an amine-based scrubbing process (Scheme 5.7) (68, 69).

Various alkylamines are used in the process, but the most common are diethanolamine (DEA), monethanolamine (MEA), methyldiethanolamine (MDEA), diisopropanolamine (DIPA), and diglycolamine (DGA). The sour gas (i.e., the gas containing H_2S and CO_2) is fed into the bottom of an absorber unit and scrubbed by the alkylamine gas stream fed in from the top. The sweetened gas free of H_2S and CO_2 is pulled off of the top of the absorber while the enriched amine stream in subsequently fed into a regenerator where the hydrogen sulfide is stripped from the amine. The H_2S-enriched gas stream is then routed into the Claus process to produce elemental sulfur (Scheme 5.8) (70).

This sulfur is then either sold or stored. If stored, the sulfur is typically poured into blocks often exceeding one million tons of yellow sulfur (67). In addition to the foregoing scrubbing process, the patent literature is replete with numerous examples of either metal chelate chemistry or oxidative methods for the removal of sour gas from petroleum feedstock (71–73). These processes involve chelation of the sulfur

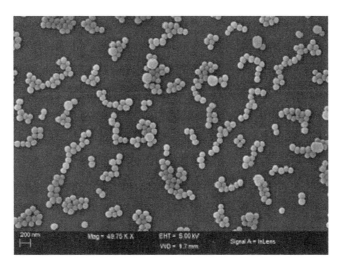

Figure 5.13   SEM micrograph of nearly monodisperse silica nanoparticles generated by modified Stöber process.

Figure 5.14   Silica nanoparticles generated by a modified Stöber process using different concentrations of water in the reaction. Larger particles appear opaque (left) in solution while smaller particles appear translucent (right).

$$RNH_2 + H_2S \longrightarrow RNH_3 + SH^-$$

Scheme 5.7   Natural gas amine-based scrubbing process (68, 69).

compounds by a reduced valence state organometallic compound or oxidation of sulfur to sulfur dioxide followed by removal of SO_2 from the gas stream.

Sulfur and sulfur-bearing gases also comprise a substantial portion of biogas. The sulfur concentration must be reduced to ppb levels to prevent poisoning of the catalyst systems. Typical approaches include adsorption onto zinc or cerium-based adsorbents at elevated temperatures (74, 75). Near room temperature, the efficiency of these catalytic adsorbents falls off precipitously. Porous activated carbon with high surface areas have also been shown to be an effective adsorbent for H_2S (76). Unfortunately, retention of H_2S on unmodified carbon adsorbents is very limited due to the weak catalytic nature of the carbon centers (75, 76). Other approaches for sulfur removal include silver (77), copper powders (78), mercury (79), Raney nickel (80), tetrabutylammonium sulfite (81), and phosphines (82).

Of the aforementioned approaches, an efficient approach for sulfur removal was demonstrated through the use of phosphine containing compounds such as triphenylphosphine (83). In this work, various environmental samples containing sulfur were reacted with triphenylphosphine and compared to other conventional methods for sulfur removals. Although sulfur removal kinetics was slightly slower than metal sulfur removal methods, it still resulted in complete removal of the sulfur and had several other advantages such as reduced toxicity.

When evaluating methods to modify silica particles such as those commonly found in conformal coating formulations, phosphines as a functionality were selected due to the many advantages mentioned earlier.

5.2.4   Phosphine Containing Alkoxysilanes

There are various approaches that may be considered for sequestration or reaction of sorbed sulfur in silicone conformal coatings. One of the most appealing methods is to incorporate a reactive species into existing conformal coatings in order to bind sorbed sulfur. In this work, we

$$H_2S + \tfrac{1}{2} O_2 \longrightarrow S + H_2O$$

Scheme 5.8   Claus process to produce elemental sulfur (70).

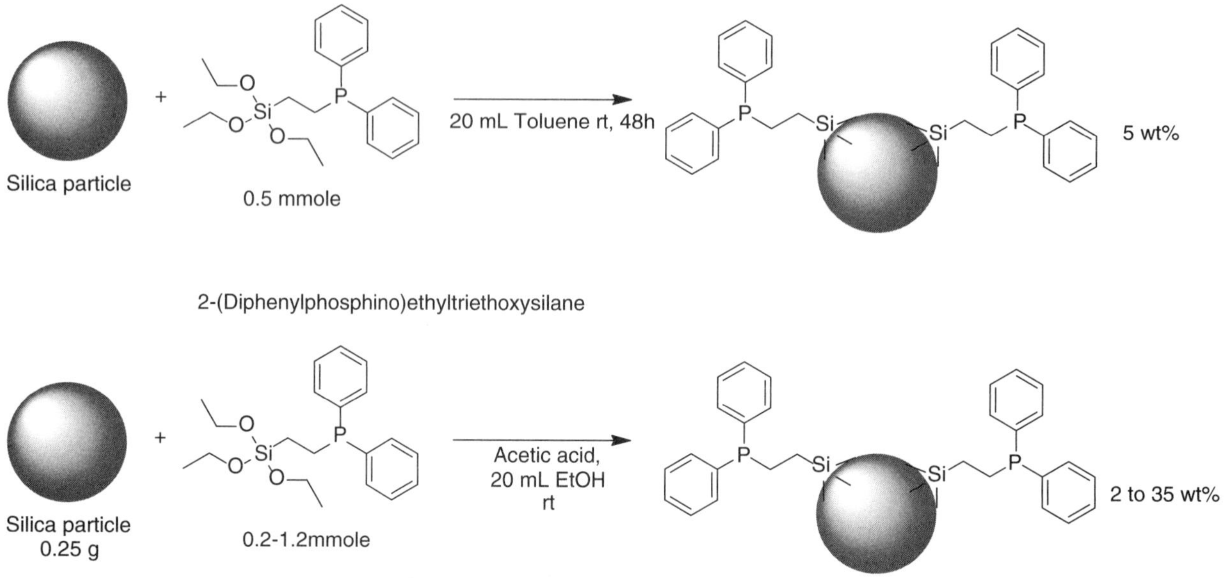

Figure 5.15   In this work, two approaches were investigated for surface modification, an approach without a catalyst (top reaction) and an acid-catalyzed approach (bottom reaction). The weight percent represents the amount of mass added to the silica particle.

explore modification of silica fillers that are common to many conformal coating formulations. Additionally, silica fillers have high surface areas which allow for increasing the concentration of phosphine functionalities.

Previous research has demonstrated that tris(diethylamino)phosphine reacts readily with various disulfides such as cyclic, benzylic, arylalkyl, and dialkyl derivatives (84). For simple dialkyl disulfides, the reaction proceeds exothermically to yield the corresponding thioethers in yields approaching 90%. The mechanism consists of an S–S bond cleavage via an SN$_2$ pathway (85) with the phosphorous compound serving as the nucleophile. A P=S bond is typically formed in one of the reaction products. Extension of this chemistry to S$_8$ suggests that phosphine-modified silica particles would be effective getters for atmospheric sulfur that is sorbed and subsequently concentrated in silicone matrices. Bifunctional phosphines can be readily immobilized on SiO$_2$ surfaces (86). Additionally, organophosphines have been used extensively for disulfide reductions (87) with great selectivity and functional group tolerance (88).

5.2.5   Phosphine Containing Alkoxysilane Surface Functionalization

A particular alkoxysilane which allows for incorporation of phosphine functionalities is $Ph_2P(CH_2)_3Si(OEt)_3$ DPPAS (2-(diphenyl-phosphino)ethyltriethoxysilane). This alkoxysilane has been investigated for surface modifications of various oxide surfaces (89). Of particular interest for this work is a silica support, which demonstrated that the pendant phosphine functionality is stable with respect to leaching. As the surface concentration of phosphine increases, the geometry of the pendant phosphine changes from a blanket configuration (where the phosphine groups bend down to lie on the silica surface) (88) to a brushlike configuration.

5.2.6   Functionalization of Silica Particles

Silica particles utilized in this work were fumed silica with a surface of ~200 m^2/g. The functionalization of silica particles with DPPAS was investigated using two approaches (Fig. 5.15). The first method utilized no catalyst and was simply dispersed in toluene. Toluene has been shown to disperse silica particles well and allow for surface modification with alkoxysilanes (89). Silica particles, toluene and DPPAS were combined and the reaction was magnetically stirred for 48 h at room temperature. The particles were isolated via centrifugation and redispersed into toluene (3 × 5000 rpm). After the last centrifugation, the solvent was decanted from the particles and the particles dried under reduced pressure at 50 °C for 48 h. Although this method allowed for ease of surface modification, it was limited as the concentration of the DPPAS could not be increased beyond 5 wt% despite increasing the DPPAS concentration. Thus, we explored the use of acid-catalyzed surface modifications. This approach is very similar to those used in commercial applications where silica particles are dispersed in ethanol with DPPAS and a catalytic amount of acetic acid (4 mol eq. to DPPAS). This reaction was conducted at room temperature for 48 h. In these reactions, the concentration of the DPPAS was varied in an effort to increase the mass addition to the silica particles. Utilizing this approach, silica particles with a mass addition ranging from 2 to 35 wt% could be achieved. The particles were isolated via centrifugation and redispersion into ethanol (3×5000 rpm). After the last centrifugation, the solvent was decanted from the particles and the particles were dried under reduced pressure at 50 °C for 48 h. At the higher mass additions, it is no longer a surface modification silicone formation. This method was used to prepare the silicone conformal coating compositions.

5.3 CORROSION RESISTANT SILICONE CONFORMAL COATING PREPARATION AND TESTING

5.3.1 Incorporation of DPPAS-Modified Silica

The DPPAS-modified silica particles as described previously were ground into a fine powder using a mortar and pestle. The ground DPPAS-modified silica particles were then added to a commercially available RTV conformal coating. This formulation was then mixed using a VMA-Getzmann Dispermat high-speed dispersion mixer, where the particle silicone formulation was mixed at 1000 rpm for 3 min. Concentrations of the modified filler in the prepared conformal coating ranged from 10 to 25 wt%. An example formulation is shown in Figure 5.16. This range is common for various silicone conformal coating formulations. Once the conformal coatings were mixed, the solutions were permitted to degas entrained air at ambient temperature for 10 min prior to being applied to resistor arrays (Fig. 5.17, right image). Once applied, the coatings were cured at ambient temperature for 24 h prior to being subjected to a sulfur-rich environment. Triplicate samples from each conformal coating were prepared.

5.3.2 Resistor Testing

To test the effectiveness of the silicone composites, standard mount thick film resistors as shown in Section 5.1.4, containing silver contacts within the resistor body, were soldered onto a PCB test card as shown in Figure 5.17 (right image). The PCB contained gold tabs allowing for

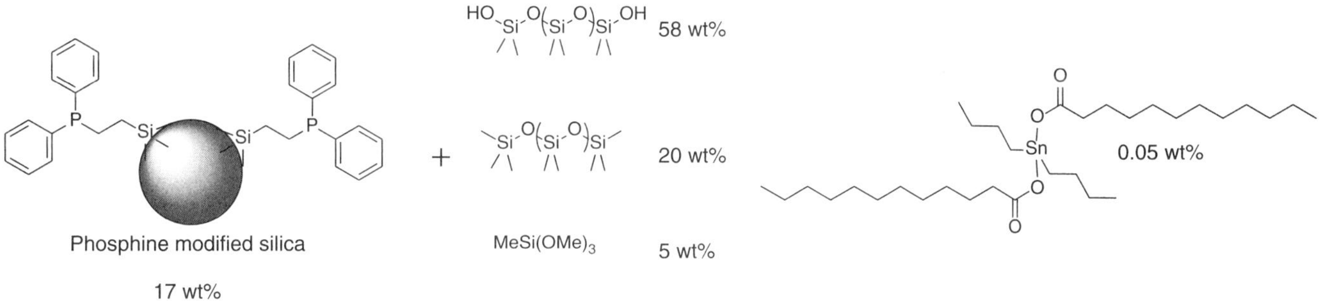

Figure 5.16 Example of formulation prepared containing phosphine-modified silica fillers. In this example, the DPPAS-modified silica was added at a concentration of 17 wt% along with the various catalyst and resins.

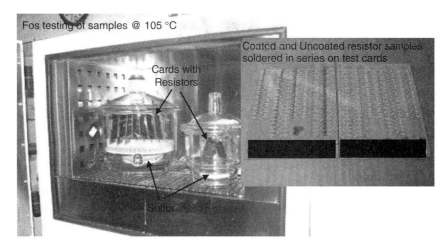

Figure 5.17 (Right) Resistor arrays which were coated with anti-corrosion silicone RTV formulations. (Left) Once the resistor arrays were coated, the printed circuit board cards were placed into a desiccator containing elemental sulfur followed by the addition to an oven at 105 °C.

easy resistance probing upon subjecting the coated resistors to corrosive environments. Once the resistor arrays were prepared, the degassed conformal coating formulation with and without the modified filler were then applied over the thick film resistors and allowed to cure at room temperature for 24 h. The DPPAS-modified silica fillers had no effect on the curing time as the conformal coating still cured within a similar time frame as the conformal coating not containing the modified filler. Once the anticorrosion conformal coatings cured, the test cards were placed into a desiccator that contained an excess of elemental sulfur. The desiccator was then placed into an oven at 105 °C (Fig. 5.17, left image). This method is known as the flowers of sulfur (FoS) method to create an environment that is sulfur rich. This method, based on a variation of ASTM B809, has been utilized as an accelerated corrosion test for the testing of various metal components in a sulfur-rich environment to induce sulfur corrosion. To measure the corrosion protection of the conformal coatings, the desiccator was removed from the oven every 24 h, allowed to cool to room temperatures followed by measuring the resistances of the thick film resistors. Once the resistances were recorded, the resistor array was placed back into the desiccator, followed by being placed back into the oven at 105 °C. This process was repeated every 24 h until all the resistors had a resistance change from 10 to 100 Ω. This indicates the conducting silver strip has begun to corrode and failure of the part.

5.3.3 Evaluation of Phosphine Binding to Sulfur

To investigate the ability of DPPAS to bind to sulfur, aluminum test bars were coated with a conformal coating formulation containing 17 wt% modified silica filler. This coated aluminum test bar was then subjected to the FoS test conditions as detailed previously. Prior to exposure, the conformal coating is an opaque white color as shown in Figure 5.18 (left image). As the conformal coating remains in the sulfur environment it progressively yellows. The yellowing of the conformal coating is observable in Figure 5.18 (right image). The aluminum test bar remained in the FoS environment for 96 h.

To identify the phosphorus–sulfur bond formation, thus the binding of sulfur and preventing diffusion to the underlying metal structure, the formulations were investigated prior and post exposure with FTIR-ATR. The phosphorus–sulfur bond is observable at approximately 610 cm^{-1}. The conformal coating formulation prior to FoS exposure has no absorbance at 610 cm^{-1} as shown in Figure 5.19 (left image). After 96 h of sulfur exposure, an absorbance at 610 cm^{-1} is observable indicating the binding of sulfur (Fig. 5.19, right image).

5.3.4 Results of Anticorrosion Conformal Coatings

The PCB which contained thick film resistors that were coated with silicone conformal coatings was probed every 24 h. The results of the conformal coating evaluation are shown in Figure 5.20. From these results, the first time to failure for any of the conformal coatings was the conformal coating containing no DPPAS-modified silica filler, which is a standard conformal coating used in the industry. This resistor had an average time to failure of approximately 48 h. After this initial failure, the time to failure of the modified silica filler conformal coating formulations increased as the concentration of the DPPAS-modified silica filler increased. Four formulations were prepared, with 10, 17, 25, and 40 wt% DPPAS-modified silica filler. The average time to failure due to sulfur corrosion of the resistor for these conformal coatings was 192 to 288 h respectively. Conformal coating formulations with no modified silica filler failed within 48 h. The amount of filler did not need to exceed 25 wt% as there was no added benefit, as this concentration is within the range for silica filers in conformal coatings. The higher loadings began to prevent conformal coating of the resistor which likely leads to no net improvement. These results demonstrate that the incorporation of phosphine-modified silica fillers can extend the time to corrosion-related failures of silver containing thick film resistors. Based upon developed acceleration standards from the modified FoS test, it is expected that the modified filler-silicone conformal coating composites will extend the resistor life several years in harsh environments with high sulfur concentrations.

5.4 CONCLUSION

In this work, we have addressed environmental sulfur corrosion and its effects on the resistor components. To generate a sulfur-rich environment, we have used FoS which accelerates the time to failure. To mitigate the sulfur corrosion, a diphenylphosphine-modified silica

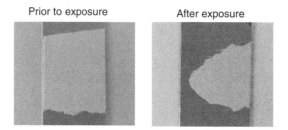

Figure 5.18   DPPAS-modified silica filler formulation coated on an aluminum test bar prior to exposure in the flowers of sulfur environment (left image) and after 96 h of flowers of sulfur exposure (right image). Phosphine-modified silica conformal coating before (left) and after (right).

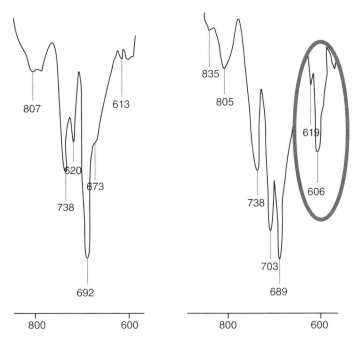

Figure 5.19 FTIR spectrum with the area blown up in the area to demonstrate the addition of phosphine–sulfur bond formation at ~610 cm^{-1}.

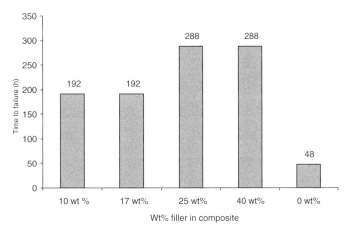

Figure 5.20 Time to failures of 10 Ω resistors coated with various conformal coating formulations.

filler was prepared and blended into a standard conformal coating. These conformal coating formations with a varying weight percentage of modified filler (5–35 wt%) were then coated onto standard thick film resistors. The time to failure was increased from 48 h to nearly 300 h. This demonstrates the ability of the phosphine-modified fillers' ability to react with and prevent the diffusion of sulfur, thus protecting the underlying metal structures from corrosion for a prolonged period of time.

5.5 EXPERIMENTAL PROCEDURE

5.5.1 Materials

Ethanol (99.5%) and acetic acid (99.7%) were purchased from Sigma-Aldrich Chemical Co.; SiO$_2$ (amorphous) and 2-(diphenylphosphino)ethyltriethoxysilane were purchased from Gelest Inc. Elemental sulfur (99.99%) was purchased from Alfa Aesar. The silicone conformal coating and resistors used in these experiments were from commercially available sources.

5.5.2   Instrumentation

Particles were blended into the silicone conformal coatings using a high sheer dispersion mixer from VMA-Getzmann. Resistance measurements were made using a Fluke multimeter 8624. Spectroscopic characterization was accomplished on a Perkin Elmer Spectrum 100 with ATR attachment. Centrifugation was done with a VWR clinical 200.

5.5.3   Modified Silica Filler Preparation

A typical procedure for preparation of modified silica filler is as follows: amorphous SiO_2 (1.0 g, 0.016 moles) was added to a 30 ml plastic bottle. To this bottle, a solution containing ethanol (15 ml), 1N acetic acid solution (0.1 g, 0.1 ml), and 2-(diphenylphosphino)ethyltriethoxysilane (0.3 g, 0.79 mmol) was added followed by a stir bar. The bottle was sealed and placed onto a magnetic stirrer. The reaction was carried out at room temperature for 24 h. The resulting modified silica powder was then purified by centrifugation at 6000 rpm at 25 °C. The solution was decanted off of the silica powder and then the silica powder was redispersed in ethanol by sonication. This process was repeated three more times.

5.5.4   Composite Preparation

Composites were prepared as follows. To a commercially available RTV conformal coating (0.8 g), mortar and pestle ground diphenylphosphine-modified silica powder (0.2 g) was added. This formulation was then mixed using a VMA-Getzmann Dispermat high-speed dispersion mixer. Concentrations of the modified filler in the prepared composites ranged from 10 to 25 wt%. After being mixed, the conformal coating solutions were permitted to degas entrapped air at an ambient temperature for 10 min prior to being applied to the resistor arrays. Once applied, the coatings were cured at an ambient temperature for 24 h prior to being subjected to the FoS environment. Triplicate samples from each conformal coating were prepared.

5.5.5   Composite Testing

To evaluate the conformal coatings effectiveness at preventing sulfur-based corrosion, a test method based on ASTM B809 was developed. To a 1 ft^3 desiccator, elemental sulfur (250 g) was added followed by resistor arrays that were coated with conformal coatings with and without the modified fillers. The desiccator was then placed into an oven at 105 °C. The resistor resistances were measured every 24 h. This was accomplished by removing the test cards from the oven, allowing them to cool to room temperature, then probing each resistor location with a Fluke multimeter.

REFERENCES

1. ITIC 2009. ITIC 2009 Global Server Hardware and Server OS Reliability Survey.

2. Persson D, Leygraf C J. Electrochem Soc. *In Situ* Infrared Reflection Absorption Spectroscopy for Studies of Atmospheric Corrosion. 1993;140:1256.

3. Roy SK, Sircar SC. Br Corros J 1978;13:191.

4. Zakipour S, Leygraf C. Br Corros J 1992;27:295.

5. Aastrup T et al. Corros Sci 2000;42:957–967.

6. Thiel PA, Madey TE. Surf Sci Rep 1987;7:211.

7. Klier K, Shen JH, Zettlemoyer AC. J Phys Chem 1973;77:1458.

8. Hall PG, Tompkins FC. Trans Faraday Soc 1962;58:1734.

9. Hultquist G, Grsj L, Lu Q, Kermark T. Corros Sci 1994;36:1459.

10. ASHRAE. 2011 Thermal Guidelines for Data Processing Environments– Expanded Data Center Classes and Usage Guidance; ASHRAE TC 9.9; American Society of Heating, Refrigerating and Air-Conditioning Engineers, Inc.; 2011.

11. Office of Pollution Prevention and Toxics, *Chemical Summary for Carbonyl Sulfide*. U.S. Environmental Protection Agency; 1994.

12. Sharma SP. J Electrochem Soc 1980;127(1):21–26.

13. Backlund P, Fjellstrom B, Hammarback S, Maijgren B. Ark Kemi 1966;26:267.

14. Drott J. Ark Kemi 1959;15:181.

15. Lorenzen JA. Proc Inst Environ Sci, Atmospheric corrosion of silver. 1971:110.

16. Rice DW, Peterson P, Rigby PB, Phipps P, Cappell RJ, Tremoureux RJ. Electrochem Soc 1981;128:275.

17. Graedel TE et al. Corros Sci 1985;25(12):1163–1180.

18. Franey JP, Kammlott GW, Graedel TE. Corros Sci 1985;25:133.

19. Graedel TE, Franey JP, Kammlott GW. Corros Sci 1983;23:1141.

20. Abbott WH. Br Corros J 1989;24(2):153–159.

21. ASHRAE. 2011 Gaseous and Particulate Contamination Guidelines For Data Centers; ASHRAE TC 9.9; American Society of Heating, Refrigerating and Air-Conditioning Engineers, Inc.; 2011.

22. Reid M, Punch J, Ryan C, Franey J, Derkits GE, Reents WD, Garfias LF. IEEE Trans Compon Packag Technol 2007;30(4):666–672.

23. Cullen D, O'Brien G. IPC Printed Circuits Expo, SMEMA Council APEX Designers Summit, Anaheim, CA; 2004. p. 23–37.

24. Veale R SMTA International Proceedings, Chicago, IL; 2005.

25. Xu C, Flemming D, Demerkin K, Derkits G, Franey J, Reents W. Apex, Los Angeles, CA; 2007.

26. Sahu AK. International Conference on Sustainable Solid Waste Management; 2007 Sep 5–7; Chennai, India; 2007.

27. Schueller R. SMTA International Proceedings; 2007 Oct; Orlando, FL; 2007.

28. Mazurkiewicz P. Proceedings of the 32nd ISTFA; 2006 Nov 12–16; Austin, TX; 2006.

29. Hillman C, Arnold J, Binfield S, Seppi J. SMTA International Conference Proceedings; Orlando, FL; 2007.

30. High-performance Buildings for High-tech Industries. 2013. Should data center owners be afraid of air-side economizer use?—A review of ASHRAE TC 9.9 white paper titled gaseous and particulate contamination guidelines for data centers. Available at http://hightech.lbl.gov/documents/data_centers/review-ashrae-tc9-9.pdf. Accessed 2013 Apr 21.

31. Purafil. 2013. Data center woes: reliability issues plague RoHS-compliant electronics. Available at http://www.purafil.com/PDFs/Technical%20Papers/Misc_Multiple%20Markets/Data_Center_Woes_Reliability_Issues_Plague_RoHS_Compliant_Electronics.pdf. Accessed 2013 Apr 21.

32. Calce Prognostics. 2013. Risks to telecommunication equipment under free air cooling conditions and their mitigation. Available at http://www.prognostics.umd.edu/calcepapers/11_Dai%20-Risks%20To%20Telecommunication%20Equipment%20Under%20Free%20Air%20Cooling%20Conditions%20And%20Their%20Mitigation.pdf. Accessed 2013 Apr 21.

33. Gouda VK, Carew JA, Riad WT. Br Corros J, Investigation of Computer Hardware Failure due to *Corrosion*, 1989;24:192.

34. ISA ANSI/ISA 71.04. Environmental conditions for process measurement and control systems: airborne contaminants. Research Triangle Park: ISA; 1985.

35. ASHRAE. *Particulate and Gaseous Contamination Guidelines for Data Centers*. Atlanta: American Society of Heating, Refrigeration and Air Conditioning Engineers, Inc.; 2009.

36. Klein LJ, Singh P, Schappert M, Griffel M, Hamann HF. 27th Annual IEEE; 2011 Mar 20–24; IEEE: 2011; SEMI-THERM; 2011.

37. Singh P, Schmidt RR and Prisco J. IBM Corp., Particulate and gaseous contamination: Effect on computer reliability and monitoring, ASHRAE 2009.

38. ISO 14644-1. Cleanrooms and associated controlled environments—Part 1: classification of air cleanliness. ISO; 1999.

39. ANSI/ASHRAE 52.2-2007. Method of testing general ventilation air-cleaning devices for removal efficiency by particle size. American Society of Heating, Refrigerating and Air-Conditioning Engineers, Inc. Atlanta, GA; 2007.

40. EPA EPA Residential Air Cleaners (Second Edition): A Summary of Available Information; EPA 402-F-08-004; 2008.

41. Blinde D, Lavoie L. Proc EOS/ESD Symp 1981;EOS-3:9.

42. Huang H, Guo X, Zhang G, Dong Z. Corros Sci 2011;53:2007.

43. Reighard, MA. 2013. Practical applications of process control in conformal coating. Available at http://www.amtest.bg/press/asymtek/Conf%20coat%20process%20control.pdf. Accessed 2013 Apr 1.

44. Lu X, Xu G. J Appl Polym Sci 1997;65:2733.

45. Ghosh MK, Mittal KL. *Polyimides: Fundamentals and Applications*. New York: Marcel Dekker, 1996.

46. Umarji GG, Ketkar SA, Phatak GJ, Giramkar VD, Mulik UP, Amalnerkar DP. Microelectron Reliab 2005;45:1903.

47. Hopf H. Angew Chem Int Ed 2008;47:9808.

48. Fulton JL, Deverman GS, Yonker CR, Grate JW, De Young J, McClain JB. Polymer 2003;44:3627.

49. Jaffe D. IEEE Trans Parts Hybrid Packag 1975;12:182.

50. Minges, ML. *Electronic Materials Handbook: Packaging*. ASM Int; 1989. p 759.

51. Gubbels F. SMT Packag 2004;4(6):1.

52. Gelest Inc. 2013. Reactive silicones: forging new polymer links. Available at http://www.gelest.com/goods/pdf/reactivesilicones.pdf. Accessed 2013 Apr 1.

53. CES Science. 2013. Polydimethylsiloxanes (PDMS). Available at http://www.silicones-science.com/chemistry/chemical-reactions-on-the-finished-silicone. Accessed 2013 Apr 1.

54. Dow Corning. 2013. Conformal coatings tutorial. Available at http://www.dowcorning.com/content/etronics/etronicscoat/etronics_cc_tutorial.asp?DCWS=Electronics&DCWSS. Accessed 2013 March 27.

55. Brinker CJ, Sherer GW. Sol-gel Science: The Physics and Chemistry of Sol-gel Processing. San Diego, CA: Academic Press, Inc.; 1990.

56. Mueller R, Mädler L, Pratsinis SE. Chem Eng Sci 2003;58(10):1969–1976.

57. Stark WJ, Pratsinis SE. Powder Technol 2002;126:103–108.

58. Ulrich GD. Combust Sci Technol 1971;4:47–57.

59. LaMer VK, Dinegar RH. J Am Chem Soc 1950;72(11):4847.

60. van Blaaderen A, Vrij A. J Colloid Interface Sci 1993;156:1–18.

61. Stöber W, Fink A, Bohn EJ. Colloid Interface Sci 1968;26:62.

62. van Helden AK, Jansen JW, Vrij AJ. Colloid Interface Sci 1981;81(2):354–368.

63. Bogush GH, Zukoski CF, Mackenzie JD, Ulrich DR, editors. Ultrasonic Processing of Advance Ceramics. New York: Wiley; 1988. p 477.

64. Tan CG, Bowen BD, Epstein NJ. Colloid Interface Sci 1987;188(1):290–293.

65. Berry BS, Susko JR. IBM J Res Dev 1977;21(2):97–208.

66. Cole M, Hedlund L, Hutt G, Kiraly T, Klein L, Nickel S, Singh P, Tofil T. 2013. Harsh environment impact on resistor reliability. Available at http://researcher.watson.ibm.com/researcher/files/us-kleinl/Resistor_Harsh_Environ_final_SMTAI2010.pdf. Accessed 2013 Apr 1.

67. Rappold TA, Lackner KS. Energy 2010;35(3):1368–1380.

68. Kohl A, Nielson R. *Gas Purification*. 5th ed. Gulf Publishing; 1997.

69. Gary JH, Handwerk GE. *Petroleum Refining Technology and Economics*. 2nd ed. Marcel Dekker, Inc.; 1984.

70. Clark PD. Sulfur and hydrogen sulfide recovery. In: *Kirk_Othmer Encyclopedia of Chemical Technology*. John Wiley & Sons. Hoboken, N.J; 2006.

71. Jeffrey GC. Process for the removal of hydrogen sulfide and optionally carbon dioxide from gaseous streams. European Patent 0,279,667 B1. 1995 Sept 20.

72. Jones TA, Snavely JES. Process for the selective removal of hydrogen sulfide from gaseous stream. European Patent 0,243,542 A1. 1987 Nov 4.

73. Parisi PJ. Removal of hydrogen sulfide. European Patent 0,581,026 B1. 1996 Oct 23.

74. Stirling D. The Sulfur Problem: Cleaning Up Industrial Feedstocks. Oxford: RCS; 2000.

75. Flytzani-Stephanopoulos M, Sakbodin M, Wang Z. Science 2006;312:1508–1510.

76. Bandosz TJ. Desulfurization on activated carbons. In: Bandosz TJ, editor. *Activated Carbon Surfaces in Environmental Remediation*. Oxford: Elsevier; 2006. p 231.

77. Buchert H, Bihler S, Ballschmiter KZ. Anal Chem 1982;313:1.

78. Blumer M. Anal Chem 1957;29:1039.

79. Goerlitz DF, Law LH. Bull Environ Contam Toxicol 1971;6:9.

80. Ahnoff M, Josefsson B. Bull Environ Contam Toxicol 1975;13:159.

81. Jensen S, Renberg L, Reutergardh L. Anal Chem 1977;49:316.

82. Jensen S, Johnels AG, Olsson M, Otterlind G. Ambio Spec 1972;1:71.

83. Anderson JT, Holwitt UJ. Anal Chem 1994;350:474.

84. Harpp DN, Gleason JG, Snyder JP. J Am Chem Soc 1968;90(15):4181–4182.

85. Dmitrenko O, Thorpe C, Bach RD. J Org Chem 2007;72(22):8298–8307.

86. Merckle C, Bluemel J. Chem Mater 2001;13:3617–3623.

87. Whitesides GM. J Org Chem 1991;56:2648.

88. Scott RPW. Silica Gel and Bonded Phases. New York: John Wiley & Sons; 1993.

89. Howarter JA, Youngblood JP. Langmuir 2006;22:11142–11147.

6

NEW NANOSCALE MATERIAL: GRAPHENE QUANTUM DOTS

DONG-ICK SON AND WON-KOOK CHOI

Interface Control Research Center, Korea Institute of Science and Technology, Sungbuk Gu, Seoul Korea

6.1 INTRODUCTION

Ever since graphene, one of new emerging advanced carbonaceous materials, was discovered in 2004, tremendous amount of research has been focused on this wonder material (a single-layered 2D honeycomb carbon allotrope) over the past few decades due to its exceptional physical, chemical, and mechanical properties igniting many exciting researches for various applications (1–6). Graphene is known to have a large surface area ($\sim$2630 m^2/g) (7, 8), high carrier transport mobility (2×10^5 cm^2/Vs at RT) (9, 10), superior mechanical strength (tensile strength, 130 GPa; Young's modulus, 1000 GPa) and flexibility (20% elongation) (11), high optical transparency (97.7%) (12), excellent thermal conductivity (5300 W/mK)/negative thermal coefficient (13–15), and chemical stability (16). Nonetheless, a zero bandgap and no saturable source–drain characteristic current in bulk graphene are obstacles in particular for the application of high-speed and high-frequency electronic nanodevices, alternative to Si (17). Besides, the application of intrinsic 2D graphene has been limited due to agglomeration and poor dispersion in common solvents. This drawback has been solved by converting 2D into 1D graphene nanoribbons (GNRs) or into 0D graphene quantum dots (GQDs) by nanolithographical tailoring and chemical synthetic approaches. Although GNRs show confined transport gaps and quantum dots (QD) associated with the geometry of the ribbons (18–22), recently, much attentions have been paid on GQDs exhibiting unusual quantum confinement and edge effects. Compared to another type of 0D carbon nanomaterials, known as quasi spherical carbon dots (C-dots), with sizes below 10 nm showing size effects (23) and mostly graphitic composition (24–27) and structural features (28), GQDs clearly possess graphene lattice inside the dots irrespective of the dot size. As the quantum confinement effect is observed convincingly in GQDs of 110 nm in diameter (29), here we discuss only the GQDs defined as graphene dots with the size less than 100 nm in diameter and less than 10 layers in thickness (Fig. 6.1a) (30) distinguished from C-dots or other graphene nanostructures (GNs). The bandgap has been observed up to $\sim$0.4 eV in GNRs (31, 32), while the energy gap of GQDs falls off as approximately $1/L$ (31, 33), where L is the average size of GQDs, and can be controlled up to $\sim$3 eV in GQDs by reducing their size (33–35), more promising for photonic nanodevices.

In chemical structure, GQDs contain carboxylic acid moieties (Fig. 6.1b) at the edge having excellent water solubility and suitability for subsequent functionalization through better surface grafting using the p–p conjugated network or surface groups. In addition, the acceptable photoluminescence (PL) quantum yield (QY), low cytotoxicity, and excellent water solubility and biocompatibility make GQDs suitable for biomedical applications such as bioimaging, protein analysis, cell cracking, isolation of biomolecules, and gene/drug delivery.

GQDs, edge-bound nanometer-size graphene pieces, can be categorized as a new class of QDs with unique electrical, optical, and chemical properties. In this perspective, recent progress in studies of nanoscale GQDs are extensively reviewed, concentrating on the synthetic methodology, origin and modulation of intrinsic properties, and energy-, environmental-, and biomedical-oriented applications of GQDs. In the first section, various synthetic methods of the GQDs are summarized by categorizing into the bottom-up and top-down approaches. The former bottom-up methods, called chemical synthesis, usually suffer from the disadvantages such as requirement of special equipment, complex processes, critical synthetic conditions, low production yield, or difficulties in control of size distributions and for mass production. On the other hand, the top-down approaches, called chemical cutting method, can control precisely the morphology and the size of the GQDs under relatively tedious synthesis procedures. Basic strategy of most top-down processes for preparing GQDs is made of forming oxygen-containing functional groups (such as epoxy, hydroxyl, carbonyl, and carboxyl) and decomposing graphene oxide (GO) into smaller sheets (36). The subsequent reduction of cleaved GO for removing the nonradiative recombination center of localized electron-hole pairs is necessary (37).

In the second section, some exciting approaches for modulating and surface modification of GQDs are introduced. Unique electronic, optical, and chemical properties of GQDs can be effectively controlled and extraordinary functionalities can be created by surface chemical treatment, doping with heteroatoms and formation of some functional groups and ligands on the GQDs for advanced device applications

Nanomaterials, Polymers and Devices: Materials Functionalization and Device Fabrication, First Edition. Edited by Eric S. W. Kong.
© 2015 John Wiley & Sons, Inc. Published 2015 by John Wiley & Sons, Inc.

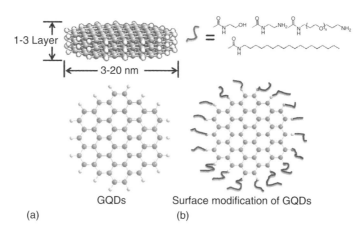

GQDs

Figure 6.1 (a) Description of GQDS and (b) conceptual structural models for edge-functionalized GQDs. (*Source*: Reproduced from Reference (30), with permission).

through hybridization with other nanomaterials. Surface functionalization with the amine group, quantum confinement and edge effect of GQDs with N-doping, alignment of GQDs on polar surface, and synthesis of semiconductor GQDs core–shell hybrid QDs are presented.

In the third section, intrinsic physical properties of GQDs are reviewed in view of electronic, magnetic, and optical properties. Origin of opening and changing in magnitude of bandgap in GQDs with the variations of size, shape, and edge environment are presented. Quantum confinement effect in GQDs is comprised of breakdown of periodic Coulomb blockade (CB) or Coulomb diamond peaks, statistical description of chaotic Dirac (neutrino) billiards for small GQDs, and long spin decoherence times caused by small spin-orbit and hyperfine interactions (38) promising for solid-state spin qubits (39, 40). Related to optical properties, in particular, the origins of PL from the GQDs are discussed and also excitation-dependent/excitation-independent and down-/upconversion of PL are presented.

Illustrations of GQDs for many applications – energy, environment, and biomedical applications – are introduced. GQD-based light-emitting diodes (LEDs), photovoltaics, fuel cells and Li-ion batteries, and lasing emission phenomenon are presented as examples for energy fields. As one representative example of environmental applications, photocatalytic system of TiO_2/GQDs is also presented. Besides intrinsic biocompatibility (low cytotoxicity) of graphene and facile biological/chemical functionalization of GO, prominent optical properties with tunable bandgap and strong excitation-dependent PL or UCPL make GQDs emerging materials in biomedical applications such as bioimaging, biosensor/labeling, and electrochemical biosensors.

6.2   CLASSIFICATION OF SYNTHETIC METHODS OF GQDs

So far, new emerging materials, GQDs have been numerously and extensively synthesized by a number of methods which are generally classified into two approaches of both chemical exfoliation methods and chemical synthesis. The first one is the cutting approach or chemical exfoliation referring to top-down methods where graphene sheets (GSs) are cut into GQDs by physical, chemical, and electrochemical processes including chemical ablation, electrochemical oxidation, nanolithography, nanotomy, ultrasonic shearing, and oxygen plasma treatment, where larger GSs are crushed into nanosized and a few layer GQDs. The other one is chemical synthesis referring to bottom-up methods involving the synthesis of graphene moieties containing a certain number of conjugated carbon atoms. This approach consists of the metal-catalyzed decomposition of fullerene (C60), stepwise organic synthesis, or solution chemistry methods during which the GQDs are formed from molecular precursors. Typically, these GQDs have surfaces rich in carboxylic acid functionalities useful for binding surface-passivation reagents. A schematic conceptual description of top-down and bottom-up methods for synthesizing GQDs is illustrated in Figure 6.2, and these two strategical approaches are listed in Table 6.1.

6.2.1   Bottom-Up Method: Chemical Synthesis

6.2.1.1   Solution Chemistry (Stepwise Organic Synthesis)   GQDs with uniform and tunable size were synthesized by stepwise solution chemistry (41). Fused graphene moieties were obtained by oxidation of polyphenylene dendrite precursors and then stabilized by attaching 2′,4′,6′-triakyl phenyl group covalently to the periphery of graphene moieties consisting of 168, 132, and 170 conjugated carbon atoms. The attached phenyl groups were distorted by the crowdedness of 3D cages created on the edge GQD and thus the alkyl chains at the 2′,6′-positions and 4′-position extend out of plane and laterally. This leads to increase of solubility by reduction of face-to-face attraction between GQDs (42) (Fig. 6.3).

Hexa-peri-hexabenzocoronene (HBC) and other large polycyclic aromatic hydrocarbons (PAHs) are generally regarded as nanoscaled fragments of graphene, but it was envisioned that fabrication of GQDs by controlled pyrolysis of large PAHs could be experimentally feasible.

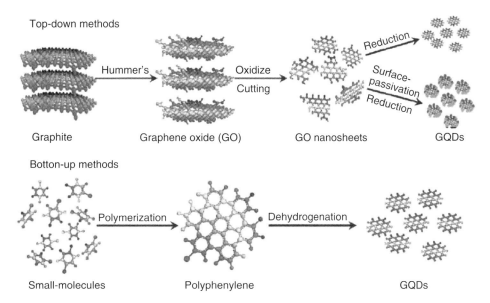

Top-down methods

Graphite → Hummer's → Graphene oxide (GO) → Oxidize / Cutting → GO nanosheets → Reduction → GQDs

Surface-passivation / Reduction → GQDs

Botton-up methods

Small-molecules → Polymerization → Polyphenylene → Dehydrogenation → GQDs

Figure 6.2 Conceptual description of the top-down and bottom-up approaches for synthesizing GQDs. (*Source*: Reproduced from Reference (30), with permission).

TABLE 6.1   A Brief Summary of the Typical Synthetic Strategies for GQDs

Synthetic Methods	Subclassification	Size of GQDs (nm)	References
Chemical synthesis (Bottom-up)	Stepwise organic synthesis (Solution chemistry)	~5	(41)
	Precursor pyrolysis	15, 1.65–21	(48, 49)
	Metal-catalyzed decomposition of C60	2.7–10	(50)
Chemical cutting method (top-down)	Chemical ablation from graphene	5–13 1.5–5	(34) (55)
	Nanolithography technique	30, 40, 80, 110, 250 50 >90	(29) (64) (65–67)
	Electrochemical synthesis	3–5 5–10 60 5–19	(71) (73) (46) (59)
	Chemical exfoliation	1–4, 4–8, 7–11 15, 18 2–7	(81) (80) (83)
	Nanotomy-assisted exfoliation	10–50	(84)
	Ultrasonic shearing of GSs	3–5	(86)
	Chemical exfoliation from metal oxide	~10	(87)

In order to synthesize the GQDs with the size of 10–100 nm using solution chemistry by a large scale, low solubility and strong intergraphene attraction of small GQDs cannot be avoidable without attaching sufficient lateral aliphatic side chains onto the edges of PAH moieties (43–45).

Liu et al. reported the synthesis of monodisperse disklike GQDs of ~60 nm diameter and 2–3 nm thickness with unsubstituted HBC as precursor. As shown in Figure 6.4, first the as-prepared GQDs through cyclodehydrogenation of HBC was pyrolyzed at 600 °C (GQD-600), 900 °C (GQD-900), and 1200 °C (GQD-1200) and then oxidized and exfoliated by a modified Hummers methods. Subsequently, aqueous solutions of the resultant GOs were heated to reflux for 48 h with oligomeric poly(ethylene glycol) diamine (PEG$_{1500N}$) and then reduced with hydrazine. PEG substituents improved the dispersibility of the GQDs. It was found that GQD-600 consisted mainly of disordered particles, while GQD-900 contained both particles and disk-shaped nanosheets. The GQD-1200 showed monodisperse disklike feature of ~60 nm diameter and 2–3 nm thickness. This solution-processable and scalable method is proved to be specific for getting monodisperse GQDs with size distribution of 10–100 nm (46).

6.2.1.2   Precursor Pyrolysis   The morphology and size of GQDs can also be controlled by carbonizing some special organic precursors by varying thermal treatment conditions. Another simple bottom-up method, the GQDs, was obtained by directly pyrolyzing a proper precursor, such as citric acid. By tuning the carbonization degree of citric acid and dispersing the carbonized products into alkaline solutions, Dong et al. obtained GQDs of ~15 nm width and 1.4 nm average height (Fig. 6.5a) (48). A complete carbonization of the precursor under prolonged heating afforded GO nanostructures consisting of sheets that were hundreds of nanometers in width and ~1 nm in height. Tang et al. reported (49) that the sole reagent glucose (sucrose or fructose)-derived GQDs were synthesized by pyrolysis using a microwave-assisted hydrothermal (MAH) method without any surface-passivation agents or inorganic additives. GQDs were prepared using 11.1 wt% glucose solution and 5 min microwave heating at 594 W and showed the average size and height of 3.4 ± 0.5 nm and 3.2 nm. The average diameter of GQDs can

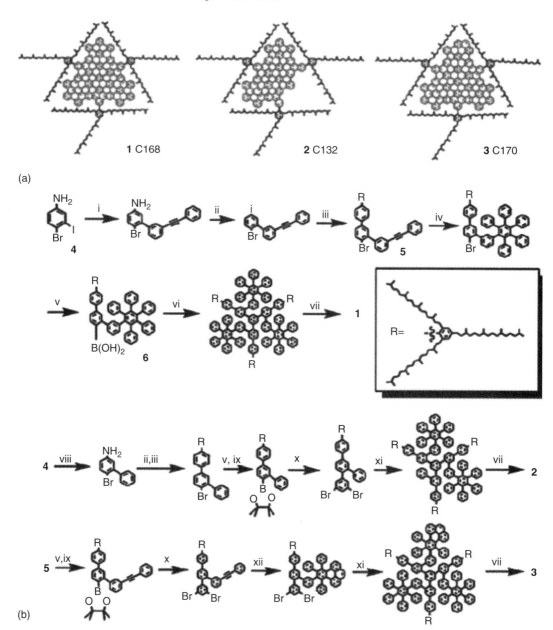

Figure 6.3   (a) Structures of three colloidal GQDs. (b) Synthesis of GQDs through a wet chemistry process. (*Source*: Reproduced from Reference (42), with permission).

be changed from 1.65 to 21 nm by changing the annealing time from 1 to 9 min. From the results, most of the carbohydrates consisting of approximately C:H:O = ~1:2:1 can be used as carbon sources to synthesize the GQDs.

6.2.1.3   *Metal-Catalyzed Cage-Opening of C60*   In order to obtain highly regular-shaped GQDs, a fabrication process incorporated with thermodynamics, as in crystal growth, would be more appropriate than solution-phase process such as defect-mediated exfoliation or cutting process of the precursor graphite flake. Lu et al. (50) first reported that atomically precise GQDs were successfully produced by Ru-catalyzed cage-opening of C60, which were similar to the formation of carbon nanotubes (CNTs), carbon onions, and graphitic domains from fullerenes (51–53). Geometrical shape of GQDs can be assembled by controlling the coverage of C60. At a high coverage of C60 ($\Theta > 0.7$ monolayer (ML)), a single layer of graphene covers a Ru(0001) surface after annealing at 1200 K for 5 min. At an intermediate coverage 0.2 ML < Θ < 0.7 ML, larger-sized and irregularly shaped GQDs are favorably formed due to the short diffusion distance between the fragmented molecules. Therefore, mean distance between C60 molecules should be longer than 15 nm ± 3 nm for the formation of GQDs. Cage-opening of C60 and transformation into GQDs on Ru(0001) surface was also revealed by tracking the evolution of a lone C60 molecule as a function of

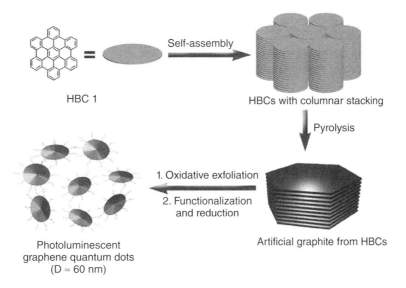

HBC 1

HBCs with columnar stacking

Pyrolysis

Self-assembly

1. Oxidative exfoliation

2. Functionalization and reduction

Photoluminescent graphene quantum dots (D ≈ 60 nm)

Artificial graphite from HBCs

Figure 6.4 Schematic synthetic process for GQDs by using HBC (1) as a precursor. (*Source*: Reproduced from Reference (42), with permission).

temperature using STM. Through annealing at 500–550 K, the C60 molecules are thermally hopped and dissociated on the terrace and move to the on-top adsorption sites. Concurrently, the Ru atom under the C60 jumps out to leave the vacancy, permitting the C60 sink lower into the surface as much as 0.5 ± 0.1Å. This kind of an adatom-vacancy mechanism has been observed in C60 on Pt(111) or predicted in C60 on Ag(111) surfaces. Decomposition of C60 on the Ru was simulated using density functional theory (DFT) assuming the model of the C60 on a Ru vacancy site (on-top vac model) and a nonvacancy site (on-top model). The stronger is the interaction of the lower hemisphere carbons with Ru atoms, the weaker is the C–C bonding in C60. This instigates the C–C bonds lengthened and creates a fault line. As temperature increases, energy generated from electron–phonon and phonon–molecular vibronic coupling (54) makes the fullerene cage rupture into two unsymmetrical hemispheres along the fault line (Fig. 6.6).

To investigate whether a C60 precursor is crucial in the formation of GQDs, the nanostructured carbons produced by two common precursors C60 and C_2H_4 are compared. At a low dosage of C_2H_4 (<1 Langmuir (L)), C adatoms are nucleated to create carbon clusters at the step edge of the substrate by dehydrogenation. On the other hand, C60 molecules get adsorbed on the terrace. The C_2H_4 with high mobility easily form large-sized and irregularly shaped graphene island; however, the C60 – with the limited mobility – leads to the formation of GQDs (Fig. 6.7).

During the fragmentation process, the transformation of GQD shape follows the trapezoid→parallelogram→triangle sequence (50) (Fig. 6.8).

6.2.2 Top-Down Methods: Chemical Cutting Methods

6.2.2.1 Hydrothermal and Solvothermal Cutting of GSs Pan et al. first reported a simple hydrothermal cutting of micrometer-sized rippled GSs into surface-functionalized GQDs (34). The hydrothermal process consists of three steps: thermal reduction of GO sheets to micrometer-sized GSs, the oxidation of chemically derived GSs in concentrated H_2SO_4 and HNO_3, and hydrothermal deoxidation of oxidized GSs (OGS). In the oxidation step, epoxy (C–O–C) and carbonyl (C=O) groups were induced at basal plane sites while carboxylic (COOH) groups at the edge or hole sites. Epoxy and carbonyl groups tend to linearly enclose sp^2 clusters with a few nanometers lateral size, as shown in Figure 6.9. These cooperatively aligned linear chains were unzipped and subsequently rupture the underlying C–C bonds and then finally removed by alkaline hydrothermal reaction, whereas relatively stable COOH were retained. During oxidation, epoxy groups were changed into energetically preferable epoxy pair and then converted to more stable carbonyl pairs. Less ordered GQDs were obtained if disordered GSs were used as a precursor and hydrothermal cutting reaction was performed under weakly alkaline (pH = 8) conditions. Their diameters and heights are ~5–13 nm and 1–2 nm, respectively. On the other hand, when high-temperature thermally reduced GO sheets were used as the precursor under strong alkaline (pH>12) conditions, well-crystallized GQDs with a lateral size ~3 nm could be obtained (55).

Another facile solvothermal method was introduced as a one-step synthetic route to produce GQDs in a large scale (56, 57). GO was first synthesized from natural graphite powder by the modified Hummers method. GO/dimethylformamide (DMF) solutions were under ultrasonication for 30 min and transferred to a poly(tetrafluoroethylene) (Teflon)-lined autoclave and heated at 200 °C for 5 h. Brown transparent suspension was segregated by consuming black precipitate, and consequently, GQDs can be obtained by evaporating the solvents. The production yield was 1.6%. The average size and height were 5.3 nm in diameter and 1.2 nm, respectively, suggesting that most of the GQDs were single or bilayered (Fig. 6.10c). The prepared GQDs exhibit strong fluorescence with green PL quantum yield (PLQY) of 11.4%. Recently, the same group demonstrated another two-step synthetic method combining a one-step solvothermal route with "top-down" cutting method. GQDs obtained by the solvothermal reaction were purified by column chromatography on silica utilizing gradient elution. The average size of GQDs prepared by two-step method was 3–5 nm, smaller than that produced by solvothermal methods (58).

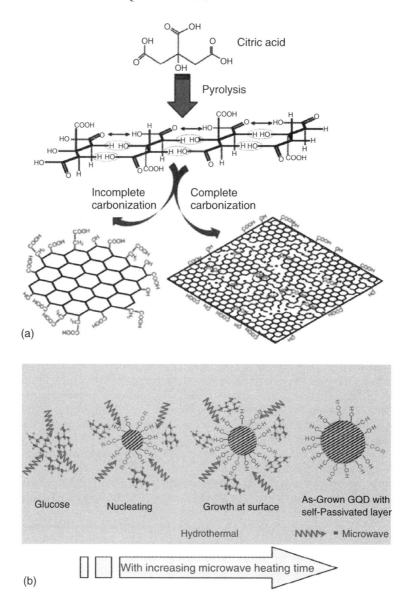

Figure 6.5 (a) A diagram for the synthesis of GQDs and GO by pyrolyzing citric acid (*Source*: Reproduced from Reference (48), with permission) and (b) preparation of GQDs by pyrolyzing glucose via MAH method. (*Source*: Reproduced from Reference (49), with permission).

In addition to hydrothermal or solvothermal methods, GQDs were also prepared by traditional hydrazine hydrate reduction of oxidized small GSs. Shen et al. (59) reported the preparation of GQDs with an oligomeric PEG$_{1500N}$ as a surface-passivation agent. GO were prepared from natural graphite powder by a modified Hummers method. PEG$_{1500N}$ was mixed with the GO solution, and the solution mixture was heated at 120 °C for 24 h. GO is reduced by hydrazine hydrate at 100 °C for 24 h. A yellow solution was separated and further dialyzed in a dialysis bag to produce GQDs with 13.3 nm average diameter, suggest that the as-prepared monodisperse GQDs are uniformly arranged. The as-synthesized GQDs were strongly fluorescent; strong blue PL was clearly shown under 365 nm and the green fluorescence was observed under a 980 nm laser. The PLQY measured using rhodamine B as a reference is 7.4%. The surface passivation can produce GQDs with higher fluorescence performance and UC properties, which will be described in detail by using the energy-level structural models of GQDs in Section 6.4.2.2 (Fig. 6.11).

Recently, the same group reported another synthetic method – GQDs surface-passivated by polyethylene glycol (GQDs-PEG) could be prepared by a one-pot hydrothermal reduction of small GO sheets and polyethylene glycol (PEG) ($M_w = 10,000$) as starting materials (60). The synthesized GQDs-PEG showed nearly monodisperse distribution with an average diameter of ~13 nm (Fig. 6.11b). By comparing the GQDs-PEG and the GQDs, the prepared GQDs-PEG show excellent blue luminescence properties; the PLQY of the GQDs-PEG with 360 nm emission was about 28.0% using rhodamine B as a reference (Fig. 6.12).

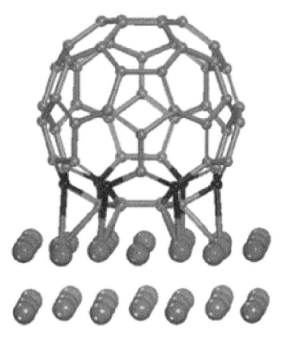

Figure 6.6 On-top_vac configuration of the C60 molecule. The single-sided arrow indicates the top-down point of view. (*Source*: Reproduced from Reference (50), with permission).

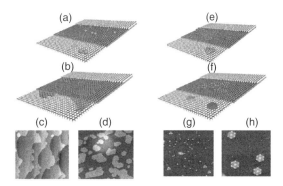

Figure 6.7 Comparison of the growth mechanism of graphene nanoislands and QDs using C_2H_4 and C60. (a) Highly mobile carbon adatoms from the dehydrogenation of C_2H_4. (b) Nucleation of the C adatom at the step edges (at <1L dose of C_2H_4). Large-sized, irregular shaped graphene islands (at <1L~<10L dose of C_2H_4). (c, d) Corresponding STM images for the growth of graphene islands from C_2H_4. (e) C60 molecules adsorb on the terrace and decompose to produce carbon clusters with restricted mobility. (f) Temperature-dependent growth of GQDs with different equilibrium shape from the aggregation of the surface-diffused carbon clusters. (g, h) Corresponding STM images for the well-dispersed triangular and hexagonal equilibrium-shaped GQDs produced from C60-derived carbon clusters. (*Source*: Reproduced from Reference (50), with permission).

6.2.2.2 *Nanolithography Technique* Ponomarenko et al. (29) microfabricated GQD-based single-electron transistor (SET) from mechanically cleaved graphene crystallites (1, 61) on top of SiO_2 (300 nm)/Si wafer. The diameter (D) of GQDs as central islands (CI) was carved as small as 10 nm utilizing a ~5 nm thick polymethylmethacrylate (PMMA) mask protecting the selected area during oxygen plasma etching and connected to side electrodes by quantum point contacts with a width of 20 nm (Fig. 6.1a). At large size (*D*>200 nm), they show periodic CB peaks as like conventional SET. The CB peaks changed into strongly nonperiodic for GQDs smaller than 100 nm due to the increase of quantum confinement. From the calculation of random spacing in CB and Coulomb diamonds, its statistics is revealed to agree well with chaotic Dirac or neutrino billiards (62). Libisch et al. (63) simulated quantum confinement effect in phase-coherent GQDs with the presence of disorder (edge roughness, charge impurities, short-ranged scatterer) with linear dimension between 10 and 40 nm, similar to that in Reference (29), and found that its statistics significantly differ from a simple Dirac billiards. Güttinger et al. (64) also fabricated SET consisting of a

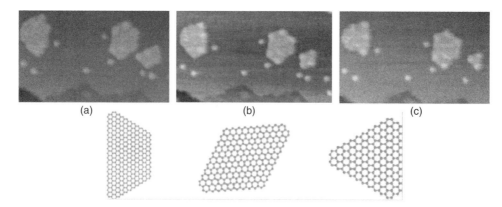

Figure 6.8   Series of STM images monitoring the transformation of trapezium-shaped GQDs to triangular-shaped GQDs at 1000 K. (a–c) The numbers in the images indicate the time lapse in seconds. Tunneling parameters: $V = 1.4$ V, $I = 0.3$ nA; image size, 25×12 nm^2. (*Source*: Reproduced from Reference (50), with permission).

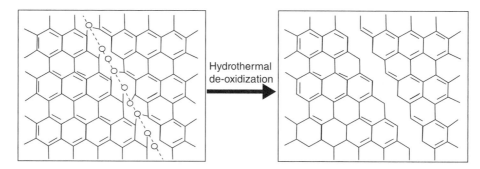

Figure 6.9   Mechanism of the hydrothermal cutting of oxidized GSs into GQDs: a mixed epoxy chain consisting of epoxy and carbonyl pair groups (left) ruptured under the hydrothermal treatment and led to a complete cut (right). (*Source*: Reprinted from Reference (55), with permission).

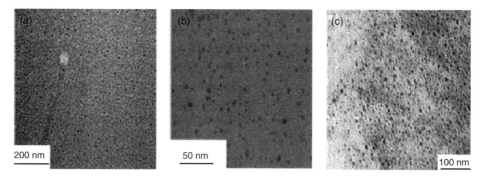

Figure 6.10   TEM images of (a) blue-luminescent (dark spot), (b) green-luminescent (dark spot) GQDs produced by hydrothermally cutting graphene sheets, (c) green-luminescent (dark spot) GQDs obtained by one-step solvothermal method. (*Source*: Reproduced from References (40, 42, 58), with permission).

50 nm wide and 80 nm long graphene GQD (Fig. 6.13a) using two successive processing steps with different 45 nm and 100 nm thick PMMA photoresist and connected to source (S) and drain (D) via two graphene constrictions with a width of 25 nm (Fig. 6.13b).

In a similar fabrication method, Wang et al. fabricated twin-dot structure consisting of smaller-sized GQDs (90 nm) and the larger dot serving as SET sensing the charge state of the nearby gate-controlled GQD. The GQD was connected by 30 nm wide tunneling barriers to source and drain contacts and coupled to a SET with a much larger diameter to form an integrated charge sensor (Fig. 6.13c) (65).

Recently, the same group further fabricated graphene double dot devices to study the transport pattern evolution in multiple electrostatic gates (Fig. 6.14). One type of the device is the gate-controlled parallel-coupled double quantum dot (PDQD) device on both single-layered and bilayered graphenes suitable for spin-based solid qubits for the quantum computation processing. The diameters of the two dots are both

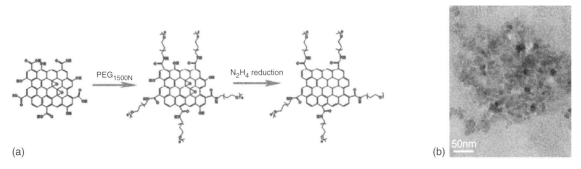

(a)   (b)

Figure 6.11 (a) Representation of GQDs containing an oligomeric PEG diamine surface-passivating agent. (b) TEM image of the GQDs. (*Source*: Reprinted from Reference (59), with permission).

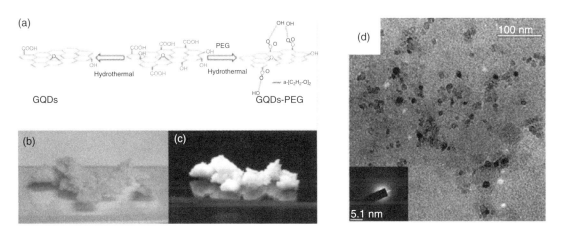

Figure 6.12 (a) Representation of the GQDs and GQDs-PEG by one-pot hydrothermal reduction. (b, c) The images of the dry GQDs-PEG under sunlight (left) and 365 nm UV lamp (right). (d)TEM images of blue-luminescent (dark spot) GQDs-PEG obtained by one-pot hydrothermal reaction. (*Source*: Reprinted from Reference (59), with permission).

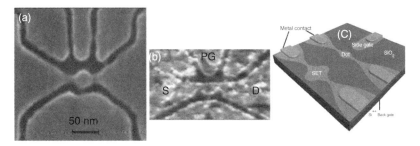

Figure 6.13 (a) A graphene-based single-electron transistor. A 30 nm GQD is connected to contact regions through narrow constrictions of 20 nm wide graphene (Reprinted from Reference (42), with permission). (b) A scanning force microscope image of an etched GQD device with source (S) and drain (D) leads and a plunger gate (PG) for electrostatic tenability (Reprinted from Reference (46), with permission). (c) A SEM image of the photo-etched sample structure consisting of the upper small Dot (GQD, 90 nm) and the bottom SET (Diameter = 180 nm). (*Source*: Reprinted from Reference (47), with permission).

100 nm, constriction between the two dots is 35 nm in width and length. The four narrow parts connecting the dot to source and drain parts have a width of 30 nm. Another one is presented in Figure 6.14a. The double QD has two isolated CI of diameter 100 nm in series, connected by 20 × 20 nm narrow constriction to source and drain contacts (S and D electrodes) and 30 × 20 nm narrow constriction with each other (Fig. 6.14b).

6.2.2.3 Electrochemical Synthesis Ever since Zhou et al. first reported the electrochemical synthesis of C-dots from MW-CNTs (68), carbon nanomaterials have been directly produced through electrochemical exfoliation by anodic oxidation and anionic intercalation of

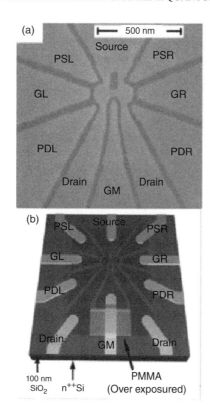

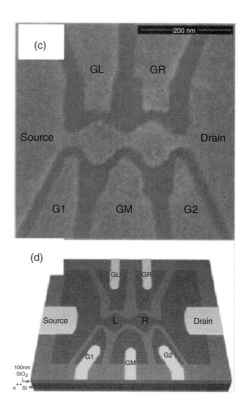

Figure 6.14   (a) SEM images of the etched parallel-coupled graphene double dot sample structure. Seven in-plane plunger gates around the dot for fine-tuning. (b) Schematic picture of the device on N-type Si (Reprinted from Reference (66), with permission). (c) SEM image of the structure of the designed multiple-gated sample. (d) Schematic of a representative device. (*Source*: Reprinted from Reference (67), with permission).

graphite electrodes in the ionic liquid (25, 69, 70). Functional GQDs were electrochemically prepared in 0.1 M phosphate buffer saline (PBS) solution by cyclic voltammetry (CV) scan within ±3 V using graphite as working electrode and subsequent filtration (71). The collected water-soluble GQDs were monodisperse and with a uniform diameter of ~3–5 nm and topographic heights of 1–2 nm (Fig. 6.15), which is advantageous of electrochemical cutting method.

In a similar electrochemical synthetic process, N-doped GQDs (N-GQDs) were prepared using the electrolyte of 0.1 M tetrabutylammonium perchlorate (TBAP) in acetonitrile, instead of PBS, and graphene films used as working electrode with typical size of ~3 cm in length, 1 cm in width, and ~30 μm in thickness with a weight of ~2 mg (72). The size of the obtained N-GQDs was well confined within 2–5 nm and their height was only ~1–5 graphene layers (Fig. 6.16).

From the XPS experiment, the formation of both pyridinic and pyrrolic N-bonding around 399 eV and 401 eV, respectively, were identified in the GQDs synthesized electrochemically from both TBAP and acetonitrile electrolyte. This incorporation of N atoms results from doping into the graphene backbone structure and also implies that the electrolyte ions and small molecules could be intercalated into graphene layers and then break the intercalated graphene layers by ion–graphene or molecule–graphene interaction at the defects or active edges under the electrochemical potentials during the electrochemical CV scanning at a relatively high potential of ± 3 V as shown in Figure 6.17a and b. Accordingly, the physical and chemical properties of the doped GQDs can be controlled by adjusting the incorporated ions based on the deliberate choice of the kinds of electrolytes in electrochemical synthesis. But the exact interaction mechanism still needs further investigation.

Recently, Zhang et al. also reported a facile electrochemical synthetic method for large scale production of GQDs with graphite rod, as an anode, inserted into 0.1 M NaOH alkaline aqueous solution, in which the O and OH radicals produced by anodic oxidation of water, can serve as electrochemical "scissors" to cut carbon nanocrystals and form oxygenated groups such as epoxy, hydroxyl, carbonyl, and carboxyl groups (73). The resulting solution from the electrolysis was reduced by adding hydrazine hydrate at room temperature. The size distribution was estimated as an average diameter of 5–10 nm and a thickness was <0.5 nm, indicating a single graphene layer. In particular, these GQDs show prominent yellow fluorescence at the wavelength of 540 nm with high quantum efficiency of 14% and excitation-independent PL. The origin of yellow luminescence is closely related to the formation of hydrazide groups $O=C-NH-NH_2$, at the binding energies of 399.7 and 400.8 eV in the N1s XPS spectrum and which is critical to reduction condition of working environment. This will be further discussed in detail in Section 6.4.2.1 (Fig. 6.18).

6.2.2.4 Chemical Exfoliation by Solution Chemistry   One of the most popular bottom-up methods to synthesize GQDs is chemical exfoliation using solution chemistry (41, 58, 74–78). Liu et al. (46) first demonstrated the preparation of GQDs using unsubstituted HBC

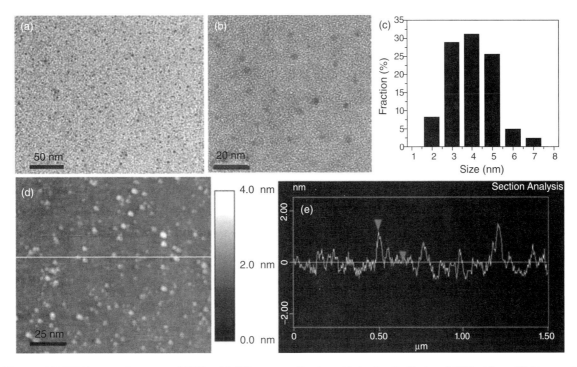

Figure 6.15 (a and b) TEM images of as-prepared GQDs with different magnifications, (c) the size distribution of GQDs, (d) an AFM image of the GQDs on Si substrate, and (e) the height profile along the line in (d). (*Source*: Reprinted from Reference (71), with permission).

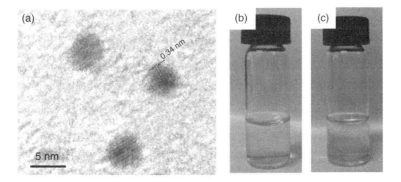

Figure 6.16 The typical high-resolution TEM images of N-GQDs and photoimages of electrochemically synthesized N-GQDs aqueous solution before (a) and after (b) one month standing. (*Source*: Reprinted from Reference (72), with permission).

as a carbon source which were made by cyclodehydrogenation of hexaphenylbenzene. Intermediate artificial graphitic column structure could be prepared by pyrolyzing HBC at 600–1200 °C for 5 h. These artificial graphites were oxidized and exfoliated with a modified Hummers method (79), and finally the aqueous solution of GO was heated to reflux for 48 h with oligomeric (ethylene glycol) diamine (PEG_{1500N}) and reduced with hydrazine. GQDs annealed at 600 °C contained disordered particles and GQDs at 900 °C consisted of both disordered particles and disk-shaped nanosheets. GQDs at 1200 °C contained homogeneous nanodisk of ~60 nm in diameter and 2–3 nm in thickness. In a similar way, Shen et al. also (59) reported a facile preparation method to obtain GQDs through hydrazine reduction of GO, first oxidized by HNO_3 and cut into small GO sheets. GO was treated with oligomeric PEG diamine (PEG_{1500N}) as a reference surface-passivation agent like Reference 46. GQDs showed the uniform distribution of average diameter 13.3 nm taken by TEM.

Dong et al. presented a one-step and high-yield simultaneous preparation of single- and multilayered GQDs from CX-72 carbon black. Carbon black was put into HNO_3 followed by refluxing for 24 h. After cooling down and centrifugation of the suspension, the GQDs1 was obtained by heating the supernatant and the GQDs2 was collected by washing the sediment and further adjusting with pH 8 with ammonia and water, respectively. GQDs1 has higher O_2 content and lower C content than GQDs2. The average sizes of GQDs1 and GQD2 were about 15 nm and 18 nm, and their topographic heights were ~0.5 nm and 1–3 nm, respectively, indicating GQDs1 are single-layered GQDs and GQDs2 are multilayered (2–6 layers) GQDs (80) (Fig. 6.19).

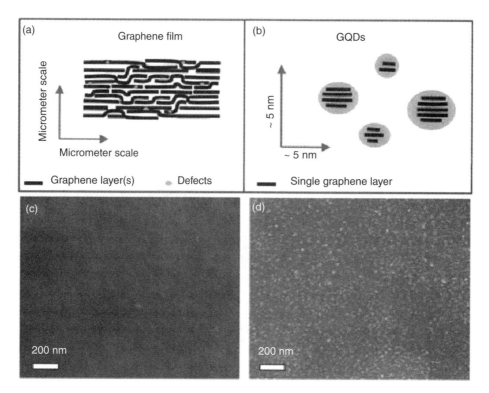

Figure 6.17   A scheme of the stacking structure of graphene layer(s) in a filtration-formed graphene film (a) and electrochemically produced GQDs (b), and the surface SEM images of the original graphene film (c) and the one after CV scan for 2000 cycles (d). (*Source*: Reprinted from Reference (74), with permission).

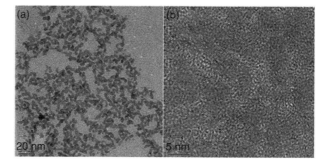

Figure 6.18   TEM images of GQDs taken from the Reference (73) in 20 nm scale and 5 nm scale. (*Source*: Reprinted from Reference (73), with permission.)

Peng et al. (81) demonstrated the synthesis of GQDs by chemical oxidation and cutting of μm-sized pitch-based carbon fibers (CF). The fiber-derived GQDs showed a narrow size between 1 and 4 nm and height distribution between 0.4 and 2 nm corresponding to 1–3 graphene layers. As shown in Fig. 6.20, the GQDs edges were predominantly parallel to zigzag orientation than armchair. The formation of zigzag edges in GQDs from CF seems to be principally related with chemical oxidation and unzipping mechanism in CNT (79) which was initiated by the lining up of formed chemical functional groups (like epoxy or carbonyl group) (82) and making graphite domains to be fractured along zigzag direction (81).

Li et al. reported a facile microwave-assisted chemical oxidation approach for the preparation of stabilizer-free greenish-yellow luminescent GQDs (gGQDs) from GO nanosheets under acidic conditions (3.2 M HNO_3 and 0.9 M H_2SO_4) for 3 h (Fig. 6.14) and blue-luminescent GQDs (bGQDs) via reducing GQDs with $NaBH_4$ for 2 h (83). The diameters of gGQDs and bGQDs were mainly distributed in the range of 2–7 nm with an average diameter of 4.5 nm and the topographic heights of GQDs were mostly between 0.5 and 2 nm with an average height of 1.2 nm, suggesting that most of GQDs were single layered or bilayered. The blueshift in PL of bGQDs might be directly caused by rather structural change after reduction than the difference in size (Fig. 6.21).

6.2.2.5 Nanotomy-Assisted Exfoliation   Mohanty et al. (84) demonstrated one of top-down methods of diamond-edge-induced nanotomy (nanoscale cutting) to produce GNs with predetermined shapes (square, rectangle, triangle, and ribbon) and tailored size. In nanotomy,

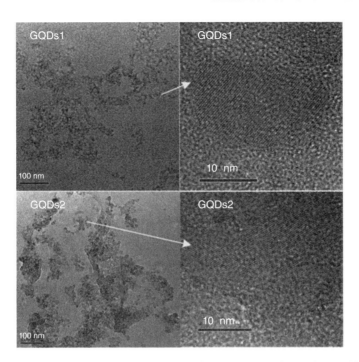

Figure 6.19   High-resolution TEM images of GQD1 and GQD2. (*Source*: Reprinted from Reference (80), with permission).

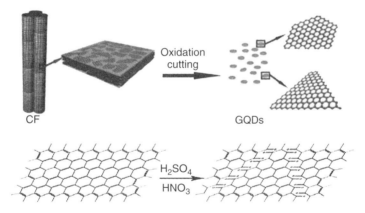

Figure 6.20   Schematics diagram of the synthesis of GQDs by chemical oxidation and cutting of μm-sized pitch-based carbon fibers (CF). (*Source*: Reprinted from Reference (81), with permission).

graphene nanoblocks (GNBs), as nanoscale graphite precursors, of controlled shape and size are prerequisite to synthesize GNs. Highly oriented pyrolytic graphite (HOPG) was cut in different directions and at controlled thickness to produce GNBs consisting of millions of columnar-stacked GNs with same dimensions and shape and consequently cleaved via C–C bond stretching and crack formation (Fig. 6.22). Raman and the high resolution transmission electron microscope (HRTEM) measurement show that the GNs had relatively smooth edges and the MD simulations and HRTEM micrographs indicate that the edges are predominantly zigzag. The GNBs were cleaved at the corresponding thickness in a single direction for getting GNRs and cleaved via two-step process in two directions at controlled width, length, and angles for GQDs and subsequently exfoliated in chlorosulfonic acid. The collected GQD_S had the size distribution of about 10–50 nm.

6.2.2.6   Ultrasonic Shearing of GSs   Li et al. reported that a one-step alkali- or acid-assisted ultrasonic treatment could be employed to synthesize monodisperse water-soluble fluorescent carbon nanoparticles (CNPs) using natural precursor, glucose as carbon source. These small CNPs are spherical and less than 5 nm in size and luminescence and up-conversion photoluminescence (UCPL) properties (85). Recently, Zhuo et al. also reported the ultrasonic synthesis of GQDs. Graphene was oxidized in concentrated H_2SO_4 and HNO_3 at room

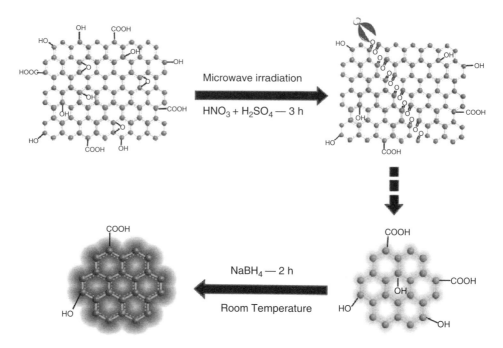

Figure 6.21   Schematic representation of the preparation path for greenish-yellow luminescent GQDs (gGQDS) (bottom-right) and blue-luminescent GQDs (bGQDS) (bottom-left). (*Source*: Reprinted from Reference (83), with permission).

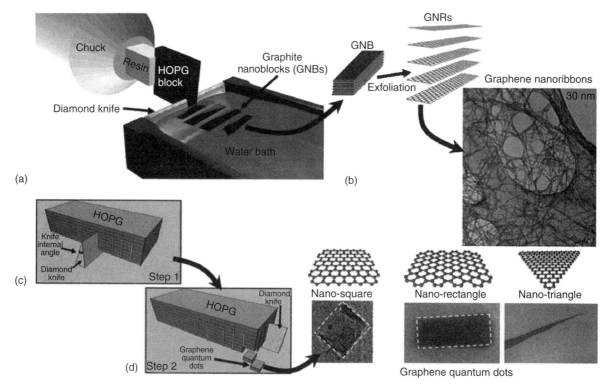

Figure 6.22   A schematic of the diamond-knife-based mechanical cleaving nanotomy process of HOPG producing (a) GNBs and (b) GNRs. (c) A sketch of the two-step nanotomy process to produce graphene nanoblocks for GQD production. (d) TEM images of square-, rectangular-, and triangular-shaped GQDs. (*Source*: Reprinted from Reference (84), with permission).

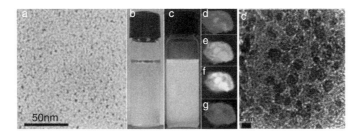

Figure 6.23   (a) TEM image of CNPs prepared from glucose with diameter less than 5 nm. (b, c) Photographs of CNPs dispersions in water with sunlight and UV (365 nm, center) illumination, respectively. (d–g) Fluorescent microscope images of CNPs under different excitation: d, e, f, and g for 360, 390, 470, and 540 nm, respectively. (f) TEM images of as-prepared GQDs. (*Source*: Reprinted from Reference (86), with permission).

temperature for 12 h. Then, the mixed solution was subsequently treated ultrasonically for 12 h. The mixture was calcined and black suspension in water solution was filtered and finally GQDs with diameters of 3–5 nm were prepared (86). GNRs with a width as narrow as 0.4 nm were also present, presumably formed from the protrudent edge of GQDs. Figure 6.23 shows the TEM images of CNPs prepared from glucose with diameter less than 5 nm and those of the as-prepared monodisperse GQDs with diameters of 3–5 nm.

6.2.2.7   *Synthesis of GQDs from Hybrid Quantum Dots*   Of late, Son et al. demonstrated a facile chemical synthesis process to form ZnO–graphene quasi core–shell structure QDs and to extract GQDs from them after removal of the ZnO core by treating with hydrochloric (HCl) acid (87). Synthetic procedure can be briefly summarized as follows and graphically depicted in Figure 6.24. As a first step, graphite oxide (GO) was obtained by ultrasonicating graphite powder with H_2SO_4 and HNO_3 mixed acid. In this reaction, chemical functional groups containing oxygen such as epoxy(C–O), hydroxyl (–OH), and carboxyl (–COOH) were induced on GO. And second, both the GO and zinc acetate dihydrate were dissolved in DMF solution. After heating the solution to 95 °C for 5 h, the formation of ZnO–graphene core–shell QDs was identified by the change of color into a white-grayish. After drying, to extract GQDs the outer layer of ZnO–graphene QDs, the ZnO–graphene QDs were treated with HCl acid and GQDs without ZnO were extracted from the solution.

From the XRD patterns (Fig. 6.25a), the occurrence of a very broad peak G(002) at $2\theta = 25.8°$ and a broad low intense peak G(100) at $2\theta = 43.5°$ strongly indicates the existence of graphene layer. Concurrently, the intense peaks of ZnO (100), (002), (101), and (102) verify the existence of ZnO core in ZnO–graphene core–shell QDs. On the other hand, after HCl treatment, no peaks related to ZnO were observed, but only a small peak at around $2\theta = 26.5°$ corresponding to graphene was found, which reveals that only outer graphene layers were exfoliated by the removal of ZnO core. Furthermore, as shown in Figure 6.25b, Raman study gives very interesting result that G-band peak is split into two symmetric lines of G^+ (1592.7 cm^{-1}) and G^- (1566.6 cm^{-1}). Because such splitting of G^- and G^+ has been usually observed in CNTs (88), it could be caused by the induced strain due to the bending of graphene outer layer surrounding the ZnO QDs. However, in the upper spectra of Figure 6.25b, only a single G-band peak at 1577.5 cm^{-1} without splitting was observed. Also, it is noteworthy that the defect-related D-band peak was greatly suppressed in comparison with G-band peak. This result indicates that detached graphene from ZnO–graphene hybrid QDs has overall good quality. The relative ratio of the I_D/I_G (= 0.1) of graphene from graphite–ZnO synthesis is much lower than any values ever reported in the previous chemical methods (89–94), implying this kind of synthetic route is one of very effective methods in producing high-quality GQDs.

Bending of outer graphene layer can be confirmed from high-resolution TEM study illustrated in Figure 6.26. The outer graphene layers over the ZnO core shown with red arrows cover well along with the surface of ZnO core particles. High-contrast image (Fig. 6.26b) of a hexagonal atomic lattice taken from the layer encircling ZnO QDs clearly unveils the existence of an ML graphene in which the measured distance between carbon atoms is of about 0.14 nm (inset of Fig. 6.26b).

6.3   SURFACE FUNCTIONALIZATION AND CONGREGATION OF GQDs

GQDs can be categorized as a new class of QDs with unique electrical, optical, and chemical properties. These extraordinary functionalities and intrinsic properties can be effectively tuned by surface chemical treatment, doping with heteroatoms, and congregation of GQDs for advanced device applications. A typical chemical treatment is likely to induce some functional groups and ligands on the GQDs surfaces which are useful for hybridization with other nano materials and will influence on the alignment of the GQDs on polar surfaces. Different surface-passivating ligands have proved with varied emission center and QY of PL and have produced some unusual properties, such as emission transformation (96).

The quantum confinement and edge effect of GQDs with the nitrogen doping was first reported by Li and coworkers. In order to dope N atoms, N-contained TBAP in acetonitrile was used as the electrolyte and N-GQDs were continuously produced by CV. As shown in Figure 6.27a and b, the mean size distribution of N-GQDs was ~2–5 nm and the corresponding atomic force microscopy (AFM) image (Fig. 6.27c) revealed a typical topographic height of 1–2.5 nm (Fig. 6.27d), suggesting that most of the N-GQDs consist of ~1–5 graphene layers (34, 97). From X-ray photoelectron spectroscopy, the N1s core-level peak was well resolved into both pyridine-like (398.5 eV) and

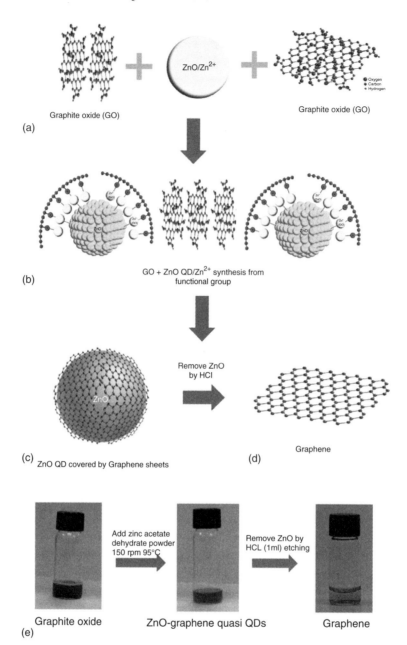

Figure 6.24 (a) Induced functional groups on GO surface after treatment with an acid. (b) Embryo ZnO QDs formed under the chemical reaction and formation of ZnO–graphene quasi core–shell QDs. The folding of graphene layers outside the ZnO inner-core QDs comes from possible chemical reactions between three kinds of functional groups (carboxyl, hydroxy, and epoxy) on GO and Zn^{2+}. (c) Schematics of ZnO–graphene quasi core–shell QDs. (d) Schematic of synthesized and unrolled graphene after removal of inner ZnO QDs by HCl. (e) Chemical process of the production of ZnO–graphene quasi QDs and graphene after removal of ZnO. (*Source*: Reprinted from Reference (87), with permission).

pyrrolic (401 eV) N atoms (98). The relative atomic ratio of N/C of N-GQDs was ~4.3% and that of the O/C was ~27% close to that of N-free GQDs. The UV–Vis absorption spectrum showed blueshift of absorption edge for N-GQDs toward ~50 nm with respect to that of N-free GQDs (Fig. 6.28). Moreover, the N-GQDs emit blue luminescence, whereas N-free GQDs shows green emission. This result seems to be similar with the blue emission due to the existence of electron-hole pairs found in the isolated sp^2-hybridized clusters within C–O matrix. Similarly, doped N atoms in the N-GQDs are assumed to contribute to the blueshift of PL spectrum due to strong electron affinity of N atoms in the N-GQDs. This assumption can be supported by the experimental observation and theoretical calculation for the N atoms within the conjugated C plane (99, 100).

Son and coworkers introduced a facile synthesizing process of a very interesting hybrid nanostructured material of an assembly of ZnO nanoparticles with GQDs (101).

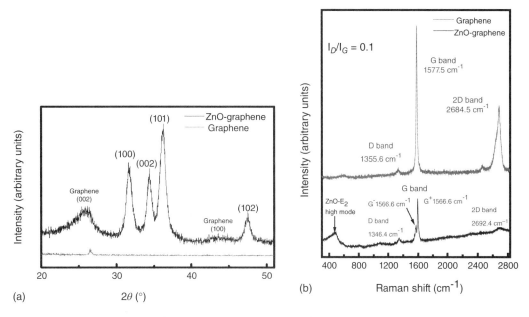

Figure 6.25 (a) XRD of pure graphene with (002) diffraction peak (top) and of the ZnO–graphene quasi core–shell QDs including graphene and ZnO QDs (bottom). (b) Raman spectrum of the graphene layers obtained by dissolving ZnO from ZnO–graphene core–shell QDs (top) by HCl and ZnO–graphene core–shell QDs (bottom). (*Source*: Reprinted from Reference (87), with permission).

GO (acid-treated graphite powders) were mixed with $Zn(CH_3COOH)$ in DMF solution and then embryo ZnO QD with average size of ∼5 nm grew first, followed by attaching of GQDs onto the ZnO inner core through $Zn–O_{epoxy}–C$ bonding as shown in Figure 6.29. Main participation of $O_{epoxy}–C$ bonding among other functional groups induced by mixture of $H_2SO_4:HCl$ acid was identified by calculating the change of density of state (DOS) of graphene using DFT. From the DFT calculation, epoxy group formed on graphene induced a splitting of the pristine lowest unoccupied molecular orbital (LUMO) level into three LUMO, LUMO + 1, and LUMO + 2 levels. Among them, LUMO and LUMO + 2 levels contain s and p orbitals and electrons belonging to s orbital participate in the charge transition to O2p valence band of ZnO. In ZnO–graphene core–shell QDs, some of near UV emission of pure ZnO QDs can be partially modulated into visible emission corresponding to the transition from modified LUMO levels of graphene to O2p valence band of ZnO QDs.

Furthermore, Hamilton et al. reported that orientations of GQDs on polar surfaces (water and mica) were controllable and aligned either parallel (face-on) to or out of plane from the surface (edge-on) (74). Three different kinds of GQDs were synthesized. The triangular core of GQDs1 with the trialkyl phenyl groups twisting the plane of the graphene has an effective area of 2.7 nm^2 and is hydrophobic. The GQDs2, with the same size of the GQDs1, were precisely made to have additional carboxylic acid groups and the GQDs3 were almost same with GQDs2, except the effective core size of 2.5 nm^2. From the measurement of the compression isotherm Π–A of GQDs 1–3 on water surface using Langmuir trough, the formation of face-on geometry was observed and which was expected due to the increased van der Waals interaction between water and the GQDs1 despite its lack of amphiphilicity. By considering the existence of carboxylic acid group in GQDs3, they are likely to be aligned out of plane from water surface. On the contrary to the case of GQDs3, it is very surprising that the GQDs3 follow a face-on geometry like the GQDs1. This result reveals that the existence of the hydrophilic functional acid groups did not greatly affect in changing the orientation of GQDs on polar surface (Fig. 6.30).

Figure 6.31 represents the typical AFM images of the GQDs 1–3 on freshly cleaved mica surface and their corresponding height profiles. The AFM images of the GQDs 1 and 2 showed very uniform film thickness of 1.4 nm, confirming the face-on geometry. On the other hand, the GQDs 3 showed a film thickness of 5 nm, indicating the edge-on geometry. The understanding of charge transport of the GQDs on different substrates by changing the orientation will be very useful for the application of electro-optical devices.

Very recently, Li and coworkers succeeded in functionalization of GQDs having the amine group by cutting GOs in a two-step process. First, the epoxy groups (C–O) are induced on the GOs and they are transformed into carboxyl groups (–COOH) through oxidation using H_2SO_4/HNO_3, resulting in the cutting of graphenes, as shown in Figure 6.32 (82). Second, subsequent reduction by N_2H_4 made graphene cut into further smaller by the removal of the bridging O atoms attaching to epoxy chains and which gives rise to the production of the GQDs (34). Afterward, oxidized GOs were reacted with diamine-terminated PEG diamine between the oxidation and reduction processes. From the literature, alkyl amine groups were attached to the oxidized GOs by the ring-opening reaction of the epoxy groups on the GOs under alkaline conditions. And then finally, the alkyl amine-functionalized GQDs (GQDs-NHR) were prepared. The formation of the functional groups in the GQDs-NHR was identified using Fourier transform infrared (FTIR) spectroscopy, as shown in Figure 6.32b. In the FTIR spectrum of the GQDs-NHR, new peaks appeared at around 1536 cm^{-1} and doublet peaks at around 2860 and 2930 cm^{-1} corresponding to N–H bending in the amine group and C–H stretching in the alkyl chains (102, 103), respectively, which strongly confirms the attachment of alkyl amine groups to the GQDs.

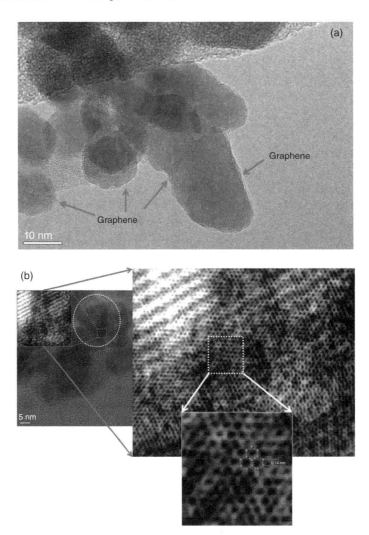

Figure 6.26 (a) HRTEM image of ZnO–graphene quasi QDs revealing the outer graphene layers. (b) HRTEM image of ZnO QD covered by graphene (left). White dot circle is an image of ZnO–graphene quasi core–shell QDs. White-squared region is the magnified image of one ZnO QD covered by a monolayer graphene. An enlarged view of the monolayer graphene is indicated by white lines. The area in the white-dotted square box is magnified to disclose the atoms more clearly: In the single layer, the white-colored atoms depict the hexagon and the hexagon center is black. (*Source*: Reprinted from Reference (87), with permission).

6.4 PHYSICAL PROPERTIES OF GQDs

6.4.1 Electronic and Magnetic Properties

Graphene is an allotrope of carbon whose structure is a single planar sheet of sp^2-bonded carbon atoms, that are densely packed in a honeycomb crystal lattice (104) and experimentally accessible (1, 2). As an unusual electronic transport phenomenon, an anomalous quantum Hall effect was experimentally measured in graphene (61, 105) and is closely linked to the unusual low-energy electronic properties. E-k relation shows a peculiar linear (or conical) dispersion for low energies around the singular K and K′ points near the six corners of the 2D hexagonal Brillouin zone (Dirac Points) as shown in Figure 6.33, leading to zero effective mass for electrons and holes. In semiempirical calculations, only the two $p_z(\pi)$ bands neglecting the σ bands separated in energy (>10 eV at Γ) are considered because the overlap between the $p_z(\pi)$ and the s or p_x and p_y orbitals is zero by symmetry (106). The equation depicting the E-k relation for the $p_z(\pi)$ electron can be expressed as

$$E^{\pm}(k_x, k_y) = \pm\gamma_0 \hbar v_F \sqrt{1 + 4\cos\frac{\sqrt{3}k_x a}{2}\cos\frac{k_y a}{2} + 4\cos^2\frac{k_y a}{2}} \tag{6.1}$$

where the Fermi velocity (effective speed of light) $v_F \equiv (\sqrt{3}/2)\gamma_0 a/\hbar \approx 10^6$ m/s is proportional to both the lattice constant (C–C distance) $a = \sqrt{3}a_{cc} (a_{cc} = 0.142$ nm$) = 0.246$ nm and to the transfer integral between first nearest-neighbor orbitals (hopping energy) $\gamma_0 \approx 2.9$–3.1 eV

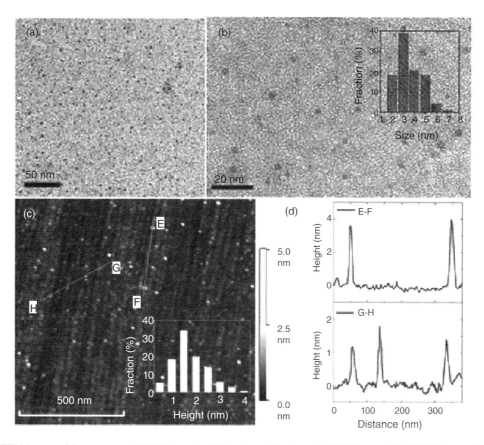

Figure 6.27   (a, b) TEM images of the as-prepared N-GQDs under different magnifications. (c) AFM image of the N-GQDs on a Si substrate. (d) Height profile along the lines in (c). The insets in (b) and (c) show the size and height distributions of N-GQDs. (*Source*: Reprinted from References (34, 97), with permission).

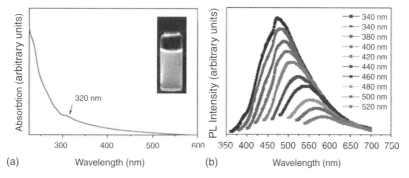

Figure 6.28   UV–Vis absorption and PL spectra of GQDs in water, respectively; inset in (a) is a photograph of GQD aqueous solution under UV irradiation (365 nm). (*Source*: Reprinted from Reference (97), with permission).

on the honeycomb lattice of carbon atoms (107). At the Brillouin zone, the energy E has a conical dependence on the 2D wave vector $\boldsymbol{k} = (k_x, k_y)$. Denoted by $\delta\boldsymbol{k} = \boldsymbol{k} - \boldsymbol{K}$, the displacement from the corner at wave vector $\boldsymbol{K}$, one has, for $|\delta k|a \ll 1$, the dispersion relation

$$|E| = h v_{\mathrm{F}} |dk| \qquad (6.2)$$

Thus, the electronic structure of graphene can be ascribed to massless relativistic carriers, called Dirac fermions, and the electronic properties of graphene are described by an equation (the Dirac equation) of relativistic quantum mechanics, even though the microscopic Hamiltonian of the carbon atoms is nonrelativistic (108, 109). Furthermore, the electronic states consist of the two sublattice components and their superposition gives an additional degree of freedom known as pseudospin (108). Two electronic processes, Andreev reflection (110) and Klein tunneling,

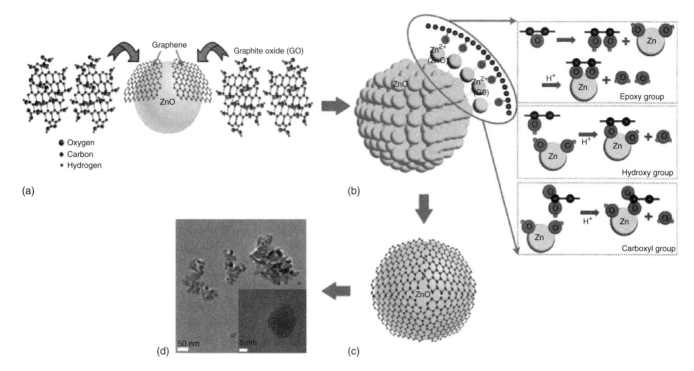

Figure 6.29 Chemical synthetic process for the ZnO–graphene quasi QDs. A schematic of (a) chemical exfoliation of graphene sheets from graphene oxide, (b) synthesis of ZnO–graphene QDs from graphene oxide and zinc acetate dehydrate, and (c) graphene-covered ZnO QDs. (d) A TEM image of the ZnO–graphene QDs. (*Source*: Reprinted from Reference (101), with permission).

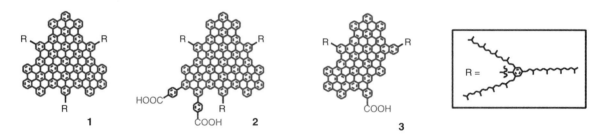

Figure 6.30 Top view (left) and side view (right) of an energy-minimized geometry of 1. The alkyl chains (marked R) form a 3D cage to separate the graphene cores (marked triangle) to avoid aggregation. The triangular core of the GQDs 1 has a length of ~2.5 nm on each edge, which with the solubilizing groups makes the overall diameter of 1–5 nm and thickness of ~1.5 nm. Both the values could vary depending on the conformation of the alkyl chain. (*Source*: Reprinted from Reference (74), with permission).

couple electron-like and hole-like states through the action of either a superconducting pair potential or an electrostatic potential. The first process is the electron-to-hole conversion at the interface with a superconductor and the second process is the tunneling through a p–n junction. The absence of backscattering, characteristic of massless Dirac fermions implies that both processes happen with unit efficiency at normal incidence. Away from normal incidence, retroreflection in the first process corresponds to negative refraction in the second process. Klein tunneling matching the wave functions of electron and positron across the potential barrier makes the graphene p–n junctions transparent by conservation of sublattice pseudospin and results from the absence of backscattering of the carriers (111–113). While graphene itself is not superconducting, it acquires superconducting properties by proximity to a superconductor. We therefore have the unique possibility to bridge the gap between relativity and superconductivity in a real material (114). Nonetheless, both gapless electronic structure and a relativistic penetration called Klein tunneling of massless Dirac fermion in graphene cannot easily permit confinement of charge carriers in graphene as in conventional 2D electron gases (115, 116).

Charge carrier confinement in graphene could be well studied by successful development of synthesizing and tailoring of quantum structure of 1D GNRs and 0D GQDs.

In lithographically patterned GNR, Han et al. observed the existence of energy gap near charge neutrality point and decrease with the increase of the ribbon width W (117). According to Sols et al., the observed energy gap in GNR through conductance measurement was

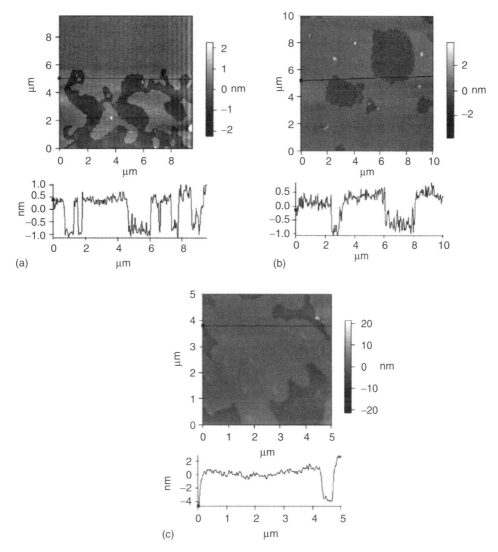

Figure 6.31 AFM images and height profiles of monolayers of GQDs (a) 1, (b) 2, and (c) 3 on mica surfaces. All the films were transferred to the substrates at a high surface pressure of 35 mN/m. The height profiles were measured along the black lines in the images and show a 1.4 nm thickness for both GQD 1 and 2 and 5 nm for 3. (*Source*: Reprinted from Reference (74), with permission).

explained to be related with CB caused by electron transport from dot to dot through graphene necks, which abruptly reduces electron conductance by increasing impedance as shown in Figure 6.34 (118, 119). This leads to temporary electron confinements and the gap is regarded as the effective charging energy of the islands.

In GNR, the formation of subband is observed and the conductance is quantized at the variation of gate voltage (V_G) at low temperature (T <100 K) and shows the dependence on the drain voltage (V_D). The conductance plateau becomes apparent for T <100 K with an asymmetry in the slope of n and p branches. The spacing ΔG between conductance plateaus is constant and decreases with increasing the length of GNR. Conductance plateau is understood by the formation of multiple 1D subband instead of CB-dominated transport as well as charge hopping by disordered edge. In GNR with width D, the bandgap $E_g = 2|\alpha|\Delta E(0 \leq |\alpha| \leq 0.5$ depends on the crystallographic orientation of the GNR, where $\Delta E = \hbar v_F \pi/W$ and gives rise to various 1D subbands. The energy spacing of 1D subbands of 30 nm wide GNRs is around 50 meV which is in good agreement with the tight-binding approximation for armchair GNRs (118). Quantum confinement in GQDs with linear dimensions between 10 and 40 nm, containing 6,000–75,000 C atoms, is explicitly simulated (121) by a third-nearest-neighbor tight-binding approximation (122) allowing for the inclusion of three kinds of disorders of edge roughness (ΔW), point defects (short range) with impurity density (N_D), and long-ranged screened Coulomb distortion due to charge impurity (scatter) (N_C) with order of $N_i = 1.8 \times 10^{-3}$ impurities/carbon (123). As shown in Figure 6.35, small weak disorder of $\Delta W = 0.6–1$ nm invokes prominent size quantization peaks with the energy separation $\Delta E = 0.1$ eV and induces interference between the cones at K and K′ which enhances confinement effects. This kind of dispersion markedly deviates from the linear dispersion $\rho \propto E$ of a massless Dirac particle. Also, the remaining of the K–K′ splitting of 12 meV even in the limit of strong disorder is believed to originate from the localization of a number of surface states at the zigzag edge. They found

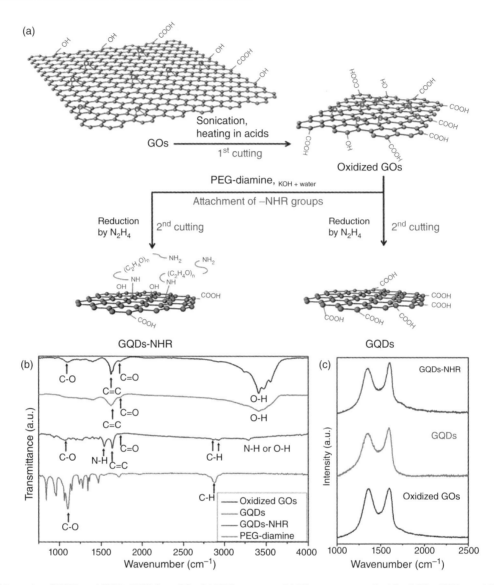

Figure 6.32 (a) Schematics of GQDs and GQDs-NHR from GOs, (b) FTIR spectra, and (c) Raman spectra of oxidized GOs, GQDs, and GQDs-NHR. (*Source*: Reprinted from Reference (82), with permission).

remarkably stable linear relations between nearest-neighbor-level spacing distribution (NNSD) parameter (β) and edge-roughness amplitude, $\beta \approx 2\Delta W$, and the density of short-range $\beta \approx 0.7\ N_D$, as well as long-range defects $\beta \approx 0.2\ N_C$. This suggests that the edge roughness of fabricated devices can be estimated by the distribution of measured CB peaks.

About energy levels of GQDs with the variations of perpendicular magnetic field to graphene plane, Güttinger et al. investigated the addition spectrum of a GQD in the vicinity of the electron-hole crossover. At all gate voltages, CB resonances with 50 nm wide dot are visible and the unique complex of the diamagnetic spectrum of a GQD are well evolved where the $n = 0$ Landau level is situated in the center of the transport gap marking the electron-hole crossover and the average peak spacing decreases with increasing magnetic field (124).

An electric-field-tunable electronic structure of 0D finite-sized (N_a, N_z) GQD denoting the number of N_a dimer lines and N_z zigzag lines was investigated using a single π-band tight-binding method in the presence of the external electric field with (125) or without (126) change of direction. The state energies, energy gap, oscillation period, and oscillation strength largely depend on the electric field strength ($F (= \gamma_o/eA) \approx 0.2$) and the field direction ($\theta = 0° -90°$ from x-axis). Such a characteristic feature originates from the anisotropy of GQD crystal. In the case of IV-type ($N_a = 9$, $N_z = 13$) GQDs, the bandgap is always zero regardless of the value of the field strength. The variations of the state energies will be directly reflected in the DOSs. In the case of the DOSs, many δ-function-like divergent peaks appeared due to the discrete feature of the electronic structures of GQDs. The peak height, numbers, and frequencies of peaks in DOS also show strong dependence of applied field strength and direction.

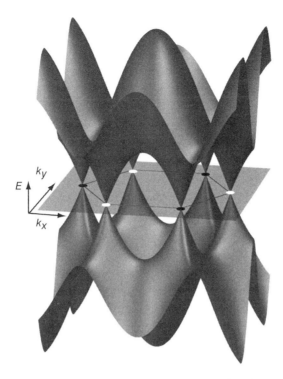

Figure 6.33 Band structure $E(k_x, k_y)$ of a carbon monolayer. The hexagonal first Brillouin zone is indicated. The conduction band ($E > 0$) and the valence band ($E < 0$) form conically shaped valleys that touch at the six corners of the Brillouin zone (called conical points, or Dirac points, or K-points). The three corners marked by a white dot are connected by reciprocal lattice vectors, so they are equivalent. (*Source*: Reprinted from Reference (106), with permission).

While uncontrolled lifting of degeneracies of the valley were previously reported (29), Recher et al. showed that in gapped single- and bilayered GQDs, the degeneracy of the valley was controllably broken by a normally applied uniform magnetic field to the graphene plane, opening up the emerging fields of valleytronics like valley filter, valves (127), or qubits (128) and spin qubits (129, 130).

The decoherence of a single-electron spin in isolated GQDs induced by hyperfine interaction with nuclear spin comes from nonuniform coupling between the spatial variation of electron wave function and nuclei located at different sites. The decoherence time is given by $\hbar N/A$ (N is the number of nuclei and A is hyperfine constant) and order of μs for a GQD which differs from the dephasing time for an ensemble of dots, $\hbar\sqrt{N}/A$ (131).

Hot carrier relaxation in GQDs is very interesting because GQDs scale-downed to a few nm show strong carrier–carrier Coulomb interaction (CB effects) and a small dielectric constant compared to bulk graphene and other compound semiconductor QDs (CdSe, InP, PbSe, etc). Mueller et al. observed the slow carrier cooling in colloidal GQDs. Based on colloidal GQDs with only sp^2 hybridization through perfect passivation of edges with H atoms, new relaxation pathways for hot carrier were investigated except Auger-like electron-hole transfer, nonadiabatic channels involving surface ligands, traps, and high vibrational modes in surface ligands. For example, since effective mass of both electrons and holes is equal near the band edge due to linear energy dispersion in GQDs, suppression from Auger-like processes followed by phonon-assisted relaxation can be excluded. The lifetimes of the hot carriers were observed as much as 100–300 ps, which are two orders of magnitude than bulk graphene. From absorption and fluorescence excitation spectra, intersystem crossing competes with internal conversion and thus both fluorescence and phosphorescence were observed simultaneously. The former can be enhanced due to the reduced singlet–triplet splitting of ~175 meV in largely conjugated GQDs. Since triplet state has longer lifetime than singlet state, triplet-state formation will enhance optical spin injection and long decoherence time is effective in spin manipulation. As illustrated in Figure 6.36a, Ponomarenko et al. (29) showed that CB peaks became strongly nonperiodic function of gate voltage (V_g) and varied in their spacing by a factor of five or more exceeding in nongraphene QDs. The distance between CB is determined by the sum of charging and confinement energies.

$\Delta E = E_c + \delta E$ and as GQDs become smaller, the average distance $<\Delta V_g>$ between CBs gradually increases and $<\Delta V_g>$ changes quicker than 1/D. The nearest-neighbor-level spacing in GQDs can be controlled by the size of Coulomb diamond. Level statistics becomes non-Poissonian for smaller GQDs and well described by Gaussian unitary distribution and follows chaotic Dirac (neutrino) billiards.

6.4.2 Optical Properties

6.4.2.1 Photoluminescence Unlike 2D graphene, finite-sized 0D GQDs are known to have a bandgap due to quantum confinement phenomenon. In the previous studies (6, 132) for polyaromatic molecules, the PL has been attributed to the electronic transition from the LUMO to the highest occupied molecular orbital (HOMO) and shows wavelength shifts toward lower energy with the increase of conjugation

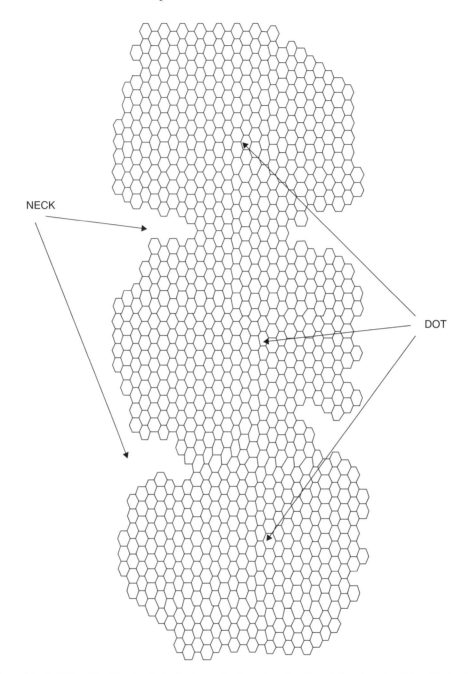

NECK

DOT

Figure 6.34 A simple model of a GNR with a disordered edge leading to the formation of necks and dots along the ribbon. Coulomb blockade occurs when the charge moves from dot to dot. (*Source*: Reprinted from Reference (118), with permission).

length. The optoelectronic properties of carbon materials are crucially determined by the π–π^* transition of the sp^2 bonding existing in the bandgap of σ–σ^* transition of the sp^3 bonding (133–135). Because PL is closely correlated with bandgap, GQDs also exhibit strong PL and thus PL behavior is highly influenced by the size, shape, and fraction of sp^2 domains of the GQDs. The strong peak generally observed at about 275 nm ($\approx$4.5 eV) in PL spectrum for GQDs is known as a π–π^* transition indicating the presence of sp^2 covalent bonding. However, the origin of strong visible light emission from GQDs ascribed to emissive surface traps and/or the edge states is still unclear.

In absorption spectra of GQDs, they show typically strong optical absorption in the UV region at 270–390 nm (34, 49) and a long extended tail into the visible range. Some studies reported the dependence (81)/independence (34, 49, 55, 56, 58) of absorption spectra on the size of GQDs; the peak positions of absorption are more closely dependent on chemical environments associated with dopants, edge structure, and defects/surface states rather than physical size of GQDs. It would be widely accepted that absorption peaks in visible range might come from the electronic transitions of n–π^* (136) in the surface states occurred with their energy levels between π and π^* states of C=C (49, 73) induced by various functional groups of C–OH, C=O, O–C=O, and C=N during the preparation of GQDs.

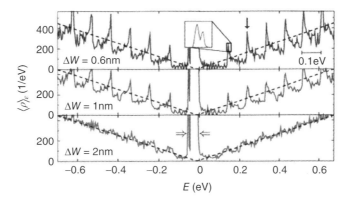

Figure 6.35 Ensemble-averaged density of states $<\rho(E)>_\xi$ of GQDs with increasing edge roughness, see different values for ΔW in the subfigures. The size of all devices is equal, $d = 20$ nm and their width $W = 16$ nm. Dashed lines indicate the averaged linear DOS for Dirac billiards. The inset shows the $K-K'$ splitting of 12 meV. (*Source*: Reprinted from Reference (121), with permission).

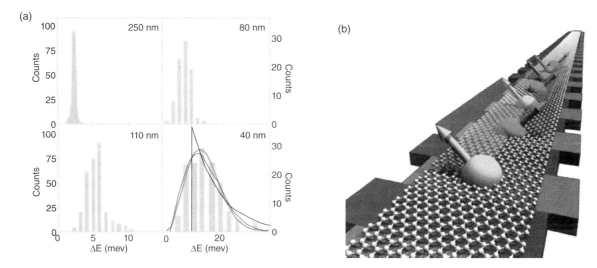

Figure 6.36 (a) Histograms of the nearest-neighbor-level spacing in GQDs of different diameters. Non-Poissonian becomes dominant for smaller GQDs (the gray, thin gray, and black curves are the best fits for the Gaussian unitary, Poisson, and Gaussian orthogonal ensembles, respectively). (b) Illustration of many spin qubits. White bars and gray bars represent QDs and barrier regions, respectively. Large coupling of the same color band denoting different spin qubits to each other via Klein tunneling. (*Source*: Reprinted from Reference (29), with permission).

Analogous to absorption spectra, PL of the GQDs is also dependent on both the physical size and shape, and chemical environment and surface functionality such as pH and solvent. The GQDs were synthesized by chemical oxidation and cutting of μm pitch-based CFs and show relatively a narrow size distribution of 1–4 nm (81). The GQDs synthesized at 120, 100, and 80 °C show blue (~400 nm), green (~500 nm), and yellow (~575 nm) emission as shown in Figure 6.37. This result suggests that different color emission originate from the difference in size. Figure 6.37c represents the relationship between the size and the energy gap which decreased from 3.90 to 2.89 eV with the increase of the size and correlates with the quantum confinement effect at lower size QDs (135). Furthermore, GQDs (3–5 nm average diameter) were prepared and possess strong green fluorescence (Fig. 6.37c) (56, 138). On the other hand, PL emission of the GQD prepared by MAH did not show size dependence (49).

Eda et al. reported the dependence of the HOMO–LUMO gap on the size of the GQDs (139, 140). Based on DFT, energy gap of $\pi-\pi^*$ decreases from 7 eV to 2 eV with the increase of the number (N) of aromatic rings from $N = 1$ to 40, as shown in Figure 6.38. Chien and coworkers (141) suggest the PL emission mechanism in GO and rGO, where the GO was synthesized using the modified Hummers method and photothermally reduced by exposing GO samples to a Xenon flash in ambient conditions. GO usually contains a large fraction of sp³ hybridized carbon atoms bonded with epoxy, hydroxyl, carboxyl, and carbonyl oxygen-functional groups, which makes it an electrically insulator. As shown in Figure 6.39a, two prominent and broad peaks, IP1 peak centered at approximately 600 nm and a small IP2 emission peak centered at roughly 470 nm, were observed. As reduction exposure times increase, the IP1 gradually decreases, but the IP2 emission increases.

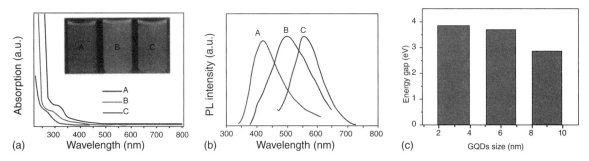

Figure 6.37   The GQDs synthesized by chemical oxidation and cutting of μm pitch-based carbon fibers (CF) at 120, 100, and 80 °C show curve A (~400 nm), curve B (~500 nm), and curve A (~575 nm) emission. (*Source*: Reprinted from Reference (81), with permission).

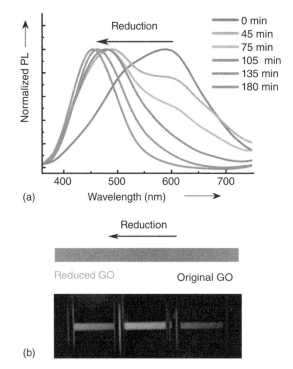

Figure 6.38   (a) Normalized PL spectra of the GO suspensions after different exposure times (0–180 min) to photothermal reduction treatment. (b) Photographs of tunable PL emission from GO at reduction times of 0 min (yellow–red), 75 min (green), and 180 min (blue). (*Source*: Reprinted from Reference (139), with permission).

As depicted in Figure 6.39b, the variation of relative intensity ratios of PL emission from two different types of electronically excited states is suggested by the result of changing the heterogeneous electronic structures of GO and rGO with variable sp^2 and sp^3 hybridizations through reduction. The number of disorder-induced defect states of GO within the π–π* gap is decreased after deoxygenation. An increased number of cluster-like states are created from the newly formed small and isolated sp^2 domains and increases the blue fluorescence at shorter wavelengths.

In addition to the size of the graphene fragments, PH and solvents in some GQD solutions influence the PL intensity of GQDs. GQDs with diameter of 5–13 nm were prepared (34) by successive oxidation and hydrothermal treatment of microscale GSs through forming a line of epoxy groups on carbon lattice and cooperative alignment inducing a rupture of the underlying C–C bonds (142) (Fig. 6.40 and b). In chemically derived GQDs, PL shows a strong blue emission centered at 430 nm (Fig. 6.40c) and may originate from free zigzag sites with a carbene-like triplet ground state described as $\sigma^1\pi^1$. This assumption is further supported by the observed pH-dependent PL. Under acidic conditions, the protonation of free zigzag sites of the GQDs by forming a reversible complex between the zigzag sites and H$^+$ makes the emissive triple carbene state broken and inactive in PL. However, under alkaline conditions, deprotonation restores the free zigzag sites and become active in PL. If pH is switched repeatedly between 13 and 1, the PL intensity varies reversibly (Fig. 6.40d). From the typical electronic transitions of triple carbenes at zigzag sites as suggested in Figure 6.40e, the two electronic transitions of 320 nm (3.86 eV) and 257 nm (4.82 eV) observed in the PLE spectra can be regarded as transitions from the σ and π orbitals (HOMOs) to the LUMO (143).

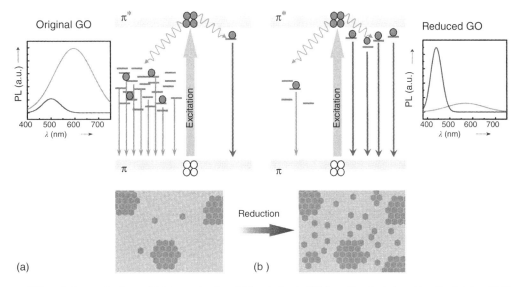

Figure 6.39 Proposed PL emission mechanisms of (a) the predominant IP1 emission in GO from disorder-induced localized states. (b) The predominant IP2 emission in rGO from confined cluster states. (*Source*: Reprinted from Reference (141), with permission).

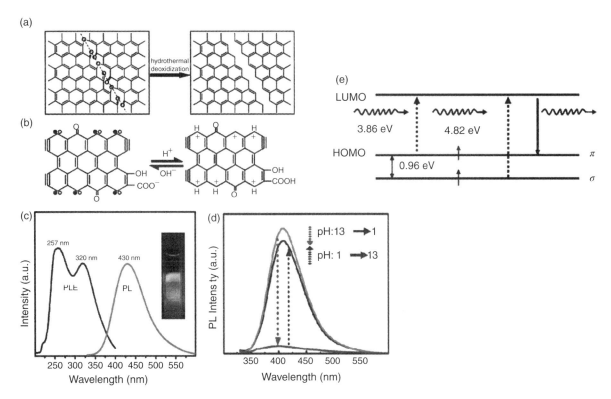

Figure 6.40 (a) Hydrothermal cutting model for oxidized GSs into GQDs. (b) Models of the GQDs in acidic (right) and alkali (left) media. Both models can be converted reversibly depending on pH. The pairing of σ and π^* localized electrons at carbene-like zigzag sites and the presence of triple bonds at the carbyne-like armchair sites are represented. (c) Blue emission from GQDs. (d) Dependence of PL on pH. (e) Typical electronic transitions of triple carbenes at zigzag sites observed in the optical spectra. (*Source*: Reprinted from References (34, 142, 143), with permission).

Zhu et al. also reported that a green emission around 515 nm of GQDs decreases in a solution of high or low pH but remains constant in a solution of pH 4–8. Moreover, in a solution of pH over 12, the PL peak shows blueshift and the full-width at half maximum (FWHM) becomes narrower. This pH-dependent PL behavior is distinct from that observed in the previous report where blue-luminescent GQDs is quenched under acidic conditions (pH = 1). Accordingly, more experimental and theoretical approaches are necessary for clear understanding of the dissimilar PL in different pHs considering different surface defects or synthetic methods of GQDs besides the emissive free zigzag sites (34).

Sun et al. passivated C-dots with organic and polymer agents of diamine-terminated oligomeric poly(ethylene glycol) $H_2NCH_2(CH_2CH_2O)_nCH_2CH_2CH_2NH_2$ (average n ~35, PEG_{1500N}) and poly(propionylethyleneimine-co-ethyleneimine) (PPEI-EI). The passivated C-dots with organic moieties attached to the surface are strongly photoluminescent both in the solution-like suspension and in the solid state and are expected to be aqueous compatible and conjugation with bioactive molecules (144). Similarly, Shen et al. demonstrated a facile hydrazine hydrate reduction of GO with surface-passivated by PEG_{1500N} method for the fabrication of GQDs to solve dependence of PL emission on pH. As-prepared GQDs-PEG showed excellent luminescence properties with PL QY of 28% at 360 nm, which was two times higher than that of pure GQDs. In an aqueous solution of neutral pH, GQDs-PEG showed bright, while the intensity of the PL peaks decreased by only about 25% under both acidic and alkaline conditions. This is because surface-passivation agent PEG would lead to a higher quantum confinement of emissive energy trapped to the GQDs surface so that the GQDs-PEG exhibit stronger PL, irrespective of degree of pH. GQDs with ~5.3 nm in diameter prepared by solvothermal route adopting DMF as a solvent show good solubility in water and most polar organic solvents due to the presence of functional groups such as –OH, epoxy/ether, CQO, and –CO–NR$_2$ originated from decomposition of DMF (138). This aqueous GQD solution recently exhibits the green emission at around 515 nm resulted from surface effect (37, 56) with the QY of 11.4% and those in the solvents of tetrahydrofuran (THF), acetone, and DMF show the blueshift of PL up to 475 nm. This solvent effect could be induced by solvent attachment or different emissive traps on the surface of GQDs (145).

Besides the factors described earlier in this section (6.4.2.1), Gokus et al. discovered that the PL intensity and thickness of graphene were directly correlated to flakes thickness (146). At first, multilayered graphene (MLG) was prepared by microcleavage of graphite on SiO_2(100 nm)/Si substrate. Through exposure to oxygen/argon (1:2) RF plasma (0.04 mbar, 10 W) for increasing time (1–6 s), single-layered graphene (SLG) was successfully obtained by layer-by-layer etching (147). Any luminescent emissions from bi- and MLGs were not observed, but the only SLG showed very pronounced PL peak with a single broad FWHM at ~700 nm (1.77 nm) as shown in Figure 6.41. These results reveal that emission from the topmost layer was quenched by subjacent untreated layers.

Recently Tetsuka et al. succeeded in optical tuning of PL emission center of GQDs by using edge-terminated GQDs with a primary amine through the effective orbital resonance of amine moieties with graphene core (148). In order to form a primary amine and alcohols by nucleophilic substitution, ammonia reacts with epoxy groups of the OGS at 70–150 °C (149, 150). Thus, self-limited sp^2 domains are extracted by ring-opening of the epoxide and simultaneously the primary amine gets bonded with a graphene edge. Initial concentration and reaction temperature determine the degree of amine functionalization and higher temperature (>120 °C) induces the dissociation of primary amine. When aqueous suspensions of amine-functionalized GQDs (af-GQDs) are irradiated by UV lamp with λ = 365 nm, emission peaks vary from 420 to 535 nm and show a redshift with the increase of amine functionalization. The emission wavelength of af-GQDs is rather dependent on the quantity of functionalized amine groups than their shape and size (Fig. 6.42). The PLQY was measured as much as 29%–19%, which decreased with the increase of functionalized amine quantity. These values are higher than ever expected in GQDs (typically <10%), which result from the smaller number of nonradiative centers like –COOH and C–O groups in af-GQDs (144).

From the XPS and FTIR, the additional functional groups can be assumed to have at least amino group, that is, –NH$_3$/–CH$_3$ pair as shown in Figure 6.43, and change the PL emission of af-GQDs from green to blue after functionalization. The adsorption of primary amine on the edge of af-GQDs will drastically change the electronic structure; degenerate HOMO levels in H-terminated graphene are lifted to higher energy due to the strong bonding with –NH$_2$ groups and consequently the bandgap get reduced. The narrowing of the bandgap decreases with the increase of the amounts of amine groups and rather independent of the size of the GQDs.

It is also very interesting that excitation-dependent or excitation-independent PL emission of GQDs in the visible range showed as the wavelength of excitation is varied. On excitation of hydrothermally prepared GQDs at λ = 320 nm (Fig. 6.44a) with a Stokes shift of 110 nm (0.99 eV), a strong PL emission was detected at 430 nm. When the excitation wavelength is increased from 320 nm to 420 nm, the PL peak shows redshift and rapid decrease of intensity, with the strongest peak intensity excited at the absorption of 320 nm. (Fig. 6.44b) (34). The blue emission is assumed to come from free zigzag edge sites with a carbene-like triplet ground state of $\sigma^1\pi^1$. Li et al. prepared oxygen-rich N-GQDs by a simple electrochemical approach. Under irradiation by a 365 nm lamp, the N-GQDs emitted intense blue luminescence (Fig. 6.44c inset) with blueshift by ca. 50 nm with respect to that of N-free GQDs of similar size (56), and like GQDs, N-GQDs show excitation-dependent PL emission (Fig. 6.44d). The PL blueshift is attributed to relatively strong electron affinity of N atoms in the N-GQDs and these excitation-dependent PL for the as-prepared N-GQDs are related to the N-doping-induced modulation of the chemical and electronic characteristics of the GQDs. Besides, the GQDs synthesized by other methods show the excitation-dependent PL and this phenomenon is explained through surface emissive traps (151) and electronic conjugation structures (72).

On the other hand, on excitation at the excitation wavelength of 360 nm, the PL spectrum for electrochemically synthesized GQDs shows the strongest peak at 540 nm (Fig. 6.45a). When excited by light with wavelengths from 340 to 410 nm, the intensity of the PL increased to the maximum and then decreased, but the position of fluorescent emission peak remained unshifted, that is, excitation independent. Optimized local structure for yellow PL of GQDs is investigated by theoretical calculation using density function theory (B3LYP/6–311++G(d,p)) and illustrated in Figure 6.45b. Absorption wavelength and the PL wavelength are calculated as ~390 nm and ~532 nm, respectively, which give evidence that the strong yellow luminescence from the GQDs was emitted by a high concentration of modified hydrazide groups chemically linked to the GQDs after absorption through the graphene π–π* and n–π* transitions. Similarly, Zhou et al. also demonstrated that the PL emission from GQDs prepared by ultrasonic reaction is not changed while the excitation wavelength is changed from 240 nm to 340 nm. This

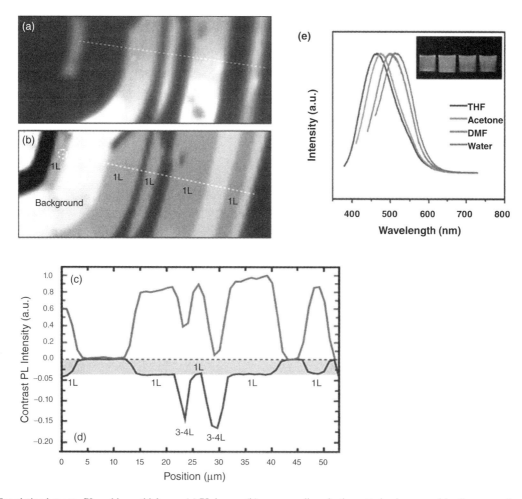

Figure 6.41 Correlation between PL and layer thickness. (a) PL image, (b) corresponding elastic scattering image, and (c, d) cross sections taken along the dashed lines in (a, b). PL is only observed from treated SLG, marked 1L. (e) Effect of solvents on the fluorescence of GQDs (at 375 nm excitation) (Inset: photograph of the four dispersions taken under UV light). (*Source*: Reprinted from Reference (146), with permission).

phenomenon is explained by the formation of ultranarrow GNRs with ~0.4 nm width during ultrasonic preparation (86). Similarly, Liu et al. observed that PL emission of GQD-1200 synthesized by unsubstituted HBC as carbon source show excitation-dependent behavior. With the variations of excitation wavelength from 320 to 480 nm, PL peak shifts from 430 to 560 nm and has a maximum green intensity at 510 nm with QY of 3.8% (46).

The GQDs – with 15 nm in size and 0.5–2.0 nm height, prepared by tuning the carbonization degree of citric acid and heating at 200 °C also showed excitation-independent blue emission at ~460 nm. This result reveals that both the size and the surface state of sp^2 clusters contained in GQDs would be uniform. On the other hand, PL from GO obtained through long annealing over 2 h depended substantially on the excitation wavelength and shifts from 450 to 542 nm with the increase of excitation wavelength from 300 to 480 nm (Fig. 6.46a and b) (47).

Tang et al. (49) observed that the GQD solution prepared by MAH method exhibits deep UV (DUV) emission at 303 nm (4.10 eV) and the broad PL peak shifts from 473 to 519 nm as the variations of the excitation from 375 to 450 nm (Fig. 6.47). The functional groups (C–OH, C=O, C–O–C, C–H) existed on the surface of GQDs form the surface energy state levels between π and π^* states of C=C, which plays as emissive traps. But it is worthwhile to note that the PL peak is not dependent on the size of GQDs, which increase from 1.65 to 21 nm with the increase of heating time from 1 to 9 min. This size-independent feature of PL is presumed to be related to the self-passivated layer of the GQDs. The QY is about 7%–11% which is slightly higher than ever reported value (4%–10%) of C-dots (34, 144).

Recently, Kim et al. reported that anomalous visible light luminescence from GQDs showed strong dependence on the size as well as shape (151). They prepared the GQDs by hydrothermal and chemical tailoring (34) of GSs obtained by thermal deoxidization of GO sheets made from natural graphite powder by a modified Hummers method. The size of GQDs (diameter; *d*) was controlled by filtering and dialysis processes and average sizes (d_a) of GQDs are accurately estimated from the HRTEM images for six kinds of samples with *d* from 5 to 35 nm at each *d* and indicated in the parentheses at the bottom of Figure 6.48. Using excitation wavelength of 325 nm, PL spectra for GQDs of 5–35 nm average sizes in deionized (DI) water are illustrated in Figure 6.49a and corresponding PL peak positions and shapes are largely varied with

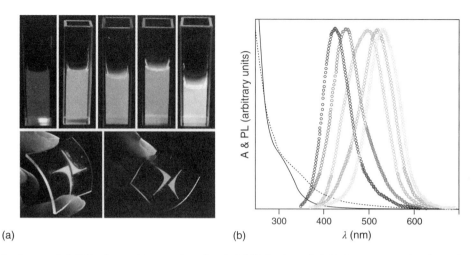

Figure 6.42 (a) Emission images of af-GQDs dispersed in water (upper) and af-GQD@polymer hybrids (bottom) under irradiation from a 365 nm UV lamp. (b) PL (color lines) and selected UV–Vis absorption spectra (black lines) of af-GQDs. (*Source*: Reprinted from Reference (148), with permission).

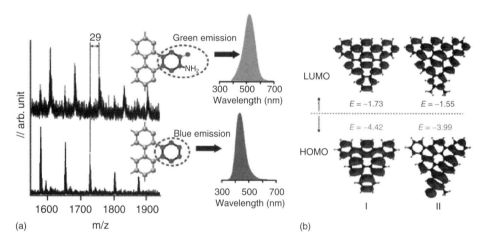

Figure 6.43 (a) Comparison of MALDI-TOF mass spectra for af-GQDs prepared at 120 °C (bottom) and 90 °C (upper). (Inset: Possible edge structures and their PL excited at 350 nm). (b) Schematic illustrations of structures used for theoretical calculations with different functional groups. I: hydrogen-terminated edge structure, II: $-NH_2/-CH_3$ pair bonded to edge (unit: eV). (*Source*: Reprinted from Reference (148), with permission).

the size of the GQDs. A sharp PL peak at 365 nm was known to stem from DI water (152). The inset of Figure 6.49a shows different colors of luminescence depending on the size of GQDs of 12–22 nm. For the GQDs of $d_a = 5$ and 12 nm, a single broad asymmetry PL peak centered at ~410 nm is observed. Especially for the GQDs of $d_a = 17$ and 22 nm, the PL spectra are resolved into two PL bands centered around 415 and 505 nm, irrespective of the excitation wavelength, possibly resulting from combination of different size/shape GQDs. For the GQDs of $d_a = 27$ and 35 nm, the PL peak shape and position restore to a single broad peak at around 410 nm and the intensity of PL peak for $d_a = 35$ nm becomes very weak.

Figure 6.49b summarizes excitation wavelength-dependent PL peak shifts for various-size GQDs as the excitation wavelength is changed from 300 to 470 nm. For all the excitation wavelengths, the size-dependent shift of PL peak is observed, except for 470 nm. For the GQDs of d_a up to 17 nm, all peak positions shift toward lower energy and the corresponding shift is maximum at 380 nm excitation up to 0.2 eV. But for the GQDs of $d_a > 17$ nm, the PL peak energy increases with increment of d_a. This unusual size dependence of luminescence is closely related to the quantum conversion efficiency (QCE); consistent with the QCE for $d_a = 17$ nm, however, the QCE no longer holds for $d_a > ~17$ nm. A similar result was found in the reconstructed Si nanoparticles (153, 154). It has been reported that after the hydrothermal treatment in the fabrication processes of GQDs, several oxygen-functional groups (OFGs) remaining at the edges of GQDs possibly modified the PL behavior of GQDs (34, 81). With the size-dependent shape/edge-state variations of GQDs for d_a of 5 to 35 nm, the peak energy of the absorption spectra monotonically decreases, whereas that of the visible PL spectra unusually shows nonmonotonic behavior, having a minimum at $d_a = ~17$ nm. The anomalous size-dependent PL behavior can be attributed to the characteristic variation feature of GQDs, that

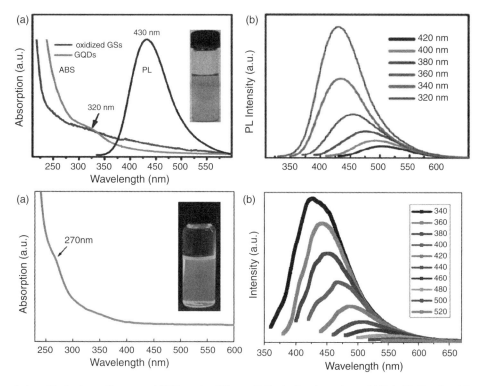

Figure 6.44 (a) PL emission of hydrothermally prepared GQDs at $\lambda = 320$ nm. (b) Excitation-dependent redshift in PL (Reprinted from Reference (34), with permission). (c) The blue emission prepared oxygen-rich, nitrogen-doped GQDs (N-GQDs) by a simple electrochemical approach. (d) Excitation-dependent PL emission of N-GQDs. (*Source*: Reprinted from Reference (56), with permission).

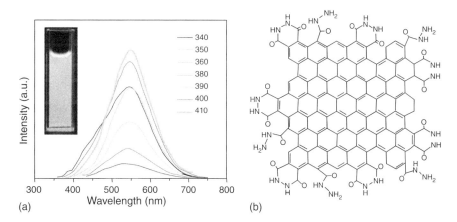

Figure 6.45 PL spectra of the GQD aqueous solution at different excitation wavelengths (a). Schematic illustration of the edge-modified GQD structure determined by theoretical calculation (b). (*Source*: Reprinted from Reference (86), with permission).

is, the circular-to-polygonal-shape and corresponding edge-state variations of GQDs at $d_a = \sim 17$ nm as the GQD size increases, as identified by HRTEM in Figure 6.48a.

Another interesting result is that the center of PL of the GQDs could be shifted by charge transfer between functional groups and GQDs. Through a two-step cutting process from GOs, GQDs functionalized with amine groups and with 1–3 layers thick and less than 5 nm in diameter were synthesized (155). For functionalized GQDs, the center of PL was redshifted as much as about 30 nm compared to the unfunctionalized GQDs. In addition, the protonation or deprotonation of the functional groups caused the change of PH and in turn led to the shift of PL emissions of the GQDs and the amine-functionalized GQDs (Fig. 6.50). All these shifts in PL emission were attributed to charge transfers between the functional groups and GQDs, which could explain the tuning of the bandgap of the GQDs.

This bandgap tuning in the GQDs through functional process was well supported by theoretical calculations from DFT. In DFT simulation, alkyl amine functional group is assumed to be an amine group ($-NH_2$) at the sp^2 clusters composed of 13 aromatic rings (Fig. 6.51a). The

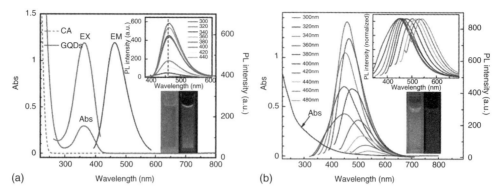

Figure 6.46 (a) UV–Vis absorption of CA and the GQDs, and PL spectra of the GQDs. (Inset: Emission spectra of the GQDs at different excitation wave wavelength (upper); (lower) Photographs of the solution of GQDs taken under visible light (left) and under 365 nm UV light (right)). (b) UV-Vis is absorption spectra and emission spectra of the GO at different excitation wave wavelength (upper). (Inset: (upper) Normalized emission spectra; (lower) Photographs of the solution of GQDs taken under visible light (left) and under 365 nm UV light (right). (*Source*: Reprinted from Reference (47) with permission).

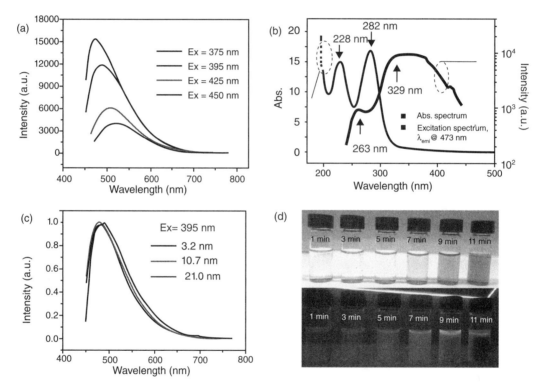

Figure 6.47 (a) The PL spectra of the GQDs excited at different excitation wavelengths. (b) The absorbance and excitation spectra of the GQD solution. (c) The normalized PL spectra of the GQDs with various sizes. (d) The GQD solutions irradiated by visible light (top) and 365 nm UV lamp (bottom). The GQDs were synthesized with 11.1 wt% glucose solution for 5 min (a and b), 5–9 min (c), 1–11 min (d) microwave heating at 595 W. (*Source*: Reprinted from Reference (49), with permission).

calculated bandgap of 13 aromatic rings is 2.508 eV which is close to that of 2.480 eV, experimentally measured for the GQDs. Because electron transfer from $-NH_2$ to the ring is prevalent, the increase of number of attached NH_2 will decrease the bandgap due to the increase of electron density in GQD–$(NH_2)_n$. The bandgap of the GQD–$(NH_2)_n$ is saturated as shown in Figure 6.51b.

For further understanding of the origin of PL from GQDs, Lingam et al. (156) systematically investigated the correlation of PL with the surface functionalization and edge configuration of GQDs by using different synthesis and annealing temperature. GQDs were synthesized with a mean diameter of ≈4 nm by modified solvothermal method and the other carbon materials; GNRs and carbon nano-onions (CNOs) were prepared for comparison. First, two different GQD samples were chemically prepared using DMF (S1-DMF) and NaOH solution (S2-NaOH) separately. As shown in Figure 6.52, both S1-DMF and S2-NaOH show nearly identical PL, exhibiting broad green emission at ~510 nm

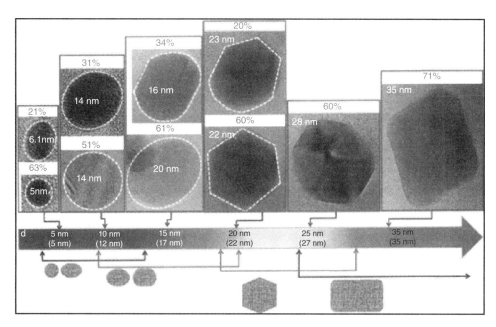

Figure 6.48   HRTEM images of GQDs corresponding populations (*p*) with the increase of average size of GQDs (The dotted line represents the region of a GQD and *p* is defined as the ratio of number of GQDs). Average sizes (d_a) of GQDs estimated from the HRTEM images at each *d* are indicated in the parentheses at the bottom. The connected arrows indicate the range of the average size in which GQDs with particular major shapes are found. (*Source*: Reprinted from the Reference (151), with permission).

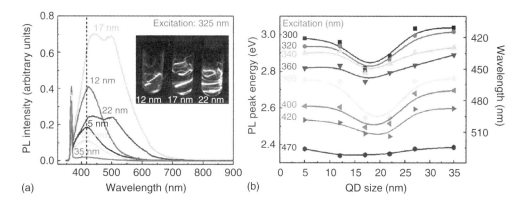

Figure 6.49   (a) Size-dependent PL spectra excited at 325 nm UV for GQDs of 5–35 nm average sizes in DI water. (Inset: Different colors of luminescence from GQDs with different average sizes of 12, 17, and 22 nm. (b) Excitation-dependent PL peak shifts on the excitation wavelength from 300 to 470 nm. (*Source*: Reprinted from the Reference (151), with permission).

and similar shifts (upshift between 400 and 460 nm and negligible shift after 460 nm) with the increment of excitation wavelength. This result reveals that the existence of different functional groups (–OF in S1-NaOH or noncovalently bonded DMF in S1-DMF) or other surface emissive defects did not have any influence on PL.

Subsequently, S1-DMF is annealed at 250, 350, and 450 °C and then the intensity of PL is observed to be reduced with increasing annealing temperature. The presence of disordered edge in S1-DMF before/after annealing is clearly identified through HRTEM; therefore, the rearrangement of edge carbons is possibly believed to decrease PL intensity. S1-DMF is further annealed in H_2 environment and large quenching of PL is observed, which implies a correlation between passivation of the edge states and PL intensity. For comparison, PL spectrum for GNR with width of 25 nm and length 200 nm is obtained and shown along with the S1-DMF annealed in H_2. Despite of the different surface morphology and size, the fact that both show the same decreasing behavior in PL intensity strongly indicates the relationship between the disordered edge state of GQDs and the origin of PL. For confirmation of this correlation, PL spectrum of cage-like structured CNOs with no free-edge carbons is taken and compared. As expected, no PL signal was observed. From the results, it is evident that the disordered edge state is responsible for the occurrence of visible PL in GQDs (Fig. 6.53).

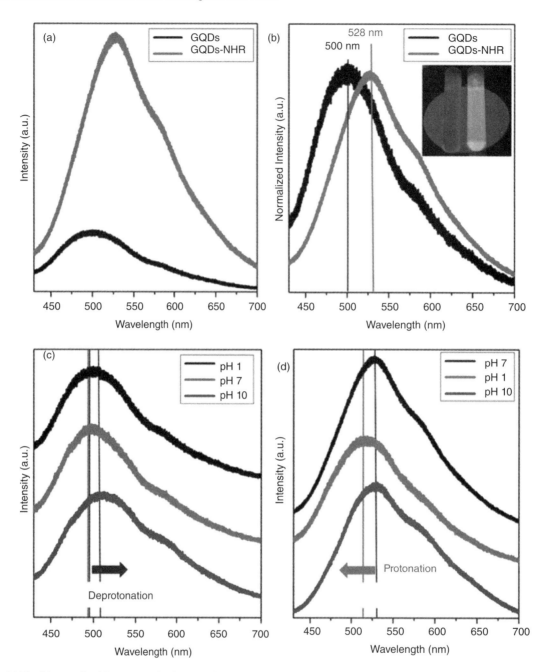

Figure 6.50 (a) PL, (b) normalized PL spectra of GQDs and GQDs-NHR in water (Inset: photograph of GQDs (left) and GQDs-NHR (right) taken under 355 nm laser excitation), (c) pH-dependent PL spectra of GQDs, and (d) GQDs-NHR. (*Source*: Reprinted from the Reference (155), with permission).

6.4.2.2 Downconversion (DC)/Upconversion (UC) Luminescence In biomedical imaging and medical diagnosis, UC fluorescence has attracted much recent attention for its many promising applications (28, 86, 156). Zhuo et al. prepared GQDs with diameters of 3–5 nm by oxidation of graphene using concentrated H_2SO_4 (10 ml) and HNO_3 (430 ml) at room temperature through subsequent ultrasonication and filtering by 0.22 μm microporous membrane and further dialyzing (3500 Da) (86). As shown in Figure 6.54, the intensity of PL spectra for the GQDs centered at ∼407 nm was invariant as excitation wavelength was varied from 240 to 340 nm, that is, excitation-independent feature which is different from the previous excitation-dependent PL behaviors (9, 34) is evident. Also, the strong downconversion (DC) PL was observed because the intensity of the PL spectra was gradually increased with increasing the excitation wavelength from 240 to 340 nm. According to the Yang group, the calculation of energy and bandgaps for GNRs using a first-principle many-electron Green's function approach within GW approximation showed self-energy corrections in the range of 0.5–3.0 eV for GNRs of 2.4–0.4 nm width. By comparison with the aforementioned calculation, the PL spectrum occurring at 407 nm (3.05 eV) corresponds to the GNR with ∼0.4 nm (3.0 eV). In preparation,

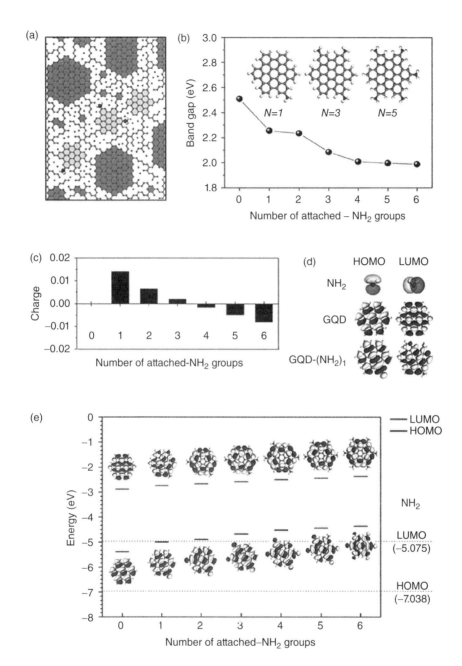

Figure 6.51 (a) Schematics of isolated sp^2 clusters with – NH_2 within the sp^3 carbon matrix. (b) Bandgap change of GQD-$(NH_2)_n$ as function of the number of attached – NH_2 groups (inset images are optimized configuration of GQD-$(NH_2)_n$), (gray, C; white, H, and dark, N_{atom}). (c) The evolution of average Mulliken charge for amino groups as a function of the number of attached –NH_2 groups. (d) HOMO and LUMO isosurface of NH_2, GQD, GQD-$(NH_2)_1$. (e) HOMO and LUMO energy levels of GQDs–$(NH_2)_n$. The dotted lines denote the HOMO and LUMO energy level of NH_2 is represented by the dotted lines. (*Source*: Reprinted from the Reference (155), with permission).

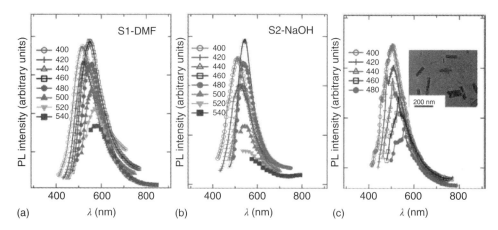

Figure 6.52   PL spectra of GQDs as a function of excitation wavelength from 400 to 540 nm (with 20 nm increments) prepared by solvothermal method in (a) DMF and (b) NaOH and (c) for GNRs suspended in DMF. The inset in (c) is the TEM image of the GNRs sample. (*Source*: Reprinted from the Reference (156), with permission).

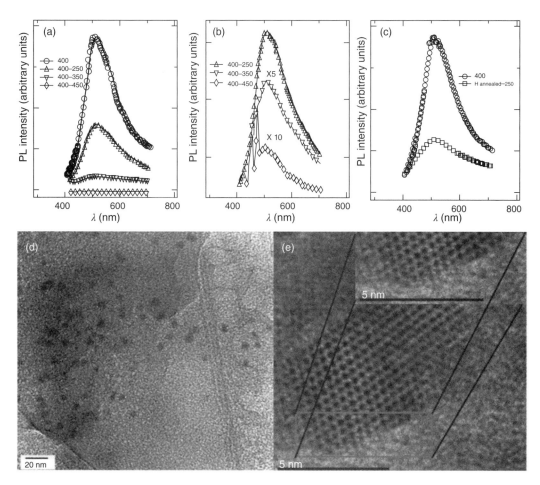

Figure 6.53   (a) PL emission with 400 nm excitation as a function of annealing temperature. (b) The magnified view of the PL spectra of annealed samples. (c) The change in the intensity of PL emission with 400 nm excitation upon annealing in H_2. The TEM and HRTEM images of GQDs post-annealed at 450 °C are shown in (d) and (e), respectively. (e) The inset shows the presence of disordered edges in GQDs. (*Source*: Reprinted from the Reference (156), with permission).

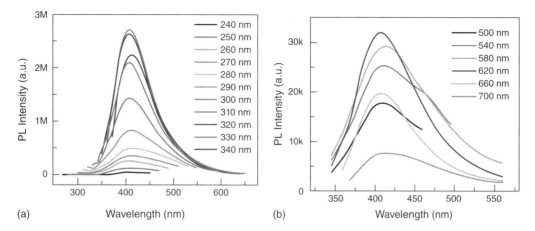

Figure 6.54 (a) PL spectra of the GQDs at different excitation wavelengths. (b) Upconverted PL spectra of the GQDs at different excitation wavelengths. (*Source*: Reprinted from the Reference (86), with permission).

the energy of ultrasound wave could cut GSs into GQDs and such tiny GNRs might originate from the protrudent edge of GQDs (157). Furthermore, UCPL of ~407 nm is clearly observed with the variation of excitation wavelength from 500 to 700 nm and also shows an excitation-independent PL behavior. In general, UCPL can be explained by the multiphoton process carbon QDs (20, 21). UC luminescence efficiency of GQDs is determined by measuring the absolute QY and QY is equal to 3.4%.

Zhu et al. also prepared GQDs (58) by two-step process: first, GO was dispersed with DMF and GO/DMF solutions were transferred in autoclave heated at 200 °C for 8 h and brown transparent GQDs were obtained. And then the GQDs were purified utilizing two gradient elutions of A (methylene chloride – MeOH) and B (water – H_2O). Under the B phase elution, the Batch 3 GQDs were obtained. When the GQDs (Batch 3) were excited from 600 to 900 nm, the PL spectrum showed a prominent peak at 427 nm and a shoulder peak at 516 nm. As the excitation wavelength increases, the peak at 427 disappears while the intensity of the peak at 516 nm greatly increases (DC) and shows excitation wavelength-dependent PL behaviors as shown in Figure 6.55. When the GQDs (Batch 3) were excited from 600 to 900 nm, intensity of PL spectrum increases and shows similarly excitation-dependent UC behavior, which can be attributed to multiphoton active process.

6.5 APPLICATIONS

6.5.1 Light-Emitting Diodes (LEDs)

Many studies about synthesis of GQDs and optical properties have been reported, whereas, relatively, there are only few reports on fabrication of LED devices utilizing GQDs as emissive layers. Mixtures of poly(2-methoxy-5-(2-ethylhexyloxy)-1,4-phenylenevinylene) (MEH-PPV) and methylene blue functionalized GQDs (MB-GQDs) with different concentrations were employed as light-emitting layer and tested in organic LEDs (OLEDs) (158). As shown in Figure 6.56a, the turn-on voltage of the device decreased from ~6 V for the pure MEH-PPV sample (trace i) to ~4 V for MEH-PPV with 1% MB-GQDs (trace ii). At a higher concentration of 3% MB-GQDs, charge trapping as well as a shortening effect occurred, possibly due to agglomeration (trace iv). The maximum light-emission intensity increased with the addition of MB-GQDs to MEH-PPV, directly reflecting the enhanced internal quantum efficiency. In mechanism, the MB-GQDs dispersed in the MEH-PPV afforded additional electrical transport paths that improved charge injection and consequently the carrier density, therefore requiring a lower turn-on voltage. But here the MB-GQDs layer is just used as electron transport layer in the OLEDs. Tang et al. showed another interesting application of GQDs as an excellent light converter, by coating a layer of GQDs onto a commercially available LED which is emitting blue light centered at 410 nm (49). After coating of GQDs, the intensity of blue emission became weak and a new broad green emission appeared at ~510 nm (Fig. 6.56b). The measured Commission International d'Eclairage (CIE) chromaticity coordinates of the LED (0.282, 0.373) reveal the white emission, which was understood as due to the mixing of the 410 nm emission from the original LED and a broad emission at 510 nm given by GQDs under the 410 nm excitation.

In the aforementioned references, GQDs are not used as emission layers for generating LED devices. Son et al. first reported that the synthesis of ZnO–GQD hybrid core–shell-type QDs and the potential use of hybrid QDs as emission layer through bandgap engineering in QD LEDs and the ZnO–graphene QD LED showed a white electroluminescence (101). Figure 6.57a shows the PL spectra of ZnO–graphene quasi core–shell QDs and ZnO QD reference sample. Compared to pure ZnO NPs, the PL spectra of the ZnO–graphene quasi QDs, excited with He–Cd ($\lambda = 325$ nm), show reduction of near UV peak and concurrent appearance of two additional peaks at the wavelength of 406 and 436 nm. The chemical bonding of GQDs to the ZnO NPs causes UV PL emission quenched up to about 71% along with the occurrence of new G–O$_{epoxy}$-related peaks. The peak around 379 nm (3.27 eV) generally appears for ZnO QDs and corresponds to the bandgap of ZnO and

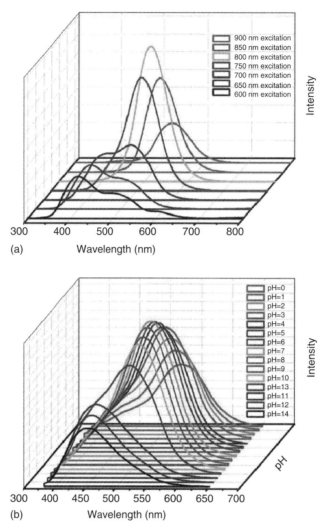

Figure 6.55   The applicable properties of the GQDs (Batch 3). (a) The upconversion PL properties of GQDs. (b) The pH-dependent PL behavior of GQDs (*Source*: Reprinted from the Reference (58), with permission).

clearly attributed to the interband transition from conduction band to valence band. In order to understand the new peaks in PL spectra, DOSs and projected DOSs of ZnO–graphene are calculated using simple model with DFT as implemented in Gaussian package (159). According to DFT result, LUMO level of pristine graphene is split into three LUMO, LUMO + 1, and LUMO + 2 levels by bridged O epoxy bonding with graphene. Among three LUMO levels, LUMO and LUMO + 2 levels consist of both s and p orbital except LUMO + 2 containing only p orbital. By selection rule, only the transition conserving the angular momentum ($\Delta l = \pm 1$) can be possible. Therefore, two graphene-related peaks in the PL of ZnO–graphene quasi QDs, as shown in schematic energy band diagram of Figure 6.57, can be ascribed to the decay of exciton from LUMO and LUMO + 2 molecular orbitals of G–O$_{epoxy}$ to the ZnO valence band mainly consisted of O2p orbital considering the aforementioned results of DFT.

Figure 6.58a represents the schematic energy band diagram of QD LED electronic structure. As shown in Figure 6.58a, two prominent electroluminescence (EL) peaks are observed at 2.89 eV and 2.74 eV. Electrons are injected from the Cs$_2$CO$_3$/Al electrode (Φ = 4.3 eV) to the outer graphene layer ((Φ = 4.4 eV) rather than to the conduction band of inner ZnO (Φ = 4.19 eV) QDs, as the energy level of the former is lower by 0.21 eV than that of the latter. This result indicates that the wavelength of excitonic emission of ZnO QD can be changed by the conjugation with graphene, opening up of a new way to tune the center of the EL of a metal-oxide semiconductor (Fig.6.58a). The modulated wavelength corresponds to the difference of energy between modified LUMO levels and valence band of ZnO. Figure 6.58b shows a photograph of the light emission from the QD LED device at applied biases of 11, 13, 15, and 17 V, where pixels with the size of 5 × 5 mm showed bluish-white color due to the combination of a series of blue (425 nm) and yellow (606 nm) emissions. At 15 V applied

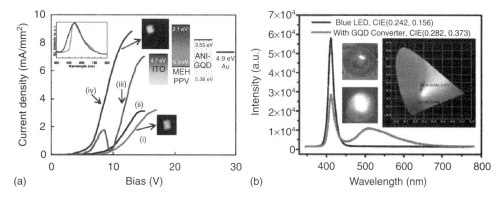

Figure 6.56 (a) Measured current density of MEH-PPV with MB-GQDs (a) 0%, (b) 0.5%, (c) 1%, and (d) 3% with the variations of the applied voltage. The inset plots the electroluminescence spectrum of the MEH-PPV (gray) and MEH-PPV/MB-GQDs (1%) (black). (Inset: the band diagram of the MEH-PPV/MB-GQDs and the brightness of MEH-PPV LED and MEH-PPV/MB-GQDs (1%) LED) (Reprinted from Reference (156), with permission). (b) PL spectra of the blue LED with and without GQDs coating. Left inset: photographs of the GQD-coated LED without (top) and with applied voltage (bottom). (*Source*: Reprinted from Reference (49), with permission).

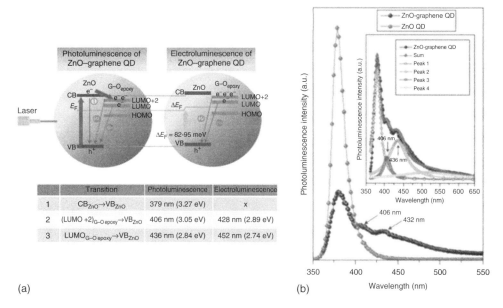

Figure 6.57 (a) PL and EL transition scheme for ZnO–graphene quasi quantum dots. Transitions 1, 2, and 3 correspond to electron transitions from the conduction band (CB) of ZnO, LUMO + 2, and LUMO levels induced by G–O $_{epoxy}$ to the valence band (VB) of ZnO, respectively. (b) Photoluminescence spectra for ZnO quantum dots and ZnO–graphene quantum dots showing quenching of photoluminescence emission. Inset: quenched photoluminescence fit with subpeaks centered at 379, 406, 436, and 550 nm, respectively. (*Source*: Reprinted from Reference (101), with permission).

bias and with optimal CIE coordinates (0.23, 0.20), the maximum brightness reached $\sim$798 cd/m^2, with an external quantum efficiency of 0.04%.

6.5.2 Photovoltaics

The colloidal GQDs show promise of applications in low-cost and high-performance photovoltaic (PV) devices owing to tunable bandgaps, high carrier mobility, high surface-to-volume ratio, and solution processability (160, 161). Li et al. applied colloidal GQDs with green luminescence as electron-acceptor materials in conjugated polymer, poly(3-hexylthiophene) (P3HT)-based thin film solar cells (Fig. 6.28a) (97). GQDs were prepared by electrochemical method and were monodispersed with a diameter of $\sim$3–5 nm and 1–2 nm height. The energy level of GQDs was approximately determined by the electrochemical method (162, 163) as 4.2–4.4 eV. In UV–Vis absorption spectrum, an absorption peak of GQD in water solution was found at 320 nm. On excitation using 365 nm, a wide and prominent PL peak was observed

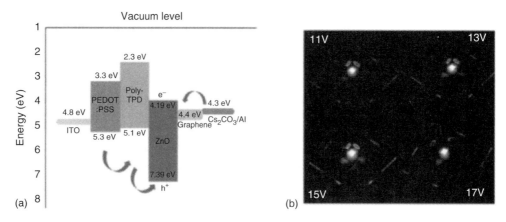

Figure 6.58 (a) Energy band diagram of the ZnO–graphene QD LED device. The pathways of holes and electrons are indicated by arrows. (b) Photograph of light emission at 11, 13, 15, and 17 V applied bias (pixel size of 5 × 5 mm). (*Source*: Reprinted from Reference (101), with permission).

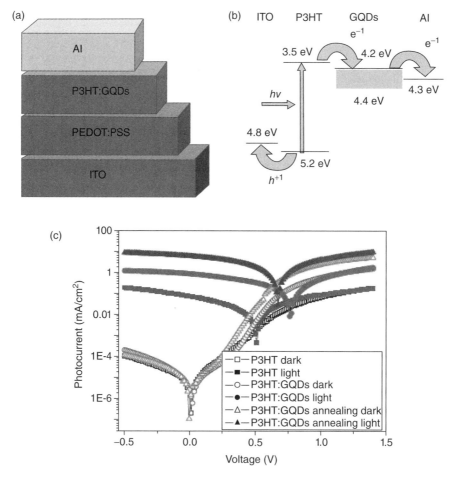

Figure 6.59 (a) Structure and (b) energy band diagram of the ITO/PEDOT:PSS/P3HT:GQDs/Al device. (c) *J–V* characteristic curves for the ITO/PEDOT:PSS/P3HT(P3HT:GQDs)/Al after annealing at 140 °C for 10 min. (*Source*: Reprinted from Reference (97), with permission).

at 473 nm and shows an excitation-dependent behavior. It is very interesting that the electrochemically synthesized GQDs radiate green luminescence with the excitation of 365 nm lamp which is different from the blue luminescence of hydrothermally prepared GQDs (34).

For comparison, P3HT and P3HT:GQDs were used as photoabsorption layer in organic PV cells, as shown in Figure 6.59a. In the case of P3HT:GQDs, the performance of organic photovoltaics (OPV) was evaluated before and after annealing at 140 °C for 10 min. In pure P3HT device, the V_{oc}, I_{sc} (mA/cm^2), fill factor (FF), and photoconversion efficiency (PCE) were 0.52 eV, 0.078, 0.23, and 0.008%, respectively. On the other hand, in the P3HT:GQDs composite device, they increased up to 0.77 eV, 0.92, 0.27, and 0.19% before annealing and to 0.67 eV, 6.33, 0.3 m, and 12.8% after annealing, respectively. The increase of V_{oc} for the pure P3HT device from 0.52 eV – corresponding to the difference of work function between Al (4.3 eV) and ITO (4.8 eV) – to 0.77 eV for the P3HT:GQDs device can be explained by the difference of about 0.8 eV between the HOMO level for P3HT of 5.2 eV (162) and the LUMO levels of 4.2–4.4 eV (165–167). In addition, GQDs can separate the excitons generated at the interface and provide charge transporting pathway to electrode, which leads to the increase of I_{sc} whereas a low I_{sc} in the P3HT device is due to poor mobility and no existence of the interface. Built-in potential at the interface of P3HT and GQDs can increase charge separation and accelerate the charge transport to electrode through high mobility media of GQDs. PCE for the P3HT:GQDs composite device was measured as 1.28%, which is better than 0.88% of OPV fabricated by solution-processable functionalized graphene (168). This result perceptibly reveals that GQDs could be a promising alternative to the widely adopted fullerene derivative (6,6)-phenyl-C$_{61}$-butyric acid methyl ester (PCBM) and C70 electron acceptors in conjunction with electron donor materials such as P3HT in organic bulk heterojunction (BHJ) solar cells.

Moreover, Gupta and his coworkers also showed that solar cell conversion efficiency of OPV used with P3HT-conjugated polymer, as electron donors, could be significantly improved when P3HT was blended with GQD acceptor rather than GS (161). The GSs were prepared by thermal oxidation of GO sheets with subsequent hydrothermal process and dialysis and further GQDs were obtained from GSs by a hydrothermal treatment (34). For the investigation of the performance of OPV, aniline-functionalized GQDs and GSs were blended with P3HT-conjugated polymer in BHJ solar cells. P3HT/ANI-GQD (GS)-based heterojunction organic solar cells were fabricated by spin-casting process on ITO/PEDOT:PSS as a bottom electrode and LiF/Al and were deposited as the top electrode. Relative ratios of ANI-GQD to P3HT were varied in the range of 0.5, 1, 3, and 5 wt%. The LUMO and HOMO levels of GQDs, determined by CV (169), were −3.55 and −5.38 eV, respectively. In P3HT/ANI-GQDs hybrid solar cell, V_{oc} was measured as 0.58–0.61 eV, which is almost coincident with the difference of 0.65 eV between the LUMO level of GQDs and work function (4.2 eV) of Al. This result is different from the V_{oc} determined by the difference between the LUMO of GQDs and HOMO level of P3HT (138). As shown in the band diagram (inset of Fig. 6.60), generated excitons in P3HT by light absorption were separated into electrons and holes by this built-in potential at the interface of P3HT and ANI-GQDs (GSs) and electrons were quickly transported through GQDs and GSs with high mobility. J_{sc} for ANI-GQDs and ANI-GSs are largely dependent on the amounts of GQDs and GSs contents and FF for ANI-GQDs is much higher (0.53) as compared to ~0.35 for ANI-GSs (170, 171). The maximum PCE = 1.14 %, V_{oc} = 0.61 V, J_{sc} = 3.51 mA/cm^2, and FF = 0.53 were obtained for 1 wt% ANI-GQDs, whereas the composite device of 10 wt% ANI-GSs showed PCE = 0.65, V_{oc} = 0.88 V, J_{sc} = 2.65 mA/cm^2, and FF = 0.28. This result indicates that GQDs can improve more efficiently the OPV characteristics than GSs. In solar cells, both charge separation and carrier transport are crucial factors to determine the PCE of the solar cells. In order to investigate the influence of surface morphology on FF, surface nanostructure was examined by an AFM. In the bottom of Figure 6.60, the AFM images over ANI-GSs represent large domains (about 100–200 nm diameters), much larger than the diffusion length of excitons (10 nm), but those for ANI-GQDs show very uniform structures. These smooth nanostructures at the interface of ANI-GQDs will easily transport generated excitons by reducing the resistance at the interface of donor/acceptor and consequently enhance the performance of the solar cells.

Yan et al. also reported that GQDs have high absorbance with the maximum at 591 nm and extend its edge around 900 nm (77). A molar extinction coefficient (ε_m) in dichloromethane was 1.0×10^5 M/cm, nearly an order of magnitude larger than that of metal complexes commonly used in dye-sensitized solar cells (172, 173). With a tight-binding model, the energy levels of HOMO and LUMO of GQDs are calculated as 5.3 and 3.8 eV below the vacuum level and relative to conduction band levels (4.3 eV) of TiO$_2$ and reduction potential (4.9 eV) of I$_3^-$/I$^-$, which make GQDs possible to use as a sensitizer in dye-sensitized solar cells made of nanocrystalline TiO$_2$ particles. Upon photoexcitation, GQDs inject an electron to TiO$_2$ and then get regenerated by accepting an electron from I$^-$. Since QY of PL and UCPL efficiency of surface-passivated GQDs-PEG are ~28% and 10 times higher than pure GDQs respectively, Shen et al. studied the photon-to-electron conversion capability of GQDs and GQDs-PEG for PEC cells (60). The GQDs-PEG photoelectrode generated a relatively small photocurrent under an 808 nm NIR laser and had higher photon-to-electron conversion capability at 365 nm UV irradiation compared with the pure GQDs. Therefore, GQDs-PEG provides a cost-effective dopant material for solar energy conversion and other PEC cells applications from the ultraviolet (UV) to the near infrared.

Furthermore, recently Son et al. presented a UV PV conversion device using core–shell-type ZnO QDs conjugated with graphene QDs. As mentioned before (174), ZnO–graphene quasi core–shell QD structures in which the inner ZnO QDs are covered with graphene QDs have been synthesized via a simple solution process method. The outer shell covering the inner ZnO QDs was identified as a single graphene (Fig. 6.61) by HRTEM and graphene was chemically connected with ZnO through Zn–O$_{epoxy}$–C (graphene) chemical bonds induced on graphene layer from acid treatment. Using spin-coating solution process, organic UV PV device with ITO/PEDOT:PSS/ZnO–graphene/Cs$_2$CO$_3$/Al was fabricated and ZnO–graphene core–shell QDs was inserted as the active layer. Figure 6.62a illustrates UV–Vis absorption spectra for ZnO QDs and ZnO–graphene core–shell-type QDs. Both ZnO and the ZnO–graphene QDs exhibited absorption in the UV region around 360 nm, but an additional absorption peak for ZnO–graphene QDs was observed at 400 nm. From the schematic energy-level diagram as shown in the inset, the further enhanced absorption at longer wavelengths around 400 nm can be ascribed to the electron transition from valence band of ZnO inner QD to LUMO level of the outer graphene shell. For clear understanding of this absorption mechanism, time-resolved PL (TRPL) for ZnO QDs and ZnO–graphene QDs was obtained and compared. Figure 6.62b represents the temporal evolution of the PL intensities at 375 and 383 nm obtained from hybrid ZnO–graphene QDs and ZnO QDs reference sample, respectively, which were measured using a time-correlated single photon counting setup. The calculated average lifetimes of ZnO–graphene core–shell QDs are 0.13 ns and 0.165 ns for the 375 and 383 nm UV emissions, respectively. It is a quite distinguishing feature that the PL lifetimes for hybrid ZnO–graphene QDs decrease to, as short as, of order of one-tenth of the 1.86 ns and 1.83 ns for the same UV emissions of reference ZnO QDs; this is indicative of the existence

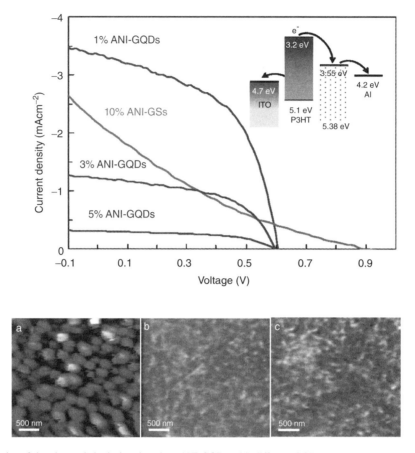

Figure 6.60 *J–V* characteristics of the photovoltaic devices based on ANI-GQDs with different GQDs content and ANI-GS (under optimized condition) annealed at 160 °C for 10 min (AM1.5G 100 mW); AFM images of (a) P3HT/ANI-GSs, (b) P3HT/ANI-GQDs, and (c) MEH-PPV/MB-GQDs. (*Source*: Reprinted from Reference (158), with permission).

of an additional high efficiency relaxation channel. This also implies that the dwelling time for excited electron in the conduction band in ZnO–graphene QDs became shorter due to quick charge transfer to LUMO level of outer graphene. This transition causes the reduction in PL intensity of UV peak, that is, it was quenched, and simultaneously another electron transition from LUMO of graphene to valence band of inner ZnO QDs occurred. The amounts of static quenching behavior were measured as large as 71% near the UV emission peak for the ZnO–graphene core–shell QDs. A faster charge separation through direct electron transfer was well supported by DFT, calculating the DOSs for graphene with or without epoxy functional group. For ZnO–graphene hybrid QDs-based UV photovoltaic cells, some characteristic values of the observed saturation current density (J_{sc}), open circuit voltage (V_{oc}), fill factor (FF), and power conversion efficiency (η) were estimated as 196.4 A/cm^2, 0.99 V, 0.24, and 2.33%, respectively (Fig. 6.63a and b). Chemically bonded GQDs with ZnO QDs represent as a good example of nanostructured material which significantly improves UV photoconversion efficiency through a fast photoinduced charge separation by absorption from valence band of inner ZnO QDs to induced LUMO level of graphene where ZnO QDs and graphene play as donor and acceptor, respectively.

6.5.3 Lasing Emission

As one of the superior luminescent materials, lasing emission phenomenon of GQD ($\leq$3 layers) was investigated and compared with that of carbon nanodots (C-dots) ($\sim$10 layers) obtained from the same functionalization process. Recently, Zhu et al. reported that lasing emission from GQDs was realized by using titanium dioxide nanoparticles as light scatterers (175). In the experiment, to utilize C-dots as a gain media of lasing emission, it was reported that either transition from $\pi \rightarrow \pi^*$ or $n \rightarrow \pi$ results from sp^2 clusters of graphene and functional group such as carboxylic group (C=O) and hydroxyl group (C–OH) attaching to edges of C-dots. GQD and C-dots were prepared by the laser irradiation using a pulsed (6 ns, 10 Hz) Nd:YAG laser ($\lambda = 1064$ nm) with a peak power density of 65 MW/cm^2 for 30 min onto GSs and graphite powder (16, 177). Emission efficiency of GQD and C-dots was compared by measuring the corresponding net optical gain, $G(\lambda)$, through the variable stripe length (VSL) method given by VSL equation as follows (178):

$$I_{tot}(I, l) = I_{sp}(l)[(\exp(G(l)L_s) - -1])/G(l) \tag{6.3}$$

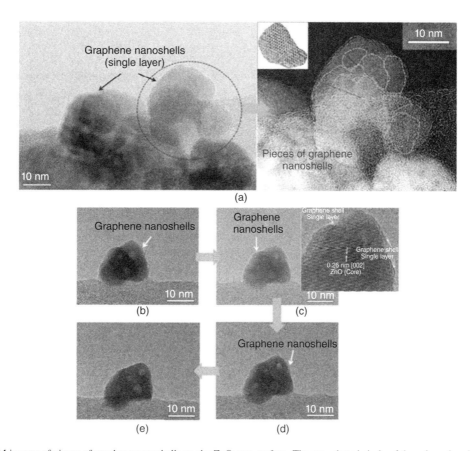

Figure 6.61 (a) TEM images of pieces of graphene nanoshells on the ZnO core surface. The gray-dotted circle of the enlarged scale image is an image of ZnO–graphene quasi core–shell QDs. The image in the white square is a magnified image of one ZnO QD covered by a monolayer graphene looking like pieces of graphene nanoshell on the ZnO core surface. (b–e) Continuous plane-view, bright-field TEM images of the ZnO–graphene quasi core–shell QDs. The ZnO petal grows along the (002) direction. (*Source*: Reprinted from Reference (174), with permission).

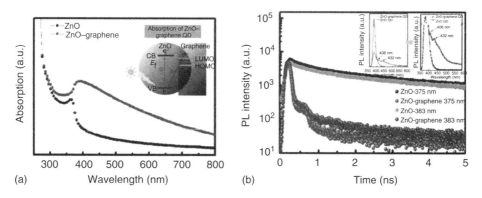

Figure 6.62 (a) UV–Vis absorption spectra for the ZnO QDs and ZnO–graphene QDs. (b) Time-resolved PL of the ZnO–graphene QDs at 375 and 383 nm. The inset of (b) is the PL spectra for ZnO QDs and ZnO–graphene QDs showing quenching of PL emission. (*Source*: Reprinted from Reference (174), with permission).

where L_s is the length of pump stripe (width of strip is fixed at 0.05 cm), I_{tot} is the total intensity measured, and I_{sp} is the spontaneous emission intensity. From the plot of the peak output intensity versus L_s, G can be calculated and then the deduced values of G versus P are plotted in the inset (top left-hand corner) of Figure 6.64. The optical gain per pump power of GQD obtained from the slope of $\partial G/\partial P$ is approximately about 9.1 cm/kW, which is five times higher than that of C-dots. Such a large optical gain of GQD was ascribed to the amount of relative ratio 1D/1G of sp^3 (1D) to sp^2 (1G). 1D/1G correspond to 0.583 for C-dots and 0.868 for GQD and indicates that C-dots have lower amounts of sp^3 carbon structures than GQDs to chemically bond with functional groups.

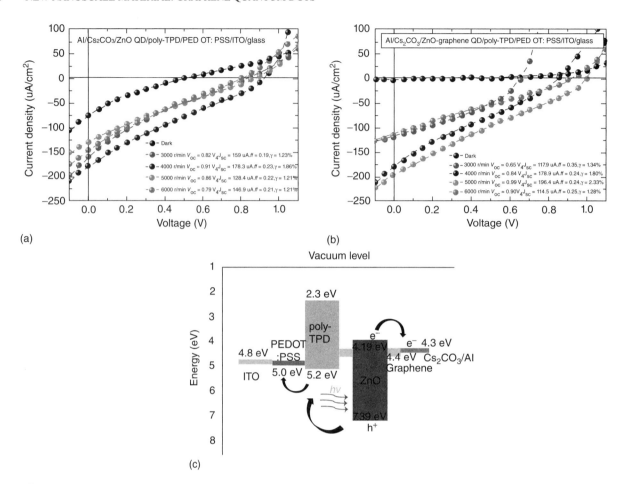

Figure 6.63   Current–voltage curves under an illumination wavelength of 365 nm for the (a) Al/Cs$_2$CO$_3$/ZnO QDs/poly-TPD/PEDOT:PSS/ITO/glass and (b) Al/Cs$_2$CO$_3$/ZnO–graphene QDs/poly-TPD/PEDOT:PSS/ITO/glass devices with active layers fabricated with various speed of spin coating. (c) Band diagram of four-layer structure UV PV cell. The pathways of the holes and the electrons are indicated by the arrows. (*Source*: Reprinted from Reference (174), with permission).

As shown in the inset of Figure 6.65, the laser cavity was made up of coaxial cylinder quartz tube (inner diameter of 1 mm and outer diameter of 2 mm as well as the length varying from 0.5 to 5 mm) and was terminated by two quartz plates. The cavity was filled with the mixture of GQD and TiO$_2$ NPs dispersed in ethanol solution. The role of TiO$_2$ NPs suppresses the lasing modes through increasing the amounts of scattering loss, but instead of that GQD amplifies the scattered light. The average size width of TiO$_2$ NPs was about 100 nm in diameter and is suitable for supporting coherent optical feedback (179, 180). Figure 6.65a shows the variations of pump threshold (P_{th}) and FWHM of the emission spectra as a function of the concentration of TiO$_2$ NPs. With the increment of concentration of TiO$_2$ NPs, P_{th} also increases as expected. However, the FWHM of the emission spectra decreases from 50 to 25 nm at the concentration of 0–0.2 mg/ml and then increases back to 50 nm for higher concentrations up to 1 mg/ml. Hence, the concentration of TiO$_2$ NPs is optimized at 0.2 mg/ml with the P_{th} of about 30–40 kW/cm^2. At the cavity length of 4 mm, the emission spectra are presented at different P_{th} of 30, 50, and 60 kW/cm^2. Lasing emission peaks with line width less than 0.3 nm are observed between 350 and 375 nm wavelength and superimposed onto a broad spontaneous emission spectra for the P_{th} larger than a threshold (slightly larger than 40 kW/cm^2) – a kink as shown in the inset.

The external power efficiency defined at $2 \times P_{th}$ was estimated as large as ~15% which is analogous to that of semiconductor lasers (~40%) and believed to be compatible with solid-state laser.

6.5.4   Bio Applications

Even though conventional CdSe or CdS and their core–shell structures QDs have been widely exploited in bioimaging, biosensor, and biomedicine, their incorporation of a heavy metal component of Cd atom, classifying as one of six elements banned by the Restriction on Hazardous Substances (RoHS), has been the obstacle for bio-related research as well as for use in living life. Since Sun et al. (181) first introduced the feasibility of GO as an efficient nanocarrier for drug delivery, biomedical applications of graphene have been widely explored from drug/gene delivery, biological sensing and imaging, antibacterial materials, to biocompatible scaffold for cell culture. Besides intrinsic biocompatibility (low cytotoxicity) of graphene and facile biological/chemical functionalization of GO, prominent optical properties with

(a)

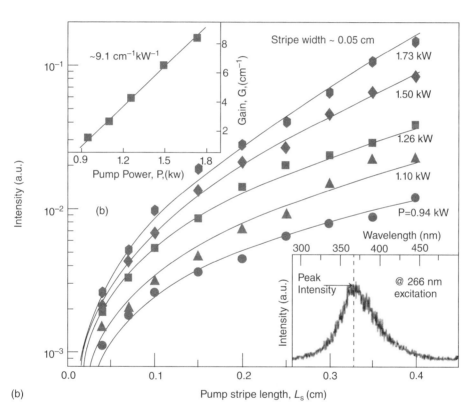

(b)

Figure 6.64 (a) Photographs of ethanol, graphene sheets, GQDs, and C-dots dispersed inside ethanol taken under the ambient illumination. (b) Output intensity versus length of pump stripe, LS, of the GQDs dispersed inside ethanol under excitation at different pump power, P, by a 266 nm laser source. (Inset: the plot of optical gain, G, versus the pump power, P (top left) and a PL spectrum of the GQDs (bottom right)). (*Source*: Reprinted from Reference (175), with permission).

tunable bandgap and strong excitation-dependent PL or UCPL make GQDs emerging materials in biomedicine with focus on drug delivery, cancer therapy, and biological imaging (182).

6.5.4.1 Cytotoxicity Cell viability on GQD was tested through mixing of GQD up to 400 μg to 150 μL of culture medium (10^4 cells) and was proved not to be weak through the MTT assay as shown in (Fig. 6.66a) (138). Using green fluorescent GQDs, bioimaging experiments were performed by the confocal fluorescence microscope as shown in Figure 6.66. Observation of the bright green area inside the cells

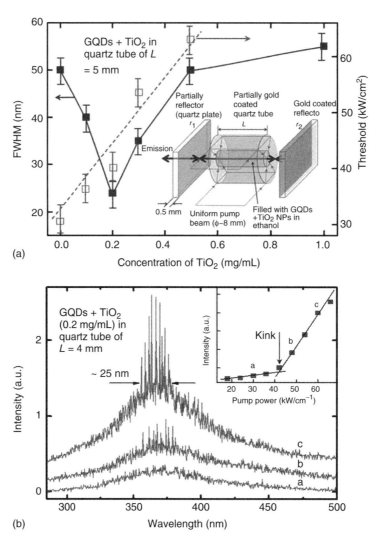

Figure 6.65   (a) Plot of pump threshold, P_{th}, and FWHM of emission spectra versus the concentration of TiO_2 NPs (Inset: Schematic diagram of the proposed laser cavity). (b) Lasing spectra of the proposed laser cavity with cavity length $L = 4$ mm under 266 nm laser excitation at RT. A kink represents the threshold of the laser. (*Source*: Reprinted from Reference (175), with permission).

indicates translocation of GQDs through the cell's membrane (405 nm excitation). The excitation-dependent PL behavior of the GQDs gives rise to impressive visible results. When the excitation light was changed to 488 nm, a green–yellow color was observed (Fig. 6.66b and c).

Peng et al. investigated that when two different human breast cancer cells MDA-MB-231 and T47D were treated with GQDs derived from CFs, no toxicity was observed in both cases at low doses of GQDS less than ~50 μg/mL compared to untreated control cells indicating least cytotoxicity of GQDs. With the increase of exposure time, cell proliferation at low doses was not influenced either (81).

6.5.4.2   *Biosensing/Bioimaging/Biolabeling*   GQDs with an average height of 0.5 nm, derived from chemical exfoliation of XC-72 carbon black, were treated with MCF-7 cell. The MCF-7 cell exhibits sharp green luminescence images, excited at 488 nm, taken by the confocal laser scanning microscopy as shown in Figure 6.67a. From the enlarged view of the section (Fig. 6.67b), it is clear that the nucleus (80) as well as the cell membrane and the cytoplasm (156, 183) could be simultaneously labeled for the first time using luminescent carbon nanomaterials.

As an another example of application of GQDs as a fluorescent probe in cell imaging, Pan et al. reported that HeLa cells cultured in an appropriate culture medium containing the GQDs displayed enhanced fluorescence around the nucleus indicating the penetration of GQDs into the cells and still emissive. The GQD-labeled HeLa cell shows unceasing PL for 10 min and can be proposed as potential substituents for dyes in the future (80).

In another *in vitro* experiment, Peng et al. (81) found that the images of human breast cancer cell T47D treated with green GQDs visualize the clear phase contrast of between T47D, nucleus stained with blue color 4',6-diamidino-2-phenylindole (DAPI), and green GQDs around each nucleus. These images suggest that GQDs are very useful for high-contrast biocell imaging (Fig. 6.68). Up to now, many studies have disclosed the biocompatibility and nontoxicity of GQDs; however, further experiments should be provided for *in vitro* and *in vivo* applications and for actual applications in the human body.

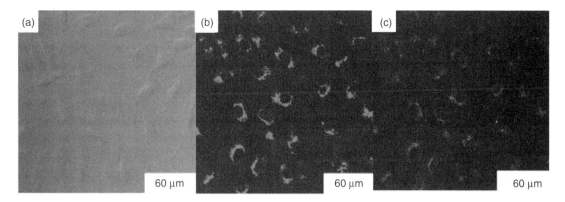

Figure 6.66 Confocal cellular imaging of GQDs; under bright field (a), 405 nm (b), and 488 nm (c). (*Source*: Reprinted from Reference (138), with permission).

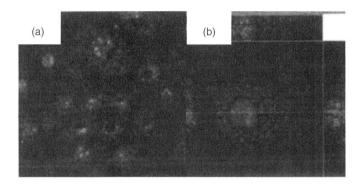

Figure 6.67 Section analysis of human breast cancer MCF-7 cells labeled with GQDs derived from carbon black. (a) Fluorescent image. (b) Enlarged fluorescent image. (*Source*: Reprinted from Reference (80), with permission).

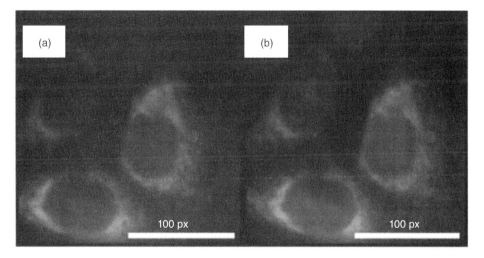

Figure 6.68 (a) Gathered green GQDs (white color) surrounding the nucleus. (b) The overlay high-contrast image of a nucleolus stained with blue (inner black part) DAPI and green (outer white part) GQDs staining. (*Source*: Reprinted from Reference (81), with permission).

6.5.4.3 Electrochemical Biosensor GQD-based platform technology for the fabrication of electrochemical biosensors was proposed to specifically recognize complementary DNA and the binding of aptamer to its target protein (184). First, the probe DNA (ssDNA) was immobilized on the pyrolytic graphite electrode surface modified by GQDs (34) and then the peak current at the modified electrode decreased, because electron transfer between electroactive $(Fe(CN)_6)^{3-/4-}$ and the electrode was inhibited by the electrostatic repulsion. When the target molecules of DNA (ssDNA-2)) or target protein such as thrombin is present in the solution, the probe ssDNA binds to the target materials to form complementary double-stranded DNA or to act as an aptamer of protein (185), which disturbs the immobilization of the probe ssDNA. The peak current of the target $(Fe(CN)_6)^{3-/4-}$ at the modified electrode increases with the concentration of the target materials and the detection limit was around 100 nm, which is lower than the previous reports (186, 187). Therefore, various electrochemical biosensors can be fabricated with this proposed platform with higher sensitivity and selectivity.

6.5.5   Photocatalyst

As aforementioned in synthesis of GQDs, the as-prepared GQDs by ultrasonic method exhibit an excitation-independent DCPL and UCPL behavior. Based upon the UC luminescence properties of GQDs, a visible light-activated rutile TiO_2/GQD and anatase TiO_2/GQD complex photocatalytic systems were fabricated (Fig. 6.69), where rutile and anatase TiO_2 nanoparticles (50 mg) were added to GQD solution (5 ml) (86). The photocatalytic degradation reaction was monitored by adding different photocatalysts with the relative concentration of 50 ml aqueous MB solution. Under the visible light irradiation (λ >420 nm), the photodecomposition efficiency was around 97% in 60 min for rutile/GQD complexes, whereas that for anatase/GQDs, it decreased only 31%. Considering the photocatalytic ability of anatase is higher than rutile, it is very surprising that photodegradation activity of rutile/GQD is three times superior to anatase. This is because the wavelength of UCPL peak in GQDs was around 407 nm (3.905 eV). This energy is larger than the bandgap of rutile TiO_2 3.0 eV (424 nm) but smaller than that of anatase 3.2 eV (388 nm). So UC light can create effectively e^-/h^+ pairs and this leads to much efficient photocatalytic behavior of rutile/GQD complex in visible light absorption. The aforementioned process is schematically illustrated in Figure 6.70.

6.5.6   Fuel Cells and Li-Ion Batteries

Qu et al. studied the feasibility of replacing high-cost Pt catalyst in fuel cells with N-doped CNT and N-graphenes to improve the resistance to time-dependent degradation and CO poisoning (98). They found that N-containing graphene (N-graphene) could act as an effective metal-free oxygen reduction reaction (ORR) catalyst associated with air-saturated 0.1 M KOH electrolyte fuel cells. As shown in Figure 6.71a, the N-graphene electrode exhibits a one-step, four-electron pathway, whereas C-graphene (pure graphene) exhibits a two-step, two-electron process, with onset at −0.45 and −0.7 eV and the steady-state current density is ~three times higher than that of the Pt/C electrode. For the testing of crossover to various fuel molecules, the Pt/C electrode on the addition of 2% methanol shows 40% reduction in chronoamperometric response, but the N-graphene electrode keeps the same value even after the addition of H_2 and glucose to methanol. On exposure to 10% CO gas, the Pt/C electrode was quickly poisoned, but the N-graphene was insensitive to CO (Fig. 6.71c). Through continuous potential cycling test, the N-graphene electrode does not show any decrease in current after 2×10^5 continuous cycles between −1.0 and 0 V.

The N-graphene shows long-term stability, tolerance to crossover, and poison effect better than the conventional Pt/C electrode and similar to that of vertically aligned N-containing CNTs (VA-NCNT). This result could be attributed to the electronegativity of N-related positive charge on adjacent C atoms, which readily attract electrons from the anode for facilitating the ORR (99). Similarly, Li et al (72). demonstrated that the graphene-supported N-GQDs with oxygen-rich functional groups (N-GQD/graphene assemblies) have superior electrocatalytic ability for the ORR to commercial Pt/C catalysts (20 wt% Pt on carbon blacks in O-saturated KOH). As shown in Figure 6.72a, the ORR onset potential was at ~−0.16 with a reduction peak at ~−0.27 V, which was comparable with those of N-GQDs and N-CNTs (100). The stable ORR current-potential curve, without characteristic peak related to CH_3OH reduction/oxidation (188), was found (Fig. 6.72b) and exhibited the remarkable tolerance to crossover effect, which was not observed in pure graphene and N-free graphene. From linear-sweep voltammetry (Fig. 6.72c), current density shows increase with the increase of the rotation rate of rotating disk electrode (RDE). The number of transferred electron (n) calculated by using the Koutecky–Levich equation (189) was derived to 3.6–4.4, suggesting the four-electron pathway for ORR.

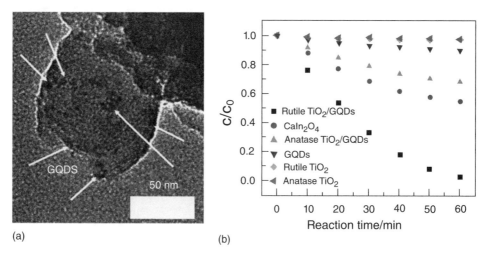

(a)   (b)

Figure 6.69   (a) TEM image of rutile TiO_2/GQD nanocomposites: The GQDs are marked with white arrows. (b) Relationship between MB concentration and reaction time for different catalysts: rutile TiO_2/GQDs, $CaIn_2O_4$, anatase TiO_2/GQDs, GQDs, rutile TiO_2 NPs, and anatase TiO_2 NPs. (*Source*: Reprinted from Reference (86), with permission).

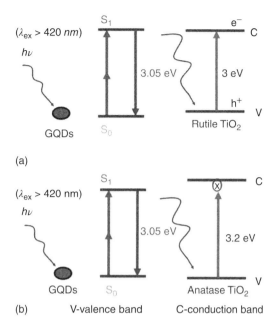

Figure 6.70 Schematic of photocatalytic process for (a) rutile TiO_2/GQD and (b) anatase TiO_2/GQD under visible light ($\lambda > 420\,nm$) irradiation. (*Source*: Reprinted from Reference (86), with permission).

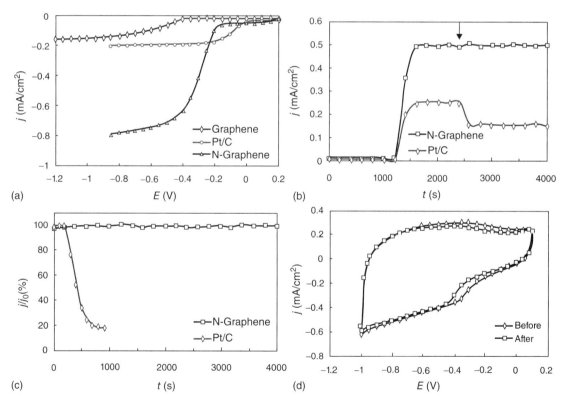

Figure 6.71 (a) RRDE voltammograms for the ORR. (b) Current density–time chronoamperometric responses. (c) Current-time chronoamperometric response of Pt/C (circle line) and N-graphene (square line) electrodes to CO. (d) Cyclic voltammograms of N-graphene electrode before (circle line) and after (square line) a continuous potentiodynamic swept for 2×10^5 cycles at RT($25\,°C$). (*Source*: Reprinted from Reference (98), with permission).

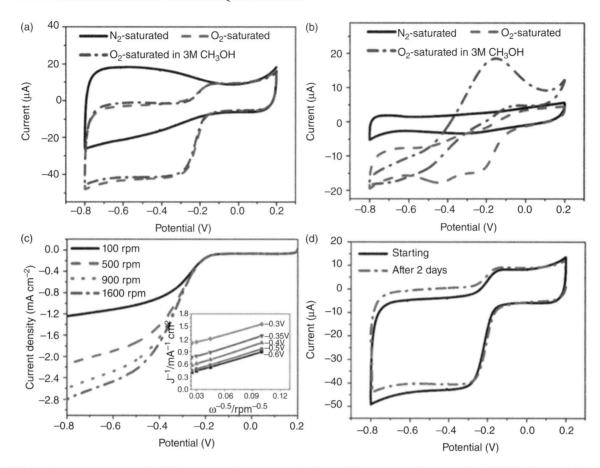

Figure 6.72 Cycle voltametries for (a) N-GQD/graphene and (b) commercial Pt/C on a GC electrode in N2-saturated 0.1 M KOH, O2-saturated 0.1 M KOH, and O2-saturated 3 M CH3OH solutions. (c) RDE curves for N-GQD/graphene in O2-saturated 0.1 M KOH with different speeds. (d) Electrochemical stability of N-GQD/graphene as determined by continuous CV in O2-saturated 0.1 M KOH. (*Source*: Reprinted from Reference (72), with permission).

6.6   PERSPECTIVES

In this chapter, we have described the recent achievements in the field of GQD-related research, focusing on the synthesis, surface chemical modification, physical properties, and various applications for energy, environment, and biology. Peculiar and unique electrical, optical, and chemical properties of GQDs will invoke more plentiful and enormous investigation to be accelerated even though the research on GQDs is still in the beginning state. Concerning the synthesis of GQDs, the bottom-up methods will be more appropriate for controlling the size and shape of the GQDs rather than the top-down method. Constrained reaction and nanoreactors which have been adopted for production of nanoparticles will be available. In addition, development of mass production method for GQDs with tailored accurate size and high quality will be highly demanded, providing the GQDs with well-defined structure and bandgap. However, because the concrete understanding of the PL mechanism for GQDs related to size, surface chemistry, and dopants is still very ambiguous, so fundamental and comprehensive understanding of optical properties on GQDs will be crucial part to make the most use of unique properties of absorption, PL, and electroluminescence of GQDs. Surface engineering will be also an essential key technology to control and improve the optical, electrical, and mechanical properties of GQDs. In the application of energy, solar cells with GQDs and with ZnO–GQDs show relatively lower efficiency (<2%) and 2.33%. For improvement of photoconversion efficiency in GQD-sensitized solar cell, GQDs demands surface engineering through conjugated bridges. Low QYs in the fluorescence device application of GQDs will be overcome by either reducing agglomeration or the use of metal-enhanced fluorescence (MEF) (190). Agglomeration among GQDs can be minimized by using dispersive matrix of QDs as ZnO QDs matrix in the form of ZnO–GQDs hybrid for a uniform emissive layer in white LED. If GQDs are located on metal@silica composite nanoparticles, fluorescence of GQDs can be enhanced or quenched by controlling the thickness of the spacer SiO_2 layer with matching the localized surface plasmon resonance (LSPR) absorption band. As seen in ZnO–GQDs and TiO_2–GQDs hybridization materials, new functionalities or enhanced features in light-emitting diodes, photovoltaics, and photocatalyst will be further expected through synthesizing hybrid system between semiconductor nanoparticles and GQDs utilizing high photoinduced charge transfer ability. In this regard, many new QD materials such as 2D *h*-BN nanosheets, BCNO nanoparticles (191), nanodiamonds (192), and Si QDs (193) will be very promising for nanoarchitecturing new kinds of dimensional hybrid materials with GQDs.

Moreover, GQDs with prominent photoluminescent property besides chemical inertness and intrinsic biocompatibility (low cytotoxicity) of GQDs have high potential as alternatives to conventional heavy metal-based hazardous QDs in biological cell imaging, clinical medicine, drug (gene) delivery, cancer therapy, and antibacterial materials. In summary, it is certain that GQDs are highly expected to become a very promising novel functional material to open new era in green chemistry, energy, environment, and biology-related fields.

REFERENCES

1. Novoselov KS, Geim AK, Morozov SV, Jiang D, Zhang Y, Dubonos SV, Grigorieva IV, Firsov AA. Electric Field Effect in Atomically Thin Carbon Films. Science 2004;306:666–669.

2. Zhang YB, Tan YW, Stormer HL, Kim P. Nature 2005;438:201–204.

3. Aleiner IL, Efetov KB. Phys Rev Lett 2006;97:236802–236804.

4. Jannik CM, Geim AK, Katsnelson MI, Novoselov KS, Booth TJ, Roth S. Nature 2007;446:60–63.

5. Schedin F, Geim AK, Morozov SV, Hill EW, Blake P, Katsnelson MI, Novoselov KS. Nat Mater 2007;6:652–655.

6. Son DI, Kim TW, Shim JH, Jung JH, Lee DU, Lee JM, Park WI, Choi WK. Nano Lett 2010;10:2441–2447.

7. Zhou K, Zhu Y, Yang X, Li C. J Chem 2010;34:2950–2955.

8. Zhou K, Zhu Y, Yang X, Li C. Electroanalysis 2011;23:862–869.

9. Shen J, Zhu Y, Zhou K, Yang X, Li C. J Mater Chem 2012;22:545–550.

10. Zhou K, Zhu Y, Yang X, Luo J, Li C, Luan S. Electrochim Acta 2010;55:3055–3060.

11. Zhou K, Zhu Y, Yang X, Jiang X, Li C. J Chem 2011;35:353–359.

12. Bae SK, Kim HK, Lee YB, Xu XF, Park JS, Zheng Y, Balakrishnan JK, Lei T, Kim HR, Song YI, Kim YJ, Kim KS, Özyilmaz B, Ahn JH, Hong BH, Iijima S. Nat Nano 2010;5:574–578.

13. Zhou K, Zhu Y, Yang X, Li C. Electroanalysis 2010;22:259–264.

14. Geim AK. Science 2009;324:1530–1534.

15. Stankovich S, Dikin DA, Dommett GHB, Kohlhaas KM, Zimney EJ, Stach EA, Piner RD, Nguyen SBT, Ruoff RS. Nature 2006;442:282–286.

16. Xie XJ, Bai H, Shi GQ, Qu LT. J Mater Chem 2011;21:2057–2059.

17. Schwierz F. Nat Nano 2010;5:487–496.

18. Han MY, Ozyilmaz B, Zhang Y, Kim P. Phys Rev Lett 2007;98:206805.

19. Todd K, Chou H, Amasha S, Goldhaber–Gordon D. Nano Lett 2009;9:416–421.

20. Stampfer C, Güttinger J, Hellmüller S, Molitor F, Ensslin K, Ihn T. Phys Rev Lett 2009;102:056403.

21. Molitor F, Jacobsen A, Stampfer C, Güttinger J, Ihn T, Ensslin K. Phys Rev B 2009;79:075426.

22. Wang M, Song EB, Lee S, Tang J, Lang M, Zeng C, Xu G, Zhou Y, Wang KL. ACS Nano 2011;5:8769–8773.

23. Baker SN, Baker GA. Angew Chem Int Ed 2010;49:6726–6744.

24. Tian L, Ghosh D, Chen W, Pradhan S, Chang X, Chen S. Chem Mater 2009;21:2803–2809.

25. Zhao Q, Zhang Z, Huang B, Peng J, Zhang M, Pang D. Chem Commun 2008;41:5116–5118.

26. Ray SC, Saha A, Jana NR, Sarkar RJ. J Phys Chem C 2009;13:18546–18551.

27. Zhou JG, Zhou XT, Li RY, Sun XL, Ding ZF, Cutler J, Sham TK. Chem Phys Lett 2009;474:320–324.

28. Li HT, He X, Kang Z, Huang H, Liu Y, Liu J, Lian S, Tsang CHA, Yang X, Lee S. Angew Chem Int Ed 2010;49:4430–4434.

29. Ponomarenko LA, Schedin F, Katsnelson MI, Yang R, Hill EW, Novoselov KS, Geim AK. Science 2008;320:356–358.

30. Shen J, Zhu Y, Yang X, Li C. Chem Commun 2012;48:3686–3699.

31. Ritter KA, Lyding JW. Nat Mater 2009;8:235–242.

32. Son Y–W, Cohen ML, Louie SG. Phys Rev Lett 2006;97:216803.

33. Zhang ZZ, Chang K, Peeters FM. Phys Rev B 2008;77:235411.

34. Pan D, Zhang J, Li Z, Wu M. Adv Mater 2010;22:734–738.

35. Güc–lü AD, Potasz P, Hawrylak P. Phys Rev B 2010;82:155445.

36. Loh KP, Bao Q, Eda G, Chhowalla M. Nat Chem 2010;2:1015–1024.

37. Mei QS, Zhang K, Guan GJ, Liu BH, Wang SH, Zhang ZP. Chem Commun 2010;46:7319–7321.

38. Min H, Hill JE, Sinitsyn NA, Sahu BR, Kleinman L, MacDonald AH. Phys Rev B 2006;74:165310.

39. Petta JR, Johnson AC, Taylor JM, Laird EA, Yacoby A, Lukin MD ; Marcus CM, Hanson MP, Gossard AC. Science 2005;309:2180–2184.

40. Koppens FHL, Buizert C, Tielrooij KJ, Vink IT, Nowack KC, Meunier T, Kouwenhoven LP, Vandersypen LMK. Nature 2006;442:766–771.

41. Yan X, Cui X, Li X. J Am Chem Soc 2010;132:5944–5945.

42. Yan X, Li B, Li L. Colloidal Graphene Quantum Dots with Well-Defined Structures. Acc Chem Res. 2013;46:2254–2262.

43. Yan X, Li LSJ. J Mater Chem 2011;21:3295–3300.

44. Simpson CD, Brand JD, Berresheim AJ, Przybilla L, R€ader HJ, Müllen K. Chem Eur J 2002;8:1424–1429.

45. Sakamoto J, Van Heijst J, Lukin O, Schlüter A. Angew Chem Int Ed 2009;48:1030–1069.

46. Liu R, Wu D, Feng X, Müllen K. J Am Chem Soc 2011;133:15221–15223.

47. Dong YQ, Shao JW, Chen CQ, Li H, Wang RX, Chi YW, Lin XM, Chen GN. Carbon 2012;50:4738–4743.

48. Dong YQ, Shao JW, Chen CQ, Li H, Wang RX, Chi YW, Lin XM, Chen GN. Carbon 2012;50:4738–4743.

49. Tang L, Ji R, Cao X, Lin J, Jiang H, Li X, Teng KS, Luk CM, Zeng S, Hao J, Lau SP. ACS Nano 2012;6:5102–5110.

50. Lu J, Yeo SE, Gan CK, Wu P, Loh KP. Nat Nano 2011;6:247–252.

51. Krishhna V, Stevens N, Koopman B, Moudgli B. Nat Nano 2010;5:330–334.

52. Cepek C, Goldoni A, Modesti S. Phys Rev B 1996;53:7466–7472.

53. Swami N, He H, Koel BE. Phys Rev B 1999;59:8283–8291.

54. Schulze G, Franke KJ, Gagliardi A, Romano G, Lin C, Da Rosa A, Niehaus TA, Frauenheim T, Di Carlo A, Pecchia A, Pascualet JI. Phys Rev Lett 2008;100:136801.

55. Pan D, Guo L, Zhang J, Xi C, Huang H, Li J, Zhang Z, Yu W, Chen Z, Li Z, Wu M. J Mater Chem 2012;12:3314–3318.

56. Li Y, Hu Y, Zhao Y, Shi G, Deng L, Hou Y, Qu L. Adv Mater 2011;23:776–780.

57. Wang J, Xin X, Lin Z. Nanoscale 2011;3:3040–3048.

58. Zhu S, Zhnag J, Liu X, Li B, Wang X, Tang S, Meng Q, Li Y, Shi C, Hu R, Ynag B. RSC Adv 2012;2:2717–2720.

59. Shen J, Zhu Y, Chen C, Yang X, Li C. Chem Commun 2011;47:2580–2582.

60. Shen J, Zhu Y, Yang X, Zong J, Zhang J, Li C. J Chem 2012;36:97–101.

61. Novoselov KS, Jiang D, Schedin F, Booth TJ, Khotkevich VV, Morozov SV, Geim AK. Proc Natl Acad Sci 2005;102:10451–10453.

62. Berry MV, Mondragon RJ. Proc R Soc London A 1987;412:53–74.

63. Libisch F, Stampfer C, Burgdörfer J. Phys Rev B 2009;79:115423.

64. Güttinger J, Stampfer C, Frey T, Ihn T, Ensslin K. Phys Status Solidi B 2009;246:2553–2557.

65. Wang LJ, Cao G, Tu T, Li HO, Zhou C, Hao XJ, Su Z, Guo GC, Jiang HW, Guo GP. Appl Phys Lett 2010;97:262113.

66. Wang LJ, Guo GP, Wei D, Cao G, Tu T, Xiao M, Guo GG, Chang AM. Appl Phys Lett 2011;99:112117.

67. Wang LJ, Li HO, Tu T, Cao G, Zhou C, Su Z, Xiao M, Guo GC, Chang AM, Guo GP. Appl Phys Lett 2012;100:022106.

68. Zhou JG, Booker C, Li RY, Zhou XT, Sham TK, Sun XL, Ding ZFJ. J Am Chem Soc 2007;129:744–745.

69. Zheng L, Chi Y, Dong Y, Lin J, Wang BJ. J Am Chem Soc 2009;131:4564–4565.

70. Lu J, Yang JX, Wang JZ, Lim A, Wang S, Loh KP. ACS Nano 2009;3:2367–2375.

71. Xie X, Qu L, Zhou C, Li J, Bai H, Shi G, Dai L. ACS Nano 2010;4:6050–6054.

72. Li Y, Zhao Y, Cheng HH, Hu Y, Shi GQ, Dai LM, Qu LT. J Am Chem Soc 2012;134:15–18.

73. Zhang M, Bai LL, Shang WH, Xie WJ, Ma H, Fu YY, Fang DC, Sun H, Fan LZ, Han M, Liu CM, Yang SH. J Mater Chem 2012;22:7461–7467.

74. Hamilton IP, Li B, Yan X, Li L-S. Nano Lett 2011;11:1524–1529.

75. Li L-S, Yan XJ. J Phys Chem Lett 2010;1:2572–2576.

76. Mueller ML, Yan X, McGuire JA, Li L-S. Nano Lett 2010;10:2679–2682.

77. Yan X, Cui X, Li B, Li L–S. Nano Lett 2010;10:1869–1873.

78. Müeller ML, Yan X, Dragnea B, Li L-S. Nano Lett 2011;11:56–60.

79. Kosynkin DV, Higginbotham AL, Sinitskii A, Lomeda JR, Dimiev A, Price BK, Tour JM. Nature 2009;458:872–876.

80. Dong Y, Chen CQ, Zheng XT, Gao LL, Cui ZM, Yang HB, Guo CX, Chi YW, Li CM. J Mater Chem 2012;22:8764–8766.

81. Peng J, Gao W, Gupta BK, Liu Z, Romero–Aburto R, Ge L, Song L, Alemany LB, Zhan X, Gao G, Vithayathil SA, Kaipparettu BA, Marti AA, Hayashi T, Zhu J, Ajayan PM. Nano Lett 2012;12:844–849.

82. Li Z, Zhang W, Luo Y, Yang J, Hou JGJ. J Am Chem Soc 2009;131:6320–6321.

83. Li LL, Ji J, Fei R, Wang CZ, Lu Q, Zhang JR, Jiang LP, Zhu JJ. Adv Funct Mater 2012;22:2971–2979.

84. Mohanty N, Moore D, Xu Z, Sreeprasad TS, Nagaraja A, Rodriguez A, Berry V. Nat Commun 2012;3:844.

85. Li HT, He XD, Liu Y, Huang H, Lian SY, Lee ST, Kang ZH. Carbon 2011;49:605–609.

86. Zhuo S, Shao M, Lee ST. ACS Nano 2012;6:1059–1064.

87. Son DI, Kwon BW, Kim HH, Park DH, Angadi B, Choi WK. Carbon 2013;59:289–296.

88. Shin HJ, Kim KK, Benayad A, Yoon SM, Park HK, Jung IS, Jin MH, Jeong HK, Kim JM, Choi JY, Lee YH. Adv Funct Mater 2009;19:1987–1992.

89. Choucair M, Thordarson P, Stride JA. Nat Nano 2009;4:30–33.

90. Fan ZJ, Kai W, Yan J, Wei T, Zhi LJ, Feng J, Ren YM, Song LP, Wei F. ACS Nano 2010;5:191–198.

91. Moon IK, Lee JH, Rodney SR, Lee HY. Nat Commun 2010;1:1–6.

92. Graupner RJ. J Raman Spectrosc 2007;38:673–683.

93. Zhou Y, Bao QL, Tang LAL, Zhong YL, Loh KP. Chem Mater 2009;21:2950–2956.

94. Yan J, Fan ZJ, Wei T, Qian WZ, Zhang M, Wei FF. Carbon 2010;48:3825–3833.

95. Kumar B, Gong H, Chow SY, Tripathy S, Hua Y. Appl Phys Lett 2006;89:071922.

96. Chen S, Liu JW, Chen ML, Chen XW, Wang JH. Chem Commun 2012;48:7637–7639.

97. Li Y, Hu Y, Zhao Y, Shi GQ, Deng LE, Hou YB, Qu LT. Adv Mater 2011;23:776–780.

98. Qu LT, Liu Y, Naek JB, Dai LM. ACS Nano 2010;4:1321–1326.

99. Gong KP, Du F, Xia ZH, Durstock M, Dai LM. Science 2009;323:760–764.

100. Wang SY, Yu DS, Dai LM. J Am Chem Soc 2011;133:5182–5185.

101. Son DI, Kwon BW, Park DH, Seo WS, Yi Y, Angadi B, Lee CL, Choi WK. Nat Nano 2012;7:465–471.

102. Cheng CF, Cheng HH, Cheng PW, Lee YJ. Macromolecules 2006;39:7583–7590.

103. Park SH, Jin SH, Jun GH, Hong SH, Jean S. Nano Res 2011;4:1129–1135.

104. Geim AK, Novoselov KS. Nat Mater 2007;6:183–191.

105. Novoselov KS, Geim AK, Morozov SV, Jiang D, Katsnelson MI, Grigorieva IV, Dubonos SV, Firsov AA. Nature 2005;438:197–200.

106. Wallace PR. Phys Rev 1947;71:622–634.

107. Avouris P, Chen Z, Perebeinos V. Nat Nano 2007;2:605–615.

108. Zhou SY, Gweon G-H, Graf J, Fedorov AV, Spataru CD, Diehl RD, Kopelevich Y, Lee D-H, Louie SG, Lanzara A. Nat Phys 2006;2:595–599.

109. Bostwick A, Ohta T, Seyller T, Horn HK, Rotenberg E. Nat Phys 2007;3:36–40.

110. Andreev AF. Sov Phys JETP 1964;19:1228–1231.

111. Allain PE, Fuchs JN. Eur Phys J B 2011;83:301–317.

112. Klein OZ. Physik 1929;53:157–165.

113. Miao F, Wijerante S, Zhang Y, Coskun U, Bao W, Lau C. Science 2007;317:1530–1533.

114. Beenakker CW. Rev Mod Phys 2008;80:1337–1356.

115. Dombey N, Calogeracos A. Phys Rep 1999;315:41–58.

116. Katsnelson MI, Novoselov KS, Geim AK. Nat Phys 2006;2:620–625.

117. Han MY, Özyilmaz B, Zhang Y, Kim P. Phys Rev Lett 2007;98:206806.

118. Sols S, Guinca F, Castro Neto AH. Phys Rev Lett 2007;99:166803.

119. Castro Neto AH, Guinea F, Peres NMR. Phys Rev B 2006;73:205408.

120. Lin Y, Perebeinos V, Chen Z, Avouris P. Phys Rev B 2008;78:61409.

121. Libisch F, Stampfer C, Burgdörfer J. Phys Rev B 2009;79:115423.

122. Reich S, Maultzsch J, Thomsen C, Orfejon P. Phys Rev B 2002;66:035412.

123. Chen J-H, Jang C, Adam S, Fuhrer MS, Williams ED, Ishigami M. Nat Phys 2008;4:377–381.

124. Güttinger J, Stampfer C, Libisch F, Frey T, Burgdörfer J, Ihn T, Ensslin K. Phys Rev Lett 2009;103:046810.

125. Chen RB, Chang CP, Lin MF. Physica E 2010;42:2812–2815.

126. Li TS, Chang SC, Chuang YC, Wu KHJ, Lin MF. Physica B 2009;404:305–309.

127. Rycerz A, Tworzydlo J, Beenakker CWJ. Nat Phys 2007;3:172–175.

128. Recher P, Trauzettel B, Rycerz A, Blanter YM, Beenakker CWJ, Morpurgo AF. Phys Rev B 2007;76:235404.

129. Recher P, Nilson J, Burkard G, Trauzettel B. Phys Rev B 2009;79:085407.

130. Trauzettel B, Bulaev D, Loss D, Burkardm G. Nat Phys 2007;3:192–196.

131. Khaetskii AV, Loss D, Glazman L. Phys Rev Lett 2002;88:186802.

132. Fox M. *Optical Properties of Solids*. Vol. 1. New York: Oxford University Press; 2010. p 227.

133. Robertson J, O'Reilly EP. Phys Rev B 1987;35:2946.

134. Mathioudakis C, Kopidakis G, Kelires PC, Patsalas P, Gioti M, Logothetidis S. Thin Solid Films 2005;482:151–155.

135. Chen CW, Robertson JJ. J Non Cryst Solids 1998;602:227–230.

136. Zhu SJ, Tang SJ, Zhang JH, Yang B. Chem Commun 2012;48:4527–4539.

137. Melenikov DV, Chelikowsky JR. Phys Rev Lett 2004;92:046802.

138. Zhu S, Zhang J, Qiao C, Tang S, Li Y, Yuan W, Li B, Tian L, Liu F, Hu R, Gao H, Wei H, Zhang H, Sun H, Yang B. Chem Commun 2011;47:6858–6860.

139. Eda G, Lin Y-Y, Mattevi C, Yamaguchi H, Chen H-A, Chen I-S, Chen C-W, Chhowalla M. Adv Mater 2010;22:505–509.

140. Li HT, He XD, Kang ZH, Huang H, Liu Y, Liu JL, Lian SY, Tsang CHA, Yang XB, Lee ST. Angew Chem Int Ed 2010;49:4430–4434.

141. Chien CT, Li SS, Lai WJ, Yeh YC, Chen HA, Chen IS, Chen LC, Chen KH, Nemoto T, Isoda S, Chen M, Fujita T, Eda G, Yamaguchi H, Chhowalla M, Chen CW. Angew Chem Int Ed 2012;51:6662–6666.

142. Li J-L, Kudin KN, McAllister MJ, Prud'homme RK, Aksay IA, Car R. Phys Rev Lett 2006;96:176101.

143. Radovic LR, Bockrath B. J Am Chem Soc 2005;127:5917–5927.

144. Sun YP, Zhou B, Lin Y, Wang W, Fernando KAS, Pathak P, Meziani MJ, Harruff BA, Wang X, Wang HF, Luo PJG, Yang H, Kose ME, Chen B, Veca LM, Xie SY. J Am Chem Soc 2006;128:7756–7757.

145. Pan D, Zhang J, Li Z, Wu C, Yan X, Wu M. Chem Commun 2010;46:3681–3683.

146. Gokus T, Nair RR, Bonettl A, Bohmier M, Lombardo A, Novoselov KS, Geim AK, Ferrari AC, Hartschuh A. ACS Nano 2009;3:3963–3968.

147. You HX, Brown NMD, Al–Assadi K. Surf Sci 1993;284:263–272.

148. Tetsuka H, Asashi R, Nagoya A, Okamoto K, Tajima I, Ohta R, Okamoto A. Adv Mater 2012;24:5333–5338.

149. Seredych M, Bandoz TJJ. J Phys Chem C 2007;111:15596–15604.

150. Lai L, Chen L, Zhan D, Sun L, Liu J, Lim SH, Poh CK, Shen Z, Lin J. Carbon 2011;49:3250–3257.

151. Kim S, Hwang SW, Kim MK, Shin DY, Shin DH, Kim CO, Yang SB, Park JH, Hwang E, Choi SK, Ko G, Sim S, Sone C, Choi HJ, Bae S, Hong BH. ACS Nano 2012;6:8203–8208.

152. Lobyshev VI, Shikhlinskaya RE, Ryzhikov BDJ. J Mol Liq 1999;82:73–81.

153. Wang X, Zhang RQ, Lee ST, Frauenheim T, Niehaus TA. Appl Phys Lett 2008;93:243120.

154. Wang X, Zhang RQ, Lee ST, Niehaus TA, Frauenheim T. Appl Phys Lett 2007;90:123116.

155. Jin SH, Kim DH, Jun GH, Hong SH, Jeon SW. ACS Nano 2013;7:1239–1245.

156. Lingam K, Podila R, Qian H, Serkiz S, Rao AM. Adv Funct Mater Evidence for Edge-State Photoluminescence in Graphene Quantum Dots 2013;23:5062–5065

157. Yang L, Park CH, Son YW, Cohen ML, Louie SG. Phys Rev Lett 2007;99:186801.

158. Gupta V, Chaudhary N, Srivastava R, Sharma GD, Bhardwaj R, Chand SJ. J Am Chem Soc 2011;133:9960–9963.

159. Frisch MJ et al. *Gaussian 03*. Wallingford (CT): Gaussian, Inc.; 2003.

160. Kamat PV. J Phys Chem B 2008;112:18737–18753.

161. Semonin OE, Luther JM, Choi S, Chen HY, Gao J, Nozik AJ, Beard MC. Science 2011;334:1530–1533.

162. De Leeuw DM, Simenon MMJ, Brown AR, Einerhand REF. Synth Met 1999;87:53–59.

163. Yu DS, Yang Y, Durstock M, Baek J-B, Dai LM. ACS Nano 2010;4:5633–5640.

164. Mandoc MM, Koster LJA, Blom PWM. Appl Phys Lett 2007;90:133504.

165. Brabec CJ, Cravino A, Meissner D, Sariciftci NS, Fromherz T, Rispens MT, Sanchez L, Hummelen JC. Adv Funct Mater 2001;1:374–380.

166. Scharber MC, Wuhlbacher D, Koppe M, Denk P, Waldauf C, Heeger AJ, Brabec CL. Adv Mater 2006;18:789–794.

167. Kymakis E, Alexandrou I, Amaratunga GAJ. J Appl Phys 2003;93:1764–1768.

168. Liu ZY, He DW, Wang YS, Wu HP, Wang JG. Sol Energy Mater Sol Cells 2010;94:1196–1200.

169. Admassie S, Ingan€as O, Mammo W, Perzon E, Andersson MR. Synth Met 2006;156:614–623.

170. Liu Z, Liu Q, Huang Y, Ma Y, Yin S, Zhang X, Sun W, Chen Y. Adv Mater 2008;20:3924–3930.

171. Liu Q, Liu X, Zhang X, Yang L, Zhang N, Pan G, Yin S, Chen Y, Wei J. Adv Funct Mater 2009;19:894–904.

172. Grätzel M. Prog Photovoltaics Res Appl 2000;8:171–185.

173. Nazeeruddin MK, Kay A, Rodicio I, Humphrybaker R, Müller E, Liska P, Vlachopoulos N, Grätzel M. J Am Chem Soc 1993;115:6382–6390.

174. Son DI, Kwon BW, Yang JD, Park DH, Seo WS, Lee H, Yi Y, Lee CL, Choi WK. Nano Res 2012;5:739–753.

175. Zhu H, Zhang W, Yu SF. Nanoscale 2013;7:1797–1802.

176. Zhang WF, Tang LB, Yu SF, Lau SP. Opt Mater Express 2012;2:490–495.

177. Zhang WF, Zhu H, Yu SF, Yang HY. Adv Mater 2012;24:2263–2267.

178. Velenta J, Pelent I, Linnros J. Appl Phys Lett 2002;81:1396–1398.

179. Chiad BT, Latif KH, Kadhim FJ, Hammed MA. Adv Mater Phys Chem 2011;1:20–25.

180. Sha WL, Liu CH, Alfano RR. Opt Lett 1994;19:1922–1924.

181. Sun X, Liu Z, Welsher K, Robinson JT, Goodwin A, Zaric S, Dai H. Nano Res 2008;1:203–212.

182. Shen H, Zhang L, Liu M, Zhang Z. Theranostics 2012;2:283–294.

183. Yu SJ, Kang MW, Chang HC, Chen KM, Yu YC. J Am Chem Soc 2005;127:17604–17605.

184. Zhao J, Chen G, Zhu L, Li G. Electrochem Commun 2011;13:31–33.

185. Li T, Fan Q, Liu T, Zhu X, Zhao J, Li G. Biosens Bioelectron 2010;25:2686–2689.

186. Centi S, Messina G, Tombelli S, Palchetti I, Mascini M. Biosens Bioelectron 2008;23:1602–1609.

187. Xiao Y, Paviov V, Niazov T, Dishon A, Kotler M, Willner I. J Am Chem Soc 2004;126:7430–7431.

188. Liu R, Wu DQ, Feng XL, Müllen K. Angew Chem Int Ed 2010;49:2565–2569.

189. Salimi R, Banks CE, Richard G, Compton RG. Phys Chem Chem Phys 2003;5:3988–3993.

190. Yang J, Zhang F, Chen Y, Qian S, Hu P, Li W, Deng Y, Fang Y, Han L, Luqman M, Zhao D. Chem Commun 2011;47:11618–11620.

191. Li W, Portehault D, Dimova R, Antonietti M. J Am Chem Soc 2011;133:7121–7127.

192. Niu K-Y, Zheng H-M, Li Z-Q, Yang J, Du X-W. Angew Chem Int Ed 2011;50:4099–4102.

193. Atkins TM, Thibert A, Larsen DS, Dey S, Browing ND, Kauzlarich SM. J Am Chem Soc 2011;133:20664–20667.

7

RECENT PROGRESS OF IRIDIUM(III) RED PHOSPHORS FOR PHOSPHORESCENT ORGANIC LIGHT-EMITTING DIODES

CHEUK-LAM HO AND WAI-YEUNG WONG
Department of Chemistry and Institute of Advanced Materials, Hong Kong Baptist University, Waterloo Road, Hong Kong, People's Republic of China

7.1 INTRODUCTION

Research in organic light-emitting diodes (OLEDs) has received much current attention because they can be utilized as high-efficiency, low-voltage, full-color, large-area flat panel displays in electronic devices, which make them perfect candidates to replace widely used liquid crystal display panels (1). OLEDs based on phosphorescent materials can significantly improve electroluminescence (EL) performance as compared with the conventional fluorescent OLEDs (2). Cyclometalated iridium(III) complexes are acquiring the mainstream position in the field of organic displays because of their highly efficient emission properties, relatively short excited-state lifetime, and excellent color tunability over the entire visible spectrum (3). In particular, extensive research efforts have been made recently for more breakthrough achievements in phosphorescent white OLEDs as they can be used for the next generation of solid-state lighting. White emission can be achieved by mixing three primary colors (red, green, and blue) or two colors from an orange emitter complemented with a blue emitter (4). To achieve highly efficient white OLEDs, there is a great demand for the efficient and bright true red color phosphorescent materials.

As compared to other colors, the design and synthesis of efficient red emitters are intrinsically more difficult, in accordance with the energy gap law (5, 6). Many red emitters suffer from poor compromise between device efficiency and color purity. The lower luminosity of red device is due to its characteristic red emission in a spectral region where the eye has poor sensitivity. Moreover, the wide bandgap host used in OLED devices and the narrow bandgap red-emitting guests have a significant difference in HOMO and/or LUMO levels between the guest and host materials. Thus, the guest molecules are thought to act as deep traps for electrons and holes in the emitting layer, causing an increase in the drive voltage of OLED. Furthermore, self-quenching or triplet–triplet annihilation by red dopant molecules is an inevitable problem in such host–guest systems especially at high doping concentrations. Therefore, from a practical standpoint, a solution to the above issues based on material's design and/or device optimization is highly needed.

This chapter provides a survey of red Ir(III) phosphorescent materials that has been used in red OLEDs in the past few years. Their structure–efficiency relationship and photophysical and electroluminescence properties will also be discussed.

7.2 IRIDIUM(III) RED DOPANTS CONTAINING VARIOUS CYCLOMETALATED LIGANDS

7.2.1 1-Phenylisoquinoline Derivatives

Although quite a number of red triplet emitters have been synthesized and developed, isoquinoline-type Ir(III) complexes, particularly 1-phenylisoquinoline-based molecules, are still the most studied one. Quinoline-based compounds have received considerable attention in optoelectronic materials due to their high electron affinities (7). The molecular design of red phosphorescent complexes with phenylisoquinoline (piq) ligand is based on the fact that the highest occupied molecular orbital (HOMO) is principally composed of a mixture of iridium d and phenyl π orbitals, while the lowest unoccupied molecular orbital (LUMO) is predominantly localized on the

Nanomaterials, Polymers and Devices: Materials Functionalization and Device Fabrication, First Edition. Edited by Eric S. W. Kong.
© 2015 John Wiley & Sons, Inc. Published 2015 by John Wiley & Sons, Inc.

π-orbitals of the piq chromophore. Greater π-electronic conjugation in isoquinoline ring would significantly lower the LUMO level and notably reduce the HOMO–LUMO energy gap. Research in the field of red phosphorescent OLEDs (PhOLEDs) started from the well-known tris(1-phenylisoquinoline)iridium(III) Ir(piq)$_3$ complex: (**1**) and Ir(piq)$_2$(acac) (**2**). The piq unit as the ligand part of these Ir(III) complexes can partially suppress the triplet–triplet annihilation and show short phosphorescent lifetime (8). Both complexes exhibit photoluminescence (PL) emission peak at around 620 nm, depending on the nature of the host used in the device structure (9). Since the initial work on **1** and **2** as red dopant in OLEDs, considerable focus was paid on **1** and **2** by developing host materials that are better matching the HOMO and LUMO energy levels in order to achieve a good exciton confinement within the emissive layer as well as by studying the effects of charge trapping, concentration of dopant, and device structure for a better control of the red OLED efficiency. We will also highlight some of the red dopant OLEDs based on **1** or in different device structures and host with attractive performance.

Kwon and coworkers reported the application of **1** in different device configurations, including multiple quantum well structure, bilayer or single layer. They illustrated that by doping **1** in bis(10-hydroxybenzo[h]quinolinato)beryllium (Bebq$_2$) matrix, a better match in the HOMO and LUMO energy levels between the guest and host systems can be achieved as Bebq$_2$ is a narrow bandgap material and works as a charge control layer with high electron mobility and very good electron-transporting characteristics. In quantum confinement approach, a multiple quantum well structure having various triplet quantum well devices from a single to five quantum wells were applied (ITO/TCTA:WO$_3$/TCTA/multiple layer (Bepq$_2$:**1**)$_{1\ to\ 5}$/Bepp$_2$/Al) (ITO, indium tin oxide; TCTA, 4,4′,4″-tri(9-carbazolyl)triphenylamine; Bepp$_2$, phenylpyridine beryllium) (10). Triplet energies in such devices are confined within the emitting layers. Among those five devices, a maximum external quantum efficiency (η_{ext}) of 14.8% was obtained with a two-quantum-well device structure, with maximum current efficiency (η_L) of 12.4 cd/A, low turn-on voltage (V_{on}) of 2.5 V, and excellent color stability with Commission Internationale de L'Eclairage (CIE) coordinates of (0.66, 0.33). Further increasing the number of quantum wells enhanced the driving voltage with a slight drop in the efficiency. The research group successfully demonstrated negligible barrier of electron transport in the red OLED by using the concept of multiple quantum well structure device and this should be very useful to future OLED display technology.

Simple bilayered device with **1** doped in the Bebq$_2$ host would be a promising way to achieve efficient and economical way for red light production (11). This device was fabricated in the configuration of ITO/NPB/Bepq$_2$:**1**/LiF/Al. Beneficial from very small exchange energy value of 0.2 eV between singlet and triplet states by using the Bepq$_2$ host, this device shows almost no barrier for the injection of charge carriers. Moderately high η_L and η_P values of 9.66 cd/A and 6.90 lm/W, respectively, were obtained. Later on, in 2010, single organic layer device using thermal evaporation technique was achieved by the same group. The key to this simplification is the direct injection of holes and electrons into the mixed host material through the opposite electrodes (12). Mixed host system with NPB:Bepq$_2$ (NPB: N,N′-di-[(1-naphthyl)-N,N′-diphenyl]-1,1′-biphenyl)-4,4′-diamine) with 1 wt% doping concentration of **1** showed the best result. Such low doping concentration device displays low V_{on} of 2.4 V, maximum η_L and η_P of 9.44 cd/A and 10.62 lm/W, respectively, with perfect CIE coordinates of (0.66, 0.33). This feature of red OLEDs paves the way to simplify the device structure and reduce the cost of red OLED manufacture.

Other researchers have improved the efficiencies of red OLEDs doped with **1** by using different host materials in the last few years. For example, deep-red **1**-based OLEDs doped in host materials 1,3,5-tris(3-(carbazol-9-yl)phenyl)-benzene, 2,4,6-tris(3-(carbazol-9-yl)phenyl)-pyridine, or 2,4,6-tris(3-(carbazol-9-yl)phenyl)-pyrimidine showed superior efficiency and suppressed efficiency roll-off (13). All the devices show very high η_{ext} of over 18% and η_P of around 19 lm/W at low current density due to the bipolar nature of the host materials.

2,7-Bis(phenylsulfonyl)-9-[4-(N,N-diphenylamino)phenyl]-9-phenylfluorene (SAF)-based red OLEDs exhibited a very low V_{on} (2.4 V) and high EL efficiencies of 15.8% and 22 lm/W, superior to those of the corresponding device incorporating with the conventional host material, 4,4′-N,N′-dicarbazolylbiphenyl (CBP; 3.2 V, 8.5%, and 8.4 lm/W) (14). The efficiencies of SAF-based red OLEDs remained high at a practical brightness of 1000 cd/m^2 (13.1%, 14.4 lm/W). Very long operational lifetime at high initial luminance of deep-red OLEDs based on **1** using double guest/host system were dedicated by Gigli (15). Device with 15 wt% of **1** was doped in NPB and bis(2-methyl-8-quinolinolato-N1,O8)-(1,1′-biphenyl-4-olato)aluminum (BAlq), which gave a maximum η_L of 9.5 cd/A and a remarkably long device lifetime of more than 2700 h at a starting luminance of 6000 cd/m^2. Host materials therefore play an important role in reducing energy barrier, emission color quality, and quantum efficiency of a device.

For **2**, comparable OLED performance as **1** has been reported in the literature. A deep-red phosphorescent **2**-based OLED hosted by a bipolar triphenylamine/oxadiazole hybrid gave an η_{ext} up to 21.6%, η_L of 15.9 cd/A, and η_P of 16.1 lm/W (16). Compared with the device using the host 3,5-di(9H-carbazol-9-yl)-N,N-diphenylaniline, the efficiencies of red OLEDs were slightly decreased but were still better than those using the common host N,N′-dicarbazolyl-3,5-benzene (mCP) (17). The maximum η_{ext} reached 19.2%. Comparable performance with η_{ext} of 19.3%, η_L of 16.4 cd/A, and η_P of 13.0 lm/W has been reported in 2012 with the OLED device structure of ITO/MoO$_3$/NPD/BBTC:**2**/BCP/Bebq$_2$/LiF/Al (BBTC, 3,6-bis-biphenyl-4-yl-9-[1,1′,4′,1″]terphenyl-4-yl-9H-carbazole; BCP, 2,9-dimethyl-4,7-diphenyl-1,10-phenanthroline) (18). This OLED showed nearly 100% internal quantum efficiency with reduced efficiency roll-off, which was attributed to the good hole mobility and low hole injection barrier of BBTC host in the emission layer. Because of the attractive results of **2**, Wei et al. announced an obvious superiority of white OLEDs using complex **2** doped in 4,4′,4″-tri(N-carbazolyl)triphenylamine (TCTA) (19). The WOLED with an optimal red dye doping concentration of 5 wt% exhibited a high color rendering index (CRI) of 89 and an η_P of 31.2 and 27.5 lm/W at the initial luminance and 100 cd/m^2, respectively. The device showed little variation of the CIE coordinates in a wide range of luminance. In addition to the host materials mentioned earlier, there are many reports on the modification of the structure of host material for **1** and **2** to improve the optical and electrical properties of related devices. Nevertheless, all these results suggest that Ir(III) dopants **1** and **2** are suitable candidates for highly efficient red and white OLEDs.

In addition to **1** and **2**, a number of isoquinoline-based Ir(III) derivatives have been synthesized and studied. The introduction of various simple substituents (e.g., *t*-Bu, F, Me, OMe) on Ir(III) complexes can not only alter the HOMO–LUMO level and emission wavelength but also minimize molecular packing and concentration quenching of luminescence. Complex **3** with 7-methyl-1-*p*-tolylisoquinoline shows a blueshift in its PL spectrum as compared to **2**, which emits a red light located at 606 nm in neat film (20). A highly efficient deep-red OLED was fabricated by doping **3** into a hole blocking material. This device shows a CIE coordinate of (0.66, 0.34) when the doping concentration is above 2%, which is very close to the National Television System Committee (NTSC) standard red point (0.66, 0.33). A maximum luminance (L_{max}) of 31,317 cd/m^2, a maximum η_L of 21.6 cd/A, and a half-lifetime of 13 h were achieved. The high efficiency can be obtained by the effective energy transfer from the BAlq host to the guest **3** and the direct recombination of electron–hole pairs on the dopants.

Liu and Wong introduced bulky and halo substituent such as F, *t*-Bu, and Me on complex **2** to reduce the self-quenching effect, which is believed to cause low device efficiency (21). Complexes **4** and **5** showed suitable red PL peak in dichloromethane at 618 and 628 nm, respectively. Red PhOLED devices incorporating indolo[3,2-*b*]carbazole/benzimidazole hybrid bipolar host material doped with the red emitters **4** and **5** revealed maximum η_{ext} of 15.6 and 15.5%, respectively. By replacing the ancillary ligand from acetylacetone in **2** to 2-acetyl-cyclohexane, **6**-doped devices with different host materials gave red emission peaks in the range of 618–636 nm (22). The best device showed L_{max} of 576 cd/cm^2. Complex **7** with free phosphine unit showed an emission signal at 600 nm, the near red range, which is about 20 nm shorter in wavelength than those of **1** and **2**. Solution processable OLED was fabricated with the structure of ITO/PEDOT:PSS/**7**/Ba/Al and redshifted emission at $\lambda_{EL} = 615$ nm was detected with CIE coordinates of (0.64, 0.34). Although the efficiencies of the devices based on **7** were not superior to that doped with **2**, it opened up a significant progress that luminescent Ir(III) units can be incorporated into organic polymer simply by phosphine coordination. This approach makes spin-coating and inkjet printing processes possible during the device fabrication, and it is conceived that we can further develop high-performance metallopolymers as electroluminescent materials or even generation of white light from spontaneous emissions from both the polymer main chain and complex **7**.

| X = H | **8** |
| X = OMe | **9** |

| X = H | **10** |
| X = OMe | **11** |

To solve the trade-off problem between device efficiency and color purity of red emitters, Wong et al. first reported the synthesis and characterization of a series of novel red-emitting Ir(III) complexes incorporating hole-transporting 9-arylcarbazole (**8–11**) (23) and triphenylamine (**12–13**) (9). These functionalized red emitters improve the charge injection and transport in **1** and **2**, help reduce the barrier height for hole injection by raising the HOMO levels, and decrease the triplet–triplet annihilation by taking advantages of the bulky carbazole and triphenylamine group. Impressive saturated red CIE coordinates of (0.68, 0.32) were reported by using both vacuum-deposited and spin-coated techniques for the doped devices made from **8** to **11**. An η_{ext} as high as 12% ph/el was achieved by **10** with η_L of 10.15 cd/A and η_P of 5.25 lm/W. The enhanced hole-transporting properties of the complexes have been demonstrated without the need of an additional hole-transporting layer in the device. While complex **12** with triphenylamine emits at 636 nm in the solution state, the emission wavelength is slightly redshifted with increasing the dendron size in **13** ($\lambda_{PL} = 641$ nm) relative to **12**. It is interesting that a substantial increase of the PL intensity and elevation of HOMO levels with increasing dendritic generation from **1** to **12** and **13** (HOMO: –4.99 eV for **12**; –4.96 eV for **13**; –5.11 for **1**). This implies that the dendritic triphenylamine structures will effectively block the self-quenching of the triplet emission core and improve the hole injection and hole-transporting properties. Highly solution processable complexes **12** and **13** were applied for the fabrication of deep-red OLEDs. They emit pure red light with λ_{PL} at around 640 nm with excellent CIE color coordinates of (0.70, 0.30). The OLEDs made from **12** (or **13**) achieved the L_{max} value of 7452 (6143) cd/m² at 17 (16) V, peak η_L of 5.82 (3.72) cd/A, η_{ext} of 11.65 (7.36)%, and η_P of 3.65 (2.29) lm/W. This class of materials with hole-transporting functionalities could provide a promising avenue for the rational design of heavy metal electrophosphors that provides superior device efficiency and color purity trade-off necessary for pure red light generation. This is actually a big step forward in the advance of high-performance true red OLEDs.

12   **13**

For small molecular materials, crystallization of their thin films may lead to the formation of excimers and exciplexes, which will decrease the device efficiency and impair the device stability. Moreover, at high doping concentration, the intermolecular interaction in thin film will lead to the self-quenching of luminescence. To address these issues, groups from Wong and Chen have adopted a 3D bulky structure and sterically hindered configuration in the cyclometalating ligand to suppress the close packing among the molecules in the solid state. They synthesized Ir(III)-based triplet emitters bearing fluorene derivatives with two sterically bulky triphenylamine in **14** (24) and spiroannulate moieties in **15** (25). The tortured geometries of these complexes not only render their highly amorphous properties but also alleviate triplet–triplet annihilation and concentration quenching in order to enhance the OLED efficiency. By taking this advantage, only slight roll-off of device efficiencies was observed with increasing current density. The pure red-emitting **14**-based device ($\lambda_{EL} = 648$ nm; CIE = 0.70, 0.30) attained L_{max} of 6471 cd/m² at 21 V, peak η_{ext} of 4.23%, η_L of 1.69 cd/A, and η_P of 0.27 lm/W. The OLED device with **15** showed an encouraging performance with peak η_{ext} of 4.6% and η_L of 7.44 cd/A; however, poor CIE coordinates of (0.644, 0.356) were obtained, which are far from the NTSC recommended red as compared to that for **14**. These pieces of work provide a possible platform to achieve anti-triplet–triplet annihilation phenomenon and relieve a bottleneck problem between red color purity and efficiency in small molecule PhOLEDs.

In general, the development of red phosphorescent Ir(III) complexes with high quantum yields (Φ_{PL}) is still difficult because these compounds are intrinsically less emissive than blue and green emitters according to the energy gap law (5, 6). Yagi et al. reported a series of rigid and extended π-conjugation framework of 1-(dibenzo[b,d]furan-4-yl)isoquinolinato-N,C3′ Ir(III) complexes (**16–18**), which yielded pure and saturated red PL emission with high Φ_{PL} (26). The Φ_{PL} of **16–18** is 0.61, 0.55, and 0.49, respectively, in toluene and much more emissive than **1** (0.26 in toluene). Although a change of the O^O ligand did not significantly affect the λ_{PL}, replacing dipivaloylmethanate (dmp) and acetylacetonate (acac) by 1,3-bis(3,4-dibutoxyphenyl)propane-1,3-dionate (bdbp) is beneficial to enhance phosphorescence Φ_{PL} and increase the solubility of the complex. Pure red EL with CIE chromaticity coordinates of (0.68, 0.31) can be obtained for **16–18** based device. The device with 0.51 mol% **16** leads to a maximum performance among the three complexes with $L_{max} = 7270$ cd/m², η_L of 3.9 cd/A, η_{ext} of 6.4%, and η_P of 1.4 lm/W. This work represents a good approach for getting highly emissive red phosphors with remarkable CIE values.

7.2.2   2-Phenylquinoline (phq) Derivatives

Nature of C^N chelates around the Ir(III) center was found to be critical for the photophysical properties of the resulting complexes. The impact of isomerization of C^N chelates from 1-phenylisoquinoline to 2-phenylquinoline was demonstrated by Suh and Kwon (27). The metal-to-ligand charge transfer (MLCT) PL emission was blueshifted for the 2-phenylquinoline derivative (complex **19**) as compared to the 1-phenylisoquinoline one (complex **2**). The former one emits an orange-red light at 581 nm with a noteworthy increase in Φ_{PL} ($\Phi_{PL} = 0.63$ in film). Methylation of the phenyl moiety of the phq ligand increases the Φ_{PL} to 0.70 and 0.76 for **20** and **21**, respectively. Dramatic increase in Φ_{PL} to 0.83 and 0.87 for **22** and **23** was observed if further methylation of the quinoline ring of the phq chelate occurred. The addition of methyl group on the C^N ligand increases the intermolecular steric interaction, resulting in reduced self-quenching effect and leading to an increment in Φ_{PL}. Moreover, after this chemical structural modification, very clear red EL spectra with much narrower full width at half maximum (FWHM) were obtained for **20–23** in the range of 609–620 nm. Impressive device efficiency data were achieved for these five red dopants. The best performance was obtained for **23** with maximum η_L of 30.1 cd/A, η_{ext} of 24.6%, and η_P of 32.0 lm/W, while the unsubstituted complex **19** got a maximum η_L of 19.1 cd/A, η_{ext} of 15.7%, and η_P of 17.6 lm/W. Changing the ancillary ligand to a 2,2,6,6-tetramethylheptane-3,5-dionate from an acac moiety enables a dramatic impact on the device performance. This piece of work successfully illustrates the orientation and substituent effects have a great impact on the intermolecular packing and efficiency of the device. In 2012, a significant enhancement in device performance based on **19** was carried out by Chang et al by wisely managing the excitons in OLED (28). Complex **19** was employed together with a green emitter [Ir(ppy)$_2$(acac)], which served as an exciton formation assistant, and excitons can be delivered to the red emitter **19**. Introduction of a green emitter gives a minimal effect on the emission spectra and keeps the λ_{EL} at 600 nm but improves the emission intensity by a factor of approximately 1.3. A high η_{ext} of 20.2% at 100 cd/m^2, corresponding to an η_L of 33.3 cd/A and an η_P of 28.0 lm/W, have been achieved without additional out-coupling enhancements. Detailed investigation on the nature of energy transfer should be carried out to gain more insight on the complex quantum mechanical processes involved for further employment of this simple technique in obtaining higher-performance PHOLEDs for displays as well as solid-state lighting applications.

The impact of changing the phenyl group to the thiophene ring on 2-phenylquinoline in **19** is not as remarkable as expected in comparison to the case of [Ir(ppy)$_2$(acac)] and only a significant bathochromic shift in the PL maximum was noted for the latter. Complex **24** with 4-methyl-2-(thiophen-2-yl)quinoline shows a very sharp emission band with FWHM of only 48 nm and relatively short wavelength emission maximum at 611 nm in the solution state as compared with the well-known deep-red Ir(III) complex **2** ($\lambda_{PL} = 620$ nm; FWHM = 58 nm). The narrow emission bandwidth of an emitter is essential for reaching high luminous efficiency. A further decrease in FWHM to 39 nm for **24** doped in a bipolar host material and a Φ_{PL} of 0.55 can benefit the dopant to get an attractive device performance. The best performance of **24**-based device shows an L_{max} of 58,688 cd/m^2 and the maximum η_{ext}, η_L, and η_P of 25.9%, 37.3 cd/A, and 32.9 lm/W, respectively. Preliminary results show that the operational lifetime of this device is estimated to be more than 2000 h at an initial luminance of 500 cd/m^2.

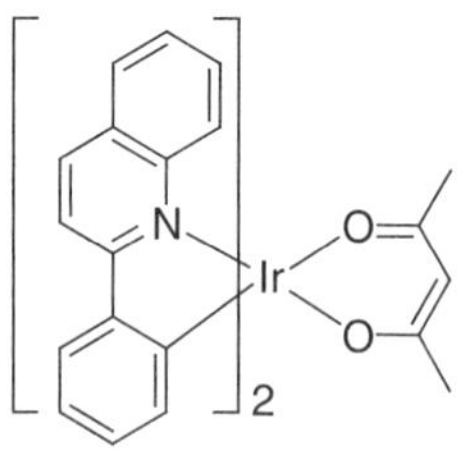

19 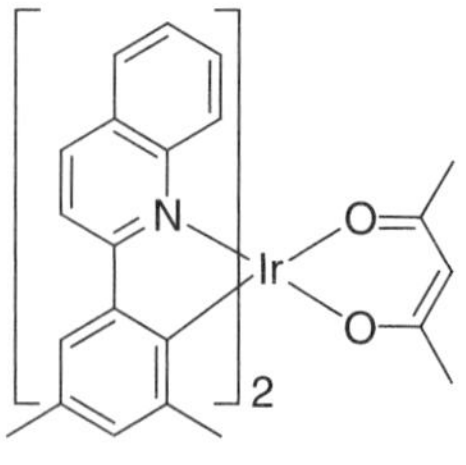   20 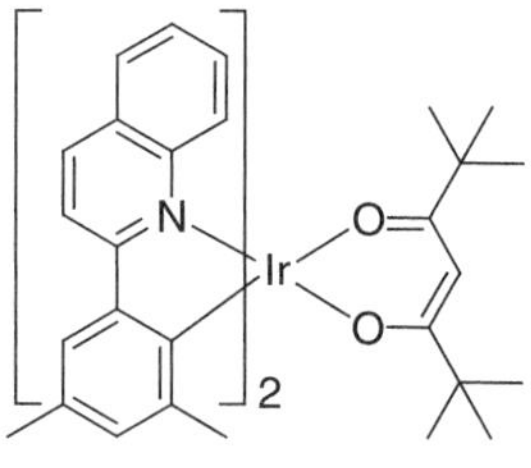   21

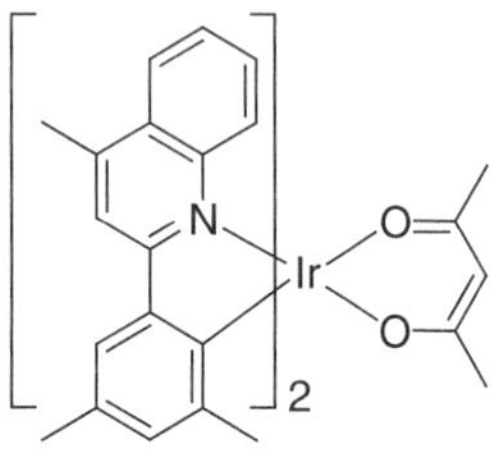

22 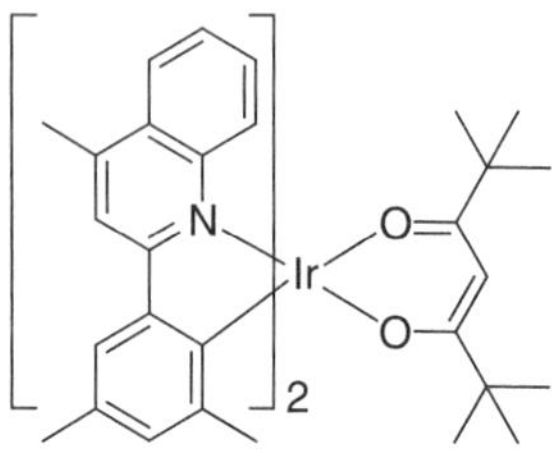   23 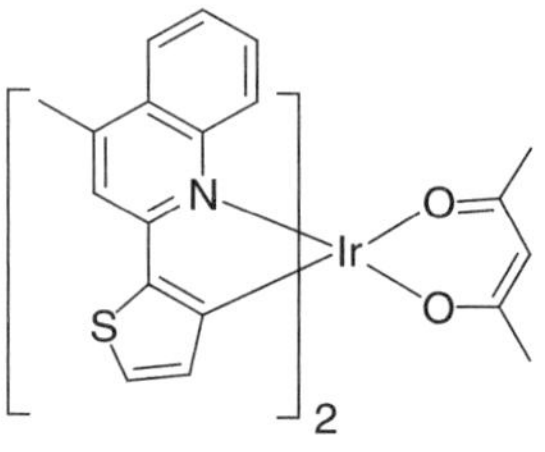   24

7.2.3   2,4-Diphenylquinoline Derivatives

Another big family of red phosphors is Ir(III) complexes based on 2,4-diphenylquinoline. From the standpoint of chemical structure, addition of a phenyl ring at the 4-position of 2-phenylquinoline does contribute to the shift of emission wavelength as the effective conjugation length increases. Therefore, 2,4-diphenylquinoline framework is able to shift the emission of Ir(III) complexes to the saturated red as the energy gap between the HOMO and LUMO levels is reduced. Complex **25** was first reported by Shu et al. (29). With respect to **19**, the 4-phenyl substituent in the 2,4-diphenylquinoline ligand leads to more reddish emission peak, which can be rationalized qualitatively by considering the decrease in the LUMO energy level that results from an increase in the π-conjugation length of the quinoline moiety induced by the 4-phenyl group. The phenyl group introduction of the quinoline ring shifts the reduction potentials of **19** (-2.41 and $-2.64\,\text{V}$) to the less negative values (-2.26 and $-2.42\,\text{V}$) in **25** because the more electron-accepting heterocyclic portion governs the reductions of this Ir(III) complex, leaving the HOMO levels unaffected. For the sake of comparison, Kim et al. systematically designed and synthesized a series of 5′-substituted derivatives of **25**, that is, Ir(III) complexes with methoxy (**26**), methyl (**27**), and fluoride (**28**) units (30). However, the Ir(III) complex with methoxy or methyl electron-donating group was not redshifted as expected. The emission wavelength of **27** was the same as that of **25**, while the λ_{EL} was even blueshifted in **26** ($\lambda_{EL} = 605\,\text{nm}$) as compared to the nonsubstituted one. The electron-withdrawing fluoro substituent causes the λ_{EL} blueshifted to $610\,\text{nm}$. The author made a calculation on the HOMO and LUMO levels on **25–28** with supportive information for the emission wavelengths of these complexes (Fig. 7.1). Although the methoxy or methyl group increased the HOMO energy level, it strongly increased the LUMO more than the HOMO and led to less MLCT mixing within their Ir(III) complex. While the fluoro substituent lowered the HOMO level of **28** intensively that caused the strongest MLCT between the Ir atom and ligand, this resulted in the best device performance of **28**. The best performance of **28** was achieved with η_L of $15.8\,\text{cd/A}$ and η_P of $12.4\,\text{lm/W}$.

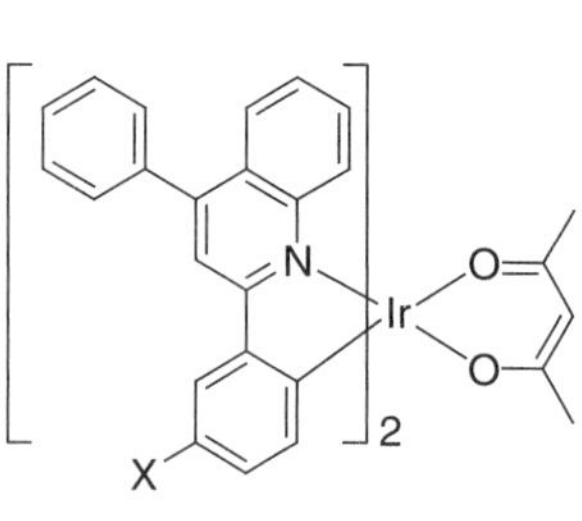

X = H   (25)
  = OCH₃   (26)
  = CH₃   (27)
  = F   (28)

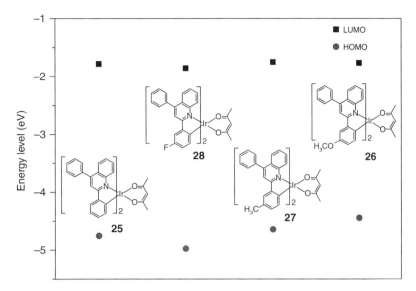

Figure 7.1 Calculated HOMO and LUMO energy levels of **25–28**.

Since the oxidation processes largely involve the Ir-aryl center whereas the reduction is generally considered to occur mainly on the heterocyclic portion of cyclometalating C^N ligand, extending the conjugated aromatic motif of the ligand framework by replacing the phenyl ring in diphenylquinoline with fluorene in **29** and **30** and naphthalene in **31** effectively elevated the HOMO level as compared to **25** with little change in the LUMO level. Likewise, a bathochromic shift in the emission wavelength in **29–31** relative to **25** can be explained by the same reason. With reference to **25**, the complex **32** with the more electron-donating ability of methylthiophene than phenyl ring, an additional 0.12 V increase in the HOMO level is revealed. However, the HOMO level elevation in **32** is not as much to the extent as in **29–31**, which implies that moving the Ir(III) complexes more reddish in emission color is more effective by extending the π-conjugation rather than replacing the phenyl ring with thiophene. Given the advantage from the presence of the long alkyl chains in fluorenyl groups of **29** and **30**, both of them are amorphous solids, which are in contrast to that of a crystalline solid in **25**. Bright-red OLED was fabricated by using **29** doped into a blend of PVK and 30 wt% of PBD. The EL emission of this device is a characteristic of **29**, with a maximum at 627 nm and CIE color coordinates of (0.68, 0.32), which satisfy the demand by NTSC. At a current density of 10.7 mA/cm^2, the η_{ext} and η_L were 10.27% and 11.0 cd/A, respectively. Even at a higher current density of 100 mA/cm^2, the device could maintain a high efficiency of 8.16%. Poorer efficiencies were reported for **30–32**.

Phenothiazine is a well-known heterocyclic compound with electron-rich sulfur and nitrogen heteroatoms. The phenothiazine group shows strong luminescence, high photoconductivity, and reversible oxidation process (31). Moreover, the nonplanar phenothiazine ring structure can restrict π-stacking aggregation, which prevents the detrimental pure singlet excitation recombination process (32). By making use of the synergistic effects between phenothiazine and phenylquinoline, Jin et al. synthesized a series of Ir(III) complexes **33–35** with different ancillary ligands of acetylacetone, picolinic acid, and picolinic acid N-oxide, respectively (33). The change of ancillary ligands moves the emission of these complexes from red to the near infrared. They emit from the predominantly ligand-centered $^3\pi-\pi^*$ excited state in the wavelength range of 626–712 nm. The HOMO level of **33** is elevated relative to **34** and **35** from −4.82 eV to −4.95 eV and −4.94 eV, respectively, while their LUMO levels are almost the same (−2.80 to −2.82 eV). Their suitable energy levels with the mixed host materials (TPD/TCTA/TPBI) (TPD, N,N'-diphenyl-N,N'-(bis(3-methylphenyl)-[1,1-biphenyl]-4,4′-diamine); TCTA, 4,4′,4″-tris(carbazol-9-yl)triphenylamine; TPBI, 5-tris[2-N-phenyl-benzimidazolyl]benzene) are essential for effectively trapping both holes and electrons in the emitting layer within the OLEDs. However, poor performances of OLED devices doped with **33–35** were noted. The maximum η_{ext} of only 0.51% in the deep-red region with CIE coordinates of (0.68, 0.30) was obtained for **35**. This may be due to the low phosphorescence yield of the dopant.

Besides, a series of carbazole-based diphenylquinoline Ir(III) complexes (**36–41**) were also reported by Jin (34–37). Systematic discussions on their structure–property relationship and efficiencies of OLEDs have been made by alternating the ancillary ligands. Because of the inherent electron-donating nature, intense luminescence, good hole-transporting properties, and high thermal stability of carbazole moiety, all these Ir(III) complexes insure the excellent thermal stabilities and good electro-optical properties. Similar to the case for the phenothiazine-based one, the use of different ancillary ligand effectively modifies the photophysical and electrochemical properties of the resulting complexes. In comparison to their physical properties, the complexes with acac ancillary ligand tend to possess a relatively lower decomposition temperature. Poorer thermal stabilities were also detected for those bearing 2-(2-methoxyethoxy)ethyl chain. In terms of PL properties, complexes with acac show more redshifts in the emission wavelength (λ_{PL} = 619 nm for **36**; 616 nm for **37**) than those with picolinic acid and picolinic acid N-oxide (λ_{PL} = 595 nm for **38–41**). The device architecture employed was ITO/PEDOT/**36–41**/BAlq/Alq$_3$/Liq/Al. Both λ_{EL} of **36** and **37** were located at 623 nm with the CIE coordinates of (0.65, 0.34), but the λ_{EL} of **38–41** shifted to 604 nm, and the emission maximum and the CIE values are insensitive to the identity of alkyl or alkoxy chain. The performance of the Ir(III) complexes containing picolinic acid N-oxide in OLEDs was remarkably higher than the other. It was believed that the electron-withdrawing nature and stronger negative inductive effect of picolinic acid N-oxide ligand could induce higher electron mobilities and result in an improved device performance. The highest L_{max} of 18,500 cd/m^2 was obtained for **41** at 215 mA/cm^2 with η_{ext} of 5.53% and η_L of 8.89 cd/A. A slightly poorer performance was obtained for **40** with η_{ext} and η_L of 4.9% and 8.36 cd/A, respectively. To boost up the efficiencies of the red OLEDs by using complexes **38** and **39**, different device configuration of ITO/PEDOT:PSS/PVK:OXD-7:TPD:**38** or **39**/cathode was adopted (37). (OXD: 1,3-bis[5-(4-*tert*-butylphenyl)-1,3,4-oxadiazol-2-yl]benzene). The use of TPD with PVK as a codopant would cause a reduction in the barrier for hole injection and can also serve as a hole-transporting layer, which allows holes to penetrate more deeply into the emittive layer. The efficiencies of the red OLEDs based on **38** and **39** were greatly enhanced and were TPD concentration dependent. All the devices displayed emission peaks at 608 nm with excellent color purity at the CIE coordinates of (0.62, 0.38). The device efficiencies increased as the TPD concentration increased and reached the maximum at 16 wt% due to the reduction of injection barrier from 0.6 to 0.3 eV in the presence of TPD. At this doping concentration, the L_{max} of the OLEDs using **38** and **39** are 8768 and 6285 cd/m^2, respectively. The OLEDs fabricated without the TPD have η_{ext}, η_L, and η_P of 6.21%, 9.15 cd/A, and 1.93 lm/W, respectively, for **38**, and were significantly increased to 10.56%, 6.42 lm/W, and 17.5 cd/A at 16 wt% of TPD. Similar observation was detected for **39** and its maximum η_{ext}, η_L, and η_P are 9.65%, 15.92 cd/A, and 4.34 lm/W, respectively. This work represents a great achievement on the research of solution-processed red-emissive OLEDs. The use of 2-pyrazinecarboxylic acid as the ancillary chelate in **42** slightly redshifted the λ_{PL} to 613 nm; however, if 5-methyl-2-pyrazinecarboxylic acid was used instead in **43**, a blueshift in λ_{PL} was obtained (λ_{PL} = 600 nm) (36). The methyl unit mainly alters the LUMO levels from −2.70 to −2.85 eV of the Ir(III) complexes without affecting their HOMO levels. CIE coordinates of (0.60, 0.39) were obtained for both devices but the more efficient electron injection and transport in **43** than **42** make the device performance of the former better. This is governed by the better matching of the LUMO gap of **43** with adjacent layers in the device. This device exhibited maximum η_{ext}, η_L, and η_P of 3.38%, 6.69 cd/A, and 1.80 lm/W, respectively.

To investigate the structure–property relationship on the ligation position of Ir(III) with C-2 and C-3 positions of the carbazole moieties, complex **44** was also prepared by the same research group, which resulted in the creation of different bandgap in a tunable manner as compared to **40** (35). The PL spectrum of **44** contains an emission peak at 668 nm. Thus, by slightly modifying the molecular structure of the carbazole–quinoline chelate, the emission peak of the Ir(III) complex was redshifted by 73 nm relative to **40**. Solution-processed PhOLEDs doped with **40** or **44** have respective peak η_{ext} of 8.74% with CIE coordinates of (0.62, 0.37) and 4.32% with deep-red CIE coordinates of (0.68, 0.28) using PVK as the host material. The combination of the high device performance and good color purity makes **44** promising candidates as deep-red phosphorescent dopant for solution-processed PhOLEDs.

Dendrimers have proved to be successful materials for solution-processable phOLEDs, in which an Ir(III) complex is surrounded by a branched shell of cyclometalating ligand to prevent self-aggregation, phase segregation, or concentration quenching of the emissive core in the solid state. By using this protection, Wang et al. reported carbazole-dendronized red dopants **45–47** by associating oligocarbazole with Ir(III) complexes by covalent bonding to form single, multifunctional dendrimers (38). The oligocarbazole dendrons not only serve as red emitter but also work as the host materials. All the dendrimers exhibited bright-red emission between 615 and 622 nm, and their emission maxima show only a small redshift of 4–11 nm as compared to **25**. Effective tuning of the intermolecular interaction can be addressed by the dendron generation and the reduced aggregation upon increasing the size of the dendrons was confirmed by the observed increase of their lifetimes. Good film-forming and charge-transporting properties of **45–47** enable them to be fabricated as OLEDs by using solution-processing technique with the nondoped configuration: ITO/PEDOT:PSS/dendrimer:TCCz-PBD/BCP/Alq/LiF/Al (TCCz N-(4-(carbazol-9-yl)phenyl)-3,6-bis(carbazol-9-yl)carbazole). A L_{max} of 1990 cd/m^2 and a maximum η_{ext} of 6.3% (4.1 cd/A, 2.4 lm/W) with the CIE coordinates of (0.67, 0.33) have been demonstrated. It should be noted that the η_{ext} of **47** is almost 4, 10, and 31 times higher than that of **46**, **45**, and **25**, which is attributed to the depressed self-quenching effect of the emissive cores upon increasing the size of the dendrons. Doped devices were also studied by Wang and coworkers as a comparison. At the brightness of 100 cd/m^2, the η_{ext} of nondoped device based on **47** at 5.0 wt% is only about 16.7% lower than that of the corresponding doped device and is very close to that of the device (5.4%) based on **25** at the optimized doping level of 2 wt%. This indicates that the nondoped device structure can be used for this kind of complexes without significant loss in efficiency. Further improvement of the device performance was carried out by reducing the thickness of the BCP layer. Pure red light with the CIE coordinates of (0.65, 0.35) for **47** was obtained, and it was found to be independent of the current density. An η_{ext} of 11.8%, an η_L of 13.0 cd/A, and an η_P of 7.2 lm/W at a brightness of 100 cd/m^2 were obtained. This molecular design strategy of combining the host and dopant to form a single multifunctional dendrimer is an efficient approach for the development of solution-processable nondoped red phosphorescent materials.

Complex **48**, an isomer of **25**, with 2,3-diphenylquinoline as the cyclometalating ligand caused a bathochromic shift of 19 nm in wavelength in its PL and EL as compared to **25** (39). The 3-phenyl group increases the probability of $\pi-\pi$ conjugation of the cyclometalating ligand and lowers the triplet energy level, thus lowering the LUMO level to result in such a shift. A red emission device with 5 wt% in dopant **48** showed $\lambda_{EL} = 620$ nm with a shoulder at about 668 nm. Its CIE coordinates changed slightly at different driving voltages (from (0.66, 0.34) at 9 V to (0.65, 0.34) at 18 V) but are still well within the red region. A V_{on} of 6 V, L_{max} of 22,040 cd/m^2 at 18 V, η_P of 2.4 lm/W, and η_L of 11.4 cd/A at 15 V were achieved, demonstrating that **48** is another dopant suitable for red OLEDs. Extending the π-electron delocalization at the quinoline ligand portion in **49** gave a longer λ_{PL} centered at 604 nm with attractive Φ_{PL} of 62% (40). Quite short phosphorescent lifetime and good emission quantum yield of **49** make it to be a promising candidate for highly efficient red OLED. Accordingly, the device based on **49** exhibited high performance with small roll-off in the corresponding device efficiency with increasing current density. It gave a maximum η_L of 17.4 cd/A and η_{ext} of 10.5%. Due to the blueshift of the λ_{EL} (604 nm) of this device, it was believed to possess great potential for high-efficiency two-element white light emission.

7.2.4 Pyridine-Based Derivatives

Instead of using fused ring in quinoline, the conjugation in pyridine-based cyclometalates of Ir(III) complexes should be extended either in the phenyl or pyridine ring to obtain red light. In complex **50**, the less electron-donating 2-position of carbazole lowers the energy of metal d orbital when it is ligated to the metal ion, giving rise to a lower energy level of the HOMO that is accompanied by a significant bathochromic shift in emission wavelength to give the red color (41). The carbazolyl ligand contribution to the excited states increases in the 2-ligated complex as compared to the 3-ligated one. The good hole injection and hole transporting properties of **50** render it to be a dual functional triplet emitter in OLEDs. The **50**-based device exhibited an L_{max} of 51,000 cd/m^2 at 15 V and respectable EL efficiencies (13.7%, 25.6 cd/A, and 19 lm/W) with the CIE coordinates of (0.60, 0.39) obtained by vacuum deposition technique. Replacing the phenyl ring at the 9-position of carbazole by a decyl chain in **51** shows a similar λ_{PL} at 594 nm; however, both of the HOMO (-4.74 eV) and LUMO (-2.31 eV) levels are elevated in **51** (42). The color of **51**-based solution-processed OLEDs with ITO/PEDOT:PSS/**51**:CzOXD/BCP/Alq$_3$/LiF/Al configuration shows a doping concentration dependence. The CIE coordinates shift from (0.64, 0.36) to (0.50, 0.36) when the doping concentration is reduced from 5 to 3 wt%. A V_{on} of 11.3 V, an L_{max} of 4894 cd/m^2, and a maximum η_L of 4.6 cd/A at 14.8 mA/cm^2 were achieved at 5 wt% with only a gentle decay in efficiencies with increasing current density.

Similar to other Ir(III) phosphors, the difference in emission energies and electrochemical properties of carbazole-based Ir(III) complexes can be rationalized by substitution approach. Bryce and coworkers reported two red-emitting homoleptic Ir(III) complexes **52** and **53** with electron-withdrawing CF$_3$ group attached at the 5- and 4-sites on the pyridine ring, respectively (43). As compared to the unsubstituted one, the CF$_3$ group shifts the oxidation potentials to more positive values, but the position of the CF$_3$ group on pyridine ring shows little dependence on the HOMO level. A better performance of solution-processed OLEDs was observed in **53** than in **52** when they are blended in the high-triplet-energy poly(9-vinylcarbazole) (PVK) polymer host. The best performance was recorded at η_L of 10.3 cd/A and η_{ext} of 5.6%. These studies provide new insights into chemically tailoring Ir complexes of carbazolylpyridine ligands, leading to red OLEDs with good color tunability appropriate for full-color display technology.

Another approach for this class of Ir(III) complex to obtain red light emission is to extend the effective electronic conjugation of the pyridine ring. The earliest example for obtaining red emission in Ir(III) complex by using this approach was achieved by coupling two fluorene units at the ortho position on the pyridine ring in **54**, and red emission located at 600 nm with a shoulder at 620 nm was observed (44). By connecting the two fluorene units, the solid-state quenching can be minimized and the electron–hole trapping on the Ir(III) complex was improved. A single active layer configuration with the PEDOT:PSS on ITO as the hole injecting bilayer electrode was employed and **54** was doped in a

PVK–PBD mixture. The emission Φ_{PL} quickly increased as the doping concentration increased since the energy was efficiently transferred from the host to the guest until the triplet emission of **54** became saturated at a very high doping concentration. The L_{max} was over 2600 cd/m^2 with maximum η_{ext}, η_L, and η_P of 5%, 7.2 cd/A, and 1.33 lm/W, respectively.

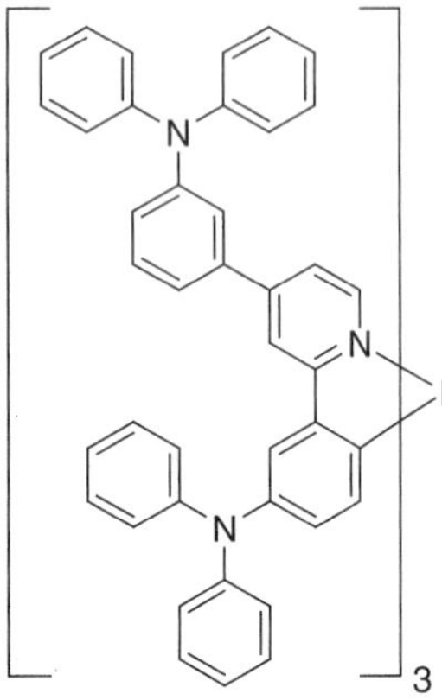

55

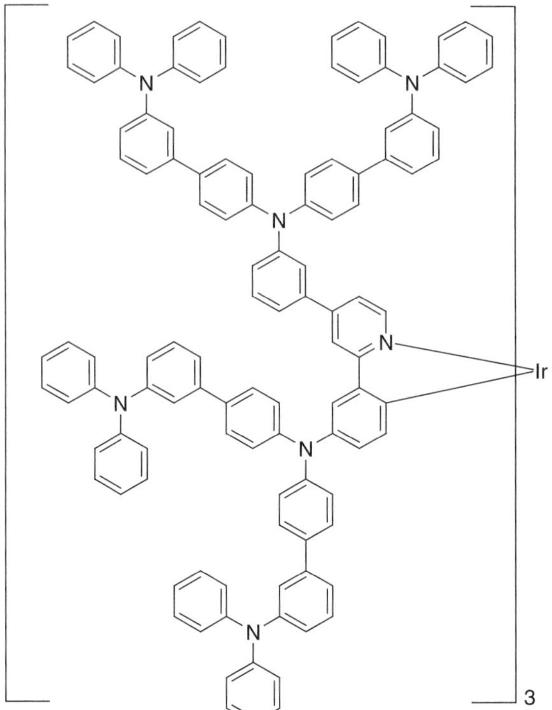

56

Yang and Wu successfully encapsulated triphenylamine and dendritic triphenylamine cores on pyridine ring peripherally, which also shift the phosphorescence peak to saturated red (45). In solution, both **55** and **56** display a structureless emission centered at 608 nm. The triphenylamine chelates in **55** and **56** serve as good antenna for the charge transfer and/or energy transfer to the emissive center and maintain the high-lying HOMO level (circa −5.2 eV) and sufficient triplet energy (circa 2.9 eV). In addition, the arylamine group directly helps to get rid of the close molecular packing in the solid state. Their high phosphorescent Φ_{PL} (Φ_{PL} = 42 and 43% for **55** and **56**, respectively) in PMMA doped films and reasonable short lifetimes benefit them to achieve high-performance OLED devices. Devices based on **55** at different doping concentrations were fabricated with a simple configuration (ITO/PEDOT:PSS/PVK:PBD:**55**/Ba/Al) and all of them exhibited bright red emission. Owing to the short phosphorescent lifetime of **55** and hence the reduced triplet–triplet annihilation effect, the device parameters were maintained high as the luminance increased. The peak η_{ext} reached 13.8% with 10 wt% level of complex **55** and turns to brighter and more efficient value at η_{ext} of 15.3% with CIE coordinates of (0.61, 0.39) at 30 wt% doping concentration. Such structural protection of the emissive core efficiently gave rise to high-performance solution-processable red OLED devices.

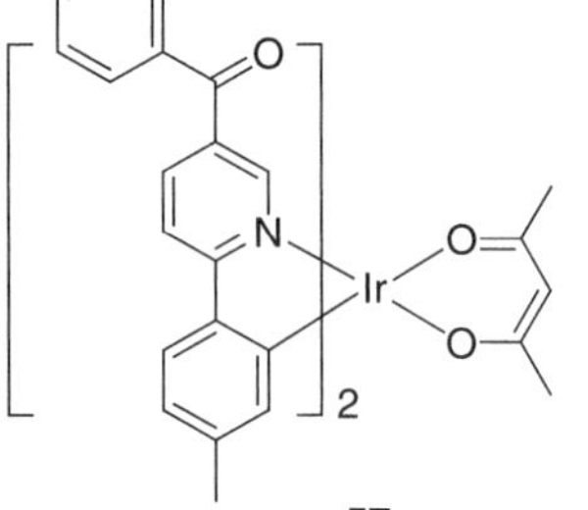

57 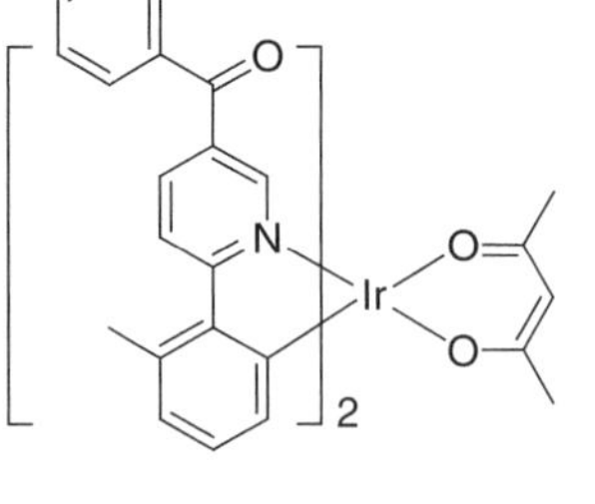   **58** 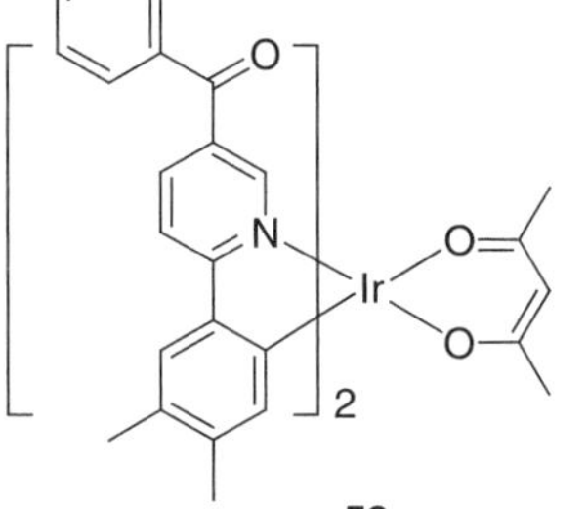   **59** 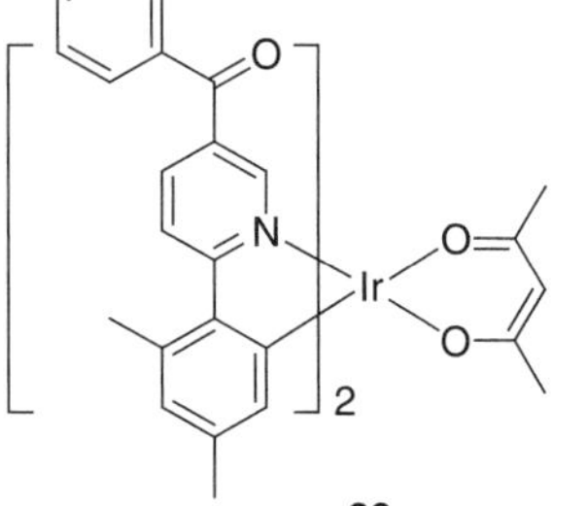   **60**

Electronic effects with inductively electron-donating or electron-withdrawing groups have to be considered for their impact on orbital energies of the Ir(III) complexes. Adding an electron-withdrawing substituent on the pyridyl ring or phenyl ring can stabilize the LUMO or HOMO level, respectively, and results in the narrowing of HOMO–LUMO gaps and makes red light emission possible. Complexes **57–60** based on 5-benzoyl-2-phenylpyridine ligands with methyl moieties can tune the emission color bathochromically as compared to [Ir(ppy)$_2$(acac)] (46). Their peak emission wavelengths ranged from 618 to 648 nm in the red region of the visible spectrum. Compared to complex **57** with a methyl group on the 4-position of the phenyl ring, the emission spectrum of **58** with a methyl group at the 6-position was

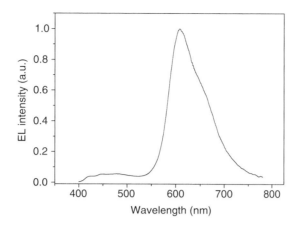

Figure 7.2 EL spectrum for **61**-based red-emitting device.

redshifted by circa 14 nm. The larger distorted angle between the phenyl and pyridine rings is caused by the 8-substituted methyl group in **58** than in **57**, leading to a small energy gap in **58** in comparison to **57**. The PL spectra of complexes **59** and **60** showed redshifted emissions by circa 15–30 nm as compared to that of complex **57** due to the increased electron density in the phenyl moiety caused by the introduction of two methyl units that destabilized the HOMO energy level of the iridium complexes. Increased methyl group addition on the phenyl chromophore did not cause any significant difference in the absorption and PL spectra. Comparing those Ir(III) complexes with one methyl group, **57** and **58**, the HOMO energy levels of the complexes with two methyl groups, **59** and **60**, were approximately 0.11–0.18 eV higher than the complexes with one methyl group. The subtle structural and electronic changes in the ligands also significantly affected the EL of the complexes. The respective CIE coordinates of the devices were (0.62, 0.38), (0.65, 0.35), (0.66, 0.34), and (0.65, 0.35) for **57–60**. In comparison with the peak η_{ext} of the devices with **57** and **58**, the value for **60** was increased by approximately 26% and 15%, respectively, at 20 mA/cm^2, while that of device **59** was decreased by 5.7% and 14%, respectively, at 20 mA/cm^2. This study suggests that structural modification of the ligands of the Ir(III) phosphors could greatly improve the EL performance of OLEDs doped with them.

In contrast to other typical Ir(III) complexes in the literature, complexes **61** with boron functionality and **62** with an electron-trapping fluorenone unit show an interesting color tuning principle (47, 48). They show obvious red shift in their PL spectra when an electron-withdrawing chromophore is introduced to the phenyl moiety of phenylpyridine. They emit at 605 and 615 nm in their solution state, respectively. The boron moiety in **61** becomes one of the major LUMO contributors that results in significant electron density transfer through the MLCT process to the B(Mes)$_2$ to stabilize the MLCT state strongly (Fig. 7.2). Similarly, electron-deficient fluorenone ring in **62** acts as an electron sink in the MLCT process leading to its red emission and their EL spectrum lies in the red region of CIE diagram (Fig. 7.3). Impressive device performance was obtained with **61** doped in PVK, which was found to have a low V_{on} of 3.7 V, an L_{max} of 16,148 cd/m^2 at 19.1 V, an η_{ext} of 9.4%, an η_L of 10.3 cd/A, and an η_p of 5.0 lm/W (Fig. 7.4). The efficiency roll-off at high current densities is not severe in this device, implying that the triplet–triplet annihilation effect is not very significant at high current density and this may be attributed to the suppression of self-quenching process by the bulky mesityl group. Reasonable OLED parameters were obtained with **62** showing a V_{on} of 5.9 V, an L_{max} of 4803 cd/m^2, a peak η_{ext} of 1.29%, an η_L of 1.48 cd/A, and an η_p of 0.49 lm/W (Fig. 7.5). These rare systems with promising electron injection and electron-transporting functionalities benefited from the electron-withdrawing moieties are vital for the optimization of future red-emitting device.

61   **62**

As pointed out earlier, a dendritic matrix around the phosphorescent dye at a molecular level can effectively control the self-quenching and thus optimize the optoelectronic properties. The site isolation effect provided by the bulky peripheral groups minimizes the undesired core–core interaction. Highly branched dendrons **63–65** integrated with rigid polyphenylene attached to 2-benzo[*b*]thiophen-2-yl-pyridyl with a hole-transporting triphenylamine surface as end groups were reported by Müllen et al. (49) The covalent linkage of inner chromophoric Ir(III) complex core and outer triphenylamine units by polyphenylene can enhance the hole capture and hole injection and improve the charge

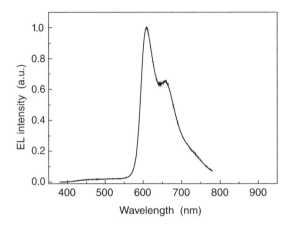

Figure 7.3   EL spectrum for **62**-based red-emitting device.

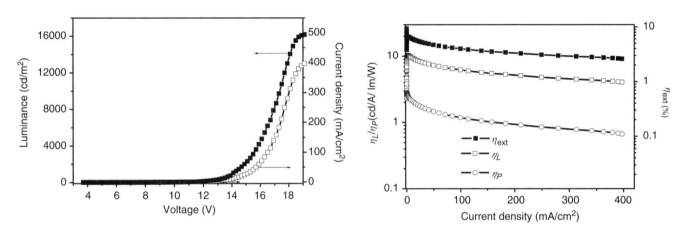

Figure 7.4   The luminance–voltage–current density curves and the efficiency versus current density relationship for **61**-based device.

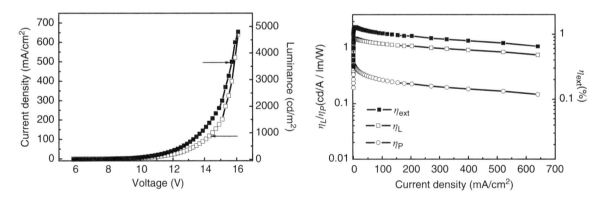

Figure 7.5   The luminance–voltage–current density curves and the efficiency versus current density relationship for **62**-based device.

recombination for light emission. The polyphenylene could participate in the electrochemical process and charge transport of the dendrimers as well as controlling the intermolecular interactions without changing the HOMO and LUMO levels with increasing the dendrimer generation. All three dendrimers possess similar PL emission bands at 621 nm with an additional shoulder at 670 nm and demonstrate similar EL peaks at 624 nm. Very close CIE coordinates to the NTSC standard for red subpixels were recorded for **63–65**. Deep-red light-emitting phosphorescent dendrimers **66–68** with carbazolyl moiety tethered to 2-benzo[*b*]thiophen-2-yl-pyridyl ligand through a nonconjugated spacer exhibited emission spectra with an identical shape and a characteristic phosphorescence emission at around 641–644 nm that are almost identical to that of the Ir(III) core (50). High encapsulation of Ir(III) unit in **66–68** by the multicarbazole dendrons ensures more efficient energy transfer within the complexes without spatial geometrical congestion, which results in increasing PL intensity with dendrimer generation. High-quality films

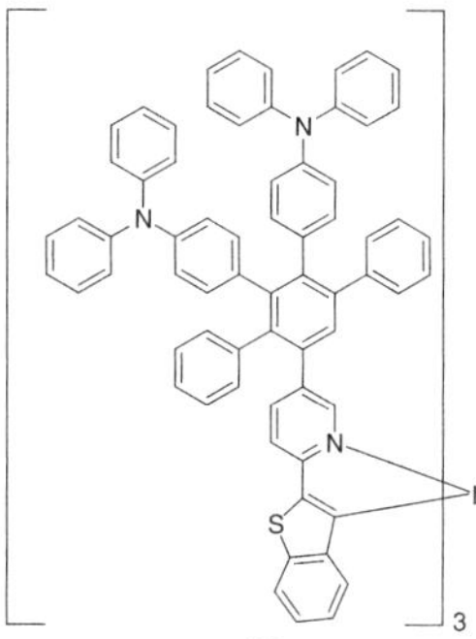

63

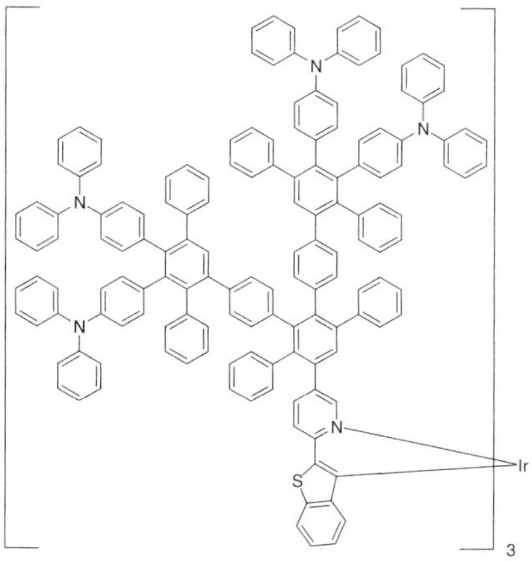

64

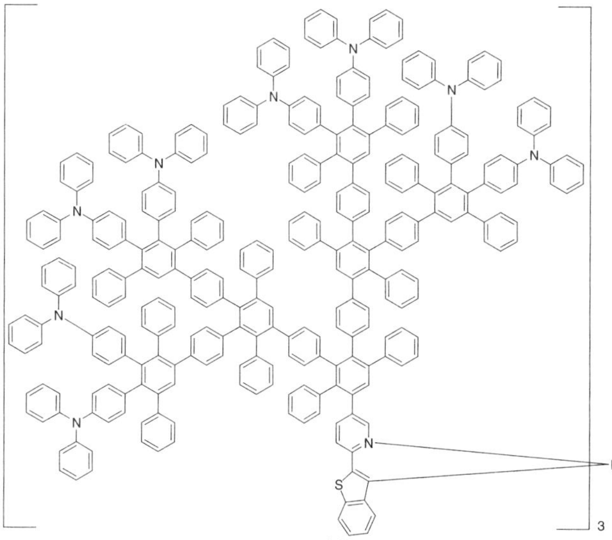

65

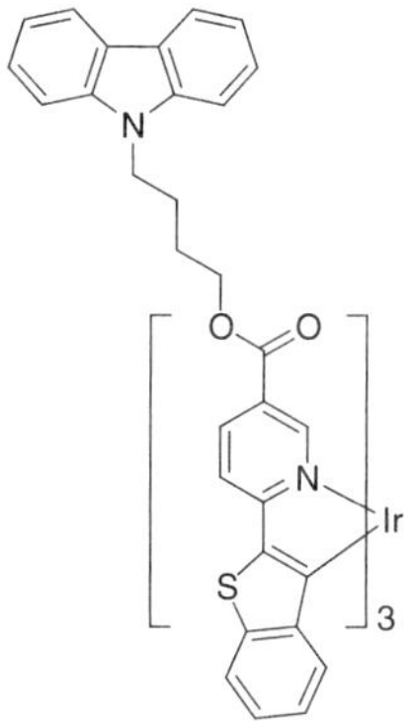

66

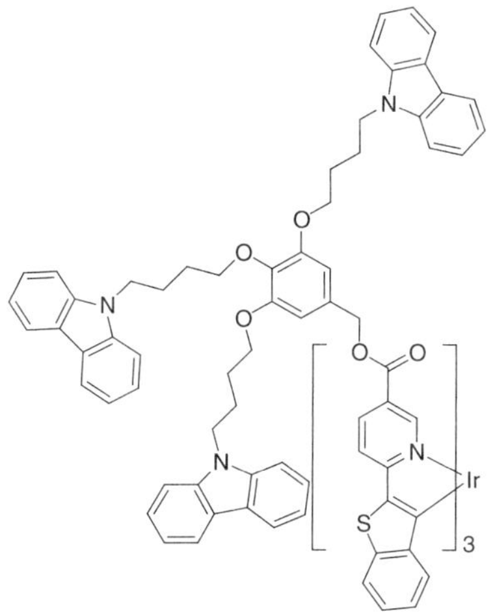

67

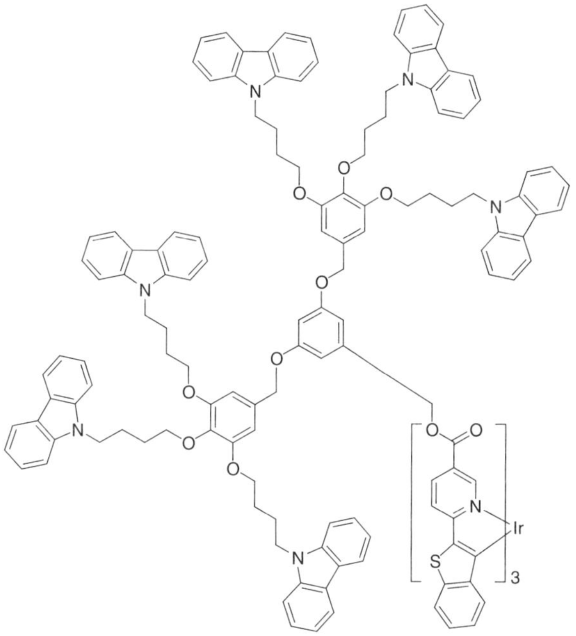

68

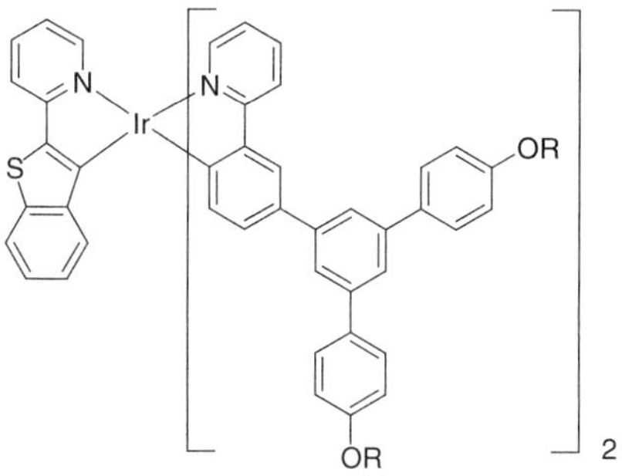

69

by spin coating for host-free OLED systems were obtained by using **66–68**. The dielectric nature with increasing size of nonconjugated spacer showed a slightly higher V_{on} in **68**. They exhibited deep-red emission in the range of 652–656 nm under forward bias voltage. At a relatively low current density, η_{ext} of **68** is higher than that of **66** but the case is reversed as current density increases. This is attributed to the better hole-transporting properties in the largest dendrimer but fast decay of efficiencies at high current density was observed. It can be conjectured that hole and electron transport becomes unbalanced, which is caused by acceleration of the hole mobility rather than the electron mobility.

Complex **69** with mixed cyclometalating ligands showed peaks at 600 and 652 nm in the 20:80 wt% **69**:CBP blend film (51). It gave a pure red EL color with CIE coordinates of (0.64, 0.36). A preliminary OLED result based on **69** was achieved with η_L of 7 cd/A and η_P of 4.0 lm/W at 100 cd/m² in a bilayer structure.

7.2.5   Other Ir(III) Red Phosphors

Besides the Ir(III) phosphors mentioned earlier, some other functionalized complexes also demonstrate red EL bands. Complex **70** with 9,9′-dimethylfluorenyl-substituted benzothiazole cyclometalating ligands showed compromised red phosphorescence (52). It was found to display red emission at 591 nm with moderate PL efficiency of 32.7% at 5 wt% **70** doped in CBP. It possesses alleviated self-quenching characteristics as Φ_{PL} remained as high as 16% at the dopant concentration of 20 wt%. Devices based on **70** at different doping ratios were also fabricated, which emit efficient red light at 600 nm (with shoulder at 653 nm) with CIE coordinates of (0.63, 0.36) and low efficiency roll-off at relatively high current density. The peak η_L and η_{ext} of 28.5 cd/A and 15.6%, respectively, were obtained at 15 wt% doping level.

70

Phthalazine compounds have a distorted heterocyclic biphenyl structure with two adjacent nitrogen atoms in one aromatic ring and had been widely used in red phosphorescent Ir(III) complexes with excellent luminescent properties. Compared with other common C^N=CH Ir(III) complexes, this C^N=N family had high thermal stabilities. This kind of complexes was easily synthesized under mild synthetic conditions. Phthalazine-based ligands could bond to the Ir atom more strongly as compared to the phenylisoquinoline one as it would create a "proximity effect." The strong bonding between metal and ligand would lead to efficient mixing of the singlet and triplet excited states and is beneficial for phosphorescence. The enlargement of the conjugated π-system in the phthalazine unit in complexes **71–77** results in their saturated red PL emission. Complex **71** emits red light peaking at 625 nm in CH_2Cl_2 solution, with a Φ_{PL} of 20% (53). The HOMO of **71** is contributed both from the Ir(III) ion and ligands and the contribution in the LUMO comes mainly from the ligand; thus, the emission of **71** is assigned to a mixture of MLCT and ligand-centered ($\pi-\pi^*$) transitions. The high thermal decomposition temperature (426 °C) of complex **71** demonstrates its potential in OLED application. The 5 wt% **71** doped device presents the best performance with η_{ext} of 8.3% and CIE coordinates of (0.69, 0.30). However, these devices suffer from severe triplet–triplet annihilation at high excitation density. A tentative lifetime of **71**-based device was tested and a lifetime of 610 h at an initial luminance of 100 cd/m² was achieved. Wang et al. reported two functionalized phenylphthalazine Ir(III) derivatives **72–73** with carbazole and diphenylamine moieties for red OLEDs (54, 55). Similarly, the incorporation of carbazole and diphenylamine units is found to improve the charge balance in the EL process, extend the π-electron delocalization of the aromatic ligand chromophore, and enhance the thermal stabilities of the compounds. With regard to the PL properties of **72–73**, they emit at 615 and 614 nm, respectively, upon irradiation with 400 nm light. The HOMO level of **73** was elevated to −5.16 eV by introducing the amine group as compared to **72** (HOMO: −5.35 eV). The solution-processed devices using **72–73** as dopant were fabricated with the structure: ITO/PEDOT:PSS/**72** or **73**:PVK+PBD/TPBI/Ba/Al. Both of the complexes emit at around 620 nm with fine vibronic structure. Higher device efficiency was obtained for **72** as it gave an L_{max} of 2948 cd/m² at a current density of 115.6 mA/cm². A maximum η_{ext} of 20.2% corresponding to an η_L of 11.3 cd/A was obtained at a current density of 0.18 mA/cm². The outstanding performance of this device can be explained by the high Φ_{PL} of **72** (Φ_{PL} = 0.46) and the reduction of nonradiative transition of the rigid ligand. The **73**-doped device at 4 wt% doping level gave a maximum η_{ext} of 13.6% corresponding to an η_L of 7.4 cd/A at a current density of 0.73 mA/cm². However, similar to **71**, both devices suffer from severe efficiency roll-off at high current density, which is probably induced by the planar structure of the cyclometalating ligand, which tends to cause aggregation of molecules at high current density and leads to excited-state intermolecular interactions. Therefore, further optimization of the devices based on **72** and **73** is needed for practical use.

Two novel Ir(III) complexes **74** and **75** with carbazole-based phthalazine-derivative were reported recently (56). They contain various functional units in order to achieve high triplet energy and good hole-transporting ability. In comparison, **75** with triphenylamine unit (λ_{PL} = 654 nm) instead of carbazole in **74** (λ_{PL} = 640 nm) led to a redshift of 14 nm in its PL profile, which suggested that the carbazole moiety in this system increased the bandgap energy. The conjugation extension of the cyclometalating ligand and the pyridine ring of 2-picolinic acid redshifted the emission as compared to **72**. Their HOMO levels are very similar ($-5.17\,eV$), which indicate that the introduction of carbazole group can regulate the energy level as the HOMO primarily resided on the Ir(III) center and 9-ethyl-9H-carbazole of the cyclometalating ligand. Their HOMO and LUMO levels were embedded between the HOMO of PVK and the LUMO of PBD, which ensure holes and electrons are effectively trapped within the emissive layer in the device. The maximum emission peak of **74** appears at 656 nm, and that of **75** at 662 nm, and both correspond to saturated red emission ((0.678, 0.291) for **74** and (0.688, 0.287) for **75**). Aggregation emission peaks were observed at 707 nm in **75**-doped OLEDs. They had a good light stability as their CIE coordinates remained almost unchanged at various doping concentration. The overall luminance of **75** was much lower than that of **74**, which indicates more balanced charge recombination and favorable electrical excitation in the EL process was achieved in the latter case. The maximum η_{ext} was 16.3% and 11.9% for **74**- and **75**-based devices, respectively. Integrating electron-rich thiophene ring with phthalazine in the homoleptic complexes **76** and **77** effectively redshifted the emission bands as compared to the structurally similar compound **72** (57). Both complexes showed saturated red emission at 648 nm with FWHM from 35 to 41 nm in the PL spectra. Changes of the substituents in **76** and **77** did not affect the effective conjugation length of 1-(thiophen-2-yl)phthalazine derivatives that resulted in an almost identical EL emission at approximately 656 nm. With the same device structure by spin-coating technique, device using **77** as the dopant has a higher efficiency and lower V_{on} than device based on **76**, which contains 2,6-dimethylphenol group instead of carbazole. **77**-based device achieved an L_{max} of 1589 cd/m^2, an η_L of 26 cd/m^2, and an η_{ext} of 10.2%. The high thermal stability, outstanding EL properties, and high PL efficiency of this class of Ir(III) materials with C^N=N structural ligand were believed to be promising for optoelectronic applications in the future.

Ir(III) complexes **78–79** with dibenzo-[f,h]quinoxaline derivatives were first reported by Cheng et al. as red emitters (58). They emit at around 610 nm with attractive Φ_{PL} of 0.53 and 0.48 for **78** and **79**, respectively. The devices based on these metal complexes with exceedingly high brightness (65,040 cd/m^2 for **78**; 59,560 cd/m^2 for **79**) and η_{ext} (11.9% for **78**; 10.4% for **79**), η_L (23.3 cd/A for **78**; 21.7 cd/A for **79**), and η_P (7.9 lm/W for **78**; 8.4 lm/W for **79**). Very recently, by applying the red phosphor **79** with device configuration ITO/MoO$_3$/NPB/TCTA/Be(PPI)$_2$ or Zn(PPI)$_2$:**79**/TPBI/LiF/Al, it further enhances the device performance significantly (Be(PPI)$_2$ and Zn(PPI)$_2$: 2-(1-phenyl-1H-phenanthro[9,10-d]imidazol-2-yl)phenol-based beryllium and zinc complexes, respectively) (59). Particularly, the Be(PPI)$_2$-based devices exhibited lower efficiency roll-off than Zn(PPI)$_2$-based one, which could be attributed to the more balanced carrier-transport characteristics of Be(PPI)$_2$. Very low V_{on} of 2.3 V was detected for both devices. The brightness of Be(PPI)$_2$-based device reached as high as 38,580 cd/m^2 with η_L of 15.9 cd/A, η_P of 19.9 lm/W, and η_{ext} of 15.1% with neglectable efficiency roll-off. These results revealed that by a suitable choice of bipolar transport material for the dopant, a remarkable improvement in efficiency of OLED device can be obtained.

The OLED performances of three heteroleptic red phosphorescent iridium(III) complexes bearing two 2-(4-fluorophenyl)-3-methyl-quinoxaline cyclometalated ligands combined with one of the following ancillary ligands, triazolylpyridine (**80**), picolinate (**81**), and acetylacetonate (**82**), were examined by Johannes et al. in 2010 (60). They emit in the spectral range of 605–628 nm in the solution state

78   **79**

from their ^{3}MLCT states, suggesting that their emission colors are influenced by the ancillary ligand. The influence of 5d-electron density of the Ir(III) center on their HOMOs led to high Φ_{PL} ($\Phi_{PL} = 0.39 - 0.42$) and short triplet lifetimes. Different ancillary ligands not only affect the emission wavelengths but also their electrochemical energy levels. The oxidation peaks are shifted to higher potentials and the first reduction peaks appear to be less negative with an increase in the π-accepting character of the ancillary ligands. The electron-withdrawing character of the fluorine substituent on these complexes leads to low-lying HOMO and LUMO energies as compared to other 2-phenylpyridyl Ir(III) complexes. Limited efficiency roll-off over a wide range of current density in the devices was demonstrated by using **80–82**. All the devices showed low V_{on} and high brightness in the range of 2.4–2.9 V and 13,252–28,019 cd/m^2, respectively. Device based on **81** showed the highest device efficiencies with η_{ext} of 12% and η_P of 14.6 lm/W. By using **82**, OLED with superior device stability with extrapolated lifetime of 58,000 h was realized. The device lifetimes are strongly correlated to the type of ancillary ligand used.

80   **81**   **82**

Ir(III) complex **83** containing 4-phenylquinazoline with hole-transporting diphenylamine unit displayed a strong red phosphorescent emission at 599 nm with a PL efficiency of 11.6% in the solution state (61). A much higher Φ_{PL} was detected in doped films with PVK:PBD blends, and it increased from 31.2% to a maximum of 54.87% on going from 1 to 4 wt-% doping concentration. Solution-casted devices based on **83** at different doping concentrations were fabricated, and complete energy transfer was achieved from the host to **83**. All the devices showed the main emission centered at 616 nm and an additional weak emission peak at about 438 nm from the host PVK–PBD at 1% and 2% doping concentrations. The electroluminescent spectral stability was very good and the CIE plots were almost identical in which the coordinates are located near (0.62, 0.38). The device at 8% doping concentration achieved an L_{max} of 7107 cd/m^2 at a current density of 78.2 mA/cm^2 and the V_{on} is 13.5 V. The maximum η_{ext} and η_L of the device at 8% doping concentration is 18.44% and 20.73 cd/A, respectively. The neglectable efficiency roll-off indicates the balance of electron and hole recombination in the host matrix, and complete energy and charge transfer from the host material to Ir(III) complex upon electrical excitation. By further optimization of the molecular and device structures with suitable energy transfer in OLEDs, promising phosphorescent devices could be achieved by using 4-phenylquinazoline-based Ir(III) complex. Employment of Ir(III) biscarbene complex **84** as the phosphorescent emitter for the EL devices gave extremely high device efficiencies (62). Basically, Ir(III) tris(carbene) complexes are known to have high triplet energy gaps and can be used as blue emitters. By using 1-(1*H*-pyrrol-2-yl)isoquinoline to form a heteroleptic Ir(III) complex **84**, it emits at 599 nm. Device based on **84** as the dopant emitter consists of the layers, ITO/NPB/TCTA/CBP:**84**/BCP/Alq/LiF/Al, which gives a red emission with the CIE coordinates of (0.60, 0.39) and reveals an extremely high η_{ext} of 24.9%, η_L of 55.4 cd/A, η_P of 43.6 lm/W, and L_{max} of 16,572 cd/m^2.

83   **84**

Using a cationic Ir(III) complex **85** as dopant, solution-processed red OLED was fabricated (63). Here, TPBI layer was inserted to suppress the exciton quenching at the cathode interface and the peak η_L of the device reached 4.2 cd/A and η_{ext} of 3.2%. The device emitted at 618 nm with CIE coordinates of (0.62, 0.38). White OLED was fabricated by codoping blue-green emitter and red-emitting complex **85** in the device structure of ITO/PEDOT:PSS/PVK:OXD-7: blue-green emitter:**85**/TPBI/Cs_2CO_3/Al. The CIE coordinates of the WOLED changed from (0.43, 0.43) to (0.35, 0.44) when the biasing voltage increased from 5.0 to 8.0 V and the device emitted at 484 and 594 nm. The peak η_L reached 16.7 cd/A, corresponding to an η_{ext} of 7.8% and an η_P of 6.8 lm/W.

85

7.3   CONCLUSION

Although red phosphor has been commercialized since 2003, there is still room for the development of new materials with highly efficient saturated red EL emission with CIE coordinates corresponding to the NTSC standard red color. Functionalization of Ir(III) cyclometalated complex with suitable ligand can afford red dopant with remarkable photophysical, thermal, and EL properties. The molecular ligand structure of the Ir(III) complex is critical in tuning the HOMO and LUMO levels of a red dopant, which in turn affects the energy difference between the host and the energies of these frontier orbitals of the dopant that has a great influence on the efficiency of the final red-emitting device. An insight into the structure–efficiency relationship of these Ir(III) cyclometalated complexes is clearly important for the future development of new high-performance red phosphorescent emitters. The saturation of triplet emissive states and triplet–triplet annihilation effect in Ir(III)-based red phosphors should be avoided so that the materials can be developed toward sought-after efficient and stable red or white OLEDs for lighting and display applications in more practical terms.

ACKNOWLEDGMENTS

W.-Y. Wong acknowledges the financial support from the National Basic Research Program of China (973 Program) (2013CB834702), Hong Kong Baptist University (FRG2/11-12/156) FRG1/13-14/053, Hong Kong Research Grants Council (HKBU203011 and HKUST2/CRF/10), and Areas of Excellence Scheme, University Grants Committee of HKSAR, China (Project No. [AoE/P-03/08]).

REFERENCES

1. Khalifa MB, Mazzeo M, Maiorano V, Mariano F, Carallo S, Melcarne A, Cingolani R, Gigli G. J Phys D Appl Phys 2008;41:155111-1–155111-3.

2. Xiao L, Chen Z, Qu B, Luo J, Kong S, Gong Q, Kido J. Adv Mater 2011;23:926–952.

3. Baldo MA, Thompson ME, Forrest SR. Nature 2000,403.750–753.

4. Zhou G-J, Wong W-Y, Suo S. J Photochem Photobiol C: Photochem Rev 2010;11:133–250.

5. Chen C-T. Chem Mater 2004;16:4389–4400.

6. Cummings SD, Eisenberg R. J Am Chem Soc 1996;118:1949–1960.

7. (a) Adachi C, Baldo MA, Forrest SR, Lamansky S, Thompson ME, Kwong RC. Appl Phys Lett 2001;78:1622–1624;(b) Coppo P, Plummer EA, Cola LD. Chem Commun 2004:1774–1775.

8. (a) Su Y-J, Huang H-L, Li C-L, Chien C-H, Tao Y-T, Chou P-T, Datta S, Liu R-S. Adv Mater 2003;15:884–888;(b) Li CL, Su YJ, Tao YT, Chou PT, Datta S, Liu RS. Adv Funct Mater 2005;15:387–395.

9. Zhou G-J, Wong W-Y, Yao B, Xie Z, Wang L. Angew Chem Int Ed 2007;46:1149–1151.

10. Park TJ, Jeon WS, Choi JW, Pode R, Jang J, Kwon JH. Appl Phys Lett 2009;95:103303-1–103303-3.

11. Park TJ, Jeon WS, Park JJ, Kim SY, Lee YK, Jang J, Kwon JH, Pode R. Appl Phys Lett 2008;92:113308-1–113308-3.

12. Jeon WS, Park TJ, Kim KH, Pode R, Jang J, Kwon JH. Org Electron 2012;11:179–183.

13. Su S-J, Cai C, Kido J. J Mater Chem 2012;22:3447–3456.

14. Hsu F-M, Chien C-H, Hsieh Y-J, Wu C-H, Shu C-F, Liu S-W, Chen C-T. J Mater Chem 2009;19:8002–8008.

15. Maiorano V, Mazzeo M, Mariano F, Khalifa MB, Carallo S, Vidalet BD, Cingolani R, Gigli G. IEEE Photonics Technol Lett 2008;20:2105–2107.

16. Tao Y, Wang Q, Ao L, Zhong C, Qin J, Yang C, Ma D. J Mater Chem 2010;20:1759–1765.

17. Cho YJ, Lee JY. Adv Mater 2011;23:4568–4572.

18. Kwak J, Lyu Y-Y, Lee H, Choi B, Char K, Lee C. J Mater Chem 2012;22:6351–6355.

19. Zhang MY, Wang FF, Wei N, Zhou PC, Peng KJ, Yu JN, Wang ZX, Wei B. Opt Express 2013:A175–A178.

20. Huang H-H, Chu S-Y, Kao P-C, Yang C-H, Sun I-W. Thin Solid Films 2009;517:3788–3791.

21. Ting H-C, Chen Y-M, You H-W, Hung W-Y, Lin S-H, Chaskar A, Chou S-H, Chi Y, Liu R-H, Wong K-T. J Mater Chem 2012;22:8399–8407.

22. Wang H, Ryu J-T, Song M, Kwon Y. Curr Appl Phys 2008;8:490–493.

23. Ho C-L, Wong W-Y, Gao Z-Q, Chen C-H, Cheah K-W, Yao B, Xie Z, Wang Q, Wang L, Yu X-M, Kwok H-S, Lin Z. Adv Funct Mater 2008;18:319–331.

24. Zhou G-J, Wong W-Y, Yai B, Xie Z, Wang L. J Mater Chem 2008;18:1799–1809.

25. Yao JH, Zhen C, Loh KP, Chen Z-K. Tetrahedron 2008;64:10814–10820.

26. Tsujimoto H, Yagi S, Asuka H, Inui Y, Ikawa S, Maeda T, Nakazumi H, Sakurai Y. J Organomet Chem 2010;695:1972–1978.

27. Kim DH, Cho NS, Oh H-Y, Yang JH, Jeon WS, Park JS, Suh MC, Kwon JH. Adv Mater 2011;23:2721–2726.

28. Chang Y-L, Puzzo DP, Wang Z, Helander MG, Qiu J, Castrucci J, Lu Z-H. Phys Status Solidi C 2012;9:2537–2540.

29. Wu F-I, Su H-J, Shu C-F, Luo L, Diau W-G, Cheng C-H, Duan J-P, Lee G-H. J Mater Chem 2005;15:1035–1042.

30. Seo JH, Lee SC, Kim YK, Kim YS. Thin Solid Films 2008;517:1346–1348.

31. Kong X, Kulkarni AP, Jenekhe SA. Macromolecules 2003;36:8992–8999.

32. Lai RY, Kong X, Jenekhe AA, Bard AJ. J Am Chem Soc 2003;125:12631–12639.

33. Park JS, Song M, Gal Y-S, Lee JW, Jin S-H. Synth Met 2011;161:213–258.

34. Lee S-J, Park J-S, Song M, Shin IA, Kim Y-I, Lee JW, Kang J-W, Gal Y-S, Kang S, Lee JY, Jung S-H, Kim H-S, Chae M-Y, Jin S-H. Adv Funct Mater 2009;19:2205–2212.

35. Song M, Park JS, Gal Y-S, Kang S, Lee JY, Lee JW, Jin S-H. J Phys Chem C 2012;116:7526–7533.

36. Song M, Park JS, Yoon M, Kim AJ, Kim YI, Gal Y-S, Lee JW, Jin S-H. J Organomet Chem 2011;696:2122–2128.

37. Song M, Park JS, Kim C-H, Im MJ, Kim JS, Gal Y-S, Kang J-W, Lee JW, Jin S-H. Org Electron 2009;10:1412–1415.

38. Ding J, Lü J, Cheng Y, Xie Z, Wang L, Jing X, Wang F. Adv Funct Mater 2008;18:2754–2762.

39. Chuang T-H, Yang C-H, Kao P-C. Inorg Chim Acta 2009;362:5017–5022.

40. Qiao J, Duan L, Tang L, He L, Wang L, Qiu Y. J Mater Chem 2009;19:6573–6580.

41. Ho C-L, Chi L-C, Hung W-Y, Chen W-J, Lin Y-C, Wu H, Mondal E, Zhou G-J, Wong K-T, Wong W-Y. J Mater Chem 2012;22:215–224.

42. Tao Y, Wang Q, Yang C, Zhang K, Wang Q, Zou T, Qin J, Ma D. J Mater Chem 2008;18:4091–4096.

43. Tavasli M, Moore TN, Zheng Y, Bryce MR, Fox MA, Griffiths GC, Jankus V, Al-Attar HA, Monkman AP. J Mater Chem 2012;22:6419–6428.

44. Gong X, Ostrowski JC, Bazan GC, Moses D, Heeger AJ. Appl Phys Lett 2002;81:3711–3713.

45. Zhu M, Li Y, Hu S, Li C, Yang C, Wu H, Qin J, Cao Y. Chem Commun 2012;48:2695–2697.

46. Lee KH, Kang HJ, Lee SJ, Kim YK, Yoon SS. Synth Met 2012;162:715–721.

47. Zhou G-J, Ho C-L, Wong W-Y, Wang Q, Ma D, Wang L, Lin Z, Marder TB, Beeby A. Adv Funct Mater 2008;18:499–511.

48. Zhou G-J, Wang Q, Wong W-Y, Ma D, Wang L, Lin Z. J Mater Chem 2009;19:1872–1883.

49. Qin T, Ding J, Baumgarten M, Wang L, Müllen K. Macromol Rapid Commun 2012;33:1036–1041.

50. Jung KM, Kim KH, Jin J-I, Cho MJ, Choi DH. J Polym Sci, Part A: Polym Chem 2008;46:7517–7533.

51. Namdas EB, Anthopoulos TD, Samuel ID, Frampton MJ, Lo S-C, Burn PL. Appl Phys Lett 2005;86:161105-1–161105-3.

52. Li M, Wang Q, Dai J, Lu Z-Y, Huang Y, Yu J-S, Luo S, Su S-J. Thin Solid Films 2012;526:231–240.

53. Mi BX, Wang PF, Gao ZQ, Lee CS, Lee ST, Hong HL, Chen XM, Wong MS, Xia PF, Cheah KW, Chen CH, Huang W. Adv Mater 2009;21:339–343.

54. Fang Y, Tong B, Hu S, Wang S, Meng Y, Peng J, Wang B. Org Electron 2009;10:618–622.

55. Fang Y, Hu S, Mang Y, Peng J, Wang B. Inorg Chim Acta 2009;362:4985–4990.

56. Mei Q, Wang L, Tian B, Tong B, Weng J, Zhang B, Jiang Y, Huang W. Dyes Pigm 2013;97:43–51.

57. Fang Y, Li Y, Wang S, Meng Y, Peng J, Wang B. Synth Met 2010;160:2231–2238.

58. Duan JP, Sun PP, Cheng CH. Adv Mater 2003;15:224–228.

59. Wang K, Zhao F, Wang C, Chen S, Chen D, Zhang H, Liu Y, Ma D, Wang Y. Adv Funct Mater 2013; 23: 2672–2680.

60. Schneidenbach D, Ammermann S, Debeaux M, Freund A, Zöllner M, Daniliuc C, Jones PG, Kowalsky W, Johannes H-H. Inorg Chem 2010;49:397–406.

61. Mei Q, Wang L, Guo Y, Weng J, Yan F, Tian B, Tong B. J Mater Chem 2012;22:6878–6884.

62. Lu K-Y, Chou H-H, Hsieh C-H, Yang Y-HO, Tsai H-R, Tsai H-Y, Hsu L-C, Chen C-Y, Chen I-C, Cheng C-H. Adv Mater 2011;23:4933–4937.

63. He L, Duan L, Qiao J, Zhang D, Wang L, Qiu Y. Org Electron 2010;11:1185–1191.

8

FOUR-WAVE MIXING AND CARRIER NONLINEARITIES IN GRAPHENE–SILICON PHOTONIC CRYSTAL CAVITIES

TINGYI GU AND CHEE W. WONG

Optical Nanostructures Laboratory, Center for Integrated Science and Engineering, Solid-State Science and Engineering, and Mechanical Engineering, Columbia University, New York, NY, USA

8.1 KERR NONLINEARITIES IN GRAPHENE–SILICON PHOTONIC CRYSTAL CAVITIES

Kerr effect is one of nonresonant electronic nonlinearities. They occur as the result of the nonlinear response of bound electrons to an applied optical field. This nonlinearity usually is not particularly large in semiconductors. But the materials having delocalized π electrons (such as graphene and two-dimensional (2D) layered materials) can have nonresonant third-order susceptibilities five orders of magnitude higher than others (1). Four-wave mixing is a Kerr nonlinear effect that can be viewed as the elastic scattering of two photons of a high power pump beam, which results in the generation of two new photons at different frequencies. Four-wave mixing in graphene attracts great attention due to its broadband operation and ultrafast response time (femtosecond scale) for applications of wavelength conversion, parametric oscillation, and amplification.

Graphene is a 2D atomic thin material with large mechanical attributes and high intrinsic mobility of the charge carries (2). The unique linear and massless band structure of graphene in a purely 2D Dirac fermionic structure has led to intense research in fields ranging from condensed matter physics to nanoscale device applications covering the electrical, thermal, mechanical, and optical domains.

8.2 EFFECTIVE KERR NONLINEARITIES IN GRAPHENE–SILICON SYSTEM

8.2.1 Calculations of Dynamic Conductivity of Graphene

The dynamic conductivity for intra- and interband optical transitions (3) can be determined from the Kubo formalism as

$$\sigma_{\text{intra}}(\omega) = \frac{je^2\mu}{\pi\hbar(\omega + j\tau^{-1})} \tag{8.1}$$

$$\sigma_{\text{inter}}(\omega) - \frac{je^2\mu}{4\pi\hbar} \ln\left(\frac{2|\mu| - \hbar\left(\omega + j\tau^{-1}\right)}{2|\mu| + \hbar(\omega + j\tau^{-1})}\right) \tag{8.2}$$

where e is the electron charge, $\hbar$ is the reduced Plank constant, ω is the radian frequency, μ is chemical potential, and τ is the relaxation time (1.2 ps for interband conductivity and 10 fs for intraband conductivity). The dynamic conductivity of intra- and interband transitions at 1560 nm is $(-0.07 \text{ to } 0.90i) \times 10^{-5}$ and $(4.15 \text{ to } 0.95i) \times 10^{-5}$, respectively, leading to the total dynamic conductivity $\sigma_{\text{total}} = \sigma_{\text{intra}} + \sigma_{\text{inter}}$ of $(4.1 \text{ to } 1.8i) \times 10^{-5}$. The real and imaginary parts of the conductivity and permittivity are calculated in Figure 8.1. Given the negative imaginary part of total conductivity, the TE mode is supported in graphene (4). The light can travel along the graphene sheet with weak damping and thus no significant loss is observed for the quasi-TE mode confined in the cavity (5). The impurity density of the 250 nm silicon membrane is $\sim 10^{11}$ cm^{-2}, slightly lower than the estimated doping density in graphene.

8.2.2 Computations of Effective Kerr Coefficient in Graphene–Silicon Cavities

Third-order nonlinearity susceptibility for graphene is reported as large as $|\chi^{(3)}| \sim 10^{-7}$ esu in the wavelength range of 760–840 nm (6). When two external beams with frequency ω_1 (pump) and ω_2 (signal) are incident on graphene, the amplitude of sheet current generated at the

Nanomaterials, Polymers and Devices: Materials Functionalization and Device Fabrication, First Edition. Edited by Eric S. W. Kong.
© 2015 John Wiley & Sons, Inc. Published 2015 by John Wiley & Sons, Inc.

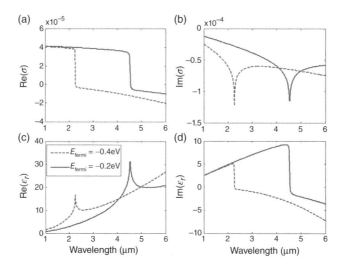

Figure 8.1 *Conductivity and permittivity of graphene in the infrared range.* (a and b) The real and imaginary parts of the total conductivity; (c and d) permittivity, with Fermi level set at −0.4 eV (dashed line) and −0.2 eV (solid line), respectively.

harmonics frequencies $(2\omega_1-\omega_2)$ is described by

$$j_e = -\frac{3}{32}\frac{e^2}{\hbar}\varepsilon_2\left(\frac{ev_F\varepsilon_1}{\hbar\omega_1\omega_2}\right)^2\frac{2\omega_1^2+2\omega_1\omega_2-\omega_2^2}{\omega_1(2\omega_1-\omega_2)} \tag{8.3}$$

where ε_1, ε_2 are the electric field amplitudes of the incident light at frequencies ω_1 and ω_2, respectively, and v_F ($= 10^6$ m/s) is the Fermi velocity of graphene. Under the condition that both ω_1 and ω_2 are close to ω, the sheet conductivity can be approximated as

$$\sigma^{(3)} = \frac{j_e}{\varepsilon_1\varepsilon_1\varepsilon_2} = -\frac{9}{32}\frac{e^2}{\hbar}\left(\frac{ev_F}{\hbar\omega^2}\right)^2 \tag{8.4}$$

Since most of the sheet current is generated in graphene, the effective nonlinear susceptibility of the whole membrane can be expressed as

$$\chi^{(3)} = \frac{\sigma^{(3)}}{\omega d} = -\frac{9}{32}\frac{e^4v_F^2}{\hbar^3c^5}\frac{\lambda^5}{d} \tag{8.5}$$

where d is the thickness of the graphene (~1 nm), λ is the wavelength, and c is the speed of light in vacuum. The calculated $\chi^{(3)}$ of a monolayer graphene is in the order of 10^{-7} esu (corresponding to a Kerr coefficient $n_2 \sim 10^{-13}$ m^2/W), at 10^5 times higher than in silicon ($\chi^{(3)} \sim 10^{-13}$ esu, $n_2 \sim 4 \times 10^{-18}$ m^2/W) (7).

Effective n_2 of the hybrid graphene–silicon cavity is then calculated for an inhomogeneous cross section weighted with respect to field distribution (8, 9) (Fig. 8.2a–f). With a baseline model without complex graphene–surface electronic interactions, the effective n_2 can be expressed as

$$\overline{n_2} = \left(\frac{\lambda_0}{2\pi}\right)^d\frac{\int n^2(r)n_2(r)(|E(r)\cdot E(r)|^2 + 2|E(r)\cdot E(r)^*|^2)d^dr}{\left(\int n^2(r)|E(r)|^2d^dr\right)^2} \tag{8.6}$$

where $E((r)$ is the complex fields in the cavity and $n(r)$ is the local refractive index (Fig. 8.2g and h). The local Kerr coefficient $n_2(r)$ is 3.8×10^{-18} m^2/W in silicon membrane and $\sim10^{-13}$ m^2/W for graphene, λ_0 is the wavelength in vacuum, and $d = 3$ is the number of dimensions. The complex electric field $E(r)$ is obtained from three-dimensional (3D) finite-difference time-domain computations of the optical cavity examined (10). The resulting field-balanced effective n_2 is calculated to be 7.7×10^{-17} m^2/W ($\chi^{(3)} \sim 10^{-12}$ esu), close to the best reported chalcogenide photonic crystal waveguides (Table 8.1) (12, 11).

The 3D Finite Difference in Time Domain Method (FDTD) method with subpixel averaging is used to calculate the real and imaginary parts of the E-field distribution for the cavity resonant mode. The spatial resolution is set at 1/30 of the lattice constant (14 nm). Time-domain coupled-mode theory, including free-carrier dispersion and dynamics and thermal time constants, is carried out with 1 ps temporal resolution.

Likewise, the effective two-photon absorption coefficient is computed in the same field-balanced approach, with a result of 2.5×10^{-11} m/W. The resulting nonlinear parameter γ ($= \omega n_2/cA_{\text{eff}}$) is derived to be 800 W/m, for an effective mode area of 0.25 μm^2 (Fig. 8.2h).

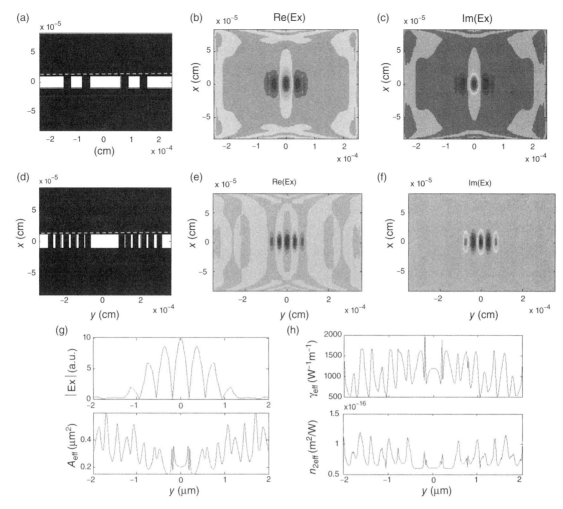

Figure 8.2 *Optical field distribution and calculation for effective Kerr nonlinearity of the hybrid graphene–silicon waveguide.* (a) The refractive index on the $x = 0$ plane, the white part is the silicon ($n = 3.45$), and the dark part is the air ($n = 1$); (b and c) real and imaginary parts of TE polarized electric field distribution on the across section in (a); (d) the refractive index on the $y = 0$ plane; (e and f) real and imaginary parts of TE polarized electric field in (d); (g) the amplitude of electric field along the cross section of waveguide (upper) and effective area (down) along the waveguide direction; and (h) nonlinear parameter γ ($= \omega n_2 / c A_{\text{eff}}$) and effective Kerr coefficient of graphene silicon waveguide.

TABLE 8.1   Field-Balanced Third-Order Nonlinear Parameters

Computed Parameters	$\overline{n_2}$ (m²/W)	$\overline{\beta_2}$ (m/W)
Graphene	10^{-13} (6)	10^{-7} (11)
Silicon	3.8×10^{-18}	8.0×10^{-12}
Monolayer graphene–silicon	7.7×10^{-17}	2.5×10^{-11}
Chalcogenide waveguide	7.0×10^{-17}	4.1×10^{-12}

8.3   DEVICE FABRICATION AND CALIBRATION

8.3.1   Graphene Growth and Transfer

Centimeter-scale graphene is grown on 25 μm thick copper foils by chemical vapor deposition of carbon. The top oxide layer of copper is firstly removed in the hydrogen atmosphere (50 mTorr, 1000 °C, 2 sccm H_2 for 15 min), and then monolayer carbon is formed on copper surface (250 mTorr, 1000 °C, 35 sccm CH_4, 2 sccm H_2 for 30 min). The growth is self-limiting once the carbon atom covers the Cu surface

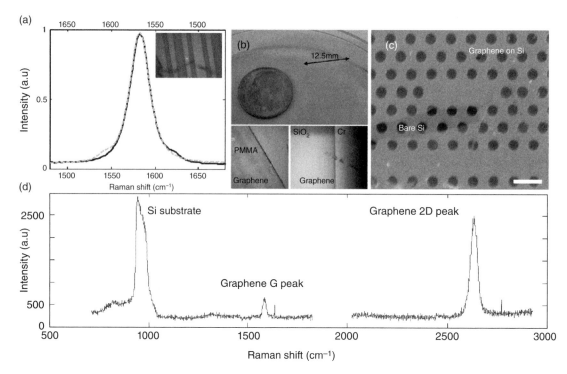

Figure 8.3 *Raman spectrum and transferred graphene samples.* (a) Raman G peak (black line) and its inverse (gray dashed line) to illustrate G peak symmetry. Inset: optical micrograph of the device with graphene transferred under Raman measurement. (b) A centimeter-scale graphene film prepared. Optical micrograph of graphene film transferred to various substrates (PMMA), air-bridged silicon membranes, silicon oxide, and partially covered metal surfaces, with graphene interface pictured. (c) Scanning electronic micrograph of example air-bridged device sample with graphene covering the whole area except the dark (exposed) region. Scale bar: 500 nm. (d) Complete Raman spectrum of the graphene-clad silicon membrane samples.

catalytic. The single layer graphene is then rapidly cooled down before being moved out of chamber. Polymethyl methacrylate (PMMA) is then spun casted onto the graphene and then the copper foil etch removed by floating the sample in FeNO$_3$ solution. After the metal is removed, graphene is transferred to a water bath before subsequent transfer onto the photonic crystal membranes. Acetone dissolves the PMMA layer, and the sample is rinsed with isopropyl alcohol and dry baked for the measurements.

8.3.2 Optical Calibration by Raman Spectrum

8.3.2.1 Raman Calibration   Confocal microscopy was used for the graphene Raman spectroscopic measurements with a 100 × (numerical aperture at 0.95) objective, pumped with a 514 nm laser.

The Raman spectra are shown in Figure 8.3d. The G and 2D band peaks are excited by a 514 nm laser and are located at 1582 cm^{-1} and 2698 cm^{-1}, respectively. The Raman spectra are homogeneous within one device and vary less than 5 cm^{-1} from sample to sample. The Lorentzian lineshapes with full width at half maximum of the G (34.9 cm^{-1}) and 2D (49.6 cm^{-1}) bands indicate the graphene monolayer (13), perhaps broadened by chemical doping and disorder. The phonon transport properties are represented by the G and 2D peak positions (varying within 1 cm^{-1} over the sample) and the intensity ratios between the G and 2D peaks (fluctuating from 1 to 1.5) (14).

Figure 8.3b and c illustrates example transfers of large-area Chemical vapor deposition (CVD) graphene into various substrates including air-bridged silicon membranes, silicon oxide, and partially covered metal surfaces. CVD grown graphene is thicker and has rough surfaces compared to exfoliated graphene, shown by the broadened 2D peak and the fluctuation of the 2D versus G peak ratio (15). The thickness of graphene is ~1 nm. Wrinkles on the graphene surface are formed during the cool down process, due to the differential thermal expansion between the copper substrate and graphene, and consistently appear only at the edges of our samples. We emphasize that at the device regions most of the devices are covered with a single unwrinkled graphene layer.

The 2D peak is observable only when the laser excitation energy (E_L) and the energy corresponding to electron–hole recombination process (E_T) follow the relation ($E_L - E_T$)/2 > E_F, where E_F is the Fermi energy of graphene. With 514 nm laser excitation, the 2D peak is located at 2698 cm^{-1} (Fig. 8.3d).

We note that wet transfer of graphene is used in these measurements. While a very thin (in the range of nanometers) residual layer of PMMA can remain on the sample after transfer, PMMA typically only has a noncentrosymmetric $\chi^{(2)}$ response with a negligible $\chi^{(3)}$ response and hence does not contribute to the enhanced four-wave mixing observations. The dopants can arise from residual absorbed molecules or ions on graphene or at the grain boundaries, during the water bath and transfer process.

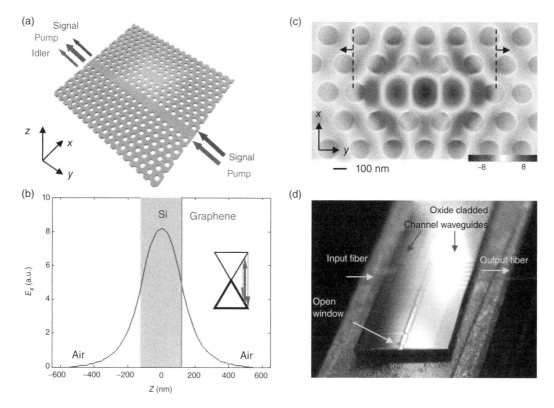

Figure 8.4 *Device layout and measurement.* (a) Structure schematic of an *L3* cavity switch formed in graphene-clad silicon membrane. (b) The electric field distribution along *z* direction simulated by FDTD method. The graphene sheet (brown line) is placed on 250 nm thick silicon membrane. Inset: schematics of graphene band diagram with photon energy of pump and converted ones (c). Top view of optic field energy distribution of an isolated S1 shifted *L3* cavity. The FDTD simulation of mode profile is superimposed on the SEM picture S1 shifted *L3* cavity with graphene cladding, with mode volume of 0.073 μm^3 and quality factor ~2.0×10^4. Scale bar: 100 nm. (d) CMOS processed integrated optical devices under test. The open window is for silicon membrane undercut and graphene cladding on the photonic crystal part.

8.3.2.2 *Optical Transparency*

The Fermi level of graphene can be modified by chemical dopants. Sulfuric acid and the nitric acid molecules are physically adsorbed on the surface of graphene without intercalations. The doping level can be well controlled beyond 1 eV, for producing graphene with high transparency (16).

With almost all of the undoped samples, the transmission was too low to ascertain good fiber–chip coupling. In addition, in the dry transfer process with solely heating and intersurface stiction, much contaminants or particles are transferred onto the pristine silicon photonic crystal chip, damaging many of the transmission features. Other than these technicalities, a 11 dB higher power should lead to similar bistability and four-wave mixing properties; however, the high power laser might introduce more background Fabry–Perot shifts (modifying the cavity lineshapes) and background noise.

8.3.3 Photonic Crystal Preparation and Transferring

The photonic crystal nanostructures are defined by 248 nm deep ultraviolet lithography in the silicon CMOS foundry onto undoped silicon-on-insulator (100) substrates. Optimized lithography and reactive ion etching was used to produce device lattice constants of 420 nm, hole radius of 124 ± 2 nm. The photonic crystal cavities and waveguides are designed and fabricated on a 250 nm silicon device thickness, followed by a buffered hydrofluoric wet etch of the 1 μm buried oxide to achieve the suspended photonic crystal nanomembranes (Fig. 8.4).

8.4 FOUR-WAVE MIXING IN PHOTONIC CRYSTAL CAVITY

The four-wave mixing measurements (Fig. 8.5) were performed on the in-plane cavity photonic crystal membranes, where a clear four-wave-mixing-generated idler intensity was observed only when graphene is present.

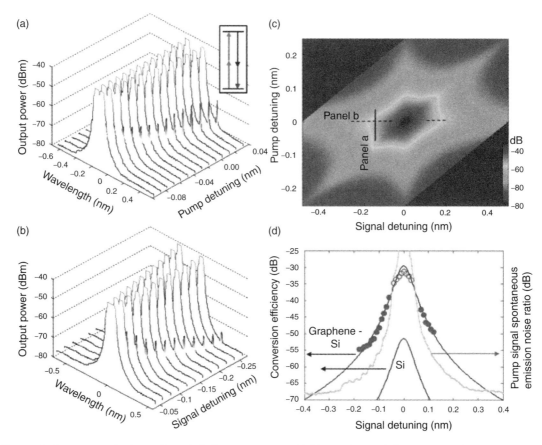

Figure 8.5 *Parametric four-wave mixing in graphene-clad silicon nanocavities.* (a) Measured transmission spectrum with signal laser fixed at −0.16 nm according to cavity resonance and pump laser detuning is scanned from −0.1 to 0.03 nm. Inset: band diagram of degenerate four-wave mixing process with pump (grey arrow), signal (black long arrow), and idler (black short arrow) lasers. (b) Measured transmission spectrum with pump laser fixed on cavity resonance and signal laser detuning is scanned from −0.04 to −0.27 nm. (c) Modeled conversion efficiency versus pump and signal detuning from the cavity resonance. (d) Observed and simulated conversion efficiencies of the cavity. Red solid dots are measured with signal detuning as in panel b, and the empty circles are obtained through pump detuning as in panel a, plus 29.5 dB (offset due to the 0.16 nm signal detuning). Solid and dashed black lines are modeled conversion efficiencies of graphene–silicon and monolithic silicon cavities, respectively. Gray dashed line (superimposed): illustrative pump/signal laser spontaneous emission noise ratio.

8.4.1 Cavity Enhancement of Light–Matter Interaction

The conversion efficiency of the single cavity is $\eta = |\gamma P_p L'|^2 \mathrm{FE}_p^4 \mathrm{FE}_s^2 \mathrm{FE}_c^2$, where FE_p, FE_s, and FE_c are the field enhancement factors of pump, signal, and idler, respectively (17). The effective length L' includes the phase mismatch and loss effects. Compared to the original cavity length (~1582.6 nm), the effective cavity length is only slightly modified by less than 1 nm. However, the spectral-dependent field enhancement factor is the square of the cavity buildup factor $\mathrm{FE}^2 = P_{\mathrm{cav}}/P_{\mathrm{wg}} = F_{\mathrm{cav}}(U/U_{\mathrm{max}})\eta_p^2$, where U/U_{max} is the normalized energy distribution with the Lorentzian lineshape. $\eta_p = 0.33$ is the correction term for the spatial misalignment between the quasi-TE mode and graphene and the optical field polarization. The field enhancement effect in the cavity is proportional to the photon mode density: $F_{\mathrm{cav}} = Q\lambda^3/(8\pi V)$ (18), where Q is the total quality factor, λ is the wavelength of the light, and V is the cavity mode volume.

The enhanced two-photon absorption and induced free-carrier absorption would produce nonlinear loss. To investigate the direct effect of two-photon absorption and free-carrier absorption on the four-wave mixing, we measure the conversion efficiency with varying input signal power as shown in Figure 8.6. Extra 4 dB loss is measured when the input signal power increases from −22 to −10 dB m, with the additional contribution from nonlinear absorption of the graphene–silicon cavity membrane.

8.4.2 Detuning Dependence

To examine only the Kerr nonlinearity, next, we performed degenerate four-wave mixing measurements on the hybrid graphene–silicon photonic crystal cavities as illustrated in Figure 8.5, with continuous-wave laser input. A lower-bound Q of 7500 was specifically chosen to allow a ~200 pm cavity linewidth, within which the highly dispersive four-wave mixing can be examined. The input pump and signal laser

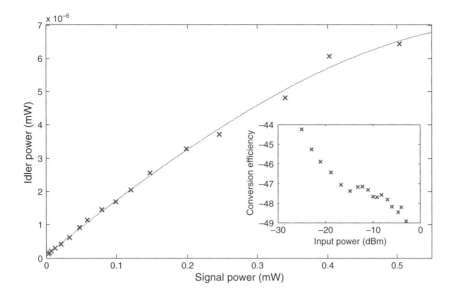

Figure 8.6 *Free-carrier absorption effects on the four-wave mixing conversion efficiency.* Measured idler power versus signal power at the transmitted port, with the pump power is fixed on the cavity resonance and the signal laser detuned by 200 pm. Experimental data (×) and quadratic fit (solid line). Inset: corresponding conversion efficiency versus signal power.

detunings are placed within this linewidth, with matched TE-like input polarization, and the powers set at 600 µW. Two example series of idler measurements are illustrated in Figure 8.5a and b, with differential pump and signal detunings, respectively. In both series, the parametric idler is clearly observed as a sideband to the cavity resonance, with the pump detuning ranging from −100 to 30 pm and the signal detuning ranging from −275 to −40 pm and from 70 to 120 pm (shown in Fig. 8.5d). For each fixed signal–cavity and pump–cavity detuning, the generated idler shows a slight intensity roll-off from linear signal (or pump) power dependence when the transmitted signal (or pump) power is greater than ∼400 µW due to increasing free-carrier absorption effects. As illustrated in Figure 8.5a and b, the converted idler wave shows a four-wave mixing 3 dB bandwidth, roughly matching the cavity linewidth when the pump laser is centered on the cavity resonance.

A theoretical four-wave mixing model with cavity field enhancement (Fig. 8.5c and d) matches with these first graphene–cavity observations. The detuning between the cavity resonance and the laser would decrease the field enhancement factor:

$$F(\lambda) = F(\lambda_{cav}) \frac{1}{1 + 4Q^2 \left(\frac{\lambda}{\lambda_{cav}} - 1 \right)^2} \tag{8.7}$$

Based on this numerical model match to the experimental observations, the observed Kerr coefficient n_2 of the graphene–silicon cavity ensemble is 4.8×10^{-17} m^2/W, an order of magnitude larger than in monolithic silicon and GaInP-related materials (19) and two orders of magnitude larger than in silicon nitride (20). The computed n_2 is at 7.7×10^{-17} m^2/W, matching well with the observed four-wave mixing derived n_2. The remaining discrepancies can arise from numerical inaccuracies or a Fermi velocity slightly smaller than the ideal values (∼10^6 m/s) in the graphene. As illustrated in Figure 8.5d for both measurement and theory, the derived conversion efficiencies are observed up to −30 dB in the unoptimized graphene–cavity, even at cavity Qs of 7500 and low pump powers of 600 µW. The highly doped graphene with Fermi level in the optical transparency region is a prerequisite to these observations. We note that for a silicon cavity without graphene, the conversion efficiencies are dramatically lower (by more than 20 dB lower), as shown in dashed black line, and even below the pump/signal laser spontaneous emission noise ratio (dotted gray line) preventing four-wave mixing observation in a single monolithic silicon photonic crystal cavity until now.

8.5   FREE-CARRIER DYNAMICS IN GRAPHENE–SILICON PHOTONIC CRYSTAL CAVITIES

Subwavelength nanostructures in monolithic material platforms have witnessed rapid advances toward chip-scale optoelectronic modulators (21–24), photoreceivers (25, 26), and high-bitrate signal processing architectures (27, 28). Coupled with ultrafast nonlinearities as a new parameter space for optical physics (29), breakthroughs such as resonant four-wave mixing (30) and parametric femtosecond pulse characterization (31, 19) have been described. Recently, graphene – with its broadband dispersionless nature and large carrier mobility – has been examined for its gate-variable optical transitions (20, 32) toward broadband electroabsorption modulators (33) and photoreceivers (34, 35), including planar microcavity-enhanced photodetectors (36, 37), as well as saturable absorption for mode locking (38). Due to

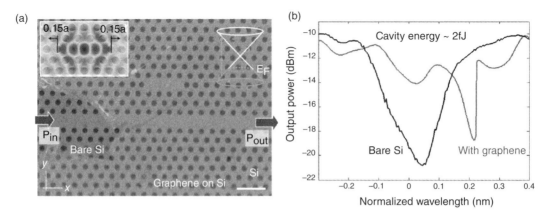

Figure 8.7 *Graphene-clad silicon photonic crystal nanostructures.* (a) Scanning electron micrograph (SEM) of tuned photonic crystal cavity, with lattice constant *a* of 420 nm. Example SEM with separated graphene monolayer on silicon for illustration. Scale bar: 500 nm. Inset: example E_z-field from finite-difference time-domain computations. Right inset: Dirac cone illustrating the highly doped Fermi level (dashed circle) allowing only two-photon transition (left arrows), while the one-photon transition (right arrow) is forbidden. (b) Fano-like lineshapes (grey line) and significantly larger redshift, compared to a control bare Si cavity sample with symmetric Lorentzian lineshapes (black line). Both spectra are measured at 0.6 mW input power and are centered to the intrinsic cavity resonances ($\lambda_{\text{cavity\_0}} = 1562.36$ nm for graphene sample and $\lambda_{\text{cavity\_0}} = 1557.72$ nm for Si sample), measured at low power (less than 100 μW input power). The intrinsic cavity quality factor is similar between the graphene and the control samples.

its linear band structure allowing interband optical transitions at all photon energies, graphene has been suggested as a material with large $\chi^{(3)}$ nonlinearities (6). In this section, we demonstrate the exceptionally high third-order nonlinear response of graphene with a wavelength-scale localized photonic crystal cavity, enabling ultralow-power optical bistable switching, self-induced regenerative oscillations, and coherent four-wave mixing at femtojoule cavity energies on the semiconductor chip platform. The structure examined is a hybrid graphene–silicon cavity (as illustrated in Fig. 8.7), achieved by rigorous transfer of monolayer large-area graphene sheet onto air-bridged silicon photonic crystal nanomembranes with minimal linear absorption and optimized optical input/output coupling. This optoelectronics demonstration is complemented with recent examinations of large-area (14, 39) graphene field-effect transistors and analog circuit designs (40) for potential large-scale silicon integration.

8.5.1   Effective Two-Photon Absorption in Graphene–Si System

With increasing input power, the transmission spectra evolve from symmetric Lorentzian to asymmetric lineshapes as illustrated in the examples of Figure 8.7. Through second-order perturbation theory (41), the two-photon absorption coefficient β_2 in monolayer graphene is estimated through the second-order interband transition probability rate per unit area as

$$\beta_2 = \frac{4\pi^2}{\varepsilon_\omega \omega^4 \hbar^3} \left(\frac{v_F e^2}{c} \right)^2 \tag{8.8}$$

where v_F is the Fermi velocity, $\hbar$ is the reduced Planck's constant, e is the electron charge, and ε_ω is the permittivity of graphene in the given frequency. At 1550 nm wavelength, β_2 is determined, through Z-scan measurements and first-principle calculations, to be in the range of ∼3000 cm/GW (41). The effective two-photon absorption coefficient of graphene on silicon is defined as

$$\overline{\beta_2} = \left(\frac{\lambda_0}{2\pi} \right)^d \frac{\int n^2(r)\beta_2(r)(|E(r) \cdot E(r)|^2 + 2|E(r) \cdot E(r)^*|^2)d^d r}{\left(\int n^2(r)|E(r)|^2 d^d r \right)^2} \tag{8.9}$$

The two-photon absorption coefficients of the hybrid cavity are calculated from 3D finite-difference time-domain field averages.

With the same CVD growth process, we also examined the dry transfer technique that controls the doping density to be low enough such that the Fermi level is within the interband optical transition region. In that case, the measured samples have a significantly increased propagation loss from ∼0 to ∼11 dB over the 120 μm length photonic crystal waveguide. The wet transfer technique significantly reduced the linear absorption, thereby allowing the various nonlinear optoelectronic measurements observed in this work.

8.5.2 Parameter Space of Nonlinear Optics in Graphene Nanophotonics

Figure 8.8 compares cavity-based switching and modulation across different platforms including silicon, III–V, and the hybrid graphene–silicon cavities examined in this work. The thermal or free-carrier plasma-based switching energy is given by $P_{0th/e} \times \tau_{th/e}$, where $P_{0th/e}$ is the threshold laser power required to shift the cavity resonance half width through thermal or free-carrier dispersion and $\tau_{th/e}$ are the thermal and free-carrier lifetimes in resonator. Note that the lifetime should be replaced by cavity photon lifetime if the latter is larger (for high Q cavity). Graphene brings about a lower switching energy due to strong two-photon absorption ($\sim$3000 cm/GW) (41). The recovery times of thermal switching (in red) are also shortened due to higher thermal conductivity in graphene, which is measured for supported graphene monolayers at 600 W/mK (42) and bounded only by the graphene-contact interface and strong interface phonon scattering.

The switching energy is inversely proportional to two-photon absorption rate (β_2). Table 8.2 summarizes the first-order estimated physical parameters from (1) coupled-mode theory and experimental data matching, (2) full 3D numerical field simulations, and (3) directly measured data. With the enhanced two-photon absorption in graphene and first-order estimates of the reduced carrier lifetimes, the switching transmission lineshape of different powers of the hybrid graphene–silicon cavity is illustrated in Figure 8.9, compared to monolithic GaAs or silicon ones.

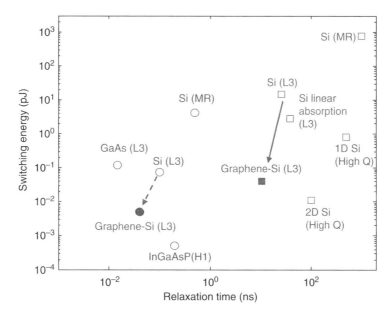

Figure 8.8 *Comparison of switching energy versus recovery time of cavity-based modulators and switches across different semiconductor material platforms.* The circles are carrier plasma-induced and the squares are thermal–optic switches with positive detuning. The arrows illustrate the operating switch energies versus recovery times, for the same material. *L3 (H1)* denotes photonic crystal *L3 (H1)* cavity; *MR* denotes microring resonator.

TABLE 8.2 Estimated Physical Parameters from Time-Dependent Coupled-Mode Theory-Experimental Matching, Three-Dimensional Numerical Field Simulations, and Measurement Data

Parameter	Symbol	GaAs (S17)	Si	Monolayer Graphene–Si
TPA coefficient	β_2 (cm/GW)	10.2	1.5(43)	25 (3D)
Kerr coefficient	n_2 (m^2/W)	1.6×10^{-17}	0.44×10^{-17} (43)	7.7×10^{-17} (3D)
Thermo-optic coeff.	dn/dT	2.48×10^{-4}	1.86×10^{-4}	
Specific heat	$c_v \rho$ (W/Km^{-3})	1.84×10^6	1.63×10^6 (cal)	
Thermal relaxation time	$\tau_{th,c}$ (ns)	8.4	12	10 (cal)
Thermal resistance	R_{th} (K/mW)	75	25 (44)	20 (cal)
FCA cross section	σ (10^{-22} m^3)	51.8	14.5	
FCD parameter	ζ (10^{-28} m^3)	50	13.4	
Carrier lifetime	τ_{fc} (ps)	8	500 (45)	200 (CMT)
Loaded Q	Q	7000	7000 (m)	
Intrinsic Q	Q_0	30,000	23,000 (m)	

CMT, nonlinear time-dependent coupled-mode theory simulation; 3D, three-dimensional numerical field calculation averages; m, measurement at low power; and cal, first-order hybrid graphene–silicon media calculations. τ_{fc} is the effective free-carrier lifetime accounting for both recombination and diffusion.

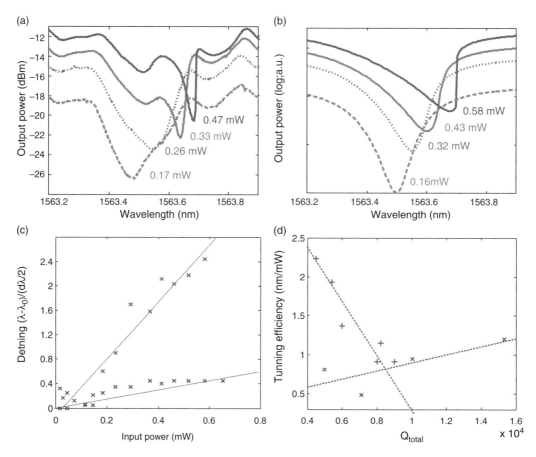

Figure 8.9 *Steady-state two-photon absorption-induced thermal nonlinearities in graphene–silicon hybrid cavities.* (a) Measured quasi-TE transmission spectra of a graphene-clad *L3* cavity with different input power levels (with extracted insertion loss from the facet of waveguides in order to be comparable to simulation in b). (b) Nonlinear coupled-mode theory simulated transmission spectra. The estimated input powers are marked in the panels. (c) Measured cavity resonance shifts versus input power, with the graphene-cad cavity samples (~4.6/mW) and the monolithic silicon control cavity sample (~0.76/mW). (d) Tuning efficiencies for graphene-clad cavity samples (*Q* in range of 4,500 ~ 9,200) and control cavity samples (*Q* in range of 4,300 ~ 15,000) for a range of cavity loaded *Q*-factors examined.

8.6   GRAPHENE THERMAL AND FREE-CARRIER NONLINEARITIES

8.6.1   Steady-State Response of Graphene–Si Cavity

8.6.1.1   Optical Bistability   We track the *L3* cavity resonance in the transmission spectra with different input powers as illustrated in Figure 8.9a. With thermal effects, the cavity resonance redshifts 1.2 nm/mW for the graphene-clad sample (*Q* ~7000) and only 0.3 nm/mW for silicon sample (similar *Q* ~7500). These sets of measurements are summarized in Figure 8.9, where the thermal redshift is sizably larger in the graphene-clad sample versus a near-identical monolithic silicon cavity. In addition, Figure 8.9d shows the tuning efficiency for a range of cavity *Q*s examined in this work – with increasing *Q*, the monolithic silicon cavity shows an increase in tuning efficiency, while the converse occurs for the graphene–silicon cavity. Figure 8.10 shows the steady-state bistable hysteresis for more detunings, including a two-cavity bistability switch. Figure 8.11a shows the temporal switching with illustrative detunings of −1.3 and 1.6 nm.

8.6.1.2   Hysteresis Loop

8.6.1.2.1   One Cavity Bistable Switch   The insertion loss seems negligible between the graphene-clad and bare sample, but this is because of the slight increased Fabry–Perot resonances after the graphene transfer. The Fabry–Perot resonances (from finite reflections in the waveguide) give rise background spectral oscillations that increase or decrease the transmission of the waveguide (Fig. 8.7). To accurately determine the additional ~1 dB or less excess loss from graphene clad over the short photonic crystal sample, we compared between transmission spectra of with graphene and without graphene. Independently, we also note that for heavily doped graphene, the linear absorption is 0.02 dB/μm (while at 0.1 dB/μm for intrinsic graphene).

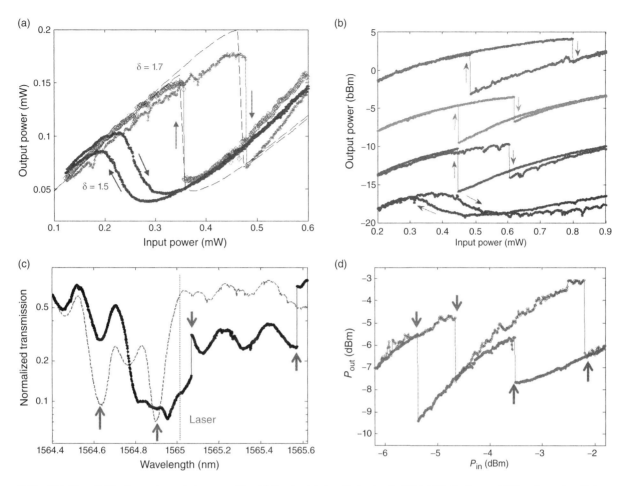

Figure 8.10 *Bistable switching in graphene-clad nanocavities.* (a) Steady-state input/output optical bistability for the quasi-TE cavity mode with laser–cavity detuning δ at 1.5 ($\lambda_{\text{laser}} = 1562.66$ nm) and 1.7 ($\lambda_{\text{laser}} = 1562.70$ nm). The dashed black line is the coupled-mode theory simulation with effective nonlinear parameters of the graphene–silicon cavity sample. (b) Measured steady-state bistability at different detunings set at 0.18, 0.23, 0.26, and 0.29 nm (from bottom to top). The plots are offset 0 dB, 2 dB, 8 dB and 15 dB for clarity. (c) Normalized transmission of two cavities with resonance separation of $\delta = 3$. The gray dashed line is measured at -16 dBm input power, and the solid black line is for 0 dBm input. (d) The transfer function for optical bistability in two-cavity system as in (c). The laser detuning is set at $\delta = 3$ to the first cavity, and $\delta = 1.5$ to the second cavity (marked as dashed line in (c)).

With coupled-mode theory, the dynamic equation for the amplitude of the resonance mode is described by coupled-mode theories. Due to the partial reflection from the facets of the waveguide, a Fabry–Perot mode is formed in the waveguide and coupled to the cavity mode (46). The interference is weak when the extinction ratio of the cavity is large ($\sim$10 dB), but the cavity resonance shift due to interference is more significant with decreased extinction ratio at high power induced by nonlinear effect. From curve fitting the transmission spectrum at low power (linear region), we obtained that waveguide facet reflectivity of 0.12 and a waveguide length of 2 mm. We fix the reflectivity and plot the transmission spectrum at five different power levels and observe maximum 0.05 nm variation of the cavity resonance shift due to different phase between waveguide and cavity at input power 0.6 mW (Fig. 8.9b).

The intrinsic $Q_{\text{intrinsic}}$ and loaded Q_{loaded} are 22,000 and 7500, respectively. The total Q obtained from the transmission spectral linewidth is related to the loaded and intrinsic Qs through the following relation: $1/Q = 1/Q_{\text{loaded}} + 1/Q_{\text{intrinsic}}$. The loaded and intrinsic Qs denote, respectively, the photon decay rates of the cavity into the coupling waveguide and into the free-space continuum. Q_{loaded} is less than 1/3 of $Q_{\text{intrinsic}}$ here (from design simulations) and thus is roughly estimated to be the same as the total Q: $Q_{\text{loaded}} = \omega_0/\Delta\omega$, where ω_0 is the cavity resonance frequency and $\Delta\omega$ is the cavity linewidth. The Qs are calibrated at low input powers and also matched with our nonlinear coupled-mode theory models. After knowing Q_{loaded}, $Q_{\text{intrinsic}}$ can be derived from $Q_{\text{intrinsic}} = Q_{\text{loaded}}/\sqrt{T}$ for a side-coupled cavity, where T is the normalized transmittance of laser power when its wavelength is set on the cavity resonance.

The cavity energy is a measure of the internal cavity energy that would be lost if the external continuous-wave laser is turned off rapidly; it is a measure used in switching and dynamical studies (47, 48). The formal definition of the intrinsic cavity energy U_c is $U_c = Q_{\text{intrinsic}}P_l/\omega$, where P_l is the power loss from the cavity and ω is the excitation frequency (49). For our side-coupled cavity in transmission, $U_c = Q_{\text{intinsic}}(1-T_{\text{min}})P_{\text{in}}/\omega$, where T_{min} is the minimum transmission and P_{in} the coupled input power, which gives the femtojoule per cavity switching energies.

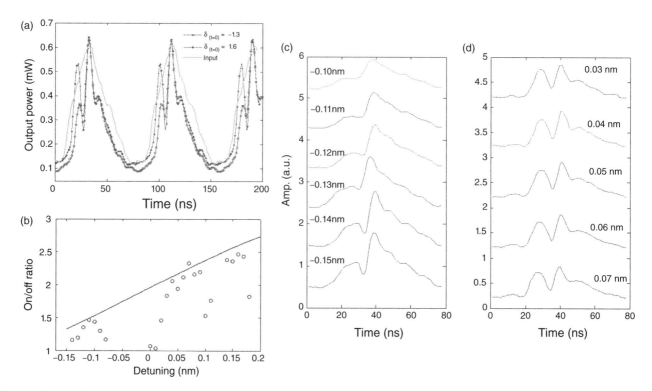

Figure 8.11 *Switching dynamics with triangular waveform drive input.* (a) The input waveform is dashed gray line. The bistable resonances are observed for both positive and negative detunings. Empty circles: $\delta(t = 0) = -1.3$ ($\lambda_{laser} = 1562.10$ nm). Solid circles: $\delta(t = 0) = 1.6$ ($\lambda_{laser} = 1562.68$ nm). Inset: schematic of high- and low-state transmissions. (b) The ratio between the on and off output intensities and different detunings. (c) The output switching dynamics with the input as in (a), at negative detunings. (d) Positive detunings.

8.6.1.2.2 Multistability in Coupled Cavity System The master equation for the dynamics of two-side-coupled-cavity system is as follows:

$$\frac{da_n}{dt} = \left[-\frac{1}{2\tau_{total,n}} + i \left(\omega_n + \Delta\omega_n - \omega_{wg} \right) \right] a_n + \kappa S_{R(n-1)} + \kappa S_{Ln} \tag{8.10}$$

$$\frac{dN}{dt} = \frac{1}{2\hbar\omega_0 \tau_{TPA}} \frac{V_{TPA}}{V_{FCA}^2} |a|^4 - \frac{N}{\tau_{fc}} \tag{8.11}$$

$$\frac{d\Delta T}{dt} = \frac{R_{th}}{\tau_{th}\tau_{FCA}} |a|^2 + \frac{\Delta T}{\tau_{th}} \tag{8.12}$$

where n is the cavity number, ω is the resonant frequency, a is the normalized cavity mode amplitude, and s is the normalized waveguide mode amplitude. $\gamma = 1/(2\zeta_t)$ is the total cavity loss rate, including the power dissipation by radiation and coupling to waveguide. ζ_t is the total cavity lifetime. $\kappa = i\exp(-i\phi/2)/\sqrt{2\tau_c}$ is the coupling coefficient between cavity and waveguide, where $\phi = \omega_{wg}n_{eff}L/c$ is the phase difference between two cavities (Fig. 8.10c and d).

8.6.2 Bistable Switching Dynamics

To verify the bistable switching dynamics, we input time-varying intensities to the graphene-clad cavity, allowing a combined cavity power–detuning sweep. Figure 8.11a shows an example of time-domain output transmission for two different initial detunings [$\delta_{(t=0)} = -1.3$ and $\delta_{(t=0)} = 1.6$] and for an illustrative triangular-waveform drive, with nanosecond resolution on an amplified photoreceiver (Fig. 8.11c). With the drive period at 77 ns, the observed thermal relaxation time is ~40 ns. Cavity resonance dips (with modulation depths ~3 dB in this example) are observed for both positive detuning (up to 0.34 nm, $\delta = 1.4$) and negative detuning (in the range from −0.15 nm [$\delta = -0.75$] to −0.10 nm [$\delta = -0.5$]). With the negative detuning and the triangular pulses, the carrier-induced (Drude) blueshifted dispersion overshoots the cavity resonance from the drive frequency and then thermally pins the cavity resonance to the laser drive frequency. Since the free-carrier lifetime of the hybrid media is about 200 ps and significantly lower than the drive pulse duration, these series of measurements are thermally dominated; the clear (attenuated) resonance dips on the intensity upsweeps (downsweeps) are due to the measurement sampling time shorter than the thermal relaxation timescale and a cooler (hotter) initial cavity temperature.

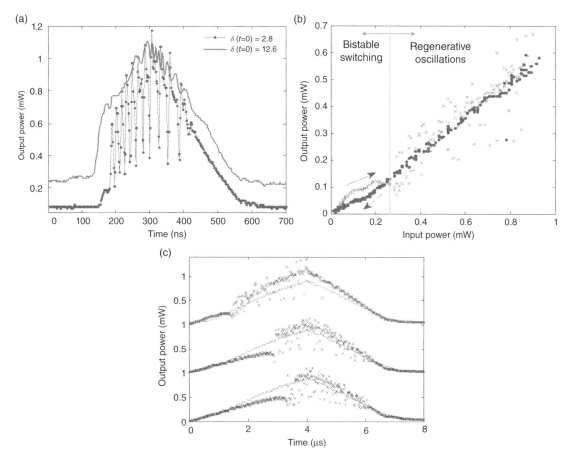

Figure 8.12 *Regenerative oscillations in graphene-clad nanocavities.* (a) Observations of temporal regenerative oscillations in the cavity for optimized detuning ($\lambda_{\text{laser}} = 1562.47$ nm). The input power is quasitriangular waveform with peak power 1.2 mW. The gray line is the reference output power, with the laser further detuned at 1.2 nm from cavity resonance ($\lambda_{\text{laser}} = 1563.56$ nm). (b) Mapping the output power versus input power with slow up (crosses) and down (circles). In the upsweep process, the cavity starts to oscillate when the input power is beyond 0.29 mW. (c) Measured regenerative oscillations at downsweep, longer temporal base widths and different detunings in graphene–silicon nanocavities. Output cavity transmission with slowly varying (7 ms) input laser intensities. The cold cavity resonance is 1562.36 nm, and the laser wavelengths from top to bottom are fixed at 1562.51, 1562.60, and 1562.62 nm. The oscillation starts when the ascending power reaches 0.29, 0.66, and 0.75 mW. Reference outputs plotted as in solid lines.

8.6.3 Regenerative Oscillation

When the input laser intensity is well above the bistability threshold, the graphene–cavity system deviates from the two-state bistable switching and becomes oscillatory as shown in Figure 8.12. Regenerative oscillation has only been suggested in a few prior studies, such as theoretically predicted in GaAs nanocavities with large Kerr nonlinearities (50) or observed in high-Q (3×10^5) silicon microdisks (51). These regenerative oscillations are formed between the competing phonon and free-carrier populations, with slow thermal redshifts ($\sim$10 ns timescales) and fast free-carrier plasma dispersion blueshifts ($\sim$200 ps timescales) in the case of graphene–silicon cavities. The self-induced oscillations across the drive laser frequency are observed at threshold cavity powers of 0.4 mW, at $\sim$9.4 ns periods in these series of measurements, which give $\sim$106 MHz modulation rates, at experimentally optimized detunings from $\delta_{(t=0)} = 0.68$ to 1.12. We emphasize that, for a monolithic silicon *L3* cavity, such regenerative pulsation has not been observed nor predicted to be observable at a relatively modest Q of 7500 and attenuated by significant nonlinear absorption. Furthermore, we show the self-induced regenerative oscillations at different detunings and also numerical model (solid line) of the two-state switching and oscillation for different detunings (Fig. 8.12c).

Figure 8.13a shows the input–output intensity cycles constructed from the temporal response measurements of a triangular-wave-modulated 1.2 mW laser with a 2 μs cycle. Clear bistability behavior is seen below the carrier oscillation threshold. The system transits to the regime of self-sustained oscillations as the power coupled into the cavity is above the threshold, by tuning the laser wavelength into cavity resonance. We show an illustrative numerical modeling in Figure 8.13: the fast free-carrier response fires the excitation pulse (blue dashed line, start cycle), and heat diffusion (red solid line) with its slower time constant determines the recovery to the quiescent state in the graphene-clad suspended silicon membrane. The beating rate between the thermal and free-carrier population is around 50 MHz, with the matched experimental data and coupled-mode theory simulation. The beating gives rise to tunable peaks in the radio-frequency spectra, which

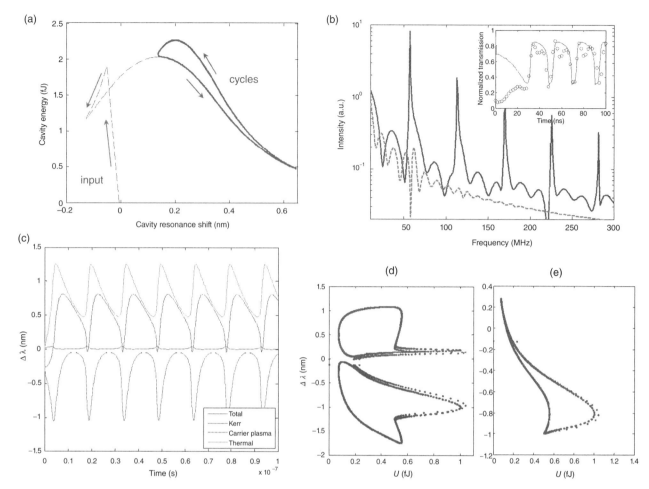

Figure 8.13 *Coupled-mode theory calculated cavity resonance oscillation.* (a) Nonlinear coupled-mode theory model of cavity transmission versus resonance shift, in the regime of regenerative oscillations. With a detuning of 0.15 nm [$\delta_{(t=0)} = 0.78$], the free-carrier density swings from 4.4 to 9.1 × 10^{17} cm^{-3}, and the increased temperature ΔT circulates between 6.6 and 9.1 K. (b) RF spectrum of output power at below (0.4 mW, gray dashed line) and above oscillation threshold (0.6 mW, blue solid line) at the same detuning $\delta_{(t=0)} = 0.78$ ($\lambda_{laser} - \lambda_{cavity} = 0.15$ nm). Inset: normalized transmission from model (curve) and (circles). (c) Cavity resonance shift due to different nonlinear dispersion versus time. (d) Thermal (lower) and free-carrier (upper) dispersion-induced cavity resonance shift versus the energy circulating in cavity. (e) The total carrier resonance dispersion versus cavity.

are absent when the input power is below the oscillation threshold (gray dashed line). We note that the model does not include a time-varying cavity quality factor, considering that the high power would usually broaden the cavity bandwidth.

The high two-photon absorption in graphene–silicon hybrid system generates large free-carrier dispersion (negative cavity resonance shift) and the thermal dispersion from the free-carrier recombination (positive cavity resonance shift). Either of the processes is demonstrated to show optical bistability in silicon (46, 52). Free-carrier dispersion dominates the carrier resonance shift initially, while the thermal dispersion dominates the long-term response (Continuous wave input as in Fig. 8.7b). In Figure 8.12a, we scan the cavity response in time domain, and the input triangular pulse width is set to make both free carrier and thermal bistability observable. It is emphasized that the role of graphene here is to enhance the nonlinear absorption, induce free-carrier population, and decrease thermal relaxation time, thermal resistance, and carrier lifetime, while the dispersion effects are similar to monolithic silicon (parameters listed in Table 8.2). The combination of the system parameter set gives the unique properties of the graphene–silicon cavity.

To further illustrate the cavity response to the laser set at red and blue detuning, the output power from cavity to step input at different detuning is simulated and shown in Figure 8.14. The switching on/off ratio is defined as the output power at time zero versus the power intensity when the cavity is switched on resonance. The on/off ratio is observed to be strongly dependent on laser detuning, both in simulation and experiment (Fig. 8.11b).

8.6.3.1 *Time-Domain Coupled-Mode Theory*
From the nonlinear coupled-mode modeling, the dynamical responses of the hybrid cavity to step inputs are shown in Figure 8.14a, illustrating the switching dynamics and regenerative oscillations. Free-carrier dispersion causes the switching on the negative-detuned laser, and the thermal nonlinearity leads to the switching on the positive side. The interplay of the

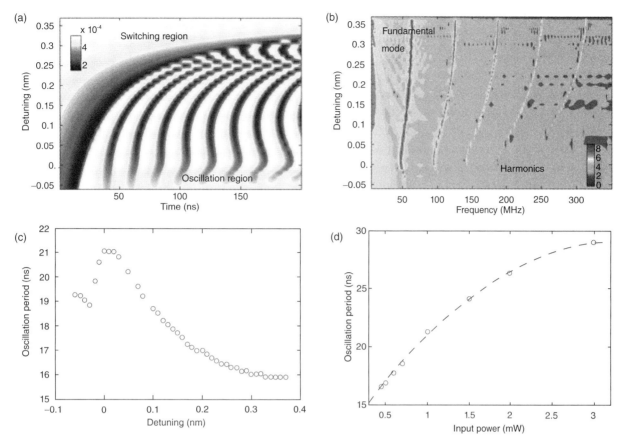

Figure 8.14 *Coupled-mode equations calculated time-domain response to a step input with a graphene-clad silicon photonic crystal L3 cavity side coupled to a photonic crystal waveguide.* (a) The output versus input powers for positive and negative detunings (laser–cavity detunings are set from −0.06 to 0.37 nm). Input laser power is set at 0.6 mW. The cavity switching dip is observed for all detunings, and regenerative oscillation exists only predominantly for positive detuning. (b) Frequency response of the cavity switching and oscillation dynamics with conditions as in (a) (in log scale). The laser detuning is set from −0.06 to 0.37 nm. (c and d) Oscillation period versus laser detunings and input powers, respectively.

free-carrier-induced cavity resonance blueshift dynamics with the thermal-induced cavity redshift time constants is observed. Figure 8.14b shows the correspondent radio-frequency spectrum. By tuning the laser wavelength, the fundamental mode can be set from 48 (zero detuning) to 55 MHz (0.3 nm detuning). The dependence of oscillation period to the detuning and input laser power is further provided in Figure 8.14c and d, respectively.

We model the nonlinear cavity transmissions with time-domain nonlinear coupled-mode theory for the temporal rate evolution of the photon, carrier density, and temperatures as described by (53)

$$\frac{da}{dt} = \left(i\left(\omega_L - \omega_0 + \Delta\omega\right) - \frac{1}{2\tau_t} \right) a + \kappa\sqrt{P_{\text{in}}} \tag{8.13}$$

$$\frac{dN}{dt} = \frac{1}{2\hbar\omega_0\tau_{\text{TPA}}} \frac{V_{\text{TPA}}}{V_{\text{FCA}}^2} |a|^4 - \frac{N}{\tau_{\text{fc}}} \tag{8.14}$$

$$\frac{d\Delta T}{dt} = \frac{R_{\text{th}}}{\tau_{\text{th}}\tau_{\text{FCA}}} |a|^2 + \frac{\Delta T}{\tau_{\text{th}}} \tag{8.15}$$

where a is the amplitude of resonance mode; N is the free-carrier density; ΔT is the cavity temperature shift; P_{in} is the power carried by incident continuous-wave laser; κ is the coupling coefficient between waveguide and cavity, adjusted by the background Fabry–Perot resonance in waveguide (46); and $\omega_L - \omega_0$ is the detuning between the laser frequency (ω_L) and cold cavity resonance (ω_0). The time-dependent cavity resonance shift is $\Delta\omega = \Delta\omega_N - \Delta\omega_T + \Delta\omega_K$, where the free-carrier dispersion is $\Delta\omega_N = \omega_0\zeta N/n$. The thermal-induced dispersion is $\Delta\omega_T = \omega_0\Delta T(dn/dT)/n$, where $\Delta\omega_K$ is the Kerr dispersion and is negligibly small compared to the thermal and free-carrier mechanisms.

The total loss rate is $1/\tau_t = 1/\tau_{in} + 1/\tau_v + 1/\tau_{lin} + 1/\tau_{TPA} + 1/\tau_{FCA}$. $1/\tau_{in}$ and $1/\tau_v$ are the loss rates into waveguide and vertical radiation into the continuum, (where $1/\tau_{in/v} = \omega/Q_{in/v}$), the linear absorption $1/\tau_{lin}$ for silicon and graphene are demonstrated to be small.

The mode volume for two-photon absorption (same as Kerr) is

$$V_{TPA/Kerr} = \frac{\left(\int n^2(r)|A(r)|^2 dr^3\right)^2}{\int_{Si} n^4(r)|A(r)|^4 dr^3} \tag{8.16}$$

The effective mode volume for free-carrier absorption is

$$V_{FCA}^2 = \frac{\left(\int n^2(r)|A(r)|^2 dr^3\right)^3}{\int_{Si} n^6(r)|A(r)|^6 dr^3} \tag{8.17}$$

The model shows remarkable match to the measured transmissions. With the first-order estimates of the thermal properties (specific heat, effective thermal resistance, and relaxation times), the carrier lifetime of the graphene-clad photonic crystal cavity is estimated to first order at 200 ps.

8.6.3.2 *Graphene–Silicon Cavity and High-Q Photonic Crystal Cavity*
Regenerative oscillations were theoretically predicted in GaAs nanocavities with large Kerr nonlinearities (50) or observed only in high-Q silicon microdisks (Q at 3×10^5) with V at $40(\lambda/n_{Si})^3$, at sub-mW power levels (51). The graphene-enhanced two-photon absorption, free-carrier, and thermal effects allow regenerative oscillations to be experimentally observable with Q^2/V values of $4.3 \times 10^7(\lambda/n)^3$ at least 50× lower, at the same power threshold levels. The regenerative oscillations with lower Qs allow higher speed and wider bandwidth operation, and are less stringent on the device nanofabrication.

8.7 CONCLUSIONS

This chapter describes graphene–silicon hybrid optoelectronic devices operating at a few femtojoule cavity recirculating energies: (i) four-wave mixing in graphene–silicon cavities, (ii) low power resonant optical bistability, and (iii) self-induced regenerative oscillations. These observations, in comparison with control measurements on solely monolithic silicon cavities, are enabled only by the dramatically large and ultrafast $\chi^{(3)}$ nonlinearities in graphene and the large Q/V ratios in wavelength-localized photonic crystal cavities. These nonlinear results demonstrate the feasibility and versatility of hybrid 2D graphene–silicon nanophotonic devices for next-generation chip-scale high-speed optical communications, radio-frequency optoelectronics, and all-optical signal processing.

REFERENCES

1. Boyd RW. *Nonlinear Optics*. San Diego (CA): Academic Press; 2002.
2. Zhang Y, Tan Y-W, Stormer HL, Kim P. Experimental observation of the quantum Hall effect and Berry's phase in graphene. Nature 2005;438:201–204.
3. Mak KF, Sfeir MY, Wu Y, Lui CH, Misewich JA, Heinz TF. Measurement of the optical conductivity of Graphene. Phys Rev Lett 2008;101:196405.
4. Bao Q, Zhang H, Wang B, Ni Z, Lim CHYX, Wang Y, Yuan Tang D, Loh KP. Broadband graphene polarizer. Nat Photonics 2011;5:411.
5. Mikhailov S, Ziegler K. New electromagnetic mode in graphene. Phys Rev Lett 2007;99:016803.
6. Hendry E, Hale PJ, Moger J, Savchenko AK. Coherent nonlinear optical response of graphene. Phys Rev Lett 2010;105:097401.
7. Dinu M, Quochi F, Garcia H. Third-order nonlinearities in silicon at telecom wavelengths. Appl Phys Lett 2003;82:2954.
8. Afshar V S, Monro TM. A full vectorial model for pulse propagation in emerging waveguides with subwavelength structures part I: Kerr nonlinearity. Opt Express 2009;17:2298–2318.
9. Gu T, Petrone N, McMillan JF, van der Zande A, Yu M, Lo GQ, Kwong DL, Hone J, Wong CW. Regenerative oscillation and four-wave mixing in graphene optoelectronics. Nat Photonics 2012;6:554.
10. Oskooi AF, Roundy D, Ibanescu M, Bermel P, Joannopoulos JD, Johnson SG. MEEP: a flexible free-software package for electromagnetic simulations by the FDTD method. Comput Phys Commun 2010;181:687.
11. Suzuki K, Hamachi Y, Baba T. Fabrication and characterization of chalcogenide glass photonic crystal waveguides. Opt Express 2009;17:22393.
12. Eggleton BJ, Luther-Davies B, Richardson K. Chalcogenide photonics. Nat Photonics 2011;5:141.
13. (a) Das A, Pisana S, Chakraborty B, Piscanec S, Saha SK, Waghmare UV, Novoselov KS, Krishnamurthy HR, Geim AK, Ferrari AC, Sood AK. Monitoring dopants by Raman scattering in an electrochemically top-gated graphene transistor. Nat Nanotechnol 2008;3:210; (b) Casiraghi C, Pisana S, Novoselov KS, Geim AK, Ferrari AC. Raman fingerprint of charged impurities in graphene. Appl Phys Lett 2007;91:233108.

14. Li X, Cai W, An J, Kim S, Nah J, Yang D, Piner R, Velamakanni A, Jung I, Tutuc E, Banerjee L, Colombo SK, Ruoff RS. Large-area synthesis of high-quality and uniform graphene films on copper foils. Science 2009;324:1312–1314.

15. Zhao W, Tan PH, Liu J, Ferrari AC. Intercalation of few-layer graphite flakes with $FeCl_3$: Raman determination of Fermi level, layer by layer decoupling, and stability. J Am Chem Soc 2011;133:5941.

16. Zhao W, Tan P, Zhang J, Liu J. Charge transfer and optical phonon mixing in few-layer graphene chemically doped with sulfuric acid. Phys Rev B 2010;82:245423.

17. Absil PP, Hryniewicz JV, Little BE, Cho PS, Wilson RA, Joneckis LG, Ho P-T. Wavelength conversion in GaAs micro-ring resonators. Opt Lett 2000;25:554.

18. Chang RK, Campillo AJ. *Optical Processes in Microcavities*. Singapore: World Scientific; 1996.

19. Pasquazi AM et al. Sub-picosecond phase-sensitive optical pulse characterization on a chip. Nat Photonics 2011;5:618.

20. Wang F et al. Gate-variable optical transitions in graphene. Science 2008;320:206–209.

21. Xu Q, Schmidt B, Pradhan S, Lipson M. Micrometre-scale silicon electro-optic modulator. Nature 2005;435:325–327.

22. Liu A et al. A high-speed silicon optical modulator based on a metal-oxide-semiconductor capacitor. Nature 2005;427:615–618.

23. Liu J et al. Waveguide-integrated, ultralow-energy GeSi electro-absorption modulators. Nat Photonics 2008;2:433–437.

24. Kuo Y-H et al. Strong quantum-confined Stark effect in germanium quantum-well structures on silicon. Nature 2005;437:1334–1336.

25. Assefa S, Xia F, Vlasov YA. Reinventing germanium avalanche photodetector for nanophotonic on-chip optical interconnects. Nature 2010;464:80–84.

26. Kang Y et al. Monolithic germanium/silicon avalanche photodiodes with 340 GHz gain–bandwidth product. Nat Photonics 2009;3:59–63.

27. Biberman A et al. First demonstration of long-haul transmission using silicon microring modulators. Opt Express 2010;18:15544–15552.

28. Pelusi M et al. Photonic-chip-based radio-frequency spectrum analyser with terahertz bandwidth. Nat Photonics 2009;3:139–143.

29. Colman P et al. Temporal solitons and pulse compression in photonic crystal waveguides. Nat Photonics 2010;4:862–868.

30. Morichetti F et al. Travelling-wave resonant four-wave mixing breaks the limits of cavity-enhanced all-optical wavelength conversion. Nat Commun 2011;2:1–8.

31. Foster MA et al. Silicon-chip-based ultrafast optical oscilloscope. Nature 2008;456:81–84.

32. Li ZQ et al. Dirac charge dynamics in graphene by infrared spectroscopy. Nat Phys 2008;4:532–535.

33. Liu M et al. A graphene-based broadband optical modulator. Nature 2011;474:64–67.

34. Mueller T, Xia F, Avouris P. Graphene photodetectors for high-speed optical communications. Nat Photonics 2010;4:297–301.

35. Xia F et al. Ultrafast graphene photodetector. Nat Nanotechnol 2009;4:839–843.

36. Engel M et al. Light-matter interaction in a microcavity-controlled graphene transistor. Nat Commun 2012;3:906.

37. Furchi M et al. Microcavity-integrated graphene photodetector. Nano Lett 2012;12:2773.

38. Sun Z et al. Graphene mode-locked ultrafast laser. ACS Nano 2010;4:803–810.

39. Bae S et al. Roll-to-roll production of 30-inch graphene films for transparent electrodes. Nat Nanotechnol 2010;5:574–578.

40. Lin Y-M et al. Wafer-scale graphene integrated circuit. Science 2011;332:1294–1297.

41. Yang H, Feng X, Wang Q, Huang H, Chen W, Wee ATS, Ji W. Giant two-photon absorption in bilayer graphene. Nano Lett 2011;11:2622.

42. Seol JH, Jo I, Moore AL, Lindsay L, Aitken ZH, Pettes MT, Li X, Yao Z, Huang R, Broido D, Mingo N, Ruoff RS, Shi L. Two-dimensional phonon transport in supported graphene. Science 2010;328:213.

43. Bristow AD, Rotenberg N, van Driel HM. Two-photon absorption and Kerr coefficients of silicon for 850–2200 nm. Appl Phys Lett 2007;90:191104.

44. Chen CJ, Zheng J, Gu T, McMillan JF, Yu M, Lo G-Q, Kwong D-L, Wong CW. Selective tuning of silicon photonic crystal nanocavities via laser-assisted local oxidation. Opt Express 2011;19:12480.

45. Barclay PE, Srinivasan K, Painter O. Nonlinear response of silicon photonic crystal micro-resonators excited via an integrated waveguide and fiber taper. Opt Express 2005;13:801.

46. Yang X, Husko C, Yu M, Kwong D-L, Wong CW. Observation of femto-joule optical bistability involving Fano resonances in high-Q/V_m silicon photonic crystal nanocavities. Appl Phys Lett 2007;91:051113.

47. Nozaki K et al. Sub-femtojoule all-optical switching using a photonic-crystal nanocavity. Nat Photonics 2010;4:477.

48. Almeida VR et al. All-optical control of light on a silicon chip. Nature 2004;431:1081.

49. Jackson JD. *Classical Electrodynamics*. New York: Wiley; 1999.

50. Armaroli A et al. Oscillatory dynamics in nanocavities with noninstantaneous Kerr response. Phys Rev A 2011;84:053816.

51. Johnson TJ, Borselli M, Painter O. Self-induced optical modulation of the transmission through a high-Q silicon microdisk resonator. Opt Express 2006;14:817–831.

52. Almeida VR, Lipson M. Optical bistability on a silicon chip. Opt Lett 2004;29:2387.

53. Haus HA. *Waves and Fields in Optoelectronics*. Englewood Cliffs (NJ):: Prentice-Hall; 1984. p 99.

9

POLYMER PHOTONIC DEVICES

ZIYANG ZHANG AND NORBERT KEIL

Fraunhofer Heinrich Hertz Institute (HHI), Berlin, Germany

9.1 INTRODUCTION

We are moving toward an era that requires that we are able to access and share information wherever we go. Web browsing, peer-to-peer file sharing, blogging, and high-definition audio/video streaming are witnessing a constant growth year over year. The average data consumption was estimated to be around 9665 petabytes (1 petabyte (PB) = 1,000,000 gigabytes) per month in 2010. This figure is projected to rise to 116,539 PB by 2015.

Wavelength division multiplexing (WDM) in passive optical networks (PONs) is one of the key access technologies to satisfy the ever-growing demand for larger Internet bandwidth due to its proven architecture and ease of scalability. A simplified schematic of this technology is illustrated in Figure 9.1. A PON is a point-to-multipoint, fiber to the premises network architecture in which unpowered optical splitters are used to enable a single optical fiber to serve multiple premises. It is a shared network, which consists of an optical line terminal (OLT) at the service provider's central office and a number of optical network units (ONUs) near end users. A PON reduces the amount of fiber and necessary central office equipment, when compared with point-to-point architectures. Using WDM technology, a number of optical carrier signals of different wavelengths from the transmitters (Tx) are carried onto a single optical fiber by a multiplexer (MUX). The incoming signals containing different wavelengths from the optical network are first decomposed by a demultiplexer (DMX) and then fed into the receivers (Rx). The ONU can well contain a tunable unit on the Tx, with which a multitude of wavelengths can be sent for various purposes.

At the backbones of the WDM PON, photonic devices with high spectral efficiency and receiver sensitivity are desired, together with the associated, preferably low-cost, processing technology. In order to scale up device functionality and reduce the footprint, an integration platform is needed to combine various components in a packaged module, which is then ready to be deployed in the optical network for plug and play. Compared to other material and integration platforms, polymers hold distinct advantages. First of all, a myriad of polymer materials have been developed in the last decades and a large range of applications have been covered. The passive functions (without electric power supply) include low-loss waveguides, optical interconnects, wavelength splitters, and (de)multiplexers, while the active functions (electric power needed) comprise high-speed electro-optic (EO) modulators, digital optical switches, and tunable optical filters and lasers (1–3).

In the early stage of polymer optical material development, poly(methylmethacrylate) (PMMA), polystyrene (PS), and epoxy resins are commonly used, among others. These materials, though relatively easy to synthesize, offer only a limited choice of refractive indexes, and often suffer from low thermal stability. Recent progress includes OE-414x and WG-101x series from Dow Corning for applications in optical interconnects in Datacom at 850 nm wavelength, ZPU and LFR series from ChemOptics for low-loss single-mode optical waveguides in Telecom at both 1310 and 1550 nm wavelengths, and EO polymer materials from GigOptix for ultrafast light switching and modulation for both Datacom and Telecom applications at 1550 nm wavelength.

To choose a proper material system, the following key issues must be considered:

1. Type of applications: For passive waveguides, the material must be transparent at the chosen wavelength window, typically around 850 nm, 1310 nm, and 1550 nm. If no other inorganic materials are used as waveguide core or cladding, the material system must be able to provide at least two different refractive indexes. For high-speed optical modulation, nonlinear polymers with decent r_{33} coefficients should be taken.

2. Method of fabrication: Different polymers exhibit large variations of chemical properties and compatibilities. Some polymers can only be structured using directly ultraviolet (UV) light illumination, while others provide more chemical resistance and can be structured

Nanomaterials, Polymers and Devices: Materials Functionalization and Device Fabrication, First Edition. Edited by Eric S. W. Kong.
© 2015 John Wiley & Sons, Inc. Published 2015 by John Wiley & Sons, Inc.

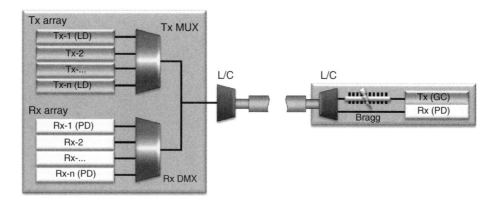

Figure 9.1   Schematic of the OLT-Tx and OLT-Rx (left) and tunable ONU (right) for the WDM PON.

using conventional photolithography with hard etching masks. There are also polymers that allow direct imprint patterning for specific applications.

3. Compatibility: Again due to the chemical varieties of polymers, they may show different surface behavior on inorganic materials and even among polymer materials themselves. Usually a specific adhesion promoter must be applied to hold polymer layers on surfaces of other materials, such as silicon and silicon dioxide. Thermal expansion and stress should be considered in thick polymer layers to avoid deformation.

4. Long-term stability: For commercial devices, the polymer material must be able to withstand harsh environmental tests such as damp heat, temperature cycling, and high optical power burn-in.

This chapter focuses on the photonic devices fabricated using the ZPU-12 series polymer material from ChemOptics. The technical advantages of using this material in the fabrication are discussed in Section 9.2, followed by the demonstration of various fabricated passive devices in Section 9.3. Thermo-optic (TO) tunable devices are further revealed in Section 9.4. The concept of hybrid photonic integration is introduced in Section 9.5. Finally in Section 9.6, the long-term reliability tests are outlined. Instead of going into depth of polymer material chemistry and device physics, the purpose of this chapter is to offer an overview of the cutting-edge polymer technology, on the individual components and device integration level that is suitable for commercial production and direct industry applications in the coming 2–3 yr.

9.2 TECHNOLOGY OVERVIEW FOR POLYMER PHOTONIC DEVICE FABRICATION

The ZPU-12 materials are mixed perfluorinated oligomer and acrylate with proprietary ingredients. The materials are optically transparent at both 1310 nm and 1550 nm telecommunication windows, at which the typical material absorption measures around 0.1 dB/cm and 0.4 dB/cm, respectively. The refractive indexes can be chosen anywhere between 1.45 and 1.48 (at 1550 nm). The material can be spin coated on a variety of substrates and then cured under UV illumination, followed by a temperature fixing at 200 °C. The "curing" is defined in polymer chemistry as a process during which the fluid polymer material toughens or hardens by cross-linking of the molecular chains, brought about by chemical additives, UV radiation, electron beam, or heat. The chemical reaction during the curving process is beyond the scope of this chapter, but this simple, fast, and low-temperature layer forming process is unique to polymers and stands out as a key advantage over complicated epitaxy process in III–V semiconductors as well as the high-temperature (>800 °C) and high-vacuum chemical vapor deposition (CVD) in silicon photonics.

9.2.1 Wafer Level Fabrication

Many novel fabrication technologies have been developed to tailor to various polymer materials. For potential industry production, however, the technology has to stay low cost and suitable for large-volume throughput. To avoid large efforts in very specialized equipment designs and to minimize the process development cost, one desirable way for polymer device fabrication is to apply similar concepts, equipment and processes that have already been well developed in the established silicon electronics industry.

Consider the buried channel waveguide structure as shown in Figure 9.2a. The cladding material index is 1.45 and the core is 1.47. To avoid waveguide geometrical birefringence, the waveguide core is kept as a square shape ($W = H$). For single-mode operations, W and H are kept below 4 μm. The transverse electric (TE) mode field E_x is shown in Figure 9.2b. As will be discussed in Section 9.4, the electrode (microheater) embedded in the polymer cladding directly below the core can provide more uniformed heating condition, and the air trenches can further help confine the heat distribution.

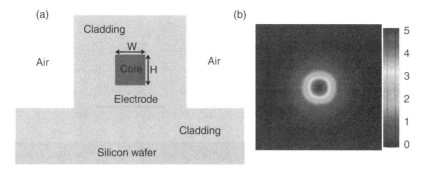

Figure 9.2 (a) Schematic of a buried channel waveguide with imbedded electrode and air trenches; (b) the TE waveguide mode profile (E_x field component).

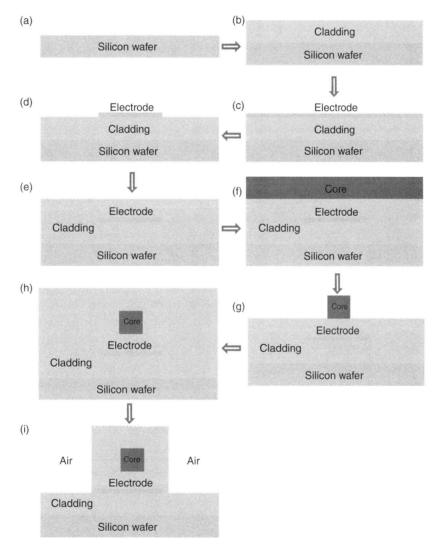

Figure 9.3 "Bottom-up" wafer fabrication process flow.

The fabrication process for the structure shown in Figure 9.2a follows a typical "bottom-up" approach in the standard wafer production, as illustrated in Figure 9.3. A silicon wafer is first primed with proper adhesion promoter. A layer of polymer cladding material is then spin coated on top and cured. The thickness of the cladding can be coarsely controlled by the spin-coating conditions such as rotation speed and time. To reach the thickness precision at the tens of nanometers level, subsequent slow-rate reactive ion etching (RIE) can be used.

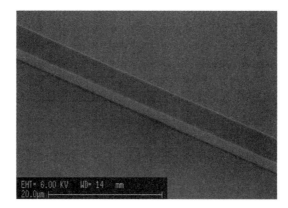

Figure 9.4   SEM image of a polymer waveguide core, etched by pure oxygen plasma.

Metal can be deposited directly on top of the polymer cladding by sputtering or thermal evaporation. A positive lithography process followed by wet chemical etching can be used to structure the metal. With subsequent spin-coating and curing processes, the metal stripes are imbedded inside the polymer.

To pattern the waveguide core, photo resist can be used as etching mask after lithography. However, since the main plasma etching chemistry for polymer etching is oxygen based, the photo resist mask will be consumed during etching. To improve the etching selectivity, defined as the ratio between the etching rates of the polymer material and the mask itself, a metal mask can be patterned on top of the waveguide core by a "lift-off" process.

During the "lift-off" process, metal is first deposited on top of the patterned photo resist, which is then washed away by an organic solver such as acetone and 1-methyl-2-pyrrolidone (NMP). Only the metal that is deposited on the openings remains. This process requires that the polymer material beneath must be chemically strong enough against the organic solver and that its chemical, physical, and optical properties should not be adversely influenced.

With pure oxygen plasma etching and metal masks such as nickel, chromium, or titanium, the etching selectivity can easily reach beyond 500. This means that deep trenches, with, for example, $50\,\mu m$ depth, can be etched in polymer by using a very thin layer of metal that is only $100\,nm$ thin. This is a clear advantage when compared to deep etching of inorganic materials, where complex and often expensive masking and etching schemes must be taken into account.

Once the waveguide core is etched, the metal mask is removed by selective wet chemistry, dry plasma etching, or ion milling. A scanning electron microscopy (SEM) image of an etched polymer waveguide core is shown in Figure 9.4. To finish the waveguide structure, another cladding layer is spin coated on top and cured. The wafer can be structured by another metal mask and deep air trenches can be etched.

9.2.2   Wafer Dicing and Facet Preparation

Usually, a wafer contains a large number of photonic components (known as die in the electronics industry, or more commonly called as chips). Fig. 9.5 shows a 4-inch wafer with ~600 chips. These components, arranged in columns and rows, must be separated before they can be further characterized, sorted, and assembled for packaging. This separation process in general is called wafer dicing. Depending on the substrate type, the wafer can either be cleaved with a diamond tool, or sawed with a rotating diamond blade. For specific applications with bristle amorphous substrate, laser cutting can be adopted to separate the chips.

Cleaving is the process of separating a crystal along its natural crystalline planes. It is achieved by creating a small notch at the edge of a wafer and then applying tensile strain to the notch. The small notch generates a stress concentration that propagates and cracks the wafer. If aligned properly, the crack will propagate along a chosen crystal plane – the path of the least resistance. The main advantage of cleaving is that the separation along natural planes leaves the cleanest edge possible due to the regularity of the materials' molecular structure. For semiconductor wafers, cleaved surfaces produce smooth mirrored surfaces that effectively reflect light. However, since most polymer materials are considered amorphous, the cracking of the substrate wafer may not tear up the polymer layer smoothly, which often results in chipping and layer delamination. The edges of the components must be aligned along the crystalline planes. Some angle misalignment during the lithography process can cause walked-off chip edges (Fig. 9.5).

Wafer sawing refers to the process by which the components are separated from a wafer mechanically with a dicing saw. During sawing, wafers are fixed on a frame with a sheet of adhesive tape. The wafer frame is then mounted on a high-precision stage, which is aligned and rotated with respect to the rotating diamond blade. As the stage moves to the blade, the wafer is fully cut, whereas the adhesive film is only slightly slit. This leaves the chips in place for further automated handling.

The main advantage of sawing is that the wafer can be cut in any lateral direction. In addition, the blade rotating speed and the stage feeding speed can be optimized to initialize an intrinsic "polishing" effect at the same time. The end facet is optically smooth enough to prevent random scattering when coupled to a light source, especially from a glass fiber, since their effective indexes are rather close, ~1.46. Therefore, the

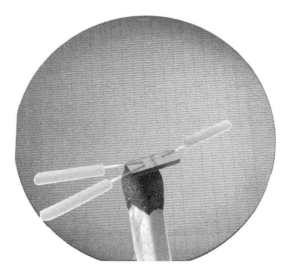

Figure 9.5   Four-inch wafer with ~600 polymer chips.

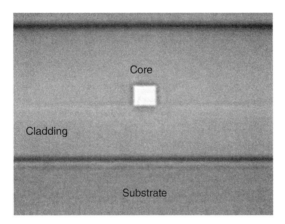

Figure 9.6   Diced facet of a polymer waveguide chip, without polishing.

time-consuming and often expensive end-facet polishing, which is common to silicon photonics, can be avoided. As an example, a photo of a diced facet from a buried single-mode polymer waveguide chip is shown in Figure 9.6. The sawing marks are barely visible.

Another advantage of the sawing process is that the cutting depth can be freely chosen. The wafer can be partially sawed in the vertical dimension, leaving a deep slot/trench without breaking the wafer or chip. A photo of such a slot is shown in Figure 9.7, where a thin film element (TFF) is already inserted. Note that the cut is through the polymer layers and partially into the silicon substrate. With dry or wet etching process, the chemistry can be well controlled to penetrate the silicon substrate without damaging the polymer layers.

By choosing blades of different thickness, the width of the slots can be varied. However, a blade with thickness less than 15 μm proves difficult to manufacture and not very durable during sawing. For narrow slots with width less than15 μm, etching is preferred. Moreover, in most cases the sawing process must be performed along the entire width of a wafer or chip, in order to guarantee a uniform cut shape. When the space is limited or the structure does not allow a complete cut-through, etching should be used to create local slots and trenches.

When the inorganic substrate such as silicon is being sawed, the destructed material leaves usually many dust particles and sometimes even bristle bits. The wafer must be thoroughly washed during and after the sawing process.

9.2.3   45° Mirror

Mirrors with 45° deflection angle are important elements in three-dimensional (3D) photonic integration. They can be used to direct the beam from the lateral wave guiding motherboard to the vertical direction so that light can be detected by a surface-illuminated photo diode photo detector (PD). They can also be used to direct the output from a vertical cavity surface emitting laser (VCSEL) to the waveguides. In silicon photonics, similar mirrors can be created by the selective wet etching of crystalline silicon. The process is often time-consuming and difficult to control. On polymer platform, such mirrors can be easily sawed by a diamond blade with a predefined 45° tip. An example is shown in

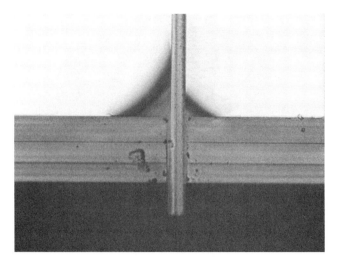

Figure 9.7 Deep slot created by sawing through the polymer layers and partially into the silicon substrate. A thin film element is inserted in the slot and fixed with index matching epoxy.

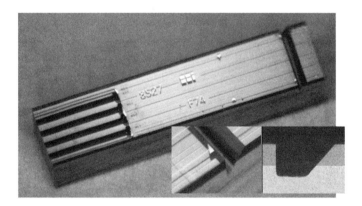

Figure 9.8 A polymer waveguide chip with a sawed 45° mirror.

Figure 9.8. The chip contains four fiber grooves, which will be explained in Section 9.3.4, and four parallel waveguides that end with a 45° facet.

The surface roughness can be minimized by the intrinsic "polishing" during sawing by choosing the right diamond concentration on the blade tip, the optimal blade rotation speed, and the feeding speed. To improve the reflectivity, the 45° surface should be coated with a thin layer of metal by directed metal evaporation.

Apart from the design and processing simplicity, such 45° deflection mirrors offer much broader bandwidth, much less polarization sensitivity, and much less beam divergence, when compared to vertical grating couplers. When the 45° groove is filled with index matching epoxy, the back reflection on the straight facet can be avoided and the beam divergence in free space is reduced. However, similar to the slot sawing process, such mirrors cover the entire width of the chip/wafer. When localized mirrors are needed, due to design or space constraints, a phase mask and subsequent shadow etching should be used.

9.3 PASSIVE POLYMER PHOTONIC DEVICES

9.3.1 Losses in Polymer Waveguides

For Telecom applications, polymer waveguides need to have low optical loss values at the three main communication wavelength windows, 850 nm, 1310 nm, and 1550 nm. For dense WDM applications, the waveguides should remain low loss in the waveguide range 1510–1620 nm, depending on the device type. Optical loss of a photonic device, or commonly known as insertion loss, comes from two main sources, that is, the device internal loss and the coupling loss with external fibers, light sources, detectors, or other elements.

The coupling loss comprises mode mismatch loss and facet scattering. Mode mismatch loss arises when two optical modes do not overlap completely due to mode size differences, lateral and angular misalignment. The "unmatched" part cannot be coupled into a guided propagating mode and eventually leaks away. In case of perfect optical alignment, the mode mismatch loss can be minimized by shaping the waveguide mode, usually with taper structures, so that its mode field matches that of the element to be connected. Facet scattering appears when the device facet is not optically smooth, due to bad sawing marks, chipping, contaminations, and so on. When the facet condition is not tolerable, a polishing process is required. Facet scattering can also cause back reflection, which appears when there is a refractive index mismatch at the optical interface. Back reflection can be minimized by filling the gap with index matching liquid or by coating the facets with antireflection (AR) layers.

The sources for internal loss include the intrinsic device loss from theory, material absorption, waveguide scattering, and polarization-dependent loss (PDL). The intrinsic device loss is derived directly from theory or numerical simulations. For example, a Mach–Zehnder interferometer (MZI) that is designed to work at the middle wavelength point between maximal and minimal interference has already an intrinsic loss of 3 dB at this wavelength from theory.

In the near infrared wavelength region, polymer material absorption comes mainly from the overtones of fundamental molecular vibrations. While C–H bonds mostly account for absorption of wavelength larger than 3000 nm, O–H bonds are the main contributor to absorption around 1400 nm. When hydrogen contents are removed through partial fluorination, the transparency of the material significantly improves. Since material absorption leads inevitably to heat generation, a unique technique called photothermal deflection spectroscopy can be applied to determine the material absorption of many polymers. As mentioned before, the ZPU-12 series exhibit a material absorption of around 0.1 dB/cm at 1310 nm and 0.4 dB/cm at 1550 nm. Though materials with higher transparency are indeed commercially available, ZPU-12 series still prove to be a good choice when material processibility, chemical and mechanical properties are concerned.

Along the waveguide path, any defects that disturb the waveguide uniformity can cause scattering. Voids, cracks, bubbles, particles, and other contaminations should be avoided at all cost during fabrication. When the polymer material is not well prepared, or the curing process is not evenly executed, there may be index inhomogeneities and optical stresses in the polymer layer across the substrate. When the waveguides pass through these index "bumps," scattering may occur.

Consider the waveguide structure in Figure 9.2a. Roughness along any of the four edges of the waveguide core can cause scattering. The sidewall roughness may result from a bad etching process. The ceiling and floor roughness may come from damaged interfaces if the core and cladding polymer materials chemically attack each other during curing. Aggressive wet chemical etching and thermally caused layer delamination can further contribute to the interface roughness.

PDL is defined as the maximal difference of attenuation between any of the two polarization states. It is, in essence, a combination of material absorption and waveguide scattering. Certain polymer materials exhibit not only high birefringence, but also the material absorption for TE- and TM-polarized light can be different. For a buried channel waveguide, TE scattering loss is mainly determined by the sidewall roughness of the core, while TM scattering loss is mainly caused by the ceiling and floor interface inhomogeneities. For the waveguide structure in Figure 9.2a, if the metal electrode is placed close enough to the waveguide core so that it disturbs the mode field, TM mode will suffer from additional light absorption in the metal layer. Sometimes, this effect is used to suppress the TM mode in the waveguide.

The material birefringence for the ZPU-12 series, that is, relative refractive index difference between the TE and TM modes, is usually less than 1%. Both polarizations appear to have similar material absorption. The PDL for the buried waveguides as in Figure 9.2a is less than 0.05 dB/cm. To study the wavelength dependence of the insertion loss, a broadband light source is launched into and collected out of the waveguide by a single-mode fiber (SMF) at both facets. Additional index matching oil should be applied to fill any possible air gaps between the fiber and the waveguide facet, in order to eliminate Fabry–Perot interferences at the interface. After normalization to the intrinsic power spectrum of the light source, the transmission spectrum for the 4.8 cm long polymer waveguide is shown in Figure 9.9. It is clear that around 1400 nm the polymer material appears to have strong absorption, indicating the presence of certain O–H bonds in the chemical contents. Around 1310 nm and 1550 nm, two loss valleys are observed, with decent 1 dB bandwidth of around 100 nm. At these two wavelength windows, low-loss photonic devices can be made.

In order to separate the internal loss, in this case the straight waveguide propagation loss, and the coupling loss from the total insertion loss, the waveguide is cut into two pieces with 1.8 and 3.0 cm length. The measurement shown in Figure 9.9 is repeated for both pieces. The insertion loss values at 1550 and 1310 nm are first extracted from the spectra and then linearly fitted. This is called "cutback" method. The results of the three waveguides of length 1.8, 3.0, and 4.8 cm are shown in Figure 9.10. The slope of the linearly fitted line indicates the waveguide propagation loss, in unit of dB/cm. At 1550 nm, the value is 0.68 dB/cm and at 1310 nm, the value is 0.45 dB/cm. Both values appear higher than the material absorption (0.4 dB/cm and 0.1 dB/cm at 1550 nm and 1310 nm, respectively). This is a direct result from waveguide scattering. It also shows that the scattering is more severe at shorter wavelength.

The cutting point on the Y-axis shows the total coupling loss from both facets. Since the waveguide is a uniform channel and the coupling with fiber is deemed identical on both sides, the single side fiber to waveguide coupling is simply half of this value, that is, 1.35 dB at 1550 nm and 1.4 dB at 1310 nm. The coupling loss can be optimized, as discussed earlier, by adding a waveguide taper that tailors the mode closer to the fiber mode field size.

9.3.2 Polymer Arrayed Waveguide Gratings

Arrayed waveguide gratings (AWGs) are very commonly used as wavelength MUX and DMX in the WDM network. They are planar optical devices based on an array of coupled waveguides. Signals with different wavelengths from one input waveguide are projected onto an array

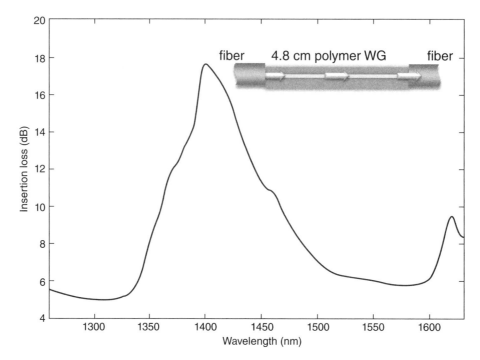

Figure 9.9 Insertion loss of a 4.8 cm long polymer waveguide.

of output waveguides, each with a specific wavelength component. The concept of AWGs was first presented by Smit (4) in 1988. An AWG is also called a waveguide grating router, a phase array, or a phasar.

Figure 9.11 shows the schematic layout of an AWG. It consists of an array of (i) input/output access waveguides, (ii) a free propagation region (FPR), and (iii) an array of curved waveguides with a fixed optical path difference between the adjacent channels.

The principle of AWG is described as follows: a beam first propagates through the access waveguide into the FPR, where it is no longer confined in the lateral dimension. The beam starts diverging the images onto the input aperture of the arrayed waveguide session. It is then coupled into all the waveguides in the array session and propagates toward the end of the array. The length of the waveguides in the array is carefully designed, so that the optical path length difference between the two adjacent waveguides equals an integer multiple of the central wavelength λ_c. For λ_c, the fields in the individual waveguides arrive at the output aperture with the same phase (or with multiple times of 2π delay/advance). Therefore, the field is reconstructed at the output aperture but reversed to the other direction. The divergent beam at the input aperture is thus transformed back into a convergent one with equal amplitude and phase distribution. The beam is further focused to form an image at the center of the output plane inside the second FPR.

As the wavelength deviates from λ_c, the propagation direction of the outgoing beam is tilted, resulting in a spatially separated focal point. If the output access waveguides are placed at proper positions along the image plane, different wavelengths are led to different output ports.

Arrayed waveguide gratings have been studied and demonstrated on many material platforms, silica, silicon on insulator (SOI), indiumphosphide (InP), and polymer, to name a few. The advantages of the polymer platform include low-cost production, fast prototyping, and the possibility to realize polarization-insensitive and temperature-insensitive devices.

Figure 9.12a shows the design of two AWGs. One is 180° rotated and combined with the other. Figure 9.12b is a microscopy photo of the intersection region from the FPR to the arrayed waveguides. Each AWG contains 12 input waveguides and 5 output waveguides. It can be used as a transmitter MUX, which combines the inputs from 12 lasers of different wavelengths into a single output and feeds into the WDM network.

The measurement results are shown in Figure 9.13. The channel spacing is 100 GHz (0.8 nm). The 1 dB bandwidth is ~0.25 nm. The insertion loss (SMF–AWG–SMF) is below 6 dB and the central wavelength cross talk is below –25 dB.

For the transmitter MUX AWG, the light sources are often coupled directly on the facet or via a short length of fiber. The polarization state is usually maintained at TE. On the receiver side, however, it is very likely that the input signals have propagated a long distance through the network and their polarization states are rendered uncertain. Therefore, polarization-insensitive DEMUX AWGs are highly desired, in order to avoid extra circuits to adjust the polarization errors electronically (Fig. 9.14).

As discussed earlier, ZPU material appears to have slight birefringence, that is, the effective indexes for the fundamental TE and TM modes are different. This will cause a shift of the spectral response for TE or TM modes, which is called polarization dispersion.

One way to minimize the polarization dispersion is to create a slot, using the method introduced in Section 9.2.2, in the middle of the phased array, and then insert a half-wave ($\lambda/2$) plate (5). Light entering the array in the TE-polarized state will be converted by the half-wave plate and travel through the second half of the array in the TM-polarized state. The TM-polarized light will similarly traverse half the array

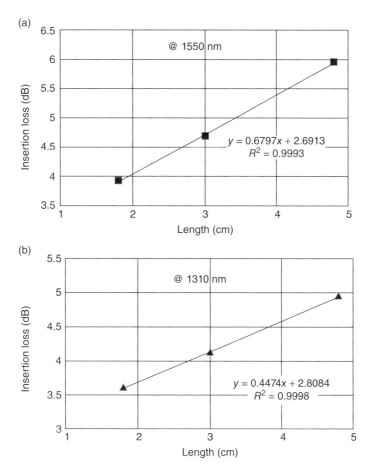

Figure 9.10 Cutback measurements at (a) 1550 nm and (b) 1310 nm for waveguides with three different lengths.

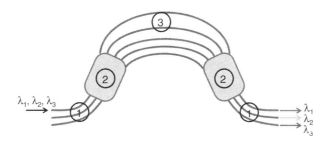

Figure 9.11 The schematic layout of an AWG as DMX.

in the TE state. As a result, both polarizations will go through the same phase variations regardless of the birefringence properties of the waveguides.

The spectral response of the DEMUX AWG is shown in Figure 9.15a, measured at TE polarization after the half-wave plate is inserted. The channel spacing measures 3.33 nm. The insertion loss is around 6 dB. Figure 9.15b shows the spectra without the half-wave plate for both polarizations. The polarization dispersion $\Delta\lambda$ is around 1.15 nm. After inserting the half-wave plate, in Figure 9.15c, $\Delta\lambda$ is dramatically reduced to below 40 pm. The excess loss introduced by the half-wave plate is less than 0.5 dB.

Since polymer materials usually have large TO effects, a temperature control unit is often necessary to stabilize the channel wavelengths of the AWGs. This requires extra power consumption and additional electronic control units. There has been much research effort in developing temperature-insensitive AWGs. One way is to use combined polymer and inorganic materials with opposite TO coefficients. The waveguide is designed such that the index change with temperature balances out to certain degree. Another method is to use a certain polymer substrate with positive coefficient of thermal expansion (CTE) that matches the optical polymer materials in the waveguides (6). The idea is to compensate

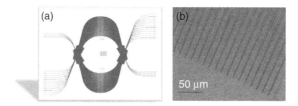

Figure 9.12   (a) Layout of two MUX AWGs; (b) microscopy photo showing the interface between the FPR and the arrayed waveguides.

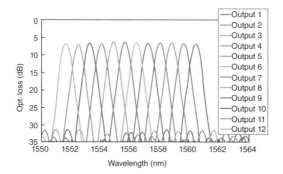

Figure 9.13   Characteristics of the transmitter MUX AWG.

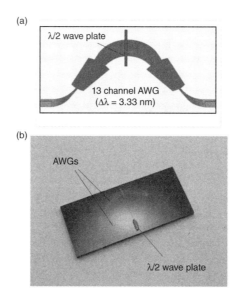

Figure 9.14   (a) The schematic layout and (b) the device photo of a receiver DEMUX AWG with a half-wave plate inserted right in the middle of the arrayed session.

the temperature-induced index change by allowing the waveguides to either physically expand or shorten themselves, so that the net optical length change is kept close to zero.

9.3.3   Multimode Interference Devices and 90° Hybrids

Multimode interference devices (MMIs) are widely used in integrated optics as compact and efficient power splitters, Mach–Zehnder switches, modulators, phase mixers, and optical hybrids. The operation of optical MMI devices is based on the self-imaging principle. The flexibility in design often offers large optical bandwidth as well as polarization insensitivity.

The self-imaging principle in uniform index slab waveguides was first suggested by Bryngdahl (7) and further explained in more detail by Ulrich (8). Self-imaging is an effect of light propagation inside a multimode waveguide, by which an input field profile is reproduced in single

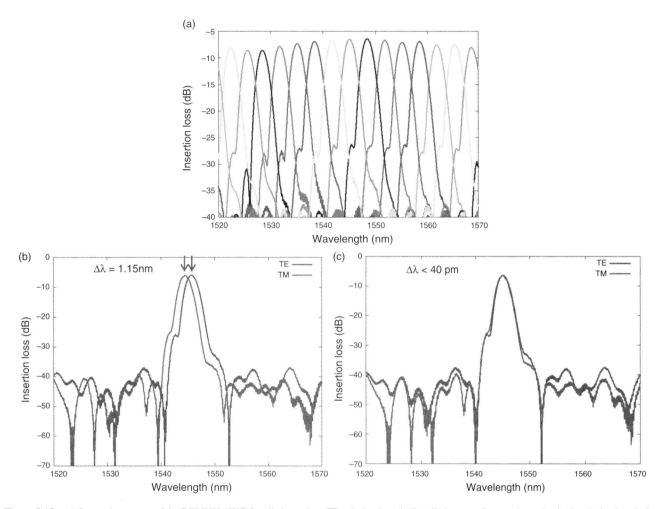

Figure 9.15 (a) Spectral response of the DEMUX AWG for all channels at TE polarization. (b) Detailed spectra for one channel at both polarizations before inserting a half-wave plate. (c) After inserting a half-wave plate, the polarization dispersion is minimized.

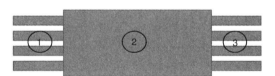

Figure 9.16 Schematic layout of an MMI. Region 1 is the input waveguide network, region 2 is the central multimode waveguide, and region 3 is the output waveguide network.

or multiple images at periodic intervals along the propagation direction of the waveguide. The schematic of an MMI is shown in Figure 9.16. The central part of an MMI is a wide waveguide designed to support a large number of modes (typically >3). In order to launch light into and collect light from the multimode waveguide, a number of access (usually single mode) waveguides are placed at both ends of the multimode session. Such devices are generally referred to as $N \times M$ MMI couplers, where N and M are the number of input and output waveguides, respectively.

A full-modal propagation analysis is probably the most comprehensive and intuitive theoretical tool to describe self-imaging phenomena in multimode waveguides (9). It not only supplies the basis for numerical simulation and design, but also provides insight into the mechanism of multimode interference. An example is shown in Figure 9.17. The device is a 4×4 MMI, in which light is injected from input port 1, propagates inside the multimode region, images itself into port 4 "focusing" nodes at the end of the multimode region, and then couples into the four output waveguides.

Since there are no strict analytical equations describing the relation between the propagation constants of the waveguide modes and their mode orders, very often, the estimated MMI multimode session length deviates from the perfect imaging length. This results in phase errors and

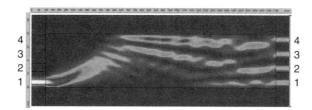

Figure 9.17 Light propagation inside a 4 × 4 MMI.

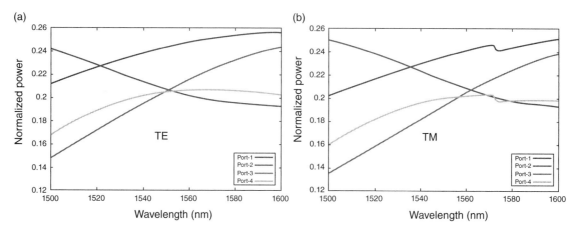

Figure 9.18 Transmission power spectra at the four outputs for both polarizations: (a) transverse electric (TE); (b) transverse magnetic (TM).

imbalance between the output modes. In addition, the coupling efficiency between the multimode session and the output access waveguides may not be 100%, due to the mismatched or misaligned mode profiles. This in turn will add to the intrinsic loss of the MMI device.

Take the 4 × 4 MMI shown in Figure 9.17, for example. Broadband light source is simulated and launched into input port 1 and the transmitted power at the four outputs is recorded. The results are plotted in Figure 9.18, for both polarizations. Before the results are analyzed, it is useful to define the imbalance (IB) of the MMI and the intrinsic loss (IL) as follows:

$$\text{IB} = -10\log\frac{\min(P_i)}{\max(P_i)} \tag{9.1}$$

$$\text{IL} = -10\log\left(\sum_{i=1}^{4} P_i\right) \tag{9.2}$$

where P_i ($i = 1, 2, 3, 4$) is the normalized transmission power fraction at output port i. At 1550 nm, the MMI appears to have IB = 0.79 dB and IL = 0.76 dB for the TE mode, while for the TM mode IB = 0.83 dB and IL = 0.86 dB.

One of the most important applications of a 4 × 4 MMI is an optical 90° hybrid, which functions as a homodyne receiver in a coherent detection system (10). Consider the diagram in Figure 9.19. The signal and local oscillator are fed in to port 1 and port 3 of the MMI, respectively. The input vector can be written as

$$E_{\text{in}} = \begin{bmatrix} E_4 \\ E_3 \\ E_2 \\ E_1 \end{bmatrix} = \begin{bmatrix} 0 \\ E_{\text{LO}} \\ 0 \\ E_{\text{Signal}} \end{bmatrix}$$

Discard the common phase term and the power penalty, the 4 × 4 MMI has an intrinsic transfer function

$$M = \frac{\sqrt{2}}{2} \begin{bmatrix} 1 & e^{j\frac{3\pi}{4}} & e^{-j\frac{\pi}{4}} & 1 \\ e^{j\frac{3\pi}{4}} & 1 & 1 & e^{-j\frac{\pi}{4}} \\ e^{-j\frac{\pi}{4}} & 1 & 1 & e^{j\frac{3\pi}{4}} \\ 1 & e^{-j\frac{\pi}{4}} & e^{j\frac{3\pi}{4}} & 1 \end{bmatrix}$$

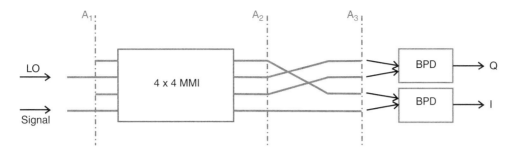

Figure 9.19 Diagram of a homodyne coherent receiver using a 4 × 4 MMI.

From plane A_1 to A_2, the intermediate vector can be written as

$$E_M = ME_{in} = \frac{\sqrt{2}}{2} \begin{bmatrix} E_{LO}e^{j\frac{3\pi}{4}} + E_{Signal} \\ E_{LO} + E_{Signal}e^{-j\frac{\pi}{4}} \\ E_{LO} + E_{Signal}e^{j\frac{3\pi}{4}} \\ E_{LO}e^{-j\frac{\pi}{4}} + E_{Signal} \end{bmatrix}$$

At plane A_3, the output ports are rearranged for balanced photo detectors (BPD). Introduce an additional 45° phase delay to the local oscillator, and the output vector can be conveniently written as

$$E_{out} = \frac{\sqrt{2}}{2} \begin{bmatrix} E_{LO} + E_{Signal}e^{-j\frac{\pi}{4}} \\ E_{LO} + E_{Signal}e^{j\frac{3\pi}{4}} \\ E_{LO}e^{j\frac{3\pi}{4}} + E_{Signal} \\ E_{LO}e^{-j\frac{\pi}{4}} + E_{Signal} \end{bmatrix} = \frac{\sqrt{2}}{2} \begin{bmatrix} E'_{LO} + E_{Signal} \\ E'_{LO} - E_{Signal} \\ jE''_{LO} + E_{Signal} \\ -jE_{LO} + E_{Signal} \end{bmatrix} = \begin{bmatrix} E_{01} \\ E_{02} \\ E_{03} \\ E_{04} \end{bmatrix} = \begin{bmatrix} I \\ Q \end{bmatrix}$$

Rewrite

$$E_{LO} = \sqrt{P_{LO}}e^{j(\omega_{LO}t+\theta_{LO})}$$

$$E_{Signal} = \sqrt{P_{Signal}}e^{j(\omega_{Signal}t+\theta_{Singal})}$$

At the BPD side, the in-phase and quadrature signals, I and Q, can be obtained separately:

$$I(t) \propto \sqrt{P_{Signal}}\sqrt{P_{LO}}\cos(\Delta\theta)$$

$$Q(t) \propto \sqrt{P_{Signal}}\sqrt{P_{LO}}\sin(\Delta\theta)$$

$$\Delta\theta = (\omega_{Signal} - \omega_{LO})t + (\theta_{Signal} - \theta_{Signal})$$

In short, via the 90° hybrid, the phase modulated and multiplexed signals are converted into intensity variations, which can be readily received and separated (I and Q) by the BPDs. The optical 90° hybrid together with the BPD array is often called a coherent receiver.

Apart from fabrication-cost advantages on the polymer platform, 45° tuning mirrors can be easily added at the output ports, which allow surface-illuminated BPDs to be mounted on top. In this way, extra mechanical support or submount is not needed for the BPDs and the assembly process is simplified. The schematic layout and the device photo are shown in Figure 9.20a and b, respectively. Details for this type of mounting techniques will be revealed in Section 9.5.1.

In order to characterize the four outputs of the MMI directly and extract the phase error information, a 1×2 delay line interferometer (DLI) is added at the input side of the MMI. Typical measurement results are illustrated in Figure 9.21. The total insertion loss is around 8 dB, including fiber–chip coupling loss and the 3 dB intrinsic loss of the optical circuits. The IB is less than 0.5 dB. The phase errors are extracted from the insertion loss measurements at each wavelength point. Typical values are around 5°.

Another advantage of coherent receivers on polymer platform is the simplicity to include a polarization control unit that allows polarization-multiplexed coherent detection scheme. The next section will reveal the functions of TFEs, with which wavelength and polarization splitting functions can be easily added wherever necessary.

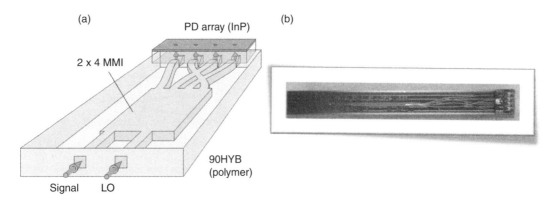

Figure 9.20   Coherent receiver based on a polymer 4 × 4 MMI and surface-illuminated BPDs. (a) Schematic layout and (b) device photo.

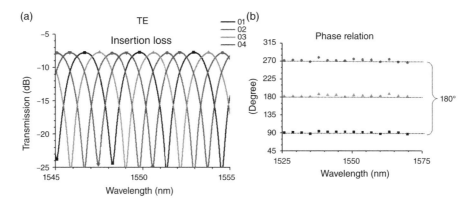

Figure 9.21   Measured results of the coherent receiver based on the polymer 4 × 4 MMI. (a) Insertion loss and (b) phase errors.

9.3.4   Fiber Grooves, Thin Film Elements, and Related Devices

9.3.4.1   U-Grooves   Historically, fiber and waveguide coupling has been challenging for planar lightwave circuits (PLCs) due to the refractive index and/or mode mismatch between fiber and waveguide. The coupling issue becomes more serious with high-index contrast PLCs, based on, for example, silicon (SOI) and III–V semiconductors. The solutions are either expensive (with nanospot size converters) or bulky (with vertical grating couplers) (11, 12). For low-index contrast PLCs, such as silica or silicon nitride (SiN$_x$) platforms, the coupling issue is less severe and can be solved by butt–joint coupling with the aid of AR coatings. However, the butt–joint coupling often suffers from poor mechanical support, and thus, an extra submount is needed to hold the fiber in place.

On the polymer platform, fiber and waveguide coupling can be achieved by means of on-chip fiber grooves, as shown in Figure 9.22, commonly known as U-grooves. The U-groove depth guarantees that once the fiber is pressed inside, the fiber core and the waveguide center are aligned vertically. Horizontally, they are defined to the respective waveguides in a self-aligned manner, thus providing virtually perfect lateral adjustment. The width of the U-groove is designed to be a few micrometers narrower than the diameter of the fiber. Since the polymer material provides certain flexibility, the fiber can be pressed inside, and the sidewalls of the U-groove clamp the fiber in place. Additionally, in an attempt to suppress back reflection, the fiber and waveguide interface can be angled at 8°. The fiber can be first cut at the same angle and then attached inside the U-groove.

The U-groove can be etched by standard oxygen RIE or inductively coupled plasma etching (ICP), utilizing a thin metal mask. Depending on the technology, the uniformity achieved across a whole 4 inch wafer can well fall into the deviation range of less than 1% from the designed values, for both the groove width and the etching depth.

With proper index matching glue and waveguide tapering for mode matching, the typical loss per facet for the U-groove-based fiber/waveguide coupling is as low as 0.25 dB over a broad spectral range from 1300 to 1600 nm. An automated "pick-and-place" fiber attachment process can also be developed.

9.3.4.2   Thin Film Elements (TFEs)   Optical filters are key components in any optical network using different operating wavelengths. The implementations of filter functions can be realized in the form of thin film filters (TFFs) or, in general, TFEs. Examples of these TFEs can be seen in Figure 9.23. Depending on the applications and spectral requirements, the thickness of the TFEs ranges between 8 and 30 μm. The lateral size can extend to hundreds of micrometers. Some examples of TFEs are shown in Figure 9.23. They are fabricated using high-precision

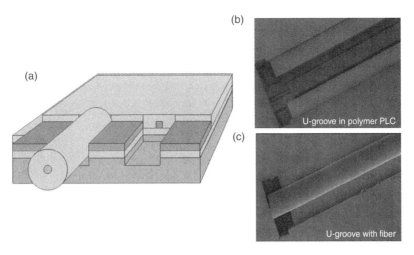

Figure 9.22 (a) Schematic layout of fiber attachment with U-groove; (b) SEM photo of an empty U-groove; (c) SEM photo of a fiber, with 8° facet cut, attached inside the U-groove.

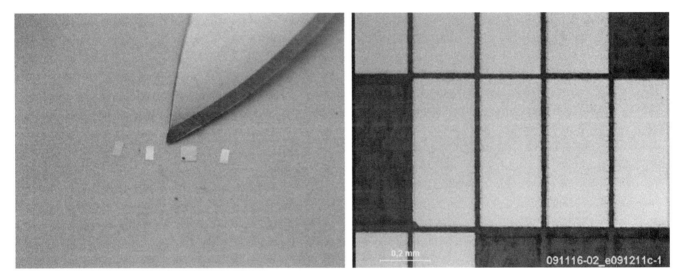

Figure 9.23 Photos of some TFEs.

ion beam sputtering process, in which at least two dielectrics of different indexes are deposited in an interleaved and often periodic manner to achieve certain dispersion characteristics. The thin films are then cut into small pieces by and excimer laser.

Thin film elements can be vertically inserted into dedicated slots crossing the PLC waveguide. The slots can be deep-etched using oxygen RIE with a thin metal mask, similar to the U-groove fabrication process. The slots can also be formed by applying a dicing saw, subject to the wafer and chip layout. The width of the slot needs to be well adjusted to the thickness of the filter platelet. The vertical tilt angle of the mounted filter and its thickness are important parameters determining the excess losses.

9.3.4.3 1 × 2 Wavelength Multiplexer The U-grooves and TFEs can be combined to realize various devices. As an example, a 1 × 2 wavelength (de)multiplexer is demonstrated. The device can be used to split and combine signals at 1310 and 1490 nm (or 1550 nm) in the optical networks. The schematic layout and the device photo are shown in Figure 9.24. The device contains three U-grooves for fiber attachment, a central slot for TFE insertion, and a few underlying polymer waveguides. The total chip made on Si substrate measures 5 mm length by 1.3 mm width. The input signals at wavelengths 1310 and 1490 nm (or 1550 nm) are launched into the input port. Using a long-wavelength-pass filter (LWPF), signal at wavelength 1490 nm/1550 nm will pass to output port 2, whereas signal at 1310 nm gets reflected by the TFE to the output port 1.

The measurement result is shown in Figure 9.25, displaying the filter performance of the LWPF. At 1310 nm, the fiber to fiber insertion loss at output port 1 is around 1.2 dB. The other channel at 1550 nm exhibits an even lower loss of 1.0 dB. The cross-talk level is below −30 dB. Across the entire wavelength range, the back reflection into the input fiber (optical return loss) is well below −50 dB. When the loss from the

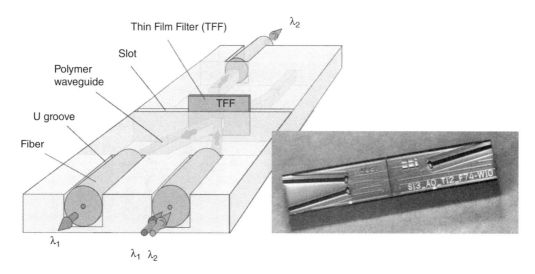

Figure 9.24   Schematic layout and device photo of a 1 × 2 wavelength (de)multiplexer.

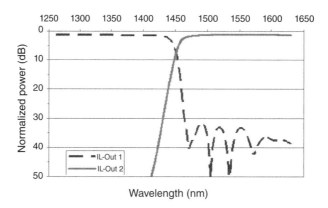

Figure 9.25   Measurement results of the 1 × 2 wavelength multiplexer.

fiber connectors is subtracted and only the net insertion loss from the input fiber to the output fiber is considered, the device itself features a loss well below 1 dB. The maximum PDL is below 0.5 dB.

9.3.4.4   1 × 1 Optical Time-Domain Reflector   Optical time-domain reflectometry (OTDR) represents a powerful means to remotely monitor the connectivity and termination status of the optical networks. It is used to identify with high spatial resolution any point of failure that may degrade transmission performance. With conventional OTDR, optical reflections of a sequence of test pulses generated from broken or disconnected fiber parts are exploited. To gain information about the fiber termination, for example, at the subscriber site, a dedicated integral optical element is required as an edge filter that is transparent for the operating wavelengths but reflects the OTDR wavelength. Low excess loss and very low device cost are the principal demands posed on such network elements, which have the potential of becoming a high-volume product when widely installed in the subscribers' terminals.

A simple and highly compact implementation of such an OTDR device can be realized on the polymer platform. The TFE is essentially a band pass filter designed to exhibit a spectral band edge at 1610 nm, which is suitable for reflecting OTDR signals at 1625 nm or above as "ping" wavelength, while letting the telecommunication wavelengths passing through.

The OTDR reflector is essentially comprised of two parts, that is, the polymer base chip and the integrated TFF, as shown in Figure 9.26. The base chip itself contains two U-grooves, a straight buried single-mode waveguide, and a slot for inserting the TFE. The chip made on Si substrate measures only 3 mm (length) by 0.7 mm (width).

The characteristics of such 1 × 1 OTDR device are shown in Figure 9.27 across the wavelength region from 1260 to 1630 nm. The filter exhibits flat transmission from 1260 to 1580 nm, and at wavelength greater than 1610 nm, the filter behaves like a 100% reflection mirror. The cross-talk level is below −23 dB, which can be further improved by fine-tuning the TFF layer deposition process. The insertion loss is below 2 dB over the 1260–1580 nm wavelength range, with a minimum value of 1.8 dB around 1550 nm. The reflected signal at 1625 nm exhibits a loss of less than 1.0 dB. The insertion loss originates mainly from the fiber and waveguide coupling, the coupling between the

(a)

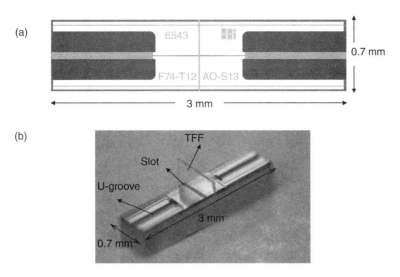

(b)

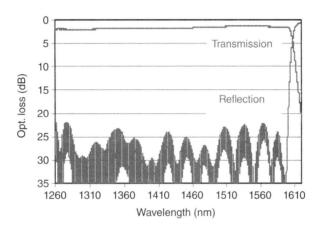

Figure 9.26 (a) Schematic layout and (b) device photo of a 1 × 1 OTDR device.

Figure 9.27 Characteristics of the 1 × 1 OTDR device.

waveguide and TFF, and the additional loss from the TFF itself. By refining the design of the slot region, the insertion loss may be further reduced. The measurements were performed using standard connectors with the reduced cladding fibers, which, on average, exhibited a loss of around 0.2 dB each. Therefore, if we consider the net insertion loss from input to output fiber, the device features a loss well below 1.5 dB.

With two of these OTDR modules, a system experiment was demonstrated in combination with a 1 km and a 10 km long fiber network. The results from an OTDR analyzer are shown in Figure 9.28. Both reflections are well detected, indicating the green ("through") status of the optical link.

9.3.4.5 TFE-Based Polarization Beam splitter polarization control is another key function with polarization diverse PLCs. In many applications, the TE and TM components need to be separated and their imbalance needs to be kept minimal. Polarization dependence in PLC is mostly caused by the waveguide geometries, the material birefringence, nonlinear process, confinement issues, and so on. Passive polarization control usually involves complicated waveguide design to compensate the different propagation constants of the TE and TM modes. This often raises a stringent challenge in the fabrication process and leads to considerable extra loss. On polymer platform, apart from using a half-wave plate to compensate TE and TM propagation differences as discussed in previous sections, a simple polarization-sensitive TFE can be adopted to split the polarizations and guide them into different channels.

The schematic architecture of the device is shown in Figure 9.29. Depending on the characteristics of the polarization splitting thin film element (PSTFE), one polarization will be reflected to output 1 and the other will pass through to output 2. The incident angle from the waveguide to the PSTFE, that is, the half-angle between the two waveguides, can be varied to optimize the splitting ratio for a desired bandwidth.

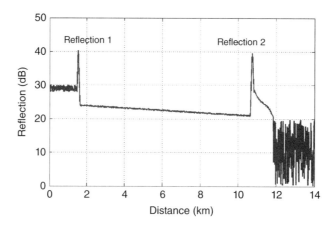

Figure 9.28 System experiment of two OTDR modules placed 9 km away from each other. Both reflections at ping wavelength are well detected, indicating a "through" state of the fiber link.

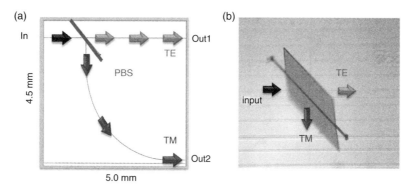

Figure 9.29 (a) Schematic layout and (b) device photo of a TEF-based polarization beam splitter on polymer platform.

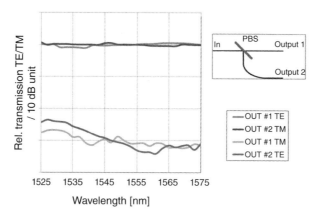

Figure 9.30 Measurement results of the TFE-based polarization beam splitter on polymer platform.

Figure 9.30 shows the spectra for TE and TM signals at the two outputs. Across a 50 nm wavelength window from 1525 to 1575 nm, the splitting ratio for the transmitted output 1 is larger than 27 dB. At the reflected output 2, the splitting remains larger than 25 dB for the same wavelength window. The insertion loss of the TFE is around 2.8 dB for both paths and both polarizations. The performance in terms of insertion loss, polarization splitting ratio, and bandwidth can be further improved by optimizing the TFE layers, as well as by carefully designing the waveguide/TFE interface so that the deflected beams are well collected.

The simplicity and high efficiency of TFE-based polarization control open up new possibilities in realizing polarization-insensitive and polarization-multiplexed devices on the polymer integration platform, such as dual-polarization coherent receivers and IQ modulators.

9.3.5   Dielectric Waveguides in Polymer

A particular strength of polymers is the possibility to deposit them on and to combine them with virtually any other material, thanks to the low process temperatures required, as well as their good surface adhesion behavior. Accordingly, there is a wide choice of useful substrate materials (commonly used is silicon), including polymer wafers, and polymer layer stacks may be formed with embedded films made of metals or any other material with desired specific physical properties, thus offering substantially enhanced design flexibility. For example, SiN_x, produced by low-temperature plasma enhanced chemical deposition (PECVD), can be incorporated as the waveguide core layer that is completely buried in polymer. The index of SiN_x ranges from 1.8 to 2.1, depending on the nitride composition, which offers a much higher index contrast when compared to polymer cores.

The buried type waveguide is composed of a polymer cladding and a SiN_x core layer, as sketched in the inset of Figure 9.31. The polymer cladding index is 1.45. In order to structure the waveguide with only conventional photolithography, the width of the waveguide should measure preferably a few micrometers. The thickness of the SiN_x core must be significantly reduced to below 200 nm for single-mode operation. For this ultraflat waveguide structures, the effective refractive index is predominantly determined by polymer cladding as well as the actual thickness of the SiN_x film.

Also observed from Figure 9.31 is that this type of waveguide appears to have very strong waveguide birefringence. The simulated intensity profiles for TE and TM modes for the waveguide with a thickness of 110 nm and width of 2.8 μm are shown in Figure 9.32a and b, respectively. Both TE and TM mode fields can be seen to spread largely into the polymer cladding at the given conditions, which is desirable for the TO tunable filters as will be discussed in Section 9.4.2. In particular for the TM mode, the confinement factor is merely 2.4% for the given SiN_x thickness. Table 9.1 summarizes key waveguide parameters for the two polarization states.

Following PECVD, the SiN_x core layer can be structured using conventional photolithography and RIE with a metal mask. After removing the mask, another spin-coating and curing process provides a 5 μm thick polymer upper cladding layer. The degradation temperature of the ZPU polymer material is 300 °C, and at this temperature, even amorphous silicon, which registers a much higher index of 3.6, can be deposited with relatively low material absorption at 1550 nm (13). Other low-temperature deposition processes, such as the ion beam sputtering process used in the TFE fabrication, can well be adopted to incorporate different dielectric materials on top of polymer cladding by a single "lift-off" process, in which no RIE is required. Typical high-index dielectric materials include TiO_2, Ta_2O_5, and Al_2O_5.

In particular, the polymer bottom cladding can be overly etched by an oxygen plasma asher to create a partially suspended dielectric membrane. The air-clad waveguide features higher index contrast and allows for more compact photonic devices. The SEM photo of such suspended SiN_x Bragg gratings is shown in Figure 9.33. In this chapter, however, only polymer-clad SiN_x waveguide is discussed as proof of concept.

To determine the waveguide propagation loss and the coupling loss to standard SMFs, waveguides were successively cut back using a dicing saw without further polishing to yield three different lengths of 4.8, 3, and 1.8 cm. Measurement data are plotted in Figure 9.34 and fitted for both polarizations. The waveguide propagation loss α, and the measured coupling losses β_M, are summarized in Table 9.2, the latter ones showing very good agreement with the respective simulated values, β_S. Since the TM mode field size is closer to that of SMF (mode

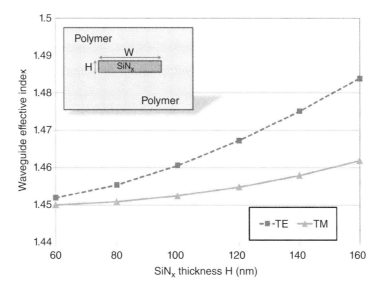

Figure 9.31   Waveguide effective index versus SiN_x layer thickness ($W = 2.8$ μm). Inset: Geometry of the buried SiN_x waveguide core in polymer cladding.

TABLE 9.1   Properties of the Fundamental TE and TM Mode

	TE	TM
Effective index n_{eff}	1.4622	1.4524
W_{1/e^2} (x direction)	2.856 μm	4.256 μm
W_{1/e^2} (y direction)	1.756 μm	3.214 μm
Confinement factor Γ	10.8%	2.4%

TABLE 9.2   Propagation Loss and Coupling Loss Comparison for Both Polarizations

	TE	TM
Propagation loss α	0.96 dB/cm	0.72 dB/cm
Measured coupling loss β_M	3.65 dB/facet	1.08 dB/facet
Simulated coupling loss β_S	3.60 dB/facet	1.05 dB/facet

field diameter ~10.4 μm), the coupling loss for TM is lower. The waveguide propagation loss for the TM mode launching conditions was measured to be 0.72 dB/cm and 0.96 dB/cm for the TE mode.

The lower value in the former case apparently reflects the fact that the TM mode expands deeper into the cladding material than the TE mode, which in turn means a lower confinement factor of the SiN$_x$ core (Table 9.1). Furthermore, the dominating field component for TE mode is E_x, which oscillates across the sidewall of the waveguide. Due to technological limitations, sidewall roughness causes more scattering and may affect the waveguide loss more severely than any inhomogeneities at the top and bottom waveguide interfaces. From the measured total optical waveguide loss a rough estimate of the optical absorption of the SiN$_x$ material itself may be deduced. Given that the actual polymer material absorption is 0.6 ± 0.05 dB/cm (as derived from loss measurements utilizing the sliding prism method) and taking into account the simulated confinement factors, SiN$_x$ absorption was deduced from the measured waveguide losses for TM and TE. As a result, absorption values range between ~2.9 dB/cm (TE, TM) in the best and 3.8 dB/cm (TE), 7 dB/cm (TM) in the worst case. By using more transparent polymer cladding materials, the waveguide propagation loss can be reduced well below the demonstrated 0.72 dB/cm.

From the well-established silica-on-silicon technology, the SiN$_x$ layer is formed by low-pressure but high-temperature (>800 °C) CVD and subsequent annealing processes in order to reduce the concentration of the loss-dominating N–H bonds. The ultralow-loss Si$_3$N$_4$ waveguide in silica has been demonstrated with propagation loss below 1 dB/m (14). This chapter shows, on the other hand, that low-loss SiN$_x$ waveguide is possible on polymer platform, using only low-temperature (<200 °C) processes. The technology remains low-cost and high-volume production compatible.

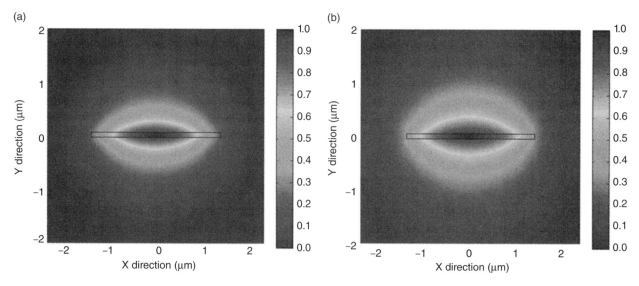

Figure 9.32   (a) Fundamental TE mode intensity profile; (b) fundamental TM mode intensity profile for the waveguide $W = 2.8$ μm and $H = 110$ nm.

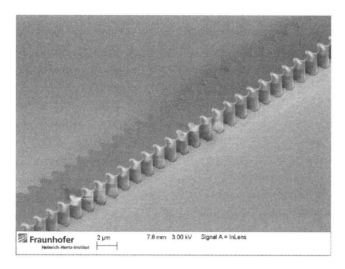

Figure 9.33 Partially air-suspended SiN$_x$ membrane Bragg grating.

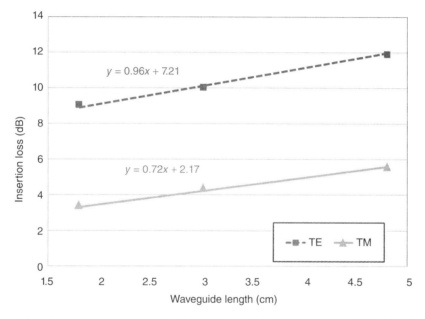

Figure 9.34 Cutback measurement results of the SiN$_x$ waveguides ($W = 2.8\,\mu$m; $H = 110\,$nm) in polymer.

9.4 THERMALLY TUNABLE POLYMER PHOTONIC DEVICES

Tunable optical filters and switches are important active components in the WDM networks. Compared to the passive components discussed earlier in this chapter, tunable filters and switches require external electrical power to alter the optical signals in a variable way. There are mainly two types of tunable devices on polymer platform. The EO polymer devices are mostly used for ultrafast modulators and switches. The EO coefficients can be much higher in specially engineered polymer materials than those in classic crystals such as lithium niobate. In addition, the electronic modulation bandwidth in EO polymers is enormous. Single channel polymer EO modulators operating at 100 Gb/s are already commercially available (15).

The second type of tunable devices is based on the TO effect. It utilizes the temperature dependence of the refractive index dn/dT to realize the tuning functionality. By applying microheaters close to the waveguide core region, the temperature gradients lead to a change of optical mode and the waveguide effective indexes, which, in turn, is used to tune the spectral responses. Compared to the EO effect, TO effect is usually very slow, with response time in the range of milliseconds, but the achievable absolute index change at certain applied voltage can be much larger. Therefore, instead of fast switching applications, the TO effect is often used for efficient tunable applications at continuous wave (CW).

TABLE 9.3 **Thermal Properties of Polymers, Silica, and Silicon**

Material	Thermo-Optic Coefficient (TOC) = dn/dT (10^{-4}K^{-1})	Thermal Expansion Coefficient (TEC) (10^{-6}K^{-1})	Thermal Conductivity (TC) $(\text{WK}^{-1}\text{m}^{-1})$
Polymers	−1 to −4	10–220	0.1–0.3
Silica	0.1	0.6	1.4
Silicon	1.8	2.5	168

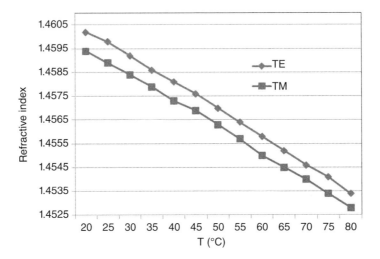

Figure 9.35 The temperature dependence of ZPU polymer refractive index for both polarizations.

Table 9.3 gives a comparison of thermal behaviors of polymers, silica, and silicon at room temperature range. It is clear that polymer materials are ideal candidates for TO devices. Due to the high thermo-optic coefficient (TOC) but at the same time low thermal conductivity, polymer TO devices will exhibit high power-conversion efficiency compared to silica and silicon devices. The TO effect in polymer is closely linked to the thermal expansion. It is usually isotropic and polarization insensitive. Figure 9.35 shows the refractive index changes for both TE and TM polarizations of a 10 μm thick layer ZPU polymer material (ZPU12-461). The index values are measured from standard M-line setup with a temperature control unit on the chip holder so that the temperature in the polymer layer can be varied uniformly. The slopes of the data points indicate a TOC of $1.14 \times 10^{-4} \text{ K}^{-1}$ for both TE- and TM-polarized light. This value is required in estimating the actual local temperature of the tunable Bragg gratings.

Consider the waveguide structure in Figure 9.2a; efficient thermal tuning can be achieved by adding a heater strip made of Au close to the waveguide core. For the waveguide mode shown in Figure 9.2b, a cladding of 5 μm is sufficient in confining the mode and isolating the optical disturbances from the metal heater itself. With the properties in Table 9.3, the thermal field distribution can be numerically simulated in order to come up with an efficient heater design. The results in Figure 9.36 show that the heater buried in the polymer cladding ~5 μm beneath the core region provides a much more uniformed heating condition, than on top of the waveguide. Deep air trenches are etched along the waveguides aiming at further improving heat confinement in the waveguide region. With the simulations of Figure 9.36, the heat source temperature is set at 100 °C, each isothermal line represents a 2 °C temperature drop, and the ambient temperature is 25 °C. For low range thermal heating, the top electrode layout can be chosen due to its fabrication simplicity. When large tuning range is desired, the buried electrode layout is preferable. In the latter case, extra openings must be etched so that the buried electrodes can be contacted by the external electrical supplies.

To understand the physics behind the simulation results in Figure 9.36, the polymer on top of the heater functions as a thermal reservoir, since it is protected by air with even lower thermal conductivity, while the polymer below the heater functions more like a transition region toward the heat sink, that is, the silicon substrate with much higher thermal conductivity. Therefore, a strong temperature gradient through the polymer layers into silicon is generated. For the buried heater, however, the polymer region above the heater tends to maintain a uniform temperature since the heat cannot escape easily into the air.

9.4.1 Polarization-Dependent Frequency Shift Controller

Coherent modulation schemes such as differential phase shift keying (DPSK) and differential quadrature phase shift keying (DQPSK) are being considered as leading solutions for the modern WDM long-haul transmission infrastructure. DLI is a key component in the D(Q)PSK

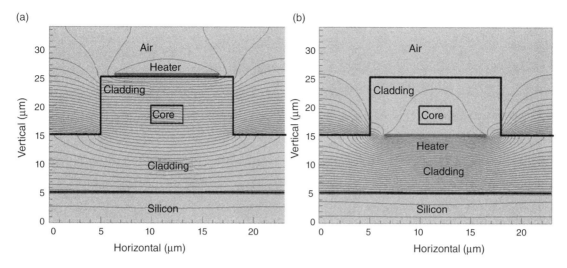

Figure 9.36 Temperature gradient distribution for (a) top electrode and (b) buried electrode layout. The heat source temperature was set at 100 °C and the ambient temperature is 25 °C. Each isothermal line represents a 2 °C temperature drop.

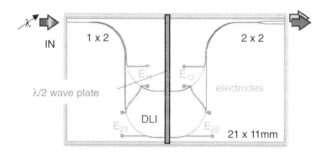

Figure 9.37 Schematic layout of a polymer-based DLI with a half-wave plate inserted and top heater electrodes to fine-tune the PDFS.

demodulator, as it converts a phase-keyed signal into an amplitude-keyed signal. In a DLI demodulator, polarization-dependent frequency shift (PDFS) should be kept minimal, if not eliminated completely.

Figure 9.37 shows a polymer DLI based on a Y-branch and a 2-by-2 MMI. The chip measures 21 mm by 11 mm. As discussed in the previous section, a half-wave plate is inserted in the middle of the symmetric device to compensate the majority of the polarization mismatch. To further fine-tune the PDFS, heater electrodes are added on top of the cladding layer. The effective electrodes located directly above the waveguides measures a width of only 10 μm. By injecting current through the electrodes, the polymer material is heated and the refractive index shifts slightly but differently for the TE and TM modes. In this way, the transmission characteristics can be controlled.

The measurement results are shown in Figure 9.38. The total insertion loss is around 7 dB, mainly due to the long waveguide, bends, and the intrinsic 3 dB loss of the MMI. Without any current, the transmission minima for the TE and TM modes exhibit a mismatch of 14 pm, which corresponds to a frequency shift of 1.7 GHz. When the current is set to be 8 mA, the PDFS is totally eliminated. Further increasing the current will reverse the PDFS.

With the help of the half-wave plate, the microheaters only need to adjust the refractive index slightly. The required temperature change for eliminating the PDFS is no larger than a few tens of kelvins. The top electrode heating scheme is sufficient in this application.

9.4.2 Tunable Bragg Grating Filters

A waveguide Bragg grating (WBG) in PLC refers to a type of waveguide with periodic perturbation along the light propagation direction. The periodic perturbation is often called a Bragg reflector, which reflects particular wavelengths of light and transmits all others. WBGs are commonly used as optical filters and optical add-drop multiplexer (OADM) in the WDM networks. They are also a crucial part of a distributed feedback (DFB) laser and external cavity tunable lasers.

The central wavelength λ_B of a WBG is given by

$$\lambda_B = 2n_{\text{eff}} \frac{\Lambda}{M} \tag{9.3}$$

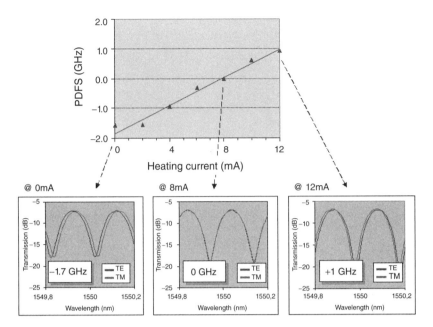

Figure 9.38 PDFS for the DLI at different heating currents.

where n_{eff} is the effective index of the waveguide, Λ is the grating period, and M is the grating order. For WBGs as back reflector filters, the grating order needs to be an odd integer. Take λ_B of 1550 nm and typical ZPU polymer waveguide effective index of 1.46, and the first order grating period Λ_1 is 530.8 nm. To make the fabrication more relaxed with standard photolithography, usually the third-order or the fifth-order gratings are taken. The periods measure 1.6 and 2.65 µm, respectively. Higher-order gratings are more loss prone because the guided mode has more chance to couple via the gratings into radiation modes.

The Bragg waveguide change with respect to temperature can be written as

$$\Delta \lambda_B = 2\frac{\Lambda}{M}\frac{\partial n_{\text{eff}}}{\partial T}\Delta T \tag{9.4}$$

Assuming the polymer core and cladding material exhibit the same TO behavior, the equation above can be rewritten as

$$\Delta \lambda_B = 2\frac{\Lambda}{M}\text{TOC}\Delta T \tag{9.5}$$

where TOC is simply the thermo-optic coefficient of the polymer material. For ZPU, it is measured to be $-1.14 \times 10^{-4}\,\text{K}^{-1}$.

The photo of a tunable WBG filter on polymer platform is shown in Figure 9.39. The waveguide structure shows a profile as shown in Figure 9.2a. It is third-order grating with a period of 1.62 µm. The heater electrode is placed only 3 µm below the waveguide core. The air trenches and the pad openings are etched at the same time. Extra plating can be applied at the pad openings to allow electric wire bonding in a module.

The measurement results are shown in Figure 9.40. Heat current has been gradually increased from 1 to 75 mA. The measured back reflection (optical return loss) peak is shifting continuously toward shorter wavelength region. A total tuning range of 48 nm, from 1569 to 1521 nm, is observed. The heating power required is ~160 mW. Increasing the heat current further brings up noticeable reflection peak power drop and at a current of 85 mA, the electrode burns out. At this point, however, the waveguide appeared to remain undamaged, concluded from the finding that the Bragg wavelength jumped back to its initial unheated position. Using Equation 9.5, the temperature change needed for this 50 nm wavelength tuning is about 390 °C. The devices have been characterized many times at these conditions, and surprisingly repeatable results have been obtained despite the extremely high local temperature inside the polymer layers. For practical long-term applications, however, the tuning range should be preferably kept below some 30 nm with this particular polymer material, so that the temperature would not exceed the polymer material degradation temperature of 300 °C.

Similar tunable WBG filters can be made using the thin SiN_x waveguide in polymer, as shown in Figure 9.41. The effective TOC of the heterogeneous waveguide structure is assumed to be the sum of the individual values of the polymer and the SiN_x material, each weighed with the respective confinement factor:

$$\Delta \lambda_B = 2\frac{\Lambda}{M}\Delta T\,[\text{TOC}_{\text{poly}}(1 - \Gamma) + \text{TOC}_{\text{SiN}_x}\Gamma] \tag{9.6}$$

Here, Γ is the mode confinement factor of the waveguide core region, as listed in Table 9.1. TOC_{poly} has been experimentally determined to be $-1.14 \times 10^{-4}\,\text{K}^{-1}$, and for $\text{TOC}_{\text{SiN}_x}$, a value of $3 \times 10^{-4}\,\text{K}^{-1}$ is assumed. Since Γ is very small, especially for the TM mode, the TO behavior

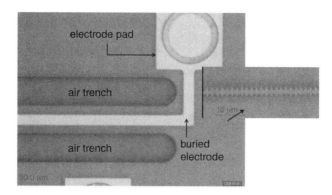

Figure 9.39 Tunable WBG filter on polymer platform.

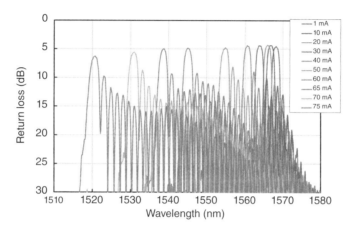

Figure 9.40 Thermal tuning of the polymer WBG.

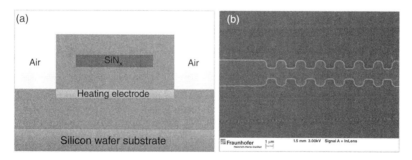

Figure 9.41 (a) Schematic layout of the tunable SiN_x WBG in polymer; (b) SEM photo of the SiN_x core.

of the waveguide is largely dominated by the polymer cladding material, and due to its negative TO coefficient, the Bragg wavelength will shift toward shorter wavelengths with increasing temperature. Because Γ_{TM} is smaller than Γ_{TE}, higher thermal tuning efficiency is expected for the TM mode.

For the TM mode, when the heater current is raised from 0 to 55 mA (corresponding to a heating power of 225 mW), a continuous shift of the center wavelength of the Bragg filter over 57 nm from 1557 to 1500 nm is obtained, as shown in Figure 9.42. The wavelength shift remains a linear relation to the heating power.

Using Equation 9.6, the temperature change in the grating region needed for this large range of tuning suggests a value of ~445 °C. Again, this large tuning range is only suitable for a lab demonstration, while for long-term stable operations, the actual temperature inside the polymer cladding must be kept below its material degradation temperature.

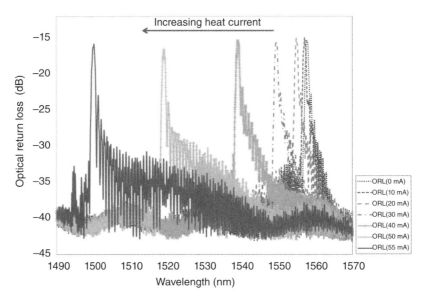

Figure 9.42   Thermal tuning of the SiN_x WBG in polymer.

Compared to the tunable polymer WGB filters, which require two compatible polymeric materials with well-adjusted refractive indexes, this type of heterogeneous structure requires only one type of polymer material. Therefore, a wider choice of useful polymers may be considered, particularly aiming at larger TOCs and stronger thermal stabilities. On the other hand, by changing the geometry of the dielectric waveguide core, mainly its thickness, the confinement factor Γ can be varied in a way that the TO effect in the core and cladding balances out. The net result is a passive temperature-insensitive filter without external temperature stabilizing units, which may find other important applications in the optical networks.

9.4.3   Grating-Assisted Directional Coupler Filter

The previously discussed WBG filters require an absolute temperature change in the waveguide region to tune the filter wavelength. Though polymer materials with higher TOC can be used, this tuning proves to be power consuming and the tuning range is eventually limited by the polymer material degradation temperature. In this section, we introduce another type of tunable filter that does not require an absolute temperature change, but rather a temperature difference between the two coupled waveguide arms. In this way, a widely tunable filter with minimal heating power consumption can be realized.

As sketched in Figure 9.41, the filter consists of two parallel heterowaveguides, one with SiN_x core and the other with polymer core. They are placed horizontally side by side with vertical centers aligned. The polymer cladding has an index of 1.45 and core of 1.47, whereas SiN_x has an index of 1.83. The polymer waveguide core has a dimension of 3.5 µm × 3.5 µm and the SiN_x core measures 2.5 µm (wide) × 135 nm (thick). Gratings are added on the SiN_x waveguide side for phase matching and wavelength selecting. When the phase matching condition is satisfied, light of certain wavelength will be coupled from one waveguide to the other. Theoretically, 100% power transfer can be realized when the waveguide propagation loss is ignored.

For simplicity's concern, we name the polymer waveguide WGP and the SiN_x waveguide WGN. np and nn are the effective indexes of WGP and WGN, respectively, with the assumption that np is smaller than nn. Λ is the period of the grating. Consider the forward coupling case, in which the transferred signal propagates in the same direction as the input signal, and the coupler central wavelength λ_0 can be derived from the phase matching condition.

$$\lambda_0 = (n_n - n_p) \cdot \Lambda \tag{9.7}$$

From Equation 9.7, the temperature derivative of the coupler can be found:

$$\frac{\partial \lambda_0}{\partial T} = \left(\frac{\partial n_n}{\partial T_n} - \frac{\partial n_p}{\partial T_p} \right) \cdot \Lambda \tag{9.8}$$

Note that we have separated the temperature response of the two waveguides. Name TOC_P and TOC_N as the TOC of WGP and WGN, respectively, and Equation 9.8 can then be rewritten as

$$\frac{\partial \lambda_0}{\partial T} = (TOC_n - TOC_p) \cdot \Lambda \tag{9.9}$$

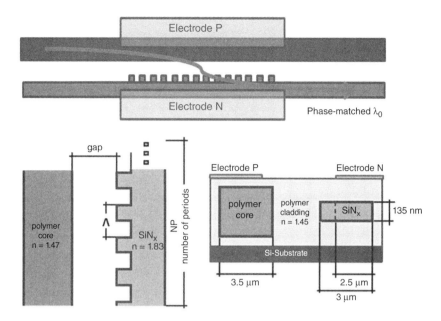

Figure 9.43 Schematic of the grating-assisted SiN_x and polymer waveguide coupler filter.

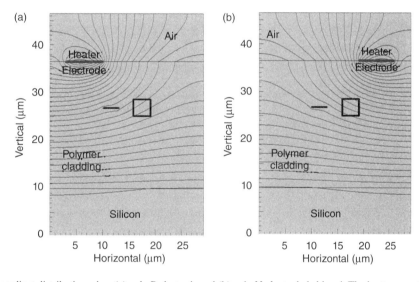

Figure 9.44 Temperature gradient distribution when (a) only P electrode and (b) only N electrode is biased. The heat source temperature was set at 100 °C and the ambient temperature is 25 °C. Each isothermal line represents a 5 °C temperature drop.

Since the polymer core and cladding are made of chemically similar materials with almost exactly the same TOC, it is convenient to assume that TOC_P is the TOC of polymer material itself, which is measured as -1.14×10^{-4} K^{-1}. TOC_n on the other hand, depends on the mode profile of WGN. In this application, WGN has a dimension of 2.5 μm (wide) × 135 nm (thick). The mode confinement factor (TE) is only 16%, that is, the majority of the mode is spread into the polymer cladding. Despite the positive value of SiN_x material TOC (on the order of $\sim 3 \times 10^{-5}$ K^{-1}), TOC_n will still have a negative sign and is estimated to be $\sim 70\%$ of TOC_p.

According to Equation 9.9, if one waveguide region has an overall temperature gradient over the other, λ_0 will either increase when WGP is "hotter" (due to the negative sign of TOC_p) or decrease when WGN is "hotter." As shown in Figure 9.43, the electrodes P and N are placed on top of the waveguides but with a small lateral offset. A temperature gradient between the two waveguides can be achieved by applying heat current through only P or N, while leaving the other electrode unbiased. Note that in this case the top electrode scheme is adopted because the buried electrode tends to provide a more uniformed heating condition, which is not desirable in this application. The thermal simulation is shown in Figure 9.44. When either P or N electrode is biased as a heat source of 100 °C, a temperature gradient of ~ 5 °C can be created between the two coupled waveguide cores.

Equation 9.9 implies a linear relation between the thermal tuning sensitivity and the grating period Λ. When Λ is very large, the central wavelength can be tuned very efficiently with minimal gradient. However, the coupler gets longer and more power is needed to generate the

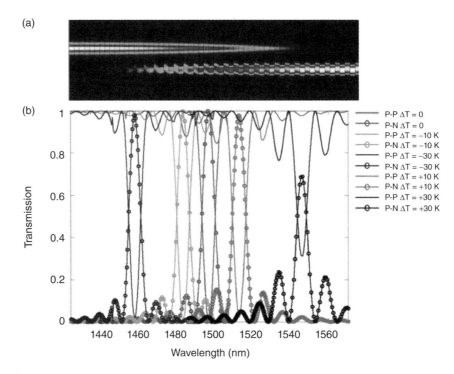

Figure 9.45 (a) Mode intensity distribution for P–N transfer at λ_0 at $\Delta T = 0$ and (b) transmission spectra at different ΔT.

temperature gradient along the waveguides. From Equation 9.7, once the central wavelength is fixed, Λ goes inversely with the effective index difference between the two waveguides. The polymer core and cladding material index contrast is very often limited to a small range due to the challenges in the material chemistry. Take ZPU series, for example; the maximal index range extends only from 1.45 to 1.48. Therefore, it is rather challenging to realize such couplers in all-polymer configuration. SiN_x, on the other hand, provides a much higher index of 1.83. The waveguide effective index can be varied effectively by changing the SiN_x core layer thickness, which offers great flexibility in the coupler design. This is the main motivation why a heterowaveguide structure with SiN_x and polymer waveguide cores is introduced.

The TO simulation results of such grating-assisted coupler filter are shown in Figure 9.45, where P–P means transmission from and out of the same polymer waveguide (drop) and P–N means transmission from polymer waveguide to SiN_x waveguide output (transfer). ΔT is positive if the temperature on the WGP side is higher. At $\Delta T = +30$ kelvin, λ_0 will increase by ~ 50 nm, whereas at $\Delta T = -30$ kelvin, λ_0 will decrease by ~ 37 nm. The unbiased mode intensity profile at λ_0 is shown in Figure 9.45a, where a 100% P–N transfer via the gratings can be seen.

Both SiN_x and polymer layers are formed at low temperature, that is, below 200 °C. Standard photolithography (320 nm) is used for structuring the waveguides. In particular, WPN can be created by a single lift-off process without subsequent etching. The misalignment of the waveguide vertical centers is kept below 5 nm by carefully controlling the polymer etching steps. The SEM photos of the grating coupler without the top cladding are shown in Figure 9.46a. The heating electrode on top of the finished device is shown in Figure 9.46b.

The grating has a perturbation depth of 0.5 μm on the side of the 2.5 μm wide WGN. The gap between the two waveguide measures ~ 2.0 μm. The grating period is 78.2 μm and the duty cycle is 50%. The total number of periods is 32, which amounts to a coupler length of ~ 2.5 mm. The electrodes are 7 μm wide. They are shifted away from the waveguide center, as shown in Figure 9.43, in order to provide a temperature gradient when singly biased.

The measurement results of such grating couplers are shown in Figure 9.47. Transmission P–P means the broadband light source is injected to the polymer input waveguide and the output from the same polymer waveguide is connected to the spectrum analyzer, whereas for P–N, the output from the coupled SiN_x waveguide is connected to the spectrum analyzer. From Figure 9.47a, a distinct wavelength drop at 1410 nm is observed for the nonbiased case, with an extinction ratio of 14 dB. The transmission loss on the P side ranges from 3–5 dB, depending on the wavelength. On the N side, higher loss is observed due to the strong SiN_x material absorption around 1400 nm wavelength.

When the P electrode is biased, λ_0 increases and at 22 mA heating current, the central wavelength shifts to 1474 nm, that is, 64 nm tuning is reached on the P side. When the N electrode is biased, the central wavelength decreases, but with a noticeably lower rate. At 24 mA (around 20 mW electric power), the wavelength drops to 1392 nm, that is, -18 nm tuning is reached on the N side. Hence, a total tuning range of 82 nm is realized.

Contrary to the theoretical estimation that TOC_n is $\sim 70\%$ of TOC_p, the wavelength tuning on the N side is much less effective. This may indicate inaccuracy in SiN_x layer thickness is causing the mode confinement factor larger than the designed value. The combined thermal optic effect for the SiN_x buried waveguide in polymer cladding is weakened, that is, TOC_n is still negative but comes with a smaller absolute

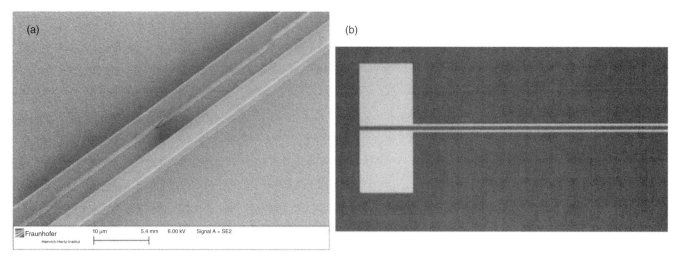

Figure 9.46   (a) SEM photo of the grating-assisted coupler (WGN/grating on top); (b) microscope image of the heating electrode.

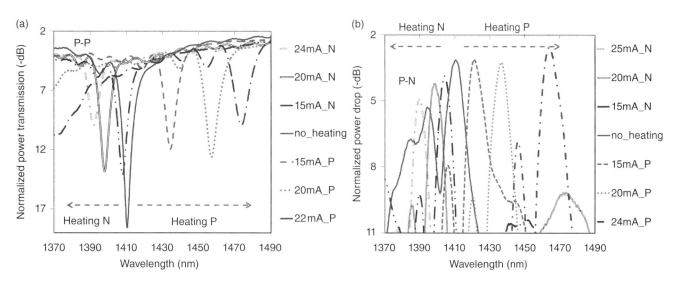

Figure 9.47   Measurement results of the tunable grating coupler: (a) P–P transmission and (b) P–N transmission.

value. Nevertheless, compared to the single waveguide-based tunable Bragg grating filters, a larger tuning range of 82 nm is achieved. The wavelength tuning can be in both directions. The heating power consumption is much reduced (from ~220 to ~20 mW). From the fabrication side, only low-temperature and standard technology is applied. The key advantages of the polymer integration platform, that is, fast prototyping and low-cost production, still remain.

9.5   HYBRID PHOTONIC INTEGRATION ON POLYMER PLATFORM

Recent progress in optical communication networks has led to the need for highly functional, small footprint, low power consumption, and low-cost optical modules. Integrating multiple optical functions into a single device is a key step for achieving the above goals. Large effort is being spent on monolithic integration technologies on both InP and silicon. However, the foundry proves to be highly expensive. The complexity in design and fabrication increases almost exponentially with the number of integrated device functionalities. Very often trade-offs in device performance need to be made in order to meet a reasonable yield. With hybrid integration technology, on the other hand, discrete elements can be made on their best suited material systems. These elements are then tested and selected prior to assembly. In this way, high freedom of device performance and yield optimization can be realized, which leads to relatively high cost-efficiency. Hybrid integration

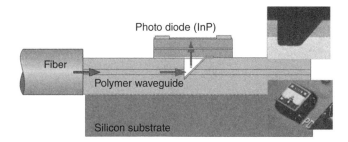

Figure 9.48   Coupling of a planar PD to the polymer optical motherboard via an integrated 45° mirror.

can offer the benefits of micro-optical packaging and allow for further miniaturization, higher degree of machine-assisted assembly, less fiber connections, and potentially higher performance and reliability. Its versatility makes it particularly useful for low/medium production volumes.

Different PLC platforms have been investigated as basis for hybrid integration including silica-on-silicon, siliconoxynitride, glass, and polymers. On the polymer platform, a broad range of passive and active optical components can be implemented that are of interest to Telecom/Datacom and optical interconnects but also to sensors, metrology, and the like. The passive elements such as TFEs and U-grooves have been covered in the previous sections. In this section, the hybrid integration techniques regarding the coupling of optoelectronic devices such as photo detectors (PDs), laser diodes (LDs), and gain chips will be revealed. In the end, some packaging activities for optical modules will be introduced.

9.5.1   Photo Diode Integration via 45° Mirrors

The optoelectronic devices such as laser diodes and photo diodes can be assembled on a polymer waveguide network via the conventional butt–joint method, in which an extra submount for the active component is necessary. The integration can also be done vertically via a 45° mirror, which eliminates the need for the extra submount and offers the possibility of semiautomatic assembly by means of a fine placer machine.

Figure 9.48 shows the vertical coupling scheme for a bottom-illuminated PD coupled to the polymer optical motherboard (PolyBoard) via an integrated 45° mirror. Also shown are the side view of the 45° mirror and the top view of a mounted PD. The 45° mirror has been introduced in the previous section. The PD is placed on top of the polymer chip, in a purely passive way, by a state-of-art bonder with positioning precision of ~1 µm. The PD is fixed by UV glue with adequate index matching to suppress back reflection. Shear force tests showed critical bonding values of up to 10 N/mm^2. The backside of the PD is AR coated against this glue, which completely fills the mirror trench. The mirror loss measured from a multimode fiber is around 1 dB. The actual mirror loss depends also on the diameter of the PD active area, its central position, and how it matches the actual beam spot. Using a 25 Gb/s capable PD with an active diameter of 20 µm, an insertion loss of only 1.8 dB was obtained at best so far, including the mirror coupling loss, fiber–waveguide coupling loss via a U-groove, and the propagation loss of a 3 mm long polymer waveguide.

The bottom-illuminated PD is designed with a full p-metal contact on its top side, which facilitates uniform current flow and serves as optical reflector for light to double-pass the absorbing layer. This is beneficial regarding achievable responsivity particularly for very high-frequency PDs, the absorption layer thickness is limited by the transit time to below 1 µm. On the other hand, the back-illuminated PD is more sensitive to beam divergence because of the longer optical path of the beam between waveguide output and the active PD region. By using adequate design methods, the effective beam diameter at the active PD area can be kept as small as ~10 µm (1/e^2). This value is well below the active diameter of PDs suitable for 25 Gb/s reception, a bit rate that is in the focus for 100 Gb/s transmission applications such as the 4 × 25 Gb/s wavelength multiplexed or DQPSK schemes.

Figure 9.49 shows the small signal frequency response of a mounted PD chip. The measurement was performed directly on the chip using a suitable RF probe head. Without extra electronics for 50 Ω impedance matching, the 3 dB bandwidth is around 25 GHz, as shown in Figure 9.49a. By applying impedance matching and some compromise to detector responsivity, the bandwidth can be well increased to 34 GHz, as shown in Figure 9.49b.

Photo diode arrays can be mounted on top of the mirror in a similar fashion. Take the 4-channel coupling sample in Figure 9.50 as an example. The optimal coupling position is firstly calculated in terms of waveguide position, cladding/core layer thickness, and the mirror opening width. The PD array (with bias-T) is then mounted on top of the 45° mirror at the proper location and fixed with UV glue. Finally, the fiber array is plugged in and glued for the measurements.

The measured responsivity mostly varies between 0.3 and 0.4 A/W, and the highest value is 0.43 A/W. Since the responsivity measured using a fiber illuminated directly on the PD is 0.65 mA/mW, the lowest insertion loss of 1.8 dB is thus obtained. The static cross talk ranges from −28 to −30 dB, for any of the individual PD elements within the array, suggesting stray light to be the limiting effect. The PDL is around 0.6 dB, in which ~0.5 dB comes from the metalized mirror itself. The optical return loss into the launching fiber (back reflection) is lower than −35 dB.

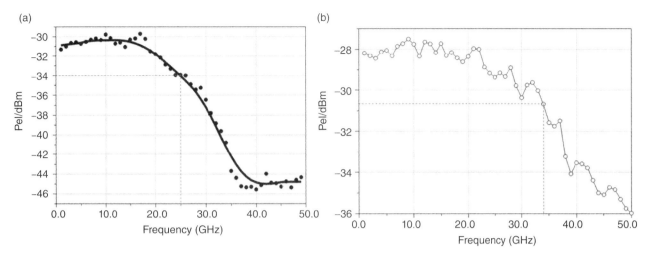

Figure 9.49 Mounted PD frequency response: (a) without 50 Ω impedance matching and (b) with 50 Ω impedance matching.

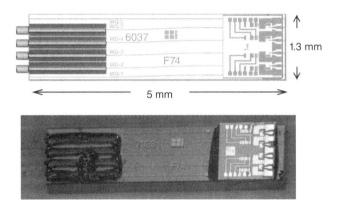

Figure 9.50 Four-channel fiber–waveguide–detector array–PolyBoard.

Another example is the coherent receiver based on a polymer 4 × 4 MMI and surface-illuminated balanced PDs. The principles of using a 4 × 4 MMI as an optical 90° hybrid have been introduced in Section 9.3.3. The balanced PD array operating at 25 Gb/s is mounted on top of the polymer chip via a 45° mirror. The measured PD responsivity is plotted in Figure 9.51. The value ranges from 0.06 to 0.07 A/W, which translates into an insertion loss range of 10 to 10.5 dB. The imbalance of the four channels is below 0.7 dB.

The last example is the implementation of an OLT receiver device (Rx), as shown in Figure 9.52a. A surface-illuminated 8-channel PD array has been vertically coupled to the output channels of a 12-channel 100 GHz (0.8 nm) polymer AWG via an integrated 45° mirror. The AWG has a Gaussian-type spectra and the 1 dB channel bandwidth is 0.25 nm. From optical characterizations, insertion loss of less than 6 dB (fiber–chip–fiber) and a central wavelength cross talk less than −30 dB have been obtained. In order to eliminate polarization dependence, a half-wave plate has been inserted in the middle of the waveguide array section. Figure 9.52b shows the effective responsivity of an 8-channel 2.5 Gb/s PD array coupled to the AWG, where all channels exhibit a uniform response of ~0.2 A/W. Compared to the balanced PDs in the coherent receiver device, the OLT-Rx PDs show a much smaller electronic bandwidth (2.5 Gb/s compared to 25 Gb/s). However, for low-speed applications, the PD sensing areas can be relative large, leading to a much improved responsivity of 0.9 A/W, compared to 0.65 A/W in the previous case. The total insertion loss of the OLT-Rx is about 5.9 ± 0.2 dB. The bandwidth of the PD arrays can be certainly increased up to 25 Gb/s at the price of a lower responsivity and some coupling constraints due to the smaller active areas of these PDs.

9.5.2 Butt–Joint Laser Diode Integration

LD and waveguide coupling is often challenging in three aspects. (i) LDs possess mostly a near-field mode field diameter of only a few micrometers. The waveguide mode must be tailored to match the laser mode size. (ii) The coupling is very sensitive to any spatial misalignment. A positioning inaccuracy of ~1 μm in the lateral dimension may lead to an extra coupling loss of more than 1 dB. (iii) Optical feedback (back scattering) must be suppressed. If some scattered light finds its way back into the laser cavity, the stable-state laser oscillation may be disturbed,

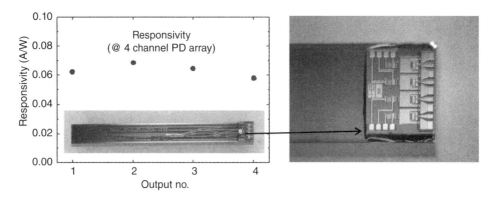

Figure 9.51   Coherent receiver based on a polymer 4 × 4 MMI and 2 pairs of surface mounted balanced PDs.

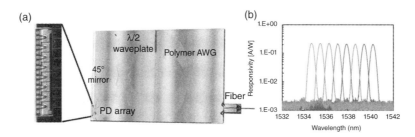

Figure 9.52   Eight-channel OLT-Rx based on a polymer AWG and 8-channel surface mounted PD array. (a) Photo of the subassembly; (b) PD responsivity measurement results.

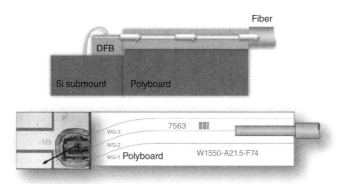

Figure 9.53   Butt–joint coupling of a curved stripe DFB laser to the polymer waveguide chip.

causing ripples in the laser spectrum. In some severe cases, the lasers can become multimode and the performance, in terms of line-width and directivity, is then seriously degraded. Therefore, the LD output is often AR coated to suppress the back scattering.

Though it is feasible to couple an LD or a VCSEL via a 45° mirror, with the help of microlenses, to the polymer waveguide, it is more straightforward to use the conventional butt–joint method, as illustrated in Figure 9.53. The DFB LD is first placed on a submount with predefined contact pads. This submount can be a silicon dice with an oxide insulating layer, or a ceramic piece for faster heat dissipation. Some high-end LDs even require a diamond submount, which offers the highest thermal conductivity among the conventional materials to help cool down the LDs in an efficient way. For high-frequency applications, the metal pads on the submount must be carefully designed as part of the impedance-matched transmission line. To facilitate the assembly process at later stage, the thickness of the submount is adjusted, by backside polishing, to reach a similar height level as the polymer board (PolyBoard).

The LD in Figure 9.53 features a curved stripe design, which proves to be more robust against external feedbacks and offers higher single-mode yield (16). The output of the laser is AR coated with respect to the polymer material. The beam shape is adjusted to obtain low divergence for optimal coupling with the buried polymer waveguide. The incident angle to the polymer waveguide is 21.6° for 1490 nm lasing wavelength. The polymer waveguide bend further eliminates the back reflection into the laser. At the output side, a single-mode fiber is

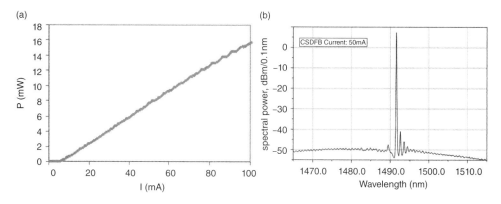

Figure 9.54 (a) CSDFB laser P–I curve at room temperature; (b) CSDFB laser spectrum. Both are measured at the fiber output in the U-groove.

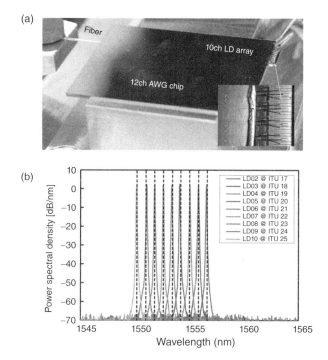

Figure 9.55 Eight-channel OLT-Tx based on a polymer AWG and 10-channel butt–joint LD array. (a) Photo of the subassembly; (b) laser spectra measured at the fiber output.

attached in the U-groove. The position of the LD is fine-tuned until the maximal output power at certain drive current is reached in the fiber. The LD is then fixed on the facet of the PolyBoard by UV glue. Thus, an LD–PolyBoard subassembly is complete.

Figure 9.54 shows the laser characteristics of the LD–PolyBoard subassembly measured from the fiber output at room temperature. The threshold current is 7.2 mA and at 100 mA the output power reaches 16 mW. The spectrum of the laser shows a distinct side mode suppression ratio (SMSR) larger than 45 dB.

LD array can be integrated to PolyBoard in a similar way. The difficulty lies in the angular adjustment between the array and the polymer chip. When a small walk-off angle exists, a significant power imbalance among the channels may occur.

An example of the LD array integration is shown in Figure 9.55a. The OLT transmitter device (Tx) consists of an InP-based 10-channel DFB LD array butt coupled to a 12-channel polymer AWG. The AWG works as a MUX and is designed with a 100 GHz (0.8 nm) channel spacing according to the ITU grid in the C-band. The laser diodes comprise a straight facet with AR coating and a tapered waveguide output that matches the polymer waveguide mode. The LDs are designed to emit at different wavelengths referring to the ITU grid. To allow fabrication tolerance as well as to compensate possible temperature drifts during laser operation, additional heater electrode is included so that the central wavelength can be tuned 1–2 nm. The direct modulation bandwidth of the LDs can go up to 10 Gb/s and the LD submount is hence designed to be high frequency compatible.

Results of the OLT-Tx device evaluation, that is, the transmission spectra measured from the butt–joint fiber, are summarized in Figure 9.55b. Driving and heater current of the LDs have been adjusted to achieve equal output power of +2 dBm at the ITU grid channels in the C-band. The sharp spectral response and ~−60 dB channel cross talk is a direct result from the superposition of the central wavelength matched laser and AWG spectra. The OLT-Tx subassembly can be packaged subsequently into an OLT-Tx MUX module.

9.5.3 Gain Chip Integration and Tunable Laser Modules

Gain chip (GC) is an irreplaceable building block as a gain medium for the construct of an external cavity laser. GC is similar to LD except the fact that the internal optical feedback loop is intentionally removed. At least one facet of a gain chip is AR coated against air or other media to eliminate internal lasing and enhance light emission. It usually employs a multiquantum well design that enhances output power and broadens the gain bandwidth.

A GC can be butt–joint coupled to a tunable polymer Bragg grating and thereupon built into an external cavity tunable laser. Schematic layouts of such structure are illustrated in Figure 9.56. The basic layout in Figure 9.56a consists of a GC with high-reflection (HR) coating at the rear facet, a tunable polymer WBG with a heater electrode, and a fiber. The grating functions as a wavelength selective reflector and together with the HR coating on the other side of the gain chip a cavity for laser oscillation is formed. When the threshold is reached, the laser emits into the fiber.

A more advanced structure in Figure 9.56b includes a curved stripe GC, which, similar to curved stripe DFB LD, offers larger optical feedback tolerance. The polymer waveguide is accordingly bent with the right coupling angle to the GC. The buried electrode with air trenches provides more efficient thermal tuning, as already discussed in the previous sections. During thermal tuning, the central wavelength of the Bragg reflector shifts toward shorter wavelength due to the thermo-induced refractive index change in the waveguide region, but at the same time, the phase reappearing condition in the laser oscillation roundtrip is inevitably altered. Since the longitudinal mode spacing of such external cavity lasers is usually narrower than the Bragg grating spectrum width, a jump in the longitudinal mode number may occur as the laser mode readjusts itself to match the phase reappearing condition. This phenomenon is called "mode hopping." When not treated properly, mode hopping can cause discontinuities in the laser P–I curve and lead to significant drop of the single-mode yield, that is, the SMSR drops. To address this issue, an extra length of phase electrode is added in order to compensate or round up the phase shift during the thermal tuning.

Some P–I curves of the laser in the advanced layout are shown in Figure 9.57. Assuming the HR coating provides ~90% broadband reflection, the optimal Bragg grating reflectivity can be derived from the rate equations, once the gain coefficient and the roundtrip loss of the cavity are known. Experimentally, a batch of such lasers with various Bragg grating reflectivity has been characterized. Thanks to the high power efficiency of the GC as well as the low coupling loss to the polymer waveguide, the Bragg grating only needs to provide a fractional reflectivity of 13% for the laser to emit an output power larger than 12 mW at a driving current of 100 mA.

The output power of the tunable laser at different heating powers has been measured. The drive current of the gain chip was chosen to amount to 50 mA. The results are summarized in Figure 9.58. A total tuning range in excess of 50 nm from 1560 to 1509 nm is achieved. The heating power needed for such a tuning range is ~161 mW. The measurements are done without applying current through the phase electrode, in order to compare the tuning power with that obtained from the stand-alone Bragg grating tuning measurement. As a result, both the tuning range and the heating power from the two measurements, as plotted in Figures 9.40 and 9.58, are in good agreement, though the starting wavelengths are not the same due to the different grating period designs.

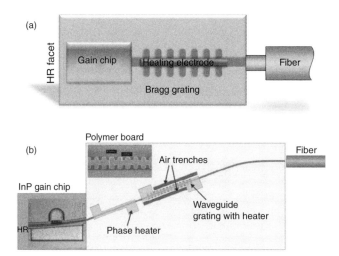

Figure 9.56 Schematic layouts of a tunable external cavity laser. (a) Basic layout: straight facet gain chip coupled to a tunable Bragg grating. (b) Advanced layout: curved stripe gain chip coupled to a tunable Bragg grating with optimized heater conditions and an additional phase electrode.

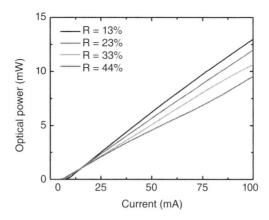

Figure 9.57 Laser P–I curve for Bragg gratings with different reflectivity.

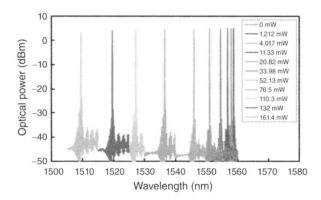

Figure 9.58 Tunable laser spectra at various heating power.

The degradation of the laser output power across such a tuning range measures less than 2 dB, indicating a flat, broadband gain spectrum from the GC, as well as good TO linearity and strong thermal stability of the polymer material, at least at the lab demonstration stage. The SMSR remains larger than 26 dB, which can be certainly improved by fine-tuning the laser spectrum via the phase electrode.

Latest research also shows that by adopting a GC design with low parasitic capacitance, shortened waveguide and Bragg grating length, and further improved facet coupling, direct laser modulation at 10 Gb/s can be achieved (17). On the other hand, with increased waveguide and Bragg grating length, a sharp laser line-width of ~500 kHz has also been demonstrated (18).

For commercial applications, the tunable laser assembly, as sketched in Figure 9.56b, needs to be placed in a sealed package with defined pins for electric contacts and fiber pigtailing. Figure 9.59 shows a typical butterfly module with integrated thermal electric cooler (TEC). The TEC is used to help dissipate the heat and stabilize the environment temperature in the module. The laser assembly is fixed on the TEC with thermally conductive epoxy. The metal pads are connected to the pins by wire-bonding. The output fiber is placed on an additional submount for mechanical support and then fed through the module port. The port is then sealed either by UV epoxy or by metal welding. In the end, a metal lid (not shown in Fig. 9.59) is welded to seal the module completely.

The packaging process may cause disturbances to the optical and electrical functions of the original assembly. Reasons may be the inappropriate temperature shock from the wire bonding process, spill of the epoxy, stretching of the coupling positions, and so on. The module needs to be checked both electrically and optically for contingencies. Very often, a set of environment tests are carried out in order to determine the reliability of the module. Details of such reliability tests will be revealed in the last section of this chapter.

9.5.4 Tunable Optical Network Unit

As introduced at the very beginning of this chapter, an ONU is a fundamental element in the WDM network. It transforms the optical signals into electronic ones that can be processed at the end user's site. It also converts the upstream electronic data from the user into optical signals through the network. A transceiver ONU usually combines both functions into one single device.

On polymer platform, such an ONU can be realized based on the 1 × 2 multiplexer discussed in Section 9.3.4.3, plus hybrid integration with a directly modulated LD and high-speed PDs. The schematic layout and the device photo are shown in Figure 9.60. On the transmitter (Tx) side, the LD emitting at 1550 nm is butt coupled to the polymer waveguide. The modulated light at 1550 nm travels to the long-wavelength-pass

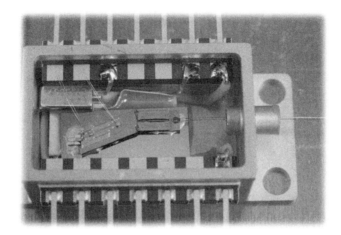

Figure 9.59 Inside view of a tunable laser module.

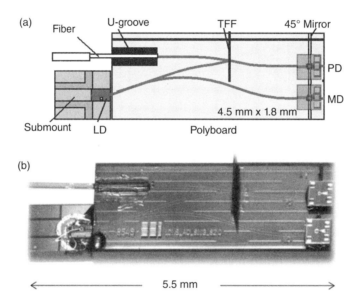

Figure 9.60 A fixed wavelength ONU with integrated LD, PDs, and TFF on PolyBoard. (a) Schematic layout; (b) device photo.

TFF, gets reflected, and transmits through the fiber into the network. About 10% of the transmitted light is tapped out, via a directional coupler, to the monitor photodiode (MD), and thus, the state of the modulator can be controlled in case of malfunction.

On the receiver (Rx) side, the incoming light from the network, at 1310 nm wavelength, enters the ONU via the fiber. It further propagates through the TFF with almost no attenuation and gets collected by the receiver PD. Both MD and PD are mounted on top of the chip and are illuminated via a 45° mirror.

Depending on the modulator bandwidth of the LD, the Tx can well operate at speed up to 40 Gb/s (19). Surface-illuminated PDs operating at 40 Gb/s are also on track to commercialization, paving way for 40 Gb/s Rx capability. For different types of applications, the bandwidth of the transmitters and the receivers can be different. For instance, the Internet end users may be more concerned about high-speed data downstream (Rx), while the broadcasting companies may require a faster upstream (Tx) into the network.

Such ONU in Figure 9.60 usually operates at fixed wavelengths due to the fact that an InP-based directly modulated lasers offer very limited range of wavelength tuning, typically only 1–2 nm. In light of the external cavity lasers with large tuning range, as discussed in Section 9.5.3, a tunable ONU can be assembled in a similar fashion by replacing the LD with a GC and by adding tunable Bragg gratings to the polymer waveguide. The layout is shown in Figure 9.61a. The TFF must be redesigned so that it reflects not only a single wavelength but a broadband of wavelengths from the tunable lasers to the fiber, while allowing the incoming signals from the fiber at other wavelengths through, in order to be received by the PD.

Though it is demonstrated in Section 9.5.3 that a single tunable laser can already cover a wavelength tuning range of ~50 nm, for long-term reliable applications, the range must be shortened to allow relaxed local heating inside the polymer layers. In order to cover telecommunication

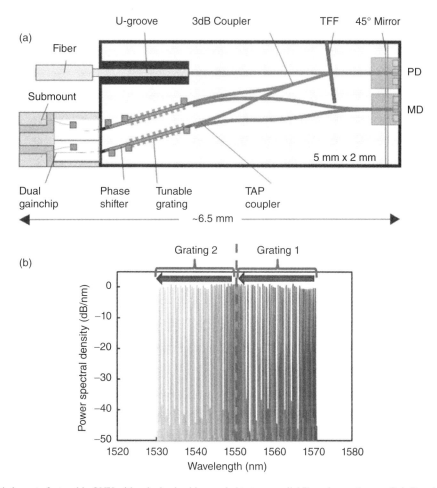

Figure 9.61 (a) Schematic layout of a tunable ONU with a dual gain chip coupled to two parallel Bragging gratings on PolyBoard. (b) Measured laser spectra from the dual-channel tunable laser.

C-band, from ~1530 nm to ~1570 nm, a dual-channel solution is taken, in which a dual-channel gain chip array is coupled to two parallel tunable Bragg gratings. The unheated Bragg wavelength of the two gratings are set 20 nm apart, for example, at 1570 nm and 1550 nm, respectively. Each grating can be tuned 20 nm to sum up the C-band. The spectra of such dual-channel lasers are shown in Figure 9.61b.

Similar to the fixed wavelength ONU, the TFF essentially determines the rules of operation. In this case, the TFF must be designed in a way that it can reflect, close 100%, all wavelengths within the C-band as Tx, but let other wavelengths through with minimal loss as Rx.

The dimension of the tunable ONU assembly is estimated to be ~6.5 mm × 2 mm. The U-groove and the 45° mirror coupling scheme both help shrink the assembly size by eliminating the need for extra submounts. The subassembly is then ready to be placed in a package for further testing.

9.6 RELIABILITY TEST

Before a packaged device can be placed in the market, the customers will mostly always ask, how reliable it is, in other words, how many years the device can function without performance degradation or complete breakdown. The company Telcordia (now part of Ericsson) has already established a set of standards defining in great detail the reliability test procedures. It is commonly accepted that the products should meet the Telcordia Generic Requirement (GR) specifications to establish confidence in quality and reliability before they can be released to the market.

Reliability can be predicted by life tests, in which a large number of the products are tested at their specified working environment. However, these tests usually take too long and the devices may miss their target market windows. Alternatively, the prediction can be determined sooner by increasing the stress on the product. Methods may include exceeding the product's maximal operating power significantly, or increasing its operating temperature above the nominal operating range, or dramatically cycling the environment conditions such as temperature and humidity. These are called accelerated life testing. Predictions by these methods take into account the number of test units and their operating

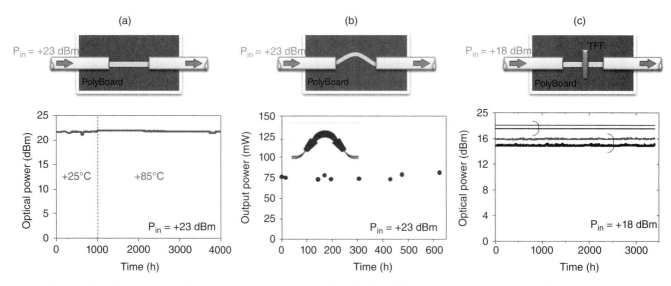

Figure 9.62 High-power optical stress test. (a) Fiber–waveguide–fiber; (b) fiber–AWG–fiber; (c) fiber–waveguide–TFF–waveguide–fiber.

hours of survival before degradation or complete failure. The data is then placed in the reliability prediction model and the expected life time of the product can be drawn.

In this chapter, the reliability tests are solely referred to the accelerated life tests. To pass these tests, certain devices need to be hermetically sealed or welded, which makes the package airtight, in order to ward off possible attacks from outside gases, water, and possible chemical contaminations. To further protect the core device in a hermetically sealed package, air can be purged and refilled with chemically inactive gas, such as N_2 and Ar.

Polymer photonic devices must go through the reliability tests and show promising results, in order to penetrate the market in a meaningful way. Typical tests include (i) optical stress tests, in which high optical power, beyond specified value on the product datasheet, is fed through the device; (ii) temperature cycling tests from −40 °C to +85 °C; and (iii) damp heat tests at 85 °C and 85% relative humidity (RH).

In Figure 9.62a, the device contains two single-mode fibers that are attached to a polymer chip via U-grooves and connected via a short distance of single-mode waveguide. The package is not hermetically sealed. Optical power of +23 dBm is monitored, rectified, and launched into the chip. The output power is constantly monitored and plotted. The test is kept at 25 °C in the first 1000 h and then elevated to 85 °C for the rest 3000 h. No degradation of the output power is seen, indicating a strong stability of the polymer waveguide as well as the fiber–chip coupling in the U-grooves.

In Figure 9.62b, the device contains two single-mode fibers that are butt–joint coupled to a polymer AWG. As discussed in Section 9.3.2, AWGs depend on the constant phase shift between adjacent waveguides in the arrayed session. If the high optical power causes some refractive index change in any of these waveguides, the constructive interference at the output waveguide will be weakened and power drop in the output fiber will be noticed. Since the AWG is temperature sensitive, a temperature control unit is added in the package to stabilize the module temperature at 25 °C. The package is again not hermetically sealed. The input power is regulated at +23 dBm and the output power shows most no variation for 650 h. Combining the results from Figure 9.62a and 9.62b, a clear indication can be seen that the polymer waveguide as well as the polymer material is stable against high optical power, in terms of not only the material loss (absorption) but also the refractive index (chemical compound).

To complete the optical stress tests, devices similar to that in Figure 9.62a are taken but with a TFF inserted in the middle. Given the fact that the properties of the multiple-layer structure in the TFF can be altered at extremely high optical power, the input power is reduced and maintained at +18 dBm. The module is not hermetically sealed and no temperature control unit is added. After 3500 h at room temperature, the device still work the same way as in the beginning of the test. The result in Figure 9.62c has shown that the technology with slot and inserted TFE can indeed lead to reliable devices against high optical power.

In Figure 9.63a, an InP PD is integrated on top of the polymer chip by 45° mirror coupling. The PD is then bond wired to the pins of the module. The input fiber is attached in the U-groove and the input power is clamped at 0 dBm (1 mW). The module is purged by N_2 gas and then hermetically sealed. The module is placed in a climatic test chamber with temperature cycling from −40 to 85 °C periodically. The photocurrent is being monitored and since the input power is 1 mW, the recorded photocurrent translates directly into the receiver responsivity. After 4000 h (360 cycles), no signs of signal degradation have been observed.

In Figure 9.63b, an InP LD is butt–joint coupled to the polymer waveguide. The bias of the LD is bond wired to the module. The output fiber is again placed in the U-groove. The package is purged with N_2 and then hermetically sealed. The module is placed in a climatic test chamber with fixed temperature of 85° and a fixed relative humidity of 85%. The drive current of the LD is kept at 50 mA and the output power is tracked from the fiber. After 1200 h, there are still no signs of degradation.

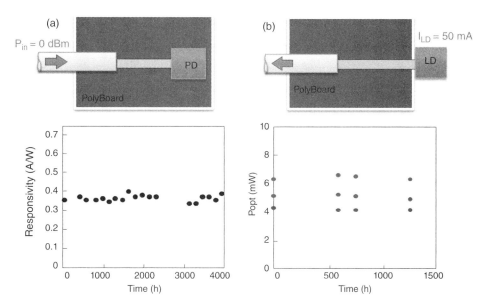

Figure 9.63 (a) Temperature cycling test: fiber–waveguide–45° mirror–PD. (b) Damp heat test: LD–waveguide–fiber.

The results in Figure 9.63 further prove that not only the polymer chips but also the hybrid integration with InP active components are reliable. With proper selection of the polymer material, optimized device fabrication, suitable hybrid integration technology, and correct packaging plans, polymer photonic devices are well on its way to being commercialized in various application fields.

REFERENCES

1. Ma H, Jen AK-Y, Dalton LR. Polymer-based optical waveguides: materials, processing, and devices. Adv Mater 1980;14(19):1339–1365.

2. Zhang Z, Mettbach N, Zawadzki C, Wang J, Schmidt D, Brinker W, Grote N, Schell M, Keil N. Polymer-based photonic toolbox: passive components, hybrid integration and polarization control. IET Optoelectron 2011;5(5):226–232.

3. Chen D, Fetterman H, Harold R, Chen A, Steier W, Dalton L, Wang W, Shi Y. Demonstration of 110 GHz electro-optic polymer modulators. Appl Phys Lett 1997;70(25):3335–3337.

4. Smit MK. New focusing and dispersive planar component based on an optical phased array. Electron Lett 1988;24:385–386.

5. Takada K, Abe M, Okamoto K. Low-cross-talk polarization-insensitive 10-GHz-spaced 128-channel arrayed-waveguide grating multiplexer–demultiplexer achieved with photosensitive phase adjustment. Opt Lett 2001;26:64–65.

6. Keil N, Yao H, Zawadzki C, Bauer J, Bauer M, Dreyer C, Schneider J. Athermal all-polymer arrayed-wave guide grating multiplexer. Electron Lett 2001;37(9):579–580.

7. Bryngdahl O. Image formation using self-imaging techniques. J Opt Soc Am 1973;63(4):416–419.

8. Ulrich R. Image formation by phase coincidences in optical waveguides. Opt Commun 1975;13(3):259–264.

9. Soldano LB, Pennings ECM. Optical multi-mode interference devices based on self-imaging: principles and applications. IEEE J Lightwave Technol 1995;13(4) 615–627.

10. Seimetz M, Weinert CM. Options, feasibility and availability of 2×4 90° hybrids for coherent optical systems. IEEE J Lightwave Technol 2006;24(3):1317–1322.

11. Almeida V, Panepucci R, Lipson M. Nanotaper for compact mode conversion. Opt Lett 2002;28:1302–1304.

12. Roelkens G, Van Thourhout D, Baets R. High efficiency grating coupler between silicon-on-insulator waveguides and perfectly vertical optical fibers. Opt Lett 2007;32:1495–1497.

13. Zhang Z, Dainese M, Wosinski L, Xiao SS, Qiu M, Swillo M, Andersson U. Optical filter based on two-dimensional photonic crystal surface-mode cavity in amorphous silicon-on-silica structure. Appl Phys Lett 2007;90:41108–41110.

14. Bauters JF, Heck MJR, John D, Dai D, Tien MC, Barton JS, Leinse A, Heideman RG, Blumenthal DJ, Bowers JE. Ultra-low-loss high-aspect-ratio Si3N4 waveguides. Opt Express 2011;19:3163–3174.

15. Dinu R, Yu G, Miller E, Chen B, Chen H, Shofman V, Wei C, Mallari J. Small form factor thin film polymer modulators for telecom applications. In: Proceedings of the Optical Fiber Communication Conference (OFC) 2012, Los Angeles, California, USA, March 3rd – 8th 2012, Paper OM3J; 2012.

16. Möhrle M, Brinker W, Wagner C, Przyrembel G, Sigmund A, Molzow WD. First complex coupled 1490 nm csdfb lasers: high yield, low feedback sensitivity, and uncooled 10 gb/s modulation. In: Proceedings of the European Conference on Optical Communications (ECOC) 2009, Vienna, Austria, September 20th – 24th, 2009, Piscataway, NJ: IEEE, 2009, ISBN: 978-1-4244-5096-1, Paper 8.1.2; 2009.

17. Klein H, Wagner C, Brinker W, Soares F, de Felipe D, Zhang Z, Zawadzki C, Keil N, Möhrle M. Hybrid InP-polymer 30 nm tunable DBR laser for 10 Gbit/s direct modulation in the C-Band. In: Proceedings of the Indium Phosphide and Related Materials Conference (IPRM), 2012, Santa Barbara, USA; 2012. p 20–21.

18. de Felipe D, Zhang Z, Soares F, Rehbein W, Brinker W, Klein HN, Zawadzki C, Möhrle M, Keil N, Grote N. Widely-tunable polymer waveguide grating laser. In: Proceedings of the European Conference on Integrated Optics (ECIO) 2012, Barcelona, Spain, April 18th – 20th, 2012, Paper TuPWGL; 2012.

19. Kreissl J, Vercesi V, Troppenz U, Gärtner T, Wenisch W, Schell M. Up to 40 Gb/s directly modulated laser operating at low driving current: buried-heterostructure passive feedback laser (bh-pfl). IEEE Photonics Technol Lett 2012;24(5):362–364.

10

LOW DIELECTRIC CONTRAST PHOTONIC CRYSTALS

JAN H. WÜLBERN[1] AND MANFRED EICH[2]

[1]*Tomographic Imaging Systems, Philips Technologie GmbH, Innovative Technologies, Research Laboratories, Hamburg, Germany*
[2]*Institute of Optical and Electronic Materials, Hamburg University of Technology, Hamburg, Germany*

10.1 INTRODUCTION

This chapter explores the possibilities to achieve full three-dimensional light confinement in resonant photonic crystal (PhC) structures (i.e., cavities) fabricated in a nonlinear optical polymer slab with air holes. At the resonant frequency, the optical field is strongly localized in the cavity volume. Refractive index changes within this volume will therefore have a considerable impact on the resonator spectral characteristics, most noteworthy the resonant frequency. Integrating the resonator structures completely into the Nonlinear Optical (NLO) polymer intrinsically leads to maximum field interaction with the NLO material and hence ensures the largest possible shift of the resonant frequency at a given amplitude of index modulation.

To observe a photonic bandgap (PBG) in a PhC consisting of a triangular lattice of air holes in a true two-dimensional geometry (not PhC slab), a minimum contrast of 1.39:1 is required (1). Since the available contrast with polymers is just slightly larger ($n_{poly} \approx 1.6$), it is clear that the obtainable bandgap size is very limited. This is especially true when PhC slabs are considered, in which case the effective index of the slab mode is always below the core material's refractive index. Additionally, in the case of PhC in slab waveguides, modes above the light line can couple to radiation modes and hence might limit the light confinement abilities of the PhC structures, necessitating a careful design of the PhC geometry. In this chapter, we explore the requirements for making low dielectric constant photonic crystals (PhCs) that can provide useful photonic processing functions.

10.2 PHOTONIC CRYSTALS

PhCs are artificial optical media, whose refractive index is varied periodically on the scale of the optical wavelength. This periodicity of dielectric interfaces causes destructive multiwave interference for certain frequency regions, prohibiting propagation of a wave inside the PhC. These frequency regions are called PBGs, in analogy to the electronic bandgaps in semiconductor materials.

PhCs are distinguished by the dimensionality of their periodicity in one-dimensional (1D), two-dimensional (2D), and three-dimensional (3D) crystals. One-dimensional crystals are essentially multilayer films, which were first studied by Lord Rayleigh in 1887 (2). However, the notion of PhCs was first coined by Yablonovitch (3) and John (4) with the introduction of 2D and 3D PhC structures. A standard introductory text to the topic of PhC is the book by Joannopoulos et al. (1).

To understand the propagation of waves in a medium with a spatially periodic dielectric constant, one starts with Maxwell's equations for time harmonic fields ($\mathbf{A}(\mathbf{r}, t) = \mathbf{A}(\mathbf{r}) \exp(-j\omega t)$):

$$\nabla \times \mathbf{H} = -j\omega \mathbf{D} + \mathbf{J} \quad \nabla \cdot \mathbf{D} = \rho \tag{10.1}$$

$$\nabla \times \mathbf{E} = j\omega B \quad \nabla \cdot B = 0$$

and the material relations

$$\mathbf{D}(\mathbf{r}) = \varepsilon_0 \varepsilon_r(\mathbf{r}) \mathbf{E}(\mathbf{r}) \quad \mathbf{B}(\mathbf{r}) = \mu_0 \mu_r(\mathbf{r}) \mathbf{H}(\mathbf{r}) \tag{10.2}$$

Nanomaterials, Polymers and Devices: Materials Functionalization and Device Fabrication, First Edition. Edited by Eric S. W. Kong.

The dielectric materials considered in this text are assumed to be free of charges ($\rho = 0$) and current ($\mathbf{J} = 0$) and the magnetic permeability is unity ($\mu_r = 1$). The relative permittivity ε_r is then related to the refractive index by $n^2 = \varepsilon_r$ and the vacuum speed of light is $c = 1/\sqrt{\varepsilon_0 \mu_0}$. Equations 10.1 and 10.2 can then be rearranged to give an expression entirely in terms of $\mathbf{H}(\mathbf{r})$, which is called the master equation

$$\nabla \times \left(\frac{1}{\varepsilon_r(\mathbf{r})} \nabla \times \mathbf{H}(\mathbf{r}) \right) = \left(\frac{\omega}{c} \right)^2 \mathbf{H}(\mathbf{r}) \tag{10.3}$$

This is an eigenvalue problem for the eigenmodes $\mathbf{H}(\mathbf{r})$ with the eigenvalue $(\omega/c)^2$. A PhC has a periodic refractive index distribution $\varepsilon(\mathbf{r}) = \varepsilon(\mathbf{r} + \mathbf{R})$, where $\mathbf{R} = N_1 \mathbf{a}_1 + N_2 \mathbf{a}_2 + N_3 \mathbf{a}_3$ is the primitive lattice vector. Thus, the Bloch–Floquet theorem can be applied to the solution of Equation 10.3. The field distribution for any given wave vector $\mathbf{k}$ consequently is a periodic function in space multiplied by $\exp(j\mathbf{k} \cdot \mathbf{r})$ and can be written in the form

$$\mathbf{H}_\mathbf{k}(\mathbf{r}) = e^{j\mathbf{k} \cdot \mathbf{r}} \mathbf{u}_\mathbf{k}(\mathbf{r}) \tag{10.4}$$

$$\mathbf{u}_\mathbf{k}(\mathbf{r}) = \mathbf{u}_\mathbf{k}(\mathbf{r} + \mathbf{R}) \tag{10.5}$$

Therefore, the analysis of the problem can be restricted to the primitive cell of the lattice. This solution is then easily expanded to the entire lattice through the application of Equation 10.5. The primitive cell is a finite domain leading to discrete eigenvalues $i = 1, 2, \ldots$. The eigenvalues $\omega_i(\mathbf{k})$ are continuous functions of $\mathbf{k}$, resulting in discrete bands when plotted versus the wave vector. In the context of PhC, the dispersion relation ω versus $\mathbf{k}$ is also called band diagram.

The master equation is scale invariant; in other words, there is no fundamental length scale. If the refractive index function is scaled in space by $\varepsilon'(\mathbf{r}) = \varepsilon(\mathbf{r}/s)$, the new mode profile is obtained by rescaling the old mode profile $\mathbf{H}'(\mathbf{r}') = \mathbf{H}(\mathbf{r}'/s)$ and similarly the new mode frequency is rescaled to $\omega' = \omega/s$. For this reason, the band diagram is typically presented in a notation where frequency $\omega a/2\pi c$ and wave vector $ka/2\pi$ are normalized to the lattice constant a.

Another important scaling property concerns the magnitude of the refractive index function. If its value is increased by a constant factor s everywhere $n'(\mathbf{r}) = sn(\mathbf{r})$, the frequency of the mode decreases by the same factor $\omega' = \omega/s$. However, the mode profiles remain unchanged.

10.2.1 Photonic Crystal Slabs

A photonic crystal slab is a slab waveguide, which has a periodic index variation in both in-plane directions. The periodicity is two-dimensional and hence it can be regarded as a 2D PhC. However, due to the slab structure, its vertical translational symmetry is broken. Therefore these structures are commonly referred to as 2.5D systems or photonic crystal slabs. PhC slabs are of technological significance, because they can be fabricated using standard planar lithography and etching processes.

Due to vertical mirror symmetry at the center of the slab, the modes can be separated into vertically even (H_z and E_{xy} components having symmetrical field distributions with respect to the mirror plane) and odd modes (antisymmetrical field distribution). Consequently, in the xy-mirror plane itself, the even mode has only H_z and E_{xy} as its nonvanishing field components and is purely transverse electrically (TE) polarized. Analogously, the odd mode is purely transverse magnetically (TM) polarized at the slab mirror plane (5). Moving away from the center of the slab, all field vector components are nonzero. Hence, the nomenclature of TE-like and TM-like modes is commonly used (1). PhC slabs with air holes in a high dielectric constant slab preferably sustain a bandgap for the TE mode. Hence, all of the following discussions consider the TE-polarized mode.

Photonic crystal slabs confine the light vertically within the slab via index guiding, a generalization of total internal reflection (TIR). Due to the in-plane periodicity, the wave vectors parallel to the plane $\mathbf{k}_\parallel$ form a band structure when plotted versus the frequency ω, which is very similar to the band structure of the true 2D case; a direct comparison of the two diagrams in displayed in Figure 10.1. The bands in the PhC slab case are shifted to higher frequencies because the effective refractive index of the slab is lower than the index of the bulk material. The

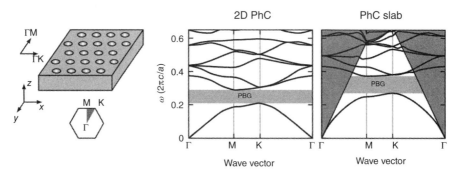

Figure 10.1   Band diagram of an infinite 2D PhC and a PhC slab that consists of a triangular lattice of air holes in Si ($n_{Si} = 3.5$) with TE-like polarization, with the irreducible Brillouin zone at the lower left.

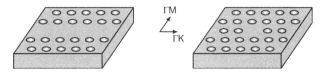

Figure 10.2 Schematic of line defect (left) and point defect (right) in a PhC slab.

eigenvalues of the cladding medium are $\omega = c_{clad}\sqrt{|\mathbf{k}_{\parallel}|^2 + k_{\perp}^2}$, where c_{clad} is the speed of light in the cladding medium $c_{clad} = c_0/n_{clad}$. When ω is plotted against $\mathbf{k}_{\parallel}$, a continuous light cone $\omega \geq c_{clad}|\mathbf{k}_{\parallel}|$ results, indicated by the darker region in Figure 10.1. Modes beneath the light cone are confined to the slab and their field decays exponentially in the vertical direction. Since modes within the light cone radiate energy to the continuum, they are also referred to as radiation modes.

When holes are omitted or shifted from their lattice positions, the periodic translational symmetry of the PhC is broken. Such deviations from the perfect PhC structure are called defects and form allowed states (defect states) within the photonic bandgap. The most prominent examples of such defects are line defects and point defects (Fig. 10.2).

Line defects are formed by omitting a row of holes along one of the lattice directions. This forms a waveguide effectively consisting of two photonic crystal mirrors, providing confinement of the mode lateral to the propagation direction. The geometric structure is periodic in the direction of propagation and hence the propagating modes are Bloch modes. In principle, for all modes below the light cone, lossless operation is possible, when neglecting absorption and scattering from geometrical imperfections of the lattice (e.g., fluctuations in hole size, position, and shape). PhC line defect waveguides have received considerable attention in the research community because their dispersion relation can be widely tuned by modifying the geometric properties of the PhC (6, 7).

A point defect is formed by omitting or changing the size of one single or multiple adjacent holes in the PhC lattice. This type of defect can define localized modes for resonant frequencies within the bandgap, as propagation is hindered in all in-plane directions by the PhC. The light is vertically confined by the TIR condition. The defect serves as a microcavity resonator. The modal volume of the cavity is typically on the order of one cubic wavelength, thus confining the optical energy very tightly. However, because the translational symmetry of the PhC is broken and the resonant mode is spatially localized, the mode now has a continuum of wave vector components, parts of which can couple to the modes in the light cone and radiate energy. Consequently, these localized modes are always lossy. Methods to reduce the vertical radiation losses in PhC slab resonators are analyzed extensively in Section 10.6. The losses of resonators are quantified by their quality factor. This quantity and its importance for resonant systems will be discussed in the next section.

10.3 CAVITIES AND RESONATORS

An optical resonator confines light at resonance frequencies, which are determined by its configuration. A resonator can be regarded as an optical transmission system with incorporated feedback, with the light repeatedly reflecting or circulating within the resonator. Optical resonators are characterized by two key parameters, related to their ability to confine the optical energy temporally and spatially. The former is quantified by the quality (Q) factor and the latter by the mode volume, V. Q is directly proportional to the storage time, a large value indicating strong temporal confinement. The mode volume measures the volume occupied by the optical mode, and a small value represents strong spatial confinement. Resonators are generally frequency selective elements and may serve as spectral analyzers and optical filters.

Assuming the energy stored in a resonator decays with the time constant τ, the optical field of this mode then decays according to

$$E(t) = E_0 e^{-j\omega_0 t} e^{-t/2\tau} \qquad (10.6)$$

where ω_0 is the resonant frequency. The frequency spectrum of the output intensity of this resonator is found by taking the Fourier transform of the previous equation

$$|FT\{E(t)\}|^2 \propto I(\omega) = I_0 \frac{1}{\left(\frac{\omega-\omega_0}{\Delta\omega}\right)^2 + 1} \qquad (10.7)$$

yielding a Lorentzian function, with a spectral line width defined by the full width at half maximum (FWHM) $\Delta\omega = 1/\tau$.

The Q factor of resonant circuits is defined as the ratio of stored energy (W) and the energy loss per cycle (P). Since, by definition, the energy decays at the rate τ, the Q factor can be related to the ratio of resonance frequency ω_0 to line width $\Delta\omega$ through

$$Q_{tot} = \frac{\omega_0 W}{P} = \omega_0 \tau = \frac{\omega_0}{\Delta\omega} \qquad (10.8)$$

Generally, multiple loss mechanisms are responsible for the resonator characteristic. These include losses at imperfect and finite-sized mirrors as well as material absorption and scattering loss inside the resonator. An individual decay time τ_i and hence the individual quality factor Q_i can be attributed to each of these dissipative mechanisms. These individual parameters are connected to the total decay constant and quality factor via

$$Q_{tot} = \omega_0 \tau = \omega_0 \left[\sum_i \frac{1}{\tau_i} \right]^{-1} = \left[\sum_i \frac{1}{Q_i} \right]^{-1}. \tag{10.9}$$

Hence, the total Q is limited by the smallest individual Q factor or, in other words, the largest loss mechanism.

A microcavity in an infinitely extended PhC slab of lossless material exhibits only vertical loss from coupling to radiation modes within the light cone. The associated quality factor is typically referred to as the vertical Q or intrinsic Q because it is an intrinsic property of the isolated cavity. In the remainder of this chapter, this quantity is expressed as $Q_\perp$. When the cavity is coupled to access waveguides, an additional loss mechanism is introduced, which is denoted by the in-plane quality factor $Q_\parallel$. The cavity acts as a filter on the transmission function from the input to the output waveguide, transmitting only the resonant frequencies.

The transmission characteristic of a resonator coupled to a waveguide can be derived from coupled mode theory, which allows us to express the transmission spectrum as a function of Q factors (a thorough derivation can be found in References 1 and 8) as

$$T(\omega) = \frac{\left(\frac{Q_{tot}}{Q_\parallel} \right)^2}{\left(\frac{\omega - \omega_0}{\Delta\omega} \right)^2 + 1}. \tag{10.10}$$

Evidently, the spectral width of the transmission is the same as for the isolated cavity. However, the peak transmission T_0 at resonance is not 100%, but determined by the ratio of total Q to in-plane Q, given by

$$T_0 = \left(\frac{Q_{tot}}{Q_\parallel} \right)^2 = \left(1 - \frac{Q_{tot}}{Q_\perp} \right)^2. \tag{10.11}$$

This means that the transmission at resonance is almost unity when $Q_\perp \gg Q_\parallel$. This is easily explicable, since the cavity mode decays much more quickly into the waveguide than into the surrounding medium.

Here, the resonator shall be used for electro-optic modulation. A variation of the refractive index within the resonator volume causes a shift of the resonance frequency $\Delta\omega_0$ and consequently a shift of the transmission spectrum, thus modulating the transmission at a given frequency ω_p with the modulation depth ΔT (see Fig. 10.3). Clearly, the larger the Q, the greater is the modulation depth for a fixed shift in resonance frequency. From this, it follows that an efficient modulator should have a Q as large as possible to exhibit a large modulation depth with small shifts in the resonance frequency. However, using resonant structures as modulators in high data rate communication schemes poses an upper limit on the permissible Q factor of the device. The photon lifetime τ should not exceed the time spacing between two data symbols, as this would lead to intersymbol interference. Targeting 100 GHz operational bandwidth results in $\tau < 10$ ps. At the telecommunication wavelength of $\lambda = 1550$ nm, this yields a maximum total Q of $Q_{tot} = 12 \cdot 10^3$.

The maximum transmission of a resonator-based modulator determines the insertion loss of the device and hence is desired to be close to unity transmission or 0 dB loss. From Equation 10.11, it is easy to find the maximum transmission of a resonant cavity once the total and intrinsic Q are known. Obviously, the optimal cavity would have $Q_\perp = \infty$; this, however, is not possible in a PhC slab nanocavity. Even in a defect surrounded by a perfect PhC lattice, the intrinsic Q can only be finite due to wave vector components that are inside the light cone. In real-world cavities, this value will be further reduced by material absorption of the dielectric slab and additional vertical scattering losses from imperfections in the geometry (e.g., fluctuations in hole size, position, and shape) of the PhC lattice.

In Figure 10.3, T_0 is plotted versus the vertical Q factor of a cavity with $Q_{tot} = 12 \cdot 10^3$, which was identified to be the upper limit of the total Q factor in a resonant EO modulator. In view of these boundary conditions, it is obvious that the overall device design has to primarily

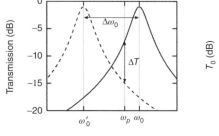
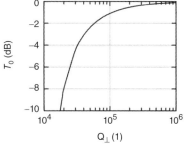

Figure 10.3 Spectral shift of the resonator transmission by $\Delta\omega_0$, resulting in a modulation depth ΔT at ω_p (left). Maximum transmission at resonance T_0 plotted against the cavity intrinsic Q factor (right).

concentrate on the maximization of the intrinsic Q factor. At the same time, this simplifies the simulation efforts, as only isolated cavities need to be considered in the first design step. The minimum intrinsic Q factor $Q_{\perp}^{min}$ required for any PhC cavity EO modulator with given maximum transmission can be found from the right-hand panel in Figure 10.3. Hence, for an insertion loss not larger than 1 dB, resonators with $Q_{\perp} > 10^5$ are a necessity.

10.4 RESONATORS IN PERIODIC RIDGE WAVEGUIDES

A PBG can be formed in a ridge waveguide that has been periodically perforated with air holes in the direction of light propagation. This geometry is comparable to one-dimensional Bragg stacks in terms of its optical properties; however, it offers a higher mechanical stability than an alternating arrangement of dielectric material and air. By introducing a defect into this periodicity, an allowed energy state is formed within the bandgap (9, 10). Such a cavity makes use of the PBG effect in one direction (x) and total internal reflection in the remaining two (y and z). Resonators with $Q = 6 \cdot 10^4$ (11) and $Q = 7.5 \cdot 10^5$ (12), respectively, have been experimentally demonstrated in high refractive index contrast systems based on silicon-on-insulator (SOI) and air-bridged silicon. We now discuss the properties of such resonator cavities in low index systems that use an NLO polymer as core material.

The basic geometry of this type of resonator is sketched in the left panel of Figure 10.4. The right-hand panel of the same figure displays the band diagram of a perfectly periodic arrangement of air holes in a low index ridge waveguide ($n = 1.54$) suspended on an ultralow index substrate ($n = 1.19$). The geometry of the ridge was chosen for single mode operation at 1.3 μm operation wavelength, with a thickness of $d = 1.35$ μm and width of $w = 0.8$ μm. The air holes have a radius of $r = 136$ nm while the lattice constant is $a = 520$ nm. This results in a relative photonic bandgap of approx. 9%, centered at 1.3 μm. The dashed line in the band diagram indicates a mode with odd symmetry, which is orthogonal to the two modes with even symmetry (solid lines) defining the band edges of the PhC.

M. Schmidt found that linearly tapering the radii of the three air holes closest to the cavity from $r = 80$ to $r = 136$ nm substantially decreases vertical scattering losses. From the finite integration technique (FIT, implemented in CST Microwave Studio (13)) simulations, total Q values exceeding 10^4 for this cavity type were obtained. Thus, they are potential candidates for electro-optic modulators, if realized in a suitable electro-optic polymer (14).

For experimental validation of the simulation results presented in Reference (14), resonator structures according to the geometric features using the same argument as in section 10.6.3.1 were fabricated. The ridge waveguides used were made from a polymer guiding layer of poly(methyl methacrylate) covalently functionalized with Disperse-Red 1 P(MMA/DR1) (15) (thickness 1.35 μm, $n = 1.54$ at 1300 nm excitation wavelength). The polymer was deposited by the standard spin coating technique on a low index substrate mesoporous silica (thickness 1 μm, $n = 1.19$ at 1300 nm excitation wavelength). To facilitate handling and ensure mechanical stability silicon wafers (thickness 0.54 mm) were used as carrier substrate. On top of the NLO polymer, a 50 nm NiCr film was deposited and subsequently a 300 nm layer of PMMA electron beam resist. The pattern was written into the top layer using standard electron beam lithography (EBL). After development of the resist, the structure was transferred into the NiCr hard mask by argon ion beam etching. The patterned NiCr layer then served as an etch mask for an electron cyclotron resonance high-density plasma etching process to define the structure in the NLO polymer layer. A detailed description of the fabrication process can be found in References 16 and 17. The holes were etched into the core material (etching depth 1.5 μm), with a slight penetration of the substrate layer.

Scanning electron micrograph (SEM) pictures of the structures are presented in Figure 10.5. The spongelike topography of the mesoporous silica substrate is clearly visible in these pictures. The high air content within the pores is responsible for the extremely low refractive index of this material.

The optical properties of these structures were characterized using the prism coupling technique. Tunable lasers in the wavelength range from 1260 to 1495 nm served as the light source. The proper plane of polarization for the TE-polarized mode was selected by a Glan polarizer, before coupling to the waveguide. The outcoupled light was detected by a standard Ge diode, with the spectrum scanned in 0.5 nm steps.

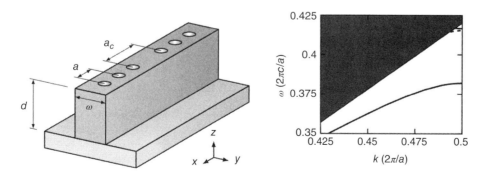

Figure 10.4 Geometry of a periodic ridge waveguide (left) and its photonic band structure (right). The PBG opens between the first two modes with even symmetry (indicated with solid lines). The dashed line represents a mode with odd symmetry.

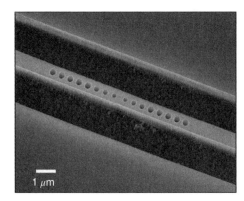
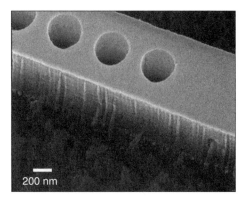

Figure 10.5   Scanning electron micrographs (SEM) displaying a PhC ridge waveguide cavity. The core is made from P(MMA/DR1), which is supported by mesoporous silica as a substrate material. The magnified image on the right emphasizes surface roughness on the etched sidewall of the waveguide. The spongelike topography of the mesoporous silica is evident in both pictures.

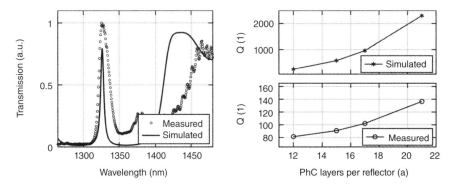

Figure 10.6   Left: Simulated (solid line) and experimental (dots) transmission spectra of a polymer PhC ridge waveguide cavity. The three innermost hole radii are linearly tapered from $r_1 = 80$ nm to $r = 136$ nm. The PBG mirrors of the resonator were 17 lattice constants in length. The simulated data was redshifted by 22 nm. Right: Simulated and experimental Q values of a polymer PhC ridge waveguide.

The transmission characteristics match the results from FIT simulations (Fig. 10.6). The relative spectral position of the dielectric band edge and the resonance are in good agreement. The air band edge lies outside the scanning range of the laser sources and hence is not visible in the transmission spectrum. However, the difference in spectral width of the resonance between simulation and experiment is significant.

The experimental quality factors of the fabricated samples with various PBG mirror lengths are compared to those obtained from FIT simulations in Figure 10.6. The quality factors found in experiments deviate significantly from the predicted simulation results. Even though the Q rises with each length increment of PBG mirrors in the experiment, the increase is much less than expected from the numeric simulations. The difference for the sample with 21 layers of PhC mirrors is more than one order of magnitude ($Q_{sim} \approx 2300$ against $Q_{exp} \approx 130$). The significantly lower Q values result from additional losses present in the system, which are not modeled by the simulation. Most noteworthy are vertical scattering losses due to fabrication imperfections. Using the relation $Q_{exp}^{-1} \approx Q_{sim}^{-1} + Q_{loss}^{-1}$, these scattering losses are found to be $Q_{loss}^{-1} = 140 \pm 15$ for all measured samples. In Figure 10.5, sidewall roughness as well as imperfect hole placement and hole shape are visible, which are limiting the value of Q_{loss}^{-1}, resulting in the low total Q factor. Such low quality factors are insufficient for application in effective broadband modulation schemes. Further substantial improvement in the fabrication technology is necessary before this type of structure in a low refractive index material is considered for the application envisioned here.

The EBL used to define the photonic crystal pattern here is essentially the same process that is used to pattern structures in high index materials, where placement accuracies below 2 nm have been achieved (18, 19). However, the etching processes and behavior of organic materials are different from semiconductors and, other than the case of photoresists, not yet as well understood (16). Consequently, we expect that the structure quality in terms of sidewall roughness and vertical uniformity could be improved with an optimized etching process. In high refractive index structures, experimental Q factors above 10^5 have been reported (11, 12) indicating that the lossy Q factor in structures with optimized fabrication processes is at least on this order of magnitude. Thus, resonators in periodically patterned low index ridge waveguides could achieve the quality factors predicted from simulations ($Q_{sim} > 10^4$ in Reference 14) if fabricated in an optimized process with significantly reduced geometrical imperfections.

10.5 OMNIDIRECTIONAL PHOTONIC BANDGAP

In this section, the existence of a complete PBG in a polymer structure is theoretically shown and experimentally proven. Kee et al. showed that polymer slabs immersed in air with a triangular array of holes can, in theory, exhibit a complete PBG for TE polarization (20). Such air-bridged structures, for which a major part of the waveguide does not rest on a solid substrate, are intrinsically mechanically unstable, and it is particularly difficult to attach a dielectric access channel to the PBG defect waveguide. In the approach presented here, the polymer core is not suspended in air; instead, an "air-like" substrate material with a refractive index close to unity is used ($n_{sub} = 1.15$ at 1300 nm). Mesoporous silica is one such material due to its air-filling fraction of 70% (21). Using this substrate instead of air offers two advantages: an additional wet etching step after structuring the waveguide core is not needed, and the PhC slab has good mechanical stability when residing on a solid substrate.

The guided-mode expansion (GME) method was used to calculate the photonic band diagram of the PhC slab structure (22). The structure is not vertically symmetric; however, following the argument in Reference 23, the modes can still be distinguished into TE- and TM-like. In order to identify the TE- and TM-like polarized modes, the field profiles of the 3D calculation were compared to the results of 2D calculations of the corresponding polarizations. The resulting band diagram is shown in Figure 10.7. This figure reveals the existence of a complete in-plane bandgap for TE-like polarized modes between 0.51 and 0.53 c/a. However, it becomes apparent that considerable parts of this bandgap are above the light line of the substrate (dashed line), which implies that modes with frequencies above the air band edge of the bandgap can couple to radiation modes propagating inside the substrate. Usually, it is desired that the PBG lie as far below the light line as possible, since otherwise the lifetime and, hence, the propagation distance of potential defect modes can be significantly reduced by this coupling mechanism. As shown previously, this problem can be avoided by etching away the PhC's underlying substrate material, thereby removing the substrate material and creating an air-bridged structure (24). Underetched waveguide structures are undesirable for the reasons mentioned earlier. We show in the following that losses incurred from vertical radiation into the continuum of substrate modes are, in fact, negligible in this experiment.

In the GME method, the intrinsic radiation losses of a mode above the light line are found and described by the imaginary part $\Im(\omega)$ of the eigenmode frequency (22). The mode is thus attenuated by $\exp(-\Im(\omega)\tau)$ per unit time τ. The propagation distance is obtained from multiplication of τ with the group velocity v_g and hence the power loss coefficient of the mode is given by $\alpha_{loss} = 2\Im(\omega)/v_g$ (25). Calculations yielded losses below 1 dB/mm for this particular geometry. Over a distance of a few tens (in this case 40) of lattice constants, the radiation losses are far below 1 dB and thus negligible as can be seen in Figure 10.8. Therefore, the additional fabrication step of etching away the substrate material underlying the PhC was omitted.

In the next step, PhC slabs were fabricated in order to experimentally confirm the existence of a complete in-plane PBG. Fabrication was carried out as described in Section 10.4. Figure 10.9 shows a SEM picture of the investigated structures. The slab waveguide was made from low index substrate mesoporous silica (thickness 1 μm) and a polymer guiding layer of P(MMA/DR1) (thickness 1.5 μm). The triangular lattice was chosen to have a lattice constant $a = 650$ nm and a hole radius of 280 nm. The bulk PhCs extended 40 lattice constants parallel to the direction of propagation and 8000 lattice constants perpendicular to it. In order to allow measurements where the light propagates in the ΓM and ΓK direction, two sets of structures were fabricated, one for each propagation direction.

To measure the transmission spectra of the structures, a measurement setup suitable to control the plane of polarization was chosen. It consisted of a white light source (100 W halogen lamp), monochromator (1/4 m, excitation wavelength range 600–2400 nm), Glan polarizer, and a Fresnel rhombus to select the desired polarization. The light was coupled in and out of the waveguide by means of prism couplers. The outcoupled light was detected by a standard Ge diode and recorded using the lock-in technique. All transmission spectra were measured with a resolution of 2 nm in a range from 1000 to 1600 nm. The spectra of the patterned waveguide were divided by the spectra of an unpatterned slab waveguide in order to obtain the transmission characteristic of the PhC without influence from the waveguide material. The results of these measurements are depicted in Figure 10.10, where the transmission spectra of the bulk PhC for each crystal orientation (ΓM and ΓK) are shown. In the ΓM direction, both the dielectric and air band edge (1400 and 1150 nm, respectively) are easy to identify. The dielectric

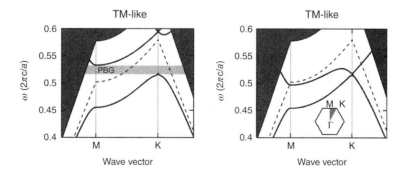

Figure 10.7 Band diagram of a triangular photonic crystal of air holes in a polymer ($n_{Poly} = 1.54$) slab waveguide suspended on an air-like mesoporous silica material ($n = 1.15$) for TE-like (left) and TM-like (right) polarizations. An omnidirectional PBG is visible for TE-like polarized modes (left). The air light cone is represented by the shaded region. The substrate light line is marked by the dashed line.

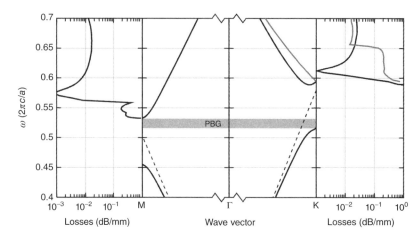

Figure 10.8   Band diagram and radiation losses of a triangular lattice photonic crystal of air holes in a polymer ($n_{Poly} = 1.54$) slab waveguide suspended on an air-like mesoporous silica material ($n = 1.15$) for TE-like modes in ΓM and ΓK directions. The losses of modes above the light line (gray line) due to coupling to radiation modes were calculated using the guided mode expansion method.

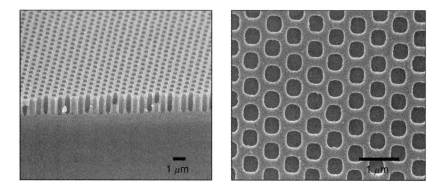

Figure 10.9   SEM displaying the cleaved edge (left) and the top view (right) of a 2D polymer triangular photonic crystal made from a P(MMA/DR1) core, with mesoporous silica as the substrate material ($a = 650$ nm, $r = 280$ nm). Holes were etched into the core material, with a slight penetration of the substrate material.

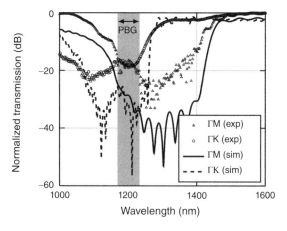

Figure 10.10   Simulated (lines) and experimental (dots) transmission spectra of a bulk 40 lattice constant triangular photonic crystal in a polymer slab waveguide. The simulated and experimental results are in good agreement. A bandgap between 1170 and 1235 nm is clearly visible.

band edge in the ΓK direction is also apparent, whereas the air band edge is outside of the measurement range. All values match the results obtained from the GME method calculations. Between 1170 and 1235 nm, a region with a signal suppression of 15 dB is clearly visible, which corresponds to the bandwidth of the omnidirectional PBG. The transient solver of CST Microwave Studio was used to calculate the expected transmission spectra. The result is plotted in Figure 10.10 and is in accordance with the experimental results.

In summary, bulk PhCs were fabricated with a triangular lattice in polymer slab waveguides on low index air-like substrates. E-beam lithography was used together with a subsequent refractive ion etching step. Band diagram calculations predicted an omnidirectional PBG for TE-like polarized modes. This prediction was confirmed by transmission measurements in the ΓM and ΓK directions of the crystal. Furthermore, the experimental results are in accordance with the results obtained from FIT calculations. For future applications, especially resonant structures and PhC waveguides realized in polymer materials, an optimized etching method needs to be developed that allows the selective underetching of the PBG structure without the destabilization of the access waveguides. This is necessary in order to lift the light line completely above the PBG and avoid losses from defect modes due to coupling to radiation modes.

10.6 MICROCAVITIES IN TWO-DIMENSIONAL LOW INDEX PHOTONIC CRYSTALS

Defects can be introduced into the PhC lattice to break its translational symmetry and hence create defect states, whose frequencies lie within the photonic bandgap (1). In a 2D PhC slab, light confinement of the microcavity is governed by distributed Bragg reflection in the slab plane and TIR perpendicular to the plane of the slab. Such resonant structures with potentially high quality (Q) factors are interesting in telecommunication applications as narrowband filters. If the resonator material is an EO polymer, the optical volume and hence the resonant frequency can be shifted by modifying the polymer's refractive index. Consequently, the wavelength filter becomes tunable and can also serve as an EO modulator.

Numerous geometries have been proposed to achieve extremely high Q ($> 10^6$) cavities with modal volumes on the order of one cubic wavelength in high index ($n \approx 3.5$) semiconductors (26). Quality factor to mode volume ratios of this magnitude lead to very strong light–matter interaction and are required to realize low threshold lasers (27), single-photon emitters (28), or compact optical buffer memories (29). High index systems are preferred for these types of applications as they intrinsically offer better light confinement properties. However, the PhC microcavities are to be integrated here into a material system with low refractive index contrast, which naturally makes their design more challenging. At the same time, the requirements on Q factor and mode volume are slightly loosened, if the PhC cavity is to be used as a filter in a wavelength division multiplex communication scheme or as a broadband EO modulator, as intended in this discussion. In the envisioned applications, mode volume is of secondary importance since a strongly enhanced light–matter interaction is not necessary. For filter applications, Q values of 12×10^3 are sufficient (see also Section 10.3), and EO modulators with $Q > 10^4$ prohibit modulation speeds in the GHz frequency range due to the long photon lifetime in the cavity (14).

In this section, three different PhC microcavity designs are presented and their geometry optimized in terms of Q factor. Finally, an analysis will be carried out to determine the intrinsic limitations of PhC microcavities in low refractive index contrast environments.

10.6.1 H1 and L3 Cavities

Among the most thoroughly studied microcavity types are the so-called H1 and L3 cavities. An H1 cavity is formed by omitting a single hole in a hexagonal lattice, whereas the L3 cavity is formed by omitting three holes in the ΓK direction (see Fig. 10.11). In silicon-based systems, Q factors of several times 10^5 have been obtained with such cavities by appropriate engineering of the lattice geometry (hole position and radius) in the immediate vicinity of the cavity (30–34).

In general, the quality factor of a cavity is determined by its losses, that is, the photon lifetime inside the cavity. If the cavity material itself is assumed lossless, the only two remaining loss mechanisms are radiation losses perpendicular to the slab plane ($Q_\perp$) and in-plane losses from energy leakage through a finite number of PhC layers ($Q_\parallel$). The total Q can be expressed as

$$\frac{1}{Q} = \frac{1}{Q_\perp} + \frac{1}{Q_\parallel} \tag{10.12}$$

If the number of surrounding PhC layers is made sufficiently large, then $Q_\parallel \to \infty$ and the total Q is determined by the radiation losses only; this value is also referred to as the intrinsic or unloaded Q factor of a cavity. Only the intrinsic quality factors ($Q_\perp$) of defect structures in low index PhC slabs are presented and discussed in this section, and hence, the subscript $\perp$ is dropped here.

The radiation losses of PhC slab microcavities arise from **k**-vector components of the cavity mode, which violate the TIR condition. This leaky region is determined by the in-plane **k**-vectors ($k_\parallel$), which satisfy the condition

$$|k_\parallel| \le k_{\text{leak}} = \frac{\omega_0}{c} \tag{10.13}$$

where ω_0 is the angular frequency of the resonant mode and c is the speed of light in the cladding material. The field distribution in k-space is connected to the optical field distribution in real space through the spatial Fourier transform. This property is exploited in the referenced works above to optimize the Q factor. By modifying the dielectric function in space, the distribution of the optical field is also altered. Hence,

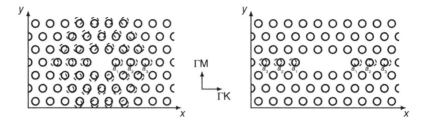

Figure 10.11 Schematic layout of the H1 (left) and L3 (right) cavities. The dashed circles indicate which holes have been shifted relative to their original position.

by proper choice of the geometry parameters, the **k**-vector components in the leaky region can be minimized. The details of this theory will be discussed thoroughly in Section 10.6.3.

In the following, the effect of lattice geometry modifications on the quality factor of H1 and L3 cavities in low index contrast PhCs are studied. Adawi et al. (35) performed such a study by shifting the innermost hole (a_1 in Fig. 10.11) of an L3 cavity. They found that the influence on the Q in a low index contrast PhC is more than an order of magnitude lower than a comparable variation in a semiconductor system with high index contrast. However, Akahane et al. showed (33) that in high index contrast PhC slabs, the structural tuning of the second and third row of holes has a strong impact ($\sim$3x) on the quality factor of a PhC cavity. Thus, the influence of the positions and geometries of the second and third nearest neighbor holes upon the PhC properties is investigated and discussed in the following.

Obviously, if one was to modify every individual hole in terms of displacement and radius, the degrees of freedom approach infinity. Here, parameter space scans were performed to study the effect of displacement of the first three hexagons of air hole layers of the H1 cavity and the displacement of the nearest three holes in ΓK direction of the L3 cavity (a_1, a_2, and a_3 in Fig. 10.11). Following the approach in Reference 33, the parameter sweeps were carried out incrementally for each parameter. Therefore, the full parameter space was not scanned, but the optimal position of the first parameter was determined before optimizing the second parameter, while keeping the first at its optimal value and so on. This incremental procedure does not necessarily yield the highest possible Q factor of the PhC cavity. However, this method was chosen to keep the computational effort within limits. A full scan of all possible parameter permutations would have required an unfeasible amount of computing time. A rigorous analytical method for optimizing the Q factor of a PhC cavity in low index slabs will be presented in Section 10.6.3.

The simulations were carried out using the transient solver of CST Microwave Studio. The bulk lattice parameters were lattice constant $a = 650$ nm, air hole radius $r = 0.34a$, slab thickness $h = a$, and slab refractive index $n = 1.6$. The structure is assumed to be air bridged; hence, the cladding and substrate index is unity. The quality factor was calculated using the FWHM criterion taken from the frequency spectrum recorded by a field probe located inside the cavity. The cavity was surrounded by 20 layers of air holes, which was found to be sufficient to eliminate the contribution of in-plane losses to the total Q. A larger simulation volume did not increase the quality factor and hence the Q is determined by radiation losses only.

The results in Figure 10.12 show that indeed the modification of the lattice geometry surrounding the cavity improves the intrinsic Q factor from ≈ 300 to almost 800 in both cases. However, this value is still one order of magnitude below the required value for efficient EO modulators.

10.6.2 Double Heterostructure Cavity

The highest reported simulated ($Q_{sim} \approx 10^9$(36)) and experimentally achieved ($Q_{exp} > 10^6$(37)) Q factors in PhC microcavities are based on photonic double heterostructures (26). In these geometries, the microcavity is formed by joining PhC defect waveguides with slightly different longitudinal lattice constants. The change in lattice constants shifts the frequency of the defect mode. As a result, a mode gap is opened between the sections of the PhC waveguides and photons with a specific frequency that can exist only in the cavity region (see Fig. 10.13). When the waveguide with the cavity lattice constant is short enough, the frequencies that photons can take in this region become quantized and a photonic microcavity is formed.

In the waveguide direction, the confinement is achieved by the mode gap effect and is not caused by the bandgap effect due to the periodic variation of the dielectric function. This leads to an exponential decay of the envelope function of the optical field outside the cavity in ΓK direction. Proper choice of lattice constants along the waveguide allows the approximation of a Gaussian envelope function for the field profile of the resonant mode and hence a greatly reduced number of **k**-vector components inside the leaky region, explaining the enormous reported Q values.

Using the same geometry parameters for the PhC lattice and simulation techniques as described in the previous section, Q factors of a double heterostructure cavity in a low refractive index contrast PhC were investigated. The lattice constant in the cavity region was chosen to be $a_{cav} = 660$ nm and $a_{cav} = 655$ nm, respectively. The Q values were computed to be $Q(a_{cav} = 660$ nm$) \approx 1500$ and with $Q(a_{cav} = 655$ nm$) \approx 2500$. Compared to the results from the H1 and L3 cavities, this is a more than twofold increase. However, in high index contrast systems, the introduction of double heterostructure cavities has led to more than one order of magnitude increase in the quality factor. In the next section, the origin of these dramatic differences in photon confinement in low and high index systems is analyzed.

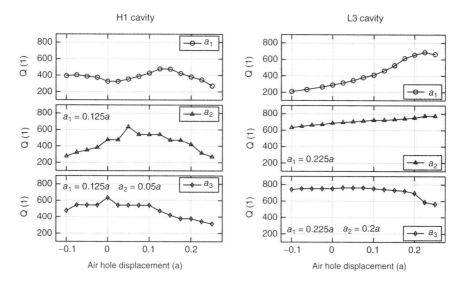

Figure 10.12   Cavity Q factors obtained numerically for H1 and L3 cavities in low index PhC.

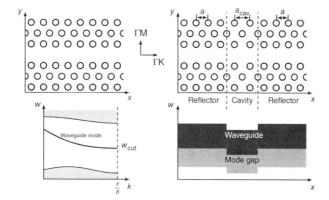

Figure 10.13   Schematic layout of a PhC waveguide (upper left) and its typical band structure (lower left). The double heterostructure is formed by changing the longitudinal lattice constant in the desired cavity region (upper right). Band structure along the ΓK direction of the waveguide (lower right). Photons with specific frequency can only exist in the cavity waveguide (37).

10.6.3   Limits of Low Index Contrast Microcavities

It is intuitively clear that the integrated optical systems with high index contrast offer better light confinement capabilities than low index systems. Therefore, the results of the previous section might not be unexpected. However, the observations do not answer the question of whether there is a general upper limit to the achievable Q factor of a PhC cavity in a low index contrast system.

In an ideal isolated PhC cavity, as in the cases described earlier, the only loss mechanism is vertical scattering determined by those in-plane **k**-vector components that violate the TIR condition given in Equation 10.13. Srinivasan and Painter therefore suggested a simple rule to minimize the optical field components in k-space to achieve high Q values in 2D PhC slab cavities (38, 39). Furthermore, Englund et al. derived an expression that relates the total radiated power from the cavity to the 2D Fourier transforms (denoted by FT_2) of the electric and magnetic field distributions in a surface just above the slab (40). They express the radiated power as

$$P \approx \frac{\eta}{2\lambda^2 k} \int_{k_{||} \leq k_{leak}} \frac{dk_x \, dk_y}{k_{||}^2} k_z \left[\frac{1}{\eta} \left| FT_2 \left\{ E_z \right\} \right|^2 + \left| FT_2 \{ H_z \} \right|^2 \right]. \tag{10.14}$$

where $\mathbf{k}_{||} = (k_x, k_y)$ and k_z are the in-plane and out-of-plane **k**-components, respectively. Furthermore, $\eta = \sqrt{\mu_0 / \varepsilon_0}$, λ is the mode wavelength in air, and E_z and H_z are the vector components of the electric and magnetic field perpendicular to the plane of the slab. In the case of TE modes, only the optical field components (E_x, E_y, H_z) are not equal to zero at the slab center; hence, for such modes, the term $|FT_2\{H_z\}|^2$ will be dominant just above the slab surface, and $|FT_2\{E_z\}|^2$ can be neglected in the analysis of the radiation loss above. Since Q is defined as the ratio of energy stored in the cavity U and energy lost per optical cycle P, that is,

$$Q \equiv \omega_0 \frac{U}{P}.$$

(10.15)

it becomes clear that the quality of a cavity can be optimized by proper engineering of the k-space distribution of H_z.

Theoretically, it is possible to start from a desired field distribution in k-space, deduce the field distribution in real space by Fourier transform and then the desired dielectric constant in space $\varepsilon(\mathbf{r})$, which satisfies these conditions. In practice, however, the rigorous approach of momentum space design is not practicable as it may lead to impractical values in the dielectric function (e.g., negative) or, even with today's technology, unachievable distributions of ε in space.

Here, the theory of momentum space design is used to investigate the feasibility of high Q cavities in low index polymers. To understand the difficulty of high Q in low index compared to high index materials, it is helpful to consider the case of a double heterostructure cavity (Section 10.6.2). Following the argument in References 40 and 41, the resonance frequency ω_0 of the cavity mode is given by the cutoff frequency of the PhC defect waveguide mode, and the maxima $\mathbf{k}_0$ of the mode distribution in k-space are given by the edge points of the first Brillouin zone of the hexagonal lattice. Hence, $k_{0x} = \pm\frac{\pi}{a}$ in ΓK and $k_{0y} = \pm\frac{2\pi}{\sqrt{3}a}$ in ΓM direction.

Figure 10.14 displays a typical field distribution in real and wave vector space of the vertical component of the magnetic field of a PhC double heterostructure resonant mode just above the slab. The k-space field profile unveils that the majority of the field components within the leaky region are oriented along the k_x axis. It is therefore reasonable to reduce the complexity of the problem to one dimension and focus on the mode shape in x direction only to understand the difficulty of achieving high Q values in low refractive index PhC cavities. To tackle the problem analytically, the envelope shape in x direction will be approximated by a Gaussian function with a width of σ. The approximation with a Gaussian envelope function is justified under the condition to find an upper estimate of Q. H_z is then given by

$$H_z(x) = \frac{1}{\sqrt{\sqrt{\pi}\sigma}} \exp\left[-\frac{1}{2}\left(\frac{x}{\sigma}\right)^2\right] \cos(k_{0x}x)$$

(10.16)

in real space and the Fourier transform FT$\{H_z\}$ gives the k-space distribution

$$\text{FT}\{H_z\} = \sqrt{\frac{\sigma}{\sqrt{\pi}}} \left(\exp\left[-\frac{1}{2}\left(\sigma\left(k - k_{0x}\right)\right)^2\right] + \exp\left[-\frac{1}{2}\left(\sigma\left(k + k_{0x}\right)\right)^2\right]\right).$$

(10.17)

The prefactor is needed to keep the mode energy, which is proportional to $\int_{-\infty}^{\infty} |H_z|^2 dx$, constant.

Figure 10.15 shows the band diagrams of a PhC defect waveguide formed by omitting one row of holes in the ΓK direction in a low index ($n = 1.6$) and a high index ($n = 3.5$) core material, respectively. From the location of the cutoff frequency of the defect mode, it is obvious that the resonant frequency of a cavity in the low index material ($\omega_0 \approx 0.48$) will be substantially higher than in the high index material ($\omega_0 \approx 0.26$), resulting in a substantially increased leaky region of wave vector components in the low index case. Figure 10.16 gives an impression of the massive increase of the integral over k within the leaky region, if the extension of the mode in x direction is assumed to be equal for the low and high index cases.

To compensate for the increased leaky region, the width of the mode profile in momentum space needs to be reduced. Figure 10.16 shows how an increase in σ leads to narrower profile in k-space, while increasing its maximum value due to conservation of energy. By evaluating the integral over the wave vector components in the leaky region, one finds that the result decreases monotonically with increasing mode

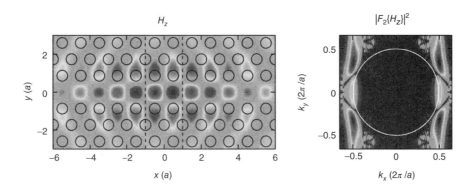

Figure 10.14 H_z field profile of the resonant mode in a double heterostructure PhC cavity in polymer (left). Fourier space representation of the H_z field component just above the slab (right). Both representations are in false color with logarithmic scaling. The leaky region is indicated by the white circle.

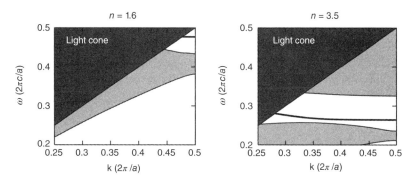

Figure 10.15 Band diagrams of W1.0 PhC defect waveguides in high index and low index materials. The defect mode is represented by the thick black line. The bulk PhC modes are indicated by the lightly shaded regions and the continuum of modes above the light line is marked by the dark shaded region.

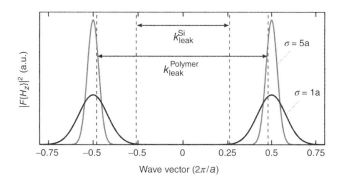

Figure 10.16 Profile of a resonant mode with Gaussian envelope function in momentum space.

width σ. Thus, theoretically, the Q value of a low index PhC cavity could indeed be driven to arbitrarily large Q factors by increasing the mode volume and maintaining a Gaussian envelope profile. Tanaka et al. (36) derived a design rule for multistep PhC heterostructures in high index waveguide cores to achieve a Gaussian envelope profile of the resonator mode with an arbitrary width in the ΓK direction. It should be noted that this method can be arbitrarily demanding on the fabrication quality and the achievable quality factor is hence principally limited by the placement accuracy of the PhC holes. Here, this concept is adapted to the case of low index PhC cores and the associated limits on the Q factor will be discussed.

10.6.3.1 Multistep Heterostructures The general geometry of a multistep heterostructure is displayed in Figure 10.17. Through modification of the properties (i.e., lattice constant, radius, waveguide width, etc.) of each PhC section PhC$_n$, the cutoff frequency is modulated along the direction of the waveguide. It will be shown in the following that by appropriate selection of the spatial cutoff frequency distribution, the desired Gaussian field envelope of the cavity mode can be achieved.

In a standard heterostructure cavity as displayed in Figure 10.13, the evanescent behavior of the field in the reflector region is determined by the imaginary part of the complex wave vector, which is denoted by $\Im(k) = q$. The field decays exponentially with

$$H \propto \exp(-qx). \tag{10.18}$$

where q is the imaginary part of the wave vector within the mode gap at the resonance frequency ω_0. However, for a Gaussian envelope the magnetic field should adhere to

$$H \propto \exp\left(-\frac{1}{2}\frac{x^2}{\sigma^2}\right). \tag{10.19}$$

From Equations 10.18 and 10.19, it becomes clear that q needs to be a linear function in the spatial coordinate x, and hence,

$$q = \frac{x}{2\sigma^2}. \tag{10.20}$$

The complex dispersion relation for the PhC waveguide can be found with the analytic continuation method (42). With this method, the real dispersion relation of the PhC mode is expanded into its complex form. Assuming the real part of the dispersion relation is known (either from the plane wave expansion or GME method), the function $f(k)$ can be fitted by a Taylor expansion of the term $(k - 0.5)$ at the Brillouin

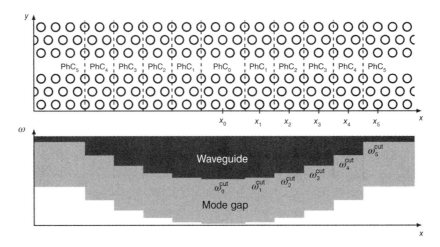

Figure 10.17   Schematic geometry (top) and band structure (bottom) of a multistep PhC heterostructure cavity.

zone boundary ($k = 0.5$). Due to the even symmetry of the dispersion relation around this point, only even order terms need to be considered in the expansion

$$\omega = \omega^{\mathrm{cut}} + C_1(k - 0.5)^2 + C_2(k - 0.5)^4 + C_3(k - 0.5)^6 + \cdots \tag{10.21}$$

Substituting $k = 0.5 - iq$ yields the dispersion in the mode gap region

$$\omega = \omega^{\mathrm{cut}} - C_1 q^2 + C_2 q^4 - C_3 q^6 + \cdots \tag{10.22}$$

The knowledge of the complex dispersion relation of the waveguide mode allows for an analytical design of the multistep heterostructure cavity. Inserting the cavity resonance frequency ω^{cav} and Equation 10.20 into the previous expression and rearrangement yields the required spatial dependency of the cutoff frequency in the reflecting section of the heterostructure

$$\omega^{\mathrm{cut}}(x) = \omega^{\mathrm{cav}} + C_1\left(\frac{x}{2\sigma^2}\right)^2 - C_2\left(\frac{x}{2\sigma^2}\right)^4 + C_3\left(\frac{x}{2\sigma^2}\right)^6 - \cdots \tag{10.23}$$

Thus, Equation 10.23 gives the required distribution of the cutoff frequency along the PhC waveguide shown in Figure 10.17.

A polymer PhC waveguide with a defect width $W = 0.7\sqrt{3}a$ yields a dispersion relation of the defect mode that is well approximated by a second-order Taylor expansion (see Fig. 10.18). This waveguide type is chosen to simplify the derivation of the necessary geometry variations. The operating frequency of a PhC is inversely proportional to its lattice constant. The cutoff frequency can therefore be tuned locally by adjusting the lattice constant in the respective PhC section according to

$$\omega^{\mathrm{cut}}(x)a(x) = \omega_0^{\mathrm{cut}}a_0. \tag{10.24}$$

Together with Equation 10.23, this gives a rule for the required spatial dependence of the lattice constant

$$a(x) = \frac{a_0}{1 + C_1 x^2/(4\sigma^4\omega_0^{\mathrm{cut}})}. \tag{10.25}$$

As long as $C_1 x^2/(4\sigma^4\omega_0^{\mathrm{cut}}) \ll 1$ is satisfied, this can be simplified to

$$a(x) = a_0(1 - C_1 x^2/(4\sigma\omega_0^{\mathrm{cut}})). \tag{10.26}$$

As indicated in Figure 10.17, the lattice parameters are changed stepwise rather than continuously, and thus, the distances x_m are approximated using

$$x_m = (2m + 0.5)a_0. \tag{10.27}$$

This yields the following expression for the lattice constant in each section:

$$a_m = a_0(1 - C_1(2m + 0.5)^2 a_0^2/(4\sigma\omega_0^{\mathrm{cut}})). \tag{10.28}$$

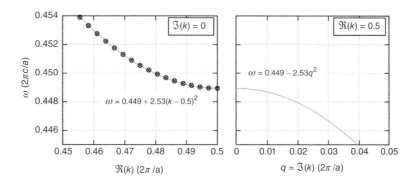

Figure 10.18 Complex dispersion relation of a polymer PhC waveguide ($W = 0.7\sqrt{3}a$). The symbols indicate the results from the band diagram calculation using the plane wave expansion method. The gray lines plot the result of a second-order polynomial fit.

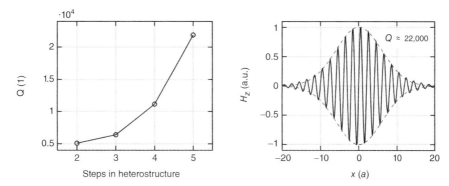

Figure 10.19 The quality factor of a multistep heterostructure increases drastically with an increase in transition steps (left). Optical field $|H_z|$ distribution in a multistep heterostructure, the dashed lines indicate a fitted Gaussian envelope function to the simulation results (right).

This design rule can be applied to fine-tune a PhC heterostructure with cavity lattice constant $a_0 = 660$ nm and bulk lattice constant $a = 650$ nm using between two and five transition steps. The results are displayed in the left panel of Figure 10.19. Using more transition steps and thus achieving a smoother approximation of $a(x)$ given in Equation 10.26 drastically increase the quality factor of the cavity. The right panel of the same figure demonstrates how well the optical field envelope follows a Gaussian envelope function. However, the incremental difference in lattice constant especially of the first transition is below 1 nm. Current nanofabrication technology allows for a placement accuracy of 2 nm or 1 nm at best, and hence, the actual fabrication of such structures is not feasible.

To determine an upper estimate of the realistically achievable quality factor, simulations were carried out under the boundary condition that the minimum step size between two PhC sections must not be smaller than either 1 nm or 2 nm. Due to the required parabolic dependence of a_m, the cavity lattice constant a_0 exceeds the value of 660 nm already at either three ($\Delta a = 2$ nm) or four ($\Delta a = 1$ nm) transition steps. This leads to a resonance frequency, which is below the bandgap of the bulk PhC and allows the optical field to couple to PhC slab modes, giving rise to undesirable in-plane losses. Hence, the maximum number of transition steps is very limited, resulting in a maximum Q of roughly 1.3×10^4 in both cases. Following the argument in Section 10.3, this value would result in an insertion loss of at least 20 dB, if the cavity is loaded to a waveguide. Clearly, this performance is not acceptable for any serious application in optical data communication.

10.6.4 Cavity Definition by Photobleaching

The refractive indices of polymers with EO chromophores can be permanently altered by photobleaching. In this process, the π-bridges in the chromophores are broken by irradiation with high-energy photons (typically in the UV), and the refractive index can be decreased up to $\Delta n = 6 \times 10^{-2}$ (43–45). The magnitude of the index change is determined by the exposure dose and can therefore be controlled by the intensity and the exposure time.

In general, the frequencies of PhC modes do not scale linearly with the refractive index of the waveguide core. However, for the variations achievable with photobleaching, which are on the order of 10^{-2}, the position of the cutoff frequency can be approximated by a linear function of the refractive index. Thus, the cutoff is described by

$$\omega^{\text{cut}}(n) = \omega^{\text{cut}}(n_0) + C_n(n - n_0). \tag{10.29}$$

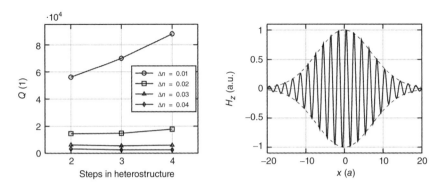

Figure 10.20 The quality factor a multistep heterostructure increases drastically with an increase in transition steps (left). Optical field in a multistep heterostructure. The dashed lines indicate a fitted Gaussian envelope function to the simulation results (right).

where C_n is a constant that depends on the particular PhC geometry. Using the same argument as in the preceding text, a condition for the spatial distribution of refractive index in x direction is found to be

$$n(x) = n_0 - C_n' \left(\frac{x}{2\sigma^2} \right)^2. \tag{10.30}$$

This theory was tested using FIT simulations completely analogous to the case discussed earlier and the simulation results are summarized in Figure 10.20. Most noteworthy is the almost one order of magnitude increase of the quality factor to $Q \approx 9 \cdot 10^4$ compared to the geometry variation optimization presented earlier. As for the previous results, the quality factor increases with a rising number of transition steps. Simulation results with more than four transition steps are not presented here because the cavity started to support a second-order mode due to the increased cavity volume. The low quality factor values for larger index variations are a result of coupling to PhC slab modes, similar to the case of unacceptably large Δa explained previously. However, it becomes clear that defining multistep PhC heterostructures using photobleaching can be a very interesting alternative to overcome fabrication limitations in the positioning accuracy of current lithography methods. The insertion loss of the loaded resonator would amount to 1.3 dB, which is acceptable for application in most optical transmission systems.

10.7 CONCLUSION

In summary, possibilities for full 3D light confinement in PhC nanocavities using low dielectric contrast materials were evaluated. Cavities in periodically structured ridge waveguides were fabricated using a polymer material as the waveguide core. The performance of these resonating cavities observed in experiments ($Q_{exp} \approx 10^2$) was more than an order of magnitude poorer than predicted from simulations. This deficiency was attributed to fabrication imperfections. Assuming that the same fabrication quality of current semiconductor technology can be achieved in polymer-based PhC ridge waveguides, cavities with loaded Q above 10^4 are feasible.

An omnidirectional PBG with gap to midgap ratio of $\approx 5\%$ for TE polarized light in a polymer slab waveguide was experimentally demonstrated. Based on these results, the achievable performance of H1, L3, and double heterostructure-type PhC nanocavities in low refractive index material was numerically studied by systematic geometry variations. For H1 and L3 type cavities, which employ light confinement by Bragg scattering in both in-plane directions, the intrinsic quality factor did not exceed 10^3. For the double heterostructure, where light confinement is based on the mode gap effect in one in-plane direction and Bragg scattering in the other, $Q_1 = 2500$ was possible. PhC nanocavities in low refractive index contrast media suffer from the intrinsic disadvantage of a resonance much closer to the air light line compared to the same structures in high dielectric constant materials. Theoretically, it is possible to achieve arbitrarily high intrinsic Q factors also in low refractive index PhC cavities by increasing the mode volume while maintaining a Gaussian envelope function of the resonant mode's optical field. However, this would require an arbitrarily fine (subnanometer) positioning accuracy for the fabrication technology. With the placement accuracy of current microstructuring technology, theoretical Q values of up to $\approx 10^4$ are possible. However, according to the argument in Section 10.2, the intrinsic Q factor should be on the order of 10^5 for an EO modulator device. To circumvent these strong requirements on placement accuracy, photobleaching of the NLO polymer material can be used to define a multistep heterostructure-type cavity. This option theoretically allows for the fabrication of PhC nanocavities with quality factors of almost 10^5. The highest reported experimental quality factors for silicon PhC slab cavities exceed 10^6 (18, 37). Therefore, it can be assumed that the simulated Q factors of 10^5 in low index PhC cavities can be achieved experimentally, if fabricated with the same quality as the high index structures. Continuing efforts are motivated by the desire to achieve compact waveguide approaches for low index materials with high levels of functionality, enabling, for example, EO modulators that can be directly integrated with Si photonics.

REFERENCES

1. Joannopoulos JD, Johnson SG, Winn JN, Meade RD. *Photonic Crystals, Molding the Flow of Light*. Princeton, (NJ): Princeton University Press; 2008.

2. Rayleigh L. On the maintenance of vibrations by forces of double frequency, and on the propagation of waves through a medium endowed with a periodic structure. Philos Mag 1887;24:145–159.

3. Yablonovitch E. Inhibited spontaneous emission in solid-state physics and electronics. Phys Rev Lett 1987;58:2059–2062.

4. John S. Strong localization of photons in certain disordered dielectric superlattices. Phys Rev Lett 1987;58:2486–2489.

5. Qiu M. Effective index method for heterostructure-slab-waveguide-based two-dimensional photonic crystals. Appl Phys Lett 2002;81:1163–1165.

6. Petrov AY, Eich M. Zero dispersion at small group velocities in photonic crystal waveguides. Appl Phys Lett 2004;85:4866–4868.

7. Petrov AY, Eich M. Dispersion compensation with photonic crystal line-defect waveguides. IEEE J Sel Areas Commun 2005;23:1396–1401.

8. Manolatou C, Khan MJ, Fan S, Villeneuve PR, et al. Coupling of modes analysis of resonant channel add-drop filters. IEEE J Quantum Electron 1999;35:1322–1331.

9. Fan S, Winn JN, Devenyi A, Chen JC, et al. Guided and defect modes in periodic dielectric waveguides. J Opt Soc Am B 1995;12:1267–1272.

10. Foresi JS, Villeneuve PR, Ferrera J, Thoen ER, et al. Photonic-bandgap microcavities in optical waveguides. Nature 1997;390:143–145.

11. Velha P, Picard E, Charvolin T, Hadji E, et al. Ultra-high Q/V fabry-perot microcavity on SOI substrate. Opt Express 2007;15:16090–16096.

12. Deotare PB, McCutcheon MW, Frank IW, Khan M, et al. High quality factor photonic crystal nanobeam cavities. Appl Phys Lett 2009;94:121106-3.

13. CST - Computer Simulation Technology. Available at www.cst.com. Accessed 2002 Feb 4.

14. Schmidt M. *Nonlinear Optical Polymeric Photonic Crystals*. Göttingen: Cuvillier; 2006.

15. Nahata A, Heinz TF. Generation of subpicosecond electrical pulses by optical rectification. Opt Lett 1998;23:867–869.

16. Hübner U, Boucher R, Morgenroth W, Schmidt M, et al. Fabrication of photonic crystal structures in polymer waveguide material. Microelectron Eng 2006;83:1138–1141.

17. Wulbern JH, Schmidt M, Hubner U, Boucher R, et al. Polymer based tunable photonic crystals. Phys Status Solidi A 2007;204:3739–3753.

18. Tanabe T, Notomi M, Kuramochi E, Shinya A, et al. Trapping and delaying photons for one nanosecond in an ultrasmall high-Q photonic-crystal nanocavity. Nat Photonics 2007;1:49–52.

19. Li J, White TP, O'Faolain L, Gomez-Iglesias A, et al. Systematic design of flat band slow light in photonic crystal waveguides. Opt Express 2008;16:6227–6232.

20. Kee CS, Han SP, Yoon KB, Choi CG, et al. Photonic band gaps and defect modes of polymer photonic crystal slabs. Appl Phys Lett 2005;86:51101.

21. Schmidt M, Böttger G, Eich M, Morgenroth W, et al. Ultralow refractive index substrates-a base for photonic crystal slab waveguides. Appl Phys Lett 2004;85:16–18.

22. Andreani LC, Gerace D. Photonic-crystal slabs with a triangular lattice of triangular holes investigated using a guided-mode expansion method. Phys Rev B 2006;73:235114–235116.

23. Qiu M. Band gap effects in asymmetric photonic crystal slabs. Phys Rev B 2002;66:33103.

24. Choi CG, Han YT, Kim JT, Schift H. Air-suspended two-dimensional polymer photonic crystal slab waveguides fabricated by nanoimprint lithography. Appl Phys Lett 2007;90:221109.

25. Gerace D, Andreani LC. Low-loss guided modes in photonic crystal waveguides. Opt Express 2005;13:4939–4951.

26. Song BS, Noda S, Asano T, Akahane Y. Ultra-high-Q photonic double-heterostructure nanocavity. Nat Mater 2005;4:207–210.

27. Painter O, Vuckovic J, Scherer A. Defect modes of a two-dimensional photonic crystal in an optically thin dielectric slab. J Opt Soc Am B 1999;16:275–285.

28. Michler P, Kiraz A, Becher C, Schoenfeld WV, et al. A quantum dot single-photon turnstile device. Science 2000;290:2282.

29. Yanik MF, Fan S. Stopping light all optically. Phys Rev Lett 2004;92:83901.

30. Akahane Y, Asano T, Song BS, Noda S. High-Q photonic nanocavity in a two-dimensional photonic crystal. Nature 2003;425:944–947.

31. Ryu HY, Notomi M, Lee YH. High-quality-factor and small-mode-volume hexapole modes in photonic-crystal-slab nanocavities. Appl Phys Lett 2003;83:4294–4296.

32. Kim GH, Lee YH, Shinya A, Notomi M. Coupling of small, low-loss hexapole mode with photonic crystal slab waveguide mode. Opt Express 2004;12:6624–6631.

33. Akahane Y, Asano T, Song BS, Noda S. Fine-tuned high-Q photonic-crystal nanocavity. Opt Express 2005;13:1202–1214.

34. Tanabe T, Shinya A, Kuramochi E, Kondo S, et al. Single point defect photonic crystal nanocavity with ultrahigh quality factor achieved by using hexapole mode. Appl Phys Lett 2007;91:21110.

35. Adawi AM, Chalcraft AR, Whittaker DM, Lidzey DG. Refractive index dependence of L3 photonic crystal nano-cavities. Opt Express 2007;15:14299–14305.

36. Tanaka Y, Asano T, Noda S. Design of photonic crystal nanocavity with Q-factor of $\sim 10^9$. J Lightwave Technol 2008;26:1532–1539.

37. Asano T, Song BS, Noda S. Analysis of the experimental Q factors (~ 1 million) of photonic crystal nanocavities. Opt Express 2006;14:1996–2002.

38. Srinivasan K, Painter O. Momentum space design of high-Q photonic crystal optical cavities. Opt Express 2002;10:670–684.

39. Srinivasan K, Barclay PE, Painter O, Chen JX, et al. Experimental demonstration of a high quality factor photonic crystal microcavity. Appl Phys Lett 2003;83:1915–1917.

40. Englund D, Fushman I, Vuckovic J. General recipe for designing photonic crystal cavities. Opt Express 2005;13:5961–5975.

41. Asano T, Song BS, Akahane Y, Noda S. Ultrahigh-Q nanocavities in two-dimensional photonic crystal slabs. IEEE J Sel Top Quantum Electron 2006;12:1123–1134.

42. Kohn W. Analytic properties of Bloch waves and Wannier functions. Phys Rev 1959;115:809–821.

43. Vydra J, Beisinghoff H, Tschudi T, Eich M. Photodecay mechanisms in side chain nonlinear optical polymethacrylates. Appl Phys Lett 1996;69:1035–1037.

44. Zhou J, Pyayt A, Dalton LR, Luo J, et al. Photobleaching fabrication of microring resonator in a chromophore-containing polymer. IEEE Photonics Technol Lett 2006;18:2221–2223.

45. Sun H, Chen A, Olbricht BC, Davies JA, et al. Microring resonators fabricated by electron beam bleaching of chromophore doped polymers. Appl Phys Lett 2008;92:193305.

11

MICRORING RESONATOR ARRAYS FOR SENSING APPLICATIONS

DANIEL PERGANDE, VANESSA ZAMORA, PETER LÜTZOW, AND HELMUT HEIDRICH

Fraunhofer Institute for Telecommunications, Heinrich Hertz Institute, Berlin, Germany

During the last decade, waveguide-based devices became more and more attractive for sensing applications in different areas, especially in chemical and biochemical detection (1, 2). In particular, integrated optical (IO) sensors exhibit extraordinary benefits such as high sensitivity in chemical as well as biochemical analysis, miniaturization and compactness, electromagnetic immunity, potential integration with electronic devices forming microsubsystems, and low-cost fabrication due to the feasibility for mass production on wafer level (3–6).

Most detection schemes operate label-free, excluding adverse effects that may be introduced by fluorescent, enzymatic, or radioactive labels. Consequently, less time-consuming preparation procedures are necessary, which have positive impacts on the time-to-final analysis and the overall costs for the individual analysis. Furthermore, a strong miniaturization allows high-level integration of complex sensor functions into smart portable to handheld devices, which should be manageable even for unskilled people.

One type of these IO devices is based on microring resonator (MRR) elements, which have received a lot of attention in recent years. Due to their resonant properties, they feature extraordinary sensitivity to respond on changes of environmental parameters. A compact and smart microsystem for diverse application can be engineered by integration with other chip-based IO, electrical, and microfluidic devices.

Accordingly, MRRs are one of the most promising classes of transducers in miniaturized IO sensors (7–9). Analyte detection is usually based on the measurement of shifts in resonance frequencies caused by local changes in the ambient refractive index. Specificity is accomplished by using biologically or chemically selective adlayers that specifically promote the accumulation of target molecules on the MRRs surface. This leads to an increase of the local refractive index related to the analyte concentration.

An ideal design of MRR devices can be obtained by considering a few important figures of merit, which will be introduced in this chapter.

As an example, highly sensitive and label-free detection of molecular species with concentrations down to the ppb level in gaseous environments has been demonstrated for MRR-based sensors (10). MRRs have been integrated with microfluidic systems and attested to be suited even for on-chip spectroscopy applications (11). MRR sensors can be operated for both reuse and single use only, as required for the specific applications and the receptor coating employed. Evidently, optical interfacing of miniaturized IO devices was challenging in the past. This problem has been solved. Optical coupling schemes with moderate alignment tolerances enable an automation of optical interfacing (12, 13). The well-engineered integration of optical circuits including MRRs will enable the development of optical transducers in labs-on-chips (LOCs).

LOCs are chip-based devices capable of analyzing a multitude of chemical or biological issues. Compared to conventional methods and equipment in laboratories, they save time, lab space, and costs (14). LOCs for biosensing contain multiple sensor elements that interact with an analyte sample flowing through small microfluidic channels. Each sensor element can be designed either to detect a specific (bio)molecule or to perform a redundant detection of the same biomolecule or a combination of both cases. Finally, the integration and parallel readout of highly sensitive and miniaturized sensor elements enable the LOCs to be used for multiparameter detection. A multiplexing approach for the LOCs will also improve the throughput as well as an on-chip referencing via control channels. For an efficient automation of such LOCs, it is obligatory to monitor and extract signal drift due to environmental influences foremost thermal effects. Processing techniques for mass fabrication, such as the complementary metal oxide semiconductor (CMOS) technology in microelectronics industry, will facilitate reliable high-volume production of chips at high quality.

Chapter 11 is organized in the following manner:

The first part introduces some basic features of the MRRs and figure of merits (FOMs) for an appropriate design of such IO sensor elements. After that, an overview is given on the most relevant material systems for the fabrication of MRR devices. Other than mere

Nanomaterials, Polymers and Devices: Materials Functionalization and Device Fabrication, First Edition. Edited by Eric S. W. Kong.
© 2015 John Wiley & Sons, Inc. Published 2015 by John Wiley & Sons, Inc.

sensor applications, some devices from optical telecommunication application are also sketched as an extended prospect of the potential of the MRRs.

In the next section, the application-relevant approaches for an implementation of MRR-based on-chip multiparameter analytics are summarized. The presented array designs and results use different optical multiplexing techniques either on chip or by using external optical devices to read out the signal of an individual MRR sensor element. As each of those MRR sensor arrays has some limitation, their use in analytics applications is limited. An appropriate multiplexing scheme is sketched to overcome such limitations. An introduction on the theory of multiplexing MRR by an electro-optical (EO) or rather a thermo-optical modulation as well as the first experimental results is given to verify the proposed concept.

In the last part of the chapter, some approaches and settings for an efficient and reliable implementation of MRR-based sensor arrays in entire sensor systems are presented. Finally, some perspectives are sketched for encouraging researchers and engineers to elaborate further innovations.

11.1   BASIC PROPERTIES OF MICRORING RESONATORS

During the last years, several textbooks dealing with optical microresonators, and especially with MRRs, have been published. Therefore, we restrict ourselves in the following to the most important theoretical considerations. Details of the mathematical description of MRR properties can be found elsewhere (15–17).

The properties of passive waveguide networks with arbitrary complexity can be calculated by using coupled mode theory. Here, the interaction between propagating modes is treated as linear superposition of abstract mode amplitudes. Therefore, an optical network can be described by a set of coupled linear equations, which can be solved analytically. Assuming single-mode waveguides, according to Figure 11.1, a simple MRR composed of a ring waveguide optically coupled to a single bus waveguide can be described by the following relations:

$$E_0 = rE_i + itE_2$$
$$E_1 = itE_i + rE_2$$
$$E_2 = ae^{-i\varphi}E_1 \tag{11.1}$$

These equations give the steady state solutions for the field amplitudes E_1, E_2 and E_0.

Here, r and t are the self- and cross-coupling coefficients, used to describe the interaction between ring and bus waveguide. The loss parameter $a = \exp(-\alpha/2L)$ describes all losses (e.g., losses due to scattering, radiation, absorption with total loss coefficient α), and $\varphi = n_{\text{eff}}2\pi/\lambda L$ is the round-trip phase, accumulated over the full length of the resonator's circumference L. In general, the effective index n_{eff} of a mode also depends on the free space wavelength λ and characterizes the phase velocity of light propagating at that particular wavelength in this particular mode inside the waveguide.

The field amplitudes, E_1 and E_2, represent the electric field strength inside the MRR and E_0 is the transmitted field strength. Assuming synchronous coupling ($r^2 + t^2 = 1$), the optical power $T(\varphi)$ transmitted past the MRR with respect to the incident power is

$$T(\varphi) = \left| \frac{E_o}{E_i} \right|$$
$$= \frac{a^2 + r^2 - 2\,a\,r\,\cos\varphi}{1 + a^2 r^2 - 2\,a\,r\,\cos\varphi} \tag{11.2}$$

The transmission spectrum is periodic in φ. Further inspection reveals that the spectrum exhibits local minima for φ being an integer multiple of 2π. The dips in the spectrum are caused by resonances inside the MRR leading to destructive interference between light inside

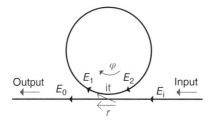

Figure 11.1   Scheme of a MRR coupled to a single bus waveguide.

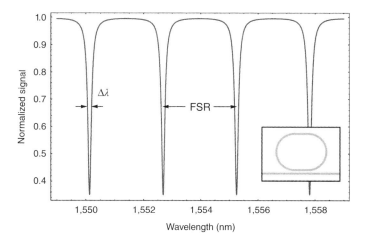

Figure 11.2 Optical spectrum of a single MRR. Resonance dips belonging to different resonance orders are separated by the free spectral range (FSR). The linewidth is defined as the full width at half maximum (FWHM) of a resonance dip. The inset shows a schematic of a racetrack-shaped MRR coupled to a bus waveguide.

the bus waveguide and light coupling back from the MRR (cf. Fig. 11.2). In more detail, the well-known resonance condition reads

$$\varphi = n_{\text{eff}} \frac{2\,\pi}{\lambda_{0,m}} L = m\,2\pi, \quad m \in \text{Integers} \tag{11.3}$$

With the help of Equation 11.3, the resonance wavelengths $\lambda_{0,m}$ can be calculated.

In the following, a few important quantities – frequently used to characterize optical resonators – will be given.

11.1.1 Free Spectral Range

The free spectral range (FSR) $\Delta\lambda$ is defined as the wavelength distance between resonances of successive resonance orders m and $(m + 1)$, respectively (cf. Fig. 11.2). From the resonance condition (11.3), it follows

$$\Delta\lambda = n_{\text{eff}} L \left(\frac{1}{m+1} - \frac{1}{m} \right) \tag{11.4}$$

For $m \gg 1$ and $\Delta\lambda \ll \lambda$ and by keeping in mind that in general n_{eff} is wavelength dependent, Equation 11.4 may be expanded to yield

$$\Delta\lambda = \frac{\lambda^2}{n_g L} \tag{11.5}$$

Here, the group index n_g of the mode with effective index n_{eff} has been introduced. The group index is defined by

$$n_g = n_{\text{eff}} - \lambda \frac{\partial n_{\text{eff}}}{\partial \lambda} \tag{11.6}$$

where the dispersion term $\frac{\partial n_{\text{eff}}}{\partial \lambda}$ defines the change of the effective index as a function of the wavelength.

The FSR carries the information about the periodicity of the spectrum. Provided resonances of successive orders are sufficiently far apart, the shape of a particular resonance of order m can be characterized by three parameters: the resonance wavelength $\lambda_{0,m}$, the linewidth $\delta\lambda$, and the extinction η.

11.1.2 Resonance Wavelength

The resonance wavelengths are calculated with the help of the resonance condition (11.3). The mth-order resonance wavelength is given by

$$\lambda_{0,m} = \frac{n_{\text{eff}} L}{m} \tag{11.7}$$

11.1.3 Linewidth

The linewidth $\delta\lambda$ of a resonance peak is usually defined as its full width at half maximum (FWHM, cf. Fig. 11.2). For the resonance dips of (11.2), the linewidth may be defined analogously as full width at $(1 + \min_\varphi T(\varphi))/2$.

An approximate expression for $\delta\lambda$ is given by

$$\delta\lambda \approx \frac{\lambda^2}{\pi\, n_g\, L}\, \frac{1 - a\, r}{\sqrt{a\, r}} \tag{11.8}$$

11.1.4 Extinction

The extinction of a resonance is defined as

$$
\begin{aligned}
\eta &= (1 - \min_\varphi T(\varphi)) \\
&= \frac{(1 - a^2)\,(1 - r^2)}{(1 - a\, r)}
\end{aligned}
\tag{11.9}
$$

An extinction of "one" corresponds to the case, where the transmission is completely suppressed on resonance. An extinction of "zero" means, that there is no change of the transmission intensity at all.

11.1.5 Lorentz Approximation

A particularly simple approximation for the line shape of a single resonance dip can be given by Taylor expansion of the cosine terms in Equation 11.2 and by introducing the aforementioned parameters $\delta\lambda$, η, and λ_0:

$$T_{\lambda_0}(\lambda) \approx 1 - \frac{\eta\,\delta\lambda^2}{\delta\lambda^2 + 4\,(\lambda - \lambda_0)} \tag{11.10}$$

This approximation has the form of a Lorentzian function, which is known to be a good approximation to the shape of atomic absorption lines as well. The approximation holds on the interval $\left[\lambda_0 - \frac{\Delta\lambda}{2}, \lambda_0 + \frac{\Delta\lambda}{2}\right]$ as long as the finesse (F) of the resonator is sufficiently large.

Here, F is defined as the ratio between FSR and linewidth:

$$F = \frac{\Delta\lambda}{\delta\lambda} \tag{11.11}$$

11.1.6 Quality Factor

Instead of the linewidth, the quality factor or Q-factor (Q) is often used to describe the quality of an MRR-based device. It is defined as the relation between the linewidth $\delta\lambda$ and the resonance wavelength λ_0:

$$Q = \frac{\lambda_0}{\delta\lambda} \tag{11.12}$$

It is connected to F via the relation

$$Q = \frac{\lambda_0}{\Delta\lambda}F \tag{11.13}$$

11.1.7 Evanescent Field Sensing with MRRs

According to Equation 11.7, the resonance wavelengths depend on the effective index n_{eff} of the waveguide mode, and in turn, n_{eff} depends on the waveguide properties. While the waveguide modes are usually centered at the waveguide core, a significant part of the fields may extend into the surrounding medium (cf. Fig. 11.3). The part of the field extending out of the core is called the evanescent field, and via the evanescent field changes in the surrounding buffer directly influence waveguide properties such as n_{eff}. Most IO sensors focus on determining changes in n_{eff} or its group velocity equivalent n_g to deduce a change in the targeted analyte. MRRs enable to determine a change in n_{eff} very precisely by means of measuring a change in the resonance wavelengths of the MRR.

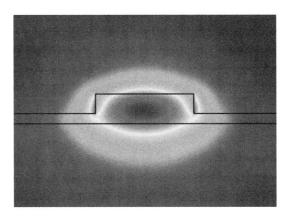

Figure 11.3 False color illustration of the electric field distribution of a waveguide mode. Black lines show the interface between regions of different refractive indices; the high refractive index region forming a waveguide is the ridgelike structure in the middle. One can clearly identify the evanescent part of the electric field leaking into the regions outside the waveguide.

11.1.8 Sensor Performance

The sensor performance depends on all aspects of the measurement system, for example, on peripheral electronics, analyte and analyte delivery, light source, data sampling, data analysis, noise sources, and intrinsic properties of the sensor MRR. A particularly important FOM is the limit of detection (LOD). It refers to the smallest change in an analyte that can be detected.

The MRR-LOD for a quantity q, measured by tracking the corresponding shift in resonance wavelength, is

$$\delta q = \frac{\delta \lambda_0}{S_q} \tag{11.14}$$

Here, $\delta \lambda_0$ is the smallest detectable change in resonance wavelength, in the following called wavelength resolution, and S_q is the sensitivity with respect to q, given by

$$
\begin{aligned}
S_q &= \frac{d\lambda_0}{dq} \\
&= \frac{d\lambda_0}{dn_{\text{eff}}} \frac{dn_{\text{eff}}}{dq} \\
&= \frac{\lambda_0}{n_g} \frac{dn_{\text{eff}}}{dq}
\end{aligned}
\tag{11.15}
$$

The first factor, (λ_0/n_g) measures the MRR's efficiency of converting changes in effective index to changes in resonance wavelength. The second factor, $(\frac{dn_{\text{eff}}}{dq})$, measures the efficiency of the waveguide geometry to convert changes in the quantity q to changes in effective index. It is called waveguide sensitivity and depends strongly on the targeted analyte as well as the cross-sectional properties of the waveguide, that is, width, height, index contrast, asymmetry, and etcetera.

Depending on the system noise and measurement bandwidth, $\delta \lambda_0$ can be larger or even much smaller than the resolution $\delta \lambda_s$ of the measured spectrum. Assuming all noise to be white Gaussian, a suitable definition of $\delta \lambda_0$ is

$$\delta \lambda_0 = \sigma_{\lambda_0} \tag{11.16}$$

Equation 11.16 is a statistical definition of $\delta \lambda_0$ and σ_{λ_0} is the variance of the resulting distribution of resonance wavelengths to be determined from an ensemble of equal measurements.

The influences of noise and resonance parameters on the wavelength resolution – when tracking the resonance wavelengths of Lorentzian-shaped resonance peaks – have been investigated phenomenologically with Monte Carlo simulations (18, 19). A general analytical methodology for calculating LODs in sensing applications, where curve fitting is used to determine the parameters of interest, was given by Hoekstra et al. (20).

The analysis of Hu et al. (4) implies that $\delta \lambda_0$ is proportional to the intensity noise level and the square root of spectral resolution and resonance linewidth. Applying the analytical formalism, presented by Hoekstra et al., it is possible to show, that $\delta \lambda_0$ is also inversely proportional to the extinction parameter.

$$\delta \lambda_0 \sim \frac{\sqrt{\delta \lambda \, \delta \lambda_s}}{\eta} \sigma_I \tag{11.17}$$

Equation 11.17 is a good approximation for high spectral resolution, that is, if $\delta\lambda \gg \delta\lambda_s$, and the Lorentzian approximation of the line shape is valid.

11.2 TECHNOLOGY OF MICRORING RESONATORS

Recent advances in microfabrication technology offer the alternative of manufacturing integrated MRR devices with a high optical quality in a great variety of natural and artificial material platforms. They are manufactured with appropriate processes to cover the requirements of current advanced applications. An excellent control of the fabrication as well as the selection of adequate materials is necessary for achieving a high confinement of light in the waveguide. Thus, the index contrast of waveguides, that is, the refractive index difference of the core and the surrounding media, plays an essential role for the performance of MRRs. There exist two ways for coupling light to the MRR: lateral and vertical coupling configurations. In the first one, the bus waveguide and the MRR are in the same plane. For vertical coupling, the bus waveguide is on top or at the bottom of the MRR. In this section, the lateral coupling MRR technologies and some commonly used material platforms from the state-of-the art will be discussed.

Several configurations for vertical coupling were recently published. Nevertheless, for all those structures the fabrication with standard planar technologies is more complex. Therefore, they are more away from practical application than the in-plane coupled devices.

11.2.1 Silicon-on-Insulator (SOI) Material

The silicon-on-insulator (SOI) platform is the favored technology utilized to design optics and electronics devices. The SOI technology emerged as an alternative platform for developing radiation devices in the military and space industry in the 1960s. An SOI platform consists of a small layer of silicon (Si) located on top of a silicon oxide layer. This buried oxide serves as a dielectric isolation layer to avoid the effects produced by the Si bulk wafer. Passive Si waveguide-based structures, such as MRRs, are promising since they offer multiple features like a high refractive index contrast, low losses at telecommunication wavelengths and compatibility with semiconductor systems. High-contrast waveguides with strong confinement allow the fabrication of small radii in order to enable a compact optical MRR-based device. Here, a brief overview of MRRs fabricated with SOI platforms will be given.

SOI waveguides are formed by a Si core surrounded by a silicon oxide bottom cladding and a top cladding (oxide or air) as illustrated in Figure 11.4. The fabrication of Si waveguides is commonly using an electron beam (e-beam) or UV lithographic process followed by a reactive ion etching (RIE) process. Both processes are well known in CMOS fabrication technology (21). Typically, fabricated Si waveguides have a width between 400 and 500 nm, and a height from 200 to 250 nm. These dimensions provide a monomode condition for one polarization (commonly TE polarization) at telecommunication wavelengths around 1550 nm.

A 5 µm radius MRR based on SOI waveguides was realized for the first time by Little et al. (22) with a Q-factor about of 250 and an FSR of 20 nm at 1550 nm. The resonance, observed as a notch in the transmission spectrum, had a depth of 15 dB. The fabrication method of such structures is based on a 200 nm thick layer of amorphous Si deposited on a 1 µm thick silicon dioxide (SiO_2) buffer layer. The strip waveguides were fabricated in polycrystalline Si via an annealing process. The bus waveguide was inversely tapered to a width of 8 µm width to increase the coupling efficient from the substrate to a lensed fiber.

Dumon et al. (23) proposed a Si racetrack MRR fabricated with deep UV lithography. A similar fabrication process has been developed for the racetrack ring with the same radius and a length of the straight coupling region of 3 µm. The optical losses were reduced significantly resulting in a Q-factor of more than 3000. A scanning electron microscope (SEM) image of a Si racetrack ring is given in Figure 11.5.

MRRs fabricated in thin SOI layers by Baehr-Jones et al. (24) achieved a Q-factor of 45,000. A Q-factor of 57,000 was obtained by adding a polymethyl methacrylate (PMMA) top cladding.

A fabrication method of Si MRRs based on a process of selective thermal oxidation of Si was demonstrated by Luo et al. (25). The process flow is sketched in Figure 11.6. First, a thermal oxide layer of 785 nm on the 500 nm SOI was grown. This layer consumed approximately 360 nm of Si. Later, the pattern was transferred on the thermally grown oxide layer with a RIE using fluorine chemistry, resulting in a thin 50 nm oxide slab. In order to strip the e-beam resist, wet thermal oxidation was used to oxidize the Si and to form the waveguides with a thin slab. Finally, the cladding was deposited with 300 nm high-temperature oxide (HTO). Furthermore, the optical confinement in the waveguides was improved by the deposition of a 1.8 µm thick layer of plasma enhanced chemical vapor deposition (PECVD) SiO_2 layer.

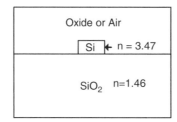

Figure 11.4 Scheme of the cross-section of a silicon waveguide based on SOI platform.

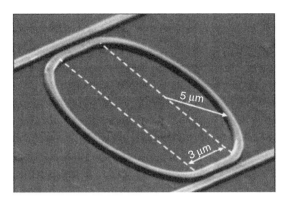

Figure 11.5 Racetrack resonator in SOI. The insolated wire width is 500 nm and the gap width is 230 nm (23).

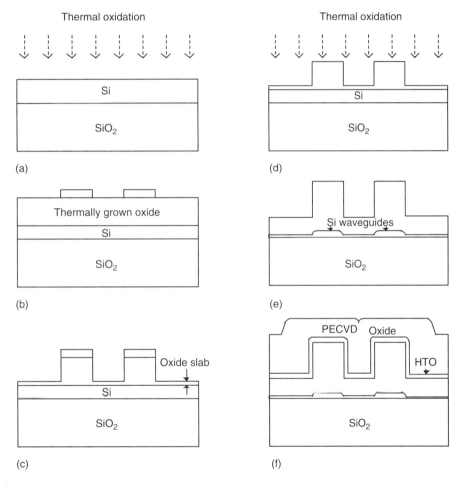

Figure 11.6 Fabrication process flow of "etchless" Si MRRs: (a) thermally grown oxide deposition on the SOI wafer. (b) Pattern transfer of the structures. (c) Etching of the thermally grown oxide, leaving a thin 50 nm oxide slab. (d) Selective wet thermal oxidation of the Si layer. (e) Structural profile of the oxidized Si waveguides. (f) Deposition of passivation layers.

Photographs of a fabricated "etchless" Si MRR and of a section in the coupling region are shown in Figure 11.7. The MRR was formed with an 800 nm wide Si waveguide supporting only the fundamental TE mode. This waveguide has an effective index of 1.6 at 1550 nm. To minimize the coupling loss, a 220 nm wide etchless Si inverse nanotaper was integrated at the input and output waveguide. By applying this fabrication method, a Q-factor of 510,000 was achieved for 50 μm radius rings (Table 11.1).

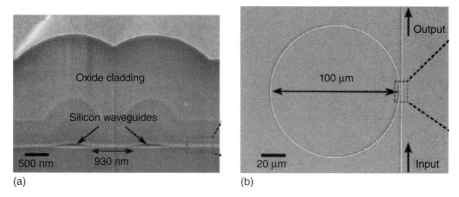

Figure 11.7   (a) Cross-sectional SEM picture of the coupling region. (b) SEM picture of the fabricated 50 μm-radius "etchless" silicon photonic MRR (25).

TABLE 11.1   Overview of Characteristics of the Presented Si-Based MRR Devices

Fabrication Process	Waveguide Dimensions (μm²)	Radius (μm)	Q-Factor/F	References
UV lithography and RIE	0.5 × 0.2	5	250	(22)
DUV lithography and RIE	0.5 × 0.22	5	3,000 (F = 28)	(23)
E-beam direct writing and ICP-RIE	0.5 × 0.12	30	45,000 57,000[b]	(24)
E-beam direct writing and RIE[a]	0.8 × 0.06	50	510,000	(25)

[a] Adding a selective thermal oxidation step.
[b] PMMA clad MRR.

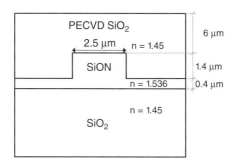

Figure 11.8   Example of a cross-section of a SiON waveguide.

11.2.2   SiN, SiON, and Si_3N_4 Materials

The silicon nitride-based material systems such as silicon mononitride (SiN), silicon oxynitride (SiON), and silicon nitride (Si_3N_4) are attractive optical materials due to their transparency in a broad range of wavelengths from 210 to 2000 nm, their refractive index of around 2.0, and optical losses below 0.2 dB/cm.

Barwicz et al. (26) realized a waveguide circuit including three coupled MRRs based on SiN material with a Q-factor of 30,000 and a FSR of 24 nm. The SiN layer was formed via low pressure chemical vapor deposition (LPCVD) exhibiting a refractive index of 2.2 at 1550 nm.

A SiON double MRR was developed by Melloni et al. with standard PECVD and etching techniques (27). The particular feature of SiON is its flexibility for adjusting the material refractive index by proper mixture of silicon oxide (refractive index: 1.45) and Si_3N_4 (refractive index: 2.0). The monomode SiON waveguide with a ridge shape is sketched in Figure 11.8. The index contrast was about 6% allowing a bending radius of 303 μm with very low radiation losses.

Alternatively, Klunder et al. (28) fabricated an MRR based on stoichiometric Si_3N_4 on SiO_2. By adding a PMMA film as a top cladding to the device, an F value of 61 has been obtained.

A further improvement in the Q-factor of Si_3N_4 MRRs could be achieved by the implementation of a temperature cycling and annealing process. Gondarenko et al. (29) used this technology to fabricate multiple MRRs with radii of approximately 20 μm at losses down to 0.12 dB/cm and intrinsic Q-factors up to 3,000,000.

Figure 11.9 Optical microscope image of the fabricated racetrack MRR with microheater elements.

TABLE 11.2 Overview of Characteristics of the Presented Silicon Nitride-Based MRR Devices

Fabrication Process	Waveguide Dimensions (µm²)	Radius (µm)	Q-factor/F	References
E-beam direct writing and RIE	1.05 × 0.44	7.3	30,000	(26)
UV lithography and RIE	2.5 × 1.8	303	26,000	(27)
UV lithography and RIE	2.5 × 0.3 (bus)	25	$F = 51$	(28)
	1 × 0.3 (MRR)		$F = 61^*$	
			*PMMA clad MRR	
E-beam direct writing and ICP-RIE	2.5 × 0.4	20	3,000,000	(29)
UV lithography and RIE	1.1 × 0.22	98	8,000	(30)

All the structures described so far were passive devices. But most of the applications require some kind of control/ modulation of the optical circumference. The sensor systems group at Fraunhofer Heinrich Hertz Institute (HHI) fabricated racetrack MRRs incorporating microheater elements in order to enable a thermo-optical modulation (30). A 220 nm thick Si_3N_4 layer ($n = 1.91$) was deposited by using ICP-PECVD. Standard lithography was used for defining the patterns from the etch masks. The structures were etched achieving a depth of 80 nm. After platinum electrodes (90 nm Pt on a 10 nm thick adhesion layer of Ti) are defined in a lift-off process, a second layer of Si_3N_4 with a thickness of 95 nm was deposited. The distance between waveguides and 10 µm width metal heaters was 4 µm to guarantee efficient heat transfer and to avoid significant light absorption from the metal. Finally, the electrical contacts were opened by removing the second Si_3N_4 layer. The waveguide had a width of 1.1 µm and supported a fundamental TE mode at 1550 nm wavelength. Racetrack rings had a radius of 98 µm and straight sections of 20 µm with a coupling gap of 0.9 µm. The MRR devices presented later in this chapter were fabricated using the presented process and a slightly modified process, respectively. Figure 11.9 shows the racetrack MRR including the microheater element. A Q-factor of about 8000 was reported with these structures, while a resonance shift of about 0.8 pm/mW heating power was obtained using the microheater elements (Table 11.2).

11.2.3 Polymer Materials

Optical polymers have been studied as an alternative material to semiconductor materials for optical integrated devices because of their capability to be fabricated using a low-cost process with a high manufacturing output. These polymers are attractive due to their particular properties such as low absorption losses, simple processing, high structural flexibility, and compatibility with semiconductor technologies. Polymeric materials have been composed with different combinations of monomers, resulting in a high control of the refractive index as well as a high structural flexibility for special purposes. Additionally, they can be constituted by dopants, such as laser dyes, rare earth ions, electro-optic chromophores, quantum dots, and etcetera. In general, polymers can be classified in different groups: thermoplastics, thermosets, and photopolymers, which can be deposited on any type of substrate. The thermo-optic coefficient of such polymers is 10 times larger than that of silica materials, enabling thermal tuning on a wider spectral range. Thus, they can be exploited for developing active optical devices based on EO and thermal–optical effects. In the following, an overview of the current techniques and polymeric materials for manufacturing MRRs will be presented.

11.2.4 Photoresist-Based and Direct Write Lithography Patterning

The structure of polymeric optical waveguides is composed by two polymers with a high optical transmission and compatibility. An index contrast between both polymers is necessary to achieve the optical confinement. The index contrast between the core and cladding polymers

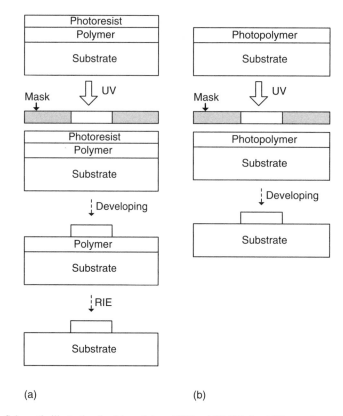

(a)   (b)

Figure 11.10   Schematic illustration for (a) mask-based UV and (b) UV direct lithography patterning processes.

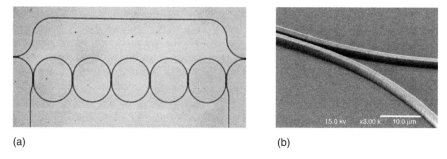

(a)   (b)

Figure 11.11   (a) Optical microscope image of the optical device, and (b) An angle-view SEM image of the coupling section between two MRRs (33).

is lower than the contrast obtained in case of semiconductor materials. However, as previously mentioned the polymer-based waveguides have unique properties. Polymer waveguide dimensions can vary between less than a micrometer and several hundred micrometers, depending on the desired purpose.

The most common technique to fabricate polymeric waveguide devices is photoresist-based lithography patterning followed by RIE, adapted from standard semiconductor processing technology. The optimization of waveguides has been particularly developed in areas of deep, smooth, and vertical sidewall etching to minimize optical losses. Another alternative method is the lithographical patterning by direct writing (UV, laser, and e-beam direct write lithography) that utilizes photosensitive polymers. This technique has the advantage of being maskless, allowing rapid prototyping, as well as the fabrication of patterns with a high grade of complexity and a small size (nanometer scale). Both processes are schematically described in Figure 11.10.

An add-drop polymer MRR was first realized by Haavisto and Pajer (31) using the laser direct writing process. The patterns were made on a layer prepared by the photopolymerization of doped PMMA on a quartz substrate. MRRs with a synthesized polymer from deuterated methacrylate and deuterated fluoromethacrylate were fabricated by Hida et al. (32). They deposited the polymer film by spin coating on a Si substrate. Later, the MRR was patterned by using UV lithography followed by RIE. Palaczi et al. (33) fabricated serially coupled racetrack MRRs using a negative novalac epoxy SU-8 ($n = 1.565$ at 1550 nm), which was coated by spinning on a 5 μm thick thermal silicon oxide layer. A 1.6 μm thick SU-8 layer was directly patterned by e-beam technique. Figure 11.11 shows the manufactured optical device and the coupling section of the racetrack MRR.

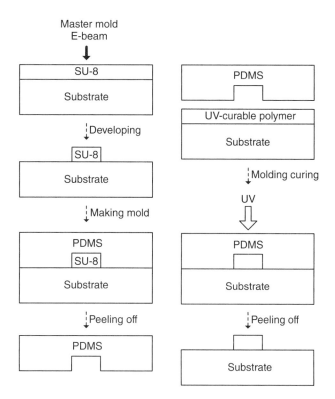

Figure 11.12   The flowchart of the soft lithography technique for MRR fabrication.

The first MRR made of benzocyclobutene (BCB) with a small radius (10 μm) was reported by Chen et al. (34). BCB is a popular polymer: cyclotene 3022-46 whose bulk index is 1.56 at 1550 nm. The undercladding was formed by spin coating a 2.5 μm thick layer of BCB on an oxidized Si substrate. A silicon oxide etch mask was deposited on the cured BCB layer. Then, the pattern was transferred from the photoresistor to the etch mask by lithography followed by an etching process. Active optical MRR-based devices have been fabricated of chromophore-containing polymers. Zhou et al. (35) and Sun et al. (36) reported an EO polymer MRR with photo and e-beam bleaching. High energy photons or electron irradiation can break bonds and decompose the organic chromophores contained in polymers, which can bleach out the color of chromophores, and therefore reduce the refractive index of polymers. Here, the used chromophore was YL124 and an EO coefficient of approximately 200 pm/V at 100 V/μm was obtained.

In order to form a 2 μm thick EO polymer film on a SiO_2 buffer layer, a mixture of PMMA and chromophore dissolved in chlorobenzene was coated on the substrate.

11.2.5   Soft Lithography and Nanoimprint Techniques

Soft lithography is a nonphotolithography technique first realized by Qin et al. (37). This technique uses an elastomeric mold/stamp to shape soft materials by casting the elastomeric material on a master fabricated via conventional or direct write lithography techniques. The molds are usually fabricated using polydimethylsiloxane (PDMS) material, because of its flexibility and poor adhesion with other materials. PDMS molds can be used to replicate the structure devices on a substrate. In general, molding methods permit feature sizes up to 10 nm. Moreover, the properties of elastomeric materials should permit a feature size down to a limit of about 1 nm (38).

SU-8 polymer MRRs was carried out by Huang et al. (39) using a soft lithography technique. Figure 11.12 shows the flowchart of the fabrication processes. The mold is fabricated via e-beam direct write lithography using a SU-8 negative e-beam resist ($n = 1.56$). The SU-8 pattern replication was made on PDMS. For the fabrication of SU-8 MRRs, the PDMS mold was used to transfer the pattern to a SU-8 UV-curable resist deposited on a SiO_2–Si substrate. The replicated MRR was cured under UV light to be solidified.

Poon et al. (40) demonstrated uncladded polystyrene (PS) and cladded SU-8 MRRs as IO filter devices. The PDMS mold fabrication is similar to the previous method. The MRRs are molded on a 3 μm-thick layer of OG-125 ($n \sim 1.45$), coated on top of a Si wafer. Afterwards, a curing process with UV light and baking is performed. For the uncladded PS MRR, 10 μL of PS diluted in toluene is deposited on the cured substrate. After patterning, the substrate was baked. In the case of a cladded SU-8 MRR, the SU-8 is UV cured followed by a deposition of the cladding – an extra layer of OG-125 with a thickness of 3 μm.

The polymer PSQ-L shows a good compatibility with soft lithography processes, and it was used for manufacturing MRRs, as proposed by Teng et al. (41). PSQ-L is an inorganic–organic hybrid polymer with good optical properties and high thermal stability. The waveguide core was made of a high refractive index polymer (PSQ-LH), while the cladding is formed by a low refractive index polymer (PSQ-LL).

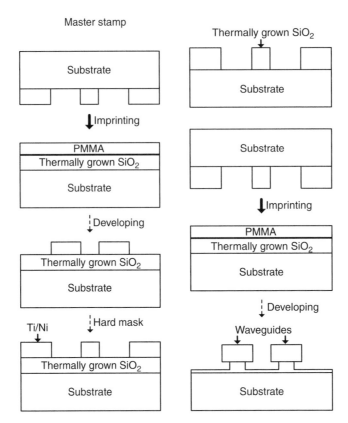

Figure 11.13   Schematic illustration of the fabrication of polymer MRRs using nanoimprint technique.

In the soft lithography process, a step is added to improve the adhesion of the PSQ-LH polymer film on the cladding layer by using oxygen plasma etching.

Moreover, CLD-1 chromophores were utilized for manufacturing a polymer ring and to achieve a wide-range spectral tuning. Poon et al. (42) proposed devices based on a polymer solution of 5.5 wt% of CLD-1/amorphous polycarbonate (APC) in trichloroethylene/dibromomethane. The structures were molded directly on a SiO_2-on-Si substrate using soft lithography replica molding. For an EO device application, a maximum wavelength shift of −8.73 nm was observed.

A further approach to form optical polymer waveguides is the nanoimprint technique. This novel technique relies on a nonconventional lithography method, which uses the deformation properties of polymeric materials, and can give a high quality for the replication of patterns with nanoscale size. The basic principle is based on the replication of structures using a hard stamp. It contains negative surface-relief structures that are imprinted into the polymeric material at controlled pressure and temperature. This process creates a controlled difference in the thickness of the polymeric material transferring the pattern into the polymeric material with a desired depth. In general, a thin residual layer of the polymeric material is left below the stamp protrusions to prevent the direct contact of the hard stamp to the substrate; thereby, it acts as a protective layer for the structures on the stamp surface.

MRRs have been performed via the nanoimprint technique by Chao and Guo (43). Briefly, the imprint process is described in Figure 11.13. As a first step, the hard stamp tool is fabricated using RIE etching and Ti/Ni liftoff processes in a thermal SiO_2 layer grown on a Si substrate, leaving on the stamp surface an inverse pattern of the ring and waveguides with a depth between 1.5–2 μm (Fig. 11.14a). Later, the fabricated stamp is used for imprinting the pattern into a thin PS layer, whose thickness ranges from 200–300 nm. It was deposited by spin coating on a thermal SiO_2–Si substrate. The final waveguide thickness is about 1.5 μm (Fig. 11.14b). After removing the PS material residues, the SiO_2 surface was treated with a hydrofluoric acid (HF) solution to create a pedestal underneath the waveguides. It provides a better optical confinement of the optical mode inside the polymer waveguides (Table 11.3).

11.3   SENSOR ARRAYS FOR MULTIPLEXED DETECTION/MULTIPARAMETER ANALYSIS

This section wants to give an overview of recent approaches in the field of MRR sensor arrays without claiming to be exhaustive. Nevertheless, the presented array designs and results are actually the most application-relevant approaches for an implementation of an MRR-based on-chip multiparameter analytics in laboratory analytics and diagnostics. It should be mentioned, that up to now such multiplexing approaches for sensing were only realized in Si or Si_3N_4 material systems.

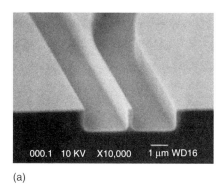 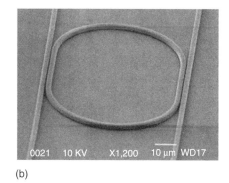

(a)   (b)

Figure 11.14 SEM images of (a) a used stamp to fabricate the PS MRR and (b) an imprinted PS racetrack MRR (43).

TABLE 11.3 Overview of Characteristics of the Presented Polymer-Based MRR Devices

Fabrication Process	Waveguide Dimensions (μm^2)	Radius (μm)	Q-factor/F	References
325 nm laser direct writing	10×1.25	4.5 cm	$F = 16$	(31)
UV lithography and RIE	7×6.4	15,900	$F = 14.8$	(32)
E-beam direct writing	2×1.6	100		(33)
UV lithography and RIE	0.9×2.5	10	13,000 ($F = 192$)	(34)
Photo- and e-beam bleaching	5×2	500	9,450	(35, 36)
Soft lithography	2×2	100	"Low"	(39)
Soft lithography	3×3	204	10,000* *PS MRR 7,100* *SU8 MRR	(40)
Soft lithography	3×2	400	42,000	(41)
Soft lithography	1.4×1.6	207	26,000 ($F = 9.3$)	(42)
Nanoimprint	1.5×2	200	300	(43)

The most straightforward approach for the realization of an MRR array was presented by University of Illinois, USA (44). A scheme of these MRR arrays is presented in Figure 11.15. Every MRR is addressed with an individual bus waveguide, which is equipped with two grating couplers for light coupling. Moving mirrors are used to scan with a laser beam and a detector in parallel along the grating couplers to sequentially read out the MRR sensors. The single MRR sensors have diameters of 30 µm and Q-factors of 43,000 and are realized by using SOI photonic wire waveguides. The MRR arrays consist of 32 individual elements and are attached to a fluidic cell made of Mylar material. The MRR arrays were characterized using salt solution with different concentrations. Bulk refractive index sensitivity (BRIS) of 163 nm/RIU and a bulk refractive index LOD of $8 \cdot 10^{-7}$ has been reported (44).

Furthermore, real multiparameter measurements were performed, and multiplexing of MRR signals during a bioanalytic measurement was demonstrated. In a first step, different MRR of the array were functionalized to enable an individual MRR to detect a specific biomolecule. All measurements were conducted in PBS buffer solutions spiked with the particular analytes. For multiplexed detection of five protein biomarkers (PSA, IL-8, AFP, CEA, TNF-alpha) the MRR arrays were functionalized using specific antibodies (45). In addition, eight nonfunctionalized reference MRRs were used to monitor environmental influences and temperature effects, respectively. Direct assays with different analyte concentrations were conducted and finally a LOD of 2 ng/ml was reported.

Parallel detection of four micro-RNA (mRNA) was demonstrated by performing a two-step assay (46). In a first step, mRNA-specific complementary DNA (cDNA) strands were attached to the particular MRR surface. Then, mRNA-cDNA-specific antibodies were given onto the MRR array, only binding to the designated combination of mRNA and cDNA. With this experiment, an LOD of 350 attomoles was found for mRNA detection.

In a further experiment, multiparameter detection was demonstrated by measuring four single-stranded DNA sequences (47). The specific recognition of the particular DNA strand was performed via specific cDNA immobilized on the MRR surface. A single nucleotide sequence detection was demonstrated by discriminating single nucleotide polymorphisms during isothermal monitoring of dissociation kinetics. Here, an LOD of 195 femtomoles was achieved for the single-stranded DNA.

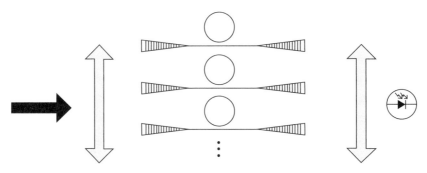

Figure 11.15 Scheme of the MRR array developed at the University of Illinois. Every MRR is addressed with an individual bus waveguide that is equipped with two grating couplers for light coupling. A laser beam (represented by the black arrow) and a detector (represented by the diode symbol) are scanned along the grating couplers to sequentially read out the MRR sensors (indicated with blue arrows).

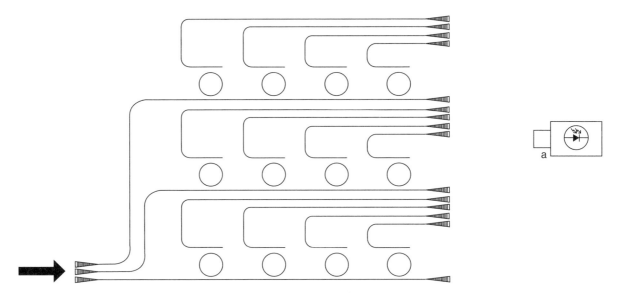

Figure 11.16 Scheme of the MRR array presented by Ghent University. A collimated laser beam (represented by the black arrow) lights three input grating couplers. Light is guided to arrays of four MRR. From every MRR light is coupled out via the drop port and guided to individual output grating couplers. Detection is realized using an infrared camera (represented by the camera symbol).

Similar to the previous MRR array design is the approach presented by Ghent University, Belgium (48). A scheme of their MRR array design is shown in Figure 11.16. A collimated laser beam with a diameter of approximately 2 mm lights three input grating couplers, each connected to a bus waveguide. Light is guided via the bus waveguides to arrays of four MRR. From every MRR light is coupled out via the drop port and guided to individual output grating couplers. At the detection side, the output signals are recorded using an infrared camera for parallel readout of the sensor signals. The single MRR sensors have diameters of 10 μm and Q-factors of 21,000. Photonic wire waveguides made of an SOI material system form the waveguide circuit. The MRR arrays consist of 12 elements and are attached to a fluidic cell made of PDMS material (48).

Parallel detection of two proteins (anti-human IgG, anti-HSA) was demonstrated. Individual MRR were functionalized by spotting complementary proteins to the sensor surface. Measurements were performed using a PBS buffer solution containing the target proteins as well as an approximately three magnitudes of order higher amount of nonspecific proteins. Nevertheless, both antibodies could be detected with a LOD of 3 ng/ml.

In another experiment, a bulk refractive index measurement was demonstrated using a digital fluidic cell (49). Single droplets of different concentrations of aqueous solutions containing sodium chloride (NaCl), glucose, and ethanol, respectively, were guided to the MRRs. A BRIS of 77 nm/RIU as well as an LOD of approximately 5×10^{-6} was reported.

The Royal Institute of Technology (KTH), Sweden, designed and fabricated an MRR array structure using a multimode interference (MMI) coupler as depicted in Figure 11.17 (50). Laser light is coupled via grating couplers to a waveguide feeding an MMI coupler. This distributes the light to seven bus waveguides, each feeding a single MRR sensor. The bus waveguides end at the cleaved edges of the optical chip. A detector array is attached to the edge leading to a parallel recording of transmitted light. The MRR structures are fabricated using slot

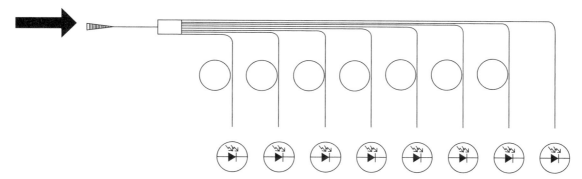

Figure 11.17   Scheme of the MRR array designed by the Royal Institute of Technology. Laser light (represented by the black arrow) is coupled via grating couplers to a waveguide leading into a MMI coupler. Here, light is distributed to seven bus waveguides feeding single MRR sensors. The bus waveguides end at the cleaved edges of the chip where a detector array (represented by diode symbols) records the transmitted light.

Figure 11.18   Scheme of the MRR array fabricated at National Research Council. Light from a laser source (represented by the black arrow) is coupled via an inverse taper mode size converter into a bus waveguide. An array of five MRR with increasing ring diameter is coupled to the bus waveguide. With a second inverse taper mode size converter, the light is coupled out of the bus waveguide and recorded (represented by the diode symbol).

waveguides in an Si_3N_4 material system (51). The diameter of an MRR is about $140\,\mu m$ exhibiting a Q-factor of 2000. The optical chip is hybrid integrated into a fluidic cartridge made of PMMA material with a sealing made of PDMS material (50).

Here, experiments for BRIS characterization of the MRR array were conducted using aqueous solutions of ethanol and methanol, respectively. A BRIS of 246 nm/RIU as well as an LOD of $5 \cdot 10^{-6}$ was demonstrated. Furthermore, surface-sensing properties were investigated by monitoring binding of anti-BSA in PBS buffer to chemical-activated MRR surface. The LOD for surface mass detection was found to be $0.9\,pg/mm^2$.

A more compact MRR array was developed by the National Research Council (NRC), Canada (51). The number of optical coupling sites is reduced to one for light feeding and one for optical readout. Light from a laser source is coupled via an inverse taper mode size converter (52) into a bus waveguide. An array of five MRRs with increasing ring diameter (from 20 to $28\,\mu m$) is coupled to the bus waveguide (Fig. 11.18). The variation of the MRR diameter leads to a variation of the FSR providing the possibility to identify the resonance peaks of a particular MRR in the optical spectrum. With a second inverse taper mode size converter, the light is coupled out of the bus waveguide and recorded with a detector. The MRRs have Q-factors of about 25,000 and are made of photonic wire waveguides in a SOI material system. The optical chip is coated with a thick SU-8 layer that is structured to create fluidic channels. A quartz top cover is used to close the channels.

Multiplexed protein detection was demonstrated with this MRR array using goat IgG and anti-rabbit IgG in a PBS buffer. The particular MRRs were functionalized using anti-goat IgG and rabbit IgG, respectively. As a result, the LOD for surface mass coverage detection is determined to be $0.3\,pg/mm^2$ (53) (Table 11.4).

11.4   MULTIPLEXING BY MODULATION OF INDIVIDUAL MRRS

The four presented approaches for the realization of an MRR-based IO sensor array use different ways of combining passive IO components, such as grating couplers or directional couplers, with a number of MRR sensor elements. In every case, the main goal is to simplify the device in order to reduce the effort of optical coupling and therefore, to lower the main barrier for commercializing an MRR-based sensor system. Nevertheless, all these approaches lack of the potential to integrate MRR sensor arrays into mobile and cost-effective systems.

In the case of the MRR array introduced by the University of Illinois, where each MRR has its own pair of grating couplers for feeding light and detection of the sensor signal, the sensor performance can be still described with Equation 11.17. However, that approach is not very cost-effective, since the number grating coupler pairs directly scales with the number of sensor MRR, resulting in a highly complex system for the readout of arrays with larger numbers of sensor elements.

The approaches introduced by Ghent University and KTH, respectively, also have a fundamental limitation for involving larger MRR arrays: If the light of a single source is split in order to feed N different MRRs, then the maximum light intensity available for each MRR is

TABLE 11.4 Overview of the Presented MRR Array Approaches from the Literature and the Demonstrated Sensor Properties

MRR Device; Material	BRIS, nm/RIU	LOD, RIU	Protein; LOD	References
32 MRR, individual readout; SOI	163	8×10^{-7}		(44)
32 MRR, individual readout; SOI			PSA, IL-8, AFP, CEA, TNF-alpha; 2 ng/ml	(45)
32 MRR, individual readout; SOI			Four mRNA; 350 amol	(46)
32 MRR, individual readout; SOI			Four single-stranded DNA; 195 fmol	(47)
3 × 4 MRR, individual readout of drop ports; SOI			Anti-human IgG, anti-HAS; 3 ng/ml	(48)
3 × 4 MRR, individual readout of drop ports; SOI	77	5×10^{-6}		(49)
7 MRR addressed by MMI coupler, individual readout; SiN slot	246	5×10^{-6}		(50)
5 MRR with increasing radius; SOI			Goat IgG, anti-rabbit IgG; 0.3 pg/mm²	(51)

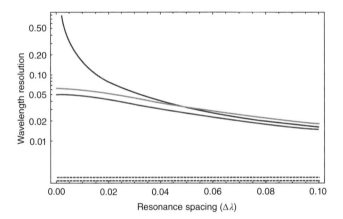

Figure 11.19 Calculated LOD with respect to the difference in resonance wavelength for two otherwise equal MRRs (solid blue line, equal MRR; solid purple and yellow line, different MRR with $\eta_1 = 1$, $\eta_2 = 0.9$; dashed blue line, single MRR $\eta_1 = 1$; dashed purple line, single MRR $\eta_2 = 0.9$).

at best $1/N$ of the source's output power. The signal-to-noise ratio (SNR) is likely to scale as $1/N$ as well. Besides, there are still N different photodetectors required.

This disadvantage can be circumvented by coupling the N MRRs to a single bus waveguide, as introduced with the device from NRC. In this case, the transmission function is simply the product of the N MRRs' single transmission functions. Here, at least N different resonance frequencies have to be determined from the combined spectrum.

But even in this case the number of elements in an MRR array is limited due to the spectral overlap between individual resonances, for example, as a result of low fabrication tolerances and high F MRRs. One may still expect the LOD to be similar to the case of a single MRR. However, if the individual MRRs' transmission spectra start to overlap, for example, due to a different response to a certain analyte, LOD will increase and become a much more complicated function of resonance parameters.

For an appropriate discussion of this problem, we will illustrate the increase of LOD due to spectral overlap by the case of two MRRs coupled to the same bus waveguide (54). Figure 11.19 shows the calculated LOD with respect to the difference in resonance wavelength for two otherwise equal MRRs with $\eta_1 = \eta_2 = 1, \delta\lambda_1 = \delta\lambda_2 = 0.1$ nm and two MRRs with different extinction parameters ($\eta_1 = 1$, $\eta_2 = 0.9$) at $\lambda_0 = 1550$ nm.

In the case of equally shaped resonances, LOD diverges with decreasing resonance spacing. For differently shaped resonances, the LOD is different for both resonances but settles for finite values. The real fitting results do not diverge at zero resonance spacing. However, a detailed look at the resulting resonance wavelength distribution at this point reveals that it becomes highly irregular, that is, the Gaussian intensity noise does not anymore translate to a Gaussian distribution of resonance wavelengths.

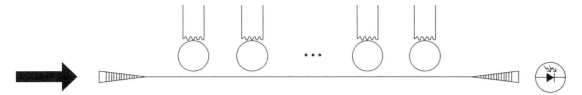

Figure 11.20 Scheme of the MRR array developed at Fraunhofer HHI. Light from a laser source (represented by the black arrow) is coupled via a grating coupler to a bus waveguide. The MRR array is coupled to the bus waveguide. The MRRs are equipped with microheater elements for modulation of the light traveling through the resonator. A second grating coupler is used to couple the light out of the bus waveguide and record the optical spectrum (represented by the diode symbol).

Nevertheless, even in the case of strongly overlapping resonances it is still possible to deduce the respective resonance wavelengths by applying appropriate fitting procedures. This is, provided that the measured spectra differ from their theoretical predictions solely due to the influence of white Gaussian intensity noise. However, other parasitic effects may influence the quality and shape of the measured spectra. Fabry–Perot oscillations between the waveguide end facets or grating couplers, other oscillatory behavior caused by mechanical oscillations or nonwhite Gaussian noise, and gradients due to the limited bandwidth of the fiber-to-chip coupling efficiency will distort the spectra and make the analysis, especially of spectra with overlapping resonances, much more complicated. Furthermore, with a larger number of coupled resonators there is usually no trivial possibility to associate the individual resonances with their respective transducer MRRs.

As result of this detailed analysis of the influence of the spectral overlap of MRR resonances, it becomes obvious that the number of MRRs in one array is rigorously limited. Even if an MRR array is well designed, as in the case of the presented device from NRC, unavoidable fabrication tolerances will lead to inaccuracies in the determination of spectral positions of resonances. Moreover, overlapping of MRR resonances during the measurement process limits the dynamic range of the MRR sensor elements.

In order to solve this problem, the sensor systems group at HHI developed an approach based on frequency modulated MRRs presented by Lützow et al. (30).

Here, every MRR is provided with a structure to periodically modulate the respective resonance wavelengths around an equilibrium value. The transmission signal for a set of discrete wavelengths is then sampled for a finite time interval and isolated at the respective modulation frequencies either using lock-in detection or using Fourier transformation to. A scheme of the MRR array is illustrated in Figure 11.20.

The recorded signal at a particular modulation frequency is proportional to the modulation amplitude impressed on the optical signal, and depends on the shape of the modulated MRR's own spectrum. The signal is large near the points of inflection of the resonance dip and vanishes at the minimum. The vanishing of the modulation amplitude allows the tracking of resonance frequencies even in dense optical spectra.

In the following, it is shown using perturbation theory that for small modulation amplitudes the modulated signal is proportional to the derivative of the nonmodulated line shape with respect to wavelength.

In Figure 11.21, the blue and red curves depict resonance dips of two independent MRRs having similar resonance frequencies. Coupling both to the same bus waveguide results in the shaded transmission spectrum due to superposition and tracking of individual resonance frequencies becomes unfeasible.

The dashed green curve in Figure 11.21 shows the lock-in signal $|F_{\text{Lock-lin}}(\lambda)|$ of the shaded spectrum $f(t, \lambda)$, where $|F_{\text{Lock-lin}}(\lambda)|$ is idealized as

$$F_{\text{Lock-lin}}(\lambda) = \frac{2}{T}\int_0^T dt \; \sin(\omega t) f(t, \lambda) \tag{11.18}$$

With the assumption that the MRR exhibiting the red spectrum is modulated by sinusoidal changing its n_{eff} with a frequency $\omega/(2\pi)$. The resonance spacing and modulation amplitude are approximately 36% and 0.7% of $\Delta\lambda$, respectively.

The combined spectrum $F^n(\lambda)$ of a group of n MRRs coupled to the same bus waveguide is the product of the spectra of the individual MRR

$$F^n(\lambda) = f(n_{\text{eff}}^{R_1}) \cdot \; \dots \; \cdot f(n_{\text{eff}}^{R_n}) \tag{11.19}$$

All waveguides are supposed to be monomode with effective index $n_{\text{eff}}^{R_1}$ to $n_{\text{eff}}^{R_n}$ for MRR 1 to n, respectively. A modulated MRR's individual resonance spectrum can be described as

$$f_{\text{mod}}(n_{\text{eff}}, \lambda) = f_{\text{mod}}(n_{\text{eff}} + \Delta n_{\text{eff}}, \lambda) \tag{11.20}$$

Assuming that the modulation induces only a small change in effective refractive index Δn_{eff}, the first two terms of a Taylor series expansion with respect to n_{eff} represent a good approximation to $f_{\text{mod}}(n_{\text{eff}}, \lambda)$:

$$f_{\text{mod}}(n_{\text{eff}}, \lambda) = f(n_{\text{eff}}, \lambda) + \frac{d}{dn_{\text{eff}}}f(n_{\text{eff}}, \lambda) \cdot \Delta n_{\text{eff}} + O(n_{\text{eff}}^2) \tag{11.21}$$

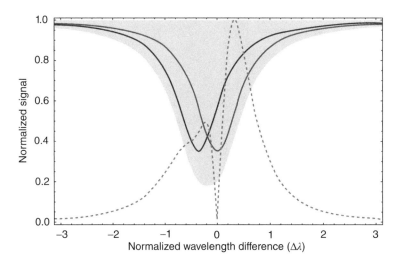

Figure 11.21 Simulated resonance dips of two MRRs with similar resonance frequencies (red and blue lines). The shaded transmission spectrum occurs, if the two MRRs are coupled to the same bus waveguide and the resonance spectra overlap. The dashed green line represents the calculated and normalized recorded signal for the MRR with the red resonance profile being modulated. Signals are plotted against the wavelength difference with respect to the resonance frequency of the modulated MRR.

With Equation 11.21, the combined spectrum with MRR R_i modulated becomes

$$F^n_{\text{mod}:R_i}(\lambda) = F^n(\lambda) + f(n^{R_1}_{\text{eff}}, \lambda) \cdot \ldots \cdot \frac{d}{dn^{R_i}_{\text{eff}}} f(n^{R_i}_{\text{eff}}, \lambda) \cdot \Delta n_{\text{eff}} \cdot \ldots \cdot f(n^{R_n}_{\text{eff}}) + O(n_{\text{eff}}^2) \tag{11.22}$$

Assuming that the modulation Δn_{eff} being a harmonic function of time, for example, $\Delta n_{\text{eff}} = \varepsilon \cos(\omega t)$ and considering that the spectrum of a MRR depends on n_{eff} only via the optical phase, spectral filtering at a frequency $\omega/(2\pi)$ (e.g., lock-in detection) gives

$$F^n_{\text{Lock-in}:R_i}(\lambda) = \varepsilon \cdot \left(-\frac{\lambda}{n_{\text{eff}}}\right) \cdot f(n^{R_1}_{\text{eff}}, \lambda) \cdot \ldots \cdot \frac{d}{d\lambda} f(n^{R_i}_{\text{eff}}, \lambda) \cdot \ldots \cdot f(n^{R_n}_{\text{eff}}) + O(n_{\text{eff}}^3) \tag{11.23}$$

$F^n_{\text{Lock-in}:R_i}(\lambda)$ is proportional to the derivative of $f(n^{R_i}_{\text{eff}}, \lambda)$ with respect to λ.

$F^n_{\text{Lock-in}:R_i}(\lambda)$ vanishes on resonance since $\frac{d}{d\lambda} f(n^{R_i}_{\text{eff}}, \lambda)$ does. In addition, $F^n_{\text{Lock-in}:R_i}(\lambda)$ changes sign on resonance of MRR i. Therefore, by taking amplitude zero crossing and phase change into account, the correct position of the resonance frequency can be identified even if the optical signal reaches zero elsewhere (55).

With Equation 11.23, it becomes obvious that the modulation of the effective refractive index of an individual MRR within an array together with an appropriate frequency filtering leads to a sensor signal, which can be analyzed independently from the resonances of other MRRs in the array. The modulation can be realized by applying a thermo-optical or EO modulation of the refractive index of the waveguide forming the MRR. The sensor elements may be modulated sequentially with the same frequency or in parallel using a distinct modulation frequency for each MRR. In the latter scheme, the distance between particular modulation frequencies has to be chosen larger than the inverse of the measurement time in order to be able to accurately resolve each spectral contribution.

In the following section, an example for the implementation of the presented modulation scheme using a SiN material system is presented.

11.5 EXPERIMENTAL DEMONSTRATION OF THE MODULATION SCHEME

The MRR arrays presented here for experimental verification of the modulation scheme are composed of either 4 or 12 MRRs, respectively. In both cases, the MRRs are coupled to a single bus waveguide. All MRRs are equipped with individual microheater electrodes for thermo-optic modulation. Figure 11.22 shows fabricated test structures in three different levels of magnification. A part of a fabricated wafer can be seen in the background of the figure. The single chips have a size of 1 cm². The left inset shows a microscope image of an exemplary MRR with metal electrodes for thermal modulation. The radius of the MRR is about 100 μm. In the SEM image in the right inset, a detailed view on waveguide and metal strip is given.

Light emitted from a tunable laser source was launched into the bus waveguides with a lensed optical fiber (LF). An in-line polarization controller was used to ensure TE-polarized optical output from the LF. The chip was stabilized in temperature during characterization. At the output side of the chip, light was collected with a high numerical aperture microscope objective and detected with a photodetector. Figure 11.23a shows the measured transmission spectrum of an array of four MRRs. The wavelength scan from 1553.3 to 1555.5 nm covers

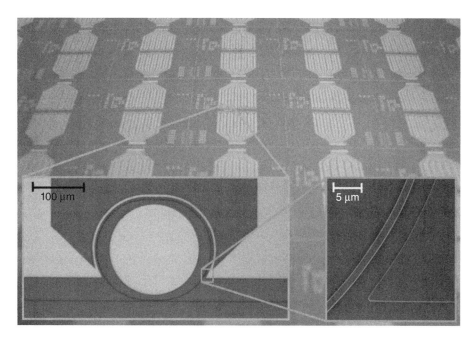

Figure 11.22   Fabricated test structures. Background: 4'' wafer with arrays of 12 MRR. Left inset: optical microscope image of an exemplary MRR. The waveguides are seen as thin lines surrounded by metal structures. Right inset: SEM image giving a more detailed view of a part of the MRR waveguide and the metal heater, respectively.

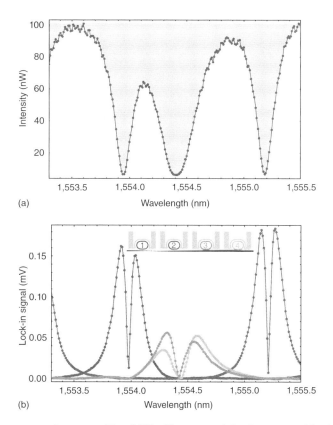

Figure 11.23   (a) Direct transmission spectrum of an array of four MRRs. The measured data is represented by the dots and lines are guides to the eye. (b) Optical spectra for the same MRRs subsequently being modulated. The measured data is represented by the colored dots. Lines are guides to the eye. The inset is a schematic of the MRR array with heating electrodes (yellow). The different colors of the MRRs refer to the colors of the respective lock-in traces.

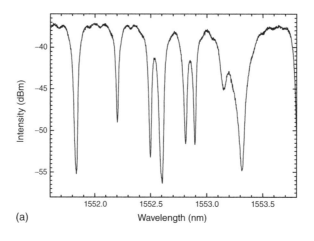

(a)

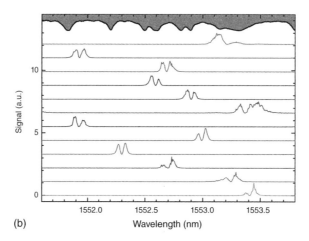

(b)

Figure 11.24   (a) Transmission spectrum of an array of 12 MRRs. (b) Recorded signals for parallel modulation and readout of the same MRR array. The different colors of the modulated MRR spectra refer to the different frequencies of the respective lock-in traces. For comparison, the nonmodulated spectrum is shown with the gray-highlighted black curve.

about 108% of the FSR. The scan resolution is 10 pm. Only three dips can be identified and the resonances are randomly distributed despite of their nominally identical layout. The spectra of two out of the four MRRs show strong overlap and cannot be separated in the direct transmission spectrum. For the two outermost resonances in the spectrum, moderate Q-factors of around 8000 are estimated.

In order to isolate the contributions from the individual MRR to the combined optical spectrum in Figure 11.23a, we applied the modulation scheme as discussed above. The optical signal was detected using a photodetector connected to a lock-in amplifier. Heating electrodes of the MRR were electrically contacted using manual probe heads and the electrical signal was supplied by a function generator. Subsequently, MRR one to four (cf. Fig. 11.23b) were modulated with a frequency of 6 kHz. More details regarding this experiment can be found elsewhere (55). A real zero crossing cannot be observed due to noise and the limited resolution of the wavelength scan, respectively. The recorded signals from MRR three and four are smaller by a factor of about 3–4 compared to the signals from MRR one and two. Due to the strong overlap between their resonance profiles, the lock-in signals from MRR three and four are partly suppressed by the near presence of the respective other MRR.

Then, we went one step further and demonstrated the parallel modulation of 12 MRRs coupled to a single bus waveguide using standard software-based Fourier analysis for readout.

Figure 11.24a shows the measured transmission spectrum of an array built up of 12 MRR. The wavelength scan from 1551.6 to 1553.8 nm covers about 110% of the FSR. The scan resolution was 5 pm. The resonances are randomly distributed and only eight dips can be identified although all the 12 MRRs exhibit a nominally identical layout. Hence, the spectra of 8 out of the 12 MRRs show strong overlap and cannot be separated.

For a fast and reliable analysis of the spectrum, the contributions of the individual MRR were isolated by applying the modulation scheme parallel to each of the MRRs adopting a software-based readout using Fourier analysis as described in the following.

The microheater elements of the MRRs were electrically contacted simultaneously using multicontact probe heads. A signal source with 12 independent outputs was used to generate the electrical modulation signals. From each output, a square wave signal with a different frequency and a stable average power of approximately 8 mW was applied to the heaters. Furthermore, the data acquisition system was triggered using the signal source to record phase-correct spectra of the MRR resonances. For each wavelength step in the spectrum, the recorded signal is frequency filtered by Fourier analysis. An individual spectrum was generated for each MRR by monitoring the absolute values of the Fourier components for each modulation frequency. The acquired traces are shown in Figure 11.24b. Dips in the signals of the modulated MRR mark the positions of resonances and the contributions of all the 12 individual MRRs can be clearly distinguished. For comparison also, the direct transmission spectrum highlighted in gray is shown. The spectra of the modulated MRR are slightly redshifted due to the average heating of the MRR caused by the applied modulation power.

11.5.1   Basic Experiments for Sensor Demonstration

For the characterization of the BRIS of modulated MRR measurements with varying concentrations of salt and glucose, respectively, were performed. Light of a tunable laser source was launched into the bus waveguide using an optical fiber and grating couplers (13) as described before. Light was collected with a photodetector and analyzed using a computer-controlled data acquisition system. Heating electrodes were electrically contacted using manual probe heads and the electrical signals were supplied by a microcontroller-based function generator. A PDMS flow cell was mounted on top of the MRR chip.

Figure 11.25a shows the response of an MRR to changes in the NaCl concentration of the solution that was flowing over the MRR. The concentration increases in 0.18% steps from 0% up to 0.9%. Since we did not apply any curve fitting, the spread of data points for constant concentration values is mostly determined by pressure variations in the fluidic system. The measured wavelength shift from DI water up to 0.9% NaCl is approximately 235 pm. We used the approximation presented in (56) to calculate the bulk refractive index of the different NaCl concentrations.

Figure 11.25b shows the response of a modulated MRR to changes in the glucose concentration of the solution that was passing by the MRR. The spread of data points for constant value of glucose concentration is consistent with the scan resolution of 10 pm since we do not apply any curve fitting algorithm at this point to improve on the resolution. However, there is a slow drift of the resonance frequencies over the measurement time that we address to changes in the microfluidic environment due to glucose deposition. The measured wavelength shifts from DI water to 1.25% glucose and from DI water to 2.5% glucose are approximately 260 pm and 540 pm, respectively. We used an effective medium approximation (57) to calculate the bulk refractive index of the respective glucose solutions.

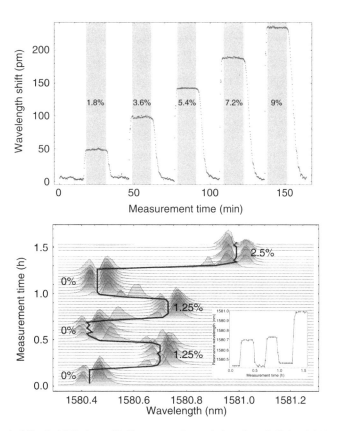

Figure 11.25   (a) Resonance wavelength shift of a MRR due to NaCl-concentration variations from 0.18% to 0.9% of an aqueous solution flowing over the MRR. (b) Response of a modulated MRR to changes in glucose concentration of an aqueous solution flowing over the MRR. The respective concentration values are displayed in the figure. The inset shows solely the extracted resonance wavelengths over the measurement time.

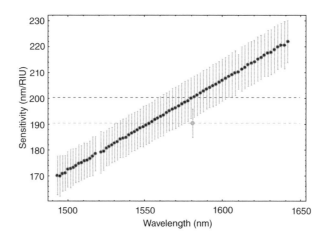

Figure 11.26 BRIS of a measured MRR over a larger wavelength range from 1490 to 1650 nm (unmodulated measurement). The green dot represents the sensitivity value of the modulated MRR. The error bars indicate twice the standard error obtained from the respective line fits.

Figure 11.26 shows that BRIS increases with increasing wavelength due to the larger evanescent field and lower group index at higher wavelengths. The BRIS value of the modulated MRR is represented by the green dot. This value of 190 nm/RIU at a wavelength of about 1581 nm differs by only 5% from the value of 200 nm/RIU obtained for the nonmodulated MRR at that wavelength that corresponds to a significant overlap of the confidence intervals. The error bars indicate twice the standard error obtained from the respective line fits.

11.6 SYSTEM INTEGRATION AND ECONOMICAL SOLUTIONS FOR OPTICAL SOURCES

Optical MRR elements are one of the most selective and manageable spectral filters to be exploited for sensor applications. They allow measuring changes in the refractive index of the surrounding medium ultimately down to $\Delta n = 10^{-8}$ by identification of the shift of the resonance peaks with accuracy down to picometer and beneath. Consequently, this challenging target has an impact on the spectral features of the light source. MRR applications require proper single-mode optical sources in order to scan MRR circuits at high spectral purity and at adequate coherence for the particular Q-factor, which in turn defines the overall optical length of interference path.

Thanks to the periodicity of the MRR resonances, it is sufficient to spread the tuning range of the light source slightly beyond the FSR of the sensing elements as the induced shift of the resonance, and not the position of the resonance in the wavelength domain, is the FOM.

Exemplarily, some impressions are given on solutions to fulfill this prerequisite without laying claim on covering everything.

11.6.1 Tuning Single-Mode Lasers

The use of single-mode laser devices would be a straightforward choice. A large amount of products at high performance are available for communication applications and are adequate for the operation of MRR circuits as well. These laser products cover the spectral range for communication technologies between 1200 and 1650 nm. Fiber-pigtailed DFB- and DBR-laser modules have DIL or Butterfly housings including an optical isolator for the suppression of reflections (typically at least up to −30 dB) to avoid mode jumping and additional noise up to loss of coherence. Furthermore, they are equipped with a temperature sensor and a Peltier element for temperature management, which is indispensable in order to guarantee the requested spectral performance. The linewidths of the components are typically in the range of 10 MHz (coherence length ≈10 m), which is adequate even for MRR applications with large Q-factor. Usually, spectral scanning is performed via variation of the injection current, which results in an increase of the emission intensity and an increase of the emission wavelength due to effective index reduction with increasing carrier injection. Thus, the use of much more expensive multisection tunable lasers with a tuning range considerably larger than 3 nm will not be necessary whenever the FSR lies in that range. All of these laser components have the advantage of large optical output power at the continuously tunable emission wavelength (typically well beyond 10 mW). The large output power offers to feed even a set of sensor circuits in parallel. Hence, the cost of a laser module will be shared by a set of sensor circuits.

Nevertheless, for certain sensor applications, this high-performance solution might be not acceptable from economic point of view.

11.6.2 Multimode Laser Sources in Combination with MRR Filter

The application of a Fabry–Perot laser (FPL) module having a multimode spectrum might be a less expensive alternative to a tunable single-mode laser module at the expense of less performance and the necessity to implement an additional MRR element operating as a mode selective filter. Nevertheless, an FPL will also require a temperature control for stable operation without mode hopping. Figure 11.27

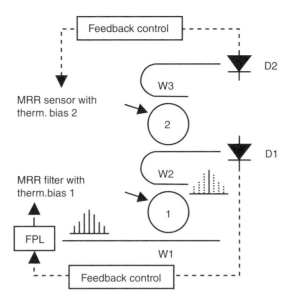

Figure 11.27   Example of an optical sensor circuit using a Fabry–Perot laser source (FPL).

gives an example for a circuit based on an FPL coupled into a single-mode waveguide (W1), which is coupled to the well-passivated MRR filter. The resonance wavelength of the MRR filter can be tuned continuously by an integrated heater at least within the bandwidth of the FSR of the FPL and MRR sensor in order to match the filter characteristic for the proper selection of one FPL mode. The waveguide, W2, coupled at the throughput port of the MRR filter is feeding a photodetector D1, which is generating the feedback signal for driving the FPL and the temperature-controlled MRR filter. Waveguide, W2, is also the input waveguide for the MRR sensor element. The shift of the resonance can be determined via the throughput port of the MRR sensor feeding a photodetector, D2. Finally, the corresponding feedback loop generates the control signal identifying the amount of induced resonance shift. Parasitic environmental influences (mechanical stress, temperature, etc.) might be avoided as these impacts are affecting both MRR elements of the circuit in parallel.

11.6.3   Broadband Sources in Combination with MRR Filter

Even less expensive than the use of a laser module would be the implementation of a super luminescence diodes (SLED). However, its use will be restricted to relatively simple MRR circuits with less performance due to their low spectral power and less coherence feature. Actually, SLED single-mode fiber-pigtailed modules are available with a 3 dB bandwidth of 50 nm for the 1550 nm wavelength window (mainly used for optical coherence tomography = OCT). The coherence length can be calculated to be in the range of 17 µm, which is beneficial for OCT application but not for MRRs. Although the output power of a fiber-pigtailed SLED module can reach 20 mW, the spectral power is only in the range of −33 dBm/pm[1]. Hence, the use of such a light source will be restricted to very small MRR elements (due to the short coherence length) with moderate resolution (due to restrictions on the quality factor) and photo diodes with low dark current have to be used in order to exploit the large contrast of the MRR elements. However, this approach might be acceptable for low-cost application of less performance, for example, by using ultracompact MRR elements with radii of a few micrometers (fabricated in SOI material) and moderate Q-factor in order to design the effective interference length well below 0.1 mm.

An example of such a circuit is sketched in Figure 11.28. The SLED is feeding a passivated MRR filter via the input waveguide W1. This first MRR generates a comb like spectrum and the overall intensity of the spectrum is detected by a photo diode, D1, at the throughput port of the MRR filter (waveguide W2). Again, the comb spectrum can be tuned to the requirements of the MRR sensor by adjusting the temperature of the MRR filter and monitoring the overall intensity at D1. The detector signal will show an incremental variation when n or $n+1$ resonances will be matched by the SLED/passband filter configuration. Waveguide W2 is feeding the MRR sensor "2." In turn, the sensor signal is recorded via waveguide W3 and photo diode D2 at the throughput port of the MRR sensor by proper shift of the bias current ("amount of tuning") to maximum output of D2 detecting the throughput intensity of a single resonance.

11.6.4   Heterodyne Detection Setup Exploiting Benefits of Electronic Signal Processing

There is another option to benefit from the potential of high-frequency electronic signal processing with its mature technology, high-volume fabrication, and its advantages to realize high-order spectral filters for the gigascale frequency domain. These advantages have been

---

[1] Data extracted from www.thorlabs.com

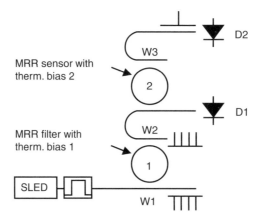

Figure 11.28 Example of an optical sensor circuit using a superluminescent light-emitting diode (SLED), followed by an optical passband filter, D1, D2: photo diodes.

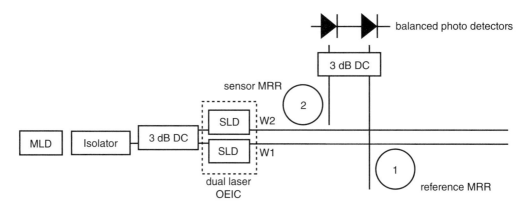

Figure 11.29 Example of an optical sensor circuit using a master laser diode that is optically isolated from an injection-locked slave laser diode (SLD) source configuration (3 dB DC: 3 dB directional coupler).

demonstrated impressively in optical heterodyne receivers as part of a coherent optical communication system (58). Here, we give an example on how to exploit coherent detection of heterodyned signals from an MRR sensor circuitry (cf. Fig. 11.29).

For such a coherent readout, a single-mode optical reference laser is used, which fits to a selected resonance of an MRR acting as a reference MRR (MRR 1). This laser is the master laser diode (MLD) for a set of two single-mode laser diodes, operating as slave laser diodes (SLDs) via injection locking. Obviously, the implementation of an optical isolator is indispensable for proper master/slave operation. In order to be able to handle temperature drifts of the SLDs in a practical manner, the SLDs have to be integrated closely on a monolithic chip (OEIC) affecting similar temperature environment of both devices.

Injection locking of this SLD by the MLD results in minimum and, to each other, synchronized phase noise/timing jitter of the SLDs and, consequently, results in optimum performance of the two coherently phase-locked laser signals. The SLD can be detuned relative to the MLD for example, up to 40 GHz @ 1550 nm wavelength without losing mode locking (59). This is a prerequisite for generating a heterodyned optical beat signal by coupling both signals in a directional coupler. For this, the lasers are feeding a passivated reference MRR 1 and a sensor MRR 2, respectively. MRR 1, the MLD module, and the SLD-OEIC are stabilized by temperature control, and a further algorithm will be used to adjust these wavelength signals to a selected resonance of MRR 1 (the feedback loops for managing these issues via control electronics are not sketched in Figure 11.29). Finally, MRR 2 and the SLDs might be tuned in such a way that detuning will be less than 40 GHz (≤ 0.32 nm @ 1550 nm center wavelength) in order to guarantee proper mode locking and not to run into challenges for the high-frequency photo diodes and the subsequent signal processing electronics. The signals of the throughput ports of both MRRs are heterodyned in a 3 dB directional coupler (both signals must have the same state of polarization) and are feeding a balanced photodetector, that is, phase noise of the lasers are reduced intrinsically. A shift of the resonances of the functionalized sensor MRR 2 will result in a variation of the beat signal of the interfering signals, which is analyzed easily and even more precise in the electrical domain as electronic filters with optimum slope performance can be realized and implanted in the electronic front end. Obviously, the range of detuning, which defines the generated optical intermediate frequency, must be controlled to the maximum bandwidth of the electronics and maximum shift of the SLD, respectively, which will be feasible by integration of an algorithm for adjustment of the MRR resonance.

11.7 PERSPECTIVES

The research and development on IO circuits (IOC) for signal processing in the domain of optical communication resulted in matured technologies and products. IO sensor devices are benefiting from more than two decades of this research, and they have become key elements in sensor systems for diverse sensing application fields such as biomedical analytics and detection of hazardous materials for safety and security purposes. The follow-up cycle of the age of information and telecommunication is labeled "postinformational technological revolution." This cycle is already running and focuses on life science including prosperous implementation of all kinds of sensors to make life more safe and comfortable. Furthermore, sensors will also improve fabrication techniques and operation of complex technical equipment by continuous process monitoring and control via a multiple of sensor elements.

During the past phase of research and development in optical communication technologies, researchers and engineers identified, for example, what is realistic and feasible for matching the tight, rigorous, and demanding specifications for optical network solutions. In optical communications, it is indispensable to keep very precise the tight spectral windows of the DWDM network with an accuracy of 1 GHz. Obviously, it is impossible to fabricate MRR-IOCs for such an application omitting tedious calibration and control of the effective optical circumference.

However, for sensing applications in general it is not necessary to match such absolute spectral frequencies. Here, the FOM is the potential of measurements of minor variations of spectral features induced by parameters of the sensor's environment to be monitored. Therefore, innovative IOC solutions are feasible and can be elaborated. For the verification of innovative sensing concepts, MRR-IOCs are among the most favorite devices for a lot of application scenarios.

As sketched out in this chapter, IOCs based on MRRs can provide a sensitive, specific, cost-effective and practicable technology platform. MRR features such as high sensitivity at low probe volumes (down picoliters to be handled corresponding to attograms to be deposited for the detection of distinct molecules) enable engineering of economical and high-performance devices. For many applications, high specificity is required which can be achieved by analyzing especially a set of differently functionalized MRR surfaces. Furthermore, the high degree of miniaturization enables economical high-volume fabrication, coming along with ongoing smart hybrid integration with microfluidic cells as well as control and signal processing electronics. The challenges and main goals are to elaborate a smart, practical, portable or handheld LOC microsystem, for example, point-of-care and safety and security applications that can be operated by unskilled people, even in (sub)tropical climate. These will encompass both analysis systems with disposable sensor elements, for example, to guarantee a sterilized environment, and sensor elements with a regeneration functionality, for example, in case of ad hoc use or equipment-implemented sensors for process monitoring and environmental monitoring, respectively.

We are just at the beginning of a new period, where OIC-based sensors will play a key role in discovering new applications and markets, respectively. In our opinion, MRR-based IOCs will become more and more important, and both the described technological approaches will be pursued depending on the requirements of the end user: (i) the one targeting high-contrast waveguide circuits for miniaturized MRRs down to a few μm radii and with large evanescent field parts based on Si photonic nanowires and (ii) the one using medium-contrast waveguide circuits for moderate MRR circumferences featuring longer interaction lengths based on, for example, SiN or polymer material. In both cases, the development can benefit from the availability of matured single-mode telecom laser sources as the readout is not restricting the application at all.

Especially for multiparameter analysis, MRR sensor arrays will be further developed in the near future following the modulation approach presented in this chapter. The identification of the individual and differently functionalized elements will be improved by implementation of EO modulation instead of thermal modulation (as described in this chapter). On the one hand, this will reduce the electric power consumption, and on the other hand, it will drastically increase the available modulation frequencies. For the Si photonic wires, a main challenge here will be to implement pn junctions without losing the performance on the MRR Q-factor. For the polymer material system, the production of reliable material with large and reproducible electro-optical coefficient is not definitely solved up to now.

In conclusion, there is a huge economic potential waiting to be covered by innovations based on MRR-operated sensor arrays as presented in this chapter.

REFERENCES

1. Passaro VMN, Troia B, La Notte M, De Leonardis F. Chemical sensors based on photonic structures. In: Wang W, editor. *Advances in Chemical Sensors.* Rijeka, Croatia: InTech; 2012.

2. Jokerst NM, Luan L, Palit S, Royal M, Dhar S, Brooke MA, Tyler T. Progress in chip-scale photonic sensing. IEEE Trans Biomed Circuits Syst 2009;3:202–211.

3. Passaro VMN, Dell'Olio F, Casamassima B, Leonardis FD. Guided-wave optical biosensors. Sensors 2007;7:508–536.

4. Armani AM, Kulkarni RP, Fraser SE, Flagan RC, Vahala KJ. Label-free, single-molecule detection with optical microcavities. Science 2007;317:783–787.

5. Wu J, Gu M. Microfluidic sensing: state of the art fabrication and detection techniques. J Biomed Opt 2011;16:080901.

6. Hunt HK, Armani AM. Label-free biological and chemical sensors. Nanoscale 2010;2:1544–1599.

7. Chao C-Y, Guo LJ. Biochemical sensors based on polymer microrings with sharp asymmetrical resonance. Appl Phys Lett 2003;83:1527–1529.

8. Chao C-Y, Guo LJ. Design and optimization of microring resonators in biochemical sensing applications. J Lightwave Technol 2006;24:1395–1402.

9. Passaro VMN, Dell'Olio F, Leonardis FD. Ammonia optical sensing by microring resonators. Sensors 2007;7:2741–2749.

10. Orghici R, Lützow P, Burgmeier J, Koch J, Heidrich H, Schade W, Welschoff N, Waldvogel S. A microring resonator sensor for sensitive detection of 1,3,5-trinitrotoluene (TNT). Sensors 2010;10:6788–6795.

11. Nitkowski A, Chen L, Lipson M. Cavity-enhanced on-chip absorption spectroscopy using microring resonators. Opt Express 2008;16:11930–11936.

12. Roelkens G, Van Thourhout D, Baets R. High efficiency silicon-on-insulator grating coupler based on a poly-silicon overlay. Opt Express 2006;14:11622–11630.

13. Maire G, Vivien L, Sattler G, Kazmierczak A, Sanchez B, Gylfason KB, Griol A, Marris-Morini D, Cassan E, Giannone D, Sohlström H, Hill D. High efficiency silicon nitride surface grating couplers. Opt Express 2008;16:328–333.

14. Janasek D, Franzke J, Manz A. Sealing and the design of miniaturized chemical-analysis systems. Nature 2006;442:374–380.

15. Chremmos I, Schwelb O, Uzunoglu N, editors. *Photonic Microresonator Research and Application*. New York: Springer; 2010.

16. Yariv A. Universal relations for coupling of optical power between microresonators and dielectric waveguides. Electron Lett 2000;36:321–322.

17. Heebner J, Grover R, Ibrahim TA, editors. *Optical Microresonators: Theory, Fabrication and Applications*. London: Springer; 2008.

18. Hu J, Sun X, Agarwal A, Kimerling LC. Design guidelines for optical resonator biochemical sensors. J Opt Soc Am B 2009;26:1032–1041.

19. White IM, Fan X. On the performance quantification of resonant refractive index sensors. Opt Express 2008;16:1020–1028.

20. Hoekstra HJWM, Lambeck PV, Uranus HP, Koster TM. Relation between noise and resolution in integrated optical refractometric sensing. Sens Actuators B 2008;134:702–710.

21. Bogaerts W, Baets R, Dumon P, Wiaux V, Beckx S, Taillaert D, Luyssaert B, Van Campenhout J, Bienstman P, Van Thourhout D. Nanophotonic waveguides in silicon-on-insulator fabricated with CMOS technology. J Lightwave Technol 2005;23:401–412.

22. Little BE, Foresi JS, Steinmeyer G, Thoen ER, Chu ST, Haus HA, Ippen EP, Kimerling LC, Greene W. Ultra-compact Si-SiO_2 microring resonator optical channel dropping filters. IEEE Photonics Technol Lett 1998;10:549–551.

23. Dumon P, Bogaerts W, Wiaux V, Wouters J, Beckx S, Van Campenhout J, Taillaert D, Luyssaert B, Bienstman P, Van Thourhout D, Baets R. Low-loss SOI photonic wires and ring resonators fabricated with deep UV lithography. IEEE Photonics Technol Lett 2004;16:1328–1330.

24. Baehr-Jones T, Hochberg M, Walker C, Scherer A. High-Q ring resonators in thin silicon-on-insulator. Appl Phys Lett 2004;85:3346–3347.

25. Luo L-W, Wiederhecker GS, Cardenas J, Poitras C, Lipson M. High quality factor etchless silicon photonic ring resonators. Opt Express 2011;19:6284–6289.

26. Barwicz T, Popovic MA, Rakich PT, Watts MR, Haus HA, Ippen EP, Smith HI. Microring-resonator-based add-drop filters in SiN: fabrication and analysis. Opt Express 2004;12:1437–1442.

27. Melloni A, Costa R, Monguzzi P, Martinelli M. Ring-resonator filters in silicon oxynitride technology for dense wavelength-division multiplexing systems. Opt Lett 2003;28:1567–1569.

28. Klunder DJW, Krioukov E, Tan FS, Van Der Veen T, Bulthuis HF, Sengo G, Otto C, Hoekstra HJWM, Driessen A. Vertically and laterally waveguide-coupled cylindrical microresonators in Si_3N_4 on SiO_2 technology. Appl Phys B 2001;73:603–608.

29. Gondarenko A, Levy JS, Lipson M. High confinement micron-scale silicon nitride high Q ring resonator. Opt Express 2009;17:11366–11370.

30. Lützow P, Pergande D, Heidrich H. Integrated optical sensor platform for multiparameter bio-chemical analysis. Opt Express 2011;19:13277–13284.

31. Haavisto J, Pajer GA. Resonance effects in low-loss ring waveguides. Opt Express 1980;5:510–512.

32. Hida Y, Imamura S, Izama T. Ring resonator composed of low loss polymer waveguides at 1.3 µm. Electron Lett 1992;28:1314–1316.

33. Palaczi GT, Huang Y, Yariv A. Polymeric Mach-Zehnder interferometer using serially coupled microring resonators. Opt Express 2003;11:2666–2671.

34. Chen WY, Grover R, Ibrahim TA, Van V, Herman WH, Ho PT. High-finesse laterally coupled single-mode benzocyclobutene microring resonators. IEEE Photonics Technol Lett 2004;16:470–472.

35. Zhou J, Pyayt A, Dalton LR, Lou J, Jen AKY, Chen A. Photobleaching fabrication of microring resonator in a chromophore-containing polymer. IEEE Photonics Technol Lett 2006;18:2221–2223.

36. Sun H, Chen A, Olbricht BC, Davies JA, Sullivan PA. Microring resonators fabricated by electron beam bleaching of chromophore doped polymers. Appl Phys Lett 2008;92:193305.

37. Qin D, Xia Y, Whitesides GM. Soft lithography for micro-nanoscale patterning. Nat Protoc 2010;5:493–502.

38. Quake SR, Scherer A. From micro- to nanofabrication with soft materials. Science 2000;290:1536–1540.

39. Huang Y, Paloczi GT, Scheuer J, Yariv A. Soft lithography replication of polymeric microring optical resonators. Opt Express 2003;11:2452–2458.

40. Poon JKS, Huang Y, Paloczi GT, Yariv A. Soft lithography replica molding of critically coupled polymer microring resonators. IEEE Photonics Technol Lett 2004;16:2496–2498.

41. Teng J, Scheerlinck S, Zhang H, Jian X, Morthier G, Beats R, Han X, Zhao M. A PSQ-L polymer microring resonator fabricated by a simple UV-Based soft-lithography process. IEEE Photonics Technol Lett 2009;21:1323–1325.

42. Poon JKS, Huang Y, Paloczi GT, Yariv A. Wide-range tuning of polymer microring resonators by the photobleaching of CLD-1 chromophores. Opt Lett 2004;29:2548–2586.

43. Chao C, Guo LJ. Polymer microring resonators fabricated by nanoimprint technique. J Vac Sci Technol B 2002;22:2862–2866.

44. Iqbal M, Gleeson MA, Spaugh B, Tybor F, Gunn WG, Hochberg M, Baehr-Jones T, Bailey RC, Gunn LC. Label-free biosensor arrays based on silicon ring resonators and high-speed optical scanning instrumentation. J Sel Top Quantum Electron 2010;16:654–661.

45. Washburn AL, Luchansky MS, Bowman AL, Bailey RC. Quantitative label-free detection of five protein biomarkers using multiplexed arrays of silicon photonic microring resonators. Anal Chem 2010;82:69–72.

46. Qavi AJ, Kindt JT, Gleeson MA, Bailey RC. Anti-DNA: RNA antibodies and silicon photonic microring resonators. Anal Chem 2011;83:5949–5956.

47. Qavi AJ, Mysz TM, Bailey RC. Isothermal discrimination of single-nucleotide polymorphisms via real-time kinetic desorption and label-free detection of DNA using silicon photonic microring resonator arrays. Anal Chem 2011;83:6827–6833.

48. De Vos K, Girones J, Claes T, De Koninck Y, Popelka S, Schacht E, Baets R, Bienstman P. Multiplexed antibody detection with an array of silicon-on-insulator microring resonators. IEEE Photonics J 2009;1:225–235.

49. Arce CL, Witters D, Puers R, Lammertyn J, Bienstman P. Silicon photonic sensors incorporated in a digital microfluidic system. Anal Bioanal Chem 2012. DOI: 10.1007/s00216-012-6319-6.

50. Carlborg CF, Gylfason KB, Kazmierczak A, Dortu F, Banuls Polo MJ, Maquieira Catala A, Kresbach GM, Sohlström H, Moh T, Vivien L, Popplewell J, Ronan G, Barrios CA, Stemmea G, van der Wijngaarta W. A packaged optical slot-waveguide ring resonator sensor array for multiplex label-free assays in labs-on-chips. Lab Chip 2010;10:281–290.

51. Barrios CA, Gylfason KB, Sánchez B, Griol A, Sohlström H, Holgado M, Casquel R. Slot-waveguide biochemical sensor. Opt Lett 2007;32:3080–3082.

52. Xu D-X, Densmore A, Delâge A, Waldron P, McKinnon R, Janz S, Lapointe J, Lopinski G, Mischki T, Post E, Cheben P, Schmid JH. Folded cavity SOI microring sensors for high sensitivity and real time measurement of biomolecular binding. Opt Express 2008;16:15137–15148.

53. Xu D-X, Vachon M, Densmore A, Ma R, Delâge A, Janz S, Lapointe J, Li Y, Lopinski G, Zhang D, Liu QY, Cheben P, Schmid JH. Label-free biosensor array based on silicon-on-insulator ring resonators addressed using a WDM approach. Opt Lett 2010;35:2771–2773.

54. Lützow P., et al. to be published.

55. Lützow P, Pergande D, Gausa D, Huscher S, Heidrich H. Microring resonator arrays for multiparameter biochemical analysis. Proc SPIE 2012;8427:84270E–84271E.

56. Quan X, Fry ES. Empirical equation for the index of refraction of seawater. Appl Opt 1995;34:3477–3480.

57. Lirtsman V, Golosovsky M, Davidov D. Infrared surface plasmon resonance technique for biological studies. J Appl Phys 2008;103:014702.

58. Nakazawa M, Kikuchi K, Miyazaki T, editors. *High spectral Density Optical Communication Technologies*. Berlin: Springer; 2010.

59. Bag B, Das A, Lu H-H, Patra AS. Injection-locked DFB laser diode in main and multiple side modes. Int J Soft Comput Eng 2011;1:4–8.

12

POLYMERS, NANOMATERIALS, AND ORGANIC PHOTOVOLTAIC DEVICES

THOMAS TROMHOLT AND FREDERIK C. KREBS

Department of Energy Conversion and Storage, Technical University of Denmark, Roskilde, Denmark

Organic photovoltaics (OPV) is a promising field, which may become one of the solutions to the challenge of moving energy production from fossil fuels to renewable energy technologies. The field has been subject to extensive research, and practical devices using OPV have been demonstrated on a large scale (1). This chapter describes the field of OPV with a focus on the materials and cell processing. A general introduction to solar cell technology and OPV is given. This is followed by an extensive discussion of the different materials applied within OPV and how they affect the overall cell performance. Additionally, the processing of OPV is described and the approaches to large-scale production, where the ambitious perspective is a production capacity of 1 GW_{peak} per day based on only abundant elements.

In 2009, the worldwide energy consumption was 12,186 million tons of oil equivalent (Mtoe). With continuous industrialization, growth in human welfare, and growth in the human population, it is clear that energy demands will only increase in the future. Consequently, from 2009 to 2010, the worldwide energy consumption increased by 5.4%. Energy is normally obtained from the combustion of fossil fuels such as coal, gas, and oil, for which there is a limited supply. At present, fossil sources of energy are being harvested at a high pace implying that in terms of present-day reserve-to-production ratio, the planet is only left with 126 years of coal, 60 years of natural gas, and 54 years of oil (2). Consequently, there is an urgent need to develop technologies that will be able to contribute to energy production and eventually take over when fossil fuel reserves are depleted. In 2009, the share of renewable energy originating from renewable energy sources was 16% of the global final energy consumption (excluding 2.8% from nuclear power) (2). Two of the currently well-exploited renewable energy sources, hydropower and wind power, contributed with 3.4% and 0.67%, respectively. The contribution from photovoltaics was rather low, since this technology has not been extensively exploited. However, a strong growth of installed capacity has been observed in recent years, where a growth of 74% from 2009 to 2010 was observed.

Photovoltaics utilize the fact that every hour, the earth receives more energy from the sun than mankind consumes in a year. Outside the atmosphere of the earth, the power density of the sunlight is 1366 W/m^2, while after passing the atmosphere impinging on the earth the power density is approximately 1000 W/m^2. Consequently, large amounts of energy are readily available to be converted into electrical energy by photovoltaics. With energy conversion efficiencies well below 100%, still only a fraction of the area of the earth needs to be covered by solar cells in order to cover the worldwide energy consumption (3).

The first observation of the photovoltaic effect, that is, generation of voltage or current in response to illumination, was made by the young physicist Becquerel in 1839 (3). However, no practical applications of this effect were developed until the arrival of the bulk semiconductors in the early 1950s. The first practical solar cell was reported by Fuller et al. based on a p–n junction of silicon where an efficiency of 6% was achieved (4). With this invention, several different applications were envisaged, and the first satellite to be powered by a solar cell was launched in 1958. Extensive development of the silicon solar cell was made with the aim of providing low-cost and stable energy generation. The silicon solar cell, referred to as a first-generation solar cell, has remained the main element for commercial solar cells ever since and efficiencies have now reached 25% (5). While the crystallinity of the silicon in different cells ranges from amorphous to monocrystalline, the latter performs better. In combination with the comparatively high efficiencies, the stability of the first-generation cell is intrinsically high since only a single-doped semiconductor makes up the solar cell. However, the production of this type of solar cell involves high temperature thermal oxidations as well as large amounts of crystalline silicon. Silicon wafers of 200–250 µm thickness are used for first-generation cells and account for around half of the cost of the total cell (6).

To overcome the problems related to material use, thin-film solar cells have been developed, where different materials are deposited onto substrates by sputtering, to obtain film thicknesses below 10 µm. The very thin films allow for creation of flexible modules if the substrate is flexible. The advantage of this is twofold: applications using flexible PV modules can easily be envisaged on curved surfaces.

Nanomaterials, Polymers and Devices: Materials Functionalization and Device Fabrication, First Edition. Edited by Eric S. W. Kong.
© 2015 John Wiley & Sons, Inc. Published 2015 by John Wiley & Sons, Inc.

Furthermore, production of the modules can be performed by roll-to-roll (R2R) processing, where a roll of substrate is rolled through different deposition steps and rerolled, which introduces tremendous advantages in production capacity and module price (7). Thin-film PV, referred to as second-generation PV, does not only use silicon as sputtering targets, but also materials such as cadmium telluride (CdTe) and copper–indium–gallium–selenide (CIGS) are used, since these materials exhibit higher light absorption than silicon. Efficiencies of thin films have reached 10.5% (silicon), 16.7% (CdTe), and 19.6% (CIGS), and thus, performance-wise, second-generation cells perform rather well (5). However, issues pertaining to reproducibility keep actual panel performance significantly lower implying that the market share in installed second-generation PV capacity is only 13% and first-generation cells are thus still dominant (2).

In recent years, different alternative paths have been taken to develop technologies that may offer a lower cost per energy, normally referred to as $/W_{peak}$, where the price is indicated at maximum light intensities at 1000 W/m^2. Both inorganic and organic technologies have been developed, which are referred to as third-generation PV. Sophisticated cell structures have been applied to second-generation cells, where, for example, multijunction cells comprising in excess of 15 different layers have been realized. With multiple junctions, different parts of the light spectrum can be harvested by the different junctions improving the overall efficiency, where the current record is 34.1% (5). Another novel technology is the dye-sensitized solar cells (DSSCs), which rely on absorption of light by a dye in an electrolytic liquid. While the first demonstration of a functional cell in 1991 by Grätzel showed an efficiency of 7.9%, the technology has only reached 11.0%. Additionally, the cells are inherently unstable due to leaking of the volatile solvent, and consequently, DSSC is currently not mature for commercialization. A third novel PV technology is the organic solar cell. With the development of semiconducting polymers dating back to 1977 by Heeger et al. (8, 9), a wide range of applications were envisaged, for example, polymer-based PV. The main advantage is that physical parameters can be tailored by synthesis of specific chemical structures. Additionally, solvent solubility can be obtained, which allows for processing of entire solar cells from solution. Consequently, extensive research efforts have been directed at developing polymer solar cells (PSCs) to a technology of commercial interest.

12.1   POLYMER SOLAR CELLS

The first demonstration of OPV was presented in 1986 by Tang who demonstrated a donor–acceptor solar cell based on evaporation of the small molecule copper phthalocyanine where power conversion efficiency (PCE) of 1% was achieved. This demonstrated the potential of OPV, and strong scientific focus was directed at this field. With the development of semiconducting polymers, polymers were envisaged as an alternative light-absorbing material for organic PV, as these could be solution processed and the properties tailored to suit specific situations. Photoinduced charge transfer was observed from conjugated polymers to fullerenes in 1992 by Sariciftci by which the donor–acceptor approach became applicable to polymers (10). In 1993, the first PSC comprising a conjugated polymer, poly[2-methoxy,5-(2′-ethylhexyloxy)-1,4-phenylenevinylene] (MEH-PPV), and C$_{60}$ was presented with a PCE of 0.04% (11). The device was a bilayer cell where the active layer of the cell consisted of a layer of polymer and a layer of C$_{60}$ deposited on top. Due to the nonpolar nature of C$_{60}$, solution processing was complicated due to low solubility, and functionalizations of fullerenes were presented by Hummelen in 1995 by which solution processing was dramatically eased (12). The development of the bulk heterojunction (BHJ) in 1995 of the active layer where the donor and the acceptor are solution processed together to form a bicontinuous network created a dramatic increase in performance to 2.9% (13). Development of alternative polymers with better absorption properties resulted, for example, in the polymer poly(3-hexylthiophene) that would become a reference within PSC for the next decade due to improved stability relative to MEH-PPV, which eased practical work significantly and provided higher efficiencies of 3.5% due to higher hole mobility (14). Further work was directed at optimizing the BHJ to allow for better charge extraction efficiency, primarily by thermal annealing, whereby efficiencies were increased to 5% (15). Recent development in polymers has included low-bandgap materials, which allow for absorbing a larger part of the solar spectrum (16). This approach proved successful as efficiencies increased rapidly from 3.2% in 2006 (17), 5.1% in 2008 (18), 7.7% in 2009, and 8.5% in 2011 (19). Additionally, state-of-the-art PSCs have not only approached but may even have exceeded the 10% barrier (5) by combining different approaches such as improved polymers and tandem structures (20).

PSC as a technology has several advantages compared to existing PV technologies. Polymer synthesis is a well-established scientific field, which allows for synthesis of a large variety of materials with different chemical moieties. Hereby, polymers are highly customizable and can be developed with different aims, such as high mobility, light absorption, high/low bandgap, and good processability. Most PV technologies rely on the use of materials based on rare elements of which there is only limited natural abundance on the earth. Consequently, a full upscaling of a given technology will prove complicated as alternative solutions to these rare elements have to be identified. Only silicon PV and PSCs are based on highly abundant elements, where the source for PSCs may be different kinds of carbon compounds as in the case of commercial plastic production. Toxicity is another issue that a PV technology such as CdTe cells is associated with, which is not an inherent issue to PSCs. Finally, the flexibility and low weight opens up for both R2R processing as well as novel application of PSCs.

Conjugated polymers are ideal for organic electronics and especially PV as they combine the properties of high optical absorptivity and hole mobility. Consequently, only layer thicknesses in the range of 100 nm are needed to achieve almost complete UV–vis light absorption, and consequently, material use is minimal. Additionally, PSCs can be processed entirely from solution as the materials can be designed with solubility in common solvents. Hereby, the need for energy consuming high temperature or vacuum steps is alleviated, which implies a highly reduced imbedded energy in each produced cell (21). Furthermore, due to the thin layers applied in PSCs, the entire cell comes flexible. Hereby, processing can be made on a R2R basis as it is also the case for thin-film solar cells on flexible substrates. As the cells can be processed during a continuous process, processing and handling is eased. Additionally, with the solution processability of PSCs, processing with conventional printing and coating techniques is possible by which very high production capacity can be achieved, where the material waste and the processing time per cell is kept at a minimum. Another beneficial side effect of the thin cells is the low weight. By this, an entire new range of applications of PSCs can be envisaged where curved surfaces may be covered, and the need for mechanical support is heavily reduced compared to conventional silicon PV panels.

While PSCs are associated with a multitude of beneficial properties in the competition with existing energy technologies, the technology has still not matured enough for a large-scale commercialization. Before such a level of maturity is needed, it has been argued that, on a module level, 10% efficiency and 10-year stability must be reached, referred to as the *10–10 goals* (22). Nevertheless, demonstration projects have been carried out where a larger number of cells have been produced and incorporated into low-power devices such as flashlights with a battery as an energy buffer (1, 23). These projects have shown that niche products can be produced with present technologies that are competitive with existing technologies. To reach the 10–10 goals, research has been focused on several different parts of the PSC. The highest efficiencies obtained have been for small, laboratory-scale cells, where glass slides with the transparent conductor indium tin oxide (ITO) have served as the substrate. The rigidity of the substrate provides a good basis for deposition of layers of high homogeneity and controlled thickness. Consequently, the highest performing PSCs with an efficiency of 8.3% have been demonstrated for tandem cells where two stacked serially connected cells increase the cell performance (20). However, when the lab-scale processes are moved to large-scale facilities and flexible substrates, the processing parameters have to be optimized, and generally lower performance is obtained on a module level, where the current record of 5.2% is held by the Japanese company Sumitomo, however, for organic small molecule solar cells (5). Consequently, higher lab-scale cell efficiencies must be reached before the module level can fulfill the 10–10 goals.

Another issue regarding the commercial potential of PSCs is the intrinsic instability of the cell components. During operation of the cell, a multitude of degradation mechanisms work in parallel to degrade the cell performance. The interfaces between the organic layers start to degrade as diffusion between the interfaces takes place. Additionally, when the light-absorbing polymer is exposed to light in the presence of oxygen, chemical reactions occur, which degrade the material and the overall cell performance. Consequently, PSC stability is presently limited to days, months, or years depending on the type of polymer, the surrounding organic layers, the substrate type, and the quality of possible encapsulation (24). However, with the increasing interest in cell stability observed in recent years, cell stability is expected to improve significantly in the coming years.

12.1.1   Principle of Operation

PSCs operation is governed by a number of steps starting with an incoming photon and ending with the extraction of an electron and a hole at the respective electrodes. All of these steps have to be optimized as the overall cell performance is governed by the rate-limiting step. A full description of the physics behind the operation of PSC is highly complex and is not within the scope of this chapter, therefore only a qualitative description is given.

The core of the PSC is the active layer, which is composed of a highly light-absorbing polymer and a so-called electron acceptor material. As light impinges on the conjugated polymer, the photons are absorbed and electrons are excited from the highest occupied molecular orbital (HOMO) to the lowest unoccupied molecular orbital (LUMO) level (Fig. 12.1a). The energy gap of this transition depends on the type of polymer, but is typically around 2 eV. Electrons with energies exceeding the bandgap are excited to higher levels if the transition is allowed but will rapidly decay to the LUMO level by thermalization. As the electron is excited, a vacancy in the HOMO level is created, which effectively acts as a particle of positive elementary charge, and is thus referred to as a hole. Two main forces work in the charge pair: Coulomb attraction and phonon collisions. In the case of an inorganic semiconductor such as Si, the dielectric constant is high ($\varepsilon_0 \approx 12$), and consequently, the Coulomb attraction falls off rapidly with the charge separation. Phonons interact with the charge pair and try to separate the charges. At room temperature, the collisions with phonons correspond to the thermal energy of 25 meV. For an electron–hole pair in Si, the binding energy is in the range of 0.01 eV, and consequently, the Coulomb well (Coulomb attraction) can be overcome and the charges separated (25). However, for semiconducting polymers, the dielectric constant is lower ($\approx$2–4), the Coulomb well is deeper, and typical binding energies are in the

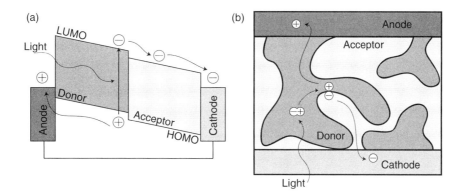

Figure 12.1   (a) Band diagram of light absorption, free charge generation, and extraction in a bilayer junction solar cell at short-circuit conditions. (b) Cross section of a solar cell comprising a bulk heterojunction.

range of 0.1–1 eV and consequently, phonon-induced charge separation is not possible. This implies that the two oppositely charged carriers effectively appear as a metastable, neutral particle, referred to as an *exciton*, which can be conceived as a hydrogenic system where the electron and the hole are in a stable orbit around each other (26).

If nothing happens to the generated exciton, it will recombine on the nanotimescale implying a loss of the photon. PSC therefore applies the approach of combining electron-donating and electron-accepting materials. By selecting a material with lower LUMO level than for the electron that is combined with the absorbing polymer, excited electrons will be transferred due to energy minimization to the electron acceptor. Hereby, the charges are physically separated from each other, the Coulomb attraction is dramatically reduced, and the charges now behave as isolated charges. Figure 12.1a shows the energy diagrams of a photoexcited electron that is transferred to the acceptor material and extracted at the cathode. Likewise, the associated hole migrates toward the anode. When the two electrodes are connected as indicated by the lower straight line, referred to as short-circuit conditions, their energy levels are leveled, which also changes the HOMO–LUMO levels across the cell. Consequently, the charges can by pure energy considerations migrate toward the electrodes and be extracted. An additional effect to charge migration is the concentration gradient close to the donor–acceptor interface, where a large number of charges are generated. As the concentration is high, the interface effectively performs as a "charge pump," which forces the generated charges away from the interface. As the electrodes are leveled, the slopes of the energy bands across the cell are maximized, and the maximum number of charge is extracted whereby the *short-circuit current* (J_{sc}) is generated. If the resistance between the two electrodes increases, a gradual energy difference between the electrodes is built up as the cathode increases and the anode decreases in energy. This energy difference increases until the point where it is energetically unfavorable for the charges to migrate to the electrodes. At this point, referred to as open-circuit condition, no charges are extracted, and the highest difference in energy, that is, voltage difference of the solar cell and the *open-circuit voltage* (V_{oc}), is obtained.

The type of exciton generated in conjugated polymers is *Frenkel excitons*, which are tightly bound to the excited atom. Consequently, only very limited diffusion of the exciton can happen, and for conjugated polymers, exciton diffusion lengths of approximately 10 nm are typically observed (27), before the charges recombine as observed by luminescence. Consequently, only excitons generated in the donor layer within 10 nm of the donor–acceptor interface contribute to the photocurrent, while all other excitons are lost.

The best configuration of donor and acceptor is consequently a system where the donor–acceptor interface is maximized, while unhindered passage to the electrodes from all interfaces is conserved. Early attempts to PSCs applied bilayer structures, where a layer of acceptor material was deposited on a layer of donor material. Hereby, percolation paths to the electrodes were ensured, while the interface area was minimal. As described earlier, a breakthrough within PSCs was the introduction of the BHJ. By coprocessing the donor and acceptor simultaneously from solution, a good compromise between interface and percolation path was obtained. Spatially, the operation of the BHJ in PSCs can thus be explained by Figure 12.1b. The BHJ is sandwiched between the two electrodes. To allow for light penetration, one of these electrodes is semitransparent, and thus, photons can pass the electrode and be absorbed in the photoactive layer, hereby generating an exciton. The exciton will diffuse in accordance with Brownian motion, and if no donor–acceptor interface is encountered within the exciton lifetime ($\approx$400 ps), recombination occurs. However, if an interface is encountered, the electron is transferred, and the charges now behave as isolated particles. The electric field, as represented by the slope of the energy levels in Figure 12.1a, forces the charges toward the appropriate electrodes. However, if no percolation path exits, the charges will be lost to recombination. Otherwise, the charges diffuse to the electrodes and are extracted, and thus contribute to the photocurrent.

12.1.2 Device Structures

Practical processing of laboratory-scale solar cells is based on glass slides onto which a thin film ($\approx$100 nm) of the transparent conductor ITO is deposited. The active layer comprising the electron donor and acceptor is processed on top of the ITO to form the BHJ. Finally, the top electrode typically applied is aluminum, which has a working function that matches the solar cell components. The first experimental work with PSC thus only applied in this glass|Active layer|Al cell stack, but the results were poor due to a low parallel resistance as the selectivity of the charge extraction of each electrode was not perfect, as many charges diffused to the wrong electrodes. As a remedy to the problem, a well-known organic conducting polymer, poly(ethylenedioxythiophene) doped with poly(styrenesulfonate) (PEDOT:PSS) was applied to the cell, which has energy levels that only allow passage of holes and not electrons (Fig. 12.2c). Hereby, well-performing cells were developed, and this layer geometry was the standard geometry for several years and is thus referred to as *normal geometry* (Fig. 12.3a).

The energy levels of the normal geometry cells are illustrated for the most studied material combination consisting of the polymer poly(3-hexylthiophene) (P3HT) as electron donor, the functionalized fullerene phenyl-C60-butyric acid methyl ester ($PC_{60}BM$), and the aforementioned hole conducting layer PEDOT:PSS in Figure 12.3b (3). The energies are determined for isolated materials and shown on an energy scale from the perspective of the electron, implying that a higher energy means a stronger electron affinity of the material. Coupling effects may occur between the materials, but this simple picture provides a reasonable basis for explaining the energy considerations of PSC operation. As electrons are absorbed by the P3HT, the transition from 5.3 to 3.0 eV is made by the electron, leaving the hole in the HOMO level. Due to energy considerations, the electron cannot enter the PEDOT:PSS layer as a higher LUMO level forbids this, and thus, it is favorable to transfer to the $PC_{60}BM$ with a LUMO level of 3.8 eV. Finally, the electron cannot transfer back into the P3HT layer, while the Al electrode with its work function of 4.1 eV receives the electron. As for the hole, the energy considerations are opposite to the electron, and energy minimization implies that the hole moves toward higher energies (i.e., ionization energies). The hole is generated on the P3HT and is situated on the HOMO level at 5.3 eV. As $PC_{60}BM$ poses a higher HOMO level, the hole does not transfer into this material, but rather into the PEDOT:PSS with a lower HOMO level of 5.2 eV and continues into the ITO electrode with a work function of 4.7 eV. This energy description shows the rectification of the cell, where a current is only generated in one direction. The overall effect is that electrons only move toward the Al electrode and holes to the ITO electrode and thus become available for external loads. From the cross section of the cell (Fig. 12.3a),

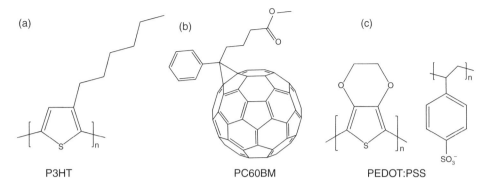

Figure 12.2   Chemical structures of the main components of normal geometry PSC. (a) P3HT, (b) PCBM, and (c) PEDOT:PSS.

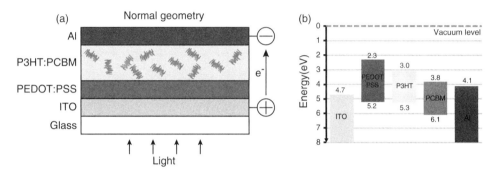

Figure 12.3   (a) Layer sequence of a normal geometry solar cell with the direction of electron transport indicated. (b) Energy diagram of the normal geometry cell relative to the vacuum level.

this implies that electron transport moves to the top electrode, which thus becomes the cathode, while the bottom ITO electrode becomes the anode.

The normal geometry cell has been the working horse for PSC research for years but suffers from several inherent limitations: the light needs to pass through the optically active PEDOT:PSS by which a significant ratio of the light (10–30%) is lost to absorption. Furthermore, as the metal electrode, which is exposed to the ambient, needs to be a relatively low-work-function electrode to accommodate charge extraction, the resistance to oxidation (i.e., removal of electrons) by oxygen is rather low. Adding to this effect is that the highly reactive excited electrons are transferred into the top electrode further increasing the top electrode reactivity. Finally, large-scale processing of low-work-function metals is at present highly cumbersome as formulation in a coatable matrix increases the rate of oxidation dramatically, and consequently, processing in inert gases is required (28).

A solution to this was gradually developed starting from year 2000, where the low-work-function top electrode was replaced by high-work-function materials such as Ag or Au (29, 30). Hereby, the ITO was the lower-work-function electrode, and the electrical field inside the cell was thus inverted. Consequently, the charges would be extracted at opposite electrodes, and new materials and layer sequences had to be introduced in the cell in order to accommodate this *inverted geometry*. While keeping the high-performance donor–acceptor combination of P3HT and PC$_{60}$BM, only moving the PEDOT:PSS toward the top electrode allows for operation of this cell geometry. However, the internal field driving the charges toward the electrodes, that is, the difference in work functions between ITO and Ag, is only 0.3 eV, and extraction problems thus arose. To increase the rectification properties of the cell, an additional charge selective layer was inserted to block holes and only conduct electrons to the ITO electrode as exemplified by ZnO (Fig. 12.2b).

A good hole-blocking layer would have a low resistance to electrons, be optically transparent and have a LUMO level that suits the cell components. Different hole-blocking layers have been applied with zinc oxide (HOMO 7.7 eV–LUMO 4.3 eV) being the most widespread. With a bandgap of 3.4 eV (360 nm), only light above this energy is absorbed. However, this part of the solar spectrum is negligible to the cell performance as photonic flux from the sun is. In inverted geometry PSC, the excited electron in P3HT (LUMO 3.0 eV) is transferred to the PC$_{60}$BM (3.8 eV) and further on to ZnO (4.3 eV) and is extracted by the ITO electrode (4.7 eV), and likewise, holes are extracted at the Ag electrode. As energy levels could be matched to high-work-function electrode, solar cells of the inverted geometry quickly matched the performance while clearly exceeding in terms of stability when compared to the normal geometry cells (31). The layer sequence of the inverted cell is thus Glass|ITO|ZnO|P3HT:PCBM|PEDOT:PSS|Ag, and electrons are driven toward the ITO electrode (Fig. 12.4b). With the change to high-work-function materials, processing electrodes by nonvacuum coating techniques were now possible as opposed to conventional thermal

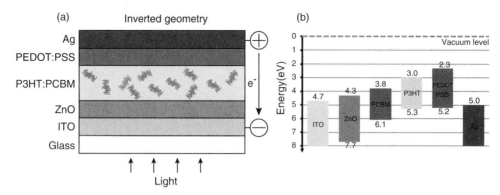

Figure 12.4   (a) Layer sequence of an inverted geometry solar cell with the direction of electron transport indicated. (b) Energy diagram of the inverted geometry cell relative to the vacuum level.

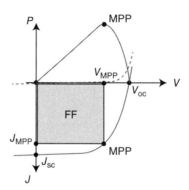

Figure 12.5   Schematic illustration of the JV (bottom part) and power (upper part) behavior of PSC where key parameters have been indicated. Dark (dashed) and light (lower solid) curves have been shown as well as a power curve (upper solid).

evaporation of, for example, Al electrodes. This has proven to be an important step as large-scale R2R processing of PSC was suddenly possible.

12.1.3   Electrical Characterization

Assessment of the performance of solar cells is made by applying a voltage across the cell and measuring the associated current. The general JV characteristics of a solar cell are that the solar cell in the darkness is a diode, implying that the current is ideally canceled in one direction, while the current is allowed to pass in the other direction. This is illustrated in Figure 12.5 as the dashed line, where the current is very low from negative bias until positive bias, where an exponential increase of the current is observed. The dark curve for a solar cell as a function of bias V can be described by

$$J_{\text{dark}}(V) = J_0 \left(e^{\frac{qV}{nk_BT}} - 1 \right).$$

where q is the elementary charge, J_0 is the saturation current, n is the ideality factor, k_B is Boltzmann's constant, and T is the Kelvin temperature.

When the solar cell is exposed to light, photo-generated carriers imply that the dark curve is shifted on the J axis, while the overall shape is still conserved (lower gray line in Fig. 12.5). When the electrodes are not connected and the cell is in open circuit, the highest voltage, V_{oc}, is obtained and the energy levels resemble the ones shown in Figure 12.3b. When the two electrodes are connected, their energy levels and a strong internal field imply high charge extraction and thus a high current, J_{sc}. The extracted power is the central parameter for the cell and is defined by $P = U \cdot J$, where U and J are the voltage and the current, respectively. Consequently, the highest power output is obtained between open circuit, where the current cancels, and short circuit, where the voltage cancels. For the JV curve shown, the associated power is shown as the upper gray line in Figure 12.5. This shows that the power increases from zero at short-circuit conditions, peaks, and decreases to zero at open-circuit conditions. The peak position is referred to as the *maximum power point* (MPP) and the associated voltage and current are the V_{MPP} and J_{MPP}. As the figure shows, these MPP voltage and currents are lower than the J_{sc} and V_{oc}, and to describe the ratio between the conservation of these parameters, the fill factor (FF) is introduced as defined by

$$\mathrm{FF} = \frac{J_{\mathrm{MPP}} \cdot V_{\mathrm{MPP}}}{J_{\mathrm{sc}} \cdot V_{\mathrm{oc}}}.$$

The FF, as normally expressed in percent, describes the degree to which follows an ideal diode behavior. The PCE of the cell describes the ratio between the electrical output power and the incoming power from the irradiation and can be expressed as

$$\eta = \mathrm{FF}\frac{J_{\mathrm{sc}} \cdot V_{\mathrm{oc}}}{P_{\mathrm{in}}}.$$

where P_{in} is the incoming solar power on the cell. The four main parameters of the solar cell (PCE, V_{oc}, J_{sc}, and FF) are used for obtaining and understanding the cell as they relate to different parts of the cell. The V_{oc} is given primarily by the potential difference between the hole energy ($E_{\mathrm{LUMO}}^{\mathrm{donor}}$) and the electron energy ($E_{\mathrm{HOMO}}^{\mathrm{acceptor}}$) in the donor–acceptor material after exciton dissociation offset by the energy it takes for the exciton to separate and can be expressed as

$$V_{\mathrm{oc}} = \frac{1}{q}E_{\mathrm{LUMO}}^{\mathrm{donor}} - E_{\mathrm{HOMO}}^{\mathrm{acceptor}} - \Delta V.$$

where q is the elementary charge and ΔV is an energy offset that ranges from 0.2 to 0.5 eV (32, 33). The J_{sc} depends to a larger degree of the quality of the processed solar cells and relates directly to the absorption of the active layer, where more photons absorbed imply a higher J_{sc}. Finally, the FF is a convoluted parameter describing primarily losses in the device by which the ideal diode curve is lost. Further description of the electrical properties of PSC can be found in the literature (34, 35).

12.2 MATERIALS FOR POLYMER SOLAR CELLS

12.2.1 Photoactive Layer: The Donor

During the last decade, a large variety of polymers have been synthesized and employed with varying success in PSC. The phenylenevinylene family (Fig. 12.6a) was the working horse during the 1990s, where device efficiencies above 3% were reached for various side chains (36, 37). However, due to a rather narrow UV–vis absorption profile and low ambient stability, alternatives were needed. A new family of polymers based on thiophene was developed, where higher efficiencies and stabilities were obtained. The higher efficiency was generally obtained by broadening the absorption window and moving it toward the peak of the solar spectrum at 700 nm. Especially P3HT demonstrated superior performance in PSC and efficiencies increased from 2.8% in 2002 to above 5% in 2008 for fully optimized devices (Fig. 12.6b) (38).

As a fundamentally different approach to increasing cell performance, focus was directed at the energy levels of the donor and the acceptor. By varying the bandgap of the polymer, two effects can be obtained: for the isolated polymer, a low bandgap implies a broad absorption window and thus high current but low voltage as the charges are not excited to higher energy states. Likewise, a high bandgap creates a large voltage but a low current. However, by combining the donor with an acceptor, the system becomes more complicated, and the matching between the two LUMO and two HOMO levels becomes important. For P3HT, the major limiting factor is the rather low V_{oc} of 0.57 V, which origins from

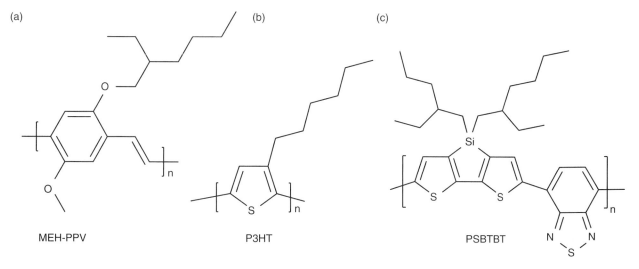

Figure 12.6 Three generations of conjugated polymers for PSC.

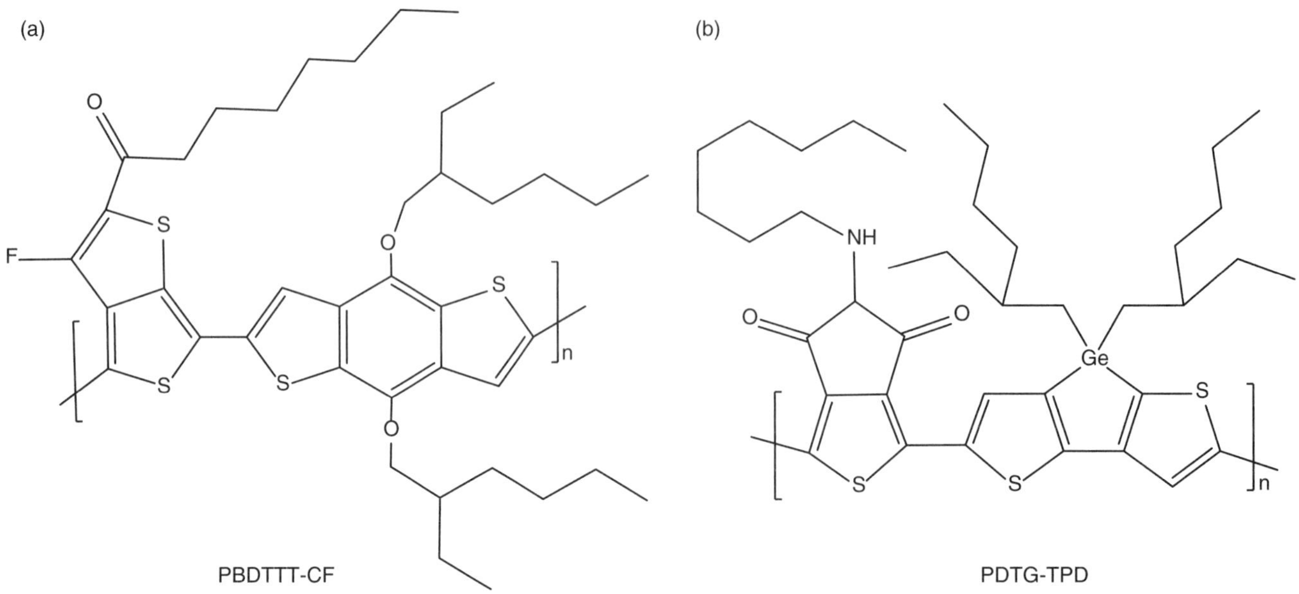

(a)

(b)

PBDTTT-CF

PDTG-TPD

Figure 12.7   Chemical structures of high-performance polymers that have been reported to yield (a) 7.7% and (b) 8.5% efficiency.

the rather poor combination of energy levels between P3HT and PCBM. As the energy levels in Figure 12.3b demonstrate, the excited electron is transferred from a LUMO level of 3.8 to 3.0 eV going from P3HT to PCBM. However, it is well established that while an energy difference between LUMO levels is needed to drive the charge separation, this difference only needs to be around 0.3 eV to separate the charges in the exciton. Consequently, a major part of the energy is lost in this charge transfer, and better energy match is needed. Minor variations in energy levels can be obtained by addition of electron-withdrawing or electron-donating substituents into the monomer. Hereby, thiophene-based polymers have been demonstrated to vary in bandgap from the original 2 eV down to 1.1 eV. A more flexible strategy is synthesizing polymers with alternating electron-rich and electron-poor units on the same backbone. By introducing push–pull effects on the electrons, the two units interact and the overall energy levels will yield a combination of the energy levels of the two individual units. This has allowed the design and synthesis of polymers with tailored energy levels. However, while polymers may be designed with optimal energy levels, issues with mobility and solubility may occur, and thus practical experience with a given polymer is still needed. By engineering the energy levels, theoretical efficiencies of 10% were predicted for donors with a HOMO level of 2.25 eV and a LUMO level of 4.0 eV (39) when blended with $PC_{60}BM$. This demonstrates the potential of aligning the energy levels for the donor and the acceptor if the optimum donor bandgap of 1.5 eV (830 nm) is reached. Consequently, donor–acceptor polymers are normally designed with a bandgap close to 1.75 eV, referred to as low-bandgap polymers, as these absorb a larger part of the solar spectrum. An example is given in Figure 12.6 where an electron-withdrawing benzothiadiazole is combined with an electron-donating silole-containing polythiophene unit with a HOMO level of 5.05 eV and a LUMO level of 3.27 eV (18). This polymer exhibits a broad UV–vis absorption peak from 650 to 750 nm, and the low bandgap combined with the optimized energy levels provides a good cell performance of 5.1%. However, theoretical upper efficiency predictions for a polymer with these energy levels and bandgap are expected to yield 8.5% efficiency, which demonstrates the deviation between theoretical predictions and the experimental values obtained for actual solar cells (38). Two present high performing polymers are shown in Figure 12.7. Chen et al. presented a donor–acceptor polymer based on a thienothiophene and a benzodithiophene unit in 2009 (Fig. 12.7). Its LUMO level of 3.45 eV matched the 3.8 eV level of PCBM well, and consequently good charger transfer was possible, while excessive energy was conserved. Due to a very broad absorption profile ranging from 300 to 750 nm, J_{sc} as high as 15.2 mA/cm² was obtained. Combined with a FF of 67% and a high V_{oc} of 0.76 V, an efficiency of 7.7% was achieved (40), which was the highest reported efficiency at that time. Recently, in December 2011, a novel polymer was reported, which gave efficiencies of 8.5%, and thus clearly beat the old record (19). The polymer was also an alternating donor–acceptor polymer based on a dithienogermole and a thienopyrrolodione unit (Fig. 12.7b), where a V_{oc} as high as 0.86 V when blended with $PC_{70}BM$ was the reason for the increased efficiency. Overall, low-bandgap polymers combine broad and low-wavelength absorption with high V_{oc}, and consequently, this approach to polymer synthesis is the most promising for materials of commercial interest. An extensive list of different low-bandgap polymers can be found in the literature (16, 41).

12.2.2   Photoactive Layer: The Acceptor

As for the polymeric donor material, electron acceptors have been extensively researched. Instead of optimizing the energy levels of the donor to the acceptor, optimization of the acceptor has also been done. Several different types of acceptors have been demonstrated, both organic and inorganic (42). A good acceptor has several properties: electron mobility on par with or exceeding the hole mobility of the donor,

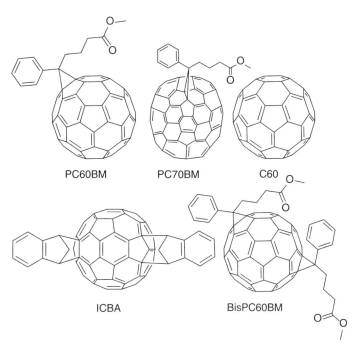

Figure 12.8 Chemical structures of C_{60} and derivatives developed for high V_{oc} PSCs.

UV–vis absorption complimentary to the donor to increase photocurrent, and a LUMO level that matches the donor to ensure a high V_{oc} while not compromising the charge transfer. Several different types of materials have been used as acceptors in BHJ solar cells such as polymers, fullerenes, carbon nanotubes, small molecules, and inorganic metal oxide particles.

The best results have been obtained with derivatives of C_{60} and C_{70} fullerenes. The fullerene is a highly symmetric molecule that exhibits excellent electron mobility with high electronegativity and the ability to stabilize negative charge and is thus an excellent choice for electron acceptor. Different derivatives of these have been developed for obtaining good solubility, morphology, but most importantly, for increasing the LUMO level to match the levels of materials such as P3HT better. Examples of five different fullerene-based electron acceptors are presented in Figure 12.8. By addition of electron-donating groups to the carbon cage, the V_{oc} can be highly increased as the LUMO gap decreases. Ranked by increasing V_{oc}, C_{60} gives 0.40 V, and increasing values are found for $PC_{70}BM$ (0.63 V), $PC_{70}BM$ (0.65 V), bisPCBM (0.73 V), and ICBA (0.87 V) (43). As the J_{sc} is conserved when going from $PC_{60}BM$ to the novel types of polymers, the same relative increase in efficiency has been obtained with a P3HT:ICBA blend where 6.5% has been reported (44). Consequently, matching the energy levels of the acceptor can be highly beneficial to the cell performance, and even higher performance increases are possible if the LUMO levels of fullerene derivatives are increased further.

Research on using carbon nanotubes (CNTs) as electron acceptors has also been carried out, as CNTs can be designed to exhibit electrical properties resembling the ones observed for the fullerenes such as high conductivity and the ability to stabilize negative charge (32). Furthermore, the high aspect ratio, which may exceed 1000, is ideal for the creation of the percolation paths to the electrode in PSC. By aligning the CNTs with an electric field, it has been anticipated that a photoactive layer consisting of highly oriented CNTs with polymer deposited on top would highly exceed the interfacial area of a conventional BHJ while keeping percolation paths to the electrode. However, in BHJ with thiophene-based polymers, cell performances are rather poor. As only semiconducting CNTs can favor electron charge transfer, minute concentrations of metallic CNTs in the blend induce short. This effect, in combination with metal catalyst impurities from the synthesis, was found to be the reason for the low performance. Consequently, PSCs applying only CNT as the acceptor have only yielded efficiencies below 2% (45).

Hybrid solar cells have been processed, where the acceptor material contains inorganic components (42, 46). Promising results have been obtained using materials as different as TiO_2, ZnO, CdSe, CdS, PbS, PbSe, SnO_2, and Si. Different formulations of these materials have been applied such as nanoparticles, nanorod, and nanowires. By controlling the shape, different morphologies can be obtained, where the best compromise between donor–acceptor interfacial area and percolation paths can be found. PPV-based polymers blended with CdSe have consequently yielded efficiencies of 2.8%, and thus, the inorganic acceptor is on par with the organic counterpart. In combination with P3HT, organic accepters have not demonstrated as promising results as performances are inferior to the $P3HT:PC_{60}BM$ cells. Actual increases in performance by application of inorganic acceptors have been reported where the inorganic acceptor is added in various concentrations of the organic acceptor. By adding ZnO nanorods to a $P3HT:PC_{60}BM$ cell to improve electron conductivity and to create a scaffold for electron extraction, the efficiency was found to increase from 1.8% to 3.9% (47). Consequently, by benefiting from the control of energy levels, shape, and aspect ratio, inorganic acceptors can advantageously be used to increase PSC performance.

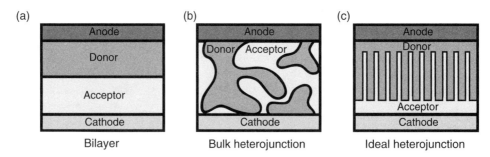

Figure 12.9 Different morphologies of the photoactive layer.

12.2.3 Photoactive Layer Morphology

While extensive focus has been directed at engineering the photoactive layer in terms of donor and acceptor materials, the morphology has been found to play an important role. In this context, morphology implies, for example, the size of the donor and acceptor domains, the crystallinity of the donor, and the vertical gradient of donor–acceptor between the electrodes. As discussed earlier, the first PSCs were processed as bilayer cells, where the donor and acceptor layers are processed individually forming two nanolayers with a horizontal interface (Fig. 12.9a). However, after the development of the BHJ (Fig. 12.9b) a dramatic increase of cell performance was achieved. However, as the morphology obtained by spin coating of a given donor–acceptor solution is not easily predicted, optimization of the BHJ has been studied extensively. The ideal photoactive layer has a phase segregated structure, where the acceptor is primarily located close to the acceptor and vice versa for the donor and the anode. A complete coverage of the electrodes of the respective component would ensure that no charges are extracted at the wrong electrode. Additionally, the donor–acceptor interface should be maximized to ensure maximum charge transfer efficiency, while percolation paths to the electrodes should be maintained. Consequently, the ideal structure would be made up of nanometer wide fingers of acceptor material traversing almost to the anode, and where the gaps are filled with donor material (Fig. 12.9c). While making the gap between the rods close to the exciton diffusion length, all excitations would result in charge transfer, and recombination of free charges would be minimized by the percolation paths of the finger structure of the donor and the accepter.

With such an ideal heterojunction, external quantum efficiencies are expected to increase significantly greatly improving the cell performance. However, with the BHJ approach, only limited control of the morphology is possible, and the properties of the ideal heterojunction are never achieved, while a rather random structure is obtained. However, the obtained morphology depends on several different parameters pertaining to the material, the processing technique, and the ambient conditions during and after film processing. Consequently, this parameter space been extensively studied in PSC processing. By variation of the weight ratio between the donor and the acceptor, highly different cell performance was obtained as the domain size was affected. Chirvase et al. demonstrated that as the relative P3HT contents was varied from 25% to 60%, efficiencies increased from 0.18%, peaked at 2.39%, and decreased to 1.94%. With the development of R2R processing techniques, the extensive practical work needed for studying a few points in this parameter space was highly reduced, as an approximately continuous gradient of P3HT from 0 to 100% was obtained during a single process. This showed that the efficiency had a symmetric bell-shaped curve as a function of P3HT content with a peak at 50% P3HT content (48).

The choice of solvent for solution processing of the active layer is another parameter that can highly influence the morphology. In early reports of PSCs, the organic solvents chlorobenzene and chloroform were typically used due to high solubility of most conjugated polymers compared to other organic solvents. With a boiling point of 131 °C and 61 °C, respectively, the solvents are rather volatile, and thus, during spin coating of the active layer, short time is left for the materials to arrange. As the BHJ is rapidly formed during the evaporation of the solvent after processing in less than 1 min, the lowest morphological energy state is not reached. By application of higher boiling point solvents such as dichlorobenzene (boiling point 181 °C), the time of evaporation of the solvent was increased significantly leaving the formation of the BHJ longer time to form. Hereby, device efficiencies were found to increase, partly due to the increase of domain sizes, and partly due to the increased crystallinity of P3HT during solvent evaporation. Consequently, for example, Mihailetchi et al. reported P3HT:PCBM performance increases from 3.1% to 3.7% by slow drying of the active layer by using dichlorobenzene (49).

While the performance of the photoactive layer depends on the donor and the acceptor materials, minor concentrations of additives have been successfully added to the solutions before film processing. Hereby, the film formation properties have been significantly changed, and improved morphologies have been obtained. Primarily alkanes with different end functionalizations have been used in small concentrations below 2 w%. They have in common that the solubility of P3HT is higher than for PCBM. Additionally, octane has a higher boiling point (270 °C) than the main solvent. During evaporation of the solvent, the main solvent evaporates first, and as the additive concentration increases, PCBM will start to form clusters and precipitate. Consequently, PCBM will form in the bottom of the layer, while the slowly evaporating additive will make the P3HT settle on the top. Consequently, the BHJ comes closer to the ideal morphology for inverted geometry cells (Fig. 12.9c) as the donor is mainly located close to the anode and vice versa for the acceptor (50). By screening the performance of different functionalized octanes, diiodooctane was found to have the most positive effect, where cell performance was increased from 3.4% to 5.1%. Another additive, octanedithiol, has also been successfully added in minor concentrations to the active layer where cell performance increased from 2.8% to 5.5% (51).

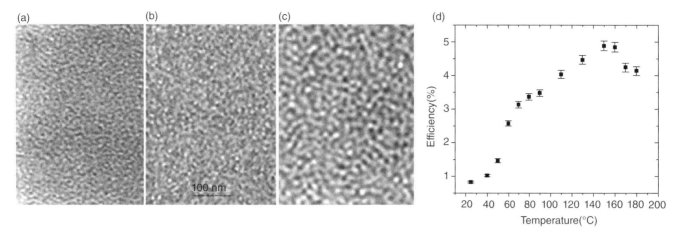

Figure 12.10 TEM images of a P3HT:PCBM photoactive layer (a) before annealing, (b) after 150 °C for 30 min annealing, and (c) 150 °C for 2 hours. (d) Solar cell efficiency after 15 min annealing at various temperatures. (*Source*: Reprinted with permission).

After processing of the active layer, the morphology can be changed, as the material is not in its morphological energy minimum. A successful approach has been by thermal annealing of the photoactive layer. When spin coating P3HT:PCBM, the optimum domain size and P3HT crystallinity cannot be obtained only by optimizing the donor–acceptor ratio, the concentration, and the solvent. However, by thermally annealing the spin-coated layer prior to deposition of new layers, the photoactive layer can be changed in a predictable way. Initially, the domain size is very small (<10 nm, Fig. 12.10a), which is too small for superior charge extraction. By thermal annealing the donor and acceptor domains will slowly grow, and crystallinity of each domain will be introduced for crystalline materials. As thermal annealing is performed, an optimum domain size is achieved (Fig. 12.10b) after which the domains grow too large (>25 nm, Fig. 12.10c), and exciton disassociation is impeded resulting in a decrease in cell performance.

As the domain size growth is a function of both time and temperature, these two parameters are normally decoupled by keeping one parameter constant while changing the other parameter. P3HT:PCBM cells have consequently been observed to increase in performance from 0.7% to 4.9% as the photoactive layer was annealed for 15 min at 25 °C and 150 °C, respectively, while above this temperature, domain sizes exceeded the optimum size and performance decreased (15). Likewise, by keeping annealing temperature on 150 °C, it was demonstrated that efficiency increased to 4.4% after 2 min and 5.0% after 60 min.

With the aim of obtaining the ideal heterojunction under controlled condition where the stochastic processes involved in forming the BHJ are avoided, different approaches have been followed. Growth of highly ordered nanostructures on surfaces has been studied extensively as such structures are interesting to energy research in general. Metal oxide nanowires such as TiO_2 and ZnO can thus be grown on ITO with high density and a height of, for example, 100 nm (46). Such a layer can function as the bottom part of an ideal morphology (Fig. 12.9c) and assuming that the electron donor can be deposited into the nanowire structure, hybrid solar cells can be obtained with ideal heterojunctions. Deposition of the subsequent layer is made by different techniques such as spin coating and physical vapor deposition or by double nanoimprint (52). Optimization of the nanowire structures for such cells is cumbersome, and problems with, for example, obtaining the optimum wire spacing exist. Additionally, the filling of the air gap with the subsequent layer is not straightforward. Consequently, using only ZnO nanowires as acceptors, cell performances do not exceed 0.5%, while up to 3% efficiency has been reached by addition of PCBM (53). Purely organic nanostructured photoactive layers have also been processed by nanoimprint. Hereby, the donor or acceptor is processed and subsequently, a patterned mold is pressed onto the layer (46). The temperature of the layer is kept above the glass temperature, whereby the material shapes around the mold to create a negative. On top of this patterned material, the second part of the active layer is then deposited. While creating a good heterojunction, this also implies that each electrode is covered with a fully covering layer of donor–acceptor using this processing approach, which reduces recombination and so on. Several different nanostructures can be processed with dimensions in the nanometer range and thus cells of predictable morphologies can be obtained. Efficiencies of P3HT:PCBM cells processed by double imprint have yielded efficiencies up to 3.25%, and consequently, imprint is an interesting approach to obtaining controlled photoactive layers.

12.2.4 Electrode Buffer Layers

To allow for good rectifying properties of the cells and to ensure that charges are extracted at the proper electrodes, buffer layers are necessary for processing high-performance PSCs. As the BHJ spans from one electrode to the other, the risk of charges being extracted at the wrong electrode is high, which may impede cell performance significantly. The two central parameters to buffer layers are the energy levels and the charge transport properties. Additional beneficial properties of buffer layers include resistance toward rough substrates and physical separation of the diffusive top electrode. Finally, some materials may be hygroscopic, and may thus act as a water scavenger, and other materials may acts as oxygen scavengers. Consequently, buffer layers contribute with a multitude of properties that allow for processing high-performance cells.

In both the normal (Fig. 12.3) and the inverted cell geometries (Fig. 12.4), the ITO can work as electrode as the work function (4.7 eV) lies reasonably close to the LUMO level of the P3HT (normal geometry) or the HOMO level of the $PC_{60}BM$ (inverted geometry). However, the interface between the neighboring material and the ITO becomes nonohmic due to the significant energy level difference severely impeding the V_{oc}. Additionally, as ITO can collect both holes and electrons due to the work function, there is no strong built-in charge selectivity of the ITO, and rectification without buffer layers becomes poor (54). Consequently, anode buffer layers have always been applied to PSC of normal geometry. Overall, the anode buffer layer material should provide good hole conduction while blocking electrons, while it should make an ohmic contact to the electrode. Additionally, from a processing perspective, it should be insoluble in the solvents that are used for processing the photoactive polymer on the anode buffer layer. Finally, it should be optically semitransparent if light needs to pass the material. The first anode buffer layer used for PSCs came was PEDOT:PSS, which had previously been applied to OLEDs for the same purpose (Fig. 12.2c). The material is a dispersion of the semiconducting thiophene-based polymer PEDOT, which is suspended in PSS providing a semitransparent layer for films of thickness in the nanoregime. PEDOT holds attractive properties as the energy levels (HOMO, 2.3 eV; LUMO, 5.2 eV) are well aligned for hole transport, while electrons are blocked. Additionally, the energy difference between the HOMO level of PEDOT and the anode work functions for normal (ITO: 4.7 eV) and inverted geometry cells (Ag: 5.0) are rather close implying ohmic contact. However, due to solubility in the same solvents as the photoactive polymers, fixation of the PEDOT was needed. This was achieved by formulating the PEDOT as a suspension in a matrix of PSS. Hereby, aqueous processing was possible, while solubility in nonpolar solvents used for photoactive processing was avoided. During the evolution of PSCs, PEDOT:PSS remains the state-of-the-art anode buffer layer and has only been slightly improved in conductivity by addition of various agents or by physical postprocessing treatments (54).

Alternative anode buffer layers have also been tested in mainly P3HT:PCBM-based solar cells. Inorganic anode buffer layers have also been researched extensively as high hole mobilities can be achieved. Additionally, a large variation in energy levels is observed for different inorganic materials why these materials may perform well as anode buffer layers. Thin layers (<10 nm) of metal oxides such as V_2O_5 (HOMO, 4.7 eV; LUMO, 2.4 eV) and M_oO_3 (HOMO, 5.3 eV; LUMO, 2.3 eV) have been applied to the ITO of normal geometry cells by thermal evaporation. When compared to the energy levels of PEDOT:PSS (HOMO, 5.2 eV; LUMO, 2.3 eV), the materials are highly similar and match the anode in PSC well. Additionally, optical transparency for thin layers combined with insolubility make inorganic materials obvious candidates for anode buffer layers. Consequently, cell performances have been comparable or even exceeded reference PEDOT:PSS-based cells (55). Nickel oxide (HOMO, 1.8 eV; LUMO, 5.0 eV) has also been applied as an anode buffer layer where pulsed laser deposition of 10 nm layers on ITO gave an overall cell performance of 5.16% (56).

Using a different approach, self-assembled monolayers have been applied to ITO to change the work function and thus the ohmic contact to the photoactive layer. Examples of this can be found in the literature where different materials are deposited on the ITO and allowed to form very thin layers (<10 nm), but with a great impact on hole extraction and wettability of the anode for subsequent processing. The work function of ITO was changed from 4.7 eV up to 5.16 eV and down to 4.35 eV and 3.65 eV, by deposition of different silyl-based agents (57). With a HOMO level of 5.3 eV for P3HT, ohmic contact was established in the case of the anode with a work function of 5.16 eV, whereby cell performance was comparable to the reference PEDOT:PSS cells. Consequently, self-assembled monolayers are interesting as they can fine-tune the work functions with even very thin layers by which optical transparency is not compromised.

Cathode buffer layers are used within PSCs to transport electrons to the electrode while blocking the holes. Many of the same demands of anode buffer layers apply to cathode buffer layers such as proper energy level alignment, high electrical conductivity, and high optical transparency. Cathode buffer layers are not always applied to normal geometry PSCs as the LUMO of PCBM (3.8 eV) is well aligned with the Al cathode (LUMO 4.1). However, for inverted geometry cells, the LUMO gap between PCBM and the ITO cathode is 0.9 eV, which implies non-ohmic contact and thus the need for a cathode buffer layer. Different materials have been applied such as metal oxides to organic materials. TiO_x and ZnO have been incorporated into inverted geometry cells where performances have been greatly increased when compared to reference devices with no cathode buffer layer (54). A 36 nm ZnO layer was deposited by chemical vapor deposition, which in a P3HT:PCBM-based cell increased cell efficiency from 0.57% to 4.18%, clearly demonstration the need for a cathode buffer layer in inverted cells (58). Different organic materials have been applied as cathode buffer layers, which are all either conjugated polymers or conjugated small molecules with well-aligned energy levels (54). The small molecule bathocuproine was thermally evaporated on top of the active layer in normal geometry cells. Hereby, an efficiency of 3.06% was obtained as compared to 1.91% for the normal cell. It was suggested that due to the low HOMO of the molecule (7.0 eV), no holes could be extracted at the cathode, which increased the efficiency.

In general, buffer layers play an important role and are needed to obtain high-performance cells. While the energy levels of a given material may match the cell components, the processing properties are central to the practical use. On a larger scale, R2R compatibility of the buffer layers is important why solution processable materials such as PEDOT:PSS for the anode and ZnO for the cathode have become highly standardized materials (59).

12.3 POLYMER SOLAR CELL PROCESSING

With the discovery of semiconducting polymers and charge transfer to fullerenes, research in PSCs was directed at increasing cell efficiencies. As the produced cells had no practical use, the size of the cells was minimal to ensure high control during processing while keeping material costs low. The normal geometry was used for the cells as only a single buffer layer was needed for the anode, and the entire cell except the photoactive layer relied on already commercially available materials. In terms of substrates, ITO-coated glass slides were used, which were well-known transparent conductors (3).

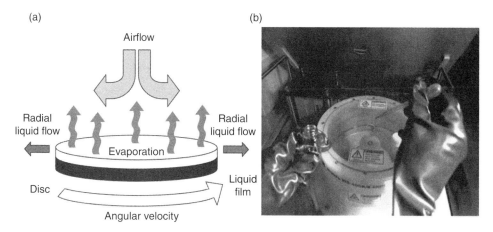

Figure 12.11   (a) Illustration of the principle behind spin coating where deposition of a droplet on a rotating substrate results in a homogeneous film. (b) Photo of spin coater and operator. (*Source*: Reprinted with permission).

The layers of this *laboratory cell* were processed from solution by *spin coating*, where the substrate is rotated in the horizontal plane onto which droplet is dropped (Fig. 12.11a). Hereby, a smooth layer with nanometer thickness is formed. Spin coating is fast, easy, and cheap, as only a spin coater is required to rotate the sample (Fig. 12.13b), and thus, no expensive equipment is needed for constructing the solar cell layer stack. Additionally, rather spatially homogeneous films can be obtained throughout the substrate. The thickness of the film can be varied by the spin coating speed, which has been one of the central optimization parameters in PSC processing (60). Spin coating has been used for driving PSC technology to the present state of maturity due to the flexibility and control of the layer processing. However, the consequence of the extensive research on the laboratory cell is that optimization takes place on a platform that has no practical and commercial interest as no direct upscaling is possible since spin coating and thermal evaporation of the electrode are not compatible with large-scale processing. As cell efficiencies are approaching 10%, focus has shifted from cell efficiency to a combination of efficiency, stability, and processability, as these are the central parameters with a commercial perspective of PSC research.

A major advantage of PSCs is that, as all layers thicknesses are in the nanometer range, the cell can be flexible, and thus processing on flexible substrates on a R2R basis is possible. This allows for processing with a new palette of layer deposition techniques implying coating and printing of the layers on a continuously rolling flexible substrate. By processing continuously, only setup of the instrument (web speed, coating head–substrate separation, etc.) has to be performed once after which the process can run without interruptions resulting in a homogeneous outcome with no upper limit to the area of the processed layer. Laboratory processing of PSCs on glass substrates is performed by successive spin coating of the different layers from orthogonal solvents. However, during spin coating, the majority of the used solution is lost as only a small ratio of the solution is actually deposited on the substrate to form the film. R2R-based coating and printing techniques are generally parsimonious as all deposited solvent is used for the film, which is an important parameter for the economical perspectives of PSCs. As in the case of laboratory cells, flexible substrates used for R2R processing are normally obtained with a transparent conductor predeposited, normally ITO. Typical substrate types are polyethyleneterephthalate (PET) and polyethylenenaphtalate (PEN) as these thin polymer substrates are highly flexible and provide good optical transparency. For processing of layers with different concentrations, surface tensions, and film-forming properties, spin coating can with its flexibility cover most solution types. However, coating and printing techniques offer less versatility and often different processing techniques are required for the different layers. As there is no single R2R processing technique that can handle all materials well, a multitude of techniques have been developed. On a research basis, processing of R2R solar cells is normally based on unrolling, layer processing, and rerolling of the substrate. This simple approach only demands for a single coating setup, which can be adjusted to the specific material to be deposited. An illustration of this approach is given in Figure 12.12a where three different steps in a R2R processes are described. As a step is carried out, the substrate is rerolled, and the setup is either modified or the web is moved to another R2R setup for subsequent processing.

As the R2R process becomes standardized and certain processing steps can be routinely done without adjustments, several processing steps can be included into the same process (Fig. 12.12b). Hereby, several different layers may be deposited simultaneously, which greatly reduces the manual handling. However, as all processes have to be performed in parallel, the web speed is the same for all processing steps, which reduces the parameter space for optimization of each step. Additionally, complete drying of a deposited layer is needed before subsequent processing steps, and thus, for example, furnaces for drying steps need to be included and synchronized with the remaining steps. However, with the integrated process approach, very high production capacity can be obtained, and the overall processing speed will only be determined by the slowest process.

Independent of the coating or printing technique, the active material is formulated as an ink, which may need different additives for optimum film-forming properties. For most coating technique, the solution is brought to the coating head by a pump, which gives control of, for example, film thickness. Multiple coating solutions can be added to the coating head with different flow rates and thus stock solutions of individual components can be used, which are only mixed during the coating. This unlocks a new range of parameters that can be studied, such as concentration

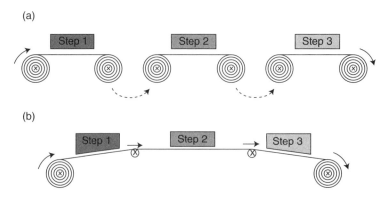

Figure 12.12   R2R processing using (a) discrete process and (b) an integrated process.

of the material in solution (adjustment of solution + extra solvent) and the effect of donor–acceptor ratio in the photoactive layer (variable flow rates of donor and acceptor stock solutions). Finally, characterization of the final solar cells normally implies either manual handling of each cell, where a cell is either manually connected to a source meter or inserted into a characterization stand. However, R2R characterization of a roll of solar cells in a fully automated setup is possible, as the cells are processed in a repeating pattern on the roll. Consequently, the manual workload associated with the laborious processing and the characterization known from laboratory solar cells is highly reduced. A comparison between laboratory and R2R solar cell processing is shown in Figure 12.13. A conventional optimization step for, for example, a novel-conjugated polymer implies processing solar cells with different donor–acceptor ratios as well as active layers of different thicknesses. Practically, this is realized by making photoactive solutions with each weight ratio (Fig. 12.13a1), manually processing the cells serially (Fig. 12.13a2), and manually performing the electrical characterization (Fig. 12.13a3). Hereby, a number of thicknesses and weight ratios can be obtained over time, which can provide a coarse impression of the maximum efficiency for the given material combination (Fig. 12.13a4). In a R2R context, such an optimization study involves processing of the active layer with a variation in the relative donor and acceptor flow rates into the coating head by which the weight ratio is continuously changed. For laboratory cells, normally less than 10 ratios are studied due to the time consumption of this work, whereas for R2R processing, the approximate continuous variation in weight ratio may provide in excess of 100 separate ratios. During a single R2R process, several ratio studies can be carried out for different thicknesses only by changing the web speed during the process. Overall, several hundreds of cells can be processed with minimal material usage as no material is wasted. Finally, the characterization is automated and thus, a high resolution of the efficiency of the optimized cell can be obtained (Fig. 12.13b3).

To process a given layer on a R2R basis, certain aspects should be considered. As the perspective with R2R coating is to produce PSCs at a volume that may cover a significant part of the worldwide electricity consumption, the selected techniques must be highly upscalable (60). Additionally, sustainable materials and solvents should be used, as for example, environmentally harmful solvents should not be applied on a commercial scale, by which the positive effect is compromised by the toxicity (61). Finally, the material chosen for PSC's processing should be abundant on the earth to such an extent that a full upscaling is possible. If all these requirements are fulfilled, the experience obtained from such R2R research is fully compatible with PSC's processing on a commercial level, and the need for reoptimizing processing parameters is alleviated.

Among the different methods for applying polymeric material to a substrate, certain techniques are proven successful. What differs between these techniques is the *dimensionality*, that is, the number of dimensions that can be controlled, where 0-dimensional implies that the entire substrate is covered, 1-dimensional that continuous lines are generated, 2-dimensional that lines can be patterned, and 3-dimensional that full control of the printing in the plane of the substrate is possible. Additionally, the process of ink transfer varies. Some techniques apply a negative (stamp, mold, etc.), which transfers ink to the substrate, and these techniques are referred to as *printing*. Alternatively, the ink can be transferred directly to the substrate by pouring, painting, spraying, and so on, normally referred to as *coating* (60). In the following, some of the central coating and printing techniques are further discussed.

12.3.1   Coating Techniques

12.3.1.1   Knife Coating   Knife or knife-over-edge coating is a coating technique mainly applied to viscous inks. It can be seen as a R2R implementation of doctor blading, where a knife is moved horizontally across a stationary substrate to distribute a deposited drop of ink. In knife coating, the knife is positioned above the web that is supported by, for example, a roller. Ink is supplied to the web before the knife and as the web passes the knife, excess ink is scraped off and a relatively homogeneous wet film thickness is obtained (Fig. 12.14a). Variation of the film thickness is controlled by the web–knife separation. The thickness of the dry film can empirically be expressed as

$$d = \frac{1}{2}\left(g\frac{c}{\rho}\right).$$

where g is the web–knife separation in cm, c is the solid material concentration in g/cm^3, and ρ is the density of the dry film in g/cm^3 (60). A continuous layer is applied to the substrate, and no patterning is thus possible. Additionally, the edges of the knife are not clearly defined,

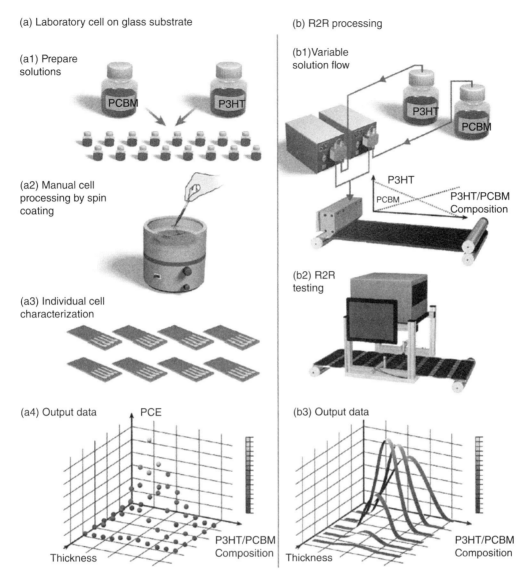

(a) Laboratory cell on glass substrate

(a1) Prepare solutions

PCBM   P3HT

(a2) Manual cell processing by spin coating

(a3) Individual cell characterization

(a4) Output data   PCE

Thickness   P3HT/PCBM Composition

(b) R2R processing

(b1) Variable solution flow

P3HT   PCBM

P3HT   P3HT/PCBM Composition
PCBM

(b2) R2R testing

(b3) Output data

Thickness   P3HT/PCBM Composition

Figure 12.13 Comparison between (a) laboratory-scale and (b) R2R processing of solar cells. Manual exploration of the parameter space of film thickness and P3HT:PCBM ratio is laborious both in terms of processing and cell characterization. R2R processing provides a highly automated approach where a large number of cells can be produced and characterized with a minimum of manual work involved. (*Source*: Reprinted with permission).

which makes knife coating a 0-dimensional coating technique. With knife coating, demonstrations of PEDOT:PSS and active layers have been made (62).

12.3.1.2 Slot Die Coating A widely used coating technique is slot die coating, which implies a coating head through which the ink is supplied to the supported web to a meniscus (Fig. 12.14a). Originally being developed for the photographic industry, this technique allows for coating materials with a large range of viscosities, which makes it a versatile technique for PSC's processing. The coating head comprises two steel sections bolted together to form an ink reservoir and a flow channel. Hereby, even wide coating heads allow for distributing the supplied ink evenly with constant pressure and velocity. The coating head can be quite complex, as they can be equipped with several ink supplies to allow for, for example, ratio experiments. Temperature control may also be added to optimize coating properties. Finally, the coating head may comprise a mask, which makes a fixed pattern across the web from the multiple exits. In contrast to knife coating, sharp edges between the mask exits can be obtained. Hereby, for example, several lines can be made in parallel, which is highly suitable for processing serially connected multicell modules (59). As only continuous depositions can be carried out, slot die coating is a 1-dimensional technique. Controlling the film thickness can be done by changing the speed of either the web or the ink supply and can be expressed by

$$d = \frac{f}{Sw}\frac{c}{\rho}$$

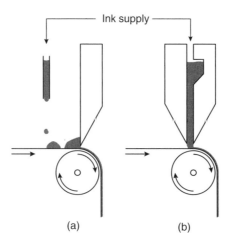

Figure 12.14 Schematic illustration of (a) knife coating and (b) slot die coating. (*Source*: Reprinted with permission).

where f is the ink flow rate in cm^3/min, S is the web speed in cm/min, and w is the width of the coating in cm. Slot die coating is a robust and flexible technique, with which it is possible to process several different components of PSCs. By changing the mask, different patterns can be obtained easily with the same coating head. Additionally, the cost of the entire coating setup can be low compared to printing setups due to the simplicity of the setup (63). These advantages have made slot die coating the most utilized PSC's processing technique, and consequently, the majority of scientific publications on R2R PSCs employ this technique for processing of low viscosity materials. Several demonstrations of slot die coating have been made of buffer layers, active layers, ZnO nanoparticles layers, and silver nanoparticles for electrodes (63, 64).

12.3.2 Printing Techniques

12.3.2.1 Gravure Printing
Gravure printing is a printing technique where a gravure cylinder transfers ink from a bath to the web (Fig. 12.15a). It is used for printing graphical products such as magazines and post cards as very high web speeds can be obtained (3). The gravure cylinder is submerged partially into an ink bath and rotated, and a knife doctor blades excessive ink off the cylinder. The gravure cylinder consists of a steel cylinder surrounded by a custom-made pattern made of cobber. Patterning of the gravure cylinder can be made with a 2-dimensional pattern of very high resolution, which is subsequently transferred to the web. The moving web is brought into contact with the gravure cylinder by an impression cylinder, which has a surface of hard rubber to allow for a good contact between gravure and web. The gravure cylinder thus works on a R2R implication of a stamp, where ink is applied to the stamp and deposited on the substrate. While a gravure pattern can be highly controlled in 2 dimensions, any new pattern demands for a new costly gravure roller. With gravure printing, low viscosity inks can be coated at even very high web speed (1–15 m/s). A disadvantage of the technique is the exposure to the ink to the ambient as the ink may stay in the ink bath for hours before the processing. The thickness of the resulting film mainly depends on the depth of the gravure pattern and the efficiency of gravure–web transfer. Gravure printing is increasing in popularity in PSC, where both active layers and cathode and anode buffer layers have been processed. Good performance has been obtained with normal geometry cells, where, for example, 2.8% was obtained for 19 cm^2 cells for gravure-printed PEDOT:PSS and P3HT:PCBM layers with thermally deposited electrodes.

12.3.2.2 Flexographic Printing
A novel R2R technique in PSC processing is flexographic printing, which is similar to gravure printing, but has additional steps in its process. A fountain roller is submerged in an ink bath and transfers ink onto an anilox roller, which is a roller with engraved cells for which the pattern will ultimately be transferred to the web (Fig. 12.15b). During rotation of the rollers, the cells are filled with ink and excess ink is doctor bladed as in the case of the gravure printing. The anilox roller transfers the ink onto a printing plate cylinder, which subsequently transfers the ink onto the moving web. The pattern thus stands out on the printing plate, and thus the transfer is different from the gravure ink-to-web transfer, which provides an alternative setting, which may perform better for some materials. Flexographic printing has been demonstrated for processing of PEDOT:PSS (65) and for surface wetting agent on the active layer to increase surface energy (23). While being a complex processing technique, high resolution features can be made, and flexographic printing may become more widely used in future PSC processing if high spatial precision gains importance.

12.3.2.3 Screen Printing
Screen printing is a printing technique that has been widely used for different applications for millennia. It is based on the application of an ink through a stencil on a surface, by which the stencil pattern is obtained. The stencil is supported by a woven mesh, which may be fixed onto a frame under tension for mechanical support. The frame with the stencil is positioned on the cell, and precise alignment is performed to deposit the electrode correctly. Ink is deposited on the mesh and distributed on the cell with a so-called squeegee by moving the squeegee across the stencil (Fig. 12.15c). The woven mesh does not allow for penetration of the ink, while ink is deposited through the pattern of the stencil and onto the cell. Screen printing offers a 2-dimensional control of the processing as the stencil can be made in any

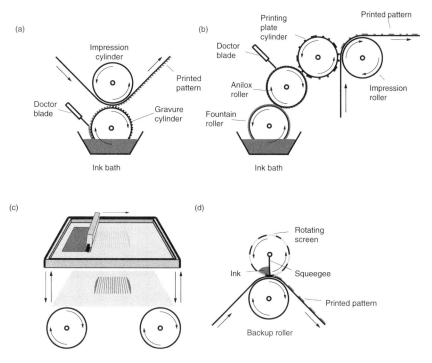

Figure 12.15 Central printing techniques for PSC: (a) Gravure, (b) flexographic, (c) screen printing, and (d) rotary screen printing. (*Source*: Reprinted with permission).

shape. Screen printing differs from other deposition techniques that wet thicknesses are obtained and that high volatility and low viscosity of the coating solution are needed. For inorganic photovoltaics, screen printing is routinely used to deposit bus bars been the different cells (60). Alternative applications include deposition of etch resists for the semiconductor industry, and text on textile fabrics. For PSCs, screen printing is not highly suitable for deposition of most buffer and active layers, as viscosities are generally low, even though successful attempt have been made (66, 67). However, silver-formulated inks are widely processed as inverted cell electrodes. Compared to thermal evaporation of the electrode, screen printing offers rapid electrode deposition while not compromising the cell performance significantly. Screen printing can applied to a R2R process where the web is moved on rolls and stopped under the screen printing frame. The frame is lowered onto the web, the ink is deposited and distributed by the squeegee, the frame is lifted off the web, and the web moves to the next cell position. While this process highly increases production capacity compared to thermal electrode evaporation, incorporation into an integrated R2R process is not possible due to the web movement that cannot be combined with other simultaneous layer coating or printing processes. However, full R2R adaption of screen printing is possible by *rotary screen printing*, where the stencil is made on a cylinder, thus working as an infinite screen (Fig. 12.15d). The ink is kept inside the rotating screen and only allowed to exit through the stencil. The web is pressed against the rotating screen, which through its rotation deposits the material. An internal squeegee removes excessive ink by which a homogeneous wet thickness is obtained. As this deposition implies a continuous movement of the web, it is fully adaptable to an integrated R2R process, and thus is rotary screen printing expected to play a role as electrode processing technique if commercialization of PSCs is realized.

12.4 POLYMER SOLAR CELL STABILITY

An inherent problem to PSC technology is that all organic materials are not chemically inert and will over time react with surrounding materials. Furthermore, the multilayer nature of the PSC implies a large number of organic layer interfaces, which will degrade due to diffusion of the organic layers. In combination with several different other degradation mechanisms, stability of PSCs is low compared to IPV, and extensive research has been directed at increasing stability as this parameter is paramount for the commercial potential of PSCs (24). Degradation of PSCs is normally divided into physical and chemical degradation. Chemical degradation includes, for example, the photochemical degradation of the active layer, diffusion of water and oxygen into the cell, degradation of the ITO, and the metal electrode. Physical degradation includes diffusion of the interfaces, evolution of the morphology over time of the active layer, and mechanical stability of cells where problems with, for example, delamination exist.

12.4.1 Degradation Mechanisms

Practical PSC stability depends on several parameters such as the components of the cell, the processing conditions (ambient/inert atmosphere), and properties of possible encapsulation. Furthermore, the stability depends highly on the degradation environment where effects such as UV content, humidity, and temperature are all known to accelerate degradation (3). The first stability studies demonstrated how unencapsulated normal geometry PPV-based cells degraded within hours (68). With the development of the inverted geometry and the more stable polyhexylthiophenes, cell stabilities increased to days for unencapsulated cells. With the development of R2R cells, encapsulation has become a standardized processing step, and stabilities have now been increased to years (24, 68).

12.4.2 Approaches to Cell Stabilization

The light absorption properties of the polymer are paramount to PSC's performance and the absorption stability is central to the discussion of PSC stability. As polymers are exposed to both light and oxygen, different degradation mechanisms take place depending on the nature of the polymer. Generally, UV illumination is very harsh to the polymers as aggressive species are generated such as hydrogen peroxide and superoxides, which will attack all organic material. Conventionally, photochemical stability of organic material is studied by monitoring the UV–vis absorbance during degradation in solar simulators, where the material is continuously exposed to illumination in the ambient. Early work on the PPV family demonstrated its low ambient stability (69), which was the major reason for low overall solar cell stability. The degradation mechanisms taking place in the ambient (photooxidation) and in inert atmosphere (photolysis) were elucidated by infrared spectroscopy (70). The instability was hereby attributed to the presence of vinylene bonds and the alkoxy substitutions. The next dominant family in PSCs, the poly(3-alkylthiophenes), presented higher photochemical stabilities as the thiophene backbone did not suffer from the same instabilities as the PPV family. Consequently, P3HT is around 50 times more stable than MEH-PPV (43), which implied significantly increased solar cell stability. Both photolysis and photooxidative degradation mechanisms were identified where the degradation was ascribed to the abstraction of the allylic hydrogen leading to chain scission and loss of absorbance (71). The recent development of donor–acceptor polymers has introduced a large variety of chemical groups with different photochemical stabilities. An extensive stability study of donor–acceptor polymers provided stability rankings of a large number of electron-withdrawing and electron-donating groups (72). Furthermore, photochemical stability of polymers was found to increase when blended with electron acceptors, which is generally ascribed to the transfer of the highly reactive, excited electron to the generally more stable electron acceptor (73). With the advent of electron acceptors with high LUMO levels, generally higher V_{oc} has been obtained for standard polymers such as P3HT, see Section 12.2.1. However, a direct consequence of an increasing LUMO level and thus a lower LUMO–LUMO energy gap was found to be decreasing blend stability as electron transfer becomes less favorable (43).

Another extensively research aspect within PSC stability is the evolution of the active layer morphology. As described in Section 12.2.2, the morphology is highly dynamic and can be affected by, for example, the choice of solvent and postprocessing annealing. Just as obtaining a good morphology is not straightforward, freezing this morphology over time adds another level of complexity. Different approaches have been followed decrease the mobility of the polymer inside the blend. Generally, the polymer side chains are included to allow for solution processing, while after processing, they only destabilize the polymer. Consequently, polymers have been developed that by thermal annealing after processing cleave off the side chains, which evaporate out of the film (74). Hereby, the polymer is not soluble anymore and thus processing of subsequent layers does not demand for orthogonal solvents. Furthermore, the highly immobile, dense structure obtained is physically stable reducing morphology evolution and thus cell stabilities increase significantly. Finally, the photochemical stability of the polymer itself increases by approximately a factor of 4 after side chain release (75). However, performances are only approaching 2% for thermocleavable polymers (76), and thus, optimization is needed to reach the 10–10 goals.

Alternative approaches to freezing the morphology have been suggested by cross-linking the polymer after processing, whereby a rigid network is created and thus the individual polymer domains are frozen spatially. By functionalization of the monomer by different groups such as oxetane, alkyl-bromide, azide, and vinyl, cross-linking takes place when the film is UV illuminated. While the photochemical stability of the polymer has been found to remain unchanged after functionalization (77), especially the cell stability in inert atmosphere is highly increased after cross-linking. Bumjoon et al. demonstrated highly stable bromine-functionalized P3HT cells, which only experienced negligible degradation after thermal annealing in the dark in inert atmosphere after 50 h annealing (78). Likewise, a bromine-functionalized high-efficiency donor–acceptor polymer demonstrated that the efficiency was conserved after functionalization and that after thermal annealing in the dark in inert atmosphere, a stable efficiency of 4.6% was obtained (79). Consequently, cross-linking holds the potential to completely freeze the photoactive layer morphology, which is important for increasing cell lifetime to the order of years.

While PSCs started out as being a highly unstable PV type, several different techniques have been developed by which lifetime has been increased. While the active layer has been highly stabilized photochemically and in terms of morphology, other degradation mechanisms such as oxygen and water diffusion are still dominant. However, with the advent of encapsulation techniques on a R2R scale, the effect of these mechanisms have been highly decreased, and R2R cells now routinely exhibit stabilities of months or even years. Overall, cell stability is expected to increase in the future as new approaches to limiting the effect of individual degradation mechanisms and the ambitious perspectives are that lifetimes increase to the order of several years for cells based on fully scalable processing techniques such as R2R.

12.5 PRESENT STATE OF POLYMER SOLAR CELLS AND PERSPECTIVES

PSC has reached a level of maturity, where high efficiencies (>10%), high stability (>1 year), and high processability have been demonstrated. As the only practically feasible approach to a large-scale PSC production is R2R processing, the performance of R2R cells is especially important. When moving from glass substrate to flexible substrate laboratory cells, practical performance drops from approximately 10% to 4% with screen-printed electrodes and generally larger active areas. However, when moving to single R2R processed cells, performance is further decreased, while on a module level ($>100\,cm^2$), performance drops to around 2% (59). The main reason for the decrease of performance with increasing active area is that R2R processing constrains the parameters by which performance can be optimized. Furthermore, larger cells are impeded by increased importance of serial resistance (80). Finally, as the active area increases, a larger area including contaminations, inhomogeneities, and so on affects the cell performance. This implies that where small cell performances typically follow a wide normal distribution, larger cells attain a more narrow efficiency distribution with lower maximum efficiency, as performance averages over a larger area.

While full commercialization of PSC is presently not viable, niche products have been demonstrated where conventional products have been shown to be replaceable with PSCs-based products. For lightning in developing rural areas without access to the electrical grid, typically kerosene-fueled lamps are used. However, a product was demonstrated where a PSC was combined with a battery and an LED (1). By replacing a kerosene lamp with the solar cell-based lamp, it was found that only 44 days of operation was needed to recover the energy invested into its production. PSCs thus at present provide the basis for niche products that are competitive with existing energy technologies and are believed to play a role in the worldwide solution to increasing future energy demands.

REFERENCES

1. Krebs FC, Nielsen TD, Fyenbo J, Wadstrøm M, Pedersen MS. Manufacture, integration and demonstration of polymer solar cells in a lamp for the "Lighting Africa" initiative. Energy Environ Sci 2010;3:512.

2. REN21 Renewables 2011 - Global Status Report; 2011.

3. Krebs F. *Polymeric Solar Cells - Materials, Design, Manufacture*. Lancaster (PA): DEStech Publications Inc.; 2010.

4. Chapin DM, Fuller CS, Pearson GL. A new silicon p-n junction photocell for converting solar radiation into electrical power. J Appl Phys 1954;25:676.

5. Green M, Emery K. Solar cell efficiency tables (version 40). Prog Photovoltaics Res Appl 2012:606–614. DOI: 10.1002/pip.

6. Bagnall D, Boreland M. Photovoltaic technologies. Energy Policy 2008;36:4390–4396.

7. El Chaar L, Lamont LA, El Zein N. Review of photovoltaic technologies. Renew Sustain Energy Rev 2011;15:2165–2175.

8. Shirakawa H, Louis EJ, MacDiarmid AG, Chiang CK, Heeger AJ. Synthesis of electrically conducting organic polymers: halogen derivatives of polyacetylene, (CH)x. J Chem Soc Chem Commun 1977:578. DOI: 10.1039/c39770000578.

9. Su W, Schrieffer J, Heeger AJ. Solitons in polyacetylene. Phys Rev Lett 1979;42:1698–1701.

10. Sariciftci NS, Smilowitz L, Heeger AJ, Wudl F. Photoinduced electron transfer from a conducting polymer to buckminsterfullerene. Science 1992;258:1474–1476.

11. Sariciftci NS et al. Semiconducting polymer-buckminsterfullerene heterojunctions: diodes, photodiodes, and photovoltaic cells. Appl Phys Lett 1993;62:585.

12. Hummelen J, Knight B. Preparation and characterization of fulleroid and methanofullerene derivatives. J Org Chem 1995;60:532–538.

13. Yu G, Gao J, Hummelen J. Polymer photovoltaic cells: enhanced efficiencies via a network of internal donor-acceptor heterojunctions. Science 1995;270:1789–1791.

14. Padinger F, Rittberger RS, Sariciftci NS. Effects of postproduction treatment on plastic solar cells. Adv Funct Mater 2003;13:85–88.

15. Ma W, Yang C, Gong X, Lee K, Heeger AJ. Thermally stable, efficient polymer solar cells with nanoscale control of the interpenetrating network morphology. Adv Funct Mater 2005;15:1617–1622.

16. Bundgaard E, Krebs F. Low band gap polymers for organic photovoltaics. Sol Energy Mater Sol Cells 2007;91:954–985.

17. Mühlbacher D et al. High photovoltaic performance of a Low-bandgap polymer. Adv Mater 2006;18:2884–2889.

18. Hou J, Chen H-Y, Zhang S, Li G, Yang Y. Synthesis, characterization, and photovoltaic properties of a low band gap polymer based on silole-containing polythiophenes and 2,1,3-benzothiadiazole. J Am Chem Soc 2008;130:16144–16145.

19. Small C et al. High-efficiency inverted dithienogermole–thienopyrrolodione-based polymer solar cells. Nat Photonics 2011:1–6. DOI: 10.1038/NPHOTON.2011.317.

20. Dou L, You J, Yang J, Chen C. Tandem polymer solar cells featuring a spectrally matched low-bandgap polymer. Nat Photonics 2012;6:180–185.

21. Espinosa N, García-Valverde R, Urbina A, Krebs FC. A life cycle analysis of polymer solar cell modules prepared using roll-to-roll methods under ambient conditions. Sol Energy Mater Sol Cells 2011;95:1293–1302.

22. Søndergaard R, Hösel M, Angmo D, Larsen-Olsen TT, Krebs FC. Roll-to-roll fabrication of polymer solar cells as the performance in terms of power conversion efficiency and operational. Mater Today 2012;15:36–49.

23. Krebs FC, Fyenbo J, Jørgensen M. Product integration of compact roll-to-roll processed polymer solar cell modules: methods and manufacture using flexographic printing, slot-die coating and rotary screen printing. J Mater Chem 2010;20:8994.

24. Jørgensen M et al. Stability of polymer solar cells. Adv Mater 2011;24:580–612.

25. Gregg BA, Hanna MC. Comparing organic to inorganic photovoltaic cells: theory, experiment, and simulation. J Appl Phys 2003;93:3605.

26. Fox M. *Optical Properties of Solids*. Oxford University Press; 2008.

27. Shaw PE, Ruseckas A, Samuel IDW. Exciton diffusion measurements in poly(3-hexylthiophene). Adv Mater 2008;20:3516–3520.

28. Malinen M, Kuusisto J-M. *Research, Development and Commercialisation Activities in Printed Intelligence.* VTT Printed Intelligence; 2010. p 66–67.

29. Arango AC, Johnson LR, Bliznyuk VN, Schlesinger Z, Carter SA, Hörhold H-H. Efficient titanium oxide/conjugated polymer photovoltaics for solar energy conversion. Adv Mater 2000;12:1689–1692.

30. Liu Y, Summers MA, Edder C, Fréchet JMJ, McGehee MD. Using resonance energy transfer to improve exciton harvesting in organic–inorganic hybrid photovoltaic cells. Adv Mater 2005;17:2960–2964.

31. Hau SK et al. Air-stable inverted flexible polymer solar cells using zinc oxide nanoparticles as an electron selective layer. Appl Phys Lett 2008;92:253301.

32. Ratier B, Nunzi J-M, Aldissi M, Kraft TM, Buncel E. Organic solar cell materials and active layer designs-improvements with carbon nanotubes: a review. Polym Int 2012;61:342–354.

33. Vandewal K, Tvingstedt K, Gadisa A, Inganäs O, Manca JV. On the origin of the open-circuit voltage of polymer-fullerene solar cells. Nat Mater 2009;8:904–909.

34. Yamamoto S, Orimo A, Ohkita H, Benten H, Ito S. Molecular understanding of the open-circuit voltage of polymer: fullerene solar cells. Adv Energy Mater 2012;2:229–237.

35. Kippelen B, Brédas J-L. Organic photovoltaics. Energy Environ Sci 2009;2:251.

36. Wienk MM et al. Efficient methano[70]fullerene/MDMO-PPV bulk heterojunction photovoltaic cells. Angew Chem Int Ed 2003;42:3371–3375.

37. Shaheen SE et al. 2.5% efficient organic plastic solar cells. Appl Phys Lett 2001;78:841.

38. Dennler G, Scharber MC, Brabec CJ. Polymer-fullerene bulk-heterojunction solar cells. Adv Mater 2009;21:1323–1338.

39. Scharber MC et al. Design rules for donors in bulk-heterojunction solar cells—towards 10 % energy-conversion efficiency. Adv Mater 2006;18:789–794.

40. Chen H et al. Polymer solar cells with enhanced open-circuit voltage and efficiency. Nat Photonics 2009;3:649–653.

41. Cheng Y-J, Yang S-H, Hsu C-S. Synthesis of conjugated polymers for organic solar cell applications. Chem Rev 2009;109:5868–5923.

42. Helgesen M, Søndergaard R, Krebs FC. Advanced materials and processes for polymer solar cell devices. J Mater Chem 2010;20:36.

43. Tromholt T, Madsen MV, Carlé JE, Helgesen M, Krebs FC. Photochemical stability of conjugated polymers, electron acceptors and blends for polymer solar cells resolved in terms of film thickness and absorbance. J Mater Chem 2012;22:7592–7601.

44. Zhao G, He Y, Li Y. 6.5% efficiency of polymer solar cells based on poly(3-hexylthiophene) and indene-C(60) bisadduct by device optimization. Adv Mater 2010;22:4355–4358.

45. Patyk RL et al. Carbon nanotube–polybithiophene photovoltaic devices with high open-circuit voltage. Phys Status Solidi RRL 2007;1:R43–R45.

46. Weickert J, Dunbar RB, Hesse HC, Wiedemann W, Schmidt-Mende L. Nanostructured organic and hybrid solar cells. Adv Mater 2011;23:1810–1828.

47. Takanezawa K, Tajima K, Hashimoto K. Efficiency enhancement of polymer photovoltaic devices hybridized with ZnO nanorod arrays by the introduction of a vanadium oxide buffer layer. Appl Phys Lett 2008;93:063308.

48. Alstrup J, Jørgensen M, Medford AJ, Krebs FC. Ultra fast and parsimonious materials screening for polymer solar cells using differentially pumped slot-die coating. ACS Appl Mater Interfaces 2010;2:2819–2827.

49. Mihailetchi VD et al. Origin of the enhanced performance in poly(3-hexylthiophene): [6,6]-phenyl C[sub 61]-butyric acid methyl ester solar cells upon slow drying of the active layer. Appl Phys Lett 2006;89:012107.

50. Yao Y, Hou J, Xu Z, Li G, Yang Y. Effects of solvent mixtures on the nanoscale phase separation in polymer solar cells. Adv Funct Mater 2008;18:1783–1789.

51. Peet J et al. Efficiency enhancement in low-bandgap polymer solar cells by processing with alkane dithiols. Nat Mater 2007;6:497–500.

52. Yang Y, Mielczarek K, Aryal M, Zakhidov A, Hu W. Nanoimprinted polymer solar cell. ACS Nano 2012;6:2877–2892.

53. Takanezawa K, Hirota K, Wei Q-S, Tajima K, Hashimoto K. Efficient charge collection with ZnO nanorod array in hybrid photovoltaic devices. J Phys Chem C 2007;111:7218–7223.

54. Po R, Carbonera C, Bernardi A, Camaioni N. The role of buffer layers in polymer solar cells. Energy Environ Sci 2011;4:285.

55. Shrotriya V, Li G, Yao Y, Chu C-W, Yang Y. Transition metal oxides as the buffer layer for polymer photovoltaic cells. Appl Phys Lett 2006;88:073508.

56. Irwin M, Buchholz D. p-Type semiconducting nickel oxide as an efficiency-enhancing anode interfacial layer in polymer bulk-heterojunction solar cells. Proc Natl Acad Sci U S A 2008;105:2783–2787.

57. Kim JS et al. Control of the electrode work function and active layer morphology via surface modification of indium tin oxide for high efficiency organic photovoltaics. Appl Phys Lett 2007;91:112111.

58. Wang J-C et al. Highly efficient flexible inverted organic solar cells using atomic layer deposited ZnO as electron selective layer. J Mater Chem 2010;20:862.

59. Krebs FC, Tromholt T, Jørgensen M. Upscaling of polymer solar cell fabrication using full roll-to-roll processing. Nanoscale 2010;2:873–886.

60. Krebs FC. Fabrication and processing of polymer solar cells: a review of printing and coating techniques. Sol Energy Mater Sol Cells 2009;93:394–412.

61. Søndergaard R, Helgesen M, Jørgensen M, Krebs FC. Fabrication of polymer solar cells using aqueous processing for all layers including the metal back electrode. Adv Energy Mater 2011;1:68–71.

62. Wengeler L, Schmidt-Hansberg B, Peters K, Scharfer P, Schabel W. Investigations on knife and slot die coating and processing of polymer nanoparticle films for hybrid polymer solar cells. Chem Eng Process 2011;50:478–482.

63. Dam HF, Krebs FC. Simple roll coater with variable coating and temperature control for printed polymer solar cells. Sol Energy Mater Sol Cells 2011;97:191–196.

64. Andersen TR et al. Aqueous processing of low-band-gap polymer solar cells using roll-to-roll methods. ACS Nano 2011;5:4188–4196.

65. Hübler A et al. Printed paper photovoltaic cells. Adv Energy Mater 2011;1:1018–1022.

66. Krebs FC, Alstrup J, Spanggaard H, Larsen K, Kold E. Production of large-area polymer solar cells by industrial silk screen printing, lifetime considerations and lamination with polyethyleneterephthalate. Sol Energy Mater Sol Cells 2004;83:293–300.

67. Shaheen SE, Radspinner R, Peyghambarian N, Jabbour GE. Fabrication of bulk heterojunction plastic solar cells by screen printing. Appl Phys Lett 2001;79:2996.

68. Frederik C. *Stability and Degradation of Organic and Polymer Solar Cells*. John Wiley & Sons, Ltd; 2012.

69. Jørgensen M, Norrman K, Krebs FC. Stability/degradation of polymer solar cells. Sol Energy Mater Sol Cells 2008;92:686–714.

70. Atreya M et al. Stability studies of poly(2-methoxy-5-(2H-ethyl hexyloxy)-p- (phenylene vinylene) [MEH-PPV]. Polym Degrad Stab 1999;65:287–296.

71. Chambon S, Rivaton A. Aging of a donor conjugated polymer: photochemical studies of the degradation of poly[2-methoxy-5-(3′,7′-dimethyloctyloxy)-1,4-phenylenevinylene]. J Polym Sci Part A Polym Chem 2007:317–331. DOI: 10.1002/pola.

72. Manceau M, Rivaton A, Gardette J-L, Guillerez S, Lemaître N. The mechanism of photo- and thermooxidation of poly(3-hexylthiophene) (P3HT) reconsidered. Polym Degrad Stab 2009;94:898–907.

73. Manceau M et al. Photochemical stability of π-conjugated polymers for polymer solar cells: a rule of thumb. J Mater Chem 2011;21:4132–4141.

74. Rivaton A et al. Light-induced degradation of the active layer of polymer-based solar cells. Polym Degrad Stab 2010;95:278–284.

75. Gevorgyan SA, Krebs FC. Bulk heterojunctions based on native polythiophene. Chem Mater 2008;20:4386–4390.

76. Manceau M, Helgesen M, Krebs FC. Thermo-cleavable polymers: materials with enhanced photochemical stability. Polym Degrad Stab 2010;95:2666–2669.

77. Helgesen M, Krebs FC. Photovoltaic performance of polymers based on dithienylthienopyrazines bearing thermocleavable benzoate esters. Macromolecules 2010;43:1253–1260.

78. Carlé JE et al. Comparative studies of photochemical cross-linking for stabilizing the bulk hetero-junction morphology in polymer solar cells. J Mater Chem 2012. DOI: 10.1039/c2jm34284g.

79. Kim BJ, Miyamoto Y, Ma B, Fréchet JMJ. Photocrosslinkable polythiophenes for efficient, thermally stable, organic photovoltaics. Adv Funct Mater 2009;19:2273–2281.

80. Choi S, Potscavage WJ, Kippelen B. Area-scaling of organic solar cells. J Appl Phys 2009;106:054507.

13

NEXT-GENERATION GaAs PHOTOVOLTAICS

GIACOMO MARIANI AND DIANA L. HUFFAKER

University of California, Los Angeles, CA, USA

13.1 INTRODUCTION

13.1.1 Unequal Energy Distribution

In 2013, NASA publicly shared a satellite image that is often referred to as a "satellite photo of Earth at night" (1). It actually represents a collection of hundreds of pictures taken from a sensor mounted onto the NASA NOAA Suomi National Polar-orbiting Partnership Satellite launched in 2011.

The sensor allows observing both atmosphere and surfaces at nighttime, constituting a map of the location of lights on Earth's surface. Every white dot can be the light of a city, fire, ship, oil well flare, or other light sources. The map (reported in Fig. 13.1) shows that the main cities are concentrated in Europe, the eastern United States, Japan, China, and India. The distribution of nighttime electricity consumption for outdoor lighting is extremely unevenly distributed, with major dark areas in South America, sub-Saharan Africa, Asia, and Australia.

13.1.2 Solar Energy

The sun is the most plentiful energy source for the earth. Wind, fossil fuel, hydro, and biomass energies exist because of sunlight. The sun delivers about 3.8×10^{26} J of energy every second, equivalent to 90 billion hydrogen bombs detonating each second.

Despite the fact that only 1 billionth of that energy (1.7×10^{17} J) falls on the surface of the earth, 1 day of sunlight is sufficient to power the whole human race needs for more than half a century. Figure 13.2 displays an increase of 40% in the global energy demand by 2030. The electricity demand is growing twice as fast, projected to be 76% more than the current usage by 2050 (2). Recent global directions witness the increasing importance of renewable sources of energy compared to conventional sources (e.g., coal, natural gas, oil, nuclear) (3). Global, domestic, and regional policies are being drafted and enacted to facilitate and foster a variety of green energy options in the quest of climate change mitigation toward grid parity with respect to the ordinary energy suppliers. The impact of solar energy installation and adoption is not only relegated to a beneficial reduction in CO_2 emission. It will create a social awareness to adopt alternative energies with a high technological, economic, and political impact. A report from the "Gigaton Throwdown Initiative" foresees that solar photovoltaics (together with other renewable energy sources) will scale up in the market more aggressively than current projections to alleviate CO_2 emissions. It is also expected that the solar energy will provide 11% of the total global electricity generated by 2050 (4).

13.1.3 Photovoltaic Technologies

Photovoltaic technology converts sunlight into electricity and is also referred to as solar electricity. Solar electricity powers everything from small calculators and remote highway signals to commercial buildings, large power plants, and satellites in space for the world's communications. Several types of solar technologies are currently available. Each technology is based on different concepts and fundamental scientific principles. Figure 13.3 summarizes the type of energy conversion, concentrated/nonconcentrated irradiation, and the resulting available technologies offered on the market. Photovoltaic solar panels represent the most commonly used solar technology for electricity

Nanomaterials, Polymers and Devices: Materials Functionalization and Device Fabrication, First Edition. Edited by Eric S. W. Kong.
© 2015 John Wiley & Sons, Inc. Published 2015 by John Wiley & Sons, Inc.

Figure 13.1 Map of light distribution during nighttime across the globe (1).

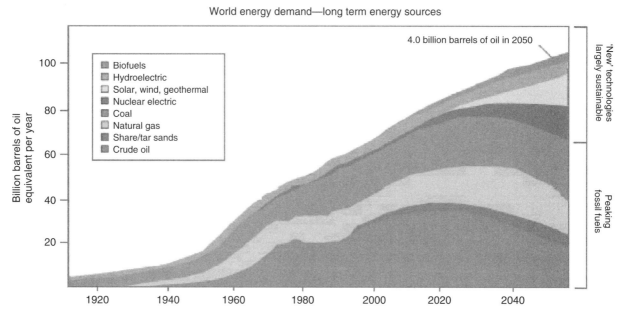

Figure 13.2 Projected world energy demand/differentiation of several sources. (*Source*: Lynn Orr, *Changing the World's Energy Systems*, Stanford University Global Climate & Energy Project (after John Edwards, American Association of Petroleum Geologists); SRI Consulting).

generation. The basic principle is to convert the solar energy (energetic photons) into useful electricity (electrons) by means of semiconductor junctions. The photovoltaic effect was first discovered in 1905 by Albert Einstein who noticed that electrons can be emitted from solids, liquids, or gases when they absorb energy from light. Semiconductor materials that exhibit such properties include monocrystalline (mc) silicon, polycrystalline (pc) silicon, amorphous silicon, cadmium telluride, copper indium selenide, and compound semiconductors.

The vast majority of photovoltaic solar panels are based on crystalline silicon. Each silicon solar cell is fabricated from wafer ingots. Solar panels based on mc-Si can achieve power conversion efficiencies between 14% and 20% (5). Pc-Si solar panels exhibit lower efficiencies but are less expensive and more resistant to irradiation degradation. The second approach is to exploit concentrated photovoltaics that uses optics, such as parabolic mirrors or polymer Fresnel lenses, to concentrate a large amount of sunlight onto a small area of semiconductor material to generate electricity. Generally, concentrated photovoltaic systems achieve the highest efficiencies (higher than 30%) among all types of solar technologies (Fig. 13.4). The amount of the semiconductor material (i.e., multijunction solar cells) is greatly reduced compared to flat plate silicon. This results in a cost-effective method to produce electricity. Some of the main limitations are the collection of diffuse radiation and the need of a precise tracking system and related moving parts that can increase both manufacturing and maintenance costs. Dye-sensitized solar

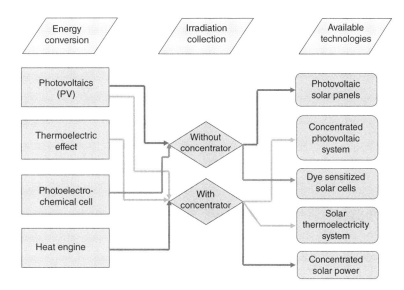

Figure 13.3 Taxonomy of solar technologies.

Figure 13.4 Overview of several solar power systems integrated on the field.

cells obtain energy from sunlight using dyes, which are abundant, cheap, and environmentally friendly. Such solar cells are semitransparent, lightweight, and flexible, offering greater possibilities and novel applications. Dye-sensitized solar cells generally consist of two sandwiched electrodes with an electrolyte solution between them. A porous network of TiO_2 nanoparticles is attached to the front surface electrical contact (anode). Organic dye molecules are adsorbed onto the surface of the TiO_2 particles. They absorb light and inject electrons into the TiO_2 particles. The electrons then diffuse through the TiO_2 particle network to the anode. Via an external circuit, they reach the back surface contact (cathode). The electrolyte solution captures the electrons at the cathode and transfers the electrons back to the dye molecules in order to complete the electrical circuit (6). The current efficiency for this technology is approximately 10%. Solar thermoelectricity produces electricity by capturing sunlight and concentrating it onto a thermoelectric generator. A thermoelectric device converts differences in temperature into useful electricity using a series of p–n semiconductor junctions. The main figure of merit to characterize thermoelectric materials is the Seebeck coefficient that represents a measure of the magnitude of an induced thermoelectric voltage in response to a temperature difference across the material. Such systems can perform in harsh environments; they are quite, extremely reliable and can be driven by low-grade heat energy. Despite that, the efficiency is still below 2%. Concentrated solar power systems utilize mirrors (i.e., heliostats) or lenses to concentrate a large area of sunlight onto a reduced area. The focused light produces heat that can drive a steam turbine connected to an electrical power generator. Steam-cooled combined cycle gas turbines can achieve efficiencies up to 60%.

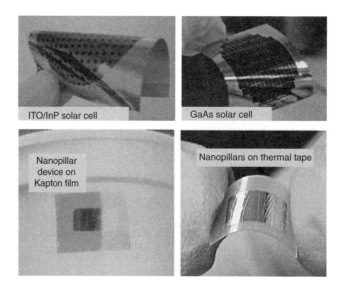

Figure 13.5 Lift-off techniques for low-cost, high-efficiency photovoltaics. From top, clockwise: lifted-off 2 inch-diameter epitaxial layer containing arrays of ITO/InP solar cells, lifted-off GaAs solar cells fabricated by combining epitaxial lift-off and cold welding onto a 50 μm-thick Kapton® sheet (7), nanopillar mesh device transferred onto thermal tape, and Kapton® film (8).

13.1.4 Increased Efficiency at Reduced Costs

One route to realize high-efficiency solar cells at low cost is to lift off the active solar cell from the native substrate after fabrication: this ensures a cyclic reuse of the semiconductor platform for consecutive devices. One of the requirements to do so is to exploit highly absorbing semiconductors where only a few microns of material is required to absorb most of the above-bandgap photons.

Direct-bandgap III–V compound semiconductors (e.g., GaAs, InP, InGaP) exhibit high optical absorption coefficients due to the direct photon transition, where the top of the valence band and the bottom of the conduction band occur at the same value of momentum. Therefore, a photon of energy E_g, where E_g is the energy bandgap, can produce an electron–hole pair in a direct-bandgap semiconductor quite easily since the electron does not need any momentum, provided by a lattice vibration called phonon. Epitaxial lift-off process enables the separation of the thin-film, single-crystal layer from the mother substrate by employing a sacrificial release layer. Such approach is attractive for photovoltaic applications by reducing the manufacturing costs through a successive reuse of the mother substrate (Fig. 13.5). In addition, semiconductor nanopillars (or nanowires) represent a promising architecture for solar energy-harvesting applications since they can drastically reduce the portion of reflected photons, enhancing the optical absorption. Releasing the nanostructures from the substrate and embedding them into mechanically flexible supports provide a viable solution for high-efficiency and low-cost photovoltaics. The high-efficiency component is maintained through the optical trapping effects and enhanced absorption sustained by the nanopillar configuration, whereas the cost-effectiveness originates from the reduced amount of material and the possibility to recycle the substrate for subsequent epitaxy.

13.1.5 Third-Generation Solar Cells

Third-generation photovoltaics aims to aggressively decrease the cost of solar energy below 1 $/W and of the second-generation solar cells to less than 0.20 $/W.

An increased efficiency results in energy at lower costs because of the smaller area necessary for a given power. Efficiency values well above 30% are required to heavily reduce the economical burden of solar technology compared to fossil-fuel-based electricity production. To achieve and surpass such efficiencies is necessary to understand the limiting mechanisms that prevent from utilizing the whole solar spectrum. There exist two principal loss mechanisms in single-junction solar cell.

The first is the thermalization of photons with energy greater than the semiconductor bandgap (Fig. 13.6, step 2), and the second is the thermalization of unutilized photons with energy less than the bandgap (Fig. 13.6, step 1).

In order to reduce both thermalization and photon loss with energy less than the bandgap, multijunction solar cells can achieve high efficiencies of up to 44.4% at 302 sun concentration (11). Multijunction solar cells can attain such high efficiency values by splitting the absorption of the broadband solar spectrum into semiconductors with different bandgaps. Higher bandgap cells can be mechanically (or monolithically) stacked on top of lower bandgap subcells. In this fashion, the solar spectrum can be optimally separated and absorbed.

For 1, 2, 3, 4, and 5 subcells, the theoretical efficiency η is 31.0% (40.8%), 42.5 (55.5%), 48.6% (63.2%), 52.5% (67.9%), and 68.2% (86.8%) for unconcentrated light (or concentrated), respectively. Typically, high-efficiency multijunction solar cells are based on III–V compound semiconductors monolithically grown by means of metal organic chemical vapor deposition (MOCVD) epitaxy. One of the main constraints for an effective material integration is to preserve the same lattice constant throughout the several bandgap stacks to avoid dislocations. Another approach to realize multijunction cells while reducing the cost per Watt is to utilize lower quality materials. Amorphous-Si (a-Si)

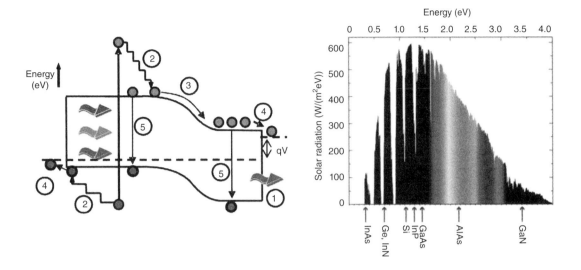

Figure 13.6 Standard loss processes in a solar cell (left) (9). Other loss mechanisms are the junction (step 3) and contact voltage loss (step 4) as well as unavoidable recombination loss (step 5). AM 1.5 solar spectrum and bandgaps of a selected number of solar cell materials (right) (10).

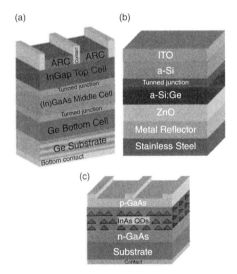

Figure 13.7 Examples of third-generation solar cells. (a) Triple-junction, (b) amorphous silicon tandem, and (c) intermediate-band solar cells.

tandem configurations combine together a-Si top cells with a:Si-Ge bottom cells with laboratory efficiencies up to 13%. Intermediate-band solar cells introduce one or more energy levels within the bandgap to absorb photons in parallel with the normal operation of a single-junction cell (Fig. 13.7). That enormously reduces the growth difficulties involved in the integration of several materials. Predicted theoretical efficiencies of 63% under concentration unveil the potential of this technology. Hot-carrier solar cells rely on carriers generated from high-energy photons (at least twice the bandgap energy) than can undergo impact ionization to create two or more carriers close to the bandgap energy. This effect has been recently demonstrated to be efficient in semiconductor quantum dots due to confinement of the process in a small volume.

13.2 OPTICAL AND ELECTRICAL PROPERTIES OF GaAs PLANAR SOLAR CELLS

13.2.1 Optical Absorption in III–V Compound Semiconductors

Compound semiconductors are formed by the interaction of elements in columns III and V of the periodic table. The binary alloys of particular interest, GaSb, InAs, GaAs, AlAs, InP, and GaP, have a crystalline zinc blende atomic structure constituted by two interpenetrating face-centered cubic sublattices displaced by a vector $\vec{r} = (a/4, a/4, a/4)$, where a is the crystalline lattice constant. A great strength of III–V

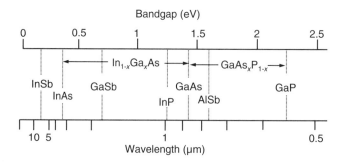

Figure 13.8   Bandgaps (in eV and μm) for typical III–V semiconductors.

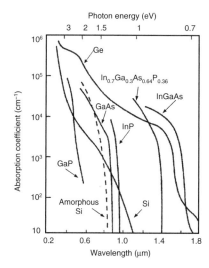

Figure 13.9   Optical absorption coefficients for common semiconductors (13).

semiconductor is the bandgap (energy difference between valence and conduction bands) spanning from ultraviolet to deep-infrared spectral regions (as schematically outlined in Fig. 13.8) (12). Besides, some of the compounds (i.e., InSb, InAs, InP, and GaAs) are direct bandgap, which means that electrons from the valence band can directly transit to the conduction band conserving momentum.

Other compounds (i.e., GaP) are characterized by an indirect bandgap, where phonon-assisted transitions of electrons from valence to conduction band are permitted. Typically, direct-bandgap semiconductors exhibit higher carrier mobilities and lower effective masses with respect to the indirect counterpart. The optical absorption coefficient dictates the penetration depth of a photon of a prescribed energy into the semiconductor material before it is absorbed.

Materials characterized by a low absorption coefficient absorb light poorly: the absorption coefficient depends on the material and also on the wavelength of light that is being absorbed. Figure 13.9 (13) shows the absorption coefficient for a variety of semiconductor materials. The plot highlights that for above-bandgap photons, the absorption coefficient is not constant but it is still strongly wavelength dependent. The probability of absorbing a photon is related to the likelihood of a photon and an electron to interact in such a way as to move from one energy band (valence) to another (conduction). For photons with energies in proximity of the bandgap, the absorption is relatively low since only those electrons directly at the valence band edge can interact with the photon to cause absorption. As the photon energy increases, not just the electrons already having energy close to that of the band gap can interact with the photon. Therefore, a larger number of electrons can interact with the photon and result in the photon being absorbed (14).

13.2.2   Window Layers in GaAs Solar Cells

GaAs solar cells are characterized by a high surface recombination due to a high density of surface states that originates from the exposure of the semiconductor surface to oxygen, which causes antisite formation of AsGa and GaAs (15). Avoiding recombination of photogenerated carriers is crucial for III–V photovoltaics with high absorption coefficients, resulting in very thin layers compared to Si-based solar cells. In fact, high surface recombination dramatically decreases the output power as well as the power conversion efficiency of GaAs solar cells. Surface recombination also increases the reverse saturation current, reducing the open-circuit voltage of the device. To mitigate the effect of surface states, higher bandgap materials are introduced to cap and avoid oxidation at the active surface of the solar cell. Such device component

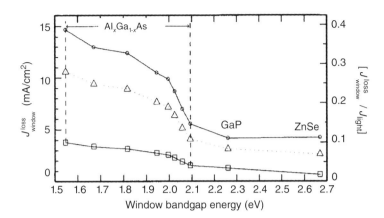

Figure 13.10   Theoretical loss of photocurrent in a GaAs solar cell due to parasitic absorption in the window passivating layer. The simulations are carried out for three different thicknesses: 100 nm (circles), 50 nm (triangles), and 10 nm (squares).

is called window layer. DeSalvo et al. analyzed the theoretical losses in the photogenerated current for a p–n GaAs photodiode due to parasitic absorption in the window layer as a function of bandgap and thickness (Fig. 13.10) (15).

It was found that for thick (100 nm) window layers, the current loss stems from the undesired optical absorption of photons that cannot reach the active junction. The current loss decreases as a function of higher window layer bandgaps. This is due to the fact that the window layers is more transparent to high-energy photons and also because it forms a higher recombination barrier in the energy band diagram to prevent more effectively annihilation of minority carriers at the surface. The majority of GaAs solar cells exploit an $Al_{0.85}Ga_{0.15}As$ passivating barrier at the top surface to successfully reduce the surface recombination. However, aluminum compounds can oxidize and they tend to degrade during air exposure, highlighting the need for encapsulants. Due to the minimal lattice mismatch, GaP and ZnSe are also identified as suitable window layers for GaAs photovoltaics. ZnSe is nearly lattice matched to GaAs with a bandgap of 2.67 eV (cutoff wavelength 465 nm), which is substantially wider compared to that of AlGaAs. Yater et al. demonstrated an 80 nm ZnSe layer grown atop a GaAs epitaxy (16). Room-temperature photoluminescence highlighted a fivefold increase in the peak signal with respect to an unpassivated GaAs sample.

13.2.3   Device Fabrication and Epitaxial Lift-Off

GaAs offers a high electron mobility and direct bandgap, and therefore, it is suitable for high-performance radio-frequency (RF) electronics and optoelectronics. Additionally, its bandgap lies at the energy for the theoretical maximum efficiency in the Shockley–Queisser approximation for single-junction solar cells (17). The epitaxial lift-off is a manufacturing technique that allows for the integration of III–V thin films/device onto arbitrary mechanical supports that can also be flexible (e.g., plastics, glass, textile, etc.). The process leaves an intact GaAs substrate behind that can be reused for growing successive epitaxial solar cells, removing a major cost hurdle from high-efficiency terrestrial solar cells.

Typical growth of GaAs solar cells is carried out using a MOCVD reactor. The entire device structure is usually grown on top of a 50 nm-thick AlAs sacrificial layer, which is selectively etched to release the solar cell from the native substrate. P-type contacts to the sample are realized by depositing Cr–Au metals, whereas n-type contacts are achieved by annealing at 380 °C aAuGe/Ni/Au metal alloy. Once fabricated, the device can be covered by a black wax, and the sacrificial epitaxial layer (i.e., AlAs for GaAs devices) can be etched away. The active solar cell is then lifted off from the native substrate and can be bonded to the new superstrate of choice. The epitaxial lift-off sequence is schematically depicted in Figure 13.11.

13.2.4   World-Record Efficiency for GaAs Solar Cells

Recently, Alta Devices® (Incorporated Company) has achieved a new world-record efficiency for a thin-film single-junction GaAs solar cell on a flexible substrate with a certified conversion efficiency of 27.6% (18, 19).

The efficiency improvement is largely due to a precise control of the dark current, which in turn results in a high open-circuit voltage. The solar cell is lifted off from the GaAs growth substrate that can be reused multiple times. The completed device is presented in Figure 13.12a. The certified efficiency is independently confirmed by the National Renewable Energy Laboratory (NREL) with a short-circuit current J_{SC} of 29.6 mA/cm², open-circuit voltage V_{OC} of 1.107 V, and a fill factor FF of up to 84.1%. The corresponding data is showed in Figure 13.12b. The device also exploits photon recycling to access such high V_{OC} values. The idea behind it is that photons absorbed in the GaAs region can be radiatively reemitted and then reabsorbed in the same GaAs region. In other words, the reemitted photons have a second chance to be absorbed and generate electron–hole pairs. Such phenomenon leads to a higher carrier density, leading to a greater quasi-Fermi level splitting and therefore increased V_{OC}. Alta Devices has currently achieved a new certified world-record efficiency for single-junction GaAs solar cells at 28.8% under 1 sun (1000 W/m²) illumination (20).

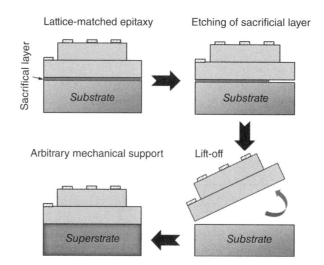

Figure 13.11   Epitaxial lift-off process for cost-effective high-efficiency solar cells.

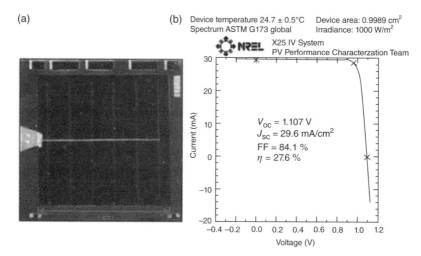

Figure 13.12   (a) Fabricated record-efficiency GaAs solar cell. (b) NREL-certified current–voltage characteristic under 1 sun illumination.

13.3   OPTICAL AND ELECTRICAL PROPERTIES OF GaAs NANOPILLAR SOLAR CELLS

13.3.1   Bottom-Up Epitaxial Growth

The first experimental demonstration of bottom-up nanowire (or nanopillar) vertical growth was reported by Wagner and Ellis in 1964 from Bell Laboratories (21). The growth method was based on a vapor–liquid–solid (VLS) mechanism; a metal (e.g., Au) catalyst particle is heated up to its liquid point in the presence of semiconductor gas precursors. The catalyst acts as a sink to collect material from the surrounding vapor sources. The metal particle supersaturates and precipitates the collected material in the form of a 1D semiconductor nanowire. The nanowire height is tuned from the amount of additional material that precipitates at the interface between the metal catalyst and the nanowire remainder (21). The VLS growth mechanism is schematically depicted in Figure 13.13.

Nanowire heterojunctions can be integrated in highly lattice-mismatched material systems since the induced strain can be coherently accommodated through lateral relaxation. III–V bulk materials are characterized by a cubic zinc blende (ZB) crystal structure. However, GaAs or InAs nanowires, for instance, exhibit a combination of ZB and wurtzite (WZ) crystal arrangements. Such effect is defined as polytypism and each ZB/WZ interface represents a rotational twin or stacking fault. Nanowire polytypism and growth direction strictly depend on the crystallographic orientation of the mother substrate. In case of GaAs, nanowires grown on <111>B GaAs substrates are vertically aligned, often polytypic with stacking faults. For <001> substrates, the nanowires are parallely aligned with no stacking fault observation. Stacking-fault-free vertical GaAs nanowires (22) are also synthesized starting from a <111>A growth substrate. Controlling polytypism represents a vibrant field of research in the nanowire community to improve the crystal quality along with electronic properties such as conductivity, carrier mobility, or junction formation. Another aspect to consider in catalyst-mediated growth is that the catalyst can

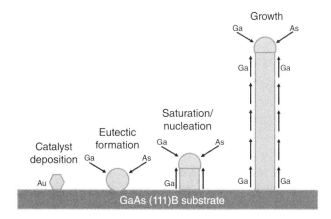

Figure 13.13   Schematic diagram showing the various steps in the VLS growth of a GaAs nanowire.

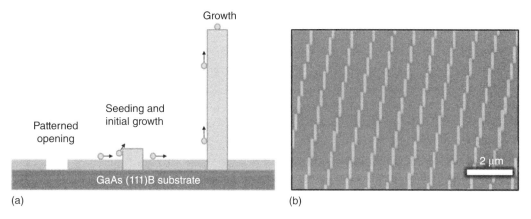

Figure 13.14   Schematic representation of selective-area nanopillar epitaxy. (a) Nanopillar growth depends on adsorption, diffusion, and hopping processes on the patterned mask and nanopillar facets. The synthesis sequence is divided into mask patterning, seeding and initial growth at the mask opening, and actual nanopillar growth. (b) Scanning electron microscope image of a selective-area GaAs nanopillar array.

incorporate into the nanowire during the crystal formation, forming deep-level traps that act as recombination centers for the charged carriers. In order to avoid this problem, selective-area epitaxy (SAE) allows for a controlled nanowire formation without requiring any catalyst (i.e., metals or semiconductor droplets) to initiate and sustain the synthesis. Patterned nanopillar formation by SAE offers a catalyst-free approach to avoid contamination, offering the ability to grow large arrays of nanostructures with lithographically defined diameters and locations. Due to the fact that the catalyst is absent, the adatom (atom that lies on a crystal surface) incorporation strictly depends on diffusion lengths and binding energies, as well as surface energetics of different crystallographic orientations (23). In SAE, an epitaxial layer nucleates in openings of a masked template (e.g., SiO_2, SiN_X, etc.) and continuously grows in height, while its lateral growth is restricted by low-energy facets.

   The wafer preparation commences with the deposition of a thin dielectric mask on an epiready semiconductor substrate (e.g., Si, GaAs, InAs, etc.). Subsequently, the nanopatterning can be achieved by nanoimprint, deep-ultraviolet, interference, or electron-beam lithography, for example. The exposed oxide areas can be wet or dry etched to uncover the underlying substrate. The nanopillar growth is generally performed by MOCVD. The growth is carried out in a vertical-flow reactor using trimethylgallium (TMGa) and tertiarybutylarsine (TBA). Metal–organics are constituted by individual metal atoms bonded to multiple organic molecules. The sequence to form catalyst-free GaAs nanopillars is presented in Figure 13.14.

13.3.2   Tuning the Optical Absorption of the Array

Nanopillar photovoltaics represents a great approach to achieve high-efficiency solar power conversion while reducing the associated manufacturing costs.

   In periodic and vertical arrangements of nanopillar arrays, the incident light field interacts with subwavelength nanostructures in a completely new manner compared to standard planar designs: while the nanopillars cover only a small percentage (5–15%) of the surface area, they exploit a principle known as resonant trapping to enhance the optical absorption by waveguiding light within the nanostructure as well as by recycling photons that bounce between adjacent nanopillars before being reflected back into the open space (24). Figure 13.15 presents a qualitative comparison of different light-coupling frameworks for planar semiconductor slabs versus a nanopillar arrangement of the same

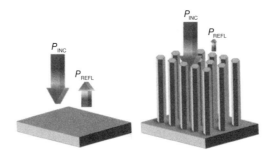

Figure 13.15   Comparison of light interaction with a planar semiconductor slab (left) and a nanopillar array of the same semiconductor (right). Resonant photon trapping and photon recycling is responsible for a dramatically reduced reflected optical power in the nanopillar array case.

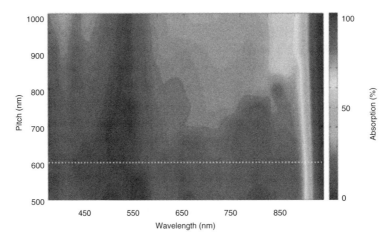

Figure 13.16   2D contour plot of broadband optical absorption with respect to increasing nanopillar pitch (from 500 nm to 1000 nm). Dark grey color indicates high absorption (conversely for lighter shade of grey color).

material. Light impinging onto a semiconductor slab is governed by Snell's law and Beer–Lambert's absorption law. With no antireflective coating or surface texturing, the portion of the reflected light is simply determined by the refractive index mismatch at the boundary of the two materials:

$$R = \left(\frac{n_1 - n_2}{n_1 + n_2} \right)^2 \tag{13.1}$$

Equation (13.1) describes the specular reflected power (R) arising from the refractive index mismatch between two heterogeneous media (n_1 and n_2). In case of the refractive index mismatch between GaAs ($n_1 = 3.9$) and air ($n_2 = 1.0$), the reflected power amounts to approximately 35% of the total incident power (24). This value can be reduced with antireflection coatings. Arrays of nanopillars are sequences of periodically arranged subwavelength structures. In such framework of objects smaller than the wavelength of light, the interaction of optical field/nanopillars cannot be analyzed with the standard geometrical optics. Full-wave simulations employing Maxwell's equations (and not approximations) are necessary to rigorously model the behavior of light once incident on nanopillars. By means of finite-difference time-domain simulations, it is possible to optimize the array geometry (i.e., diameter, center-to-center pitch, tiling pattern, etc.) to maximize the optical absorption while minimizing material usage.

Mariani et al. studied the GaAs nanopillar-array absorption as a function of pitch and tiling pattern, providing useful design guidelines in patterned geometries (25). The pitch dependence is summarized in Figure 13.16. The plot highlights that with highly packed arrays the absorption is enhanced: optical-resonance-guided modes increase the electromagnetic field that arises from the interaction of incident light with high-Q leaky modes. For larger pitch values, the optical field becomes less and less confined, causing a severe drop in the absorption of the array (e.g., pitch > 700 nm).

In Figure 13.17, four different tiling patterns (triangular, square, rectangular, parallelogram) are compared with respect to GaAs planar substrates. The areas for the unit cell in all four patterns are identical (600 nm × 600 nm), as a result, the filling ratio is kept constant, and only the effect of tiling pattern is taken into account in this comparison (25). The simulations show an optical absorption fairly independent with respect to different tiling patterns. Slight variations in the absorption can be attributed to the differences in the symmetry of the packing arrangements.

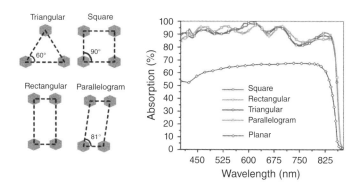

Figure 13.17 Wavelength-dependent absorption as a function of different tiling patterns (pitch at 600 nm).

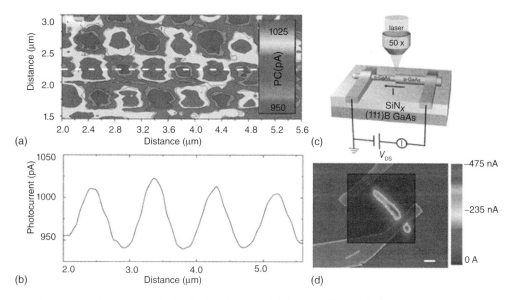

Figure 13.18 Scanning photocurrent microscopy applied to single and arrays of GaAs nanopillars. (a) SPCM map of a nanopillars array with 600 nm pitch. Red is indicative of maximum photocurrent, while blue is minimum photocurrent output. (b) Corresponding scan line. (c) Setup for single-nanowire SPCM and (d) resulting SPCM plot.

13.3.3 Mapping the Photocurrent Profile at the Nanoscale

Scanning photocurrent microscopy (SPCM) is a potent technique to analyze the charge transport in 1D semiconductors. In a conventional SPCM setup, a diffraction-limited laser spot is raster scanned at the surface of an electronic device, while the photocurrent is recorded as a function of the illumination position. SPCM constitutes an essential tool to analyze charge carrier transport as a function of space in the device: it can identify recombination processes before extraction at the contacts, quality of the p–n junction, and uniformity in the photocurrent spatial distribution across the whole solar cell. It represents an effective characterization method to study the operation of p–n junctions at the nanoscale for photovoltaic applications where both optical and electrical processes are equally important. Mariani et al. reported on an SPCM image (26) that maps the photocurrent profile of an array of nanopillars to verify that each nanostructured p–n junction indeed functions as an individual solar cell. The sample is mounted onto an X–Y piezotranslational stage with a step resolution of 50 nm. A green HeNe laser ($\lambda = 544$ nm) is used to illuminate the structure, and zero-bias photoconductivity is measured using a data acquisition board and a preamplifier. Figure 13.18a and b displays the acquired photocurrent map along with the corresponding scan line. Because of the laser spot size and the close proximity of adjacent nanopillars, there is a nonzero photocurrent even when the laser beam is centered among four nanopillars. Despite that, the photocurrent maximum peaks are consistent with the 600 nm spacing of the patterned growth. Such measurement highlights inherent differences from nanopillar to nanopillar arising, for instance, from contact inhomogeneities or dopant variations in the GaAs crystal. The overall solar cell performance is the result of a collective optoelectronic behavior of thousands of nanopillars connected in parallel so mapping spatially the photocurrent pattern is highly required. Gutsche et al. recently demonstrated successful SPCM measurements for individual radial nanowire diodes (27). The setup is schematically represented in Figure 13.18c. The photocurrent is collected as a function of the scanned position, with a laser illumination at 532 nm and an estimated focused spot diameter of 1.2 µm. The map (Fig. 13.18d) shows that the region

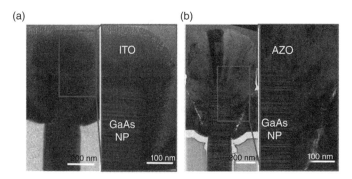

Figure 13.19   Cross-sectional transmission electron microscope images of a GaAs nanopillar coated with (a) ITO and (b) AZO.

(blue color) with the highest photocurrent is where the complete radial p–i–n junction is intact. The rest of the nanowire has been etched to solely contact the n-GaAs core. No appreciable photocurrent is recorded in this region. The photocurrent profile highlights a strong response in the vicinity of the electrical contact to the p-type shell, witnessing a lowered series resistance due to the p-shell.

13.3.4   Transparent Electrical Contact Limitations and Comparisons

Electrical contacts are as important as the active semiconductor region in solar cells in order to extract up to the terminals the maximum amount of photocarriers available. One of the major challenges in contacting vertical, three-dimensional structures like nanopillars is that standard metal contacts cannot be utilized. This is due to the fact that each nanopillar requires to be contacted individually, while preserving an efficient light coupling. In order to achieve that, the electrical contacts have to satisfy a (1) high optical transparency to maximize photogeneration rates into the semiconductor and (2) form an ohmic interface with contact resistance comparable to that of metal electrodes. Transparent conducting oxides (TCOs) are doped metal oxide thin films predominantly used in optoelectronic devices, for example, flat panel displays and photovoltaics. A TCO is a wide bandgap semiconductor that has a high free-electron concentration in its conduction band. This peculiarity is due to either defects in the material or extrinsic dopants, the impurity levels that lie near the conduction band edge. The most common composites used in TCOs are tin oxide, indium oxide (ITO), indium tin oxide, zinc oxide, and aluminum zinc oxide (AZO). It exists a general trade-off between electrical conductivity and optical transmittance. In the ideal case, both figures of merit will be as high as possible.

RF sputtering is a type of deposition utilized to dispense TCO contacts on top of semiconductor surfaces. It can be imagined as a highly energetic plasma that ionizes atoms from a preinstalled target (TCO material source). The atoms are sputtered (or literally sprayed) onto the required surface. This type of deposition is able to achieve high conformal TCO profiles onto high-aspect ratio structures such as nanopillars as presented in Figure 13.19.

For this reason, RF sputtering is the mainstream deposition method used to realize transparent electrical contacts in nanopillar-based solar cells. The electrical properties of the TCO/semiconductor interface mainly depend on the preparation of the nanopillar surface, sputtering conditions of the TCO, and postdeposition thermal annealing treatments.

GaAs nanopillar photovoltaic devices are characterized and compared through dark/light current–voltage measurements and external quantum efficiency (EQE) from 500 to 1000 nm wavelength range. Under AM 1.5 G illumination, samples processed with AZO contact exhibit a $V_{OC} = 0.2$ V, $J_{SC} = 8.1$ mA/cm^2, and FF = 36% with a PCE = 0.58% (power conversion efficiency). On the other hand, ITO devices showed a $V_{OC} = 0.39$ V, $J_{SC} = 17.6$ mA/cm^2, and FF = 37% with a PCE = 2.54% (26). The data is presented in Figure 13.20a and b. The nearly fivefold increase in the PCE is attributed to a more favorable contact interface offered by the ITO TCO. Additionally, the polarity of the TCO is important to determine if the minority carriers will encounter a barrier before being extracted. A p-type TCO electrode in contact with an n-doped GaAs will most certainly impede an ohmic electronic transport at the contact interface. Figure 13.20c elucidates on a degraded EQE response once AZO is utilized. The electrical analysis of the two contacts highlighted a highly resistive (627 Ω/sq) AZO thin film, with p-type polarity from Hall measurements. The ITO cathode presented a sheet resistance of 15 Ω/sq with n-type polarity. Considering that both TCO films contacted an n-type shell in the p–n junction of the solar cell, the extraction of the photogenerated carriers is greatly fostered when ITO is exploited as top electrode, translating into higher PCEs.

13.4   ADVANCED TOPICS FOR HIGH-EFFICIENCY NANOPILLAR PHOTOVOLTAICS

13.4.1   Direct Dopant Mapping in Semiconductor Nanopillars

Semiconductor solar cells are typically based on p–n junctions. Each region in the junction is doped with impurity atoms to turn the intrinsic material into a p-type or n-type semiconductor. Dopant incorporation at the nanoscale drastically differs from bulk devices, presenting many challenges in the degree of control. The most conventional characterization techniques (e.g., capacitance–voltage doping profiling) exhibit limitations due to the difference in nanopillar geometry with respect to planar designs as well as inherent uncertainties (e.g., gate

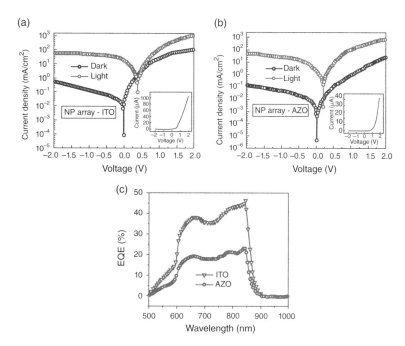

Figure 13.20 Comparison of device performance utilizing different transparent contact materials in the solar cell. (a) Current density–voltage characteristic for a GaAs nanopillar solar cell fabricated with ITO or (b) AZO as top transparent electrode; (c) external quantum efficiency for both contacts.

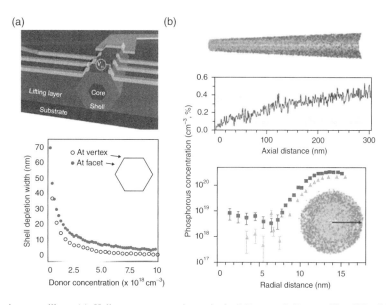

Figure 13.21 Doping profiling in nanopillars. (a) Hall measurements in a single InP core–shell nanopillar (28). (b) Side-view cross section of a phosphorous-doped germanium nanowire and corresponding radial dopant mapping.

capacitance in a field-effect transistor). Despite that, Storm et al. recently analyzed experimentally doping levels by means of Hall-effect-based measurements (28). The group characterized the shell of core–shell InP nanopillars. The technique is able to spatially resolve and quantitatively determine the carrier concentration and mobility of the nanopillar shell. The spatial mapping is carried out by topologically defining three separate pairs of contacts along the nanopillar body. Longitudinal electrical current is flown while a magnetic field is applied. In this fashion, a Hall voltage can be measured and directly correlated to the doping concentration at that particular probing region. Figure 13.21a describes the measurement setup and the measured shell depletion width as a function of donor concentration. The graph shows that a doping-dependent electrical shell thickness has to be taken into account at low donor concentrations (28).

Another viable technique to map out the dopant profile within the nanowire crystal is a local electrode atom probe tomography. The method destructively analyzes a portion of the nanopillar by knocking off atoms one by one by means of a highly ionizing electric field. For each atom,

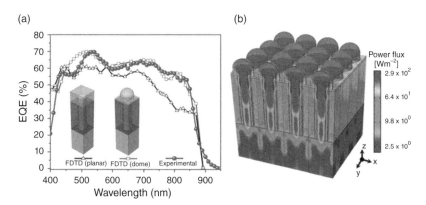

Figure 13.22   Light management analysis (30). (a) External quantum efficiency measurements (blue dots) of the solar cells from 400 to 950 nm. Analytical simulations are carried out to analyze the impact of a planar (black triangles) and a dome-shaped ITO layer (red squares) on the final optical coupling performance. (b) Integrated-AM 1.5 G optical power flux within the periodic structure. Each ITO dome acts as a subwavelength nanolens.

the relative time of flight is recorded and correlated back to the material species on the periodic table. From such measurements, it can be inferred that in Ge nanowires the core is substantially underdoped (9×10^{17} cm^{-3}), whereas the shell is highly overdoped (1.95×10^{20} cm^{-3}) as reported in Figure 13.21b (29).

13.4.2   Efficient Light Management

As discussed in Section 13.3.2, the geometrical design of a patterned nanopillar array is essential to enhance the optical absorption that stems out from light trapping effects. However, the fabricated device structure often differs from the starting bare nanopillar arrays, including several dielectric layers and particular morphologies of transparent contacts that can modify the photonic coupling into the solar cell itself. Mariani et al. reported on dome-shaped ITO top electrode in GaAs solar cells that functions as a two-dimensional periodic array of subwavelength lenses that focus the local density of optical states within the nanopillar active volume (30). The presence of domes is the result of a conformal sputtering deposition of ITO on top of the nanostructure tips. Such domes display an appreciable optical focusing effect, which is broadband. Finite-difference time-domain simulations are successfully correlated with EQE experimental curves as shown in Figure 13.22a. Several resonance peaks are identified and attributed to Mie resonance effects generated in the dome contact and funneled into each nanopillar.

Figure 13.22b presents a power flux density profile of a 4-by-4 nanopillar array under 1 sun illumination conditions. The figure is directly calculated as active absorbed power from the Poynting vector in the electromagnetic propagation. A large portion of power density is concentrated into the nanopillar, resulting in a higher density of optical states. Additionally, the photonic coupling of the incoming light field with the nanopillar array is vastly dependent on the shape of the ITO contact; dome morphology is more advantageous compared to a planar configuration.

13.4.3   Passivation of Nanopillar Surfaces

Semiconductor nanopillar devices are characterized by a high surface-to-volume ratio, resulting in optoelectronic properties strictly related to surface treatments. The GaAs nanopillar facets are characterized by dangling bonds, which become chemically and electronically stable after air exposure. This, in turn, creates Ga–O and As–O defects that are notoriously known to form surface states that pin the Fermi level at energy levels that are located near the center of the bandgap of GaAs. For this reason, several research groups endeavored both *ex situ* and *in situ* surface treatments to mitigate the deleterious effect caused by surface states. Typically, the sidewall surfaces of GaAs nanopillars can be chemically passivated using ammonium sulfide $(NH_4)_2S$ aqueous solution. As a result, the passivation is expected to reduce surface recombination rates and surface depletion in the nanostructures (31). Such treatment has shown to increase the nanowire conductivity over two orders of magnitude with a reported residual surface trap density of $N_{SS} = 7 \times 10^{11}$ cm^{-2} compared to untreated samples ($N_{SS} = 3 \times 10^{12}$ cm^{-2}) (31). Despite that, sulfur-based passivation only provides a short-term stability due to the weak covalent bonds formed by sulfur atoms. A long-term alternative is represented by capping the GaAs nanowire with a high-bandgap material (e.g., AlGaAs, InGaP, or GaP) that acts as a barrier to avoid carrier recombination at the surface. Electron-beam-induced current measurements at metal–semiconductor single-nanowire Schottky junctions extrapolated minority carrier diffusion lengths in the order of 30 nm for a bare GaAs nanopillar, whereas in AlGaAs-passivated GaAs nanowires were increased up to 180 nm (32). Recently, *in situ* epitaxial passivation has been demonstrated for vertically standing arrays of GaAs nanopillars. Lattice-matched InGaP passivating shell has been grown onto arrays of GaAs nanopillars, with a net PCE improvement of over 600% from 1.02% to 6.63% efficiency. The surface passivation decreased the surface state density from 5×10^{12} cm^{-2} eV^{-1} to 7×10^{10} cm^{-2} eV^{-1} (25), justifying the increase in performance of the solar cells. Photoelectrical measurements highlight the need for *in situ* passivation treatment to substantially enhance the EQE and PCE in GaAs nanowire-based photovoltaics.

13.5 CONCLUSIONS AND FUTURE DIRECTIONS

GaAs photovoltaics are currently marginalized in the fan of solar cell solutions due to the high manufacturing costs mostly related to expensive substrates. GaAs junctions are generally sandwiched as middle absorber in triple-junction solar cells, widely utilized in concentrated photovoltaics. One stratagem to lower the "dollar-per-watt" figure is to employ the epitaxial lift-off technique to reuse the mother substrate in subsequent epitaxial growths. The intriguing aspect in adopting vertical semiconductor nanostructures is to trap light more efficiently than planar solar cells, and with less material. The ultimate goal is to harvest the energy of the sun at a better quality and at a cheaper cost since less absorbing material is required. This new class of devices can outperform the current solar cell generation only if each aspect of the active junction is carefully analyzed and studied. Future directions in this regard will require to deeply focus on the formation of the doped junction at the nanoscale, on novel transparent ohmic contacts that can challenge interface conductivity of typical electrical metal contacts, while exploring new routes to suppress the detrimental effect generated by surface state charges.

REFERENCES

1. Available at http://geology.com/articles/satellite-photo-earth-at-night.shtml Accessed 2014 Nov 14.
2. Available at http://www.energyandcapital.com/articles/trickle-down-enernomics/1488 Accessed 2014 Nov 14.
3. Sheikh N, Kocaoglu F. Technology management in the energy smart world (PICMET). 2011 Proceedings of PICMET '11; 2011.
4. International Energy Agency. Energy Technology Roadmaps: Charting a low-carbon energy revolution. Available at http://www.indiaenvironmentportal.org.in/files/Synthesis_Roadmap.pdf Accessed 2014 Nov 14.
5. Chu Y, Meisen P. Global Energy Network Institute; 2011.
6. Available at www.exeger.com/dye-sensitized-solar-cells Accessed 2014 Nov 14.
7. Service R. Science 2013;339:6115.
8. Lee CH, Kim D, Zheng X. Proc Natl Acad Sci U S A 2010;107:22.
9. Conibeer G. Mater Today 2007;10:42.
10. Brown G, Wu J. Laser Photonics Rev 2009;4:394–405.
11. Available at http://optics.org/news/4/6/44 Accessed 2014 Nov 14.
12. Wieder H. J Vac Sci Technol 1971;8:210.
13. Shur M. *Physics of Semiconductor Physics*. Englewood Cliffs (NJ): Prentice Hall; 1990.
14. Available at www.pveducation.org Accessed 2014 Nov 14.
15. DeSalvo G, Barnett A. IEEE Trans Electron Dev 1993;40:705.
16. Yater J, Landis G, Bailey S, Olsen L, Addis F. 25th Photovoltaic Specialist Conference; 1996.
17. Schockley W, Queisser H. J Appl Phys 1961;32:510.
18. Kayes B, Nie H, Twist R, Spruytte S, Reinhardt F, Kizilyalli I, Higashi G. 37th IEEE Photovoltaic Specialists Conference; 2011.
19. Li E, Chaparala P. National Renewable Energy Laboratory PV Module Reliability Workshop; 2013.
20. Green M, Emery K, Hishikawa Y, Warta W, Dunlop E. Prog Photovoltaics Res Appl 2013;21:827.
21. Wagner R, Ellis W. Appl Phys Lett 1964;4:89.
22. Wacaser B, Deppert K, Karlsson L, Samuelson L, Seifert W. J Cryst Growth 2006;287:504.
23. Shapiro J, Wong P-S, Scofield A, Tu C, Senanayake P, Mariani G, Liang B, Huffaker D. Appl Phys Lett 2010;97:243102.
24. Mariani G, Huffaker D. SPIE Defense, Security and Sensing, Volume 8725; 2013. p 872510.
25. Mariani G, Scofield A, Hung C-H, Huffaker D. Nat Commun 2013;4:1497.
26. Mariani G, Wong P-S, Katzenmeyer A, Leonard F, Shapiro J, Huffaker D. Nano Lett 2011;11:2490.
27. Gutsche C, Lysov A, Braam D, Regolin I, Keller G, Li Z, Geller M, Spasova M, Prost W, Tegude F. Adv Funct Mater 2012;22:929.
28. Storm K, Halvardsson F, Heurlin M, Lindgren D, Gustafsson A, Wu P, Monemar B, Samuelson L. Nat Nanotechnol 2012;7:718.
29. Perea D, Hemesath E, Schwalbach E, Lensch-Falk J, Voorhees P, Lauhon L. Nat Nanotechnol 2009;4:315.
30. Mariani G, Zhou Z, Scofield A, Huffaker D. Nano Lett 2013;13:1632.
31. Tajik N, Chia C, LaPierre R. Appl Phys Lett 2012;100:203122.
32. Chang C, Chi C, Yao M, Huang N, Chen C, Theiss J, Bushmaker A, LaLumodiere S, Yeh T, Povinelli M, Zhou C, Dapkus P, Cronin S. Nano Lett 2012;12:4484.

14

NANOCRYSTALS, LAYER-BY-LAYER ASSEMBLY, AND PHOTOVOLTAIC DEVICES

JACEK J. JASIENIAK[1], BRANDON I. MacDONALD[2], AND PAUL MULVANEY[3]

[1]*Materials Science and Engineering, CSIRO, Clayton, Victoria, Australia*
[2]*QD Vision, Lexington, MA, USA*
[3]*School of Chemistry and Bio21 Institute, University of Melbourne, Parkville, Victoria, Australia*

14.1 INTRODUCTION

Since the advent of the hot-injection method for the preparation of II–VI semiconductor nanocrystals by Murray et al. in 1993 (1), researchers have rapidly optimized the necessary conditions for the preparation of high-quality, monodisperse nanocrystals. By selection of the appropriate chemical and physical conditions needed for control of the nucleation and growth kinetics, size (2–5), crystal structure selection (6, 7), and surface/bulk atomic composition (8–10) have all been realized. These collective achievements are a sign of maturity, and this, in turn, has shifted the focus to the application of nanocrystals in a wide variety of technologies (11).

One of the most exciting potential applications for semiconductor nanocrystals is solar energy conversion. Originally proposed in the late 1970s, early attempts to harness sunlight with colloidal materials were unsuccessful, largely because of the poor crystallinity of the materials (12). However, over the last 20 years, the improvement in the understanding of nanoscale electronics and optics has rekindled interest in the exploitation of quantum effects for efficient light harvesting and energy storage. In general, three major types of nanocrystal solar cells have been identified: (i) hybrid organic–inorganic (or pure inorganic) bulk heterojunction (BHJ) (13, 14), (ii) organic or inorganic nanocrystal-sensitized photovoltaic devices (15–17), and (iii) inorganic nanocrystalline or "sintered" thin-film devices (18, 19). Examples of these are depicted in Figure 14.1.

Each type of solar cell presents its own advantages and disadvantages. Sensitized solar cells, for instance, are a proven technology, and solar conversion efficiencies above 12% have been reported (20). However, recent progress to push their efficiencies above this level has been very slow, which is a major concern, given that the theoretical limit for single-cell devices is ~20% (21). Recent reports of hot-carrier extraction (22) and multiexciton generation (23, 24) in quantum-confined systems have stimulated interest in these solar cell structures. This is because carrier extraction appears to occur on subpicosecond timescales at type II interfaces. The predicted increase in efficiency stemming from these so-called "3rd-generation" photovoltaic mechanisms (25, 26) provides great incentive to continue research on these device structures.

Hybrid and inorganic bulk heterojunctions are an elegant adaptation from the organic community (27, 28). They can be fabricated readily on plastic substrates due to the low processing temperatures (<150 °C) required, and in principle, such systems offer a theoretical efficiency above 20% (29). However, the requirement to form interpenetrating networks at the nanoscale has been a severe challenge for the reproducible fabrication of devices (13). This has limited the overall efficiencies for such devices to ~5% (14), which is substantially lower than the ~9.2% reported for single-junction, pure organic BHJ counterparts (30).

Conversely, inorganic thin-film-based photovoltaic devices are much more closely aligned to existing technologies, which typically employ techniques such as sputtering or closed-space sublimation. The basic solar cell consists of nanocrystals less than ~15 nm in diameter, which are chemically passivated in such a manner as to ensure high electrical coupling. Examples of these include PbSe Schottky (31) and PbSe/ZnO bilayer excitonic (32) devices. To date, maximum efficiencies from such devices have approached 6% (33). Two of the major drawbacks that may ultimately limit the performance of such devices are (i) the poor electronic properties of nanocrystals compared to their bulk analogues and (ii) the large surface-to-volume ratios of nanocrystals, which enhances surface recombination of charge carriers.

Both of these drawbacks are overcome in a second type of inorganic solar cell device structure, which is based on bulk nanocrystalline materials. In these "sintered" solar cells, nanoparticles are deposited as thin films and are subsequently chemically and thermally treated to yield bulk nanocrystalline materials. To date, research into this variety of solar cell has yielded devices with efficiencies above 7%

Nanomaterials, Polymers and Devices: Materials Functionalization and Device Fabrication, First Edition. Edited by Eric S. W. Kong.
© 2015 John Wiley & Sons, Inc. Published 2015 by John Wiley & Sons, Inc.

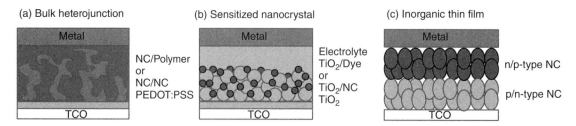

Figure 14.1   A depiction of the three major classes of solar cell architectures that have the potential to employ nanocrystals. Within this figure, TCO and NC refer to transparent conductive oxide and nanocrystal, respectively.

using CdTe (34), $CuIn_xGa_{1-x}Se_2$ (35), and $Cu_2ZnSnSe_4$ (36) thin-film layers. The theoretical maximum efficiency of a single-junction, bulk semiconductor solar cell is given by the Shockley–Queisser limit, which, for materials with a bandgap of 1.1–1.5 eV, is above 30%. This not only suggests that there is plenty of opportunity for further improvement, but also highlights the attractiveness of this class of solar cells for development through low-cost, solution-based processing.

This chapter is focused on the development of such solar cells, with particular reference to one of the most successful inorganic thin-film solar cell materials, CdTe. We begin by introducing the reader to some of the fundamental concepts associated with the synthesis of high-quality, inorganic semiconductor nanocrystals. This is followed by a detailed description of how such nanocrystals can be utilized to grow high-quality, inorganic thin films that are suitable for electronic applications. Finally, we discuss the application of such inorganic layers to the fabrication of bilayer and compositionally graded, inorganic thin-film solar cells. It is hoped that the reader will not only gain a comprehensive understanding of the fabrication of nanocrystal solar cells, but will also be able to apply the concepts to the fabrication of any nanocrystal-based optoelectronic technology.

14.2   NUCLEATION AND GROWTH OF COLLOIDAL NANOCRYSTALS

We start by considering the challenges of producing large amounts of monodisperse nanocrystals. These challenges relate to control of the nucleation and growth kinetics of nanocrystals. Despite a century of investigation, nucleation is still a poorly understood phenomenon. This is particularly true for nanocrystal systems in which nucleation is typically rapid and strongly coupled to both growth and Ostwald ripening. The most widely accepted models for colloid nucleation include classical homogeneous equilibrium nucleation theory (37), steady-state nucleation theory (38), and the more recent atomistic nucleation theory (39). Many researchers are currently studying dynamic systems in which nucleation and growth theories are being presented simultaneously within coupled frameworks (40, 41). Numerical approaches such as the Kampmann and Wagner model enable prediction of actual particle size distributions, which can be directly compared to experiment (42). Such models are indeed a necessary step for a more complete understanding of nanocrystalline colloid systems, particularly when nucleation and growth occur simultaneously. Here, we present a brief overview of (1) the nucleation process from a classical steady-state viewpoint and (2) the associated growth of the formed nanocrystals under reaction- and diffusion-limited conditions. These concepts are critical for understanding how to grow colloidal nanocrystals of an appropriate size and distribution, which in turn is fundamental to the formulation of nanocrystal dispersions appropriate for optoelectronic applications.

14.2.1   Thermodynamics of Nucleation

One practical definition of nucleation is that it is a process whereby small, unstable (subcritical) nuclei are dynamically transformed into metastable(supercritical) nanoparticles through the continuous addition of free supersaturated monomers. Classical nucleation theory assumes that nuclei must overcome an energetic barrier due to the energy cost of generating a newly forming surface. The free energy (ΔG) needed to create a nucleus with a size r is given by

$$\Delta G(r) = \frac{4}{3}\pi r^3 \Delta G_v + 4\pi r^2 \gamma \tag{14.1}$$

where ΔG_v is the change in the Gibbs free energy per unit volume (J/m^3) and γ (J/m^2) is the surface energy. The first part of this equation accounts for the favorable free energy of forming a defined volume of material, while the second part accounts for the (positive) surface free energy. Assuming that surface energy is independent of size, in Figure 14.2 a representative plot of the overall free energy is shown as a function of particle size. It is evident that at small particle sizes the free energy of the system increases. This arises because the surface component of the Gibbs free energy dominates over the thermodynamically favorable volume component. As the size increases, this volume component begins to make the major contribution to the barrier to nucleation. The barrier maximum occurs when $d\Delta G/dr = 0$, corresponding to the critical radius (r^*).

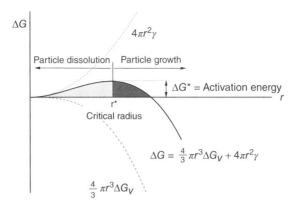

Figure 14.2 A schematic of the Gibbs free energy of nucleation.

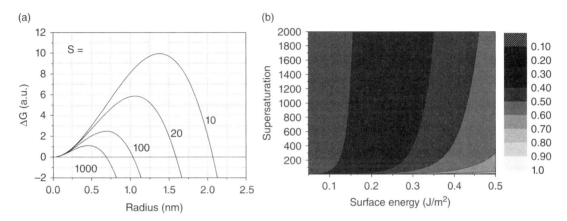

Figure 14.3 (a) The effect of supersaturation on the overall Gibbs free energy to nucleation of CdSe for a fixed surface energy of 0.2 J m^{-2} at a temperature of 473 K. (b) A contour plot showing the effect of surface energy and supersaturation on the relative magnitude of the activation barrier to nucleation. For ease of visualization, we have plotted $\log_{10}\Delta G^*$.

It can be readily shown that the critical radius is related to the surface tension exhibited by the particle surface in a given reaction medium, temperature, and supersaturation (S) of the system, through

$$r^* = -\frac{2\gamma}{\Delta G_v} = \frac{2\gamma V_m}{RT \ln S} \tag{14.2}$$

$$\Delta G^* = \frac{16\pi\gamma^3}{(3\Delta G_v)^2} = \frac{16\pi\gamma^3 V_m^2}{3R^2 T^2 (\ln S)^2} = \frac{4\pi\gamma}{3} r^{*2} \tag{14.3}$$

where V_m is the molar volume, S is the supersaturation, and $\Delta G_v = -RT \ln S/V_m$ is used to simplify these equations. Equation 14.2 is a form of the Gibbs–Thomson equation.

From Equations 14.1 and 14.3, it is evident that the thermodynamic stability of a small crystal is strongly size dependent. For radii smaller than r^*, the crystal will tend to dissolve, while those that are larger than r^* will be thermodynamically stable and be able to grow into larger particles. Thus, one of the key factors in nucleation is the precise control of the critical radius. It is important to note that the value of the surface free energies of crystals is difficult to determine and, consequently, only approximate values for r^* can be calculated. Moreover, the surface free energy is strongly affected by chemisorption, which further complicates any quantitative predictions.

In Figure 14.3(a), we demonstrate that for a fixed surface energy, a variation in supersaturation has a drastic effect on both r^* and ΔG^*. More specifically, both parameters are found to decrease with increasing supersaturation. Analogous results are observed for changes in surface energy and a fixed supersaturation. It should be noted that a decrease (increase) in r^*, regardless of the cause, always leads to a reduction (increase) in ΔG^*. In Figure 14.3(b), we show a normalized plot of log (ΔG^*) as a function of both γ and S. It is evident from this figure that both surface energy and supersaturation have dominant effects on the barrier to nucleation for supersaturation values of S less than ~100. At higher supersaturation values however, the surface energy provides the major contribution to ΔG^*. In nanocrystal systems, nucleation typically occurs at supersaturation values ranging between 200–2000, and consequently, the surface free energy term becomes the dominant one. This is one reason why surface ligands have such a dominant effect on the nucleation rates in the synthesis of colloidal nanocrystals.

14.2.2   Classical Homogeneous Nucleation

The nucleation of nanocrystals in a solution is a complex function of the solvent composition, the concentrations of ligands, and the chemical form of the precursors used to generate the active monomers. Existing nucleation models do not include ligand effects, the particle charge, or the role of the electrical double layer (39, 43). Generally, they are based on gas-phase dynamics. According to the steady-state model of Becker and Döring (39), nucleation occurs as a consecutive series of bimolecular reactions between monomers and clusters of a given size. Under such conditions, the steady-state nucleation rate (J_0) is given by

$$J_0 = \omega^* \Gamma C \exp\left(-\frac{\Delta G^*}{kT}\right)$$ (14.4)

where C is the free monomer concentration in solution, ω^* is the frequency of monomer attachment to a critical nucleus, ΔG^* is the barrier to nucleation as given by Equation 14.3, and

$$\Gamma = \frac{V_m \Delta G_v^2}{8\pi\sqrt{\gamma^3 kT}}$$ (14.5)

is the Zeldovich factor, which accounts for the thermal fluctuations in the free energy ΔG^* (43).

For a process that involves the diffusion of precursors followed by a kinetic reaction with the surface, the frequency factor can be considered to be the product of a diffusive flux (f^*), the surface area, and a kinetic factor associated with a defined energy barrier (ΔG_{desolv}) (39):

$$\omega^* = 4\pi r^{*2} f^* \exp\left(-\frac{\Delta G_{\text{desolv}}}{kT}\right)$$ (14.6)

The diffusive flux accounts for the transfer of the monomer from the bulk solution to the surface. One can easily show that $f^* = DC/r^*$ (where D is the diffusion coefficient of the monomer in m²/s) by solving the appropriate diffusion problem (43). The kinetic factor encompasses a whole series of steps associated with the attachment step. For example, the rate-determining step may be the breaking of chemical bonds in the precursor molecules to form the active monomer, not; it may reflect the energy to desolvate the monomer (i.e., break monomer–solvent bonds or monomer–ligand bonds), or it may be related to the detachment of surface ligands and adsorbates to create a vacant, surface adsorption site. The potential energy profile for such a process is depicted in Figure 14.4 (44). This schematic shows that species which complex monomers and/or surface sites more strongly will, exhibit higher (ΔG_{desolv}) values and, consequently, lower nucleation rates.

Overall, Equations 14.4–14.6 can be combined and simplified to give the overall steady-state rate of homogeneous nucleation in units of mol/m³:

$$J_0 = N_A \sqrt{\frac{kT}{\gamma}} S^2 \log(S) C_{\text{flat}}^0 D \exp\left(-\frac{\Delta G_{\text{desolv}}}{kT}\right) \exp\left(\frac{16\pi\gamma^3 V_m^2}{3R^2 T^2 (\ln S)^2}\right)$$ (14.7)

where N_A is Avogadro's number and the simplification $C = C_{\text{flat}}^0 S$ has been introduced. C_{flat}^0 is the concentration of the monomer in equilibrium with a flat crystal.

In Figure 14.5(a), J_0 is plotted as a function of the supersaturation for a fixed value of ΔG_{desolv} and for surface energies between 0.18 mJ/m² and 0.22 mJ/m². It is immediately evident that below a critical supersaturation value, nucleation is almost negligible. As the supersaturation increases above this threshold, the rate of nucleation abruptly increases to a point where it can be considered "explosive" in nature. From this figure we observe that only small variations in surface energy are required to have a drastic effect on the critical supersaturation value. Although not shown here, analogous results are obtained for a variation in C_{flat}^0, that is, the overall solubility of the crystal.

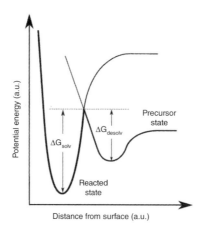

Figure 14.4   A schematic showing the relative potential energy profile of precursors reacting with a surface. For completeness, we have included the desolvation (ΔG_{desolv}) and solvation activation barriers (ΔG_{solv}).

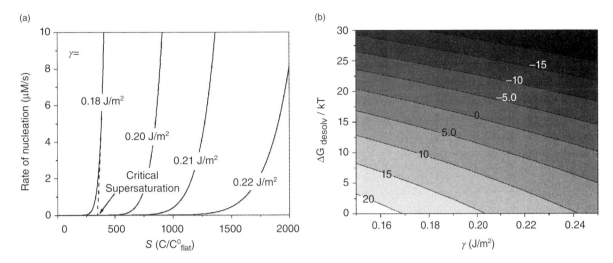

Figure 14.5 (a) The rate of homogeneous nucleation under steady-state conditions for a fixed value of $\Delta G_{\mathrm{desolv}}/kT = 10$. (b) A contour plot of $\log_{10}(J_0)$, where J_0 is in particles $\mathrm{m}^{-3}\,\mathrm{s}^{-1}$ as a function of γ and $\Delta G_{\mathrm{desolv}}/kT$ for a fixed $S = 100$. The remaining parameters are $T = 500\,\mathrm{K}$, $D = 1 \times 10^{-9}\,\mathrm{m}^2\,\mathrm{s}^{-1}$, $V_m = 3.3 \times 10^{-5}\,\mathrm{m}^3\,\mathrm{mol}^{-1}$, and $C_{\mathrm{flat}}^0 = 0.01\,\mathrm{mol}\,\mathrm{m}^{-3}$.

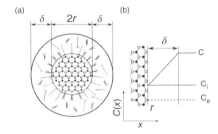

Figure 14.6 (a) A schematic of a nanocrystal with radius r that is surrounded by a ligand sphere of thickness δ. (b) This diffusive barrier creates a monomer concentration profile extending from the nanocrystal boundary.

Both γ and C_{flat}^0 clearly affect the thermodynamics of nucleation; the desolvation factor, however, introduces an additional kinetic factor, which affects only the rate. In Figure 14.5b, the variation of $\log_{10}(J_0)$ is shown as a function of changing γ and $\Delta G_{\mathrm{desolv}}$ for a fixed supersaturation value of 100. It is evident that an increase in $\Delta G_{\mathrm{desolv}}$ acts to significantly reduce the nucleation rate for a given critical size. As already discussed, the value of γ is also found to have a significant impact on the nucleation rate; however, any changes in this factor will concurrently act to alter the nucleating particle size. One can thus see that an interplay between thermodynamics and kinetics is always occurring for a given nucleation event. It is important to note that nucleation is always occurring concurrently with growth. As such, a burst nucleation event coupled with growth acts to rapidly deplete the monomer concentration from the system. The resulting decrease in supersaturation suppresses any further nucleation from occurring. Growth and Ostwald ripening at this point become the two major mechanisms of evolution for a colloidal ensemble. Both of these processes will now be briefly discussed.

14.2.3 Diffusion- and Reaction-Limited Growth

The growth of any crystal involves the attachment of monomers from the bulk solution onto one of its surface sites. The rate-determining step may be either (i) the diffusion of the monomer from the solution to the nanocrystal surface or (ii) the adsorption of that monomer to the surface. We summarize this overall picture in Figure 14.6 for a given nanocrystal of radius r that is overcoated by a monolayer of stabilizer. C is the bulk reservoir concentration of the monomer, C_i is the monomer concentration at the particle/solution interface, and C_e is the size-dependent, equilibrium monomer concentration (given by the Gibbs–Thomson equation) (45).

Assuming that there is no convection and that the crystal is composed of a single, uncharged monomer, the overall flux of the monomer under diffusion (J^{diff})- and reaction(J^{react})-limited conditions are given by

$$J^{\mathrm{diff}} = 4\pi r D(C - C_i) \tag{14.8}$$

$$J^{\mathrm{react}} = 4\pi r^2 (k_g C_i - k_d) \tag{14.9}$$

Here, D is the diffusion coefficient of the monomer in solution, k_g is the rate constant (m/s^1) for attachment of the monomers to the surface, while k_d (mol/m^2/s^1) is the rate of desorption. If we assume the size dependence of these rate constants is accounted for by the Gibbs–Thomson equation, then the reaction-limited flux becomes

$$J^{\text{react}} = 4\pi r^2 \left(k_g^{\text{flat}} C_i \exp\left(-\alpha \frac{2\gamma V_m}{rRT} \right) - k_d^{\text{flat}} \exp\left(\beta \frac{2\gamma V_m}{rRT} \right) \right) \tag{14.10}$$

where $k_{g(d)}^{\text{flat}}$ is the rate constant for a flat interface, α and β are the transfer coefficients ($\alpha + \beta = 1$), and the subscript $g(d)$ denotes the growth (dissolution) component.

Under steady-state conditions, the amounts of monomer passing through the diffusion layer and also undergoing reaction-mediated growth are equal, that is, $J^{\text{diff}} = J^{\text{react}}$. Since the flux of the monomer to the surface is directly related to the rate of growth of the spherical nanocrystal by

$$J = \left(\frac{4\pi r^2}{V_m} \right) \frac{dr}{dt} \tag{14.11}$$

the size-dependent growth rate is found to be (46)

$$\frac{dr}{dt} = DV_m C_0^{\text{flat}} \left\{ \frac{\dfrac{C}{C_0^{\text{flat}}} - \exp\left(\dfrac{2\gamma V_m}{rRT} \right)}{r + \dfrac{D}{k_g^{\text{flat}}} \exp\left(\alpha \dfrac{2\gamma V_m}{rRT} \right)} \right\} \tag{14.12}$$

or in terms of dimensionless units,

$$\frac{d\tilde{r}}{d\tau} = \frac{S - \exp\left(\dfrac{1}{\tilde{r}} \right)}{\tilde{r} + K \exp\left(\dfrac{\alpha}{\tilde{r}} \right)} \tag{14.13}$$

with the dimensionless radius

$$\tilde{r} = \frac{RT}{2\gamma V_m} r \tag{14.14}$$

the dimensionless time

$$\tau = \frac{R^2 T^2 D C_{\text{flat}}^0}{4\gamma^2 V_m} \tag{14.15}$$

and the Damköhler number

$$K = \frac{RTD}{2\gamma V_m k_g^{\text{flat}}} \tag{14.16}$$

In these equations, $C_0^{\text{flat}} = k_d^{\text{flat}}/k_g^{\text{flat}}$. The Damköhler number reflects the overall ratio between mass transfer (here diffusion limit) and reaction-limited rates. Thus, a K value of 100 corresponds to a strongly reaction-limited process, while a value of 0.01 corresponds to a process that is almost purely diffusion limited.

In Figure 14.7, the size-dependent, steady-state growth rates predicted by Equation 14.13 are displayed for physically reasonable variations in (a) Damköhler number, (b) supersaturation, and (c) surface energy. In all cases, the zero-growth rate by definition occurs at r = r*

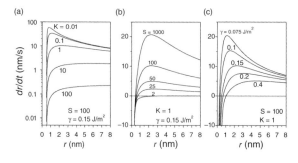

Figure 14.7   The size-dependent steady-state growth rate as determined in Equation 14.13 for physically reasonable variations in (a) Damköhler number, (b) supersaturation, and (c) surface energy. In all figures, the remaining constants were fixed at $T = 500\,\text{K}$, $D = 1 \times 10^{-12}\,\text{m}^2\,\text{s}^{-1}$, $V_m = 3.3 \times 10^{-5}\,\text{m}^3\,\text{mol}^{-1}$, and $C_{\text{flat}}^0 = 0.02\,\text{mol}\,\text{m}^{-3}$.

(see Eq.14.2). It is immediately evident from Figure 14.7a that a decrease in K causes a large increase in the growth rate for all particle sizes and a pronounced maximum to the size-dependent growth rate. As we will discuss in more detail shortly, this latter factor enables narrow size distributions to be achieved during the growth of nanocrystals. Notably, as the Damköhler number does not vary the thermodynamics of the system, r^* is found to be constant. In Figure 14.7b, we see that an increase in supersaturation gradually increases the growth rate and decreases r^*. This occurs simply because the growth rate depends on the monomer concentration, whereas the desorption of monomers from the crystal is assumed to be independent of the monomer concentration in the solution. Finally, in Figure 14.7c, we demonstrate that a decrease in surface energy also has a drastic effect by (i) increasing the growth rate and the maximum rate of growth and (ii) reducing r^*. These effects are synergistic, that is, both lead to faster particle growth and indeed faster nucleation, but are significantly more drastic than the effects caused by variations in supersaturation. Although the effects of temperature are not shown here, it is evident from Equation 14.2 that a decrease in temperature has the same qualitative effect on the growth rates as does an increase in surface energy, that is, an overall reduction in the rates of nucleation and growth.

For a typical ensemble of nanocrystals, the growth rate of a nanocrystal of any given size will depend on where in the distribution the critical radius is located at that moment (i.e., the zero-growth point). Nanocrystals above r^* will evidently grow, and those below r^* will dissolve. This dynamic process is commonly called Ostwald ripening. Any given nanocrystal that grows will consume the monomer from the solution, which in turn will reduce the overall supersaturation and simultaneously increase r^*. In dynamic systems where particles are growing and dissolving, all particles and parameters such as r^* and S are coupled. If one thus considers a system that has nucleated and is growing under steady-state conditions, then r^* will always be located on the small-particle-side of the distribution maximum (47). As the distribution continues to evolve, the nature of the growth mechanism will ultimately determine the final particle distribution.

In the case of diffusion-limited conditions (or extremely low γ), the growth rate steeply climbs smaller than to a maximum above r^* and then gradually decreases for larger particle sizes. One can gauge that as larger particles years possess smaller growth rates, overall, the distribution will rapidly narrow – a condition, which has been termed "focusing" (48). The temporal consumption of the monomer by growing particles gradually decreases the supersaturation and, consequently, the overall growth rates. The growth rates of particles larger than that at the growth rate maximum now show little variation, while the slope of the rate between r^* and the maximum is also significantly reduced. Both of these effects cause particles above the maximum to grow at an almost uniform rate while causing those between r^* and the maximum to grow progressively slower. A broadening of the size distribution ("defocusing") is thus expected, with the eventual steady-state profile being similar to that predicted by the classical Lifshitz–Slyozov–Wagner theory (46, 48).

It is often assumed that particle growth is a diffusion-limited process (46, 48). The early work on nanocrystal nucleation and growth was carried out in glass matrices (49). In such highly viscous media, the mean size of the particles obeys a $t^{1/3}$ rate law, which is the hallmark for diffusion-limited growth (50). Of course, in fluids where the solvent viscosity may be a million times lower, monomer diffusion is facile and it is the displacement of ligands or decomposition of precursors that is rate determining. The reaction kinetics in liquids are usually reaction-limited. If the surfaces were not protected by strongly adsorbed ligands then particle coagulation would be facile. It is also generally assumed that "focusing" is characteristic of diffusion-limited growth conditions(46, 48). However, van Embden has recently shown that focusing is also possible under reaction-limited growth, though the effect is not quite as pronounced as at low Damköhler numbers (51).

14.2.4 Nucleation and Growth of Metal Chalcogenide Nanocrystals

The most successful route to the synthesis of nanocrystal ensembles has been under hot-injection, lyophobic synthetic conditions (1). To ensure narrow size distributions in such reactions, stabilizers must be chosen such that they (i) provide sufficient steric stabilization to avoid particle–particle aggregation, (ii) coordinate the precursor in solution in order to control the monomer activity, and (iii) bind to the nanocrystal surface to control the nucleation and growth dynamics.

Sufficient stabilization is achieved by adsorbing polymer or organic layers to the nanocrystal surface (52). Practically, the component of the ligand or polymer that acts to provide this stabilization is in the form of an alkyl group with 6 or more carbon groups, and it is tethered to the nanocrystal through direct adduct formation with the surface. The exact nature of the anchoring group depends strongly on the surface state to be passivated, the required adduct bond strength, and the system of interest. Typically, these moieties are, however, limited to either carboxylate, amine, thiol, or phosphine (and its oxidized derivatives) functional groups. Only through the judicious selection of suitable capping agents can the optimal conditions for nucleation and growth of crystallites be determined.

To exemplify these concepts, in Figure 14.8, we show how the concentration of oleic acid (OA) within a high-temperature reaction mixture can be used to control the particle size of CdS nanocrystals (53) – a system that was first studied by Yu and Peng (54). In these reactions, cadmium oxide is heated within a mixture of 1-octadecene and OA. The typical concentration of the cadmium in these reactions is ~20 mM. Following degassing at moderate temperatures, the reaction is heated under nitrogen gas to ~300 °C to form cadmium oleate, which is the precursor for nanocrystal formation. In order to induce nucleation and growth of CdS, a solution of sulfur dissolved in 1-octadecene is injected to give a final cadmium–sulfur ratio of ~1:1.25. The extent of nanocrystallite growth is controlled by heating the reaction mixture at 240 °C for the desired time.

It is evident from Figure 14.8a that under the aforementioned experimental conditions, the addition of higher OA concentrations with respect to the base concentration of cadmium acts to increase the particle size at all times during the reaction. As seen in Figure 14.8b, a concomitant decrease in the particle concentration is also observed. These findings are consistent with the fact that increasing the oleic concentration acts to decrease the supersaturation of the system by increasing the solubility of the monomer in the solution. A reduced

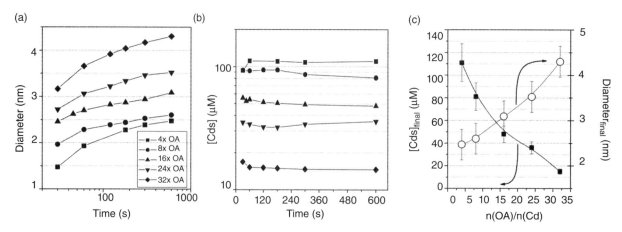

Figure 14.8   The temporal evolution of CdS size (a) and concentration (b) of samples in which the oleic acid–cadmium ratio is modified in the reaction mixture. (c) A summary of particle size and concentration following 10 min of growth. The errors for the respective measurements are of the order of 0.3 nm for the size and 15% of the total concentration. A common legend is included in (a). (Reprinted with permission from Reference 53).

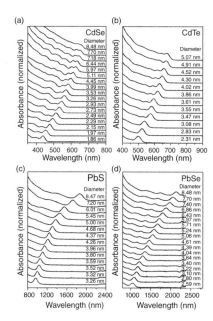

Figure 14.9   Absorption spectra of size-tunable (a) CdSe, (b) CdTe, (c) PbS, and (d) PbSe nanocrystals prepared under high-temperature nucleation and growth conditions. (Reprinted with permission from Reference 55. Copyright 2011 American Chemical Society).

supersaturation causes an increase in the critical radius for nucleation (see Fig. 14.2) and, consequently, a reduced nucleation rate (see Fig. 14.3). For a fixed concentration of monomer in the initial reaction mixture, the extra ligand leads to a smaller number of nuclei and ultimately to the growth of larger nanocrystals. One may also argue that higher OA concentrations lower the interfacial surface energy. However, contrary to what is observed, this would increase the rate of nucleation, although a small decrease in surface free energy may lead to a narrower size distribution (see Fig. 14.7).

From this case study on CdS, it is clear that even the concentration of OA has a drastic influence on both the nucleation and growth stages. This highlights the difficulty of quantitatively understanding and predicting the evolution of any colloidal ensemble when parameters such as concentration, temperature, and precursor decomposition rates are all considered. Nevertheless, protocols for the synthesis of highly monodisperse metal chalcogenide nanocrystals have been developed. In Figure 14.9, we show absorption spectra demonstrating the size tunability for a number of archetypal inorganic systems – each of which has been successfully utilized for solar applications (55). Notably, as the sizes of these nanocrystals are below the effective Bohr radii of their respective bulk material, clear evidence of quantum confinement is exhibited.

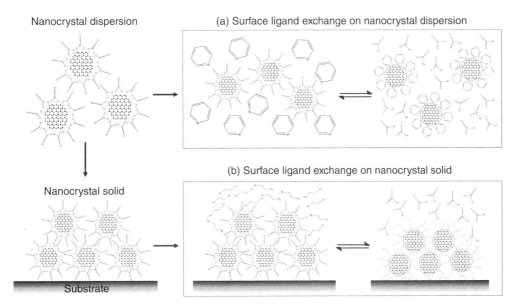

Figure 14.10 A general depiction of the two main strategies involved in performing a surface ligand exchange on colloidal nanocrystals. The most conventional approach is to perform a dynamic ligand exchange directly in solution. An emerging approach is to deposit the nanocrystal as a solid film and do a subsequent chemical treatment step by exposing the nanocrystal film to an appropriate ligand present in either a vapor or a liquid phase.

14.3 LAYER-BY-LAYER ASSEMBLY OF SINTERED NANOCRYSTAL THIN FILMS

Colloidal nanocrystals are attractive materials for optoelectronic devices because of their ease of synthesis and their size-tunable optical and electronic properties and because they are "solution processable." However, processing these nanocrystals requires not just highly concentrated and colloidally stable dispersions but also tailoring of the rheological properties of the resultant inks to optimize their surface tension, density, viscosity, and wetting properties. The optimization depends on the deposition technique being employed. In this section, we critically assess such factors with respect to developing high-quality "sintered" CdTe and $CdSe_xTe_{1-x}$ thin films.

14.3.1 Surface Chemistry of Nanocrystal Thin Films

There are two general strategies for controlling the surface chemistry of nanocrystals for electronic devices: (i) direct solution-based surface ligand exchange and (ii) manipulation of the surface chemistry after the nanocrystals have been deposited as a thin film. Both of these strategies are depicted in Figure 14.10.

In the first approach, dispersed nanoparticles are treated with a ligand that is capable of displacing the existing surface ligands. This ligand exchange process is dynamic and is usually modeled using a Langmuir isotherm (56). To a first approximation, Pearson's hard and soft acid and base (HSAB) theory can be used as a guide for gauging the relative interactions of ligands with a particular surface species. For example, in the case of CdSe nanocrystals, Cd^{2+} is considered a soft acid, and HSAB suggests that soft bases such as alkyl thiolates should bind more strongly than hard bases such as alkyl amines to surface cadmium species. Measurements of the relative binding constants of alkyl thiols and alkyl amines on CdSe nanocrystals are consistent with this prediction (56). Thiols, which notably chemisorb to the surface as a thiolates, possess an equilibrium constant that is 2–3 times higher than for amines. In order to maximize the extent of surface ligand exchange, the exchange process is often repeated multiple times with intermediate precipitation, centrifugation, and redispersion steps.

The advantage of a dynamic surface exchange is that the nanocrystals remain dispersed in solution. This is critical for effectively depositing nanocrystals within polymer/nanocrystal or sol–gel/nanocrystal composites, which are useful within bulk-heterojunction solar cells (13) or lasers (57).

The necessity to retain colloidal stability during ligand exchange imposes strong restrictions on the nature of the ligands that can be employed. Generally, the ligands need to be long enough to offset van der Waals interactions between the nanocrystals; however, this drastically limits the interparticle electrical conductivity. For this reason, films of such particles tend to be very strong insulators. This restriction has recently been overcome by the group of Talapin through the implementation of pure chalcogenides (S^{2-}) or metal chalcogenides (e.g., $Sn_2S_6^{2-}$) as direct surface passivants (58, 59). The high dielectric constants of the passivants and their close to "atomic thickness" ensure there is sufficient interparticle electronic coupling for electronic applications.

The second approach toward modifying the surface chemistry of nanocrystals is arguably simpler – it also enables strong electronic coupling between particles, and it does not require colloidal stability to be maintained during surface treatment. Within this approach, as-deposited films of nanocrystallites are exposed to either a vapor or liquid phase containing a ligand that adsorbs strongly to the surface

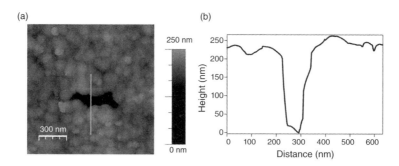

Figure 14.11   (a) Tapping mode AFM image of a single-layer ∼260 nm CdTe film that was CdCl$_2$ treated and annealed at 350 °C for 1 min. (b) A plot of relative height versus distance for the blue trace shown in (a). The ∼250 nm depth of the feature shown agrees with the thickness of the film as measured by profilometry. (Reprinted with permission from Reference 34. Copyright 2011 American Chemical Society).

and displaces the existing ligands. The only requirements of this exchange process are that the nanocrystals must not possess solubility in the liquid phase containing the ligands nor gain solubility after surface exchange. As a result, multifunctional ligands or ligands that greatly enhance particle–particle interactions can be used.

As an example, by treating as-deposited films of OA-capped nanocrystals with dilute solutions of short-chain amines or dithiols, a nearly quantitative removal of the OA-capping ligand can be achieved (60, 61). This treatment brings the nanocrystals into close contact, greatly enhancing charge transport between the nanocrystals and allowing the fabrication of PbS(e)-based devices such as solar cells (31), field-effect transistors (62), and photodetectors (63).

This approach has recently been extended by the Sargent group to include the use of organic halide salts, such as cetyltrimethylammonium bromide (CTAB), to achieve inorganic atomic ligand passivation of PbS (33). Such a treatment resulted in a 200-fold increase in electron mobility relative to dithiol-capped nanocrystals and resulted in solar cells with efficiencies as high as 6 %. A similar approach by Zhang and coworkers found that treatment of both Pb- and Cd-chalcogenide nanocrystals with ammonium sulfide ((NH$_4$)$_2$S) resulted in the displacement of the original surfactant ligands and the formation of a monolayer thick, metal sulfide shell (64).

In practice, a combination of both surface modifications strategies is used to deposit thin films of nanocrystals that are useful for electronic applications. As part of this process, the initially bulky, long-chained ligands stemming from the original synthetic protocols are replaced by more compact, short-chained ligands. Then, following deposition of such a dispersion as a thin film, a secondary chemical treatment step is applied. Depending on the ligand selection at each of the surface exchange steps, nanocrystal thin films with tunable degrees of electronic coupling and mechanical properties can be fabricated.

14.3.2   Layer-by-Layer Assembly of Nanocrystal Thin Films

To obtain optimal performance in solution-processed electronic devices, it is important that the constituent thin films are smooth, uniform, and defect-free over large areas. This is difficult to achieve in nanocrystal films that have been chemically treated, particularly with ligands that have significantly altered the original interparticle spacing, because the induced stresses in the film eventually result in macroscopic defect formation (65). Likewise, if nanocrystal thin films are heated to a point where either the surface ligands thermalize or the nanocrystals undergo large-scale grain growth, defects inevitably result (66). Fortunately, such structural problems can be overcome through the use of layer-by-layer (LbL) nanocrystal assembly (34).

The LbL assembly method is a versatile bottom-up technique for assembling thin films (67). It consists of a series of solution-based deposition steps through which a thin film can be assembled with a high degree of spatial and compositional control. The technique was originally designed as a method for depositing films of oppositely charged polyelectrolytes (68). It has since been adapted for many other applications including the assembly of three-dimensional periodic structures (69), organized films of metal nanoparticles (70), nanoparticle–polyelectrolyte composites (71), and, more recently, graphene-based transparent electrodes (72).

The LbL process is appealing because multiple deposition cycles eventually fill any voids, vacancies, and defects. It also gives control over the film thickness, which might not be achievable in a single deposition cycle. In doing so, short-circuit pathways are effectively passivated, and high-quality, nanocrystal-based transistors (65), LEDs (73), and solar cells (34, 66) can be readily fabricated.

To exemplify the necessity for LbL assembly, in Figure 14.11, we show a typical topographic atomic force microscopy (AFM) profile of a CdTe nanocrystal thin film, with original nanocrystal diameters of ∼4 nm, which has been chemically treated with cadmium chloride (CdCl$_2$) and annealed at 350 °C. It is clear that this grain growth and sintering process leads to defect formation in the CdTe films. In this instance, a pinhole approximately 300 nm long and 100 nm across can be observed. The depth of this pinhole is determined to be ∼250 nm, which is in agreement with the total thickness of the film as determined via profilometry. Such structural defects are common for any nanocrystal thin film that has undergone shrinkage during processing.

To a first approximation, the simplest approach to minimize stress and, therefore, defect formation is to maximize the volume fraction that the inorganic material occupies within the film. For ligand-free nanoparticles of spherical, cylindrical, and cubic shapes, it can be shown that the maximum volume fraction occupied under cubic close packing is 74%, 79%, and 100%, respectively, and is independent of crystal size

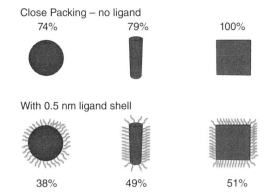

Figure 14.12 A schematic showing the volume fraction occupied by cubic close-packed nanocrystals of different shapes, with and without a 0.5 nm ligand shell.

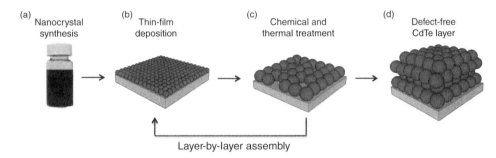

Figure 14.13 A schematic representation of the layer-by-layer assembly process used for the fabrication of defect-free sintered CdTe nanocrystal films. (a) CdTe nanocrystals are synthesized and dispersed in a solvent suitable for multilayer deposition. (b) The nanocrystals are deposited as a thin film using a suitable solution-processing technique such as spin coating. (c) The CdTe film is first exposed to a chemical treatment consisting of exposure to a solution of CdCl$_2$ followed by thermal annealing at elevated temperature. This leads to nanocrystal sintering and defect formation. (d) This process is repeated a number of times until the desired CdTe thickness is achieved.

(Fig. 14.12). This immediately suggests that under these ideal conditions cubic nanoparticles would result in no shrinkage, while cylindrical particles and spherical nanoparticles would experience significant and similar shrinkage.

In reality, nanocrystals must be stabilized using ligands to maintain their colloidal stability. The volume fraction occupied by the ligands can be considerable. If one considers CdTe nanocrystals of 2 nm radius, cylinders of the same radius with a typical aspect ratio of 8:1, and cubes with vertices of 4 nm in length, all passivated by ligands with a length of 0.5 nm (typical for small molecules such as pyridine), the volume fractions now become 38%, 49%, and 51%, respectively. Thus, the effect of ligands is to greatly reduce the volume fraction of spheres relative to cylinders, while causing the volume fraction of cylinders to become similar to that of cubes.

This simple analysis suggests that it is significantly more difficult to develop defect-free thin films with spherical nanocrystals than with cylindrical or cubic particles. Supporting this assumption, studies made on sintered CdTe nanorod thin films have shown limited structural defect formation (19, 74), unlike the results presented in Figure 14.11 on spherical CdTe nanocrystals. However, synthesizing well-defined, nonspherical nanocrystals is generally difficult. This is especially true for novel materials, whose synthetic procedures are not yet well understood. Therefore, there is a need for the development of a generic route for obtaining high-quality sintered films, regardless of the crystal morphology.

To overcome the problem of defect formation in sintered CdTe films, LbL assembly process is adopted where chemical and/or thermal treatment steps are performed following each deposition step (34). In this way, the defects that do form can be overcoated or filled in by subsequent depositions. A similar process has been utilized for PbS(e)-based photovoltaic devices (31).

The LbL assembly process is outlined in Figure 14.13. First, the as-synthesized CdTe nanocrystals are washed to remove excess precursors and subjected to ligand exchange if necessary. A variety of ligands have been found to be suitable for the LbL process including OA, pyridine, 5-amino-1-pentanol, and short-chain amines. Once the desired surface functionalization has been obtained, the nanocrystals are dispersed in a suitable solvent. Again, a variety of solvents including 1-propanol, chloroform, chlorobenzene, 1,2-dichlorobenzene, pyridine, and mixtures of pyridine/1-propanol have been used to generate high-quality LbL films. For the remainder of this chapter, we discuss the LbL assembly of cadmium chalcogenide nanocrystals, which have been functionalized with pyridine ligands and dispersed in a 1:1 pyridine/1-propanol (v/v) solvent mixture.

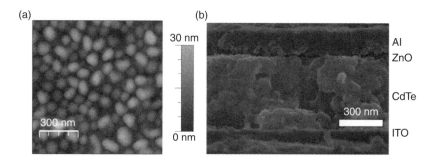

Figure 14.14 (a) Tapping mode AFM image of a layer-by-layer-deposited CdTe film. The film consists of four CdTe layers that have been treated with CdCl$_2$ and annealed at 350 °C for 30 s following each deposition. (b) A cross-sectional SEM image of a layer-by-layer-deposited CdTe film deposited on an ITO substrate. The CdTe has been overcoated with a ~60 nm thick ZnO layer and 100 nm Al to make a complete solar cell. (Reprinted with permission from Reference 34. Copyright 2011 American Chemical Society).

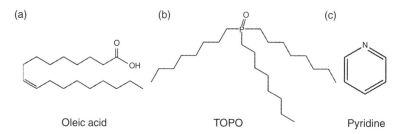

Figure 14.15 The chemical structures of oleic acid (OA), trioctylphosphine oxide (TOPO), and pyridine.

Once the CdTe film has been deposited, it is annealed briefly at 150 °C to remove any residual solvent and to ensure that the film is well adhered to the substrate. If desired, the CdTe can then be exposed to a CdCl$_2$ chemical treatment step in order to modify the surface chemistry and to promote crystal growth upon annealing (34). Although some defect formation may occur at this stage, it is generally not until the film is annealed and grain growth occurs that large-scale defects are formed. Once the film has been annealed at elevated temperature for the desired time it is then cooled to room temperature. This process of deposition, chemical treatment, and thermal treatment is repeated a number of times to obtain the desired CdTe thickness and to eliminate any large-scale defects.

The efficiency of the LbL process can be demonstrated through AFM imaging. Shown in Figure 14.14a is an AFM image of an LbL-deposited CdTe film that consists of four layers, each approximately 100 nm thick that were CdCl$_2$ treated and then annealed at 350 °C. The image shows a uniform and tightly packed particle film, free of any large-scale defects. This result is typical of LbL-deposited CdTe films indicating that this process is a suitable method for creating homogeneous, pinhole-free, sintered nanocrystal thin films.

While AFM provides information about the surface of a thin film, scanning electron microscopy (SEM) can be used to obtain cross-sectional images, allowing the inner structure of the film to be determined. From the SEM image of an LbL-deposited CdTe film shown in Figure 14.14b, it is evident that the film is uniform across the entire thickness of the CdTe layer, with no visible defects or obvious gradient in grain sizes.

14.3.3   Influence of Surface Chemistry on CdTe Nanocrystal Grain Growth

As we will later show, to achieve good performance in sintered nanocrystal devices it is necessary that the grain size within the film reaches bulk proportions. To examine the role of surface chemistry on CdTe nanocrystal grain growth, thin films of CdTe nanocrystals coated with the commonly used organic ligands oleic acid (OA), trioctylphosphine oxide (TOPO), and pyridine were compared (75). The chemical structures of these ligands are shown in Figure 14.15. Grain growth occurs upon annealing and was monitored by UV-visible absorption measurements. With increasing crystal size the absorption onset red shifts as the effects of quantum confinement are diminished. This red shift continues until the crystals attain bulk optical properties, beyond which the absorption onset and optical bandgap remain unchanged.

For CdTe nanocrystals coated with native OA ligands (Fig. 14.16a) and TOPO (Fig. 14.16b), limited crystal growth was found up to temperatures of 400 °C. It was not until the samples were annealed at a temperature of 450 °C that the absorption onset approached that of bulk CdTe (76).

While both OA and TOPO are high boiling point and relatively strong surface-binding ligands, pyridine is significantly more volatile and coordinates only very weakly to CdTe surfaces. Despite this, thermal annealing of pyridine-coated CdTe nanocrystal films was found to induce significant grain growth only at a slightly lower temperature of 400 °C compared to both of the aforementioned surface chemistries. This may be due to the fact that quantitative surface exchange is not achieved with pyridine (77, 78), so that the resulting films behave similarly to

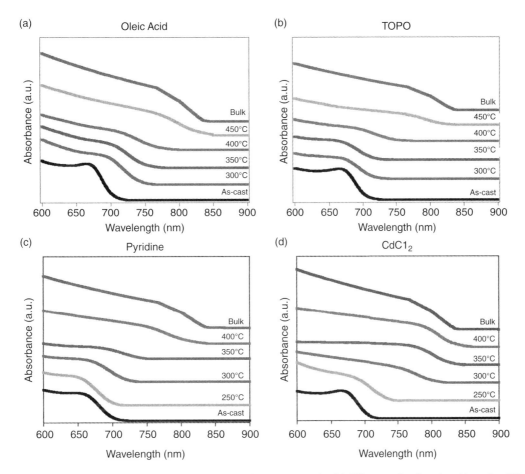

Figure 14.16 (a–c) Absorption spectra of ~100 nm thick films made from CdTe nanocrystals with different surface ligands and heated at different temperatures for 2 min. All as-synthesized nanocrystals were coated with oleic acid and subsequently exchanged with either TOPO or pyridine prior to deposition. (d) Absorption spectra for films that have been treated with CdCl$_2$ before annealing. The bulk CdTe absorption spectrum shown in all graphs is calculated from bulk optical constants and is shown for reference. Spectra have been offset for clarity. (Reprinted with permission from Reference 75. Copyright 2014 Elsevier.)

OA-capped nanocrystals (Fig. 14.16c). As the films are annealed in air, surface oxidation may also play a role in stabilizing the nanocrystals against growth (79).

For vacuum-deposited CdTe, it is well established that a CdCl$_2$ treatment can promote recrystallization and grain growth upon annealing, provided the as-deposited crystals are in the submicron size regime (80). This approach was extended to nanocrystal films by simply dipping pyridine-capped CdTe nanocrystal thin films into a saturated solution of CdCl$_2$ in methanol for 10 s and then rinsing any excess with 1-propanol (34). As shown in Figure 14.16d, a ~5 nm absorbance red shift was induced by this chemical treatment. This is likely due to a slight increase in nanocrystal size resulting from the formation of a Cd-rich surface layer. Thermal annealing of the treated particles exhibited a clear signature of grain growth at temperatures as low as 300 °C. It is evident that CdCl$_2$ treatment drastically accelerates grain growth of CdTe nanocrystal films and also significantly lowers the necessary temperature for grain growth.

This finding highlights the importance of selecting an inorganic passivant (or dopant) for stimulating low-temperature grain growth and recrystallization. For CdTe, this passivant is conveniently provided by CdCl$_2$. For copper indium gallium selenide (CIGS), it has been shown that the presence of sodium (81) or antimony (82) has similar effects.

14.3.4   Fourier Transform Infrared Spectroscopy of CdTe Nanocrystals Thin Films

The role of the CdCl$_2$ treatment in the crystal growth process can be partially understood through the use of Fourier transform infrared (FTIR) spectroscopy. FTIR is commonly used to probe the surface chemistry of nanocrystals as it is sensitive to the structure and binding modes of organic passivants (83). Films of pyridine-capped CdTe nanocrystals that had not been chemically treated with CdCl$_2$ show characteristic C–H peaks in the aliphatic region corresponding to pyridine and residual OA-capping ligands. Although these peaks decrease steadily with increasing annealing temperature, it is not until 400 °C that they disappear entirely (Fig. 14.17). This serves as an indication that the ligands remain bound to the nanocrystal surface up to temperatures well beyond their solution boiling point, thus preventing large-scale crystal growth from occurring.

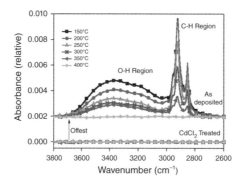

Figure 14.17 FTIR spectra of CdTe thin films annealed at different temperatures with and without $CdCl_2$ treatment. Spectra of the untreated films have been offset for clarity. (Reprinted with permission from Reference (34). Copyright 2011 American Chemical Society).

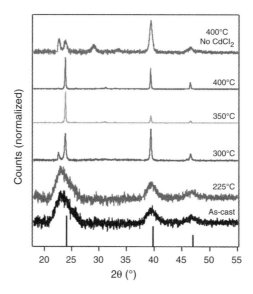

Figure 14.18 X-ray diffraction patterns for CdTe thin films that have been treated with $CdCl_2$ and annealed at various temperatures for 2 min. Also shown are the XRD patterns for a film that has been annealed at 400 °C without $CdCl_2$ treatment and the standard reference values for cubic-phase CdTe (JCPDS \#75-2086). (Reprinted with permission from Reference 75. Copyright 2014 Elsevier.)

In contrast, the spectra of $CdCl_2$-treated films contained no detectable C–H contributions, even at room temperature. This confirms that the $CdCl_2$ treatment effectively strips all the organic ligands from the nanocrystal surface and presumably leaves an atomically passivated chloride ion-capped surface. It can, therefore, be concluded that chemical treatment steps that induce inorganic surface passivations can reduce the energy barrier to crystal growth as well as the carbon contamination levels within the final nanocrystalline films.

14.3.5 X-Ray Diffraction of Sintered CdTe Nanocrystal Thin Films

14.3.5.1 Influence of Annealing Temperature Absorbance measurements provide a measure of crystallite diameter in the quantum size regime. However, they are unable to provide such information once the crystal has attained bulk optical properties. In this size range, characterization techniques such as X-ray diffraction (XRD), AFM, and SEM are the preferred methods of size determination.

XRD results of CdTe nanocrystal thin films subjected to different chemical and heating treatments are shown in Figure 14.18. From these, it is apparent that at annealing temperatures of 300 °C and above, $CdCl_2$-treated CdTe samples possess a crystal phase that is unequivocally cubic, the most stable form of bulk CdTe (84). The average CdTe crystallite size has been determined using the Scherrer equation (85); a summary of these values are included in Table 14.1.

For an as-cast film, which has not been $CdCl_2$ treated or annealed, broad diffraction peaks characteristic of small nanocrystals are observed. The observed peaks are consistent with cubic-phase CdTe, and peak analysis yields an average crystal diameter of 4 nm, in good agreement with absorption results. For a film that has been treated with $CdCl_2$ and annealed at 225 °C, virtually no change is observed in the XRD pattern, indicative of little to no crystal growth. When the annealing temperature is elevated to 300 °C, a greater than 10-fold increase in

TABLE 14.1 Average Crystal Diameter of CdTe Nanocrystal Thin Films That Have Been Annealed at Different Temperatures for 2 min

Annealing Temperature (°C)	Average Crystal Diameter (nm)
As-cast	4
225	5
300	42
350	67
400	94
400 (no CdCl$_2$)	20

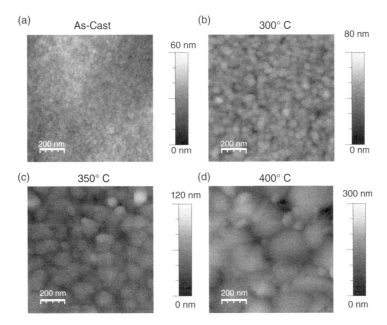

Figure 14.19 (a) Tapping mode AFM image of an as-cast CdTe film. (b–d) AFM images of films that have been treated with CdCl$_2$ and annealed for 2 min at 300 °C, 350 °C, 400 °C, respectively. (*Source*: Reprinted with permission from Reference 75. Copyright 2014 Elsevier).

crystal diameter is observed with the average crystal size reaching 42 nm. As expected, at higher annealing temperatures, the extent of crystal growth is enhanced, with average diameters of 67 nm at 350 °C and 94 nm at 400 °C. Notably, accurate approximations of the crystallite size for higher temperatures could not be determined as these exceeded the size limitation placed on the validity of the Scherrer equation. In comparison, the film that had not been chemically treated and annealed at 400 °C showed only limited growth, with an average crystal diameter of only 20 nm. This finding is consistent with simple absorption measurements.

In Figure 14.19, topographic AFM images of CdTe films treated with CdCl$_2$ and annealed at different temperatures are shown. The as-cast film consists of tightly packed, uniform grains. For films that have been annealed at temperatures of 300 °C and above, there is clear evidence of increased grain growth with increasing annealing temperature. The crystallites remain uniform and tightly packed for the films annealed at 300 °C and 350 °C, with minimal void space separating adjacent grains. At 400 °C, the formation of void space between crystals becomes evident, even when the film is deposited in an LbL fashion. This is likely to increase the likelihood of device failure due to shorting if very thin layers are utilized. Nevertheless, AFM also reveals the presence of small grains located primarily in the grain boundary regions. A similar finding has been previously reported for bulk CdTe films treated with CdCl$_2$ and attributed to the ongoing recrystallization process (80).

14.3.5.2 Influence of Annealing Time While annealing temperature is a strong determinant of the extent of crystal growth achieved within a sintered nanocrystal layer, it is also important to understand the temporal evolution of crystal size at a given temperature. XRD measurements were performed on CdTe films treated with CdCl$_2$ and annealed at 350 °C for different lengths of time (75). As discussed earlier, as-deposited films are comprised of 4 nm crystallites that exhibit cubic crystal structure. Annealing for only 5 s was sufficient to greatly reduce the width of the CdTe diffraction peaks, indicative of rapid and facile crystal growth (Fig. 14.20a). The average crystal size in this film was determined to be 17 nm (Table 14.2). For such a short annealing time, it is unlikely that the substrate will reach the 350 °C temperature set point, and only

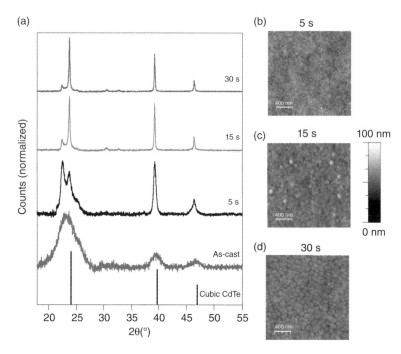

Figure 14.20 (a) X-ray diffraction patterns for an as-cast CdTe thin film as well as those that have been treated with $CdCl_2$ and annealed at 350 °C for 5, 15, and 30 s. Also shown are the standard reference values for cubic-phase CdTe (JCPDS \#75-2086). The XRD patterns have been normalized to the peak of greatest intensity and offset for clarity. (b–e) Tapping mode AFM images of CdTe nanocrystal thin films that have been treated with $CdCl_2$ and annealed at 350 °C for 5, 15, and 30 s, respectively. (*Source*: Reprinted with permission from Reference 75. Copyright 2014 Elsevier).

TABLE 14.2 Average CdTe Nanocrystal Diameter as a Function as Annealing Time at 350 °C as Determined by XRD and AFM

Annealing Time at 350 °C	Average Crystal Diameter (XRD) (nm)	Average Crystal Diameter (AFM) (nm)
As-cast	4	<5
5 s	17	32
15 s	53	66
30 s	68	87
60 s	64	88
120 s	66	82

a very limited opportunity for crystal growth is presented. With further annealing, the CdTe crystal size continues to increase up to times of 30 s, at which point an average crystal size of 68 nm is determined. Beyond this time the crystal size essentially saturated.

A similar trend was observed based on AFM results – annealing times greater than 30 s resulted in little change to the grain size or surface roughness (Fig. 14.20b–d). It should be noted that in AFM imaging it is the geometrical grain diameter and not necessarily the crystallite size that is being measured. As a result, these values are typically larger than those determined by XRD. Nevertheless, the general trend in crystal size versus annealing time is the same for both techniques. The causes of the saturation in crystal size are most likely related to the formation of surface oxides such as CdO, $CdTeO_3$, and TeO_2, whose formation at the expense of CdTe upon annealing in air is well documented (86–88).

14.3.6 LbL Assembly of Sintered $CdSe_xTe_{1-x}$ Nanocrystal Thin Films

The development of defect-free, neat layers is a critical first step toward any thin-film electronic application. A natural extension to this development is the formation of compositionally graded structures. In traditional CdTe solar cells, high-temperature annealing induces diffusion between the n-type CdS and p-type CdTe layers (81). This results in a compositionally smooth junction, necessary for optimum device performance.

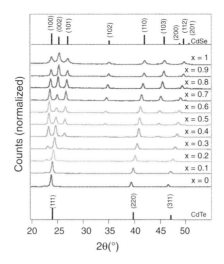

Figure 14.21   (a) X-ray diffraction patterns for ~100 nm thick $CdSe_xTe_{1-x}$ thin films that have been treated with $CdCl_2$ and annealed at 350 °C for 30 s. For reference, we have included diffraction patterns for bulk CdSe and CdTe in the native hexagonal and cubic phases, respectively. (*Source*: Reprinted with permission from Reference 92. Copyright 2011 American Chemical Society).

The incorporation of alloyed layers into solar cells also allows for the fabrication of compositionally graded structures, with tunable bandgap and energy-level alignment. Such compositional engineering is also critical for obtaining high-efficiency CIGS devices, where a three-stage evaporation process is employed to vary the Ga/(In + Ga) ratio throughout the film (89, 90).

An advantage of the LbL process is that it provides a means to directly create such gradients during the deposition. Moreover, through the use of nanocrystals as building blocks with subsequent chemical and thermal treatment steps, mixtures of separately synthesized nanocrystals can be utilized to facilitate their formation. To exemplify this concept, here, we describe a method for fabricating $CdSe_xTe_{1-x}$ alloys by inducing grain growth and sintering of thin films containing mixtures of CdTe and CdSe nanocrystals.

It is possible to directly synthesize $CdSe_xTe_{1-x}$ nanocrystals with tunable composition (8); however, it is generally accepted that it is difficult to reproducibly control the exact stoichiometry in such alloyed systems (91). For this reason, the use of binary components provides a more facile approach for developing various alloys. The structural, optical, and electronic properties of these $CdSe_xTe_{1-x}$ materials will now be explored, with details of their incorporation into both single- and graded-composition thin films, and solar cells provided in later parts of this chapter.

Using ~5 nm diameter pyridine-coated CdTe and CdSe nanocrystals as starting materials, blending controlled quantities of nanocrystals in solution enabled the deposition of compositionally tunable sintered $CdSe_xTe_{1-x}$ films (92). These were prepared using an analogous $CdCl_2$ treatment to the one described for neat CdTe and then annealing at 350 °C in air. Grazing angle XRD showed that neat CdTe and CdSe films exist in their thermodynamically most stable cubic and hexagonal crystal phases, respectively (Fig. 14.21) (84). From the XRD patterns of the Te-rich alloys, it is clear that all of the scattering contributions associated with the CdTe phase shift toward higher 2θ values with increasing x. This is consistent with the formation of an alloyed cubic phase that possesses an increasingly smaller lattice constant. For $x = 0.3$ and above, however, the emergence of a secondary phase characteristic of a hexagonal crystal structure is observed. The overlap of the (111) peak of the cubic phase with the (101) peak of the hexagonal phase for the most Se-rich samples prevents us from unequivocally identifying the upper boundary of this dual-phase region. These observations of a biphasic system are in agreement with previous reports on bulk $CdSe_xTe_{1-x}$ alloys (95, 96).

The formation of $CdSe_xTe_{1-x}$ alloys is accompanied by large-scale grain growth and nanocrystal sintering. Analogous to neat CdTe thin films, this process creates large structural defects in the thin films. Again, this problem can be overcome through the use of LbL processing. AFM images of LbL-deposited $CdSe_xTe_{1-x}$ show that nanocrystalline layers with RMS roughness of between 8.6 nm (CdTe) and 4.2 nm (CdSe) are obtained (Fig. 14.22a–d). Having started with ~5 nm nanoparticles, there is clear evidence of grain growth following the chemical and thermal treatment steps; however, the extent of grain growth does decrease with increasing Se content, with the average crystal size decreasing from 38 nm for CdTe to 17 nm for CdSe, as determined from XRD using the Scherrer equation. This is consistent with AFM images that exhibit the same trend toward smaller grains with increasing Se content (Fig. 14.22e). We attribute this effect, at least partly, to the larger lattice energy of CdSe (749 kcal/mol) compared to CdTe (680 kcal/mol) (97). Since the surface energies for both CdSe and CdTe are similar (98), this would cause the CdSe to exhibit a higher activation energy for recrystallization relative to CdTe (99), thus reducing its size under identical annealing conditions. We also note that much like CdTe, CdSe is also known to oxidize upon annealing in air (88). It is expected that the formation of such oxide species will further restrict grain growth from occurring (79). Therefore, the intrinsic differences between the rates of recrystallization and oxidation across the investigated composition range are believed to cause the observed variation of crystallite sizes.

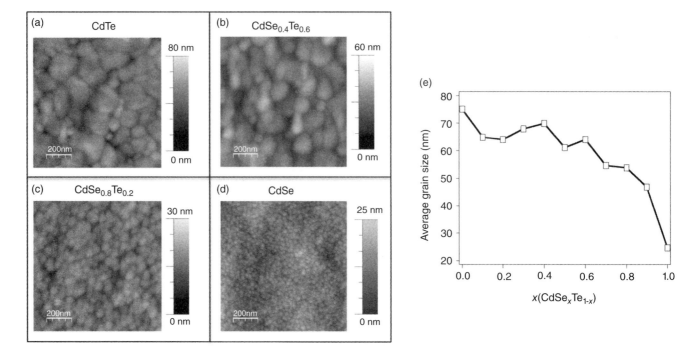

Figure 14.22 (a–d) Tapping mode AFM images of CdTe, $CdSe_{0.4}Te_{0.6}$, $CdSe_{0.2}Te_{0.8}$, and CdSe, respectively. For all samples, two layers were deposited, with $CdCl_2$ treatment and thermal annealing step performed after each layer. The total film thickness was ~200 nm. (e) Average grain size as determined from AFM images as a function of $CdSe_xTe_{1-x}$ composition. (*Source*: Reprinted with permission from Reference 92. Copyright 2011 American Chemical Society).

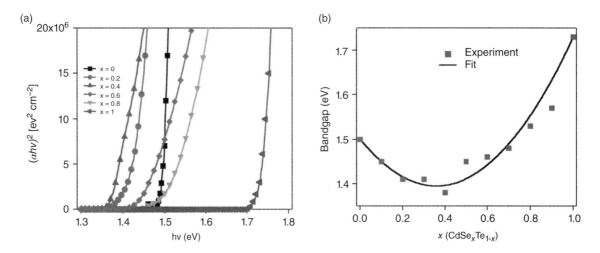

Figure 14.23 (a) Selected plots of $(\alpha h v)^2$ versus hv as a function of $CdSe_xTe_{1-x}$ composition. Measurements were made on 200 nm thick films that had been $CdCl_2$ treated and annealed at 350 °C for 30 s. The linear onset indicates that the materials possess a direct bandgap. (b) Bandgap of $CdSe_xTe_{1-x}$ alloy films as a function of composition (red squares) as well as a fit to the standard bowing equation. Experimental values were determined by extrapolating the linear portion of $(\alpha h v)^2$ versus hv plots to zero. (*Source*: Reprinted with permission from Reference 92. Copyright 2011 American Chemical Society).

14.3.7 Optical Properties of Sintered $CdSe_xTe_{1-x}$ Nanocrystal Thin Films

To investigate the effects of alloy formation on the optical properties of $CdSe_xTe_{1-x}$ thin films, UV-visible absorption measurements were performed (Fig. 14.23a) (92). From these spectra, the value of the optical bandgap, along with its nature, direct or indirect, can be ascertained through Tauc's relation:

$$(\alpha h v)^\gamma = A(hv - E_g) \tag{14.17}$$

where α is the absorption coefficient, hv is the photon energy, A is a constant, E_g is the bandgap, and γ is a number that characterizes the transition process (100). A value of $\gamma = 2$ corresponds to allowed direct transitions, while $\gamma = 1/2$ corresponds to allowed indirect transitions.

The value of γ can be determined from a plot of $(\alpha h v)^{\gamma}$ versus photon energy. For $CdSe_xTe_{1-x}$ films, a linear onset is observed when γ equals 2, confirming that this material possesses a direct bandgap, consistent with previous results (101).

The value of the bandgap can be determined by extrapolating the linear region of the plots in Figure 14.23(a) to the x-axis. Plotting the bandgap as a function of alloy composition reveals clear optical bowing (Fig. 14.23(b)), that is, a parabolic variation of the bandgap with alloy composition. For materials that exhibit optical bowing, the bandgap of the alloyed material is often smaller than either of its binary components. This effect has been described by Van Vechten and is a result of the disorder created by the presence of multiple anions (or cations) in the crystal lattice (94). Optical bowing has been commonly observed in $CdSe_xTe_{1-x}$ (102, 103), as well as many other semiconducting alloys (104, 105). Thus, by tuning the $CdSe_xTe_{1-x}$ film composition, it is possible to vary the bandgap from a minimum of 1.38 eV (absorption onset ~899 nm) for $x = 0.4$ to 1.73 eV (absorption onset ~717 nm) for $x = 1$. As will be shown in the following, this bandgap tuning can be exploited as a means to tune the spectral response of sintered nanocrystal solar cells.

The relationship between the optical bandgap and alloy composition can be expressed using the standard bowing equation

$$E_g(x) = bx(1 - x) + (E_{g,CdTe} - E_{g,CdSe})x - E_{g,CdTe} \tag{14.18}$$

where b is the optical bowing parameter and $E_{g,CdTe}$ and $E_{g,CdSe}$ are the bulk bandgaps of CdSe and CdTe, respectively. Fitting the $CdSe_xTe_{1-x}$ results to this equation yields a bowing parameter of 0.81 eV. While this parameter has no physical significance, it is in good agreement with literature values for the $CdSe_xTe_{1-x}$ system, which vary from 0.59–0.91 eV (106, 107).

14.3.8 Electronic Structure of Sintered $CdSe_xTe_{1-x}$ Nanocrystal Thin Films

To determine the electronic structure of the $CdSe_xTe_{1-x}$ films, photoelectron spectroscopy in air (PESA) was used. In PESA, the yield of photoemitted electrons is measured as a function of incident photon energy. From the cubic root of this photoemission yield, the minimum ionization energy of inorganic semiconductor samples can be determined (55). Typical PESA plots for $CdSe_xTe_{1-x}$ films are shown in Figure 14.24a. Extrapolation of these curves to the baseline value provides an estimation of the ionization energy of a sample, which for an intrinsic semiconductor is equivalent to the valence band edge. These results show that the valence band edge steadily shifts to higher energies with increasing Se content, from 5.23 eV for CdTe to 5.92 eV for CdSe.

The corresponding conduction band edges of the samples can be approximated by adding the optical bandgap, as determined previously, to the ionization energy, with the results shown in Fig. 14.24B. In contrast to the nearly linear variation observed in ionization energy with changing composition, the conduction band edge initially shifts to higher energies with increasing x before leveling off when $x \geq 0.4$. These results are consistent with theoretical studies of $CdSe_xTe_{1-x}$, which found that the majority of filled states near the valence band edge are located on the anionic atoms, in particular Te, while states near the conduction band edge are predominantly located on the Cd atoms (107). In the films examined here, it is only the ratio [Se]/[Te] that is changing, while [Cd] is constant. It is, therefore, unsurprising that varying the alloy composition has a stronger affect on the valence band energy levels.

14.3.9 Vertically Graded Sintered $CdSe_xTe_{1-x}$ Nanocrystal Thin Films

We now consider the fabrication of vertically graded $CdSe_xTe_{1-x}$ nanocrystal thin films. These films were prepared on a glass/ITO substrate utilizing graded $CdSe_xTe_{1-x}$ layers. As part of this LbL process, subsequent depositions of a nanocrystalline ZnO layer and a thermally evaporated aluminum layer were performed in order to fabricate complete PV devices as shown in Figure 14.25.

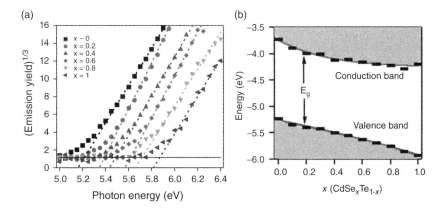

Figure 14.24 (a) PESA spectra of films with varying $CdSe_xTe_{1-x}$ composition. (b) Valence band maxima and conduction band minima as a function of alloy composition. Valence band levels were determined directly from PESA, while conduction band levels were approximated by combining the valence band levels with the experimentally determined optical bandgaps. Solid red lines are a guide to the eye. (*Source*: Reprinted with permission from Reference 92. Copyright 2011 American Chemical Society).

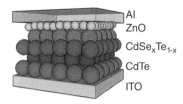

Figure 14.25   A schematic of a completed solar cell device that possesses a vertically graded $CdSe_xTe_{1-x}$ compositional structure.

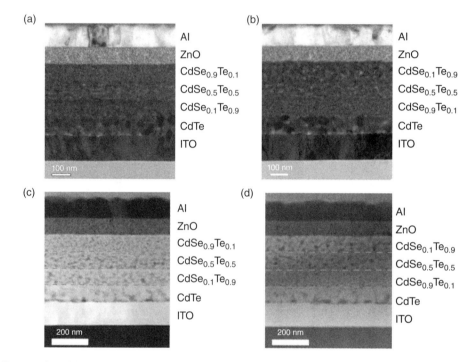

Figure 14.26   (a–b) Cross-sectional bright-field TEM images of forward- and reverse-graded solar cells, respectively. (c–d) Cross-sectional dark-field STEM images of the same devices. The dashed green lines serve as a guide to the eye. (*Source*: Reprinted with permission from Reference 92. Copyright 2011 American Chemical Society).

Two compositional gradings were investigated, one in which the selenium content was increased closer toward the ZnO layer and the other where it was decreased. These devices were respectively termed "forward" and "reverse," in recognition of their expected electronic structure. In both instances, the initial layer was comprised of neat CdTe.

Structural and compositional characteristics of such graded structures were obtained through cross-sectional device imaging using a focused ion beam (FIB) and both bright- and dark-field transmission electron microscopy (TEM). The TEM images for both device structures clearly establish the existence of the individual $CdSe_xTe_{1-x}$ layers, along with the nanocrystalline ZnO layer, and ITO and Al contacts (Fig. 14.26). It can also be seen that each layer conformally coats the underlying layer, thereby reducing the amount of void space within the device structure. The marked contrast between the ZnO and $CdSe_xTe_{1-x}$ layers and the sharpness of this interface suggests that widespread interdiffusion is not occurring between these materials.

The compositions of the graded structures were measured using energy-dispersive X-ray spectroscopy (EDS) and are presented in Figure 14.27. In both the forward- and reverse-graded devices, the cadmium concentration is relatively constant across the entirety of the $CdSe_xTe_{1-x}$ layers. A slightly Cd-rich composition is observed throughout, approximately 55%. It is possible that this excess Cd is found in the crystal bulk; however, a more likely explanation is that a Cd-rich surface layer is created by the $CdCl_2$ treatment. Of greater interest is the variation of chalcogenide content across the device as this provides information on the extent of interdiffusion between the $CdSe_xTe_{1-x}$ layers. On examining the chalcogenide content of the forward-graded structure, we see that both the Se and Te concentrations vary in a nearly stepwise manner.

This important result not only proves the existence of compositional grading in these multilayered thin films but also confirms that limited interlayer diffusion occurs at the annealing temperatures used here (350 °C). Notably, for device structures that require controlled interlayer diffusion, higher temperatures should easily facilitate such outcomes. Overall, based on the results presented earlier for $CdSe_xTe_{1-x}$,

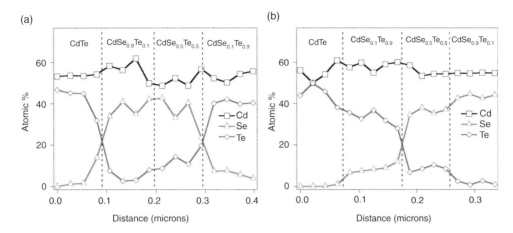

Figure 14.27 (a) Cross-sectional EDS profile of the CdTe/CdSe$_x$Te$_{1-x}$ layers in a forward-graded device. (b) Cross-sectional EDS profile of the CdTe/CdSe$_x$Te$_{1-x}$ layers in a reverse-graded device. Dashed lines indicate the boundaries between individual device layers. The nominal composition of each layer, as determined prior to film deposition and annealing, is provided. (*Source*: Reprinted with permission from Reference 92. Copyright 2011 American Chemical Society).

it is clear that LbL assembly can be exploited for the fabrication of inorganic thin films with controlled structural, optical, and electrical properties.

14.4 SINTERED NANOCRYSTAL SOLAR CELLS USING LbL ASSEMBLY

In the previous two sections, we have described how (i) high-quality nanocrystals can be synthesized, (ii) the surface chemistry of such nanocrystals can be modified, and (iii), through the use LbL assembly, high-quality sintered nanocrystal thin films can be deposited. These critical steps can be applied to develop numerous technologies, however in this chapter we will focus on their use in solar cells.

By virtue of the nanocrystalline structure of the thin films prepared at the temperatures used for fabrication of our CdSe$_x$Te$_{1-x}$ thin films, the resulting devices vary substantially from that of their bulk analogues. Nevertheless, promising devices can still be fabricated because of a relatively high doping density and a high internal efficiency of charge separation and collection. We begin this section by taking a look at conventional CdTe-based solar cell device architectures, before progressing to see how LbL assembly enables colloidal particle to be used in such devices.

14.4.1 CdTe Solar Cell Device Structures

Early investigations into CdTe solar cells focused on p–n homojunctions, which reached efficiencies up to 11% (108). It has long since been established that higher performances can be obtained using a heterojunction configuration (79). While CdTe can be doped either p or n type, it has been found that the best performing devices consist of a p–CdTe absorber layer paired with a wide bandgap n-type semiconductor, commonly referred to as the "window" layer. At the interface between these two materials, a charge-separating p–n junction will form. The requirements for a good window layer in a conventional CdTe solar cell are high transparency in the visible region to minimize parasitic absorption losses, reasonably high doping density to ensure that the bulk of the depletion region is located within the CdTe, and the ability to form a high-quality junction with CdTe in order to minimize interfacial recombination losses.

The two different configurations typically used for fabricating CdTe heterojunction solar cells, commonly termed superstrate and substrate, are depicted in Figure 14.28a–b. In the superstrate configuration, a transparent conducting oxide (TCO) is first deposited onto a glass support structure. The most widely used TCO materials are SnO$_2$:F (FTO), In$_2$O$_3$:SnO$_2$ (ITO), ZnO:Al (AZO), and Cd$_2$SnO$_4$ (109). This is generally followed by a high-resistivity buffer layer. The purpose of this layer is to prevent the CdTe from coming into direct contact with the TCO via any cracks or pinholes in the thin window layer. The device is then completed through the sequential deposition of the window layer, CdTe, and back contact. In the substrate configuration this process is reversed, with device fabrication beginning from the back contact and finishing with the TCO. This structure has the advantage of not requiring the incident light to pass through the glass support structure.

The device configuration used throughout this work is slightly different from either of the aforementioned cases and can be considered an "inverted superstrate" geometry (Fig.14.28c). As with a conventional superstrate device, the device is built on top of a glass/TCO support and completed with back-contact deposition. However, the order of the semiconductor depositions is reversed, with the CdTe deposited first, followed by the n-type layer. In this geometry, the incident light first strikes the CdTe layer. This means that the "window" layer in this structure is no longer required to be highly transparent. As a consequence, this layer can be made thicker, minimizing the likelihood of pinhole formation and thus eliminating the need for a high-resistivity buffer layer. This section will examine inverted superstrate devices using a CdTe/ZnO structure, with a particular focus on the CdTe layer.

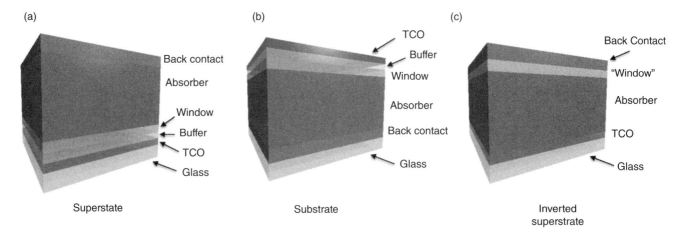

Figure 14.28 Schematic representations of superstrate, substrate, and inverted substrate CdTe solar cell device configurations. The change in the window layer color from yellow to gray represents the use of ZnO in place of the more conventional CdS. In all cases, the order of deposition is from bottom to top.

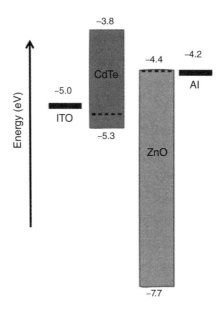

Figure 14.29 The flat-band electronic energy levels with respect to vacuum ($E_{vac} = 0$ eV) for all components of an ITO/CdTe/ZnO/Al solar cell. The dashed lines indicate the approximate Fermi levels of the given materials.

14.4.2 The CdTe/ZnO Heterojunction

It has long been known that CdTe/ZnO heterojunctions can be used to produce efficient photovoltaic devices. Using monocrystalline CdTe and ZnO deposited by spray pyrolysis, power conversion efficiencies as high as 8.8% have been reported (110). More recently, it has been established that films of ZnO nanocrystals can be used as effective charge transport layers in solution-processed solar cells (111) and LEDs (93). This is due in part to the high levels of n-type doping obtained in ZnO nanocrystals that have been exposed to ultraviolet light, making the films highly conductive (112). Meanwhile, the deep valence band of ZnO (−7.7 eV) makes it an effective hole-blocking material. When paired with CdTe, it is expected that an abrupt p–n junction will be formed with an energetic structure that is suitable for PV applications (Fig. 14.29). Finally, the synthesis of ZnO nanocrystals can be readily achieved using inexpensive precursors and mild reaction temperatures. All of these factors make the CdTe/ZnO heterojunction a promising device architecture for use in sintered nanocrystal solar cells.

14.4.3 Single-Layer Sintered CdTe/ZnO Nanocrystal Solar Cells

As we have discussed in Section 14.3, thin films of sintered CdTe nanocrystals that are deposited as a single layer exhibit significant structural defects. When such layers are incorporated into an inverted superstrate solar cell configuration with the architecture ITO/CdTe/ZnO(60 nm)/Al,

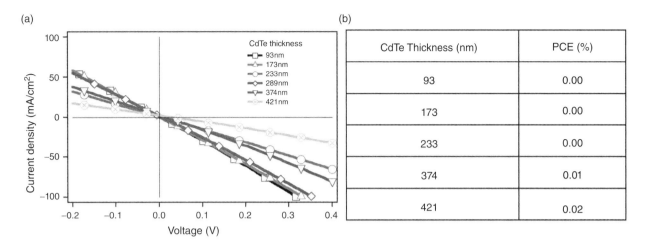

Figure 14.30 (a) Current–voltage curves for ITO/CdTe/ZnO(60 nm)/Al solar cells with varying CdTe thickness. In these devices, a single deposition and annealing treatment of the CdTe layer was performed. (b) Power conversion efficiencies of the solar cells in (a) as a function of CdTe thickness. All devices failed due to shorting.

the existence of these defects becomes immediately evident from the current–voltage curves (Fig. 14.30). Even though such devices were prepared with varying CdTe layer thicknesses, each exhibited negligible photovoltaic performance as a result of shorting. This shorting behavior was observed for all devices in which a single annealing treatment was performed, including attempts at CdTe/CdSe devices and CdTe-only Schottky devices. From these results, it is clear that the single-layer approach is not suitable for the fabrication of sintered nanocrystal solar cells using spherical nanocrystals. Instead, an LbL deposition method must be applied.

14.4.4 LbL Assembly of Sintered CdTe/ZnO Nanocrystal Solar Cells

LbL assembly of CdTe thin films was carried out as described in the previous section, with $CdCl_2$ chemical and thermal annealing steps performed following every deposition. It was found that a total of four CdTe layers were required in order to obtain reproducible solar cell performance, regardless of the thickness of each individual layer. After the CdTe layering was completed, a 50–100 nm thick colloidal ZnO layer was deposited, also through spin coating, and a 100 nm thick aluminum back electrode was thermally evaporated. A schematic representation of the LbL process for the fabrication of the inverted superstrate ITO/CdTe/ZnO/Al solar cells is provided in Figure 14.31.

14.4.4.1 Influence of CdTe Grain Size We have shown that annealing at higher temperatures increases the average CdTe grain size within the $CdCl_2$-treated CdTe nanocrystal films. To understand how this influences solar cell performance, ITO/CdTe (400 nm)/ZnO (60 nm)/Al (100 nm) devices were fabricated with the thermal annealing temperature of the CdTe varied between 150 and 450 °C.

The photovoltaic performance in such solar cells is presented in Figure 14.32 and summarized in Table 14.3. The results are based on optimum annealing times for each temperature, which will be examined in depth later in this chapter. When the CdTe layers were annealed at temperatures too low to induce significant grain growth (<300 °C), the devices exhibited poor performance, primarily due to low short-circuit current densities. At higher annealing temperatures the average grain size was increased to >40 nm; this was accompanied by a significant enhancement in device photocurrent and overall performance. For devices annealed in the temperature range 300–400 °C, efficiencies of >5% were obtained. Maximum efficiency was obtained at a temperature of 350 °C, where a total area power conversion efficiency of 6.9% was recorded. When the temperature was further increased to 400 °C, the performance dropped to 5.2%, while at 450 °C, the devices failed completely due to shorting. In the latter case, the films were thinner and visibly much rougher following annealing, indicating possible CdTe sublimation or extreme oxidation. Additionally, the grain sizes were >100 nm (verified by SEM).

These results build on previous studies by the groups of Alivisatos (19) and Carter (74) in the area of solution-processed CdTe nanocrystal devices. They also demonstrate that in LbL-assembled solar cells, lower annealing temperatures are required to achieve good device performance relative to single-layer CdTe nanorod devices. Indeed, it has been explicitly stated by Ju et al. that performance in CdTe/CdSe nanorod devices drops substantially at annealing temperatures below 380 °C (113).

The grain-size-dependent behavior in these systems can be partly explained by the light absorption properties of such films. In devices with very small CdTe crystallites, the total absorption is lower because of the blue shift in absorption onset as a result of quantum confinement effects as well as a lower-absorption cross section. However, this only accounts for a small portion of the differences in device performance. A greater contribution is accounted for by considering the exciton binding energy of small crystallites. In bulk CdTe, the exciton radius is 6.8 nm and the binding energy is just 10 meV, so charge carrier separation is very facile at room temperature. In a strongly confined system, the enhanced electron–hole Coulomb interaction induces an increased exciton binding energy, E_b, which prevents effective charge separation

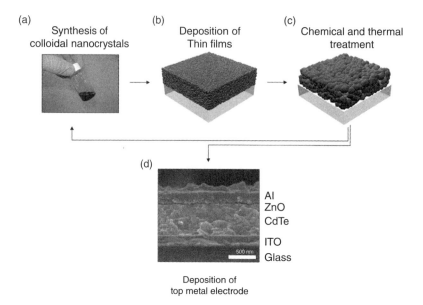

Figure 14.31 A schematic representation of the layer-by-layer assembly process used in the fabrication of ITO/CdTe/ZnO/Al solar cells. (a) As-synthesized nanocrystals are purified and their surface chemistry modified as desired. (b) The nanocrystals are deposited as a thin film onto an ITO-coated glass substrate by spin coating. (c) The thin film is exposed to a chemical treatment (typically $CdCl_2$), followed by thermal annealing to induce crystal growth via nanocrystal sintering. These steps are repeated until the desired device architecture is obtained. (d) A scanning electron microscope (SEM) image of a typical polycrystalline CdTe/ZnO solar cell fabricated using the LbL approach. (*Source*: Reprinted (adapted) with permission from Reference 34. Copyright 2011 American Chemical Society).

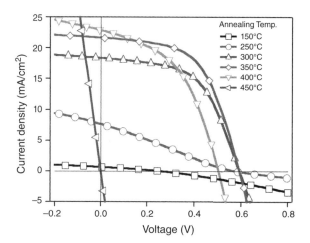

Figure 14.32 Current–voltage curves for layer-by-layer ITO (125 nm)/CdTe(400 nm)/ZnO(60 nm)/Al(100 nm) solar cells with different CdTe annealing temperatures. All cells consist of four CdTe layers, with $CdCl_2$ and annealing steps being performed after each layer for an optimized time. (*Source*: Reprinted with permission from Reference 34. Copyright 2011 American Chemical Society).

from occurring. The value of E_b can be estimated from the Coulombic component of the Brus equation (114) as

$$E_b = \frac{1.8e^2}{4\pi\varepsilon_r\varepsilon_0 r} \tag{14.19}$$

where e is the elementary charge, ε_0 the vacuum permittivity, ε_r the relative dielectric constant ($\varepsilon_{r,CdTe} \sim 10.3$), and r the crystal radius. For smaller crystal sizes, such as those found in CdTe films annealed at $<300\,°C$, the exciton binding energy is larger than the thermal energy, kT, and is in a range comparable to those reported for organic semiconducting materials (115, 116). When the crystallite diameter is increased to ~20 nm, the exciton binding energy decreases below that of the thermal energy attained at room temperature. This enables excitons to

TABLE 14.3 Average Crystal Diameter as Determined by XRD, and Performance Characteristics of ITO(125 nm)/CdTe(400 nm)/ZnO(60 nm)/Al(100 nm) Solar Cells Fabricated at Different CdTe Annealing Temperatures (34). All Values Have Been Corrected for Spectral Mismatch

Annealing Temperature (°C)	Average Crystal Diameter (nm)	J_{sc} (mA/cm^2)	V_{oc} (V)	FF	Efficiency (%)
150	4	0.7	0.25	0.30	0.1
250	5	7.4	0.57	0.25	1.1
300	42	17.6	0.59	0.55	5.8
350	67	20.7	0.59	0.56	6.9
400	94	21.9	0.50	0.37	5.2
450	>100	0.0	0.00	0.00	0.0

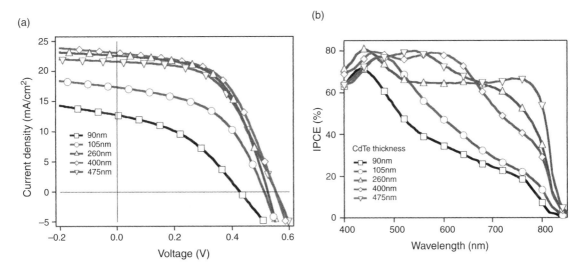

Figure 14.33 (a) Current–voltage curves for ITO(125 nm)/CdTe/ZnO(60 nm)/Al(100 nm) solar cells with different CdTe thicknesses under AM1.5 illumination. (b) IPCE curves for the same cells. (*Source*: Reprinted with permission from Reference 34. Copyright 2011 American Chemical Society).

dissociate into free charges without the need for a charge-separation mechanism, such as an external electric field, a type II interface, or a surface trap state. Additionally, increased particle size also reduces the surface-to-volume fraction and the number of grain boundaries present in the film, further decreasing the likelihood of charge recombination (117). Accordingly, only when the CdTe layers are annealed above the bulk-transition temperature are appreciable solar cell power conversion efficiencies observed.

14.4.4.2 Influence of CdTe Absorber Layer Thickness   In addition to the processing conditions, absorber layer thickness is one of the most critical parameters affecting thin film solar cell performance. Due to its high absorption coefficient of $\sim 10^5$cm^{-1} at energies greater than the bandgap, CdTe is an ideal material for making very thin solar cells (79). In a typical vacuum-deposited cell, the thickness of the CdTe layer is on the order of 3–10 μm (118). However, this thickness range is cause for concern due to the relative scarcity of Te. As highlighted by both Fthenakis and Zweibel, the thickness of the CdTe layer will have to be decreased to <1 μm in order for TW-scale energy production from CdTe solar cells to be achievable based on the current, annual, and global supply of tellurium (119, 120).

Recent work by the Compaan group has shown promising results in regard to reducing the CdTe thickness to submicron levels. Using a magnetron sputtering process, this group has manufactured cells with a conventional geometry of FTO/CdS/CdTe/(Cu + Au) (121). By optimizing the processing conditions, device efficiencies as high as 12% for 1.1 μm and 9.7% for 0.5 μm CdTe thicknesses were attained (122). These promising results show that Te supply will not necessarily be the limiting factor in CdTe solar energy production.

To study the effects of absorber layer thickness on nanocrystal-based solar cells, ITO/CdTe/ZnO(60 nm)/Al(100 nm) devices were fabricated for CdTe thicknesses varying from 90 to 475 nm (34). Current–voltage curves obtained with these devices are shown in Figure 14.33a and summarized in Table 14.4. Functioning devices were obtained with as little as 90 nm of CdTe. However, the short-circuit current density in this device, as well as that with 105 nm CdTe were limited due to incomplete absorption of the incident light. This is clearly seen in the incident photon-to-current efficiency (IPCE) spectra, which exhibit a depressed response across nearly the entire visible spectrum (Fig. 14.33b). Nevertheless, the fact that functioning devices were obtained with such thin layers is a testament to the effectiveness of the LbL process.

TABLE 14.4 Performance Characteristics of ITO(125 nm)/CdTe/ZnO(60 nm)/Al(100 nm) Solar Cells with Varying CdTe Thickness

CdTe Thickness	J_{sc} (mA/cm^2)	V_{oc} (V)	FF	Efficiency (%)
90 nm	12.4	0.43	0.40	2.1
105 nm	16.9	0.51	0.50	4.3
260 nm	22.0	0.52	0.55	6.3
400 nm	21.9	0.57	0.54	6.7
475 nm	20.7	0.59	0.56	6.9

Only when the CdTe thickness is increased to more than 250 nm are high values of J_{sc} and device efficiencies of > 6% obtained. The J_{sc} values of 21–22 mA/cm^2 obtained in these cells are not far from those of the highest efficiency CdTe cells, which are on the order of ~25 mA/cm^2 (123). Notably, the devices presented here do not have an antireflection coating, which typically increases the J_{sc} of a device by a relative ~5% (124). These thicker devices show peak IPCE values of >80% and broadband conversion efficiencies of >60% for wavelengths of less than 700 nm.

These devices also show that open-circuit voltage decrease with decreasing CdTe thickness. This is consistent with previously reported experimental work and device simulations, where the decrease in V_{oc} has been attributed to increased back-contact recombination and device shunting (121, 122, 125). From these results it is clear that high-efficiency devices can be obtained with minimal CdTe consumption. Moreover, the similarity between the device performance of conventionally prepared CdTe solar cells and those presented here for equivalent thicknesses confirms the great promise offered by solution processing for developing solar cell technologies.

14.4.4.3 Optical Modeling and Internal Quantum Efficiency In order to understand the efficiency of the photocurrent generation process in solar cells, it is vital to gauge how well-absorbed photons are converted to free carriers and then collected. This figure of merit is defined as the internal quantum efficiency (IQE). IQE is the ratio of the number of charge carriers collected within a solar cell to the number of photons that are absorbed within the device. Typically, this is accomplished through the use of conventional reflectance and transmittance measurements, which allow the determination of the wavelength-dependent device absorptance. The IQE of a device is then calculated as

$$IQE = \frac{IPCE}{(1 - T - R)} = \frac{IPCE}{A} \tag{14.20}$$

where T is the transmittance, R the reflectance, and A the absorptance, which under the assumption of minimal scattering is equal to $1 - T - R$. By integrating the product of the absorptance values and the equivalent current density of the AM1.5 spectrum, the maximum photocurrent attainable under standard conditions can be determined. However, when performed on multilayered thin-film solar cells, these measurements implicitly include parasitic contributions such as optical interference and absorption by adjunct layers (including the substrate and electrodes), resulting in an overestimation of the true active layer absorptance (126). In order to remove these parasitic contributions, it is useful to model the system using a scattering matrix method. By using this approach, it is possible to model the entire multilayered device structure and explicitly calculate the wavelength dependence of parasitic contributions (127).

To confirm the applicability of such modeling to the multilayered solar cells described previously, the calculated reflectance spectra of completed devices to that determined experimentally for a variety of CdTe thicknesses were compared (75) (Fig. 14.34). For all these devices the experimentally determined reflectance was in good agreement with that predicted, especially at wavelengths below 700 nm. The discrepancy at longer wavelengths is most likely due to the fact that the modeling calculations assume perfectly smooth interfaces with no scattering or intermixing between layers. A comparison of the experimentally determined absorptance for the entire solar cell to that of the extracted CdTe layer shows excellent agreement for thicker CdTe layers, with the standard reflection and transmission measurements only slightly overestimating the true absorptance at wavelengths below the CdTe optical bandgap (~845 nm), while at higher wavelengths this overestimation significantly worsens. For the thinnest device with 70 nm of CdTe there is a much greater divergence between the two curves, as the relative parasitic contribution to the absorptance becomes much higher. Overall though, the excellent agreement between the experimental and modeled results across the entire thickness range examined here demonstrates the effectiveness of this approach in predicting the absorption properties of complex, multilayered nanocrystal devices.

From the calculated spectra, the internal quantum efficiencies based on the absorptance of the CdTe layer alone were calculated (128). A comparison of the wavelength-dependent IQE for solar cells with 260 nm and 475 nm CdTe thicknesses shows good agreement between the two calculation methods, with the IPCE/A result typically 2–4% higher (Fig. 14.35a–b). Both devices show exceptionally high internal quantum efficiencies across the CdTe absorption range, with values in excess of 80%. For the 475 nm device, IQE values approaching unity are obtained in the region centered at 550 nm. These results are in good agreement with the high short-circuit current densities measured in these devices indicating minimal recombination losses.

From the scattering matrix model it is possible to determine the maximum achievable J_{sc} for a given CdTe thickness. This can be accomplished through integration of the calculated absorptance with the AM1.5 spectrum and by assuming that 100% of the absorbed photons in the CdTe layer are collected as photocurrent. A comparison of experimentally determined J_{sc} values as a function of CdTe thickness to the

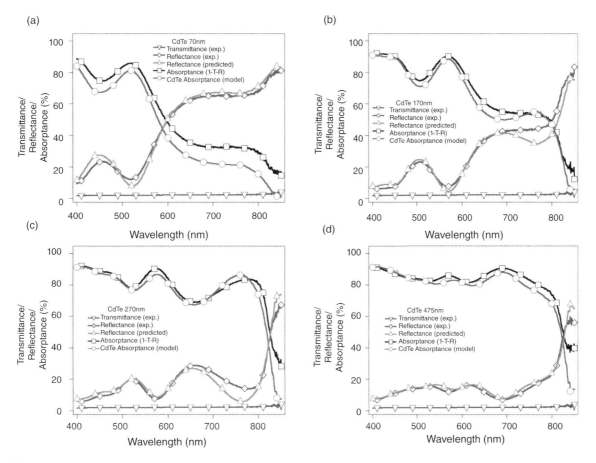

Figure 14.34 Experimentally determined transmittance, reflectance, and absorptance values of glass/ITO (145 nm)/CdTe/ZnO(60 nm)/Al (100 nm) solar cells with varying CdTe thickness. The reflectance values are compared to those calculated using a scattering matrix model based on bulk material parameters. In comparison to the measured absorptance, we also present the calculated CdTe absorptance, from which parasitic contributions have been removed. The CdTe thickness was determined using regression analysis of the calculated reflectance data as well as by profilometry. These values were found to agree to within ±5%. (*Source*: Reprinted with permission from Reference 75. Copyright 2014 Elsevier).

expected values from the scattering matrix model is shown in Figure 14.35(c). The error bars represent the 2.5% variation in measured device area from the nominal value of 0.205 cm^2. Good agreement between the experimental and predicted values is observed across the entire CdTe thickness examined. As the model assumes 100% of the absorbed light within the CdTe layer in the device is converted to photocurrent, we are able to determine the overall IQE of our devices by

$$IQE = \frac{J_{SC,exp}}{J_{SC,calc}} \tag{14.21}$$

For all thicknesses, total IQE values of >75% are obtained, with many devices possessing an IQE of more than 90%. Again, this implies that under short-circuit conditions nearly all photons absorbed within the device are collected as photocurrent.

14.4.4.4 Depleted Heterojunction Mechanism   To understand the origin of these exceptionally high J_{sc} and IQE values, it is important to understand the charge-separation mechanism in these solar cells. In general, high-efficiency polycrystalline thin-film solar cells are fabricated with either a p–n or p–i–n device configuration (117). In a p–n-type cell, when two oppositely doped semiconducting materials are brought into contact, Fermi-level equilibration occurs through an exchange of charge carriers, leaving behind fixed ions. This creates a depletion region that is free of mobile charge carriers and generates a built-in electric field whose distribution within the cell is dependent on the relative doping densities of the p- and n-type materials (Fig. 14.36a). Photons absorbed within the depletion region experience electric field drift and are quickly swept toward their respective electrodes. Photons that are absorbed outside of the depletion region, in the quasineutral region, do not experience this drift and rely on charge separation by diffusion alone, making recombination more likely.

In a p–i–n-type cell, a lightly doped intrinsic (i-type) semiconductor is sandwiched between more highly doped p- and n-type materials (Fig. 14.36b). This extends the width of the electric field and is vital for effectively collecting photocarriers in systems where the diffusion coefficients are small (129). For the CdTe/ZnO system, initial inspection of the band structure would lead to the expectation of a p–n junction. This would naturally arise due to the p-type conductivity of CdTe when treated with CdCl$_2$ and annealed (109). Meanwhile, ultraviolet

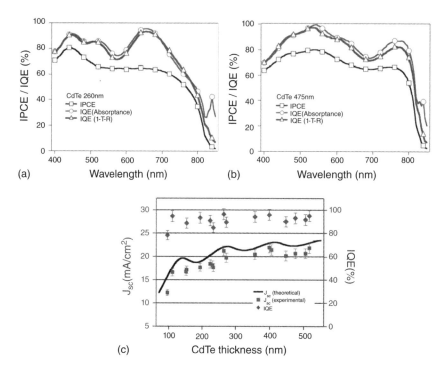

Figure 14.35 A comparison of external quantum efficiency (IPCE) and internal quantum efficiency for ITO (125 nm)/CdTe/ZnO (60 nm)/Al (100 nm) solar cells with CdTe thicknesses of 260 nm (a) and 475 (b). IQE spectra were determined both from the CdTe absorptance calculated from scattering matrix modeling (red curves) and from experimentally determined values of transmittance and reflectance (blue curves). (c) Experimentally measured J_{sc} values plotted as a function of CdTe thickness (red squares) compared to the predicted J_{sc} from modeling assuming that 100% of absorbed photons within the CdTe layer are converted to photocurrent. From the ratio of the predicted and experimental values, the total IQE under short-circuit conditions is determined (blue diamonds). (*Source*: Reprinted with permission from Reference 34. Copyright 2011 American Chemical Society).

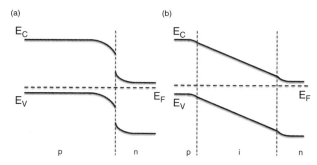

Figure 14.36   (a) Band profile of a p–n solar cell at equilibrium. (b) Band profile of a p–i–n solar cell at equilibrium.

irradiation of ZnO nanocrystal films leads to desorption of surface-bound oxygen or carbon dioxide, which in turn creates free electron densities of approximately 1×10^{19} cm^{-3} and renders the ZnO highly n-type (130). However, as we will now show, this system operates more like a p–i–n junction due to extended depletion of the CdTe layer.

14.4.4.5 Capacitance–Voltage Measurements   Capacitance–voltage (CV) analysis is a useful approach for determining the electronic properties of solar cells. It is able to identify the presence of a depletion layer stemming from a p–n or p–i–n junction and an estimate of the depletion layer width. It is also able to provide information about bulk doping densities and the built-in field (129). This information can be ascertained from a plot of the square of the inverse capacitance as a function of the applied voltage. For an abrupt one-sided junction, where the doping density on one side of the junction is much higher than the other, the capacitance per unit area is defined as

$$C = \sqrt{\frac{q\varepsilon_r\varepsilon_0 N}{2}} \left(V_{bi} - V - \frac{2kT}{q} \right)^{-\frac{1}{2}}$$

(14.22)

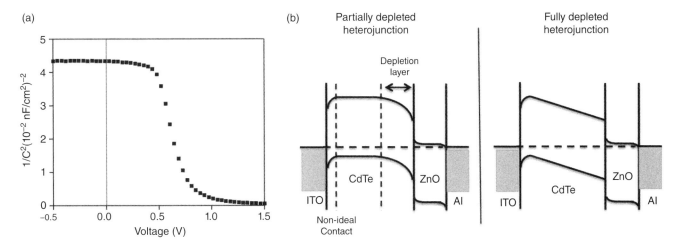

Figure 14.37   (a) Inverse capacitance squared versus voltage plot of an ITO(125nm)/CdTe(200nm)/ZnO(60nm)/Al(100nm) solar cell. (b) Schematic representations of partially and fully depleted abrupt heterojunctions based on an ITO/CdTe/ZnO/Al device architecture.

where N is the doping density, V_{bi} is the built-in voltage, and V is the applied bias. This equation is applicable to the CdTe/ZnO system as it is expected that the doping density of ZnO will be several orders of magnitude greater than that of CdTe. Rearrangement of this equation leads to

$$\frac{d\left(\frac{1}{C^2}\right)}{dV} = -\frac{2}{q\varepsilon_r\varepsilon_0 N} \tag{14.23}$$

From this equation, the doping density can be easily extracted. This further enables the depletion layer width (W_D) to be determined using the expression

$$W_D = \sqrt{\frac{2\varepsilon_r\varepsilon_0}{qN}\left(V_{bi} - \frac{2kT}{q}\right)} \tag{14.24}$$

The $1/C^2$ versus V plot for a device with 200 nm thick CdTe layer is provided in Figure 14.37(a). From the slope of this plot, a CdTe doping density of 1.1×10^{16} cm^{-3} is obtained. This is higher than the 10^{14}–10^{-15} cm^{-3} typically reported in CdTe (131, 132). It is generally thought that the p-type doping density in CdTe is limited by defect-compensation effects (133). However, values up to 1.5×10^{16} cm^{-3} have been reported previously for bulk CdTe deposited by closed-space vapor transport (134) and electrodeposition (135). For nanocrystalline CdTe devices, even higher values of $\sim 7 \times 10^{16}$ cm^{-3} have been reported, where it was hypothesized that residual phosphorus from the TOPO ligands acts as a p-type dopant (136). Although the CdTe films here do not utilize phosphorus-containing ligands, it is possible that sintered nanocrystalline materials have inherently higher defect densities as a result of their small size. It is also worth noting that the purity of the nanocrystal precursors used in this work is relatively low, which may allow for the introduction of extrinsic impurity doping.

From Figure 14.37a, it is also clear that the value of $1/C^2$ is largely independent of the applied voltage at biases below ~ 0.3 V. This is consistent with the CdTe being fully depleted of charge carriers (137) and corresponds to a calculated depletion layer width of ~ 245 nm for this device. In this case, the built-in electric field will extend across the entire CdTe thickness (Fig. 14.37b), providing a strong driving force for charge separation. Indeed, for such a system, it is expected that very high J_{sc} and IQE values would be observed under short-circuit conditions, exactly as has been observed. For devices where the CdTe layers are thicker, a significant fraction of the CdTe will remain depleted, minimizing the reliance on diffusion contributions.

The CV behavior of these multilayered, nanocrystalline solar cells was also examined using SCAPS software (138). Simulations of our system were performed using the energy levels provided in Figure 14.29, a CdTe trap density of 2×10^{14} cm^{-3} and a ZnO donor density of 1×10^{19} cm^{-3}. To obtain comparable bulk capacitance values the CdTe thickness was fixed at 183 nm, slightly lower than the experimental value of 200 nm. We have found that this deviation is within the experimental error of using profilometry to determine film thickness.

Results for varying CdTe bulk acceptor densities exhibit excellent agreement with experimental results for an acceptor density of 1.5×10^{16} cm^{-3}, very similar to the calculated value of 1.1×10^{16} cm^{-3} (Fig. 14.38). From these results, it can be seen that for CdTe doping densities of up to $\sim 2.5 \times 10^{16}$ cm^{-3}, it is expected that full depletion will occur at 0 V applied bias.

In these simulations, the low trap density ensures that the main depletion mechanisms stem from the CdTe/ZnO junction and the ITO/CdTe interface. In reality, the polycrystalline nature of this system would result in large surface trap densities, which would act to deplete each particle and lead to formation of a Schottky barrier extending from the grain boundaries.[117] Under these circumstances, depletion would arise from both grain boundary and junction contributions. It was found that this additional mechanism does not substantially alter the trend observed in the $1/C^2$ versus voltage plot. Thus, while the bulk doping density may be varied somewhat from that determined previously, the fully depleted heterojunction picture remains valid.

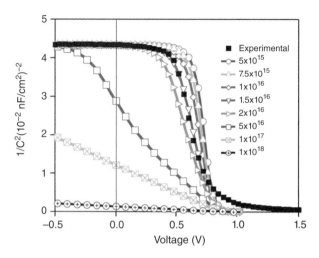

Figure 14.38   Simulated inverse capacitance squared versus voltage plots for ITO/CdTe(183 nm)/ZnO (60 nm)/Al solar cells with varying CdTe acceptor densities. Also shown for comparison is the experimentally determined data. (*Source*: Reprinted with permission from Reference 34. Copyright 2011 American Chemical Society).

If we assume no surface state pinning, the built-in voltage of a fully depleted heterojunction is governed by the Fermi level of the photocathode and the heavily doped n-type material with an Ohmic contact. The work function of cadmium chloride-treated ITO, as determined by PESA, lies between 4.8 and 5.0 eV, depending on the exact processing conditions. Using the literature value of 4.4 eV for the electron affinity of ZnO (138) and considering the high donor doping density, the expected built-in voltage would be of the order of 0.4–0.6 V. This prediction is in good agreement with our experimental device results, further supporting the fully depleted heterojunction model.

14.4.5   LbL Assembly of Sintered $CdSe_xTe_{1-x}$ Nanocrystal Solar Cells

In the previous sections, we have focused on the development of simple, effective bilayer CdTe/ZnO solar cells through the use of LbL assembly. We now utilize this approach for the fabrication of solar cells that contain the $CdSe_xTe_{1-x}$ alloys presented in Section 14.3.6 and demonstrate how device performance can be tuned using compositional gradient engineering.

14.4.5.1 $CdSe_xTe_{1-x}$ Solar Cell Performance   Solar cells were fabricated using a variation of the LbL assembly method as outlined in Figure 14.31 with a structure that has been previously depicted in Figure 14.25. In general, a complete device consisted of a single CdTe layer, approximately 100 nm thick; three $CdSe_xTe_{1-x}$ layers, with a total thickness of ~300 nm; and a nanocrystalline ZnO layer. Using a CdTe buffer layer between the alloy layers and the underlying ITO substrate served two purposes. First, it ensured an identical rectifying interface with the ITO for all devices. This enabled interfacial effects between the semiconducting layers and ITO to be minimized, which in turn permitted the role of the alloy layers on solar cell performance to be isolated. Second, based on the energy levels presented in Figure 14.24b, the CdTe acts as an electron blocking layer, ensuring that only holes from the alloy layers are able to be transported through to the ITO. Meanwhile, the highly n-type ZnO nanocrystal layer acts as an effective hole-blocking layer and enhances charge separation within the active layer by creating an extended depletion region within the device.

Current–voltage curves for selected $CdSe_xTe_{1-x}$/ZnO devices are shown in Figure 14.39a, with full performance characteristics summarized in Table 14.5. The highest device performance in this series was obtained for the CdTe-only device, with a power conversion efficiency of 7.3%. As the Se content within the device increases, a concomitant decline in overall performance was observed. For devices made using alloy layer compositions of $x = 0.1$ and 0.2, the performance and the efficiency are only slightly lower at 7.1% and 7.0%, respectively. As the Se content is further increased, the decline in device performance is accelerated, with the lowest efficiency for an alloy-containing device being 2.5% for $x = 0.9$.

A closer examination of the device characteristics shows that this decrease in performance is primarily due to lower J_{sc} values. As the Se content increases from $x = 0$ to 0.9, the J_{sc} steadily declined to approximately half the value of the CdTe-only device. For solar cells with very thin absorber layers, such as those examined here, the short-circuit current is strongly dependent on the optical properties of the absorber material, with lower-absorption cross sections leading to incomplete light absorption and consequently smaller photocurrent. Comparing the absorption characteristics of $CdSe_xTe_{1-x}$ devices shows that with increasing Se content, total light absorption within the device decreases (Fig. 14.39b). This decreased absorbance largely explains the decline in short-circuit current. This result is promising as it suggests that device performance is not inherently limited by the incorporation of alloy layers and that through optimization of alloy layer thickness higher efficiencies should be possible.

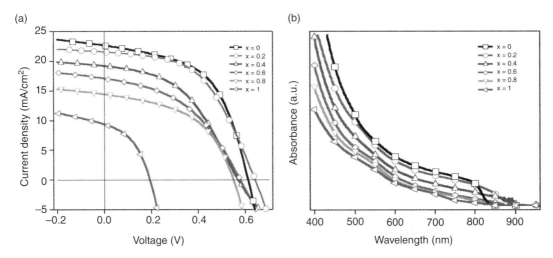

Figure 14.39 (a) Current–voltage curves for $CdSe_xTe_{1-x}$ solar cells. The structure of the device stack was ITO/CdTe ($\sim$100 nm)/$CdSe_xTe_{1-x}$($\sim$300 nm)/ZnO (60 nm)/Al (100 nm). All CdTe and $CdSe_xTe_{1-x}$ layers were treated with $CdCl_2$ and annealed at 350 °C for 30 s in air. The ZnO layer was annealed in air at 300 °C for 2 min. (b) Absorbance spectra for the same devices. Devices with higher Se content exhibit lower total absorption. (*Source*: Reprinted with permission from Reference 92. Copyright 2011 American Chemical Society).

TABLE 14.5 Performance Characteristics of Solar Cells with Varying Alloy Composition. All Cells Consist of an ITO Front Electrode and 100 nm Thick Al Back Electrode (92). Current–Voltage Measurements Were Taken under 100 mW/cm² AM1.5 Conditions and Have Been Corrected for Spectral Mismatch

Device Structure	J_{sc} (mA/cm²)	V_{oc} (V)	FF	Efficiency (%)
CdTe(400 nm)/ZnO	22.6	0.61	0.53	7.3
CdTe (100 nm)/$CdSe_{0.1}Te_{0.9}$ (300 nm)/ZnO	20.3	0.61	0.57	7.1
CdTe (100 nm)/$CdSe_{0.2}Te_{0.8}$ (300 nm)/ZnO	31.6	0.64	0.51	7.0
CdTe (100 nm)/$CdSe_{0.3}Te_{0.7}$ (300 nm)/ZnO	19.5	0.63	0.47	5.8
CdTe (100 nm)/$CdSe_{0.4}Te_{0.6}$ (300 nm)/ZnO	18.7	0.57	0.46	4.9
CdTe (100 nm)/$CdSe_{0.5}Te_{0.5}$ (300 nm)/ZnO	16.0	0.56	0.49	4.4
CdTe (100 nm)/$CdSe_{0.6}Te_{0.4}$ (300 nm)/ZnO	16.7	0.58	0.44	4.2
CdTe (100 nm)/$CdSe_{0.7}Te_{0.3}$ (300 nm)/ZnO	15.2	0.59	0.54	4.8
CdTe (100 nm)/$CdSe_{0.8}Te_{0.2}$ (300 nm)/ZnO	14.0	0.55	0.49	3.8
CdTe (100 nm)/$CdSe_{0.9}Te_{0.1}$ (300 nm)/ZnO	11.9	0.49	0.42	2.5
CdTe (100 nm)/CdSe (300 nm)/ZnO	9.4	0.19	0.41	0.7

In contrast to J_{sc}, the open-circuit voltage of these devices is essentially independent of the alloy layer composition. Indeed, for even the most Se-rich alloy device, the V_{oc} retains greater than 80% of the value for the CdTe-only device. This result is consistent with the absorbing layers of the device being fully depleted under short-circuit conditions and the built-in voltage of the device being fixed by the work function difference between the ITO and ZnO layers.

14.4.5.2 Spectral Response To examine the effects of optical bowing on the spectral response of $CdSe_xTe_{1-x}$-containing solar cells, incident IPCE measurements were performed. All device compositions show a broad spectral response extended across the visible range and are convoluted with thin-film interference effects (Fig. 14.40). Of particular interest in these devices is the wavelength at which the spectral response reaches zero as this can provide insights into the contribution of alloy layers to the device photocurrent. For CdTe-only devices, the IPCE approaches zero near the bulk CdTe absorption edge of $\sim$860 nm, while for alloyed devices, values as high as 30% are obtained at this wavelength, with responses extending into the near-IR region to as far as $\sim$900 nm. This clearly demonstrates that the $CdSe_xTe_{1-x}$ layers are actively contributing to the device photocurrent and that through this approach a degree of control over spectral response can be obtained.

14.4.5.3 Compositionally Graded Solar Cells In Section 14.3.9, the use of sintered $CdSe_xTe_{1-x}$ nanocrystal layers to form compositionally graded thin films termed "forward" and "reverse" were described. In Figure 14.41(a), we present the corresponding flat-band energy levels

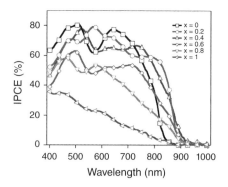

Figure 14.40 (a) Incident photon-to-current efficiency (IPCE) curves for ITO/CdTe (~100 nm)/CdSe$_x$Te$_{1-x}$ (~300 nm)/ZnO (60 nm)/Al (100 nm) devices. The oscillations observed in the photoresponse spectra of these solar cells are a direct result of thin-film interference effects. (*Source*: Reprinted with permission from Reference 92. Copyright 2011 American Chemical Society).

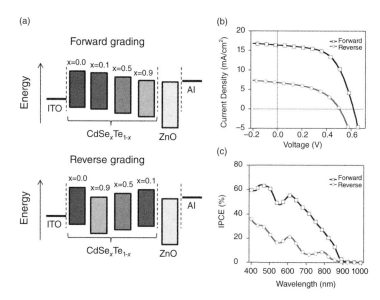

Figure 14.41 (a) Flat-band energy diagrams for forward- and reverse-graded nanocrystal solar cells. (b) Current–voltage curves of graded cells under 1 sun illumination. In both devices, the thickness of each individual CdSe$_x$Te$_{1-x}$ layer was ~100 nm for a total absorber layer thickness of ~400 nm. The thickness of the ZnO layer was 60 nm. (c) IPCE spectra for the same graded devices. The forward-graded structure exhibits superior performance across the entire spectral range. (*Source*: Reprinted with permission from Reference 92. Copyright 2011 American Chemical Society).

of these device structures as determined through PESA. In the forward-graded cell, the energy levels are chosen such that an energy-level cascade will be formed for both electrons and holes, promoting charge separation and transport through the device. In the reverse-graded device, band alignment is unfavorable as a number of energy-level barriers are introduced.

The current–voltage characteristics of these graded nanocrystal devices are presented in Figure 14.41b and summarized in Table 14.6. In the forward-graded cell, a power conversion efficiency of 5.4% is obtained. Examination of the performance parameters reveals that the open-circuit voltage and fill factor of this cell are comparable to those of the highest efficiency CdTe/ZnO devices. The only major limitation in the performance of this device is a relatively low value of J_{sc}, 15.5 mA/cm². As explained earlier, this is directly due to the reduced absorption of the CdSe$_x$Te$_{1-x}$ layers compared to pure CdTe films.

In comparison, the reverse-graded cell exhibits inferior performance across all device parameters. The resulting PCE of 1.5% is substantially lower than both the forward-graded cell and any of the nongraded, alloy-containing devices. This observation is consistent with poor charge transport through the cell due to the unfavorable band structure. Based on the energy levels presented in (Fig. 14.41a), this cell would not be expected to deliver any meaningful photovoltaic performance. However, this simple diagram ignores effects such as band bending, Fermi-level pinning, and the presence of trap states, which will make the true band structure of a device much more complex. Evidently, this does allow for partial extraction of photogenerated carriers from within the device.

TABLE 14.6   Performance Characteristics of "Forward"- and "Reverse"-Graded Nanocrystal Solar Cells (92). All Measurements Were Taken under AM1.5 Illumination and Have Been Corrected for Spectral Mismatch

Device Structure	J_{sc} (mA/cm^2)	V_{oc} (V)	FF	Efficiency (%)
Forward	15.5	0.62	0.56	5.4
Reverse	6.5	0.50	0.46	1.5

To further examine the performance difference between the graded structures, their IPCE spectra have been examined (Fig. 14.41c). For both devices an extended NIR response relative to CdTe is observed, once again confirming the photocurrent contribution of the alloy layers. While the shapes of both IPCE spectra are nearly the same, the forward-graded device exhibits a superior response across the entire spectral range. As the absorption profiles of the devices are nearly identical, light absorption differences can be excluded as the cause of this effect. From these results, it is clear that through the use of compositional device grading, it is possible to either promote or hinder charge transport through the cell. Through further optimization of the grading profiles and layer thicknesses, it is possible that the performance of these graded devices may exceed that of single composition cells.

14.5   CONCLUSIONS

With continuing research efforts in the general area of colloidal nanocrystals, an ensemble of new and exciting technologies will be developed. Some of these will undoubtedly harness the quantum size effects in semiconductor nanocrystals, while others will simply utilize the elegant simplicity that nanocrystal inks present as a means to fabricate low-cost devices. In this chapter, we have tried to highlight the importance of controlling both nanocrystal nucleation and growth, which is the starting point for creating well-defined materials for thin-film devices. Moreover, a more profound understanding of nanocrystal surface chemistry is imperative for the optimization of structural, optical, and electronic properties in nanoparticulate thin films. In this chapter, we have presented each of these steps in a detailed and practical manner with a focus on the development of $CdSe_xTe_{1-x}$-based sintered solar cells. Importantly, all the discussions with reference to such solar cells can be extended to many other colloidal nanocrystal systems and directly applied to most emerging nanocrystal-based electronic application.

REFERENCES

1. Murray CB, Norris DJ, Bawendi MG. Synthesis and characterization of nearly monodisperse CdE (E = S, Se, Te) semiconductor nanocrystallites. J Am Chem Soc 1993;115:8706–8715.

2. van Embden J, Mulvaney P. Nucleation and growth of CdSe nanocrystals in binary ligand system. Langmuir 2005;21:10226–10233.

3. Milliron DJ, Hughes SM, Cui Y, Manna L, Li J, Wang L-W, Alivisatos AP. Colloidal nanocrystal heterostructures with linear and branched topology. Nature 2004;430:190–195.

4. Manna L, Milliron DJ, Meisel A, Scher EC, Alivisatos AP. Controlled growth of tetrapod-branched inorganic nanocrystals. Nat Mater 2003;2:382–385.

5. Peng X, Manna L, Yang W, Wickham J, Scher E, Kadavanich A, Allvisatos AP. Shape control of CdSe nanocrystals. Nature 2000;404:59–61.

6. Mohamed MB, Tonti D, Al-Salman A, Chemseddine A, Chergui M. Synthesis of high quality zinc blende CdSe nanocrystals. J Phys Chem B 2005;109:10533–10537.

7. Jasieniak J, Bullen C, van Embden J, Mulvaney P. Phosphine-free synthesis of CdSe nanocrystals. J Phys Chem B 2005;109:20665–20668.

8. Bailey RE, Nie S. Alloyed semiconductor quantum dots: tuning the optical properties without changing the particle size. J Am Chem Soc 2003;125:7100–7106.

9. Zhong X, Feng Y, Knoll W, Han M. Alloyed $Zn_xCd_{1-x}S$ nanocrystals with highly narrow luminescence spectral width. J Am Chem Soc 2003;125:13559–13563.

10. Jasieniak J, Mulvaney P. From Cd-rich to Se-rich the manipulation of CdSe nanocrystal surface stoichiometry. J Am Chem Soc 2007;129:2841–2848.

11. Talapin DV, Lee J-S, Kovalenko MV, Shevchenko EV. Prospects of colloidal nanocrystals for electronic and optoelectronic applications. Chem Rev 2009;110:389–458.

12. Graetzel M. Artificial photosynthesis: water cleavage into hydrogen and oxygen by visible light. Acc Chem Res 1981;14:376–384.

13. Huynh WU, Dittmer JJ, Alivisatos AP. Hybrid nanorod-polymer solar cells. Science 2002;295:2425–2427.

14. Rath AK, Bernechea M, Martinez L, de Arquer FPG, Osmond J, Konstantatos G. Solution-processed inorganic bulk nano-heterojunctions and their application to solar cells. Nat Photonics 2012;6:529–534.

15. O'Regan B, Gratzel M. A low-cost, high-efficiency solar cell based on dye-sensitized colloidal TiO_2 films. Nature 1991;353:737–740.

16. Plass R, Pelet S, Krueger J, Grätzel M, Bach U. Quantum Dot sensitization of organic–inorganic hybrid solar cells. J Phys Chem B 2002;106:7578–7580.

17. Lee Y-L, Lo Y-S. Highly efficient quantum-dot-sensitized solar cell based on co-sensitization of CdS/CdSe. Adv Funct Mater 2009;19:604–609.

18. Wu Y, Wadia C, Ma W, Sadtler B, Alivisatos AP. Synthesis and photovoltaic application of copper(I) sulfide nanocrystals. Nano Lett 2008;8:2551–2555.

19. Gur I, Fromer NA, Geier ML, Alivisatos AP. Air-stable All-inorganic nanocrystal solar cells processed from solution. Science 2005;310:462–465.

20. Yella A, Lee H-W, Tsao HN, Yi C, Chandiran AK, Nazeeruddin MK, Diau EW-G, Yeh C-Y, Zakeeruddin SM, Grätzel M. Porphyrin-sensitized solar cells with cobalt (II/III)–based redox electrolyte exceed 12 percent efficiency. Science 2011;334:629–634.

21. Snaith HJ. Estimating the maximum attainable efficiency in dye-sensitized solar cells. Adv Funct Mater 2010;20:13–19.

22. Tisdale WA, Williams KJ, Timp BA, Norris DJ, Aydil ES, Zhu X-Y. Hot-electron transfer from semiconductor nanocrystals. Science 2010;328:1543–1547.

23. Schaller RD, Klimov VI. High efficiency carrier multiplication in PbSe nanocrystals: implications for solar energy conversion. Phys Rev Lett 2004;92:186601.

24. McGuire JA, Joo J, Pietryga JM, Schaller RD, Klimov VI. New aspects of carrier multiplication in semiconductor nanocrystals. Acc Chem Res 2008;41:1810–1819.

25. Green MA. Third generation photovoltaics: ultra-high conversion efficiency at low cost. Prog. Photovoltaics Res. Appl. 2001;9:123–135.

26. Hanna MC, Nozik AJ. Solar conversion efficiency of photovoltaic and photoelectrolysis cells with carrier multiplication absorbers. J Appl Phys 2006;100:074510–074518.

27. Yu G, Gao J, Hummelen JC, Wudl F, Heeger AJ. Polymer photovoltaic cells: enhanced efficiencies via a network of internal donor-acceptor heterojunctions. Science 1995;270:1789–1791.

28. Halls JJM, Walsh CA, Greenham NC, Marseglia EA, Friend RH, Moratti SC, Holmes AB. Efficient photodiodes from interpenetrating polymer networks. Nature 1995;376:498–500.

29. Koster LJA, Shaheen SE, Hummelen JC. Pathways to a new efficiency regime for organic solar cells. Adv Energy Mater 2012;2:1246–1253.

30. He Z, Zhong C, Su S, Xu M, Wu H, Cao Y. Enhanced power-conversion efficiency in polymer solar cells using an inverted device structure. Nat Photonics 2012;6:591–595.

31. Luther JM, Law M, Beard MC, Song Q, Reese MO, Ellingson RJ, Nozik AJ. Schottky solar cells based on colloidal nanocrystal films. Nano Lett 2008;8:3488–3492.

32. Choi JJ, Lim Y-F, Santiago-Berrios MEB, Oh M, Hyun B-R, Sun L, Bartnik AC, Goedhart A, Malliaras GG, Abruña HCD, et al. PbSe nanocrystal excitonic solar cells. Nano Lett 2009;9:3749–3755.

33. Tang J, Kemp KW, Hoogland S, Jeong KS, Liu H, Levina L, Furukawa M, Wang X, Debnath R, Cha D, et al. Colloidal-quantum-dot photovoltaics using atomic-ligand passivation. Nat Mater 2011;10:765–771.

34. Jasieniak J, MacDonald BI, Watkins SE, Mulvaney P. Solution-processed sintered nanocrystal solar cells via layer-by-layer assembly. Nano Lett 2011;11:2856–2864.

35. Guo Q, Ford GM, Agrawal R, Hillhouse HW. Ink formulation and low-temperature incorporation of sodium to yield 12% efficient Cu(In,Ga)(S,Se)2 solar cells from sulfide nanocrystal inks. Prog. Photovoltaics Res. Appl. 2013;21:64–71.

36. Guo Q, Ford GM, Yang W-C, Walker BC, Stach EA, Hillhouse HW, Agrawal R. Fabrication of 7.2% efficient CZTSSe solar cells using CZTS nanocrystals. J Am Chem Soc 2010;132:17384–17386.

37. Volmer M, Weber A. Keimbildung in übersättigten Gebilden. Z Phys Chem 1926;119:277.

38. Becker R, Döring W. Kinetische Behandlung der Keimbildung in übersättigten Dämpfen. Ann Phys 1935;24:719–752.

39. Markov IV. *Crystal Growth For Beginners: Fundamentals of Nucleation, Crystal Growth, and Epitaxy*. Singapore: World Scientific Publishing Co. Pte. Ltd.; 1995.

40. Robson JD. Model the evolution of particle size distribution during nucleation, growth and coarsening. J Mater Sci Technol 2004;20:441–448.

41. Myhr OR, Grong Ø. Modelling of non-isothermal transformations in alloy containing a particle distribution. Acta Mater 2000;48:1605–1615.

42. Kapman R, Wagner R. *Materials Science and Technology*. Vol. 5. Weinheim: Wiley-VCH; 1991.

43. Kashchiev D, van Rosmalen GM. Review: nucleation in solutions revisited. Cryst. Res. Technol. 2003;38:555–574.

44. Atkins P, de Paula H. *Atkin's Physical Chemistry*. 7th ed. Oxford: Oxford University Press, Inc.; 2002.

45. Sugimoto T. Preparation of monodispersed colloidal particles. Adv Colloid Interface Sci 1987;28:65–108.

46. Talapin DV, Rogach AL, Haase M, Weller H. Evolution of an ensemble of nanoparticles in a colloidal solution: theoretical study. J Phys Chem B 2001;105:12278–12285.

47. Talapin DV, Rogach AL, Shevchenko EV, Kornowski A, Haase M, Weller H. Dynamic distribution of growth rates within the ensembles of colloidal II-VI and III-V semiconductor nanocrystals as a factor governing the photoluminescence efficiency. J Am Chem Soc 2002;124:5782.

48. Peng X, Wickham J, Alivisatos AP. Kinetics of II-VI and III-V colloidal semiconductor nanocrystal growth: "focusing" of size distributions. J Am Chem Soc 1998;120:5343–5344.

49. Woggon U. *Optical Properties of Semiconductor Quantum Dots*. Heidelberg: Springer-Verlag; 1996. p 252.

50. Gaponenko SV. *Optical Properties of Semiconductor Nanocrystals*. Cambridge: Cambridge University Press; 1998. p 245.

51. van Embden J, Sader JE, Davidson M, Mulvaney P. Evolution of colloidal nanocrystals: theory and modeling of their nucleation and growth. J Phys Chem C 2009;113:16342–16355.

52. Cao G. *Nanostructures & Nanomaterials: Synthesis, Properties & Applications*. London: Imperial College Press; 2004. p 433.

53. Jasieniak JJ. *Synthesis and Application of II-VI Semiconductor Quantum Dots*. Melbourne, Australia: University of Melbourne; 2008.

54. Yu WW, Peng X. Formation of high-quality CdS and other II-VI semiconductor nanocrystals in noncoordinating solvents: tunable reactivity of monomers. Angew Chem Int Ed 2002;41:2368–2371.

55. Jasieniak J, Califano M, Watkins SE. Size-dependent valence and conduction band-edge energies of semiconductor nanocrystals. ACS Nano 2011;5:5888–5902.

56. Bullen C, Mulvaney P. The effects of chemisorption on the luminescence of CdSe quantum dots. Langmuir 2006;22:3007–3013.

57. Signorini R, Fortunati I, Todescato F, Gardin S, Bozio R, Jasieniak JJ, Martucci A, Della Giustina G, Brusatin G, Guglielmi M. Facile production of up-converted quantum dot lasers. Nanoscale 2011;3:4109–4113.

58. Nag A, Kovalenko MV, Lee J-S, Liu W, Spokoyny B, Talapin DV. Metal-free inorganic ligands for colloidal nanocrystals: S2–, HS–, Se2–, HSe–, Te2–, HTe–, TeS32–, OH–, and NH2– as surface ligands. J Am Chem Soc 2011;133:10612–10620.

59. Kovalenko MV, Scheele M, Talapin DV. Colloidal nanocrystals with molecular metal chalcogenide surface ligands. Science 2009;324:1417–1420.

60. Law M, Luther JM, Song O, Hughes BK, Perkins CL, Nozik AJ. Structural, optical, and electrical properties of PbSe nanocrystal solids treated thermally or with simple amines. J Am Chem Soc 2008;130:5974–5985.

61. Luther JM, Law M, Song Q, Perkins CL, Beard MC, Nozik AJ. Structural, optical and electrical properties of self-assembled films of PbSe nanocrystals treated with 1,2-ethanedithiol. ACS Nano 2008;2:271–280.

62. Klem EJD, Shukla H, Hinds S, MacNeil DD, Levina L, Sargent EH. Impact of dithiol treatment and air annealing on the conductivity, mobility, and hole density in PbS colloidal quantum dot solids. Appl Phys Lett 2008;92:212105.

63. Konstantatos G, Howard I, Fischer A, Hoogland S, Clifford J, Klem E, Levina L, Sargent EH. Ultrasensitive solution-cast quantum dot photodetectors. Nature 2006;442:180–183.

64. Zhang HT, Hu B, Sun LF, Hovden R, Wise FW, Muller DA, Robinson RD. Surfactant ligand removal and rational fabrication of inorganically connected quantum dots. Nano Lett 2011;11:5356–5361.

65. Talapin DV, Murray CB. PbSe nanocrystal solids for n- and p-channel thin film field-effect transistors. Science 2005;310:86–89.

66. Guo Q, Ford GM, Hillhouse HW, Agrawal R. Sulfide nanocrystal inks for dense Cu(in(1-x)Ga(x))(S(1-y)Se(y))(2) absorber films and their photovoltaic performance. Nano Lett 2009;9:3060–3065.

67. Ariga K, Hill JP, Ji QM. Layer-by-layer assembly as a versatile bottom-up nanofabrication technique for exploratory research and realistic application. Phys Chem Chem Phys 2007;9:2319–2340.

68. Lvov Y, Decher G, Mohwald H. Assembly, structural characterization, and thermal-behavior of layer-by-layer deposited ultrathin films of poly(vinyl sulfate) and poly(allylamine). Langmuir 1993;9:481–486.

69. Ho KM, Chan CT, Soukoulis CM, Biswas R, Sigalas M. Photonic band-gaps in 3-dimensions - new layer-by-layer periodic structures. Solid State Commun 1994;89:413–416.

70. Chapman R, Mulvaney P. Electro-optical shifts in silver nanoparticle films. Chem Phys Lett 2001;349:358–362.

71. Kotov NA, Dekany I, Fendler JH. Layer-by-layer self-assembly of polyelectrolyte-semiconductor nanoparticle composite films. J Phys Chem 1995;99:13065–13069.

72. Bae S, Kim H, Lee Y, Xu XF, Park JS, Zheng Y, Balakrishnan J, Lei T, Kim HR, Song YI, et al. Roll-to-roll production of 30-inch graphene films for transparent electrodes. Nat Nanotechnol 2010;5:574–578.

73. Sun L, Choi JJ, Stachnik D, Bartnik AC, Hyun B-R, Malliaras GG, Hanrath T, Wise FW. Bright infrared quantum-dot light-emitting diodes through inter-dot spacing control. Nat Nanotechnol 2012;7:369–373.

74. Olson JD, Rodriguez YW, Yang LD, Alers GB, Carter SA. CdTe Schottky diodes from colloidal nanocrystals. Appl Phys Lett 2010;96:242103.

75. MacDonald BI, Gengenbach TR, Watkins SE, Mulvaney P, Jasieniak JJ. Solution-processing of ultra-thin CdTe/ZnO nanocrystal solar cells. Thin Solid Films 2014;5581:365–373.

76. Adachi S, Kimura T, Suzuki N. Optical-properties of CdTe - experimental and modeling. J Appl Phys 1993;74:3435–3441.

77. Lokteva I, Radychev N, Witt F, Borchert H, Parisi J, Kolny-Olesiak J. Surface treatment of CdSe nanoparticles for application in hybrid solar cells: the effect of multiple ligand exchange with pyridine. J Phys Chem C 2010;114:12784–12791.

78. Kuno M, Lee JK, Dabbousi BO, Mikulec FV, Bawendi MG. The band edge luminescence of surface modified CdSe nanocrystallites: probing the luminescing state. J Chem Phys 1997;106:9869–9882.

79. McCandless BE, Sites JR. Cadmium telluride solar cells. In: *Handbook of Photovoltaic Science and Engineering*. 2nd ed. West Sussex, United Kingdom: John Wiley and Sons, Ltd; 2011.

80. Moutinho HR, Al-Jassim MM, Levi DH, Dippo PC, Kazmerski LL. Effects of $CdCl_2$ treatment on the recrystallization and electro-optical properties of CdTe thin films. J Vac Sci Technol A 1998;16:1251–1257.

81. Romeo A, Terheggen M, Abou-Ras D, Bätzner DL, Haug FJ, Kälin M, Rudmann D, Tiwari AN. Development of thin-film Cu(In,Ga)Se$_2$ and CdTe solar cells. Prog Photovoltaics Res Appl 2004;12:93–111.

82. Yuan M, Mitzi DB, Liu W, Kellock AJ, Chey SJ, Deline VR. Optimization of CIGS-based PV device through antimony doping. Chem Mater 2009;22:285–287.

83. Fafarman AT, Koh W-K, Diroll BT, Kim DK, Ko D-K, Oh SJ, Ye X, Doan-Nguyen V, Crump MR, Reifsnyder DC, et al. Thiocyanate-capped nanocrystal colloids: vibrational reporter of surface chemistry and solution-based route to enhanced coupling in nanocrystal solids. J Am Chem Soc 2011;133:15753–15761.

84. McGill TC, Sotomayor Torres CM, Gebhardt W. *Growth and Optical Properties of Wide-Gap II-VI Low-Dimensional Semiconductors*. New York: Plenum Press; 1989. p 349.

85. Scherrer P. Estimation of the size and structure of colloidal particles by Rontgen rays. Nachr Ges Wiss Gottingen 1918;26:96–100.

86. McCandless BE, Hegedus SS, Birkmire RW, Cunningham D. Correlation of surface phases with electrical behavior in thin-film CdTe devices. Thin Solid Films 2003;431:249–256.

87. Werthen JG, Haring JP, Bube RH. Correlation between cadmium telluride surface oxidation and metal junctions. J Appl Phys 1983;54:1159–1161.

88. Wang F, Schwartzman A, Fahrenbruch AL, Sinclair R, Bube RH, Stahle CM. Kinetics and oxide composition for thermal-oxidation of cadmium telluride. J Appl Phys 1987;62:1469–1476.

89. Gabor AM, Tuttle JR, Albin DS, Contreras MA, Noufi R, Hermann AM. High efficiency $CuIn_xGa_{1-x}Se_2$ solar-cells made from $(In_xGa1-x)_2Se_3$ precursor films. Appl Phys Lett 1994;65:198–200.

90. Contreras MA, Tuttle J, Gabor A, Tennant A, Ramanathan K, Asher S, Franz A, Keane J, Wang L, Noufi R. High efficiency graded bandgap thin-film polycrystalline Cu(In,Ga)Se-2-based solar cells. Sol Energy Mater Sol Cells 1996;41–42:231–246.

91. Xie R, Rutherford M, Peng X. Formation of high-quality I-III-VI semiconductor nanocrystals by tuning relative reactivity of cationic precursors. J Am Chem Soc 2009;131:5691–5697.

92. MacDonald BI, Martucci A, Rubanov S, Watkins SE, Mulvaney P, Jasieniak JJ. Layer-by-layer assembly of sintered $CdSe_xTe_{1-x}$ nanocrystal solar cells. ACS Nano 2012;6:5995–6004.

93. Qian L, Zheng Y, Xue J, Holloway PH. Stable and efficient quantum-dot light-emitting diodes based on solution-processed multilayer structures. Nat Photonics 2011;5:543–548.

94. Van Vechten JA, Bergstresser TK. Electronic structures of semiconductor alloys. Phys Rev B 1970;1:3351–3358.

95. More PD, Shahane GS, Deshmukh LP, Bhosale PN. Spectro-structural characterisation of CdSe1-xTex alloyed thin films. Mater Chem Phys 2003;80:48–54.

96. Muthukumarasamy N, Balasundaraprabhu R, Jayakumar S, Kannan MD. Investigations on structural phase transition in hot wall deposited CdSexTe1-x thin films. Mater Chem Phys 2007;102:86–91.

97. Zhao X, Wanga X, Lin H, Wang Z. Relationships between lattice energy and electronic polarizability of ANB8-N crystals. Opt Commun 2010;283:1668–1673.

98. Adachi S. *Properties of Group-IV, III-V and II-VI Semiconductors*. Wiltshire, Great Britain: John Wiley and Sons, Ltd.; 2005.

99. Bailey JE, Hirsch PB. The recrystallization process in some polycrystalline metals. Proc R Soc Lond A Math Phys Sci 1962;267:11–30.

100. Tauc J, Menth A. States in the gap of chalcogenide glasses. J Non Cryst Solids 1972;8–10:569–585.

101. Muthukumarasamy N, Balasundaraprabhu R, Jayakumar S, Kannan MD, Ramanathaswamy P. Compositional dependence of optical properties of hot wall deposited CdSexTe1-x thin films. Phys Status Solidi A 2004;201:2312–2318.

102. Feng ZC, Becla P, Kim LS, Perkowitz S, Feng YP, Poon HC, Williams KP, Pitt GD. Raman, infrared, photoluminescence and theoretical-studies of the II-VI-VI ternary CdSeTe. J Cryst Growth 1994;138:239–243.

103. Hannachi L, Bouarissa N. Electronic structure and optical properties of CdSe(x)Te(1-x) mixed crystals. Superlattices Microstruct 2008;44:794–801.

104. Benamar E, Rami M, Fahoume M, Chraibi F, Ennaoui A. Electrodeposition and characterization of CdSexTe1-x semiconducting thin films. Solid State Sci 1999;1:301–310.

105. Wu J, Walukiewicz W, Yu KM, Ager JW, Li SX, Haller EE, Lu H, Schaff WJ. Universal bandgap bowing in group-III nitride alloys. Solid State Commun 2003;127:411–414.

106. Islam R, Banerjee HD, Rao DR. Structural and optical-properties of $CdSe_xTe_{1-x}$ thin-films grown by electron-beam evaporation. Thin Solid Films 1995;266:215–218.

107. Tit N, Obaidat IM, Alawadhi H. Origins of bandgap bowing in compound-semiconductor common-cation ternary alloys. J Phys Condens Matter 2009;21:075802.

108. Cohen-Solal G, Lincot D, Barbe M. High efficiency shallow p^+nn^+ cadmium telluride solar cells. In: Bloss WH, Grassi G, editors. *Fourth E.C. Photovoltaic Solar Energy Conference*. Netherlands: Springer; 1982. p 621–626.

109. Bosio A, Romeo N, Mazzamuto S, Canevari V. Polycrystalline CdTe thin films for photovoltaic applications. Prog Cryst Growth Charact Mater 2006;52:247–279.

110. Aranovich JA, Golmayo D, Fahrenbruch AL, Bube RH. Photovoltaic properties of ZnO/CdTe heterojunctions prepared by spray pyrolysis. J Appl Phys 1980;51:4260–4268.

111. Small CE, Chen S, Subbiah J, Amb CM, Tsang S-W, Lai T-H, Reynolds JR, So F. High-efficiency inverted dithienogermole-thienopyrrolodione-based polymer solar cells. Nat Photonics 2012;6:115–120.

112. Beek WJE, Wienk MM, Kemerink M, Yang XN, Janssen RAJ. Hybrid zinc oxide conjugated polymer bulk heterojunction solar cells. J Phys Chem B 2005;109:9505–9516.

113. Ju T, Yang L, Carter S. Thickness dependence study of inorganic CdTe/CdSe solar cells fabricated from colloidal nanoparticle solutions. J Appl Phys 2010;107:104311.

114. Brus LE. Electron–electron and electron–hole interactions in small semiconductor crystallites: the size dependence of the lowest excited electronic state. J Chem Phys 1984;80:4403–4409.

115. Hill IG, Kahn A, Soos ZG, Pascal RA. Charge-separation energy in films of pi-conjugated organic molecules. Chem Phys Lett 2000;327:181–188.

116. Alvarado SF, Seidler PF, Lidzey DG, Bradley DDC. Direct determination of the exciton binding energy of conjugated polymers using a scanning tunneling microscope. Phys Rev Lett 1998;81:1082–1085.

117. Nelson J. *The Physics of Solar Cells*. London: Imperial College Press; 2003.

118. Miles RW, Zoppi G, Forbes I. Inorganic photovoltaic cells. Mater Today 2007;10:20–27.

119. Fthenakis V. Sustainability of photovoltaics: the case for thin-film solar cells. Renewable Sustainable Energy Rev 2009;13:2746–2750.

120. Zweibel K. The impact of tellurium supply on cadmium telluride photovoltaics. Science 2010;328:699–701.

121. Gupta A, Parikh V, Compaan AD. High efficiency ultra-thin sputtered CdTe solar cells. Sol Energy Mater Sol Cells 2006;90:2263–2271.

122. Plotnikov V, Liu X, Paudel N, Kwon D, Wieland KA, Compaan AD. Thin-film CdTe cells: reducing the CdTe. Thin Solid Films 2011;519:7134–7137.

123. Britt J, Ferekides C. Thin-film CdS/CdTe solar cell with 15.8% efficiency. Appl Phys Lett 1993;62:2851–2852.

124. Roshko VY, Kosyachenko LA, Grushko EV. Theoretical analysis of optical losses in CdS/CdTe solar cells. Acta Phys Pol A 2011;120:954–956.

125. Amin N, Sopian K, Konagai M. Numerical modeling of CdS/CdTe and CdS/CdTe/ZnTe solar cells as a function of CdTe thickness. Sol Energy Mater Sol Cells 2007;91:1202–1208.

126. Burkhard GF, Hoke ET, McGehee MD. Accounting for interference, scattering, and electrode absorption to make accurate internal quantum efficiency measurements in organic and other thin solar cells. Adv Mater 2010;22:3293–3297.

127. Pettersson LAA, Roman LS, Inganas O. Modeling photocurrent action spectra of photovoltaic devices based on organic thin films. J Appl Phys 1999;86:487–496.

128. Semonin OE, Luther JM, Choi S, Chen H-Y, Gao J, Nozik AJ, Beard MC. Peak external photocurrent quantum efficiency exceeding 100% via MEG in a quantum dot solar cell. Science 2011;334:1530–1533.

129. Sze SM, Ng KK. *Physics of Semiconductor Devices*. 3rd ed. Hoboken (NJ): John Wiley and Sons, Inc.; 2007.

130. Lakhwani G, Roijmans RFH, Kronemeijer AJ, Gilot J, Janssen RAJ, Meskers SCJ. Probing charge carrier density in a layer of photodoped ZnO nanoparticles by spectroscopic ellipsometry. J Phys Chem C 2010;114:14804–14810.

131. Okamoto T, Yamada A, Konagai M. Characterization of highly efficient CdTe thin film solar cells by the capacitance-voltage profiling technique. Jpn J Appl Phys Part 1 2000;39:2587–2588.

132. Reislohner U, Hadrich M, Lorenz N, Metzner H, Witthuhn W. Doping profiles in CdTe/CdS thin film solar cells. Thin Solid Films 2007;515:6175–6178.

133. Desnica UV. Doping limits in II-VI compounds - challenges, problems and solutions. Prog Cryst Growth Charact Mater 1998;36:291–357.

134. Bube RH, Fahrenbruch AL, Sinclair R, Anthony TC, Fortmann C, Huber W, Lee CT, Thorpe T, Yamashita T. Cadmium telluride films and solar-cells. IEEE Trans Electron Devices 1984;31:528–538.

135. Das SK. Characterisation of $CdCl_2$ treated electrodeposited CdS/CdTe thin film solar cell. Sol Energy Mater Sol Cells 1993;29:277–287.

136. Olson JD, Gray GP, Carter SA. Optimizing hybrid photovoltaics through annealing and ligand choice. Sol Energy Mater Sol Cells 2009;93:519–523.

137. Barkhouse DAR, Debnath R, Kramer IJ, Zhitomirsky D, Pattantyus-Abraham AG, Levina L, Etgar L, Grätzel M, Sargent EH. Depleted bulk heterojunction colloidal quantum dot photovoltaics. Adv Mater 2011;23:3134–3138.

138. Burgelman M, Nollet P, Degrave S. Modelling polycrystalline semiconductor solar cells. Thin Solid Films 2000;361:527–532.

139. Ellmer K, Klein A, Rech B, editors. *Transparent Conductive Zinc Oxide: Basics and Applications in Thin Film Solar Cells*. Heidelberg, Germany: Springer; 2008.

15

NANOSTRUCTURED CONDUCTORS FOR FLEXIBLE ELECTRONICS

JONGHWA PARK, SEHEE AHN, AND HYUNHYUB KO

School of Nano-Bioscience and Chemical Engineering, Ulsan National Institute Science and Technology (UNIST), Ulsan, Republic of Korea

15.1 INTRODUCTION

Mechanically flexible and high-performance electronic devices offer a great promise to expand current, rigid electronics to the novel application areas such as handheld, rollable displays and solar cells; human-interactive, skin-attachable sensors and devices; and shape-reconfigurable electronic devices. For the realization of bendable and stretchable electronic devices, there exist challenges in the design of active and passive components, which should be compatible with the external stress and strain imposed on the flexible devices. In particular, electrically conductive materials, which are typically used in passive components in the device, are getting increasing attention as key materials to affect the final performance of flexible electronics.

The main application areas of conductive materials in flexible electronics include transparent electrodes, stretchable conductors, and pressure-sensitive films (Fig. 15.1). Flexible and transparent electrodes are currently under great attention for applications in flexible solar cells, displays, and touch screens. Stretchable conductors enable the coverage of arbitrary surfaces and the elasticity in the movable parts, which may find applications in robotic skins, actuators, and joints; prosthetic devices; and wearable electronics. While the first two abovementioned application areas require the constant conductivity under mechanical stress, pressure-sensitive films need the sensitive change of conductivity in response to external stimuli such as pressure. The pressure-sensitive films have potential applications in electronic skins, pressure and strain sensors, and biomedical applications. Here, we introduce different types of nanostructured conductors based on carbon nanotubes (CNTs), graphenes, metal nanowires, and hybrid nanostructures for applications in transparent electrodes, stretchable electrodes, and pressure-sensitive conductors.

15.2 TRANSPARENT ELECTRODES

For the last decades, transparent electrodes have played an important role in many optoelectronic devices such as solar cells (1–3), liquid-crystal displays (LCDs) (4, 5), organic light-emitting diodes (OLEDs) (6–8), touch panels (9–11), e-papers (12, 13), and smart windows (14). Metal-doped oxide, especially indium tin oxide (ITO), has been mainly used as transparent electrodes due to its high transparency (~90%) and low sheet resistance (15–20 Ω/sq) (15, 16). However, the brittleness of ITO has limited its applications in flexible and stretchable electronic devices. The increasing interest and demand of flexible electronics has fueled the search for novel transparent electrodes with high flexibility and low cost (17). Transparent electrodes for solar cells and displays require a low sheet resistance of <10 Ω/sq and a high transmission of >90%. Likewise, the transparent electrodes for touch screen and smart window also should satisfy the sheet resistance of <500 Ω/sq and the high transmission of >90%.

Conducting polymers such as polyaniline (PANI) (18) and poly(3,4-ethylenedioxythiophene)-poly(styrene sulfonate) (PEDOT:PSS) (19) and conductive nanomaterials including CNTs (9, 20–22), graphenes (10, 23, 24), metal nanowires (25–27), metal nanogrids (25), and hybrid nanostructures (28–30) show great promise as flexible and transparent electrodes. Figure 15.2 shows the recent literature values of transmission at visible light as a function of sheet resistance for various transparent electrodes based on ITO films, graphene, single-walled CNTs, metal films, Ag nanowires, and PEDOT films (31).

Nanomaterials, Polymers and Devices: Materials Functionalization and Device Fabrication, First Edition. Edited by Eric S. W. Kong.
© 2015 John Wiley & Sons, Inc. Published 2015 by John Wiley & Sons, Inc.

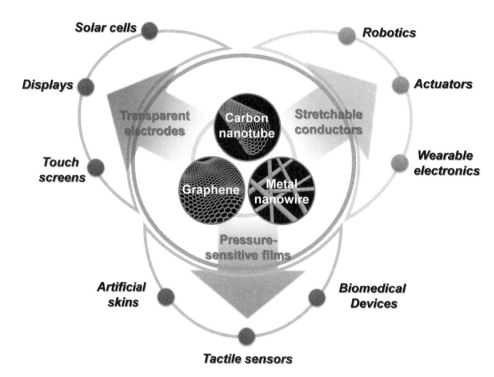

Figure 15.1   Nanostructured conductors based on carbon nanotubes, graphenes, metal nanowires, and hybrid nanostructures for applications in transparent electrodes, stretchable conductors, and pressure-sensitive films.

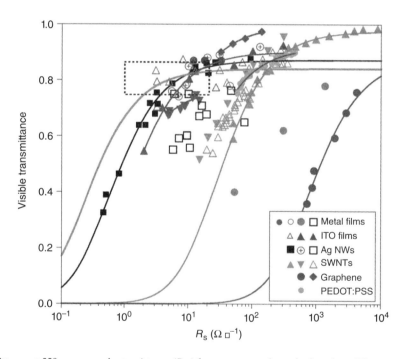

Figure 15.2   *Optical transmittance at 550 nm versus sheet resistance (R_{sh}) for transparent electrodes based on different nanostructured conductors*: metal films based on nickel and gold, the metal oxide film of ZnO:Al/Au/ZnO:Al trilayers, sputtered ITO films, silver nanowire films, single-walled carbon nanotube films, graphene films, and PEDOT:PSS films. The dotted rectangle is the application target region with high transmittance and low sheet resistance for the transparent electrodes. (*Source*: Reprinted with permission from Reference 31. Copyright (2012) Nature.)

15.2.1   Carbon Nanotubes

Depending on the number of cylindrical shell, CNTs are sorted as single-walled carbon nanotubes (SWNTs) and multiwalled carbon nanotubes (MWNTs). Among different types of CNTs, SWNTs are predominantly used as transparent electrodes because of the high optical transparency in addition to the high electrical conductivity. The fabrication of transparent electrodes using CNTs is typically performed using solution-based assembly of random networks of CNTs, which enables large-scale production via roll-to-roll process, the biggest advantage as compared to the vacuum deposition of metal oxide transparent electrodes (Fig. 15.3a) (17). One of the first transparent electrodes based on CNT random network films is demonstrated by using vacuum filtration methods (6, 22, 32–34). This method, by simply filtering CNT solution through filter membranes, enables an easy control of the film thickness with inexpensive and simple processes. In addition, the filtrated CNT films can be easily transferred to various flexible or stretchable substrates via dissolving filter membranes (22) or contact printing with elastomeric stamp (6). However, the disadvantages of this method include the limited scalability, small sample size, and low throughput.

To overcome these limitations, a spray coating method was introduced (36, 39, 40). In this method, CNT solution is sprayed using an air brush over a substrate that is heated to a certain temperature to provide quick evaporation of solvent, resulting in uniform formation of CNT networks without any stains from the coffee ring effects. Mayer rod coating is another method that can be easily explored in the laboratory scale to produce tens of centimeter-size transparent electrodes (41). In this coating process, a Mayer rod wrapped around with a wire is used to trap CNT solution between the rod and the substrate, and is drawn in one direction for the uniform coating of CNT films with desired thickness. The critical part in the Mayer rod coating is the control of solution rheology and wetting properties. Triton X-100 was used to increase the viscosity of surfactant-stabilized CNT solution, rendering uniform formation of CNT network films (41). Recently, the high-molecular-weight sodium carboxymethyl cellulose (CMC) has the ability to disperse CNTs in water and at the same time can increase the solution viscosity appropriate for the Mayer rod coating (35). In addition, after the Mayer rod coating process, the CMC can be easily removed by washing with HNO_3, rendering further modification of various functional polymers (Fig. 15.3b). Roll-to-roll process with capabilities of large-scale and high-throughput production is the most practical way to fabricate industrial-scale transparent electrodes. However, one of the challenges is the control of the rheology of CNT solution, the wetting, and the adhesion on the desired substrates.

According to the chirality-dependent energy gap calculation, one-third of SWNTs show metallic and two-third show semiconducting behaviors (42). Therefore, the as-synthesized CNTs, without further purification steps to separate metallic SWNTs (m-SWNTs) from semiconducting ones (s-SWNTs), typically have the problem of contact resistance between junctions of m- and s-SWNTs when they are assembled into the thin film structures of transparent electrodes (43). In addition, the surfactants used in the stable dispersion of CNTs frequently cause the increase of contact resistance between CNTs in the network films. The contact resistance in the junction of CNTs is the main bottleneck of CNT applications in transparent electrodes.

One method of decreasing junction resistance between metallic and semiconducting CNTs is the doping of semiconducting CNTs. Because the CNTs in the air environments are p-doped due to the presence of oxygen, the CNTs are typically p-doped in the fabrication of transparent electrodes. Typical dopant materials for the p-doping of CNTs include HNO_3 (33, 36, 44), $SOCl_2$ (36, 44), $FeCl_3$ (45), and $AuCl_3$ (46). The immersion of CNTs in HNO_3 solution is the most common method to remove surfactants and at the same time to p-dope the CNTs, lowering the contact resistance in the network films by a factor of 2–3 (Fig. 15.3c) (36). However, one of the critical problems of these dopants is the rapid increase of electrical resistance when the films are exposed to air environments because these dopants are not stable to moisture, heat, and chemicals. To overcome this issue, PEDOT:PSS capping layers were used to prevent the desorption of HNO_3 and consequently increase the doping stability in the air. Recently, MoOx was introduced as a stable dopant to CNTs and graphenes, where the sheet resistance of CNT networks decreased by a factor of 2 with the MoO_3 coating (Fig. 15.3d) (37). The sheet resistance even further decreased by a factor of 5–7 when the CNT films are annealed (450–500 °C for 3 h) in Ar environments. The other example of stable doping agent is graphene oxide nanosheets, which have the ability to p-dope CNT films due to the presence of the electron-withdrawing functional groups on the graphene sheet (Fig. 15.3e) (38). The p-doping of CNT films by simple coating of graphene nanosheets resulted in 60% reduction of sheet resistance of transparent electrodes (Fig. 15.3f).

15.2.2   Metal Nanowires and Nanogrids

The flexibility of carbon-based transparent electrodes is advantageous in flexible electronics, but their relatively low conductivity still imposes a limitation on their use in optoelectronic devices. With their high conductivity and optical transparency when the thickness is below tens of nanometers, metals are another promising candidate as highly conductive and transparent electrodes. Ultrathin metal films showed performances of 40–80% and 30–1250 Ω/sq for nickel (Ni) (47) and 75% and 4.5–9 Ω/sq for silver (Ag) and gold (Au) (48). Metal nanogrids can improve the optical transparency with empty space between metal strips while maintaining the electrical conductivity via increasing the thickness of metal strips (Fig. 15.4a) (49). Theoretical prediction showed that sheet resistance of 0.8 Ω/sq at 90% transmission is achievable with a silver nanogrid (50). Among various methods, nanoimprint lithography was introduced for the efficient and large-scale production of metal nanogrids on rigid and flexible substrates (51). Various metals including Au, Ag, Cu, and Al can be imprinted by using this method. The roll-to-roll nanoimprinting technique is already in use at industry producing processing transparent and flexible electrodes with 0.4–1 Ω/sq at 85% transmission.

The metal nanowire with their high electrical conductivity is a strong candidate for future transparent electrodes to replace ITO electrodes. Most of the solution assembly techniques used in the fabrication of CNT films can be adopted in the preparation of a random network of metal nanowires, rendering the possibility of low-cost and large-area processing. For example, the Mayer rod coating method was used to coat silver nanowire (AgNW) networks on flexible substrates with controllable thickness (Fig. 15.4b) (26). In this work, 20 Ω/sq at 80% transmittance

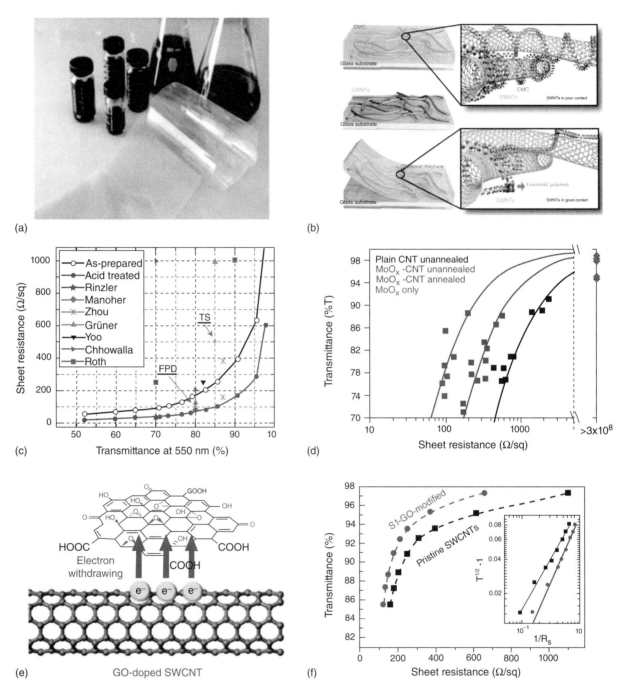

Figure 15.3 *Transparent electrodes based on carbon nanotube networks.* (a) Large-area fabrication of transparent electrodes based on roll-to-roll deposition of CNT ink. (Reprinted with permission from Reference 17. Copyright (2011) John Wiley & Sons, Inc.) (b) Networks of single-walled carbon nanotube (SWNT) wrapped with high-molecular-weight CMC fabricated by Mayer rod coating. CMC increases the viscosity of SWNT solution for the uniform Mayer rod coating. (Reprinted with permission from Reference 35. Copyright (2012) American Chemical Society.) (c) A sheet resistance reduction of spray-coated CNT films by immersing in HNO_3 solution due to the p-type doping and surfactant removal. (Reprinted with permission from Reference 36. Copyright (2007) American Chemical Society.) (d) Transparent CNT films doped with air-stable MoOx dopants. (Reprinted with permission from Reference 37. Copyright (2012) American Chemical Society.) (e) Graphene oxide doping of SWNT films. The SWNTs lose the electron easily (p-doping) due to the electron-withdrawing groups of graphene oxide. (f) The sheet resistance of CNT films doped with graphene oxide. (*Source*: Reprinted with permission from Reference 38. Copyright (2012) Royal Society of Chemistry).

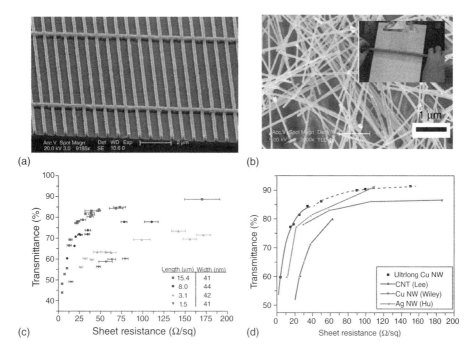

Figure 15.4 *Transparent electrodes based on metal nanowires and nanogrids.* (a) An SEM image of the silver nanogrid fabricated by nanoimprinting lithography. (Reprinted with permission from Reference 49. Copyright (2007) John Wiley & Sons, Inc.) (b) A silver nanowire (AgNW) network films on flexible substrate by a Mayer rod coating. (Reprinted with permission from Reference 26. Copyright (2010) American Chemical Society.) (c) The transmittance and sheet resistance of AgNW network films depending on the length and diameter of AgNWs. (Reprinted with permission from Reference 52. Copyright (2012) Royal Society of Chemistry.) (d) The transmittance and sheet resistance of the ultralong copper nanowire networks in comparison with the literature values from carbon nanotube, short copper nanowire, and silver nanowire. (*Source*: Reprinted with permission from Reference 53. Copyright (2012) American Chemical Society).

was achieved. The diameter and length of nanowires significantly affect the conductivity and transparency of Ag NW electrodes. The increase of NW length improves the electrical connectivity and conductivity of the network films. Figure 15.4c shows that the transmittance of Ag NW films increases with the increase of NW length at a given sheet resistance (52). On the other hand, the increase of NW diameter causes the increase of surface roughness, which results in device shorting for thin active layers and also a higher haze factor due to the light scattering (17). While the haze is not desirable for display applications, it improves the short circuit current of solar cells due to improved light absorption (25).

Although the Ag NWs has a great promise for the applications in transparent electrodes, it is still expensive and the price is similar to ITO. Recently, transparent electrodes based on copper nanowires have emerged as a low-cost transparent electrodes because copper is 100 times cheaper and more abundant than silver. Large-scale (>1 g) synthesis of copper NWs has been demonstrated by a simple solution method with dimensions of ~90 nm in diameter and <10 µm in length (54). The copper nanowire films showed a sheet resistance of 15 Ω/sq at 65% transmittance. The low transmission in this film was attributed to the low aspect ratio of copper nanowires and NW aggregates in the films. Recently, ultralong single-crystalline copper NWs have been synthesized with excellent dispersibility, rendering efficient solution processing of Cu NWs for the fabrication of transparent electrodes (53). The transparent electrodes based on ultralong Cu NWs showed significantly improved performances with transmission of 90% at sheet resistance of 90 Ω/sq as compared to the short Cu and Ag NWs (Fig. 15.4d).

The junction resistance in random network of metal NWs is one of the limiting factors for high-performance transparent electrodes. Typically, insulating organic ligands are used for the synthesis and dispersion of NWs in solution. Such insulating ligands in the NW junctions after the solution-based deposition of NW networks should be removed for the decreased contact resistance. For example, annealing of Ag NW films at 200 °C can effectively remove poly(vinyl pyrrolidone) (PVP) and partially fuse NWs together, resulting in one order of decrease of sheet resistance (3). Mechanical pressing method was used to decrease the junction resistance and also the surface roughness (26). Recently, plasmonic welding, which uses large field enhancements and local heating at the NW junctions, has been introduced as an effective way to remove ligands and weld together Ag NWs selectively at the junctions (55). A short exposure (~2 min) of Ag NW films to broadband light lowered the film resistance by a factor of 1000.

The high aspect ratio of metal nanowires in network films leads to the desired electrical conductivity of the films at a low percolation threshold due to the decrease of the number of junction points, which also results in high transmittance (56). Electrospinning method was successfully applied to the fabrication of ultralong metal nanowires and their network films (57). In this work, transparent electrodes with 50 Ω/sq at 90% transmittance were achieved based on copper nanowires with aspect ratio up to 10^5. Further improvement of performances has been achieved through fabrication of metal nanotrough network using an electrospinning method (Fig. 15.5a) (58). Here, the concave shape of metal nanotrough reduces the electromagnetic cross sections, resulting in the increase of light transmission as compared to the

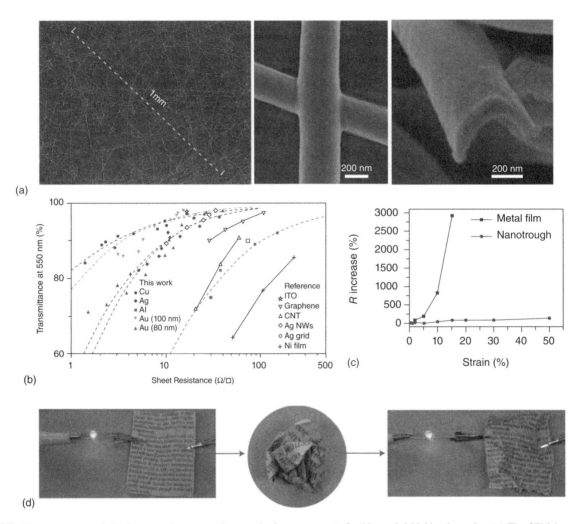

Figure 15.5 *Electrospinning of ultralong metal nanotrough networks for transparent, flexible, and foldable electrodes.* (a) The SEM images of gold nanotrough networks (left), the crossed junction of gold nanotrough (middle), and the concave shape of the gold nanotrough (right). (b) The transmittance and sheet resistance of metal nanotrough networks made of copper, silver, aluminum, and gold. The copper nanotrough network shows the best performance with 2 Ω/sq of sheet resistance at ∼90% transmittance. (c) The change in resistance versus uniaxial strain for gold nanotrough network fabricated on PDMS substrate. (d) Photograph of gold nanotrough network on a paper. The conductive paper maintains its function even after repeated crushing and unfolding the paper. (*Source*: Reprinted with permission from Reference 58. Copyright (2013) Nature).

flat nanostripes. The transparent electrodes showed sheet resistances of 2, 8, and 10 Ω/sq at 90% transmission for copper, gold, and silver nanotrough, respectively (Fig. 15.5b). The stretchable properties of nanotrough networks are shown in Figure 15.5c, where the uniaxial strain of 50% on nanotrough webs increased the sheet resistance by ∼40%. The strong mechanical properties are demonstrated in Figure 15.5d, where even after the crushing of transparent electrodes on the paper, the electrical conductivity is maintained.

15.2.3 Graphene and Hybrid Nanostructures

A monolayer graphene is a zero-bandgap semiconductor with a layer thickness of 0.34 nm. These properties endow a graphene with high electron mobility (∼15,000 cm^2/Vs) and optical transparency (59). The optical transmittance of a monolayer graphene was measured to be 97.7%, which is also proved by a theory predicting that a graphene absorbs incident white light by $\pi\alpha \approx 2.3\%$, where α is the fine-structure constant (60). Therefore, few-layer graphenes are expected to have an optical transmittance of $T = 100 - 2.3N$ (%), where N is the number of graphene layers. Due to the unique electronic structures of graphene, the delocalized charge carriers can travel without scattering, rendering high conductivities. Depending on the number of layers, the sheet resistance of highly doped graphene can be estimated as $R_{sh} = 62.4/N$ (Ω/sq) (61). The experimental demonstration of large-scale, transparent graphene electrodes was achieved by using chemical vapor deposition to grow 30 inch scale graphene sheet on copper foils (10). After etching the copper foil and transferring four layers of graphene onto flexible substrates, this work showed transparent electrodes with 30 Ω/sq at 90% transmission.

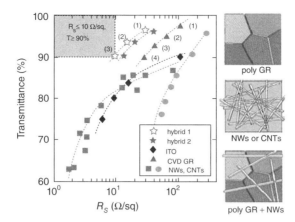

Figure 15.6 *Transmittance and sheet resistance of hybrid nanostructures based on polycrystalline graphene and silver nanowires.* The silver nanowires cross over the resistive grain boundaries and bridge the neighboring grains of polycrystalline graphene, resulting in the decrease of sheet resistance. (*Source*: Reprinted with permission from Reference 62. Copyright (2011) American Chemical Society).

Most of graphene films grown from CVD methods are polycrystalline containing numerous resistive grain boundaries, which limit the conductivity of graphene films. To improve the performance of graphene-based transparent electrodes, recently, hybrid nanostructures combined with the advantages of the other conducting nanomaterials have been investigated. Hybrid nanostructures based on polycrystalline graphene film and a subpercolating network of silver nanowires showed a great promise to overcome problems related to the highly resistive grain boundaries. In this approach, the silver nanowires improved the conductive pathways via crossing over the resistive grain boundaries and bridging the neighboring grains, which resulted in transparent electrodes with 20 Ω/sq at 90% transmission (Fig. 15.6) (62).

15.3 STRETCHABLE CONDUCTORS

The electronic devices, which maintain their high device performances while they are stretched, bent, and twist, would enable a new area of applications, for which the current, rigid electronics cannot be used. The stretchable electronics enable the coverage of arbitrary surfaces and the elasticity in the movable parts, which may find applications in robotic skins, actuators, and joints, prosthetic devices, and wearable electronics. One of the key challenges in the realization of stretchable electronics is the development of stretchable electrodes, which maintain their electrical connections between active components in the devices while they are mechanically stretched. Recent development of stretchable active and passive components in the devices showed great promise in applications like artificial skin (63), stretchable solar cell (64), strain sensors (65), conformal integrated circuits (66), eyeball digital camera (67), and stretchable inorganic light-emitting diode (68, 69).

15.3.1 Materials for Stretchable Conductors

The early form of stretchable conductors are based on the conductive rubber composites, which have been fabricated by simply filling conductive materials such as metal particles (70), CNTs (71–73), and carbon nanoparticles (74) into a rubbery matrix. This approach combines the advantages of the excellent electrical conductivity of fillers and the mechanical elasticity of rubbers, enabling conductive rubber composites with high conductivity, flexibility, and even stretchability. However, there is a trade-off between the conductivity and elasticity because of the decreased elasticity of composite rubbers with the increase of filler contents. Printable elastic conductors were demonstrated based on ionic liquid-stabilized SWNTs uniformly dispersed in a fluorinated rubber with the conductivity of 57 S/cm at 38% tensile strain (71). In this work, the use of ionic liquid enabled the uniform dispersion of SWNTs without the sacrifice of the mechanical flexibility of the rubber matrix. Here, the stretchability up to 134% with reasonable electrical conductivity (6 S/cm) was obtained via design of net-shaped structure through the perforation of nanotube films and subsequent filling with PDMS rubber. In this approach, the net-shaped structure provides additional stretchability. The other approach is based on the use of ionic liquid and jet-milling process, which enables the uniform dispersion of long and fine SWNTs in rubber matrix. Here, the dispersion of long SWNTs provides conductive paste with the appropriate viscosity for printing process, enabling printable elastic conductors with the conductivity of 102 S/cm at 15.8 wt% SWNTs and the stretchability of 118% at 1.4 wt% SWNTs (75). Figure 15.7a shows the elastic SWNT/PDMS composite patterns on flexible substrates, which were fabricated by screen printing of composite pastes. Because the SWNT content affects both the conductivity and stretchability, these properties can be precisely tuned depending on the SWNT content. Figure 15.7b shows the stretchability and conductivity of elastic conductors as a function of SWNT content.

The hybrid mixtures of CNTs with metal particles benefit the high electrical conductivity of metals and the stretchability of CNT networks. Figure 15.7c shows a hybrid nanostructure composed of MWNTs, Ag nanoparticles, and Ag microflakes (76). Here, the CNTs provide conductive networks bridging neighboring silver flakes. The role of Ag nanoparticles self-assembled on the CNT surface is to improve the contact interface with the Ag flakes. Figure 15.7d shows the conductivity of the hybrid Ag–MWNT film under tensile strain for different

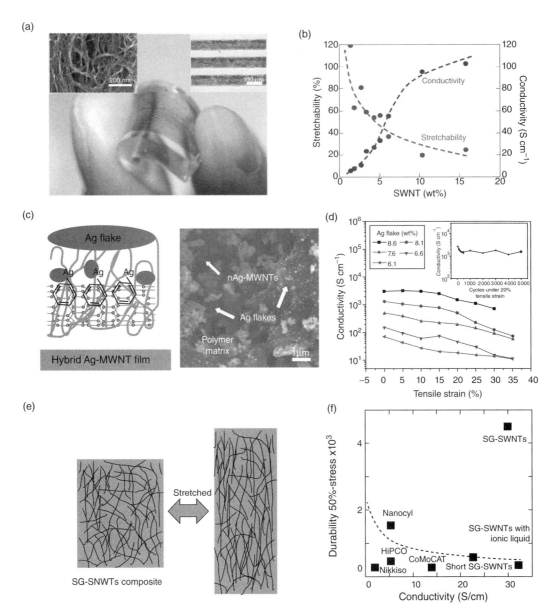

Figure 15.7 *Stretchable conductor based on composites of conductive materials and elastomer.* (a) Line-patterned printable SWNTs/fluorinated copolymer elastic conductor on PDMS substrate. The insets show SEM image of uniformly dispersed SWNTs in the rubber and AFM image of line width of 100 μm. (b) A relationship between stretchability and conductivity as a function of wt% of SWNTs. (Reprinted with permission from Reference 75. Copyright (2009) Nature.) (c) An illustration and a SEM image of hybrid Ag–MWNT composite film. (d) Conductivity of hybrid Ag–MWNT composite film under tensile strain for different wt% of Ag flake. (Reprinted with permission from Reference 76. Copyright (2010) Nature.) (e) An illustration of SWNT network maintaining conductive network by elongation along the strain direction. (f) Durability (the number of cycle at decreased 50% strain from initial state) and conductivity of stretchable conductors for different types of CNT. (*Source*: Reprinted with permission from Reference 77. Copyright (2011) American Chemical Society).

contents of silver flakes. First, we can observe that the conductivity of the hybrid Ag–MWNT film increase with the increase of content of Ag flake. With the increase of tensile strain, the composite films maintain a relatively good conductivity with a slow decrease of conductivity. A high conductivity of 706 S/cm at 30% strain was obtained at silver flake concentration of 8.60 wt%. In addition, the mechanical durability of the films was demonstrated by showing a minimal conductivity change even after 5000 cycles, as shown in the inset of Figure 15.7d.

For applications in stretchable electronics, the mechanical durability against strain cycles is a significant factor to be considered in addition to the conductivity. Ultralong SWNTs, grown by water-assisted chemical vapor deposition method, were successfully used as conductive fillers for the fabrication of mechanically durable and highly conductive elastomeric composite (77). The key design principle in this approach is that the long SWNTs with a high aspect ratio (>10⁶) and a small diameter (~3 nm) maintain the electrical networks by elongation along the strain direction (Fig. 15.7e). As can be seen in Figure 15.7f, most of the other conductive fillers with small aspect ratio cannot be elongated enough when a large strain is applied, resulting in the decrease of mechanical durability. We also observe that although the high loading of CNTs provides the higher conductivity, the increase of stiffness leads to the decreased mechanical durability.

The uniform dispersion of CNTs in the elastomers without damages on the CNTs is a critical challenge in the solution-based mixing processes. The other way to fabricate composite elastic conductors is based on the preconfiguration and arrangement of conductive network nanostructures and subsequent transfer onto or filling with elastic rubbers (72, 73, 78, 79). The advantages of these approaches include the precise control of the desired CNT network structures before or even after hybridizing with the elastic rubbers and the minimal damages on the CNTs. For example, well-aligned CNTs fabricated by drawing method from CNT forests could be embedded in the PDMS rubbers while maintaining the original aligned structures, resulting in the stretchable conductors with minimal change in conductivity with 100% strains (72). While the stretchable conductors based on the unidirectional CNT arrays show anisotropic electrical conduction, stretchable conductors with isotropic electrical conduction can also be demonstrated by using cross-stacked aligned CNT films. The stretchable conductors with cross-stacked CNT arrays showed a stable electrical conductivity under a strain of ~35% (79).

15.3.2 Structure Engineering for Stretchable Conductor

15.3.2.1 Wavy or Buckled Structures Although the hybrid structures based on conductive fillers and elastic rubbers showed great performances for stretchable conductors, the stretchability is intrinsically limited by the elastic modulus of the hybrid conductors. The structural design of the hybrid conductors, for example, wavy shapes, buckles, or herringbone configurations of the conductors, can provide a large stretchability because the deformation of the applied strain can be accommodated through the amplitude and wavelength changes in the structured conductors (80, 81). The fabrication procedures for the buckling of conductive films on elastic rubbers are shown in Figure 15.8a. First, the conductive films are coated or laminated on the prestrained elastomeric substrate such as PDMS or Ecoflex. Next, the prestrained elastomeric substrate is released, resulting in the spontaneous formation of periodically buckled patterns. Here, the conductive films buckle or winkle to relieve the stress during the release of prestrained PDMS. Figure 15.8b shows an example of buckled conductors based on a CNT film on a PDMS substrate (82). In this work, the significant improvement of stretchability by the buckled configuration was achieved, as can be seen in Figure 15.8c. As compared to the significant resistance change over 10% strain for plain CNT films, the buckled CNT films showed a minimal resistance change until 40% strain. Graphene is another good candidate as conductive materials in buckled conductors owing to its two dimensionality, high flexibility, conductivity, and transparency.

Other than CNTs, metal nanoparticles (83) and nanowires (84) have also been employed in the buckled conductive films. However, nanoparticles or short nanowires with low aspect ratios have limited performances in stretchable conductors. The high-aspect-ratio metal nanowire films in buckled geometry are expected to significantly improve the stretchability and conductivity of stretchable conductors. Figure 15.8d shows a stretchable conductor based on an ultralong silver nanowire buckled film on a superelastic substrate. The stretchable conductors showed stretchability over 460% without a notable resistance change. This high performance can be attributed to the fact that the ultralong Ag nanowires can provide conductive networks without rupture or disconnection even under a large strain level as compared to those conductors based on short Ag nanowires and nanoparticles.

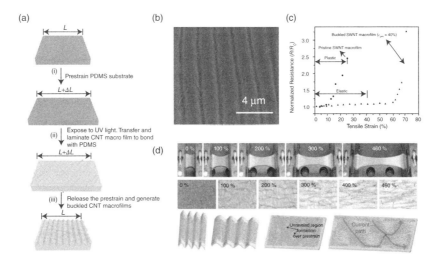

Figure 15.8 *Wavy or buckled stretchable conductors.* (a) An illustration of fabrication procedures of wavy or buckled conductive films on elastic substrate for stretchable conductors. (b) An SEM image of the uniformly buckled SWNT film. (c) Changes of the normalized resistance (R/R_0) under tensile strain for pristine SWNT film and buckled SWNT film. (Reprinted with permission from Reference 82. Copyright (2009) John Wiley & Sons, Inc.) (d) Stretchable conductors based on long Ag nanowires on Ecoflex. The macroscopic (top row) and microscopic (middle row) images show the deformation of Ag nanowire networks on Ecoflex stretched from 0% to 460%. The illustration (bottom row) shows a stable connection of long and buckled Ag nanowire networks maintaining conductive networks under stretched condition. (*Source:* Reprinted with permission from Reference 69. Copyright (2012) John Wiley & Sons, Inc).

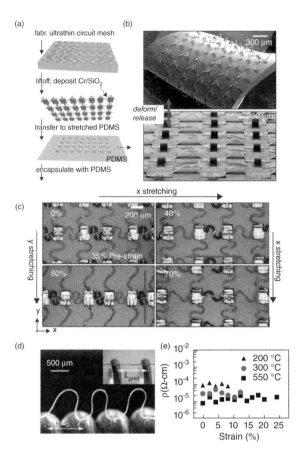

Figure 15.9 *Stretchable conductors based on wavy metal interconnects.* (a) A schematic fabrication procedure of stretchable circuits on an elastomeric substrate based on the buckled interconnects. (b) SEM images of an array of CMOS inverter under twisting (above) and released state (below). Other classes of deformations such as shearing, bending, and stretching can be accommodated by moving up and down of the thin, narrow metal interconnect bridges. (c) Optical images of an array of noncoplanar CMOS inverters with serpentine bridges under stretching in the *x* and *y* directions. (Reprinted with permission from Reference 85. Copyright (2008) by the National Academy of Sciences of the USA.) (d) An optical image of stretchable silver arches formed by releasing straight silver interconnects printed on prestrained spring. (e) Electrical resistance of the silver arch interconnects as a function of strain for different annealing temperatures. (*Source:* Reprinted with permission from Reference 86. Copyright (2009) American Association for the Advancement of Science).

The other approach of stretchable electrodes by a structure configuration is based on the stretchable interconnects (80, 81). In this strategy, the wavy interconnects bridging the neighboring active devices accommodate the stress by deformation. Figure 15.9a shows the fabrication procedures for noncoplanar mesh structures, where the ultrathin circuit mesh is transferred to stretched PDMS and the subsequent release of stretched PDMS results in the noncoplanar stretchable complementary metal–oxide–semiconductor (CMOS) integrated circuits. In these device structures, the wavy metal interconnects deform to accommodate various types of stresses (85). The stretchable circuits can be stretched to large tensile strains up to 100%. In addition, due to the deformability in any direction, complex twisting, shearing, and other types of deformation can be accommodated without any disconnection between the adjacent device islands as shown in Figure 15.9b. While this design strategy based on wavy interconnects needs enough free space between the rigid device islands, serpentine-shaped bridges can be used to expand the deformability without sacrificing the device density in the chip. Figure 15.9c shows serpentine-shaped bridges interconnecting device islands. The serpentine-shaped interconnects are capable of compensating the strain up to ~140%.

The wavy interconnects can also be fabricated by direct printing of metal nanoparticle inks on the desired location of the devices and the subsequent annealing. Figure 15.9d shows a stretchable wavy interconnects fabricated by direct printing of silver nanoparticle ink onto a prestrained spring and the subsequent releasing to form arch-shaped silver interconnects (86). Here, the resistivity and robustness of metal interconnects depend on the annealing temperature. As shown in Figure 15.9e, the resistivity decreases and the maximum strain increases with the increase of annealing temperature. The best condition was achieved with annealing temperature of 550 °C, which shows 25% stretchability.

15.3.2.2 Net-Shaped Nanostructures Net-shaped structures are capable of improving the stretchability by accommodating the stress through the deformation of the net shape (74). Similarly, porous structures can be also utilized to fabricate stretchable conductors, where the pores deform to accommodate the stress. As shown in Figure 15.10a–c, porous PDMS films, which were made by applying a pressurized

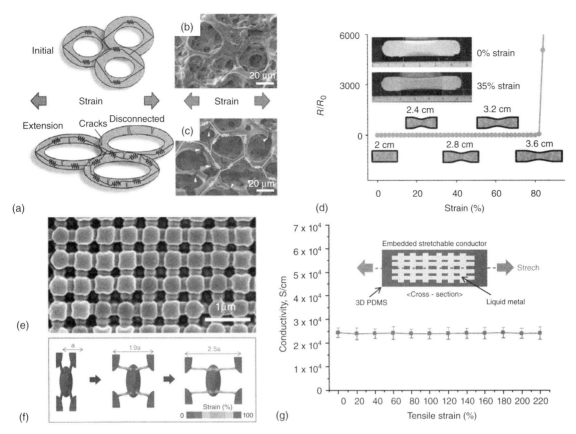

Figure 15.10 *Net-shaped stretchable conductors.* (a) Schematic illustration of electrical connections of porous PDMS films coated with Ti/Au metal before stretching (upper) and after 20,000 stretching cycles under an applied strain of 30% along the direction indicated by the arrow (below). After stretching cycles, although cracks are formed at edges of metal coated pores, the electric pathway remains through the interconnected parts. (b) An SEM image of metal-coated porous PDMS under tensile strain of 0%. (c) An SEM image of metal-coated porous PDMS under tensile strain of 30%. (d) Normalized resistance as a function of strain of porous PDMS conductors. The maximum stretching capability is 80%. The insets are optical images of porous electrode under stretching 0 and 35%. (Reprinted with permission from Reference 87. Copyright (2012) Nature.) (e) An SEM image of stretchable conductors based on net-shaped 3D PDMS. (f) Strain distribution in 3D PDMS under tensile strain simulated by finite element analysis (FEA). The deformations are accommodated by rotating and stretching of bridging elements. (g) Conductivity of 3D stretchable conductor as a function of tensile strain. The stretchable conductor shows an excellent conductivity at large strain due to the ability of the fluid-phase liquid metal (EGaIn) that conforms to any deformation of structure. (*Source*: Reprinted with permission from Reference 68. Copyright (2012) Nature).

steam to uncured PDMS films, were uniformly coated with metals to fabricate stretchable conductors (87). When a tensile strain is applied to the conductors, the porous conductors retain the electrical connectivity through the deformation of pores and partial breakage of some electrical pathways. The stable porous networks with electrical pathways support strains up to 80% without any significant degradation of conductivity (Fig. 15.10d).

When the tensile strain is applied to the irregular porous structures, the uneven stress distribution will result in a stress concentration on the local area and the breakage of electrical connectivity at that point. The 3D regular array of net-shaped conductors can support uniform distribution of the applied stress, extending the stretchability. Figure 15.10e shows 3D regular nanostructured conductors with a giant stretchability (68). The design strategy is based on the fabrication of the 3D net-shaped PDMS nanostructures and the injection of liquid-phase eutectic gallium–indium (EGaIn) into 3D PDMS nanostructures. In this configuration, the liquid-phase EGaIn has the ability to follow arbitrary shape changes of the stretched 3D PDMS nanostructures, thus maintaining the conduction pathways throughout the porous network. In addition, the bridges in the net-shaped structures stretch to accommodate the deformation under the applied stress with minimal strain on the neighboring ellipsoids (Fig. 15.10f). As can be seen in Figure 15.10g, these stretchable conductors showed a stretchability of over 200% with high electrical conductivity (~24,100 S/cm).

15.3.3 Applications

The stretchable wavy-shape interconnects showed great promise in various applications of electronic devices. One of the challenges in the realization of flexible and wearable electronics is the development of flexible energy storage devices. Recently, flexible supercapacitors using buckled thin films of CNTs (82) or CNT-soaked fibrous textiles (88) have been reported. Thin, flexible Li-ion batteries have been demonstrated by using a single layer of commercial paper as separators and free-standing CNT films as current collectors (89). Stretchable batteries will be the key component in the development and practical realization of stretchable electronics. As shown in Figure 15.11a, stretchable serpentine interconnects were utilized in the fabrication of stretchable batteries, which can provide sufficient powers to operate light-emitting diodes (90). The stretchable batteries maintained capacity density of 1.1 mA h/cm^2 while being stretched up to 300%.

For applications in displays, stretchable serpentine interconnects have been applied to stretchable LEDs (Fig. 15.11b), where uniform emissions of LEDs were retained under stretching of ~48% (91). Skin-like electronic devices have the ability to attach on the human skin to monitor electrical activity from the heart, brain, and skeletal muscles. Figure 15.11c shows skin-like electronic membranes incorporating electronics, sensors, and communication devices interconnected with stretchable electrodes. Because of the appropriate thickness and elastic moduli, the skin-like devices can be conformally attached onto the surface of the skin (66).

In addition to the flexibility of conventional photovoltaic solar cells (80, 92), stretchability is in critical need for the development of photovoltaic devices with the ability of conformal contacting onto arbitrary surfaces. The transparent electrodes and active layers on top of buckled PDMS films enabled the fabrication of stretchable photovoltaic solar cells, which can accommodate strains up to 27% (93). Recently, ultrathin (<2 μm) and flexible organic solar cells attached on prestrained elastomer were introduced to demonstrate stretchable solar cells (Fig. 15.11d) (1). The extreme bendability of the solar cells was demonstrated by the 3D deformation on top of 1.5 mm diameter plastic tube and by wrapping around a human hair with a radius of 35 μm. Furthermore, the solar cells showed an excellent power conversion efficiency of 4.2% with a specific weight of 10 W/g, demonstrating the high-performance, ultralight, and stretchable solar cells.

15.4 PRESSURE-SENSITIVE FILMS

The pressure-sensitive films have recently attracted great attentions for applications in electronic skins, pressure and strain sensors, wearable electronics, prosthetic limbs, remote surgery, and a wide range of biomedical applications. For applications in pressure-sensitive electronic skins, pressure-sensitive, conductive rubber films can be used independently or these conductive rubber films are integrated with the active matrix of transistor arrays. While stretchable conductors maintain the electrical conductivity during the stretching cycles, pressure or stain sensitive films change the electrical resistance in response to pressure or strain variation applied to the films. Therefore, the most of design strategies of materials and structures used in stretchable conductors can be employed to fabricate pressure or strain sensitive films through a modification of configuration or structure. Here, we introduce recent achievements in the fabrication of pressure or strain sensitive films based on novel materials and structural concepts with high sensitivity.

15.4.1 Materials and Structures for Pressure- or Strain-Sensitive Films

Traditionally, carbon particles such as carbon black (94) and graphite powder (95) are used as conductive fillers in the elastomeric matrix for the fabrication of piezoresistive films. Here, the applied mechanical stress leads to the morphology changes of conductive fillers within the matrix, resulting in the modulation of conductive networks and the electrical conductivity. In this materials configuration, the amounts of conductive fillers are controlled close to the percolation threshold of fillers, resulting in highly sensitive piezoresistive films.

CNTs are ideal candidates for conductive fillers in pressure- or strain-sensitive films stems due to their outstanding electrical properties with high mobility, current-carrying capacitance, and mechanical properties. The one-dimensional structures of CNTs with high aspect ratios provide a great advantage to manipulate the structural configuration of conductive networks to maximize the sensitivity. CNT-based strain-sensitive films were recently reported based on the hybridization of the vertical MWNT forests and the polyurethane films (96). A 3D accordion-like structure of a conductive CNT network held together with an elastic polymer binder allows it to have high stretchability with high electrical conductivity. This composite provides highly reproducible changes in resistivity on stretching for strains up to 40%.

The contact resistance change in side-by-side packed CNTs can be utilized in strain sensors. For example, the reversible fracture formation of conductive materials under the applied strain was utilized in strain sensing with conductive films (65, 97). Figure 15.12a shows aligned SWNT films used as strain-sensitive conductors. When strain is applied to CNT films, the CNT films fracture into gaps and islands and bundles bridging the gaps. The fracture in the CNT films results in the increase of film resistance. The strain-sensitive films based on the film fractures efficiently support the high sensitivity of the film with the applied strains up to ~280% (Fig. 15.12b). A similar operation principle about switching the fracture and bridging under strain is also realized in graphene-woven fabrics (97, 98).

In addition to the conductive nanomaterials, the piezoelectric nanomaterials are also good candidates for flexible strain sensors. Figure 15.12c shows stretchable strain sensors based on ZnO nanowire/polystyrene nanofiber hybrid structures attached on a PDMS film (99). As can be seen in Figure 15.12d, the relative change in resistance increases monotonously over 5000% when the strain changes from 0 to 50%. Here, the network of polystyrene nanofibers enables the high stretchability and the ZnO nanowires provide the electrical contact points. When the hybrid film is elongated, the number of contact points or the electrical pathways decreases, resulting in an increase of electrical resistance.

The variation of contact area by the applied force can be also used in pressure- or strain-sensitive films (100, 101). An electrical resistance change of two interlocked arrays of high-aspect-ratio nanofibers was reported for applications in highly sensitive strain sensors (Fig. 15.12e). In this work, a small displacement of interlocking pillars in response to an external loading can be precisely converted into a resistance change,

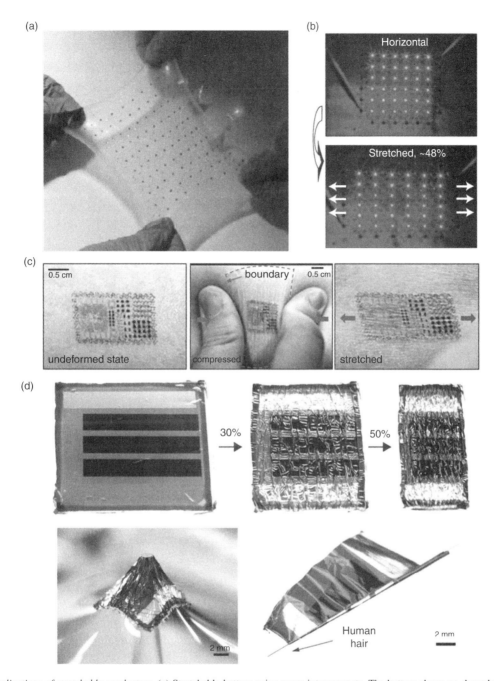

Figure 15.11 *Applications of stretchable conductors.* (a) Stretchable battery using wavy interconnects. The battery shows no degradation of brightness of LEDs with biaxial stretching of 300%. (Reprinted with permission from Reference 90. Copyright (2013) Nature.) (b) Inorganic LED arrays on stretchable substrate based on serpentine bridge interconnect. The LED shows constant brightness even at 48% stretching. (Reprinted with permission from Reference 91. Copyright (2010) Nature.) (c) Stretchable electrodes for epidermal electronics on skin. Initial state (left), compressed state (middle), and stretched state (right) of epidermal electronic devices. (Reprinted with permission from Reference 66. Copyright (2011) American Association for the Advancement of Science.) (d) Ultralight and stretchable organic solar cell. The solar cell has a high stretchability and an extreme conformability showing conformal deformation on top of tube with 1.5 mm tip diameter and flexibility by wrapping around a human hair with a radius of 35 μm. (*Source*: Reprinted with permission from Reference 64. Copyright (2012) Nature.)

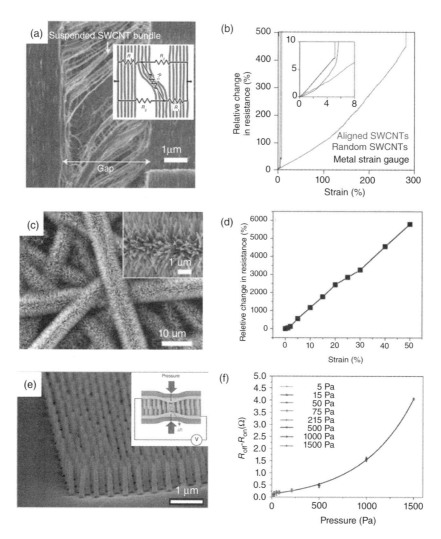

Figure 15.12 *Pressure- or strain-sensitive conductive films.* (a) SEM image of the suspended SWCNT bundled films. The inset shows a simple model of basic circuit of the fractured film. (b) Relative change in resistance under strain for the well-aligned SWCNT film (right), randomly aligned SWCNT film (middle), and metal thin film (left). (Reprinted with permission from Reference 65. Copyright (2011) Nature.) (c) An SEM image of ZnO nanowire/polystyrene nanofiber hybrid network structures. The inset shows a high resolution SEM image of ZnO NWs. (d) The relative change in resistance under strain up to 50% for ZnO NWs/polystyrene nanofiber hybrid structure on an elastomeric substrate. (Reprinted with permission from Reference 99. Copyright (2011) John Wiley & Sons, Inc.) (e) SEM image of Pt-coated polyurethane nanofibers used in strain sensor. The inset shows a schematic illustration of interlocking nanofibers under pressure. (f) Change of electrical resistance ($R_{off} - R_{on}$) under different mechanical loading of pressure. (*Source:* Reprinted with permission from Reference 100. Copyright (2012) Nature).

as illustrated in the inset of Figure 15.12f. This design of strain-sensitive films was successfully applied to detect various types of stresses such as pressure, shear, and torsion.

15.4.2 Applications

The flexible pressure- or strain-sensitive films have a wide range of applications such as artificial electronic skins, flexible strain sensors, remote surgical devices, and skin-attachable health monitoring. The applications of electronic skins in robotic hands require the detection of local pressure distribution for the dexterous handling of objects. Figure 15.13a shows local pressure detection capability of artificial electronic skins (63). Here, the pressure-sensitive, conductive rubber films were laminated and integrated with transistor arrays, where parallel semiconducting nanowire arrays were used as channel materials. When the external pressure distribution corresponding to letter "C" is applied on the electronic skin, the pressure-sensitive film, which is the outmost layer of the electronic skin, detects the local pressure distribution and changes the local conductance. Subsequently, the local conductance distribution leads to the modulation of transistor characteristics and thus the different output signals on each pixel.

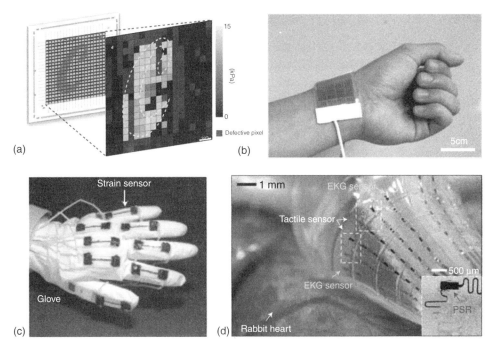

Figure 15.13 *Applications of pressure- or strain-sensitive conductive films.* (a) Artificial electronic skins with pressure-sensitive pixel arrays. The front image shows the local pressure distribution applied on the electronic skin by the local pressure of letter "C," which corresponds to a normal pressure of 15 kPa. (Reprinted with permission from Reference 63. Copyright (2010) Nature.) (b) Multiplex, real-time, skin-attachable strain sensors attached on the artery of the wrist. (Reprinted with permission from Reference 100. Copyright (2012) Nature.) (c) Flexible strain sensors for human-motion detection. SWCNT strain sensors assembled on a glove detect the fine finger movements. (Reprinted with permission from Reference 65. Copyright (2011) Nature.) (d) Tactile sensors for multifunctional balloon-catheter devices. The inset shows a magnified image of tactile sensor. (*Source*: Reprinted with permission from Reference [102]. Copyright (2011) Nature).

The other application area of pressure-sensitive film is health monitoring devices. As can be seen in Figure 15.13b, the piezoresistive films based on interlocked arrays of high-aspect-ratio nanofibers were utilized as skin-attachable sensors to detect heartbeat signals (100). For robotic applications, the strain-sensitive conductive films based on CNTs can be utilized to monitor the fine motion of robotic fingers (Fig. 15.13c) (65). One of the emerging application areas of pressure-sensitive films is the biomedical devices. Figure 15.13d shows a multifunctional balloon catheter integrated with various functional devices including sensors, actuators, and semiconducting devices (102). Here, the tactile sensors have a critical role in the detection of external forces applied to heart tissue. The tactile sensors in surgical tools enable the simultaneous monitoring of mechanical interactions with the tissue during the surgery, which minimize the damage on the tissue.

15.5 CONCLUDING REMARKS

In this chapter, we have presented different types of nanostructured conductors for applications in flexible electronics. In particular, we have reviewed recent developments of materials and architectures in nanostructured conductors for application areas including transparent electrodes, stretchable conductors, and pressure-sensitive films, areas that attract increasing attention recently as key components in the development of high-performance flexible electronics. The critical challenge is the fabrication of these electrodes and conductors with high performance using low-cost and large-area processes. Here, we introduced CNTs, graphenes, metal nanowires, and their hybrid nanostructures as the main materials for the fabrication of nanostructured conductors. All of these materials have the potential to be processed by using low-cost and large-scale processes such as roll-to-roll coating and composite manufacturing.

REFERENCES

1. Rowell MW, Topinka MA, McGehee MD, Prall HJ, Dennler G, Sariciftci NS, Hu LB, Gruner G. Appl Phys Lett 2006;88(23):233506–233506-3.

2. Wang X, Zhi LJ, Mullen K. Nano Lett 2008;8(1):323–327.

3. Lee JY, Connor ST, Cui Y, Peumans P. Nano Lett 2008;8(2):689–692.

4. King RCY, Roussel F. Appl Phys A 2007;86(2):159–163.

5. Blake P, Brimicombe PD, Nair RR, Booth TJ, Jiang D, Schedin F, Ponomarenko LA, Morozov SV, Gleeson HF, Hill EW, Geim AK, Novoselov KS. Nano Lett 2008;8(6):1704–1708.

6. Zhang DH, Ryu K, Liu XL, Polikarpov E, Ly J, Tompson ME, Zhou CW. Nano Lett 2006;6(9):1880–1886.

7. Lewis J, Grego S, Chalamala B, Vick E, Temple D. Appl Phys Lett 2004;85(16):3450–3452.

8. Aguirre CM, Auvray S, Pigeon S, Izquierdo R, Desjardins P, Martel R. Appl Phys Lett 2006;88(18):183104–183104-3.

9. Hecht DS, Thomas D, Hu LB, Ladous C, Lam T, Park Y, Irvin G, Drzaic P. J Soc Inf Disp 2009;17(11):941–946.

10. Bae S, Kim H, Lee Y, Xu XF, Park JS, Zheng Y, Balakrishnan J, Lei T, Kim HR, Song YI, Kim YJ, Kim KS, Ozyilmaz B, Ahn JH, Hong BH, Iijima S. Nat Nanotechnol 2010;5(8):574–578.

11. Madaria AR, Kumar A, Zhou CW. Nanotechnology 2011;22(24):245201.

12. Park YB, Hu L, Gruner G, Irvin G, Drzaic P. 37.4: late-news paper: integration of carbon nanotube transparent electrodes into display applications. SID Symposium Digest of Technical Papers. Wiley Online Library; 2008. p 537–540.

13. De S, Higgins TM, Lyons PE, Doherty EM, Nirmalraj PN, Blau WJ, Boland JJ, Coleman JN. ACS Nano 2009;3(7):1767–1774.

14. Hu L, Gruner G, Li D, Kaner RB, Cech J. J Appl Phys 2007;101(1):016102–016102-3.

15. Lewis BG, Paine DC. MRS Bull 2000;25(8):22–27.

16. Minami T. Thin Solid Films 2008;516(7):1314–1321.

17. Hecht DS, Hu LB, Irvin G. Adv Mater 2011;23(13):1482–1513.

18. Yang Y, Heeger AJ. Appl Phys Lett 1994;64(10):1245–1247.

19. Ha YH, Nikolov N, Pollack SK, Mastrangelo J, Martin BD, Shashidhar R. Adv Funct Mater 2004;14(6):615–622.

20. Artukovic E, Kaempgen M, Hecht DS, Roth S, GrUner G. Nano Lett 2005;5(4):757–760.

21. Hu LB, Hecht DS, Gruner G. Chem Rev 2010;110(10):5790–5844.

22. Wu ZC, Chen ZH, Du X, Logan JM, Sippel J, Nikolou M, Kamaras K, Reynolds JR, Tanner DB, Hebard AF, Rinzler AG. Science 2004;305(5688):1273–1276.

23. Wassei JK, Kaner RB. Mater Today 2010;13(3):52–59.

24. Morozov SV, Novoselov KS, Katsnelson MI, Schedin F, Elias DC, Jaszczak JA, Geim AK. Phys Rev Lett 2008;100(1):016602.

25. Hu LB, Wu H, Cui Y. MRS Bull 2011;36(10):760–765.

26. Hu LB, Kim HS, Lee JY, Peumans P, Cui Y. ACS Nano 2010;4(5):2955–2963.

27. Rathmell AR, Wiley BJ. Adv Mater 2011;23(41):4798–4803.

28. Tung VC, Chen L-M, Allen MJ, Wassei JK, Nelson K, Kaner RB, Yang Y. Nano Lett 2009;9(5):1949–1955.

29. Tokuno T, Nogi M, Jiu J, Suganuma K. Nanoscale Res Lett 2012;7(1):1–7.

30. Yin ZY, Wu SX, Zhou XZ, Huang X, Zhang QC, Boey F, Zhang H. Small 2010;6(2):307–312.

31. Soukoulis CM, Wegener M. Nat Photonics 2011;5(9):523–530.

32. Green AA, Hersam MC. Nano Lett 2008;8(5):1417–1422.

33. Jackson R, Domercq B, Jain R, Kippelen B, Graham S. Adv Funct Mater 2008;18(17):2548–2554.

34. Hecht DS, Heintz AM, Lee R, Hu LB, Moore B, Cucksey C, Risser S. Nanotechnology 2011;22(7):075201.

35. Li X, Gittleson F, Carmo M, Sekol RC, Taylor AD. ACS Nano 2012;6(2):1347–1356.

36. Geng HZ, Kim KK, So KP, Lee YS, Chang Y, Lee YH. J Am Chem Soc 2007;129(25):7758–7759.

37. Hellstrom SL, Vosgueritchian M, Stoltenberg RM, Irfan I, Hammock M, Wang YB, Jia C, Guo X, Gao Y, Bao Z. Nano Lett 2012;12(7):3574–3580.

38. Han JT, Kim JS, Jo SB, Kim SH, Kim JS, Kang B, Jeong HJ, Jeong SY, Lee G-W, Cho K. Nanoscale 2012;4(24):7735–7742.

39. Kaempgen M, Duesberg GS, Roth S. Appl Surf Sci 2005;252(2):425–429.

40. Tenent RC, Barnes TM, Bergeson JD, Ferguson AJ, To B, Gedvilas LM, Heben MJ, Blackburn JL. Adv Mater 2009;21(31):3210–3216.

41. Dan B, Irvin GC, Pasquali M. ACS Nano 2009;3(4):835–843.

42. Lu X, Chen ZF. Chem Rev 2005;105(10):3643–3696.

43. Fuhrer MS, Nygard J, Shih L, Forero M, Yoon YG, Mazzoni MSC, Choi HJ, Ihm J, Louie SG, Zettl A, McEuen PL. Science 2000;288(5465):494–497.

44. Parekh BB, Fanchini G, Eda G, Chhowalla M. Appl Phys Lett 2007;90(12):121913.

45. Liu X, Pichler T, Knupfer M, Fink J, Kataura H. Phys Rev B 2004;70(20):205405.

46. Kim SM, Kim KK, Jo YW, Park MH, Chae SJ, Duong DL, Yang CW, Kong J, Lee YH. ACS Nano 2011;5(2):1236–1242.

47. Ghosh DS, Martinez L, Giurgola S, Vergani P, Pruneri V. Opt Lett 2009;34(3):325–327.

48. Meiss J, Riede MK, Leo K. J Appl Phys 2009;105(6):063108.

49. Kang MG, Guo LJ. Adv Mater 2007;19(10):1391–1396.

50. Catrysse PB, Fan SH. Nano Lett 2010;10(8):2944–2949.

51. Kang MG, Kim MS, Kim J, Guo LJ. Adv Mater 2008;20(23):4408–4413.

52. Bergin SM, Chen YH, Rathmell AR, Charbonneau P, Li ZY, Wiley BJ. Nanoscale 2012;4(6):1996–2004.

53. Zhang DQ, Wang RR, Wen MC, Weng D, Cui X, Sun J, Li HX, Lu YF. J Am Chem Soc 2012;134(35):14283–14286.

54. Rathmell AR, Bergin SM, Hua YL, Li ZY, Wiley BJ. Adv Mater 2010;22(32):3558–3563.

55. Garnett EC, Cai WS, Cha JJ, Mahmood F, Connor ST, Christoforo MG, Cui Y, McGehee MD, Brongersma ML. Nat Mater 2012;11(3):241–249.

56. De S, Coleman JN. MRS Bull 2011;36(10):774–781.

57. Wu H, Hu LB, Rowell MW, Kong DS, Cha JJ, McDonough JR, Zhu J, Yang YA, McGehee MD, Cui Y. Nano Lett 2010;10(10):4242–4248.

58. Wu H, Kong D, Ruan Z, Hsu P-C, Wang S, Yu Z, Carney TJ, Hu L, Fan S, Cui Y. Nat Nanotechnol 2013;8(6):421–425.

59. Geim AK, Novoselov KS. Nat Mater 2007;6(3):183–191.

60. Nair RR, Blake P, Grigorenko AN, Novoselov KS, Booth TJ, Stauber T, Peres NMR, Geim AK. Science 2008;320(5881):1308.

61. Wu JB, Agrawal M, Becerril HA, Bao ZN, Liu ZF, Chen YS, Peumans P. ACS Nano 2010;4(1):43–48.

62. Jeong C, Nair P, Khan M, Lundstrom M, Alam MA. Nano Lett 2011;11(11):5020–5025.

63. Takei K, Takahashi T, Ho JC, Ko H, Gillies AG, Leu PW, Fearing RS, Javey A. Nat Mater 2010;9(10):821–826.

64. Kaltenbrunner M, White MS, Głowacki ED, Sekitani T, Someya T, Sariciftci NS, Bauer S. Nat Commun 2012;3:770.

65. Yamada T, Hayamizu Y, Yamamoto Y, Yomogida Y, Izadi-Najafabadi A, Futaba DN, Hata K. Nat Nanotechnol 2011;6(5):296–301.

66. Kim D-H, Lu N, Ma R, Kim Y-S, Kim R-H, Wang S, Wu J, Won SM, Tao H, Islam A. Science 2011;333(6044):838–843.

67. Ko HC, Stoykovich MP, Song J, Malyarchuk V, Choi WM, Yu C-J, Geddes JB III, Xiao J, Wang S, Huang Y. Nature 2008;454(7205):748–753.

68. Park J, Wang S, Li M, Ahn C, Hyun JK, Kim DS, Rogers JA, Huang Y, Jeon S. Nat Commun 2012;3:916.

69. Lee P, Lee J, Lee H, Yeo J, Hong S, Nam KH, Lee D, Lee SS, Ko SH. Adv Mater 2012;24(25):3326–3332.

70. Bowden N, Brittain S, Evans AG, Hutchinson JW, Whitesides GM. Nature 1998;393(6681):146–149.

71. Sekitani T, Noguchi Y, Hata K, Fukushima T, Aida T, Someya T. Science 2008;321(5895):1468–1472.

72. Zhang Y, Sheehan CJ, Zhai J, Zou G, Luo H, Xiong J, Zhu Y, Jia Q. Adv Mater 2010;22(28):3027–3031.

73. Cai L, Li J, Luan P, Dong H, Zhao D, Zhang Q, Zhang X, Tu M, Zeng Q, Zhou W. Adv Funct Mater 2012;22(24):5238–5244.

74. Someya T, Kato Y, Sekitani T, Iba S, Noguchi Y, Murase Y, Kawaguchi H, Sakurai T. Proc Natl Acad Sci U S A 2005;102(35):12321–12325.

75. Sekitani T, Nakajima H, Maeda H, Fukushima T, Aida T, Hata K, Someya T. Nat Mater 2009;8(6):494–499.

76. Chun K-Y, Oh Y, Rho J, Ahn J-H, Kim Y-J, Choi HR, Baik S. Nat Nanotechnol 2010;5(12):853–857.

77. Ata S, Kobashi K, Yumura M, Hata K. Nano Lett 2012;12(6):2710–2716.

78. Chen M, Tao T, Zhang L, Gao W, Li C. Chem Commun 2013;49(16):1612–1614.

79. Liu K, Sun Y, Liu P, Lin X, Fan S, Jiang K. Adv Funct Mater 2011;21(14):2721–2728.

80. Rogers JA, Someya T, Huang YG. Science 2010;327(5973):1603–1607.

81. Kim DH, Xiao JL, Song JZ, Huang YG, Rogers JA. Adv Mater 2010;22(19):2108–2124.

82. Yu C, Masarapu C, Rong J, Wei B, Jiang H. Adv Mater 2009;21(47):4793–4797.

83. Hyun DC, Park M, Park C, Kim B, Xia Y, Hur JH, Kim JM, Park JJ, Jeong U. Adv Mater 2011;23(26):2946–2950.

84. Xu F, Zhu Y. Adv Mater 2012;24(37):5117–5122.

85. Kim D-H, Song J, Choi WM, Kim H-S, Kim R-H, Liu Z, Huang YY, Hwang K-C, Zhang Y-W, Rogers JA. Proc Natl Acad Sci U S A 2008;105(48):18675–18680.

86. Ahn BY, Duoss EB, Motala MJ, Guo X, Park S-I, Xiong Y, Yoon J, Nuzzo RG, Rogers JA, Lewis JA. Science 2009;323(5921):1590–1593.

87. Jeong GS, Baek D-H, Jung HC, Song JH, Moon JH, Hong SW, Kim IY, Lee S-H. Nat Commun 2012;3:977.

88. Hu L, Pasta M, Mantia FL, Cui L, Jeong S, Deshazer HD, Choi JW, Han SM, Cui Y. Nano Lett 2010;10(2):708–714.

89. Hu L, Wu H, La Mantia F, Yang Y, Cui Y. ACS Nano 2010;4(10):5843–5848.

90. Xu S, Zhang Y, Cho J, Lee J, Huang X, Jia L, Fan JA, Su Y, Su J, Zhang H. Nat Commun 2013;4:1543.

91. Kim R-H, Kim D-H, Xiao J, Kim BH, Park S-I, Panilaitis B, Ghaffari R, Yao J, Li M, Liu Z. Nat Mater 2010;9(11):929–937.

92. Yang L, Zhang T, Zhou H, Price SC, Wiley BJ, You W. ACS Appl Mater Interfaces 2011;3(10):4075–4084.

93. Lipomi DJ, Tee BCK, Vosgueritchian M, Bao Z. Adv Mater 2011;23(15):1771–1775.

94. Wang L, Ding T, Wang P. IEEE Sens J 2009;9(9):1130–1135.

95. Someya T, Sekitani T, Iba S, Kato Y, Kawaguchi H, Sakurai T. Proc Natl Acad Sci U S A 2004;101(27):9966–9970.

96. Shin MK, Oh J, Lima M, Kozlov ME, Kim SJ, Baughman RH. Adv Mater 2010;22(24):2663–2667.

97. Li X, Zhang R, Yu W, Wang K, Wei J, Wu D, Cao A, Li Z, Cheng Y, Zheng Q. Sci Rep 2012;2:870.

98. Li X, Sun P, Fan L, Zhu M, Wang K, Zhong M, Wei J, Wu D, Cheng Y, Zhu H. Sci Rep 2012;2:395–402.

99. Xiao X, Yuan L, Zhong J, Ding T, Liu Y, Cai Z, Rong Y, Han H, Zhou J, Wang ZL. Adv Mater 2011;23(45):5440–5444.

100. Pang C, Lee G-Y, Kim T-I, Kim SM, Kim HN, Ahn S-H, Suh K-Y. Nat Mater 2012;11(9):795–801.

101. Engel J, Chen J, Chen N, Pandya S, Liu C. Multi-walled carbon nanotube filled conductive elastomers: materials and application to micro transducers. 19th IEEE International Conference on Micro Electro Mechanical Systems. MEMS 2006 Istanbul; 2006. IEEE. p 246–249.

102. Kim D-H, Lu N, Ghaffari R, Kim Y-S, Lee SP, Xu L, Wu J, Kim R-H, Song J, Liu Z. Nat Mater 2011;10(4):316–323.

16

GRAPHENE, NANOTUBE, AND NANOWIRE-BASED ELECTRONICS

Xi Liu[1], Xiaoling Shi[2], Lei Liao[3], Zhiyong Fan[1], and Johnny C. Ho[2]

[1]*Department of Electronic and Computer Engineering, Hong Kong University of Science and Technology, Kowloon, Hong Kong SAR, China*
[2]*Department of Physics and Materials Science, City University of Hong Kong, Kowloon, Hong Kong SAR, China*
[3]*Department of Physics and Key Laboratory of Artificial Micro- and Nano-Structures of Ministry of Education, Wuhan University, Wuhan, China*

16.1 GRAPHENE

The carbon-based nanomaterials, such as graphene and carbon nanotubes, are emerging as an exciting category of new materials for future electronics due to their superior electronic properties. Graphene and carbon nanotubes share a variety of common characteristics in view of their fundamental electronic properties (1, 2). Since the discovery of carbon nanotubes in 1991 (3), the worldwide intensive research has brought about a lot of exciting breakthroughs and prototype manifestations of diverse electronic circuits and devices. However, the practical application of carbon nanotubes has been baffled due to the in competence of controlling their electronic properties and the requirements of unconventional assembly procedures to align nanotubes into ordered arrays for highly integrated circuits. Fortunately, after graphene was discovered in 2004 (4–7), the carbon-based nanomaterials have received a renewed interest for future electronics due to their two-dimension geometry of graphene that could be utilized to resemble conventional planar electronics and can be readily processed to highly integrated circuits over large scale (8–10). In this section, the opportunities, challenges, as well as the recent advances in the development of high-speed graphene transistors and circuits are reviewed.

Graphene, with a single atomic sheet of sp^2-bonded carbon (Fig. 16.1a), is a unique material from many aspects (4–7, 14). It is probably the thinnest material, and meanwhile owning excellent chemical stability and impermeability, optical transparency, mechanical strength, and electrical and thermal conductivity. Ever since its discovery merely almost 10 years ago, graphene has been one of the most interesting materials for both fundamental studies and large amounts of potential applications. Graphene is a semimetal material with zero bandgap. Its valence and conduction bands are cone shaped and meet at the K points of the Brillouin zone (Fig. 16.1b). Precisely so, bulk graphene-based devices cannot be switched off, which limits the achievable on–off current ratios but does not rule out analogue radio frequency (RF) device applications.

Graphene shows a unique combination of multiple characteristics to guarantee an exciting material for the next generation electronics with unprecedented performance. Particularly, graphene owns the highest carrier mobilities (up to 200,000 cm^2/Vs) among any known material at room temperature (Fig. 16.1c) (6, 15–17), which is not only ∼100 times greater than that of silicon, but also about 10 times better than the state-of-the-art high-mobility group III–V semiconductors materials (12). In particular, it is worth noting that these high carrier mobility can be achieved at relative high carrier concentration (e.g., ∼10^{12} cm^{-2}). Although some semiconductors exhibit room-temperature mobility as high as ∼77,000 cm^2/Vs (e.g., InSb), these values are usually quoted for undoped bulk semiconductors.

Carrier mobility represents the charge transport speed under a given electrical field, and is an important parameter of charge transport characteristics of a semiconductor material and traditional semiconductor devices. The extraordinarily high carrier mobility in graphene makes it an attractive material for ultrahigh speed electronics (6, 16, 17). However, the carrier mobility only describes the charge transport speed in low electrical fields. With the reducing gate lengths in modern field-effect transistors (FETs) and increasing electrical field in the channel, the steady-state carrier velocity starts to saturate. At this saturation point, the intrinsic carrier mobility becomes less relevant to device performance, and the carrier saturation velocity becomes another important parameter of the charge transport characteristics. In this regard, graphene also exhibits exceptional carrier saturation velocity characteristics (Fig. 16.1d). The maximum carrier velocity is expected to reach around 4.5×10^7 cm/s in graphene, while only 2×10^7 cm/s for GaAs and 1×10^7 cm/s for silicon (18–20). Moreover, when in high fields the velocity in graphene does not drop as severely as those in III–V semiconductors. Therefore, graphene also shows significant features over

Nanomaterials, Polymers and Devices: Materials Functionalization and Device Fabrication, First Edition. Edited by Eric S. W. Kong.
© 2015 John Wiley & Sons, Inc. Published 2015 by John Wiley & Sons, Inc.

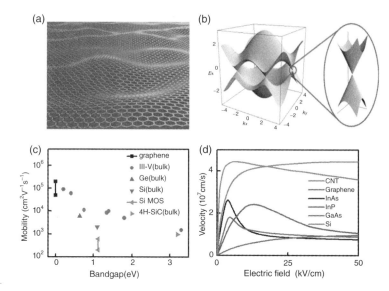

Figure 16.1 The physical properties of graphene. (a) A schematic showing corrugated graphene as a single atomic sheet of sp^2-bonded carbon. (b) The K points of the Brillouin zone of graphene. (Adapted from Reference 11.) (c) Electron mobility in low electric fields versus bandgap for different materials at room temperature. The mobility data relates to undoped material except for the Si MOS data (2000–100 cm²/Vs). (d) Electron drift velocity versus electric field for common semiconductors, carbon nanotubes, and large-area graphene (simulation). (*Source*: Adapted from References 12, 11, 13).

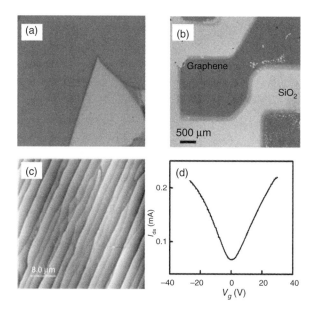

Figure 16.2 The growth of graphene. (a) Optical image of mechanical peeled graphene. (Adapted from Reference 25.) (b) Optical image of the CVD-grown graphene transferred from the Ni substrate onto SiO_2/Si substrate. (Adapted from Reference 26.) (c) AFM image of graphene on 6H–SiC (0001) with a nominal thickness of 1.2 monolayer formed by annealing in Ar. (*Source*: Adapted from Reference 27). (d) Typical transfer characteristics (I_{ds}-V_g) curves of a back-gated graphene FET on 100 nm SiO_2/Si.

conventional semiconductors in regard to high-field transport (12). In addition, graphene also exhibits many other satisfactory properties, for example, single atomic thickness, high thermal conductivity (5000 W/mK) (21, 22), large critical current densities ($\sim$3 × 10⁹ A/cm²) (23) and exceptional mechanical strength (24), all of which make graphene an attractive and ideal material for high-speed rigid/flexible electronics.

In the past several years, various methods have been developed for fabrication, growth, and synthesis of graphene and its derivatives (Fig. 16.2). Due to weak van der Waals forces between the adjacent graphene layers, the pristine graphene can be obtained by mechanical exfoliation of graphite using adhesive tapes (Fig. 16.2a) (5). This approach can produce the best quality graphene with the highest mobility up to 200,000 cm²/Vs at room temperature, but often with sizes only several tens of micrometers. Therefore, it is still adequate for fundamental investigation of graphene-based materials and devices, but not very practical for large-scale applications. On the other hand, large-area

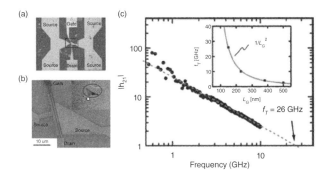

Figure 16.3 Electrical properties of a first graphene RF transistor. (a) Optical image of the device layout with ground–signal–ground accesses for the drain and the gate. (b) (False color) SEM image of the graphene channel and contacts. The inset shows the optical image of the as-deposited graphene flake (circled area) prior to the formation of electrodes. (c) Small signal current gain |h21| as a function of frequency of a graphene FET with L_G= 150 nm, showing a cutoff frequency at 26 GHz. The dashed line corresponds to the ideal $1/f$ dependence for |h21|. Inset: the maximum f_T as a function gate length for the four GFETs measured with a $1/L_G^2$ dependence. (*Source*: Adapted from Reference 40).

single- or a few-layer graphene can now be readily grown on metal foil (e.g., Ni or Cu) through chemical vapor deposition (CVD) method and then transferred onto various substrates (Fig. 16.2b) (28–32, 26, 33). This CVD-grown graphene reported to date usually exhibits highly variable mobility values ranging from 800 to 16,000 cm²/Vs. In addition, large-area continuous graphene can also be obtained by high-temperature annealing of carbon-containing substrates, for example, SiC (Fig. 16.2c), namely, carbon segregation process. The graphene obtained through this method is often called epitaxial-grown graphene, and typically own a mobility of around 1000–18,100 cm²/Vs (4, 34, 35). Moreover, large quantity of graphene flakes can also be produced via solution chemical exfoliation process, which yet usually have relatively poor quality and is not commonly used for high-performance electronic devices (36, 37).

For fundamental investigations, most graphene devices are fabricated on silicon/silicon oxide substrate due to the convenience in visualizing single-layer graphene on the substrate (38, 39). To minimize the substrate influence and investigate the intrinsic electrical properties of graphene, suspended graphene devices have also been fabricated and studied (16, 17). A typical high-performance graphene exhibits a conspicuous ambipolar electric field effect (Fig. 16.2d), for instance, charge carriers can be tuned continuously between electrons and holes with the carrier concentrations as high as 10^{13} cm⁻² and the carrier mobilities exceeds 20,000 cm²/Vs even at ambient conditions (17).

The IBM group reported the first comprehensive experimental studies on the high-frequency response of top-gated graphene transistors with variable gate lengths in 2009 (40). The small signal current gain (the ratio of small-signal drain and gate currents) of the graphene transistors derived from S-parameter measurement was found decreasing with increasing frequency and following the ideal $1/f$ dependence expected for conventional FETs, from which the f_T can be extrapolated to the point where the current again reduced to unity. The resulted f_T values exhibit a strong gate voltage dependence and is proportional to the DC transconductance. The peak cutoff frequency was found to be inversely proportional to the square of the gate length, and for a gate length of 150 nm, a peak f_T as high as 26 GHz was obtained (Fig. 16.3). This work represented a significant step toward the actualization of graphene-based high-frequency electronics. The frequency response of these initial devices is mainly restricted by two factors. First, the dielectric integration approach uses an oxidative functionalization layer that causes a severe degradation of the electronic properties of graphene, with a carrier mobility as low as 400 cm²/Vs. Second, there is a large gap between source and gate or gate and drain electrodes, in which the graphene is not modulated by the gate and function as a series access resistance. Through using an improved dielectric integration approach with a thin layer of aluminum nucleation layer, the IBM group were able to achieve improved carrier mobilities up to 2700 cm²/Vs. In addition, the use of a dual-gate configuration allows them to investigate the impact of access resistance on RF performance. In this dual gate structure with both local top- and global back-gate, the access resistance of the graphene transistor is modulated by the back-gate through electrostatic doping. By varying the back-gate voltage, the access resistance can be reduced by more than a half, leading to a fourfold increase of transconductance. Consequently, a significantly improved intrinsic cutoff frequency of 50 GHz is achieved in a 350-nm-gate graphene transistor (41).

Taking a further step, IBM group also first presented graphene FETs fabricated on a 2 inch epitaxial graphene on SiC wafer (Fig. 16.4a), with a cutoff frequency as high as 100 GHz (42). In this case, graphene was epitaxially grown on the Si face of a semi-insulating SiC wafer by thermal annealing at 1450°C and exhibited an electron carrier density of ~3 × 10^{12} cm⁻² and a Hall mobility between 1000 and 1500 cm²/Vs. In order to fabricate the top-gate stack without excessive damage to the graphene, an interfacial polymer layer made of a derivative of polyhydroxystyrene was spin coated on the graphene before atomic layer deposition of a 10 nm thick HfO₂ insulating layer. Arrays of top-gated FETs were fabricated with various gate lengths (L_G) down to 240 nm. The drain current I_{ds} of graphene FETs measured as a function of V_g exhibited n-type behavior with a rather negative Dirac point below $V_g < -3.5$ V (Fig. 16.4b). This value corresponds to a rather high electron density (>4 × 10^{12} cm⁻²) in the graphene channel at a zero gate bias state and is advantageous for achieving low series resistance of graphene FETs. As a result, the device transconductance, g_m, defined by dI_{ds}/dV_g, is nearly constant over a wide V_g range in the on state (right axis in Fig. 16.4b). The output characteristics show no clear current saturation up to a drain biases of 2 V or before device breakdown, which is different from the conventional Si FETs due to the absence of a bandgap in graphene (Fig. 16.4c). The current gain |h21| exhibits the expected $1/f$ frequency dependence with the extrapolated f_T, which reaches as high as 100 GHz for 240 nm device operating at a drain bias of 2.5 V

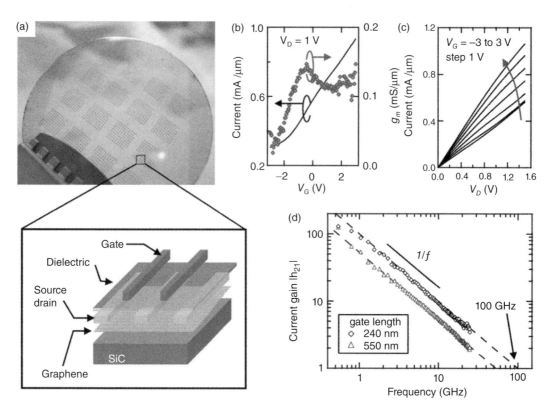

Figure 16.4 Graphene transistors with 100 GHz f_T. (a) Image of devices fabricated on a 2 inch graphene wafer and schematic cross-sectional view of a top-gated graphene FET. (b) The transfer characteristics and transconductance of a graphene FET (gate length $L_G = 240$ nm) as a function of gate voltage at drain bias of 1 V. (c) The output characteristics of a graphene FET ($L_G = 240$ nm) for various gate voltages. (d) Measured small-signal current gain |h21| as a function of frequency f for a 240 nm gate ($\Diamond$) and a 550 nm gate (Δ) graphene FET at $V_D = 2.5$ V, with the extrapolated cutoff frequencies of 53 GHz and 100 GHz for the 550 nm and 240 nm devices, respectively. (*Source*: Adapted from Reference 42).

(Fig. 16.4d). The high-frequency performance of these epitaxial graphene transistors exceeds that of state-of-the-art silicon transistors of the same gate length. Similarly sized Si devices are limited to 30 GHz operation. Assuming these devices can be scaled, they will undoubtedly present a significant speed increase over the conventional silicon devices (43–45).

Earlier last year, the IBM group also reported a systematic study of top-gated RF transistors based on CVD-grown graphene with gate lengths scaled down to 40 nm, and demonstrated the highest cutoff frequencies up to 155 GHz in CVD graphene transistors (46).

Despite the significant efforts, the RF performance of graphene transistors described earlier is still far from the potential that graphene could achieve, which is mainly limited by two adverse factors in the device fabrication process. The first limitation is related to the severe mobility degradation that resulted from graphene–dielectric integration process that introduces substantial defects into pristine graphene lattices. Extensive efforts have been devoted to developing unconventional and innovative dielectric integration approaches to mitigate potential damages to graphene (43, 45, 47–55). To deal with this dilemma, Liao et al. have recently developed a new strategy to integrate high-quality high-k dielectrics with graphene using a physical assembly approach without introducing any appreciable defects into the graphene lattice, and demonstrated the top-gated graphene transistors with the highest carrier mobility exceeding 20,000 cm²/Vs (23, 25, 56–59). The second limitation of the top-gated graphene transistors reported to date is the large access resistance due to nonoptimum alignment of the source-drain and gate electrodes, which can have particularly unfavorable influence on the short channel devices. When decreasing channel length, an increasing demand is necessary for a more precise device fabrication process. In the state-of-the-art silicon MOSFET technology, a self-aligned gate structure is used to ensure that the edges of the source, drain, and gate electrodes are precisely positioned to ensure no overlapping or significant gaps exist between them. However, these conventional dielectric integration and device fabrication processes used in silicon technology cannot be readily applied for graphene-based electronics, as many of them would severely damage the monolayer of graphene lattices, thus degrading the graphene performance.

To address these challenges, Liao et al. have recently developed a self-aligned approach to fabricate graphene transistors with ultrashort gate length using a metal/oxide core/shell NW as the top gate, which is integrated on graphene through a physical assembly process at room temperature (56, 57, 60). In brief, a single layer of graphene is first transferred onto a highly resistive Si/SiO₂ substrate. The metal/oxide core/shell NWs are then aligned on top of the graphene through a physical assembly process, followed by an electron-beam lithography process and metallization process to define the source, drain, and gate electrodes. A thin layer of Pt (5–10 nm) was then deposited on graphene

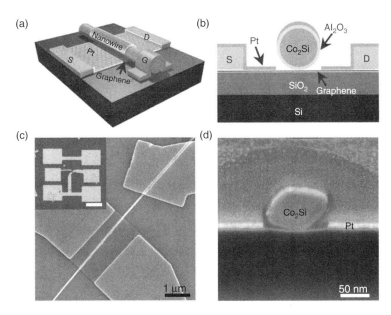

Figure 16.5 Graphene transistors with a self-aligned NW gate. (a) Schematic illustration of the three-dimensional view of the device layout. (b) Schematic illustration of the cross-sectional view of the device. In this device, the Co_2Si/Al_2O_3 core/shell NW defines the channel length, with the 5 nm Al_2O_3 shell functioning as the gate dielectrics, and the metallic Co_2Si core functioning as the self-integrated local gate, and the self-aligned Pt thin film pads as the source and drain electrodes. (c) An SEM image of a graphene transistor with a self-aligned NW gate. The inset shows an optical microscope image of the overall device layout. (d) The cross-sectional SEM image of a typical device shows that the self-aligned Pt thin film source and drain electrodes are well separated by the NW gate and precisely positioned next to the NW gate. (*Source*: Adapted from Reference 60).

across the NW, in which the NW separates the Pt thin film into two isolated pads that form the self-aligned source and drain electrodes (Fig. 16.5a and b).

Using the self-aligned approach, Liao et al. have recently demonstrated a graphene transistor with the record-high cutoff frequency up to 300 GHz (60). The graphene transistors were fabricated using a Co_2Si/Al_2O_3 core/shell NW as the self-aligned gate. In this device, the Co_2Si/Al_2O_3 core/shell NW defines the channel length, with the 5 nm Al_2O_3 shell functioning as the gate dielectrics, and the metallic Co_2Si core functioning as the local top-gate, and self-aligned Pt thin film pads as the source and drain electrodes. Figure 16.5c shows the scanning electron microscopy (SEM) image of a self-aligned graphene transistor and an optical microscope image of the overall device layout (Fig. 16.5c, inset). The cross-sectional SEM image of a typical device shows the self-aligned Pt thin film source and drain electrodes are well separated by the NW gate and precisely positioned next to the NW gate (Fig. 16.5d), clearly demonstrating that the self-alignment process can be used to effectively integrate graphene with a NW gate and nearly perfectly positioned source and drain electrodes.

The formation of the self-aligned source and drain electrodes allows precise positioning of the source–drain edges with the gate edges, and thus substantially reduces the access resistance and improve the graphene transistor performance. Figure 16.6a shows the I_{ds}–V_{ds} output characteristics under various gate voltages. The device can deliver a maximum scaled on-current of 3.32 mA/μm at $V_{ds} = -1$ V and $V_{TG} = -1$ V. The I_{ds}–V_{TG} transfer curve recorded for a self-aligned device shows a current modulation of ~42% (Fig. 16.6b), which is comparable with the typical values observed in long channel graphene transistors, suggesting that the access resistance in this ultrashort channel device is largely removed through the self-alignment process. The transconductance $g_m = dI_{ds}/dV_{TG}$ can be extracted from the I_{ds}–V_{TG} curve. Significantly, the transconductance of the self-aligned device is about 60 times larger than that of the non-self-aligned devices, showing an exceptionally high value of scaled transconductance of 1.27 mS/μm at $V_{ds} = -1$ V (Fig. 16.6c).

The aforementioned discussions clearly demonstrate that the self-aligned NW gate can allow achieving graphene transistors with unprecedented DC performance to promise excellent RF performance. The RF characteristics of the self-aligned transistors were characterized with on-chip microwave measurements. The small signal current gain |h21| derived from the S-parameters measurement displays the typical $1/f$ frequency dependence expected for an ideal FET (Fig. 16.6d). Extrapolating the curve to unit gain yields a cutoff frequency $f_T = 300$ GHz at $V_{TG} = 1$ V and a drain bias $V_{ds} = -1$ V. It should be noted that the S-parameter measurement was conducted by following a standard de-embedding procedure to exclude the parasitic capacitance, resistance, and inductance. Due to the extremely small dimension of the active device and relatively large ratio between the parasitic and gate capacitance, a ~10% error is expected for this extraction. Importantly, the f_T value extracted from S-parameter measurement is also consistent with the projected value (323 GHz) using the well-known relation $f_{T,intrinsic} = g_m/(2\pi C_g)$ established for conventional FETs (61). The speed of this self-aligned graphene device represents the highest in all graphene devices reported to date. Furthermore, it is about two times faster than that of the best silicon MOSFETs of comparable channel length (e.g., about 150 GHz for a 150 nm Si-MOSFET), and comparable to those of the very best InP high electron mobility transistors

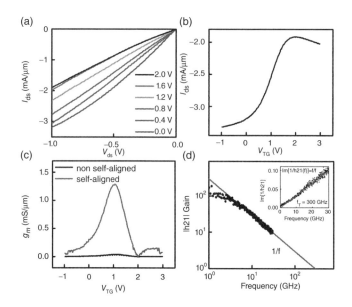

Figure 16.6   Electrical properties of the self-aligned graphene transistors. (a) The I_{ds}–V_{ds} output characteristics at various gate voltages (V_{TG} = 0.0, 0.4, 0.8, 1.2, 1.6, and 2.0 V) for the self-aligned device. (b) I_{ds}–V_{TG} transfer characteristics at V_{ds} = −1 V. (c) Transconductance at V_{ds} = −1 V as a function of top-gate voltage V_{TG} before (black) and after (red) the deposition of the self-aligned Pt source-drain electrodes, highlighting the self-alignment process increases the peak transconductance by a factor of >60. (d) Small-signal current gain |h21| versus frequency f at V_{ds} = −1 V for a device with gate length = 144 nm. (*Source:* Adapted from Reference 60).

(HEMT) and GaAs metamorphic HEMT with similar channel lengths (62). This study clearly demonstrates the exciting potential of graphene as a new electronic material for RF electronics.

Furthermore, it has also been achieved recently that the self-aligned approach can be scaled by integrating dielectrophoresis assembled NWs array on top of CVD graphene (Fig. 16.7a) (59). Briefly, single-layer graphene is first grown by CVD approach and transferred onto a glass substrate. The graphene is then patterned into isolated graphene blocks. Subsequently, pairs of electrodes are defined across each graphene block, and a dielectrophresis assembly process is then used to precisely settle the NWs array on top of each patterned graphene block. The dielectrophresis method is a self-limiting process allowing assembly of only a single NW on each pair of electrodes with high yield by controlling the hydrodynamic and electric field forces (63, 64). Finally, the self-aligned graphene transistors are fabricated using the same approaches described earlier.

Figure 16.7b shows electrical performances of a 170 nm long channel CVD graphene transistor with a NW gate on glass. The I_{ds}–V_{ds} output characteristics show that this device can deliver a significant on-current of about 1.26 mA/μm at V_{ds} = −1 V and V_g = 0 V. Significantly, a peak transconductance of 0.36 mS/μm can be obtained at V_{ds} = −1 V (V_g = 1.5 V) (Fig. 16.7c), representing the highest value achieved in CVD-grown graphene transistors at the time. The small signal current gain |h21| obtained at V_{ds} = −1 V displays the typical 1/f frequency dependence with an extrapolated f_T value of 72 GHz (Fig. 16.7d).

The fabrication of self-aligned CVD-grown graphene transistors on glass brings another feature to enable excellent extrinsic cutoff frequency. Although extraordinarily high intrinsic f_T up to 300 GHz has been reported from various types of graphene and/or device fabrication approaches, the extrinsic cutoff frequency of these graphene transistors is typically limited to only ~10 GHz or even less (41, 42, 46, 60, 65). This dramatic difference between intrinsic and extrinsic f_T is primarily due to the large ratio between parasitic pad capacitance and gate capacitance. Importantly, the fabrication of graphene transistors on insulating glass substrate minimizes the parasitic pad capacitance to allow unprecedented extrinsic f_T. Indeed, the measurement the above self-aligned device without the de-embedding procedures shows an extrinsic f_T value of 55 GHz (Fig. 16.7d), representing the highest extrinsic f_T achieved in any graphene transistors reported to date.

The achievement of exceptionally high cutoff frequency (particularly the extrinsic cutoff frequency) in graphene transistors is significant and can be allowed to configure graphene transistor–based RF circuits, for instance, frequency multipliers and signal mixers operating in the gigahertz regime. Frequency multiplication is an important signal generation technique, where a signal of frequency f_0 is introduced to a nonlinear element to generate harmonics at higher frequencies, which is of great importance for all major areas of digital and analog communications, radio astronomy, and terahertz sensing. The conventional frequency multipliers at terahertz frequencies are usually based on GaAs Schottky diodes, which offer relatively high efficiencies but lack signal amplification (66, 67). FET-based frequency multipliers can offer signal gain at the cost of efficiency (66, 67). In both cases, the spectral purity of the generated signal is usually very poor, and additional components are often needed to filter the undesired frequencies.

The unique properties of graphene can allow for completely novel devices that are impossible for other materials, for example, nonlinear devices for full-wave signal rectification and frequency multiplication (68–70). Graphene FETs typically exhibits ambipolar characteristics with a "V"-shaped current–gate-voltage characteristic: the drain current is based on hole conduction when the gate voltage is below the Dirac

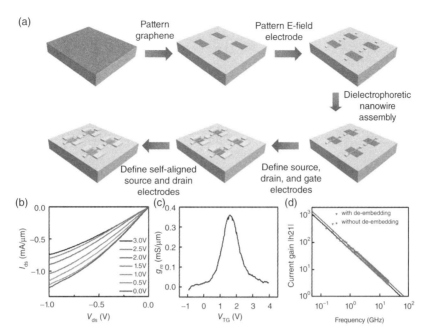

Figure 16.7 Scalable fabrication of self-aligned graphene transistors. (a) Schematic illustration of the scalable fabrication of the top-gated graphene transistors with self-aligned NW gate. (b) I_{ds}–V_{ds} output characteristics of the device at various gate voltages. (c) Transconductance at $V_{ds} = -1$ V as a function of top-gate voltage V_{TG}. (d) The small-signal current gain |h21| as a function of frequency f at $V_{ds} = -1$ V, highlighting an intrinsic cutoff frequency of 72 GHz, and a record high extrinsic cutoff frequency of 55 GHz without de-embedding process. (*Source*: Adapted from Reference 59).

point, while at higher voltages, electron conduction dominates (Fig. 16.8a). Because of this unique characteristics, full-wave rectification can be obtained at the drain contact of the device, where the output drain current alternates between electron and hole transport only if the gate electrode of graphene FET is biased at Dirac point with a superposed sinusoidal signal of frequency (Fig. 16.8b). In this way, the graphene transistors allows full-wave rectification to be realized with a much simpler circuit and can hence greatly simplify RF communication circuits that require rectifiers. Additionally, such ambipolar graphene transistors can also function as a frequency doubler where the output signal has a fundamental frequency doubling that of the input signal. Wang et al. first reported a graphene transistor–based frequency doubler (71). In their study, a 10 kHz input signal was applied to the device gate, clear frequency multiplication is observed in the output signal, where the fundamental frequency is 20 kHz. This frequency-doubler device shows high spectral purity in the output RF signal, where 94% of the output RF energy is at the fundamental frequency (20 kHz). These graphene frequency doublers offer an alternative approach to the existing complicated frequency-doubling circuit, and more important to achieve a high output spectral purity without any filtering elements.

However, the operation frequency of these initial graphene RF circuit is merely up to 20 MHz due to the severe limitation of the low extrinsic f_T of the graphene transistors. The achievement of high extrinsic f_T in self-aligned device on glass substrate described earlier can readily enable the configuration of graphene transistor–based RF circuits, operating in the gigahertz regime (59). To this end, Liao et al. have recently configured the self-aligned graphene transistors into RF frequency doublers and signal mixers. Importantly, with the excellent extrinsic f_T, the frequency doubler configured using the self-aligned graphene transistors shows a clear doubling function. With the input signal frequency at 1.05 GHz, a doubling frequency of 2.1 GHz is clearly observed at the output (Fig. 16.8d). Spectrum analysis shows that the frequency-doubler device exhibits a high spectral purity in the output RF signal, with 90% of the output RF energy at the doubling frequency (2.1 GHz) (Fig. 16.8e). This study demonstrates for the first time that a single graphene transistor–based frequency doubler can be operated in the gigahertz regime with high output spectral purity.

The availability of graphene transistors with high extrinsic cutoff frequency can also allow the configuration of signal mixers operating in the gigahertz regime. Mixers are electrical circuits used for frequency conversion and are critical components in modern RF communication systems. Liao et al. have further configured a single graphene transistor–based RF mixer (Fig. 16.8c) and performed two-tone measurements. To configure a graphene transistor–based mixer, the graphene transistor is modulated by two signals (the RF input f_{RF} and local oscillator f_{LO}) through a power combiner, and produces a drain current that contains the mixed frequencies – the intermediate frequency f_{IF} corresponding to the sum ($f_{RF} + f_{LO}$) and the difference ($f_{RF} - f_{LO}$). Our recent study shows that a single graphene transistor on glass substrate can be configured into an RF mixer operating in the range up to 10 GHz regime. Figure 16.8f shows the output spectrum of the RF mixer with the RF input $f_{RF} = 6.72$ GHz and local oscillator $f_{LO} = 2.98$ GHz. The RF mixing function is clearly seen at intermediate frequency (IF) $f_{IF} = f_{RF} - f_{LO}$ = 3.74 GHz and $f_{IF} = f_{RF} + f_{LO}$ = 9.70 GHz. It should be noted that these frequency doubling or mixing circuits do not show an apparent conversion gain, but with a relative large conversion loss (~−30 dB at 1 GHz), which is also the case in previous reports of graphene or carbon nanotube–based mixers operating at much lower frequencies (~−35 to 43 dB at 20 MHz) under similar power input. The loss can be largely attributed to the undesirable impedance mismatch and the nonideal testing setup.

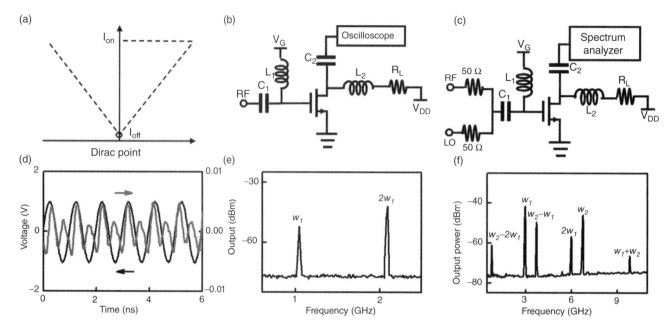

Figure 16.8   Graphene frequency doubler and mixer. (a) Linear approximation of the "V"-shaped transfer characteristic of ambipolar graphene transistors. (b) Graphene-based circuit for frequency multiplication of the input signal. With the device being biased at Dirac point, electrons (e$^-$) and holes (h$^+$) conduct alternatively in neighboring half-cycles of the output signal. (Adapted from Reference 59.) (c) The circuit diagram of a graphene transistor-based RF mixer. (Adapted from Reference 59.) (d) Measured input (red) and output (black) signals of the frequency-doubling circuit made of a self-aligned graphene transistor on glass. The input frequency is 1.05 GHz and the output frequency is 2.10 GHz. (e) Output spectrum with single RF input f_{RF} = 1.05 GHz. The frequency doubling is clearly visible. The signal power at $2f_{RF}$ = 2.1 GHz is about 10 dB higher than the signal power at f_{RF} = 1.05 GHz without filtering. (f) Output spectrum of graphene transistor–based RF mixer with LO input W_1 = 2.98 GHz and RF input W_2 = 6.72 GHz at equal power. The presence of strong signal power at $W_2 - W_1$ and $W_1 + W_2$ clearly demonstrates mixing operations up to nearly 10 GHz. (*Source*: Adapted from Reference 59).

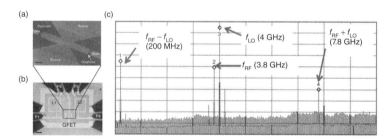

Figure 16.9   Integrated graphene RF mixer. (a) SEM image of a top-gated, dual-channel graphene transistor used in an integrated RF mixer. The gate length is 550 nm and the m. (b)□m. Scale bar, 2 □ total channel width, including both channels, is 30 m. (c) A□Optical image of an on-chip integrated graphene mixer. Scale bar, 100 snapshot of output spectrum, between 0 and 10 GHz, of the mixer with f_{RF} = 3.8 GHz and f_{LO} = 4 GHz. Each x and y division corresponds to 1 GHz and 10 dBm, respectively. The input RF power was adjusted to 0 dBm, so that the output spectrum power measured the actual loss (gain) with respect to the RF input. The frequency mixing was visible with two peaks observed at frequencies of 200 MHz and 7.8 GHz with signal power of −27 and −52 dBm, respectively. (*Source*: Adapted from Reference 72).

Meanwhile, the IBM scientists have created the first integrated RF mixer that integrates an epitaxial graphene transistor on SiC with other components (e.g., resistors, capacitors, or inductors) on chip (Fig. 16.9a and b) (72). Figure 16.9c shows the output frequency spectrum of the graphene mixer with input signals f_{RF} = 3.8 GHz and f_{LO} = 4 GHz. The two tones observed at frequencies f_{IF} of 200 MHz and $f_{RF} + f_{LO}$ of 7.8 GHz distinctly demonstrate the expected frequency mixing function. This integrated RF mixer also shows a relatively large conversion loss of about −27 dB, which can be largely attributed to the relatively poor performance of graphene transistors (intrinsic f_T ~9 GHz) used in the integrated circuits.

With a unique combination of many desirable properties, graphene has emerged as an attractive material for analog RF electronics. Extraordinary progress has been made to demonstrate the exciting potential of graphene-based devices for RF applications. Graphene transistors with the highest intrinsic cutoff frequency up to 300 GHz have been achieved on silicon/silicon oxide substrate and the highest extrinsic cutoff frequency up to 55 GHz have been achieved on glass substrate. Recent studies have also demonstrated that graphene transistors can be used to construct RF circuits such as frequency multipliers or signal mixers, but with an operating frequency typically limited within 10 GHz regime, with relatively large loss.

The fabrication of high-speed graphene transistors is of significant challenge since the conventional fabrication process often introduces undesired damage to graphene lattice to degrade its electronic performance or results in nonideal device geometry with excessive parasitic capacitances or serial resistances. Significant efforts have made in designing novel device architectures and developing innovative material integration and device fabrication approaches to mitigate these challenges. To date, graphene transistors with the highest intrinsic cutoff frequency up to 300 GHz have been achieved on silicon/silicon oxide substrate and the highest extrinsic cutoff frequency up to 55 GHz have been achieved on glass substrate.

16.2 CARBON NANOTUBE (CNT)

16.2.1 Physics of Carbon Nanotubes

In the close relationship to graphene, carbon nanotubes (CNTs) have an array of unique structural, mechanical, and electronic properties, such as the extraordinary aspect (length-to-diameter) ratio, chemical inertness, excellent mechanical strength while low density, high thermal stability and conductivity, one-dimensional (1D) ballistic electron transport, and biocompatibility.

16.2.1.1 Electrophysics In general, the electrical properties of CNTs can be derived similarly from the electronic structures of graphene while being different due to their carbon atom arrangement – helicity. Single-walled carbon nanotubes (SWNTs) can be either metallic or semiconducting depending on the (n, m) wrapping vector. Theoretically, SWNTs possess several unique properties. For example, a metallic SWNT can have an electrical current density more than 1000 times greater than the one of copper. As a 1D structure, the electron transport in SWNTs can take place through quantum effects and propagate along the tube axis.

16.2.1.1.1 Band Structure of Single-Walled Carbon Nanotubes The distinct electrical properties of SWNTs mainly come from the sp^2 hybridization of constituent carbon atoms, which can be discussed starting from the 2D energy dispersion of graphene. Graphene is a semimetal whose valence and conduction bands degenerate at only six **K** corners, defining the first Brillouin zone as shown in Figure 16.10a. These six points can be observed and determined from the Fermi surface of the graphene sheet. After rolling this 2D sheet into a 1D SWNT, the wave vector **K** becomes quantized due to the periodic boundary conditions: $\mathbf{K}\cdot\mathbf{C} = 2\pi q$, where q is an integer and **C** is the chiral vector. By this constraint, only a set of discrete states are allowed. As shown in Figure 16.10, if the wave vector passes through the **K** points, the nanotube will be metallic (Fig. 16.10b) and if not, it will be semiconducting (Fig. 16.10c). It is clear that the (9, 0) tube contains a **K** point but (10, 0) tube has none. However, the curvature of the tubes induces the mixing of π/π^* bonding and π/π^* antibonding orbitals on carbon (73); then the wave vector shifts away from the **K** point so as to produce small gaps in $(n, 0)$ and (n, m) metallic tubes with the bandgap depending inversely on the square of the tube diameter (74). On the other hand, armchair (n, n) tubes are purely metallic because there is no such shift (75). As a result, $(n, 0)$ or chiral (n, m) SWNTs are small bandgap metallic tubes when $(nm)/3$ is an integer and otherwise semiconducting.

The density of states (DOS) of SWNTs is then shown in Figure 16.11. The left plot (Fig. 16.11a) is the DOS of a metallic SWNT and the right plot (Fig. 16.11b) is the DOS of a semiconducting tube. Near the Fermi level EF, the possibility of electron occupation for a metallic SWNT is nonzero, while it is zero for a semiconducting tube. The DOS plots of both kinds of tubes are symmetric with very high density mirrored spikes in both conduction and valence bands, called von Hove singularities (VHS), which result from the 1D structure and the curvature of the SWNTs.

16.2.1.1.2 Electrical Transport Properties of Single-Walled Carbon Nanotubes The unique electronic transport properties of SWNTs come from the confinement of electrons in their 1D structure. Metallic SWNTs can hold current densities up to 109 A/cm^2 and this large value is more than 100 times larger than the ones of metal such as Cu. The strong carbon–carbon bonds and the small scattering effects both contribute to this property. Because of this property, SWNTs are highly resistive to the electron migration, which is attractive to many electronic and electrical applications. In addition, metallic SWNTs have a Fermi velocity similar to the ones of metal and higher hole-mobility than the one of silicon. Also, a semiconducting SWNT FET behaves like a p-type transistor because oxygen adsorption on the tubes induces the p-channel characteristics. Specifically, the tube conductance will decrease as the electric gate bias increases. The transistor on-stage is then operated at the negative gate bias and the off-stage is at the positive bias.

16.2.1.2 Optical Physics The optical properties of SWNTs are also a consequence of the 1D confinement of their electronic structures (77). This unique electronic structure is a result of the VHS of SWNTs. All the optical techniques used to characterize SWNTs are based on this property. Such optical methods, including absorption, Raman, infrared spectroscopy, and others, are simple, quick, and nondestructive. More importantly, individual nanotubes can be characterized using such methods. The basis of all these techniques can be understood by the so-called Kataura plot as shown in Figure 16.12 (79). Each point in the plot shows one optical transition energy, E_{ii}, which determines the energy of the light absorption by the nanotube, for a specific (n, m) nanotube with a diameter, dt.

16.2.1.2.1 Raman Spectroscopy of Single-Walled Carbon Nanotubes Raman spectroscopy is widely used to characterize SWNTs. The method is simple, sensitive, and requires no sample preparation. As shown in Figure 16.13, the Raman spectra of SWNTs have many

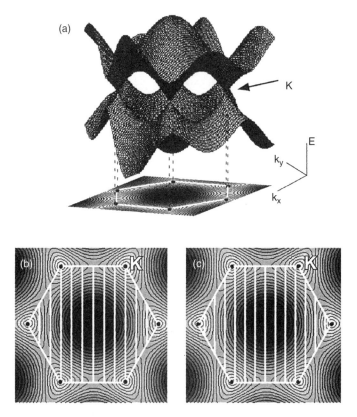

Figure 16.10   Band structure of graphene and its relationship to the SWNTs. (a) Three-dimensional view of the graphene π/π^* bands and their 2D projection. (b) Example of the allowed 1D subbands for a metallic tube. Schematic depicts (9, 0) configuration. (c) Example of the quantized 1D subbands for a semiconducting tube. Schematic depicts (10, 0) configuration. The white hexagon defines the first Brilluion zone of graphene, and the black dots in the corners are the graphene K points. (*Source*: Reprinted with permission from Reference 73, Copyright (2000) American Chemical Society).

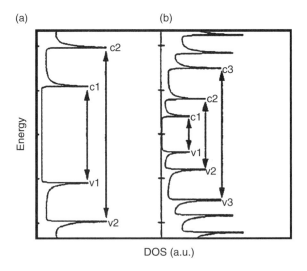

Figure 16.11   Schematic diagram of electronic density of states of SWNTs. (a) Metallic and (b) semiconducting SWNTs. (*Source*: Reprinted with permission from Reference 76, Copyright (2001) American Chemical Society).

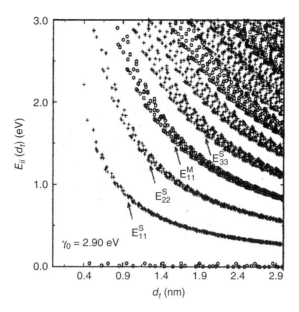

Figure 16.12 Kataura plot. Calculated energy separation, E_{ii}, between van Hove singularities for (n, m) nanotube versus the nanotube diameter. (*Source*: Reprinted with permission from Reference 78, Copyright (2002) American Chemical Society).

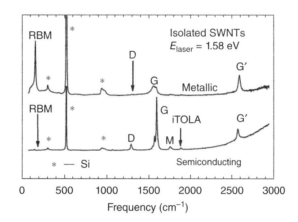

Figure 16.13 Raman spectra from a metallic and a semiconducting SWNTs. (*Source*: Reprinted with permission from Reference 80, Copyright (2005) American Institute of Physics).

characteristic features. Herein, Liao et al. discuss three of these characteristics – the radial breathing mode (RBM), D mode, and G mode – because these three will be used in the next sessions and are the primary modes by which nanotube samples are typically characterized. The Raman signal is strongly dependent on the excitation laser wavelength (energy) and in particular, matching the excited energy to the optical transition energy can greatly enhance the Raman signal, termed as resonance Raman spectroscopy (RRS). In addition, the Raman intensity is also highly influenced by the orientation between the excited light polarization and the tube direction. Only a tube with the axis along the polarized direction can result in the strongest intensity.

The RBM is one of the most important regions in SWNT Raman spectra. The RBM peak position and intensity can be used to determine the diameter, (n, m), and other properties of SWNTs. In this chapter, Liao et al. demonstrate the relationship between the diameter and the RBM frequency, which is relevant in the following sessions. Calculations from the RBM are one of the most accurate methods to measure the diameter of a SWNT. Numerous experiments and theoretical calculations have concluded that the diameter of a SWNT is inversely proportional to its RBM peak frequency, although many different analytical equations have been published in the literature (80). The RBM phenomenon mainly comes from the radial vibration of the tube and is dependent on the environment around the tubes such as the substrate, surfactant, outer-force, and so on. On the other hand, the G mode is derived from the G band in graphite at a single peak of $1582\,\mathrm{cm^{-1}}$ and the G band in a SWNT is in the range between 1565 and $1590\,\mathrm{cm^{-1}}$ because of the tube curvature. The G band also has very important information about the tube type, with metallic and semiconducting tubes having different line-shapes as shown in Figure 16.13. A metallic tube generally

TABLE 16.1   Young's Modulus for CNTs

Nanotube No	Length (μm)	Outer Diameter (nm)	Inner Diameter (nm)	Young's Modulus (TPa)
1	1.17	5.6	2.3	1.06
2	3.11	7.3	2.0	0.91
3	5.81	24.8	6.6	0.59
4	2.65	11.9	2.0	1.06
5	1.73	7.0	2.3	2.58
6	1.53	6.6	2.3	3.11
7	2.04	7.0	3.0	1.91
8	1.43	6.6	3.3	4.15
9	0.66	7.0	3.3	0.42
10	1.32	9.9	3.0	0.40
11	5.10	8.4	1.0	3.70

Source: Reprinted with permission from Reference 81, Copyright (1996) Nature Publishing Group.

has a broader shoulder and this feature can be used to distinguish the tube type. The D mode generally appears in the range between 1350 and 1370 cm^{-1} while it represents the disordered sp^2 bonding such as sp^1 and sp^3 carbon. Amorphous carbon and multiwalled carbon nanotubes (MWNTs) often show the strong D band, and the D band can indicate the defect and impurity level of SWNT samples.

16.2.1.3 Mechanical Physics   At the same time, carbon nanotubes have exceptional mechanical properties because of their atomic structures and chemical bonding. The carbon–carbon bond is the strongest bond known in nature and the cylindrical structures of CNTs can further enhance the mechanical strength. Some individual nanotubes are listed in Table 16.1 with their Young's modulus, showing that the average modulus value of CNTs is 1.8 TPa. However, the mechanical properties are also dependent on the structural quality; thus, high-quality tubes are desired for the mechanical applications.

16.2.2   Synthesis of Carbon Nanotubes

16.2.2.1 Arc-Discharge and Laser Ablation   Arc discharge and laser ablation were the first methods utilized to fabricate CNTs and SWNTs. CNT research then became a hot topic after Iijima had synthesized CNTs by arc discharge method at NEC Lab in Japan. In the arc-discharge scheme, carbon atoms are evaporated by plasma of helium gas ignited by high currents passing through the opposing carbon anode and cathode. From then, arc-discharge has been developed into an excellent method for producing both high-quality multiwalled nanotubes and single-walled nanotubes. MWNTs can be obtained by controlling the growth conditions such as the pressure of inert gas in the discharge chamber and the arcing current. In 1992, a breakthrough in MWNT growth by arc-discharge was first made by Ebbesen and Ajayan who achieved the growth and purification of high-quality MWNTs at the gram level (82). Importantly, the synthesized MWNTs have lengths on the order of ten microns and diameters in the range of 5–30 nm. The nanotubes are typically bound together by strong van der Waals interactions and form tight bundles. Also, the MWNTs are very straight, indicative of their high crystallinity. For as-grown materials, there are few defects such as pentagons or heptagons existing on the sidewalls of the nanotubes. The by-product of the arc-discharge growth process is multilayered graphitic particles in polyhedron shapes. This way, the purification of MWNTs can be achieved by heating the as-grown materials in an oxygen environment to oxidize away the graphitic particles (82). The polyhedron graphitic particles exhibit higher oxidation rate than MWNTs; nevertheless, the oxidation purification process also removes an appreciable amount of nanotubes.

For the growth of single-walled tubes, a metal catalyst is needed in the arc-discharge system. The first success in producing substantial amounts of SWNTs by arc-discharge was achieved by Bethune and coworkers in 1993 (83). They used a carbon anode containing a small percentage of cobalt catalyst in the discharge experiment, and found abundant SWNTs generated in the soot material. The optimization of SWNT growth by arc-discharge was achieved by Journet and coworkers using a carbon anode containing 1.0 at% of yttrium and 4.2 at% of nickel as catalyst (84). At the same time, the growth of high-quality SWNTs at the 1–10 g scale was achieved by Smalley and coworkers using a laser ablation (laser oven) method (85). This method utilized intense laser pulses to ablate a carbon target containing 0.5 at% of nickel and cobalt. The target was placed in a tube furnace heated to 1200 °C. During the laser ablation, a flow of inert gas was passed through the growth chamber to carry the grown nanotubes downstream to be collected on a cold finger. The produced SWNTs are mostly in the form of ropes consisting of tens of individual nanotubes closely packed into hexagonal crystals via van der Waals interactions.

In the SWNT growth by both arc-discharge and laser ablation, typical by-products include fullerenes, graphitic polyhedrons with enclosed metal particles, and amorphous carbon in the form of particles or overcoating on the sidewalls of nanotubes. A purification process for SWNT materials has been developed by Smalley and coworkers (86) and is now widely used by many researchers. This method involves refluxing the as-grown SWNTs in a nitric acid solution for an extended period of time, oxidizing away amorphous carbon species, and removing some of the metal catalyst species.

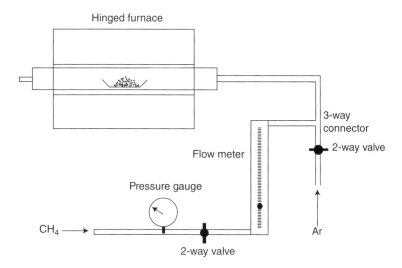

Figure 16.14 Schematic experimental setup for methane CVD synthesis of SWNTs. (*Source*: Reprinted with permission from Reference 87, Copyright (1999) American Chemical Society).

16.2.2.2 Chemical Vapor Deposition CVD methods have been successfully developed for making carbon fiber, filament, and nanotube materials since more than 10–20 years ago (87–92). A schematic experimental setup for the CVD growth of nanotubes is depicted in Figure 16.14. The growth process involves heating a catalyst material to high temperatures in a tube furnace and flowing a hydrocarbon gas through the tube reactor for a period of time. Materials grown over the catalyst are collected upon cooling the system to room temperature. The key parameters in the nanotube CVD growth are the hydrocarbon flow conditions, catalyst choices, and growth temperature. The active catalytic species are typically transition-metal nanoparticles formed on a support material such as alumina. In general, the CVD nanotube growth mechanism includes the dissociation of hydrocarbon molecules catalyzed by the transition metal, and dissolution and saturation of carbon atoms in the metal nanoparticle (93).

16.2.3 CNT Characterization

16.2.3.1 Raman Spectroscopy Raman spectroscopy is one of the most powerful tools for the characterization of carbon nanotubes. The most characteristic features are summarized as following: (i) a low-frequency peak $<200 \, cm^{-1}$ (or bunch of peaks for polydispersed samples when resonating conditions are met), characteristic of SWNTs assigned to a "breathing" mode of the tubes, whose frequency depends essentially on the diameter of the tube (RBM: radial breathing mode); (ii) a large structure ($1340 \, cm^{-1}$) assigned to residual ill-organized graphite, the so-called D-line (D: disorder); (iii) a high-frequency bunch (between 1500 and $1600 \, cm^{-1}$) called G band, also characteristic of nanotubes, corresponding to a splitting of the E_{2g} stretching mode of graphite (94). This bunch could be superimposed with the G-line of residual graphite (95); (iv) a second-order mode observed between 2450 and $2650 \, cm^{-1}$ assigned to the first overtone of the D mode and often called G mode; (v) a combination mode of the D and G modes between 2775 and $2950 \, cm^{-1}$. The RBM is directly dependent on the nanotube diameters through the relation, $\omega_{RBM} = A/d + B$, where the parameters A and B are determined experimentally (96). However, the frequency of RBM is not related to the chiral angle θ of the nanotubes (97–99). As an example, for an isolated SWNT located on the Si/SiO$_2$ substrate, the experimental value of A is found to be $248 \, nm \, cm^{-1}$ and $B = 0 \, cm^{-1}$ (100). Using a force constant model with interactions to the fourth neighbor, Bandow et al. calculated the value of $A = 224 \, nm \, cm^{-1}$ (101). For SWNT bundles, the tube/tube interactions lead to values of $A = 234 \, nm \, cm^{-1}$ and $B = 10 \, cm^{-1}$ (102). A calculated value of RBM parameters ($A = 232 \, nm \, cm^{-1}$ and $B = 6.5 \, cm^{-1}$) is also given by Alvarez et al. using the crystal of identical and infinite nanotubes and taking into account the van der Waals interactions between tubes with a Lennard–Jones potential (103). Summarized in Bandow et al. (104), other values of parameter A are given according to the models used: zone folding method ($A = 223 \, nm \, cm^{-1}$) (105), force constant model ($A = 218 \, nm \, cm^{-1}$) (97), local density approximation ($A = 234 \, nm \, cm^{-1}$) (98), pseudopotential density functional theory ($A = 236 \, nm \, cm^{-1}$) (99), and elastic deformation model ($A = 227 \, nm \, cm^{-1}$) (104). Table 16.2 illustrates the dependence of the RBM on the nanotube diameters. The direct environment of the nanotubes causes a modification of the RBM mode as well. Thus, the correction due to the bundling (tube–tube interaction) is an up-shift of about $14 \, cm^{-1}$ for (9,9) nanotubes (107). The tube–tube interaction is not the only one to be taken into account, the interaction between graphene sheets in the case of MWNTs also contributes to the RBM modification (104). This phenomenological relation is only valid in a weak range of size ($1 \, nm < d < 2 \, nm$). Especially, with nanotube diameters below 1 nm, a chirality dependence of ω_{RBM} appears due to the distortion of nanotube lattice: the relation is no longer valid in this range of diameters. Moreover, with diameters greater than 2 nm, the intensity of RBM is weak (96).

On the other hand, SWNTs (diameters and metallic/semiconducting features) can be characterized by studying the G band line shape although the information obtained is less accurate than that from RBM. The Raman line shape differs between metallic and semiconductor nanotubes, which then allow the distinguishment between the two types. The line shape of G band is composed of up to six peaks but only the

TABLE 16.2 Observed versus Predicted Raman Radial Breathing-Mode Frequencies for Metallic and Semiconducting CNTs

Metallic	d_t (nm)	ω_{RBM} (cm^{-1})	α	Predicted
(15, 0)	1.1909	199.1	1.2	200.2
(10, 7)	1.1749	203.6	0.2	202.7
(13, 1)	1.0740	221.1	0.8	220.6
(9, 6)	1.0382	227.5	0.0	227.8
(7, 7) or (11, 2)	0.9626	245.2	0.6	244.7
(12, 0)	0.9527	247.9	0.8	247.1
(8, 5)	0.9017	261.4	1.1	260.4
Semiconducting	d_t (nm)	ω_{RBM} (cm^{-1})	α	Predicted
(10, 3)	0.9360	255.4	0.8	251.3
(11, 1)	0.9156	255.9	1.3	256.6
(7, 6)	0.8251	283.5	0.9	283.4
(6, 5)	0.7573	309.1	1.0	307.6
(6, 4)	0.6921	336.9	1.3	335.4
(9, 4)	0.7276	316.5	0.0	319.7

Source: Reprinted with permission from Reference 106, Copyright (2003) American Chemical Society.

two most intense peaks are useful for analysis (96). The lower and higher frequency components (ω_G^- and ω_G^+) are, respectively, associated with the atomic displacements along circumferential direction and along the tube axis (100). For semiconducting nanotubes, the profile of the two components ω_G^- and ω_G^+ is narrow (108) and fits to a Lorentzian line shape (96, 100). However, for metallic tubes, the profile of the lower frequency component ω_G^- is broad, asymmetric, and can be described by a Breit–Wigner–Fano (BWF) line shape (96, 100, 108). There is no dependence of the frequency ω_G^+ (about 1591 cm^{-1}) on the nanotube diameters contrary to ω_G^-.

16.2.3.2 Transmission Electron Microscopy High-resolution transmission electron microscopy (HRTEM) is another powerful instrumental technique which can reveal the diameters of the single-walled and multiwalled CNTs, the number of walls, and the distance between the walls (87, 109–113). In addition, the electron diffraction mode of transmission electron microscopy (TEM) helps to identify the nature of the cap on top of the CNT, which is usually composed of the metal catalyst (114, 115). In TEM, a thin solid specimen ($\leq$200 nm thick) is bombarded in vacuum with a highly focused and monoenergetic beam of electrons. The beam is of sufficient energy to propagate through the specimen. A series of electromagnetic lenses then magnifies this transmitted electron signal and diffracted electrons are observed in the form of a diffraction pattern beneath the specimen. This information is used to determine the atomic structure of the material in the samples. For TEM study, the nanotubes are dispersed in dimethylformamide using tip ultrasonification. The samples are then prepared by placing a droplet of the suspension onto a carbon coated grid and dried in air. Figure 16.15 displays typical TEM images of SWNTs grown from discrete nanoparticles, showing particle–nanotube relationships.

16.2.4 Applications of Carbon Nanotubes

Since their discovery by Iijima in 1991, CNTs have attracted great interest in chemistry, physics, electronics, and materials science. Because of their amazing properties, CNTs show a wide range of current and potential applications such as FETs (116), sensors (117), light emitters (118), logic circuits (119), and so on.

16.2.4.1 Carbon Nanotube–Based FETs Silicon transistors can be shrunk to nanosize but challenging to less than 10 nm. The tiny CNT with a diameter around 1 nm has been demonstrated to be the choice to speed up the nanoscale electronics revolution. Researchers have previously demonstrated excellent FETs (120–123) using individual SWNTs. Figure 16.16a and b depicts the transfer (I_{DS}–V_{GS}) and output (I_{DS}–V_{DS}) characteristics of the state-of-the-art individual SWNT-FET with self-aligned source/drain contacts and the near-ballistic transport (122). Impressive performance with subthreshold slope (SS) of 110 mV/dec, on-state conductance of $0.5 \times 4e^2/h$ and saturation current up to 25 μA per tube (diameter~1.7 nm) has been achieved in devices with channel lengths down to 50 nm (122). Better SS of ~70 mV/dec, which is close to the theoretically limit of 60 mV/dec, has also been achieved in transistors with slightly longer channel lengths (500 nm) (121). More recently, SWNT-FETs with sub-10 nm channel lengths have been demonstrated (123). Such devices exhibit an impressive SS of 94 mV/dec, current on–off ratio of 10^4, and on-current density of 2.41 mA/μm, which outperform silicon FETs with comparable channel length.

16.2.4.2 Carbon Nanotube Sensor In addition, carbon nanotubes are also effective sensing elements because of their 1D electronic structure, all their atoms located on the surface, and their high aspect ratio. The aforementioned characteristics combined with the unique

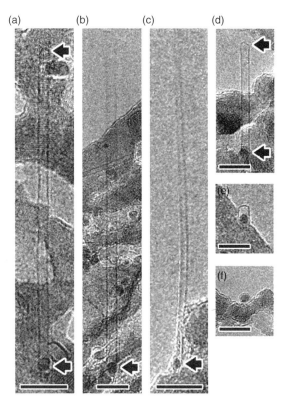

Figure 16.15 TEM images of SWNTs grown from discrete nanoparticles, showing particle-nanotube relationships. Scale bars: 10 nm. (a–d) SWNTs grown from discrete nanoparticles (dark dots at the bottom of the images). The arrows point to the ends of the nanotubes. The ends extending out of the membrane in (b) and (c) are not imaged due to thermal vibration. The background roughness reflects the TEM grid morphology. (e) Image of an ultrashort (~4 nm) nanotube capsule grown from a ~2 nm nanoparticle. (f) Image of a nanoparticle surrounded by a single graphitic shell. (*Source*: Reprinted with permission from Reference 109, Copyright (2002) American Chemical Society).

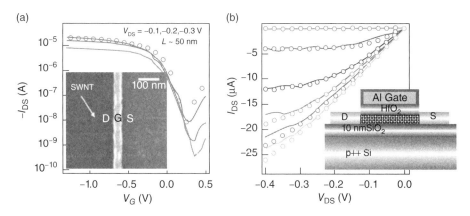

Figure 16.16 State-of-the-art individual SWNT transistors. (a) I_{DS}–V_{GS} characteristics of a self-aligned ballistic SWNT-FET with a channel length of 50 nm. Inset: SEM image of the device. (b) Experimental (solid line) and simulated (open circle) I_{DS}–V_{DS} characteristics of the same device shown in panel (a). Inset: Schematic of the device. (*Source*: Reprinted with permission from Reference 122, Copyright (2004) American Chemical Society).

electrical, electrochemical, and optical properties provide potential for detecting low or ultralow chemical gas concentration, especially toxic gas in the ambient environment. One of the first sensors based on CNTs to be demonstrated is an individual semiconducting SWNT gas sensor (117). The conductance changed by orders of magnitude upon the gas exposure. The sensing mechanism is still a controversy. In brief, researchers argued between a Fermi level shift of the tube and the Schottky barrier modulation of the tube-electrode contact in the gas exposure after several extensive investigations. Even if the mechanism is unclear, many methods have been suggested to improve the sensitivity and selectivity of CNT gas sensors. One of them is binding specific ligands to CNTs through which the selectivity and sensitivity can be greatly improved by utilizing the affinity of the ligand to the gas molecules as shown in Figure 16.17 (124).

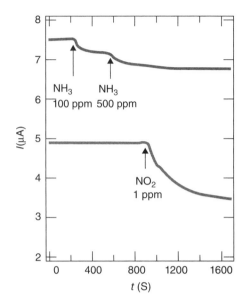

Figure 16.17   Red (top) curve: a device coated with Nafion exhibits response to 100 and 500 ppm of NH_3 in air, and no response when 1 ppm of NH_3 is introduced to the environment. Blue (bottom) curve: a PEI-coated device exhibits no response to 100 and 500 ppm of NH_3 and large conductance decrease for the exposure to 1 ppm of NO_2. (*Source*: Reprinted with permission from Reference 124, Copyright (2003) American Chemical Society).

Besides the gas sensor, the glucose sensor as shown in Figure 16.18 (125, 126), the DNA sensor (127–129), and other biosensors have also been fabricated accordingly. The aforementioned sensors are all based on FET devices of CNTs. A novel capacitance sensor has as well been developed (130) with a transparent SWNT network with the electrodes patterned on the substrate as the sensing element. This sensor is highly sensitive, fast, and reversible. The capacitance is changed either up or down depending on the analyte.

16.2.4.3   Carbon Nanotube–Based Solar Cells   In photovoltaic (PV) fields, Shockley and Queisser used detailed balance calculations to derive a fundamental limit to the power conversion efficiency (PCE) of a single junction solar cell (131). Calculations have since been refined giving a limit of ~33%; however, this is based on the assumption of one photon producing one exciton. SWNTs have been shown to exhibit multiple exciton generation (MEG) from a single photon (132). In this process, a photon having energy equal to n multiples of the bandgap is absorbed and produces n excitons. However, a complete solar cell device exploiting this property, and thus allowing violation of the Shockley Quiesser limit, is yet to be realized. Currently, the literature contains many cases of research utilizing carbon nanotubes in solar cell devices. For example, nanotubes have been integrated into a variety of organic photovoltaics (OPVs) (133). In any case, in these systems the carbon nanotubes are not so much responsible for the exciton generation upon absorption of light but rather, by introducing carbon nanotubes into these polymeric systems, exciton dissociation or electron transport is enhanced within the materials. There has also been much work in recent years incorporating carbon nanotubes into photoelectrochemical cells (PECs) either on their own (134) or as elements in donor–acceptor hybrids in conjunction with fullerenes (135), fullerenes/P3HT (136), fullerenes/porphyrin (137), porphyrins (138), pyrenes (139), polythiophenes (140) and other photoactive polymers (141), phthalocyanines (142), PAMAM dendrons (143), quantum dots (144), and more. There is also a plethora of fundamental work investigating the photoinduced charge transfer processes between carbon nanotubes and other species. It must be noted that the boundary between OPVs and PECs is blurred since photo-electrochemistry is also a method of characterizing new OPV systems. Similarly, the implementation of carbon nanotubes in dye-sensitized solar cells (DSCs) has included the replacement for the platinum catalyst counterelectrode (145), improvement of the electrical properties of titania (146), or replacement of the titania scaffold (147).

At the same time, nanotube-silicon heterojunction (NSH) solar cells are a recent photovoltaic system which has been reported to utilize carbon nanotubes in the photocurrent generation. A typical device has architecture similar to that of a single junction crystalline silicon solar cell with the exception that the emitter layer is replaced by a thin film of single-, double-, or multiwalled carbon nanotubes (SWNTs, DWNTs, or MWNTs). Wadhwa et al. (148) report a novel method of improving NSH solar cells through the electronic junction control of SWNT/n-Si devices by the use of a gate potential applied to the junction via the ionic liquid electrolyte 1-ethyl-3-methylimidazolium bis(trifluoromethylsulphonyl)imide (EMI-BTI) (Fig. 16.19). The device exhibits a PCE of 8.5% which can be dynamically and reversibly adjusted to between 4 and 11% by the electronic gating. Jia et al. (149) has achieved the highest efficiency so far with a PCE of 13.8% by the *in-situ* doping of nanotube films with 0.5 M HNO_3. The untreated devices can only demonstrate a PCE of 6.2% and the improvement with doping is primarily due to an increase in J_{SC} from 27 to 36 mA/cm^2 coupled with an increase in the FF from 0.47 to 0.72.

16.2.4.4   Carbon Nanotube Logic Circuit   Mechanically flexible logic gates such as inverter, 2-input NAND, and NOR have also been demonstrated using SWNT thin film transistors (TFTs) as shown in Figure 16.20 (150). The voltage transfer characteristics (VTC) of a diode-loaded inverter are presented in Figure 16.20b. Respectable voltage gain of ~30 at $V_{DD} = 5$ V, symmetric input/output behavior (i.e.,

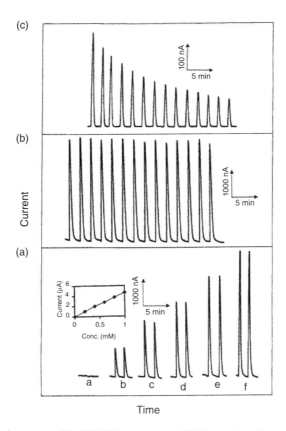

Time

Figure 16.18 (a) Flow injection amperometric response of the CNT/Teflon detector to NADH solutions of increasing concentrations (0.2–1.0 mM, b–f), along with the response to the blank solution (a) and the resulting calibration plot (inset). (b, c) Flow injection response to 14 successive injections of a 1.0 mM NADH solution at the MWCNT/Teflon and bare glassy-carbon electrodes, respectively. (*Source*: Reprinted with permission from Reference 125, Copyright (2003) American Chemical Society).

single V_{DD} operation), and rail-to-rail swing have been successfully achieved. The aforementioned characteristics enable cascading multiple stages of logic blocks for the larger-scale integration, where the output of the preceding logic block needs to be able to drive the ensuing logic block directly. Moreover, owing to the excellent mechanical flexibility of the carbon nanotube networks, the fabricated inverter circuit is extremely bendable and exhibit outstanding mechanical reliability. The measured inverter VTC shows the minimal performance change under various curvature radii (down to 1.27 mm) (Fig. 16.20c) and repeated bending tests up to 2000 cycles (Fig. 16.20d). Similarly, 2-input NAND (Fig. 16.20e) and NOR (Fig. 16.20f) logic gates have as well been demonstrated with the diode load. Both circuits are operated with a V_{DD} of 5 V and input voltages of 5 and 0 V are treated as logic "1"and "0," respectively. For the NAND gate, the output is "1" when either one of the two inputs is "0", while for the NOR, the output is "0" when either one of the two inputs is "1." The measured output characteristics confirm the accurate functioning of the fabricated circuits based on CNTs. All these demonstration have further illustrated the technological potency of CNTs for various high-value applications.

16.3 NANOWIRE

16.3.1 Nanowire Growth

A nanowire (NW) is a nanostructure, with the diameter of the order of a nanometer (10^{-9} m). Alternatively, NWs can be defined as structures that have a thickness or diameter constrained to tens of nanometers or less and an unconstrained length. At these scales, quantum mechanical effects are important, hence such wires are also known as "quantum wires."

Typical NWs exhibit aspect ratios (length-to-width ratio) of 1000 or more. As such they are often referred to as one-dimensional (1D) materials. Nanowires have many interesting properties that are not seen in bulk or 3D materials. This is because electrons in NWs are quantum confined laterally and thus occupy energy levels that are different from the traditional continuum of energy levels or bands found in bulk materials. Peculiar features of this quantum confinement exhibited by certain NWs manifest themselves in discrete values of the electrical conductance. Such discrete values arise from a quantum mechanical restraint on the number of electrons that can travel through the wire at

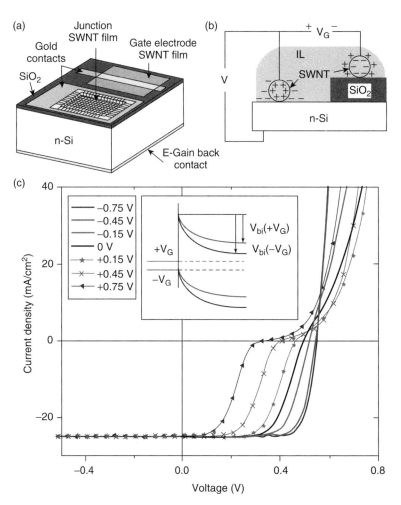

Figure 16.19 (a) Schematic illustration of electrically gated NSH solar cells; (b) illustration of the charge distribution during the gated operation; (c) *J–V* characteristics of the device with varying gate voltage. Inset shows the change of built-in potential, V_{bi}, due to more or less silicon band-bending with different magnitudes of the gate voltage. (*Source*: Reprinted with permission from Reference 148, Copyright (2010) American Chemical Society).

the nanometer scale. These discrete values are often referred to as the quantum of conductance and are integer multiples of

$$\frac{2e^2}{h} \approx 77.41 \ \mu S$$

They are inverse of the well-known resistance unit h/e^2, which is roughly equal to 25812.8 ohms, and referred to as the von Klitzing constant R_K (after Klaus von Klitzing, the discoverer of exact quantization). Since 1990, a fixed conventional value R_{K-90} is accepted. Many different types of NWs exist, including metallic (e.g., Ni, Pt, Au), semiconducting (e.g., Si, InP, GaN, etc.), insulating (e.g., SiO_2, TiO_2), and molecular NWs, which are composed of repeating molecular units either organic (e.g., DNA) or inorganic (e.g., Mo_6S_9-xIx). There are many applications where NWs may become important in electronic, optoelectronic and nanoelectromechanical devices, as additives in advanced composites, for metallic interconnects in nanoscale quantum devices, as field-emitters and as leads for biomolecular nanosensors.

Generally, there are two basic approaches to synthesizing NWs: bottom-up and top-down. A bottom-up approach synthesizes the NW by combining constituent adatoms. Most synthesis techniques use a bottom-up approach, in which there are vapor phase growth and solution phase growth. A top-down approach reduces a large piece of material to small pieces, by various means of solid phase fabrication methods such as lithography or electrophoresis, which will not be discussed in this chapter.

16.3.1.1 Vapor Phase Growth High-temperature vapor phase growth assisted by a thermal furnace is a straightforward approach that controls the reaction between metal vapor source and oxygen gas (151). In order to control the diameter, aspect ratio, and crystallinity, diverse techniques have been exploited including thermal CVD, direct thermal evaporation (152), pulse-laser-deposition (PLD) (153), and metal–organic chemical vapor deposition (MOCVD) (154), and so on. These growth methods are based on three mechanisms: vapor–solid (VS), vapor–liquid–solid (VLS), and vapor–solid–solid (VSS).

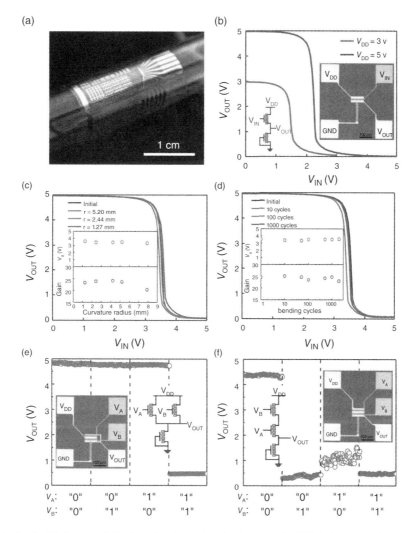

Figure 16.20 (a) Photograph of a flexible integrated circuit made with semiconducting nanotube TFTs being wrapped on a test tube with a curvature radius of 5 mm. (b) VTC of a flexible nanotube inverter measured with V_{DD} of 3 or 5 V. (c) Inverter VTC measured at various curvature radii. Inset: Inverter threshold voltage and gain as a function of curvature radius, showing the minimal performance change even when bent down to 1.27 mm radius. (d) The inverter VTC is measured after various numbers of bending cycles, indicating the good reliability after 2000 cycles. (e, f) Output characteristics of the diode-loaded 2-input NAND (e) and NOR (f) logic gates. (*Source*: Reprinted with permission from Reference 150, Copyright (2012) American Chemical Society).

16.3.1.1.1 Vapor–Solid Growth Vapor solid process occurs in many catalyst-free growth processes (155–157). It is a commonly observed phenomenon but still lacks fundamental understanding. Quite a few experimental and theoretical works have proposed that the minimization of surface-free energy primarily governs the VS process (158–160). Under high-temperature condition, source materials are vaporized and then directly condensed on the substrate placed in the low-temperature region. Once the condensation process happens, the initially condensed molecules form seed crystals serving as the nucleation sites. As a result, they facilitate directional growth to minimize the surface energy. This self-catalytic growth associated with many thermodynamic parameters is a rather complicated process that needs quantitative modeling. Examples of VS synthesized NWs include ZnO, SnO_2, In_2O_3, CdO, Sn, and Sb_2Te_3. Figure 16.21 illustratesa typical thermal CVD setup consisting of a horizontal quartz tube and a resistive heating furnace. Source material is placed inside the quartz tube; another substrate (SiO_2, sapphire, etc.) deposited with catalyst nanoparticles is placed at downstream for nanostructures growth.

16.3.1.1.2 Vapor–Liquid–Solid growth Vapor–liquid–solid mechanism was first proposed by Wagner and Ellis in 1964 (161) while observing the growth of Si whisker (162) In essence, VLS is a catalyst-assisted growth process which uses metal nanoclusters or nanoparticles as the nucleation seeds. These nucleation seeds determine the interfacial energy, growth direction, and diameter of NW. Therefore, proper choice of catalyst is critical. In the case of growing NW, VLS process is initiated by the formation of liquid alloy droplet which contains both catalyst and source material. Precipitation occurs when the liquid droplet becomes supersaturated with the source material. Under the flow of oxygen, NWs crystals are formed and grow outward from the nanoclusters (163). Simply turning off the source can adjust the final length of

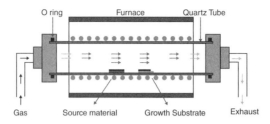

Figure 16.21   A schematic illustration of a thermal furnace synthesis system that is used in vapor phase growth methods including CVD, thermal evaporation, and PLD. (*Source*: Reprinted with permission from Reference 151, Copyright 2006, Elsevier B.V).

(a)   (b)

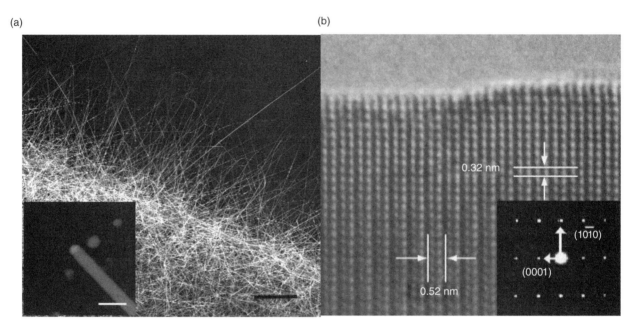

Figure 16.22   (a) An SEM image shows the as-grown ZnO NWs Inset: a ZnO NW with diameter of 50 nm terminated with an Au nanoparticle. (scale bar is 100 nm) (b) High-resolution TEM image of a ZnO NW with lattice spacing 0.52 nm. The inset SAED indicates that the growth direction is [0001] along the c axis. (*Source*: Reprinted with permission from Reference 176, Copyright 2004, American Institute of Physics).

the NWs. Normally the resulting crystal is grown along one particular crystallographic orientation which corresponds to the minimum atomic stacking energy, leading to NW formation. This type of growth is epitaxial, resulting in high crystalline quality. Wu et al. have provided direct evidence of VLS growth by means of real-time in situ transmission electron microscope observations (164). This work describes a vivid dynamic insight and elucidates the understanding of such microscopic chemical process.

A majority of oxide NWs has been synthesized via this catalyst-assisted mechanism, such as ZnO (165), MgO (166), CdO (167), TiO_2 (168), SnO_2 (169), In_2O_3 (170), and Ga_2O_3 (171). Several approaches have been developed based on the VLS mechanism. As an example, thermal CVD synthesis process utilizes a thermal furnace to vaporize the metal source, then proper amount of oxygen gas is introduced through mass flow controller. In fact, metal and oxygen vapor can be supplied via different ways, such as carbothermal or hydrogen reduction of metal oxide source material (172, 173) and flowing water vapor instead of oxygen (174, 175). For example, Figure 16.22a shows the scanning electron micrographs of ZnO NWs synthesized by directly evaporating pure Zn powder at 700 °C in low oxygen concentration (<2%) using Au nanoparticles as catalyst. Figure 16.22b is a HRTEM image and the inset selected-area electron diffraction (SAED) pattern indicates single crystalline growth along [0001] direction. The growth direction determined in Q1D system is usually along the crystal plane which is the most stable and owns the lowest surface free energy.

Figure 16.23 illustrates another example for Si and Ge NW growth by VLS method. The schematic in Figure 16.23a shows silane molecules reacting at the Au surface through a hydrogen decomposition reaction sequence that introduces atomic Si into the Au nanodot (Fig. 16.23b). The temperature is held near the eutectic temperature of 363 °C and as the composition approaches the eutectic composition of 19 at% Si (Fig. 16.23c) melting occurs. Upon further addition of Si, the Si concentration in the liquid Au moves to the right of the liquidus line and the supersaturation becomes sufficient for crystallization of Si at the liquid–solid interface. The liquid drop size defines the liquid–solid interface and thus the area available for Si crystal growth. This maintains the NW growth at a fixed diameter as long as Au is not lost from the nanodrop.

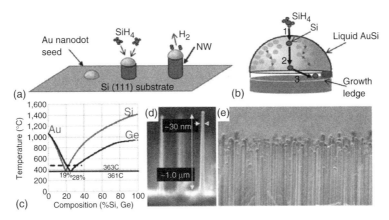

Figure 16.23 Vapor–liquid–solid (VLS) growth mechanism for NW synthesis. (a) Schematic of VLS Si NW growth from a liquid Au catalyst seed which floats on top of the NW as it grows and defines the growth diameter. (b) Enlarged view illustrating the three kinetic steps for NW growth: (i) silane decomposition at the vapor–liquid interface, (ii) Si atom diffusion through the AuSi liquid, and (iii) NW crystallization by Si incorporation into a step at the growing liquid–solid interface of the NW. (c) Binary phase diagrams for Au–Si and Au–Ge overlaid to illustrate their similar eutectic melt properties; dashed line indicates a typical growth temperature. (d and e) SEM images of Si0.1Ge0.9 alloy NWs grown epitaxially from a Si(111) surface. (*Source*: Reprinted with permission from Reference 177, Copyright 2010, Springer).

Since Si and Ge exhibit similar eutectic behavior with Au (see Fig. 16.23c), Si, Ge, and SiGe alloy NWs can all be readily grown by this same approach. Figure 16.23d illustrates a $Si_{0.1}Ge_{0.9}$ alloy NW. In general, Ge NW growth is carried out at ~275 to 375 °C and Si at 450–550 °C, since Ge precursors are more easily decomposed than Si precursors in the VLS process due to the greater reactivity of germane compared to silane. An important advantage of low growth temperatures is that conventional vapor–solid (VS) CVD growth on the NW sidewalls and Si substrate is minimized during NW VLS growth. Simultaneous sidewall growth reduces the maximum aspect ratios that can be achieved and results in a linear tapering of the NWs. Tapering can also occur as a result of the loss of Au atoms from the NW tip if there is significant Au surface diffusion along the Si sidewall during growth, however this usually occurs only at ultrahigh vacuums and low pressures and is not a consideration for usual CVD reactor growth conditions.

It is interesting to observe that several dynamical process that can affect growth. Adsorption occurs from the fluid (whether gaseous, liquid, or supercritical) phase. Adsorption might be molecular or dissociative and may either occur (vii) on the NW (viii) on the particle, or (ix) on the substrate. A natural way for the catalytic particle to direct material to the growth interface is if the sticking coefficient (probability of adsorption) is higher on the particle and vanishingly small elsewhere. Diffusion of adatoms will occur (i) across the substrate (if the sticking probability is not negligible), (ii) across the particle, and (iii) along the sidewalls. Diffusion across the substrate and along the sidewalls must be rapid and cannot lead to nucleation events. Nucleation of the NW anywhere other than on the particle must be suppressed so that growth only occurs at the particle/NW interface and so that sidewalls do not grow independently of the axial growth. There may be (vi) diffusion of material through the catalytic particle in addition to (ii) diffusion along its surface. Substrate atoms might also be mobile. They might (v) enter the particle directly or else (iv) surface diffusion along the substrate can deliver them to the surface of the particle. Not shown in the diagram is that atoms from the catalytic particle might also be mobile and diffuse along the sidewalls and the substrate.

Four major distinctions in the growth process are illustrated in Figure 16.24: root versus float growth and multiprong versus single-prong growth. The particle may either end up at the bottom (root growth) or top (float growth) of the NW. In multiprong growth, as in Figure 16.24c, more than one NW grows from a single particle. In this case, the radius of the NW r_w must be less than the radius of the catalytic particle r_p. In single-prong growth, there is a one-to-one correspondence between particles and NWs. A natural means to exercise control over the NW diameter in single-prong growth would be if the NW radius determines this value and $r_w \approx r_p$. Here it should be mentioned that in single-prong growth, $r_w \approx r_p$ is usually observed but that the catalyst particle sometimes is significantly larger and occasionally is somewhat smaller than the NW radius. In multiprong growth, r_w is not determined directly by r_p but must be related to other structural factors such as the curvature of the growth interface and lattice matching between the catalytic particle and the NW.

16.3.1.1.3 Vapor–Solid–Solid Growth A majority of efforts have employed the VSS approach at growth temperatures above the metal–semiconductor eutectic. Sub-eutectic vapor–solid–solid (VSS) growth has received less attention but also possible and may provide advantages including reduced processing temperatures and more abrupt heterojunctions. VSS growth has been reported for $TiSi_2$-catalyzed Si NWs, SiO_x-catalyzed InAs and Au-catalyzed InAs and GaAs NWs. In general, the mechanism of VSS and VLS is almost the same, as shown in Figure 16.25, except that (i) the phase of the catalyst is solid, (ii) the growth temperature can be below the bulk eutectic point for VSS growth, and (iii) therefore the growth kinetics and the requirements for catalyst stability are different.

Figure 16.26 shows the schematic of the difference in NW VSS and VLS growth, when the melting point suppression occurs as a result of the catalyst supersaturation, leading to VLS growth well below the eutectic temperature. If sufficient supersaturation is not maintained, the

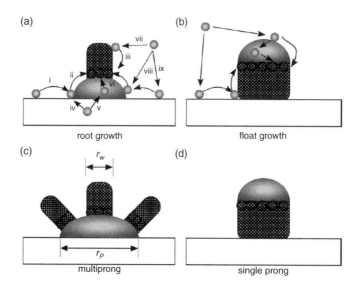

Figure 16.24   The processes that occur during catalytic growth. (a) In root growth, the particle stays at the bottom of the NW. (b) In float growth, the particle remains at the top of the NW. (c) In multiple prong growth, more than one NW grows from one particle and the NWs must necessarily have a smaller radius than the particle. (d) In single-prong growth, one NW corresponds to one particle. One of the surest signs of this mode is that the particle and NW have very similar radii. (*Source*: Reprinted with permission from Reference 178, Copyright 2006, Elsevier).

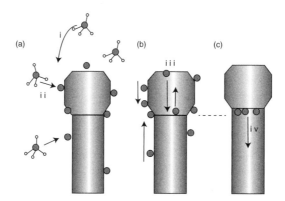

Figure 16.25   Schematic of VSS NW growth. (a) Delivery of the precursor to the NW surface (i) and decomposition of this precursor (ii). (b) Atoms are delivered to the metal catalyst/NW interface by surface or bulk diffusion processes (iii). (c) Si or Ge atoms are incorporated at the growth interface (metal catalyst/NW interface) leading to anisotropic growth (iv). (*Source*: Reprinted with permission from Reference 179, Copyright 2009, Royal Society of Chemistry).

particle solidifies and VSS growth proceeds. In similar, both VLS and VSS growth, the interface between the semiconductor and the catalyst provides a preferential interface for the incorporation of adatoms into the growing crystal.

16.3.1.2   Solution Phase Growth   Solution-phase synthesis refers to techniques that grow NWs in solution. The major advantages of the solution-based technique (in aqueous or nonhydrolytic media) for synthesizing anomaterials are high yield, low cost, and easy fabrication. The solution-based technique has been demonstrated as a promising alternative approach for mass production of metal, semiconductor, and oxide nanomaterias with excellent controls of the shape and composition with high reproducibility. In particular, this technique is able to assemble nanocrystals with other functional materials to form hybrid nanostructures with multiple functions with great potential for applications in nanoelectronic and biological systems. To develop strategies that can guide and confine the growth direction to form NWs, researchers have used various approaches which could be grouped into template-assisted method and template-free method (151). In addition, solution-based techniques and electrospinning are also classified in this category.

16.3.1.2.1   Template-Assisted Synthesis   Large-area patterning of metal oxide NWs arrays assisted by template have been achieved (180). By using periodic structured template, for example, anodic aluminum oxide (AAO), molecular sieves, and polymer membranes, nanostructures can grow inside the confined channels. For instance, AAO membranes have embedded hexagonally ordered nanochannels. They are prepared by the anodization of pure aluminum in acidic solution (181). These pores can be filled to form vertical NWs arrays through electrodeposition

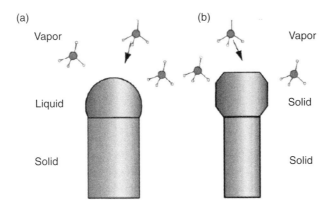

Figure 16.26 Schematic depicting the relevant phases in (a) the vapor–liquid–solid growth mechanism and (b) the vapor–solid–solid growth mechanism. (*Source*: Reprinted with permission from Reference 179, Copyright 2009, Royal Society of Chemistry).

or sol–gel deposition methods. Because the diameter of these nanochannels and the interchannel distance are easily controlled by the anodization voltage, it provides a convenient way to manipulate the aspect ratio and the area density of NWs.

ELECTROCHEMICAL DEPOSITION Electrochemical deposition has been widely used to fabricate metallic NWs in porous structures. It was found that it is also a convenient method to synthesize metal oxide nanostructures. Actually, there are both direct and indirect approaches to fabricate metal oxides NWs using electrodeposition. For the direct method, by carefully choosing the electrolyte, ZnO (182), Fe_2O_3 (183), Cu_2O (184), and NiO (185) NWs have been successfully synthesized; for an indirect approach, Chen et al. (186) deposited tin metal into AAO and then thermally annealed it for 10 h to obtain SnO_2 NWs embedded in the template. In addition, ZnO NWs had also been obtained by this method (187).

SOL–GEL DEPOSITION In general, sol–gel process is associated with a gel composed of sol particles. As the first step, colloidal (sol) suspension of the desired particles is prepared from the solution of precursor molecules. An AAO template is then immersed into the sol suspension, so that the sol will aggregate on the AAO template surface. With an appropriate deposition time, sol particles can fill into the channels and form structures with high aspect ratio. The final product will be obtained after a thermal treatment to remove the gel. Sol–gel method has been utilized to obtain ZnO (188) by soaking AAO into zinc nitrate solution mixed with urea and kept at 80 °C for 24–48 h followed by thermal heating. MnO_2 (189), ZrO_2 (190), TiO_2 (191), and a number of multicompound oxide nanorods (NRs) (192, 193) had been synthesized based on similar processes.

16.3.1.2.2 Template-Free Methods Instead of plating nanomaterials inside a template, much research effort is triggered to develop new techniques to direct NW's growth in liquid environment. Several methods are described in the following, including surfactant method, sonochemistry, hydrothermal, solution-based techniques, and electrospinning.

SURFACTANT-ASSISTED GROWTH Surfactant-promoted anisotropic NW crystal growth has been considered as a convenient way to synthesize oxide NWs. This anisotropic growth is often carried out in a microemulsion system composed of three phases: oil phase, surfactant phase, and aqueous phase. In the emulsion system, these surfactants serve as microreactors to confine the crystal growth. To acquire desired materials, one needs to prudently select the species of precursor and surfactants, and also set the other parameters such as temperature, pH value, and concentration of the reactants. As a result, surfactant-assisted system is a trial-and-error based procedure which requires much endeavor to choose proper capping agents and reaction environment. By using this process, Xu et al. had synthesized ZnO (194), SnO_2 (195), and NiO (196) nanorods. Reports on lead oxide (PbO_2) (197), chromate ($PbCrO_4$, $CuCrO_4$, $BaCrO_4$) (198), and cerium oxide (CeO_2) (199) nanorods have also been published.

SONOCHEMICAL METHOD Sonochemical method uses ultrasonic wave to acoustically agitate or alter the reaction environment, thus modifies the crystal growth. The sonication process is based on the acoustic cavitation phenomenon which involves the formation, growth, and collapse of many bubbles in the aqueous solution (200). Extreme reaction conditions can be created at localized spots. Assisted by the extreme conditions, for instance, at temperature greater than 5000 K, pressure larger than 500 atm, and cooling rate higher than 10^{10} K/s, nanostructures of metal oxides can be formed via chemical reactions. Kumar et al. have synthesized magnetite (Fe_3O_4) nanorods in early days by ultrasonically irradiating aqueous iron acetate in the presence of beta-cyclodextrin which serves as a size-stabilizer (201). Hu et al. later demonstrated that linked ZnO rods can be fabricated by ultrasonic irradiation under ambient conditions and assisted by microwave heating (202). Recently, nanocomposite materials have been grown by applying this technique; Gao et al. synthesized and characterized ZnO nanorod/CdS nanoparticle (core/shell) composites (203). Rare earth metal oxides nanomaterials, such as europium oxide (Eu_2O_3) nanorods (204) and cerium oxide (CeO_2) nanotubes (205), have also been obtained via this method.

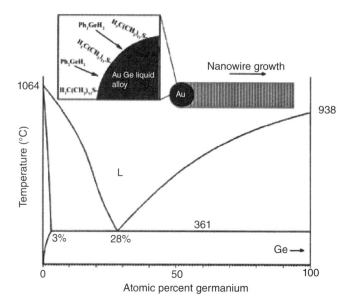

Figure 16.27 Schematic of NW growth. The organometallic precursor thermolytically degrades in the supercrital reaction environment. Above the eutectic temperature, the Ge atoms dissolve in the Au nanocrystals to form a liquid AuGe seed droplet. The seed droplet reaches the eutectic composition, Au0.72Ge0.28, and the continued supply of Ge results in the production of Ge NWs. (*Source*: Reprinted with permission from Reference 219, Copyright 2003, WILEY-VCH).

HYDROTHERMAL Hydrothermal process has been carried out to produce crystalline structures since the 1970s. This process begins with aqueous mixture of soluble metal salt (metal and/or metal–organic) of the precursor materials. Usually the mixed solution is placed in an autoclave under elevated temperature and relatively high pressure conditions. Typically, the temperature ranges between 100 °C and 300 °C and the pressure exceeds 1 atm. Many works have been reported to synthesize ZnO nanorods by using wet-chemical hydrothermal approaches (206–208). Through this technique, other nanomaterials have also been produced, such as CuO (209), cadmium orthosilicate (210), Ga_2O_3 (211), MnO_2 nanotubes (212), perovskite manganites (Fe_3O_4) (213), CeO2 (214), TiO_2 (215), and In_2O_3 (216).

SOLUTION-BASED TECHNIQUES Korgel et al. proposed that NW growth can not only take place in gaseous environments, but also in liquid media (217). The supercritical fluid–liquid–solid growth method (217, 218) can be used to synthesize semiconductor NWs, for example, Si and Ge. By using metal nanocrystals as seeds (219), Si and Ge organometallic precursors are fed into a reactor filled with a supercritical organic solvent, such as toluene. Thermolysis results in degradation of the precursor, allowing release of Si or Ge, and dissolution into the metal nanocrystals. As more of the semiconductor solute is added from the supercritical phase (due to a concentration gradient), a solid crystallite precipitates, and a NW grows uniaxially from the nanocrystal seed. One method utilizes highly pressurized supercritical organic fluids enriched with a liquid silicon precursor, such as diphenylsilane and metal catalyst particles, as indicated in Figure 16.27. At reaction temperatures above the metal–silicon eutectic, the silicon precursor decomposes and silicon forms an alloy with gold. Analogously to the VLS mechanism, the alloy droplet in this supercriticalfluid–liquid–solid (SFLS) method starts to precipitate a silicon NW once the alloy gets supersaturated with silicon (217, 219, 220). Crystalline NWs with diameters as low as 5 nm and several micrometers in length have been fabricated using this approach. Similar to the VSS mechanism, silicon-NW growth via a solid catalyst particle has also been demonstrated for the solution-based method. Micrometer-long NWs were synthesized at a temperature of merely 500 °C using copper particles as catalysts (221). Another high-yield silicon-NW production method is the so-called solution–liquid–solid (SLS) method. Here, the growth environment is not a supercritical liquid, but an organic solvent at atmospheric pressure, and the production of micrometer-long crystalline wires, 25 nm in diameter, has been demonstrated (222). The SLS method probably represents the most cost-effective NW-production method, as it can be realized without high-priced equipment.

ELECTROSPINNING Electrospinning is a simple and versatile technique for fabricating uniform ultrafine fibers with diameters ranging from several micrometres down to a few nanometres. During electrospinning, a thin charged jet is formed when the electrostatic force generated by a high operating voltage overcomes the surface tension of the polymer droplet. The jet is accelerated toward the grounded collector and produces ultrathin fibers in the form of nonwoven mats (223, 224). In the past decade, besides polymeric materials, inorganic fibers (225, 226) also can be fabricated by electrospinning and calcinating. The electrospun fibers have been developed as a good potential candidate in many fields, such as tissue engineering, drug release, nanosensors, energy applications, biochips, and catalyst supports (226). However, the random orientation of fibrous mats fabricated by the conventional electrospinning may limit the potential applications of electrospun fibers, especially in the fields of electronics, photonics, PVs, and actuators, which need direct, fast charge transfer or regular, uniform structures. To solve this problem, a variety of strategies have been proposed by many research groups, such as pair electrodes collection (Fig. 16.28),

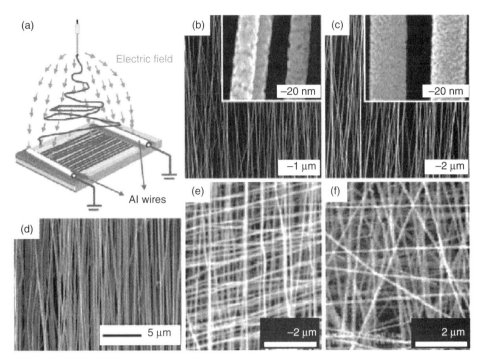

Figure 16.28   (a) Schematic illustration of the gapping method setup for electrospinning. The collector contains two conductive Al wires separated by a gap. (Reprinted from Reference 227, Copyright 2009, with permission from the American Chemical Society.) (b–d) SEM images of uniaxially aligned semiconductor nanofibers made of (b) TiO2, (c) Sb-doped SnO_2, (Reprinted from Reference 228, Copyright 2003, with permission from the American Chemical Society) and (d) $BaTiO_3$. The insets show enlarged SEM images of these semiconductor fibers. (*Source*: Reprinted from Reference 228, Copyright 2006, with permission from Elsevier Ltd). (e and f) SEM images of cross-aligned nanofibers network structures made of (e) TiO_2 and (f) $Li_4Ti_5O_{12}$.

rotating drum or disc collection, auxiliary electric or magnetic electrospinning, double spinning, electroconductive template collection, and so on.

Near-field electrospinning (229–234) is another powerful way to fabricate orderly nanofibres on flat substrates and realize patterned deposition of nanofibers. In near-field electrospinning, because of the short electrode-to-collector distance (500 μm–10 mm), bending instability and splitting of the charged jet in electrospinning are avoided, so a straight-line jet between the spinneret and the collector can be utilized to direct-write an orderly nanofiber (Fig. 16.29a). Particularly, by putting the collector on an X–Y motion stage, the movement of the collector can be controlled in the preprogrammed track via a host computer. When the collector speed matches the ejection speed of the jet stream, a mechanical drag force is generated on the suspended nanofiber from the collector and straight-line nanofibers without helical structure can be obtained. And various patterns can be constructed by adjusting the collector speed and track. For example, Chang et al. (231) have deposited solid nanofibres with orderly patterns over large areas via continuous near-field electrospinning, various complex patterns of polymer fibers have been fabricated (Fig. 16.29b–e). Also, Chang et al. (233) have demonstrated piezoelectric poly(vinylidene fluoride) nanogenerators directly written onto a flexible plastic substrate using near-field electrospinning. The energy conversion efficiency is found to be as high as 21.8% with an average of 12.5%, which is much greater than that of typical power generators made from poly(vinylidene fluoride) thin films (0.5–4%). And it could be the basis for an integrated power source in nanodevices and wireless sensors or new self-powered textile by direct-writing nanofibers onto a large area cloth to boost the total power output for portable electronics. Moreover, Rinaldi et al. (236) have used the nearfield electrospinning to grow well-aligned TiO_2 nanofibers with a diameter of about 200–400 nm on a planar silicon dioxide substrate (Fig. 16.29f–g). The SEM image shows the presence of microcrystallites, which can increase the gas or tension sensing sensitivity. Compared with other electrospinning techniques, the advantage of this technique is a precise control of the deposition location and pattern of electrospun nanofibers. The weakness is that the assembly efficiency is relatively low.

16.3.2   Nanowire-Based Electronic and Optoelectronic Devices

In this section, the structure and properties of NWs and their interrelationship will be discussed. The discovery and investigation of nanostructures were spurred on by advances in various characterization and microscopy techniques that enable materials characterization to take place at smaller and smaller length scales, reaching length scales down to individual atoms. For applications, characterization of the NW structural properties is especially important so that a reproducible relationship between their desired functionality and their geometrical and structural characteristics can be established. Due to the enhanced surface to volume ratio in NWs, their properties may depend sensitively on their surface condition and geometrical configuration. Even NWs made of the same material may possess dissimilar properties due to

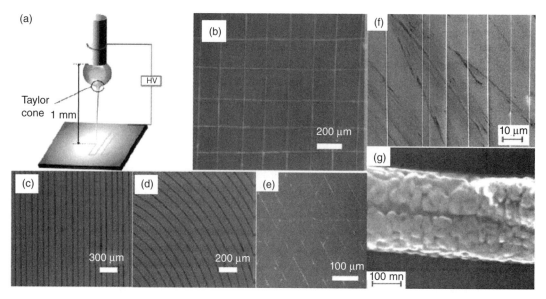

Figure 16.29 (a) A schematic of low-voltage near-field electrospinning showing the solution droplet, Taylor cone, and the jet stretched by the electric field and patterned onto a substrate, (Reprinted with permission from Reference 230, Copyright 2011, American Chemical Society.) (b) Optical image of a grid pattern with controlled 200 mm spacing. (c) Parallel fibers with controlled 100 mm spacing. (d) Arc pattern with controlled 100 mm spacing, (Reprinted with permission from Reference 235, Copyright 2010, Elsevier Ltd.) (e) Optical image of a triangular pattern, (Reprinted with permission from Reference 231, Copyright 2008, American Institute of Physics.) (f) and (g) SEM images of aligned TiO2 nanofibers at low (f) and high (g) magnifications. (*Source*: Reprinted with permission from Reference 236, Copyright 2007, Elsevier Ltd).

differences in their crystal phase, crystalline size, surface conditions, and aspect ratios, which depend on the synthesis methods and conditions used in their preparation.

16.3.2.1 *Nanowire-Based Field-Effect Transistors (FETs)*

16.3.2.1.1 Single Nanowire FETs and Electrical Properties NWs have been fabricated into FETs to serve as the fundamental building blocks of electronic devices such as logic gate, computing circuits, photodetectors, chemical sensors, and so on. Various metal oxides including ZnO (176), Fe_2O_3 (237), In_2O_3 (238), SnO_2 (153), Ga_2O_3 (171), V_2O_5 (171), and CdO (167) have been configured to FET. In brief, the fabrication process can be described as following. NWs are first dispersed in a solvent, usually isopropanol alcohol or ethanol to form a suspension phase, and then deposited onto a SiO_2/Si substrate. The bottom substrate underneath the SiO_2 layer is degenerately doped (p++ or n++), serving as the back gate. Photolithography or electron-beam lithography is utilized to define the contact electrode pattern. Assuming a cylindrical wire of radius r and length L, the capacitance per unit length with respect to the back gate may be simply represented as

$$\frac{C}{L} = \frac{2\pi\varepsilon\varepsilon_0}{\ln(2h/r)} \tag{16.1}$$

where ε is the dielectric constant of the gate oxide and h is the thickness of the oxide layer. From a well-defined transfer characteristics, one can estimate the Q1D carrier concentration and mobility using two simple relations:

Carrier concentration:

$$n = \frac{V_g(th)}{e} \times \frac{C}{L} \tag{16.2}$$

Carrier mobility:

$$\mu_e = \frac{dI}{dV_g} \times \frac{L^2}{CV_{ds}} \tag{16.3}$$

where V_g (th) is the gate threshold voltage at which the carriers in the channel are completely depleted, dI/dV_g denotes the transconductance.

Zinc oxide NWs have been intensively studied and fabricated into FETs for electrical transport measurements. Figure 16.30a shows the drain–source I_{DS}–V_{DS} characteristics at different gate bias of a ZnO FET contacted by Ti/Au electrodes, exhibiting high conductance, excellent gate dependence, and high on/off ratio. It is worth to note that the CVD-grown ZnO nanostructures are single crystalline, rendering them with superior electrical property to polycrystalline thin film. For instance, an electron field-effect mobility of 7 cm²/Vs is considered high for ZnO thin film transistors (240). However, Park et al. have reported an electron mobility of 1000 cm²/Vs after coating the NWs with polyimide passivation layer to reduce the electron scattering and trapping at surface (241). Lately, it has been discovered that after coating the

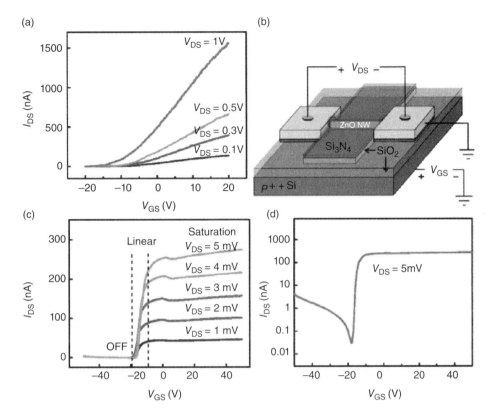

Figure 16.30 Transport measurements present (a) I_{DS}–V_{GS} curves of a ZnO NW FET without surface treatments showing typical n-type semiconducting behavior. (b) Schematic of surface passivated ZnO NW FET with SiO_2/Si_3N_4 bilayer covering the NW channel. (c) I_{DS}–V_{GS} of a surface treated NW FET exhibits significantly enhanced on–off ratio and transconductance. (d) Semilog plot demonstrates a 10-fold reduction in the subthreshold swing. At large negative gate voltages, band bending gives rise to hole conduction. (*Source*: Reprinted permission from Reference 239, Copyright 2006, American Institute of Physics).

ZnO NW with a layer of SiO_2 followed by Si_3N_4 to passivate the surface states, field-effect mobility is dramatically improved and exceeds 4000 cm²/Vs (as described in Fig. 16.30). These results indicate that the ZnO NWs device has exceptional potential in high-speed electronics application (41).

In addition, C.M. Lieber et al. (242) used homogeneous doped NWs FETs as a prototypical example to represent key building blocks for a variety of electronic devices and, moreover, studies of FETs enable the evaluation of the typical performance of NWs devices as compared with their corresponding planar complements. The representative NW materials including Si, Ge, and GaN have been shown to be prepared with complementary n- and p-type doping. For example, studies of NWFETs fabricated from boron- or phosphine-doped Si NWs have depicted that the devices are turned on when the gate voltage becomes more negative/positive, which is the characteristic of p-/n- channel FETs (Fig. 16.31a and b). Notably, analysis of these results has illustrated that the doped Si, Ge, and GaN NWFETs can exhibit performance comparable to the best values prepared for their planar counterparts. Studies have also shown the high electron mobility of InAs NWFETs with a wrap-all-around gate structure. This finding is noteworthy since the NWFETs are fabricated using nonconventional methods (e.g., solution assembly), which opens up opportunities in the areas not possible with traditional single-crystal wafer-based electronics.

The high performance of these homogeneous NW devices has been further confirmed by low-temperature measurements. For example, proximity-induced superconductivity has been realized in InAs NWs contacted with Al-based superconductor electrodes (246). All these results indicate Schottky barrier-free contacts are formed between NWs and metals, suggesting the phase-coherence length for electron propagation in these NWs up to hundreds of nanometers. C.M. Lieber et al.have also shown that molecular-scale Si NW devices (247) configured as single-electron transistors display single period Coulomb blockade oscillations (Fig. 16.31c) and coherent transport through single NW "islands" for lengths up to 400 nm. This result demonstrates that Si NWs are expected as a clean system without much structural/dopant variation on this length scale. In contrast, lithographically defined Si NWs have more significant structural and/or dopant fluctuations and thus yield a reduced length scale (e.g., at least one magnitude smaller) for electronically distinct regions (248). Notably, coherent transport has been observed in molecular scale Si NWs down to the last few charges (Fig. 16.31d) (247), which further depicts the high quality of the NW material, long carrier mean free-paths, and the potential to serve as a unique building block for both low- and room-temperature applications.

16.3.2.1.2 Crossed Nanowire Structured FETs NW building blocks as well as more complex and complicated NW device architectures can open up new opportunities apart from conventional paradigms. Specifically, the crossed NW architecture reported in 2001 (248, 245) is a

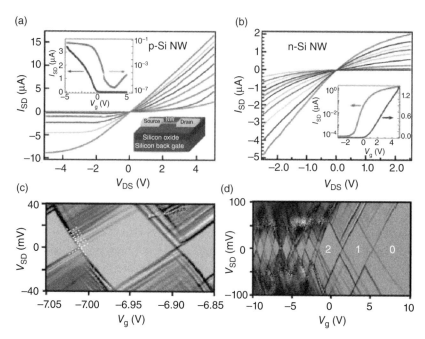

Figure 16.31 Si NWFETs: family of source-to-drain current versus source-to-drain voltage (I_{ds}–V_{ds}) plots for a representative (a) 20 nm p-Si NW array device (channel length of 1 μm; from red to pink, V_g = −5 V to 3 V); and (b) 20 nm n-Si NW device (channel length of 2 μm; from yellow to red, V_g = −5 V to 5 V) in a standard back-gated NWFET geometry as depicted. Insets in (a) and (b) are source-to-drain current versus gate-voltage (I_{ds}–V_g) curves recorded for NWFETs plotted on linear (blue) and log (red) scales at V_{ds} = −1 V and 1 V, respectively. (c) dI/dV_{sd}-V_{sd}-V_g data recorded at 1.5 K on a 3 nm diameter Si NW device in the Coulomb blockade regime with 100 nm channel length. Dark lines (peaks in dI/dV_{sd}) running parallel to the edges of the diamonds correspond to individual excited states and are highlighted by white dashed lines. (d) dI/dV_{sd}–V_{sd}–V_g data for a Si NW device with diameter of 3 nm and channel length of ~50 nm at 4.2 K. The carriers are completely depleted for V_g > 5.5 V, below which carriers are added consecutively to the dot and the first three carriers are labeled as 0, 1, and 2, respectively. (*Source*: Reprinted with permission from References 243–245. 2004 Wiley-VCH and 2004, 2005 American Chemical Society, respectively).

clear illustrative example as the key device properties are typically defined by the assembly of two NW components instead of by lithography. This way, the dimensions of the crossed NW device are only limited by the NW diameters, which lead the architecture readily scalable for high integration density and, depending on the selection of NWs, the structure can give a variety of critical device components, including transistors and diodes (248, 245). For instance, crossed NWFETs can be configured from one NW as the active channel while the second crossed NW can be performed as the gate electrode separated by a thin SiO$_2$ dielectric shell on the Si NW surface, with the gate located on the surface of one or both of the crossed NWs (245). This concept was first introduced utilizing Si NWs as the channel and GaN NWs as the gate electrodes, including the integration of multiple crossed NWFETs on a single Si NW channel to illustrate both NOR logic-gate structures (Fig. 16.32a) and basic computation (245). What's more, C.M. Lieber et al. extended this idea of crossed NWFETs to further establish a general approach uniquely addressing a large array of NW devices. Discriminating chemical modification is also employed to differentiate specific cross points in a four-by-four crossed Si NWFET array (Fig. 16.32b), therefore enabling the selective addressing of the four individual outputs (Fig. 16.32c). More importantly, these results can deliver a proof-of-concept in which the assembled crossed NW arrays can be served as the basis for addressable integrated nanosystems with the signals restored at the nanoscale.

16.3.2.1.3 Axial Nanowire Heterostructure FETs The device integration at the nanoscale can also be performed during the NW synthesis simply by varying the composition and/or doping in the axial elongation, in which the resulting axial junctions can give controlled nanoscale device function without the need for lithography. A typical example is a GaAs/GaP compositionally modulated axial heterostructures (Fig. 16.33a) (250). Since GaAs is a direct bandgap semiconductor while GaP has an indirect bandgap, these NW heterostructures can be patterned synthetically and emit light as nanoscale barcodes. At the same time, p–n junctions formed within individual NWs can also be constructed in a similar way. Forward-biased n-InP/p-InP single NW devices can be performed as nanoscale LEDs with the intense light emission at the p–n interface as depicted in Figure 16.33b.

Moreover, C.M. Lieber et al. have also employed this key concept of the composition modulation to define functional devices in several other directions, such as electronic and optoelectronic applications. First, the selective transformation of Si NWs into metallic NiSi NWs and NiSi/Si NW heterostructures can be achieved by the thermal annealing of as-made Si NWs with Ni (Fig. 16.33c) (251). Notably, this method yielded the first demonstration of atomically sharp metal–semiconductor interfaces between single metallic (NiSi) and semiconductor (Si) NWs. In these heterostructures, Si NWFET source–drain contacts can now be defined by the metallic NiSi NW regions, which act as excellent ohmic contacts at room temperature (Fig. 16.33d), and therefore provide an integrated solution for nanoscale contacts and interconnects.

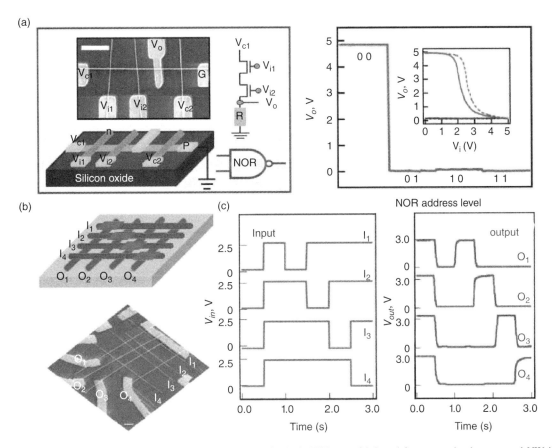

Figure 16.32 Crossed NW electronic FET devices. (a, left) Schematic of a logic NOR gate fabricated from a one-by-three crossed NW junction array employing one Si NW and three GaN NWs; insets show a representative scanning electron image of the device (scale bar, 1 μm) and symbolic electronic circuit. (a, right) Output voltage versus the four possible logic address level inputs; inset is the V_o–V_i relation, where the solid and dashed red (blue) lines correspond to V_o–V_{i1} and V_o–V_{i2} when the other input is 0 (1). (b) Schematic and scanning electron image of a four-by-four crossed Si NW array address decoder, with four horizontal NWs (I_1 to I_4) functioned as inputs and four vertical NWs (O_1 to O_4) taken as signal outputs. The four diagonal cross points in the array were chemically modified (green rectangles) to differentiate their responses from the input gate lines. Scale bar, 1 μm. (c) Real-time monitoring of the V_g inputs (blue) and signal outputs (red) for the four-by-four decoder. (*Source*: Reprinted with permission from References 245, 249. 2001 and 2003 American Association for the Advancement of Science, respectively).

The modulation of axial doping has also been shown for Si NWs (252), therefore providing another approach to introduce complex function at the initial stage of building block synthesis. For example, the pure axial growth of n^+-$(n$-$n^+)_N$ Si NWs with essential properties, such as the number, size, and period of the differentially doped regions, can be defined in a controllable manner during synthesis (Fig. 16.34a and b) (252). The synthetic modulation of dopant concentration can also be exploited for various types of nanoelectronic devices and circuits. Specifically, arrays of modulation doped NWs as depicted in Figure 16.34c can be employed to create address decoders (Fig. 16.34d). An important point of this method is that lithography is used entirely to define a regular array of microscale gate wires and is not required to form a specific address code in the nanoscale as reported previously (249). As a result, it offers the possibility to break lithography barriers in the ultra-dense arrays.

At the same time, the synthetic control of the size and separation of modulation-doped regions can be utilized to define quantum dot (QD) structures, in which the band offset induced by the variations in dopant concentration can generate potential barriers confining the QD (252). Modulation-doped Si NWs with the structural form of n^+-n_1-n_{QD}^+-n_2-n_{QD}^+-n_1-n^+ (Fig. 16.34e, left) exhibit a single Coulomb oscillation period consistent with two weakly coupled QDs when the barrier n_2 is large and, since this barrier is decreased (through synthesis), the tunneling conductance between QDs is then increased (Fig. 16.34e, right) (252). All these evidently illustrate the potency of encoding functional information into NWs during the synthesis, and this concept will be critical for defining unique electronic and optoelectronic device capabilities in NWs as compared with their lithographically patterned structures.

16.3.2.1.4   Radial Nanowire Heterostructure FETs   On the other hand, radial composition and doping modulation in the NW structures can facilitate another approach to further enhance the device performance as well as to enable new functionalities through the synthesis versus lithography. With the aim to push for the performance limits of NWFETs, a one-dimensional hole gas system based on an undoped epitaxial

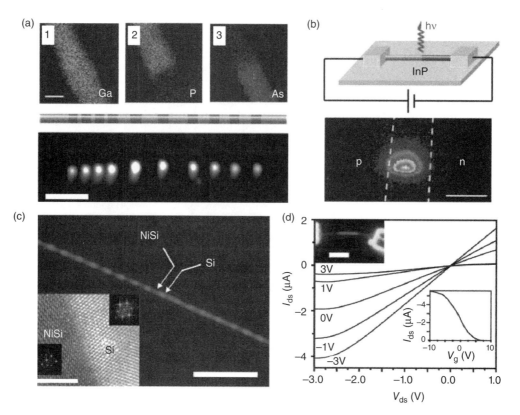

Figure 16.33   Axial NW heterostructures. (a, top 1–3) TEM elemental EDS mapping of a single GaAs/GaP NW heterojunction, illustrating the uniform spatial distribution of Ga (gray), P (red), and As (blue) at the junction. Scale bar, 20 nm. (a, bottom) Schematic and photoluminescence (PL) image of a 21-layer, (GaP/GaAs) 10 GaP, NW superlattice. The 10 bright regions correspond to GaAs (blue, direct bandgap) regions, while the dark segments are originated from the GaP (red, indirect bandgap) regions. (b) Schematic of a modulation-doped InP NW LED and the image of emission from the device. Dashed white lines represent the edges of the electrodes. Scale bar, 3 μm. (c) Dark-field optical image of a single NiSi/Si NW superlattice heterostructure. The bright green segments correspond to Si and the dark segments present the NiSi regions; scale bar is 10 μm. Inset shows a high-resolution TEM image of the atomically abrupt interface between the NiSi and Si; scale bar is 5 nm. (d) I_{ds}–V_{ds} curves of a NiSi/p-Si/NiSi heterojunction NWFET constructed employing a 30 nm diameter p-type Si NW; upper inset is a dark-field optical image of the same NW device describing that the contacts are made to the metallic NiSi regions only. Scale bar, 3 μm. Lower inset is the I_{ds}–V_g obtained with $V_{ds} = -3$ V. (*Source*: Reprinted with permission from References 250, 251. 2002 and 2004 Nature Publishing Group, respectively).

Ge/Si core/shell structure has been designed and experimentally realized (Fig. 16.35a) (253, 254). In this work, the valence band offset of ~500 meV between Ge and Si at the heterostructure interface can perform as a confinement potential for the quantum well. When the Fermi level lies below the valance band edge of the Ge core, free holes can gather in the Ge channel. Low-temperature electrical transport characterization have further illustrated the distinct conductance plateaus corresponding to transport through the first four subbands in the Ge/Si NW (Fig. 16.35b), in which the subband spacings (Fig. 16.35c), $\Delta E_{1,2} = 25$ mV and $\Delta E_{2,3} = 30$ mV, are in the good agreement with theoretical calculations (253). It is also noted that the conductance displays little temperature dependence here, consistent with the computations of minimized backscattering in this one-dimensional system, demonstrating that transport is ballistic even at room temperature.

In this regard, the unique transport properties of Ge/Si core/shell NW heterostructures make them the ideal building blocks for high-performance NWFETs and potential alternatives to planar metal-oxide-semiconductor field-effect transistors (MOSFETs). Importantly, Ge/Si NW devices can be demonstrated with scaled transconductance (3.3 mS/μm) and on-current (2.1 mA/μm) values, which are three to four times better than state-of-the-art MOSFETs and the highest ones obtained in NWFETs (Fig. 16.35d) (254). Another significant benchmark of transistor performance is the intrinsic delay, $\tau = CV/I$, in which C is the gate capacitance, V is the power supply voltage, and I is on-current. The data again illustrate an obvious speed advantage at a given channel length, L, for the Ge/Si NWFETs versus Si p-MOSFETs (Fig. 16.35e). Overall, these results confirm for the first time a true performance benefit of NWs, represent the best performance achieved to date in NWFET devices, and act as a benchmark for future development.

Moreover, the versatility of this band-structure engineering in creating NW carrier gases has been further investigated by the accomplishment of an electron gas in dopant-free GaN/AlN/AlGaN radial NW heterostructures (255). Achieving both hole and electron gases is essential since they are required to enable high-performance complementary nanoelectronics as well as to assess the fundamental properties of both one-dimensional electron and hole gases. In brief, the designed NW structure consists of an intrinsic GaN core and sequentially

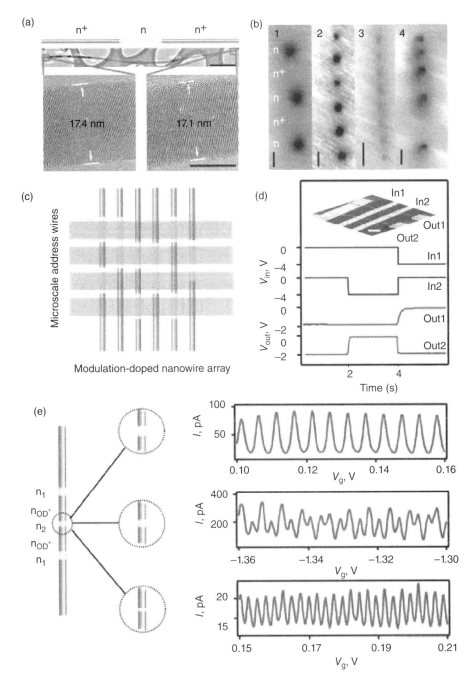

Figure 16.34 Modulation-doped Si NWs and their corresponding applications. (a, top) Schematic and low-resolution TEM image of an n^+–n–n^+ modulation-doped Si NW. Scale bar, 500 nm. (a, bottom) High-resolution TEM images recorded at the two ends of the NW showing the deficiency of radial coating; scale bar is 10 nm. (b) Scanning gate microscopy images (1–4) of n^+-$(n$-$n^+)_N$ NWs recorded with a tip voltage of −9 V and V_{sd} = 1 V. The dark regions denote the reduced conductance corresponding to lightly doped NW segments. Scale bars, 1 μm. (c) Schematic of lithography independent address decoder based on modulation-doped NW array, in which microscale address wires and modulation-doped NWs act as inputs and outputs, accordingly. (d) Plots of input (blue) and output (red) voltages for the two-by-two decoder configured employing two modulation-doped Si NWs as outputs (Out1 and Out2) and two Au metal lines deposited over a uniform Si_3N_4 dielectric as inputs (In1 and In2). (e, left) Schematics of a coupled double-QD structure in modulation-doped Si NW, in which the n^+ QD structure is confined by two barriers from the n-type regions. The width of n_2 region between the two n^+ QDs is adjustable. (e, right) I–V_g data recorded at 1.5 K on three double-QD NW devices with the n^2 sections grown for 15 s, 10 s, and 5 s (top to bottom) showing different coupling. (*Source*: Reprinted with permission from Reference 252. 2005 American Association for the Advancement of Science).

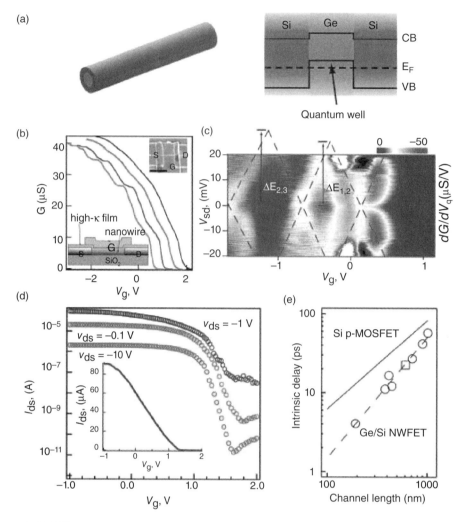

Figure 16.35 Ge/Si core–shell NWFETs. (a) Schematic illustration of an undoped Ge/Si core–shell NW and the corresponding band diagram displaying the formation of a hole gas in the Ge quantum well confined by the epitaxial Si shell, where CB denotes the conduction band and VB denotes the valence band. The dashed line indicates the Fermi level, E_F. (b) G-Vg recorded at different temperatures on a 400 nm long top-gated device; the red, blue, green, and black curves correspond to temperatures of 5, 10, 50, and 100 K, accordingly. Insets show a schematic and scanning electron image of a top-gated NWFET; scale bar is 500 nm. (c) Transconductance dG/dV_g as a function of V_{sd} and V_g. Dashed lines guide and indicate the evolution of conductance modes with V_{sd} and V_g. The vertical arrows highlight values of subband spacings $\Delta E_{1,2}$ and $\Delta E_{2,3}$, respectively. (d) I_{ds}–V_g data for a Ge/Si NWFET (190 nm channel length, 4 nm HfO$_2$ dielectric) with blue, red, and green data points corresponding to V_{ds} values of -1 V, -0.1 V, and -0.01 V, accordingly; inset gives the linear scale plot of I_{ds}–V_g measured at $V_{ds} = -1$ V. (e) Intrinsic delay, τ, versus channel length for seven different Ge/Si NW devices with HfO$_2$ dielectric (open circle) and ZrO$_2$ dielectric (open square). (*Source*: Reproduced with permission from References 253, 254, Copyright 2005, National Academy of Sciences USA and Nature Publishing Group, respectively).

deposited undoped AlN and AlGaN shells (Fig. 16.36a), in which the epitaxial AlN interlayer is employed to reduce the alloy scattering from the AlGaN outer shell and to enable a larger conduction band discontinuity for the enhanced electron confinement (255). More importantly, temperature-dependent transport results indeed confirm the accumulation of an electron gas in these undoped GaN/AlN/Al$_{0.25}$Ga$_{0.75}$N NWs, and give an intrinsic electron mobility of 3100 cm^2/Vs at room temperature which reaches 21,000 cm^2/Vs at 5 K (Fig. 16.36b). This enhanced mobility at low temperature is consistent with minimized phonon scattering of the electron gas. The room-temperature value is comparable to the value reported in planar GaN/AlGaN heterostructures, and substantially higher than the ones obtained in n-type GaN NWs. In addition, top-gated FETs fabricated with these NW radial heterostructures display the scaled transconductance (420 mS/μm) and subthreshold slope (68 mV/dec) values (Fig. 16.36c) which are significantly improved than the previously reported n-channel NWFETs. In any case, all these affirm to the functional and operational advantage of developing more complex building blocks in radial heterostructures.

16.3.2.1.5 Vertical Nanowire FETs   In order to increase the integration density of nanoscale devices and fully utilize the scaling advantage, intense efforts have been made to build vertical FETs. A vertically surround-gate NW FET was first fabricated by Ng et al. (256). In this

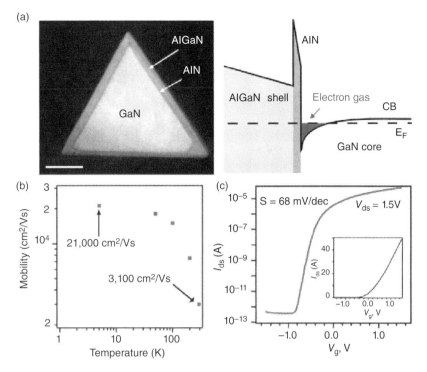

Figure 16.36 GaN/AlN/AlGaN radial heterostructured NWFETs. (a, left) Cross-sectional, high-angle annular dark-field scanning TEM image of a GaN/AlN/AlGaN radial NW heterostructure; scale bar is 50 nm. (a, right) Band diagram illustration of a dopant-free GaN/AlN/AlGaN NW depicting the formation of an electron gas (red region) at the core-shell interface confined by the epitaxial AlN/AlGaN shells. (b) Plot of the intrinsic electron mobility of a GaN/AlN/Al$_{0.25}$Ga$_{0.75}$N NWFET as a function of temperature, in which the values were obtained after the correction for contact resistance. (c) Logarithmic scale I_{ds}–V_g curve recorded at $V_{ds} = 1.5$ V, on a top-gated GaN/AlN/Al$_{0.25}$Ga$_{0.75}$N NWFET (channel length 1 µm, 6 nm ZrO$_2$ dielectric); inset display the linear scale plot of the same data. (*Source*: Reproduced with permission from Reference 255, Copyright 2006, American Chemical Society).

work, the positions of NWs were controlled via lithographic patterning technique. Vertical aligned ZnO NWs were observed to grow from lithographically patterned Au spots. These vertical NWs were then surrounded with SiO$_2$ and Cr which functions as the gate oxide and gate electrode, respectively, as illustrated in Figure 16.37a. Figure 16.37b shows the drain current versus absolute deviation of gate voltage (V_{GS}) from threshold voltage (V_{th}) for both n-channel and p-channel vertical surrounding gate VSG-FETs. The n-channel VSG-FET shows a linear dependence while the p-channel shows strong nonlinearity. This is because that in the n-channel, the variation of gate-induced charge involves essentially electrons that are mobile in the channel; whereas in the p-channel, the gate-induced charge involves both holes and ionized impurities in the depletion region, and the hole concentration governs conduction and increases with the gate deviation.

The same group has also demonstrated another type of vertical NW FET based on aligned In$_2$O$_3$ NWs (257). Instead of using conductive SiC substrate, direct electrical contact is made by a self-assembled underlying In$_2$O$_3$ buffer layer. This buffer layer was formed during the synthesis process right on top of the nonconductive sapphire substrate. This depletion mode n-type In$_2$O$_3$ NW vertical FET architecture uses a top-gate configuration which places the gate dielectric capping on the Pt electrode. These successes in fabricating vertical nano devices can lead to the integration of electronic and optoelectronic devices with high packing density, design flexibility, and function modularity.

16.3.2.2 Gas Sensors

It is well established that the electrical conductivity of conducting nanomaterials, such as semiconductor NWs (176, 258–261), metallic NWs (262–266), and conducting polymers NWs (267–269), with NW structure is affected by exposure to various organic and inorganic gases (270, 271). This has led to the investigation, by a lot of research groups, of the new type of nanostructures as building blocks for gas sensors. In this section, metallic and conducting polymers NWs gas sensors will not be discussed.

The advantage of using NWs for chemical sensing is manifold. With a large surface-to-volume ratio and a Debye length comparable to the NW radius, the electronic property of the NWs is strongly influenced by surface processes, yielding superior sensitivity than their thin film counterparts. In order to achieve maximum sensitivity, thin film gas sensors are often operated at elevated temperature (272, 273). This indicates that a single sensing device needs to incorporate temperature control unit, which will certainly increase the complexity of sensor design and power consumption. Fortunately, NW-based gas sensors have demonstrated significantly higher sensitivity at room temperature. For example, room temperature NO$_2$ sensing with ZnO NW shows more than 50% conductance change under an exposure of 0.6 ppm NO$_2$ (259); in contrast, NO$_2$ sensor made of doped ZnO thin film demonstrates less than 2% conductance change when exposed to an even higher concentration of NO$_2$ (1.5 ppm) (272). This encouraging result manifests the potential of building room temperature operating highly sensitive

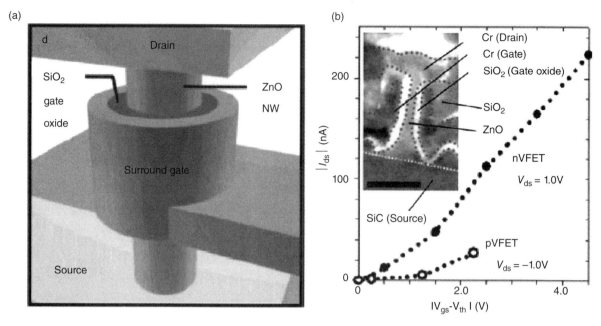

Figure 16.37 (a) A 3D schematic illustrating the critical components of a VSG-FET. (b) *I–V* characteristics for two n- and p-channel VSG-FETs. The inset shows a cross-sectional image of a VSG-FET with a channel length of about 200 nm. (*Source*: Reproduced with permission from Reference 256, Copyright 2004, American Chemical Society).

gas sensors. In addition, NWs configured as field-effect transistors have demonstrated that a transverse electric field can effectively tune the sensing behavior of the system (259, 274).

At a given temperature, the conductance of the semiconducting NWs is defined as

$$G = \frac{ne\mu\pi D^2}{4L} \tag{16.4}$$

where n is the initial carrier concentration, e the electronic charge, μ the mobility of the electrons, and D and L are the diameter and length of the NW channel, respectively. During gas sensing, the change in conductance (ΔG) of the semiconducting NWs will result from change in carrier concentration Δn_s according to (274, 275)

$$\Delta G = \frac{\Delta n_s e\mu\pi D^2}{4L} \tag{16.5}$$

Hence, the sensitivity of sensors can be defined as

$$\frac{\Delta G}{G} = \frac{\Delta n_s}{n_0} \tag{16.6}$$

The dependence of the sensitivity on the changes in carrier concentration Δn_s is linear (276), which indicates a more measurable change in carrier concentration could improve the sensing performance. Concerning selectivity and sensitivity, improvements could be obtained either by modifying the surface with catalyst particles or by modifying the intrinsic properties of semiconducting NWs by plasma treatment (275, 277).

The sensing mechanism of semiconductor is mainly governed by the fact that the oxygen vacancies on the oxide surfaces are electrically and chemically active. In this case, two kinds of sensing responses have been observed. (i) Upon adsorption of charge accepting molecules, such as NO_2 and O_2, at the vacancy sites, electrons are withdrawn and effectively depleted from the conduction band, leading to a reduction of conductivity. (ii) On the other hand, in an oxygen-rich environment, gas molecules such as CO and H_2 could react with the surface adsorbed oxygen and consequently release the captured electrons back to the channel, resulting in an increase in conductance. Conclusively, if one categorizes such redox sensing response into reducing and oxidizing, which manifests in an increase and a decrease in the channel conductance, the sensing responses can be represented using two examples:

$$\text{Reducing response}: \ CO + O \rightarrow CO_2 + e^-$$

$$\text{Oxidizing response}: NO + e^- \rightarrow NO^-$$

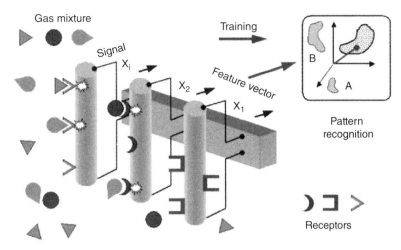

Figure 16.38 A proposed NW-based "electronic nose." The NW surfaces may be functionalized with molecule-selective receptors. The operation is based on molecular selective bonding, signal transduction, and chemical detection through complex pattern recognition. (*Source*: Reproduced with permission from Reference 274, Copyright 2004, Annual Reviews).

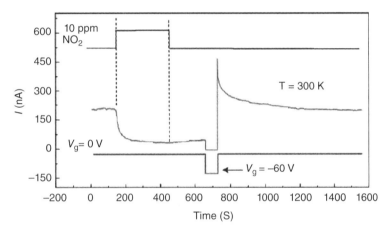

Figure 16.39 Response of ZnO NW FET exposed to 10 ppm NO_2 gas then return to pure Ar. After applying a gate voltage pulse with 60 s duration, the conductance is turned off followed by a current surge. In about 4 min, the conductance recovers to the initial level. (*Source*: Reproduced with permission from Reference 259, Copyright 2005, American Institute of Physics).

which results in the change of the electrical conductance of the semiconductors, and therefore the conductivity of the NWs, $\Delta G \sim \Delta n_{CO}$, increases monotonically with CO concentration.

Selectivity has always been a major hurdle for solid-state gas sensors. There have been a number of approaches developed to improve the selectivity of gas sensors, including doping metal impurities (278), using impedance measurement (279), modulating operating temperature (280), surface coating (124), and so on. In the frame work of studying metal oxide NW field-effect sensors, it is discovered that the transverse electrical field induced by back gate can be utilized to distinguish different chemical gases (259, 281). Ultimate goal is to develop an "electronic nose" system mimicking the mammalian olfactory system by assembling multicomponent sensing modules integrated with signal processing and pattern recognition functions. This concept is illustrated in Figure 16.38 (274).

Bulk and thin film metal oxides have been used for sensing gas species such as CO, CO_2, CH_4, C_2H_5OH, C_3H_8, H_2, H_2S, NH_3, NO, NO_2, O_2, O_3, SO_2, acetone, humidity, and so on. (282, 283). In the past few years, a surge of research effort focuses on the chemical sensing based on semiconductor NWs, such as zinc oxide (ZnO), tin oxide (SnO_2), indium oxide (In_2O_3), aluminum oxide (Al_2O_3), gallium oxide (Ga_2O_3), tungsten oxide (WO_3), vanadiumoxide (V_2O_5), Copper oxide (CuO), titanium oxide (TiO_2), and silicon (Si).

Oxygen (O_2), ozone (O_3), and nitrogen dioxide (NO_2) are oxidizing gases that were found to be easily detected by ZnO. Pearton et al. proposed ZnO nanorods ozone sensors (284). Because of the higher oxidizing ability of O_3, nanorods sensing response to ozone is pronounced in either N_2 or O_2 ambient. Fan et al. reported O_2 and NO_2 oxidizing sensing with field-effect transistors constructed of individual ZnO NWs (176, 259). The conductance of the NW FET decreases with the presence of O_2 and NO_2 (Fig. 16.39). In addition, it was observed that the sensitivity can be tuned by the back gate potential, that is, above the gate threshold voltage of FET, sensitivity increases with decreasing

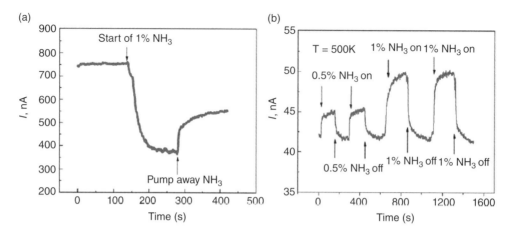

Figure 16.40   (a) ZnO NW conductance decreases in the presence of NH_3 at room temperature (b) Conductance increases when exposed to NH_3 at 500 K. (*Source*: Reproduced with permission from Reference 258, Copyright 2006, IEEE).

gate voltage. This suggests that the gate voltage could be used to adjust the sensitivity range. Moreover, a gate-refreshing mechanism was proposed (259). As demonstrated in Figure 16.39, the conductance of NW can be electrically recovered by applying a negative gate voltage much larger than the threshold. The applied gate voltage weakens the bond and facilitates the desorption of the absorbed gas molecules, thus recovering the conductance of the sensor to the original level. This establishes an efficient method to refresh sensors at room temperature, as the refreshing procedure is normally performed by using UV illumination or thermal heating. Furthermore, it was shown that from the gate-refresh voltage and the temporal response of the electrical conductance, different chemical adsorbants can be distinguished. These results provide an exciting prospect of a novel handheld sensor which utilizes the gate voltage as a knob to control sensitivity, refreshability as well as distinguishability.

Ammonia (NH_3) sensing shows temperature-dependent behavior. Figure 16.40a shows oxidizing sensing response of a single ZnO NW to NH_3 at room temperature, demonstrated by a conductance decrease. However, at 500 K, the conductance shows an increase upon exposure to NH_3 (Fig. 16.40b). These phenomena are attributed to the fact that the chemical potentials of the NW and the ammonia adsorbants lie closely to each other, and the temperature-induced chemical potential level shift results in the change of relative position of the chemical potentials, causing a reversal in the charge transfer direction (258). Similar phenomena have also been observed in tungsten oxide NH_3 sensing (285).

Carbon monoxide (CO) is known as a reducing gas. CO sensing was carried out at 500 K in synthetic air (20% oxygen). Admittance of CO into the test chamber immediately increased the NW conductance. This is caused by the reaction between CO and the surface adsorbed O_2, which releases the electrons withdrawn by O_2 adsorption back to the conduction channel (258).

Hydrogen (H_2) detection and storage are drawing increasing attention due to the demand of solid oxide fuel cell which uses H_2 as a source. Pioneering works have been done by Pearton et al. on hydrogen sensing by utilizing treated and untreated ZnO nanorods. Pt (286), Pd (287)-coated, and pristine ZnO nanorods have been examined, showing reducing response. It is worth to note that hydrogen introduces a shallow donor state in ZnO which may also contribute to the conductance increase. Furthermore, surface functionalized nanorods have shown higher H_2 sensitivity down to 10 ppm and faster response time than pristine rods at room temperature.

Hydrogen sulfide (H_2S) sensing by using hydrothermally prepared ZnO nanorods has been tested (288, 289). Because of the low H–SH bond energy, H_2S is easy to dissociate and can be readily detected at room temperature, showing a conductance increase. Wang et al. (289) has demonstrated the sensitivity down to 50 ppb of H_2S at room temperature.

Ethanol (C_2H_5OH, or EtOH) is a commonly used flammable chemical. Compared to H_2S, because of the moderately higher bond energies in C_2H_5OH, reducing sensing response is only pronounced at elevated temperature (289). Wan et al. fabricated ZnO NW ethanol sensors using microelectromechanical system (MEMS) technology (290). In this work, NWs were placed between Pt interdigitating electrodes. Under an operation temperature of 300 °C, the resistance of NWs significantly decreased upon exposure to ethanol vapor. In addition, humidity (H_2O) sensing has been tested on Cd-doped ZnO NWs (291). It is suggested that water molecules are dissociated at the oxygen vacancies, thus providing protons as charge carriers for the hopping transport in low-humidity environment. In contrast, when at high humidity, water condensed on the NW surface aids the electrolytic conduction of protonic transport and further increases the conductance.

A number of studies on In_2O_3 NW sensing of ethanol, NH_3, and NO_2 have also been reported. Ethanol sensor using In_2O_3 NWs prepared by carbonthermal synthesis was fabricated by Chu et al. (292). It shows an increase in channel conductance. On the other hand, Li et al. have investigated oxidizing gas sensing of NO_2 and NH_3 (293). In this work, detection of NO_2 down to 5 ppb levels at room temperature has been achieved by using a mesh of multiple-wires (261), as shown in Figure 16.41. As mentioned previously, NH_3 demonstrates oxidizing sensing at room temperature and reducing sensing at high temperature. Similar to this temperature-dependent change in the sensing response, Li et al. manipulated the carrier concentration and thus the chemical potential by carefully adjusting oxygen partial pressure during the CVD growth, and observed the reverse sensing behavior (294), that is, lower (higher) carrier concentration corresponding to lower (higher) chemical potential yielding reducing (oxidizing) sensing performance.

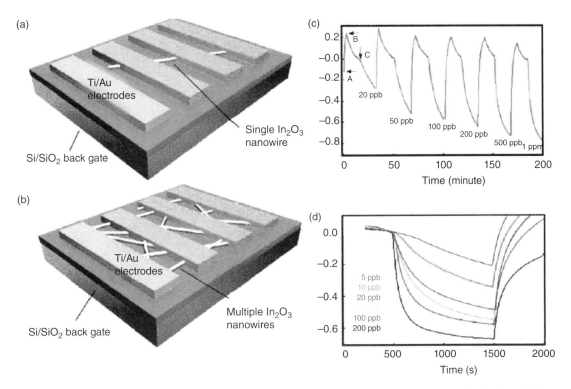

Figure 16.41 Schematic illustration of (a) a single NW and (b) a multiple NW transistor structure, where Ti/Au are deposited on NW-decorated Si/SiO₂ substrate as drain and source electrodes; (c and d) sensing response of the single-NW device and the multiple-NW device to NO₂ diluted in air, respectively. (*Source*: Reproduced with permission from Reference 261, Copyright 2004, American Chemical Society).

Although single NWs have characteristic response to different chemical species, in order to achieve fast identification of the composition and concentration of mixed species, it is more practical to build a sensing device with arrays of NWs made of different materials. Due to the difference on chemical compositions and surface states, NWs made of different materials may have different response to different species. On the other hand, the response of NW to specific specie can be also modified via surface decoration/modification. For example, Pd decorated NW surface leads to improved H_2 sensing property (295). The ultimate objective of fabricating NW array based gas sensor is to realize "Electronic Nose," which is composed of large number of sensors. In fact, realization of this concept with metal oxide NWs has been demonstrated by Kolmakov and coworkers (296). In this work, single SnO_2, Ni-doped SnO_2, In_2O_3, and TiO_2 mesowires are integrated on the same chip to form an array of chemoresistors, as shown in Figure 16.42a. Note that these structures were actually randomly placed on the four different regions of the same Si/SiO₂ wafer for device fabrication (296). Figure 16.42b shows the response of these nano/mesowires to H_2 pulses after low pressure O_2 exposure. Similar responses were observed in the case of carbon with an exception of In_2O_3 mesowire. At the chosen range of concentrations, CO did not cause a measurable change in conductance in this mesowire. In addition, the In_2O_3 whisker response to hydrogen was slower than that for any of the other structures. In this work, though both H_2 and CO increase the conductance of the chemiresitors, and the selectivity of the individual nanostructure is not sufficient to discriminate between them, gas distinguishability can be achieved using pattern recognition. Specifically, as shown in Figure 16.42c, signal patterns are depicted for these three sensing elements as radial plots: each radial beam shows the signal of one single sensing element normalized to its maximum value and the radial plots of H_2 and CO gas responses over the chemiresistor array are prominently different for two gases leading to feasible pattern recognition.

The concept-proof smart sensing with multiple NWs has confirmed that an "Electronic Nose" can only be realized via heterogeneous integration of NWs. In addition, this integration has to be more or less deterministic, namely, specific type of NWs have to be placed at the defined location, therefore, arrays of sensor elements can be easily fabricated at low cost. Recently, a scalable contact printing technique has been developed to achieve heterogeneous integration of semiconductor NWs for sensor applications (297). As shown in Figure 16.43a, this method involves the directional sliding of the NW growth substrate with randomly aligned NWs on top of a receiver substrate. During this process, NWs are effectively combed (i.e., aligned) by the directional shear force, and are eventually detached from the growth substrate and transferred to the target substrate which has lithographically defined patterns. This NW transfer method has been proven highly generic for a number of NW materials and also compatible with various type of substrates, including Si/SiO₂, glass, plastic, and paper. More importantly, all-NW integrated sensing circuits have been fabricated using two different types of NWs. Therefore, this technique provides an ideal technological platform for building NW arrays toward selective and smart chemical sensing, as illustrated in Figure 16.43b.

16.3.2.3 *Optical Properties Characteristic of Nanowires*
A wide range of optical techniques are available for the characterization of NWs to distinguish their properties from those of their parent bulk materials. Some differences in properties relate to geometric differences, such as

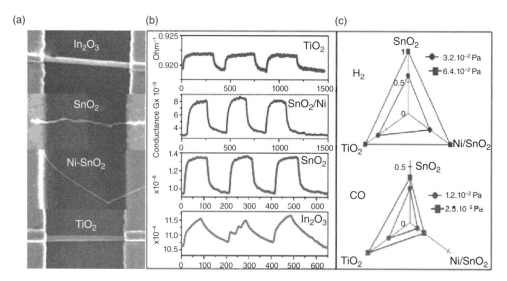

Figure 16.42   (a) SEM images of single In_2O_3, SnO_2, Ni-doped SnO_2, and TiO_2 mesowires integrated on the same chip. (b) The response of the array of the chemiresistors to hydrogen pulses with partial pressure of 6.4×10^{-2} Pa. (c) The response of a three-chemiresistor array to H_2 (top) and CO (bottom) inputs, normalized by maximum value.

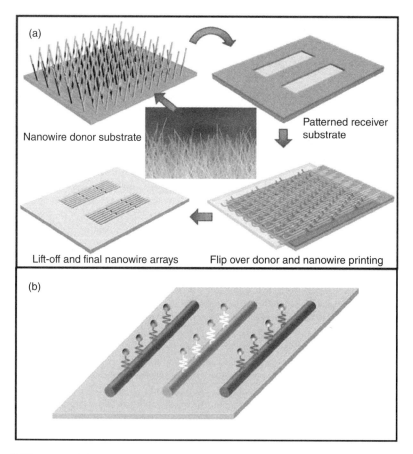

Figure 16.43   (a) Schematic of NW contact printing process. (*Source*: Reprinted with permission from Reference 297, Copyright 2008, American Chemical Society). (b) Schematic of heterogeneous NW integration for smart sensing application.

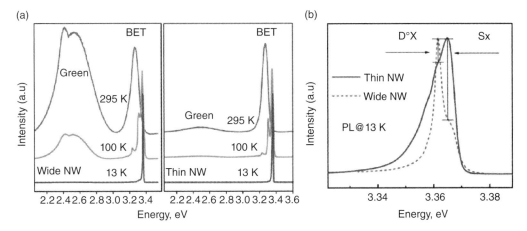

Figure 16.44 (a) PL spectra of wide (left) and thin (right) ZnO NWs at different temperatures show that the bulk defect states (Green) are more pronounced in wide NWs. Intensity is normalized with respect to the band-edge transition peak. (b) A close-up view of band-edge transition (BET) of thin and wide NWs at 13 K, indicating the blue shift in thinner wires due to the prevailing surface bound exciton emission. (*Source*: Reprinted with permission from Reference 302, Copyright 2007, American Institute of Physics).

the small diameter size and the large length to diameter ratio (also called the aspect ratio), while others focus on quantum confinement issues. The optical and transport properties are closely related to the inherent native defects for the undoped NWs and extrinsic impurities for the doped NWs. Many spectroscopy techniques such as photoluminescence (PL) (298), cathodoluminescence (CL) (299), and electroluminescence (EL) (300) have been employed to probe the optical and electronic properties of NWs. PL measurement is one of the most sensitive tools to investigate radiative defects in the material. Emission techniques probe the NWs directly and the effect of the host material does not have to be considered. This characterization method has been used to study many properties of NWs, such as the optical gap behavior, oxygen vacancies in ZnO NWs (182), strain in Si NWs (301), and quantum confinement effects in InP NWs (250).

While the size approaches fundamental length scale, optical properties of metal oxide Q1D materials begin to show size effects. Chang et al. (302) and Shalish (303) et al. compared the ratio of the deep level/near band edge of different-sized ZnO NW and found that the ratio is directly related to the wire diameter/length ratio. The temperature-dependent PL spectra of ZnO NWs with different diameters (d) are shown in Figure 16.44a (302). The green luminescence band at ~2.5 eV, originated from bulk defects, is much stronger for wide NW ($d > 100$ nm) than the thinner wires ($d < 30$ nm). Figure 16.44b shows the zoomed-in view of the band-edge transition (BET) emission at 13 K for different diameters (302). The emission from wide NWs shows a sharp peak at 3.362 eV arising from the donor bound excitons and a minor peak at 3.366 eV as a result of surface bound exciton emission. The emission peak at 3.366 eV becomes prevailing in the thinner wires. This blue shift is a result of finite size effect, due to the merging of the surface bound exciton emission that becomes dominant in smaller diameter NWs.

Figure 16.45 shows the photoluminescence of InP NWs as a function of wire diameter, thereby providing direct information on the effective bandgap. As the wire diameter of an InP NW is decreased, it becomes smaller than the bulk exciton diameter of 19 nm, quantum confinement effects set in, and the bandgap is increased. This results in an increase in the PL peak energy because of the stronger electron–hole Coulomb binding energy within the quantum-confined NWs as the wire radius gets smaller than the effective Bohr radius for the exciton for bulk InP. Since the smaller the effective mass, the larger are the quantum confinement effects. When the shift in the peak energy as a function of NW diameter (Fig. 16.45a) is analyzed using an effective mass model, the reduced effective mass of the exciton is deduced to be $0.052m_0$, which agrees quite well with the literature value of $0.065m_0$ for bulk InP. Although the linewidths of the PL peak for the small diameter NWs (10 nm) are smaller at low temperature (7 K), the observation of strong quantum confinement and bandgap tenability effects at room temperature are significant for photonics applications of NWs.

The resolution of photoluminescence (PL) optical imaging of a NW is, in general, limited by the wavelength of light. However, when a sample is placed very close to the detector, the light is not given a chance to diffract, and so samples much smaller than the wavelength of light can be resolved. This technique is known as near-field scanning optical microscopy (NSOM) and has been used successfully (304) to image NWs. For example, Figure 16.46 shows the topographical (a) and PL (b) NSOM images of a single ZnO NW.

16.3.2.4 Photodetectors NWs photoconductors are probably the simplest configuration of NW-based photodetectors (305). In general, single NWs or NW mashes (either randomly distributed, or aligned along a preferential direction) are drop-casted on an insulating substrate, and external bias is applied between two as-fabricated top metal electrodes by photolithography and evaporation processes. Upon illumination with incident light irradiation object to the NWs axis, the electrical conductivity increases, thus providing light-sensing capabilities. The unique properties of individual NWs or vertical arrays of NWs photodetectors, such as light polarization sensitivity, light absorption enhancement, and internal photoconductive gain, could be exploited for the realization of efficient and highly integrated devices such as optical interconnects, optical switches, or image sensors. Both the device geometry and the intrinsic material properties (such as charge carrier density and mobility)

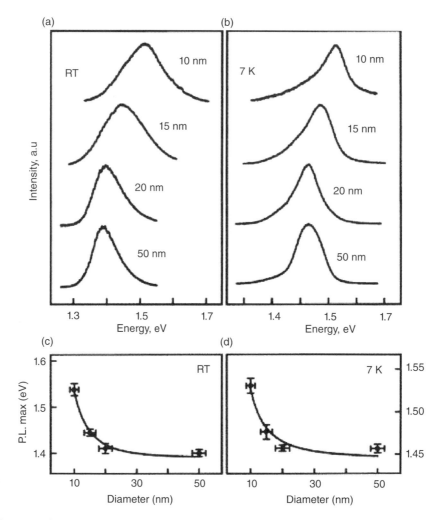

Figure 16.45   Photoluminescence of InP NWs of varying diameters at 7 K (b and d) and room temperature (a and c) showing quantum confinement effects of the exciton for wire diameters less than 20 nm. (*Source*: Repinted from Reference 251, Copyright © 2002, Nature Publishing Group).

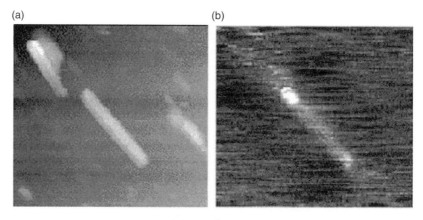

Figure 16.46   (a) Topographical and (b) photoluminescence (PL) near-field scanning optical microscopy (NSOM) images of a single ZnO NW waveguide. (*Source*: Reproduced from Reference 304. Copyright © 2001, American Chemical Society).

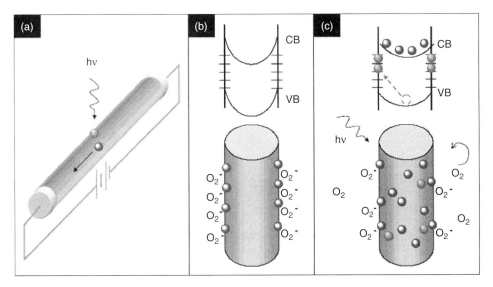

Figure 16.47 (a) Photoconduction in NW photodetectors. (a) Schematic of a NW photoconductor. Upon illumination with photon energy above E_g, electron–hole pairs are generated and holes are readily trapped at the surface. Under an applied electric field, the unpaired electrons are collected at the anode, which leads to the increase in conductivity. (b and c) Trapping and photoconduction mechanism in ZnO NWs: the top drawing in (b) shows the schematic of the energy band diagrams of a NW in dark, indicating band-bending and surface trap states. VB and CB are the valence and conduction band, respectively. The bottom drawing shows oxygen molecules adsorbed at the NW surface that capture the free electron present in the n-type semiconductor forming a low-conductivity depletion layer near the surface. (c) Under UV illumination, photogenerated holes migrate to the surface and are trapped, leaving behind unpaired electrons in the NW that contribute to the photocurrent. In ZnO NWs, the lifetime of the unpaired electrons is further increased by oxygen molecules desorption from the surface when holes neutralize the oxygen ions. (*Source*: Reproduced from Reference 306, Copyright © 2007, American Chemical Society).

play an essential role in determining the overall photoresponse of NWs photoconductors. In this section, different categories of NW-based photodetectors will be discussed according to the photogeneration mechanism.

16.3.2.4.1 Semiconductor Nanowire Photodetectors It is commonly accepted that photoconductivity (PC) in semiconductor NWs is dominated by a charge-trapping mechanism mediated by oxygen molecules adsorption and desorption on the NWs surface (Fig. 16.47) (306): when in dark environment, oxygen molecules are adsorbed onto the NWs surfaces and capture the free electrons present in the semiconductor materials, and thus a low-conductivity depletion layer will be formed close to the NW surface:

$$O_2(g) + e^- \rightarrow O_2^-(ad)$$

Upon illumination at photon energies above E_g, electron–hole pairs will be photogenerated [$h\nu \rightarrow e^- + h^+$]; holes migrate to the surface along the potential slope produced by band-bending and discharge the negatively charged adsorbed oxygen ions, and consequently oxygen molecules are photo-desorbed from the NW surface:

$$h^+ + O_2^-(ad) \rightarrow O_2(g)$$

The unpaired electrons are either collected at the cathode or recombined with holes generated when oxygen molecules are re-adsorbed and ionized at the surface. By elongating the photocarrier lifetime, this mechanism further improves the NW photoresponse and leads to extremely high photoconductive gain. This also causes saturation of the photoresponse at high illumination intensity due to the reduction of the number of available holetraps and accordingly to the shortening of the carrier lifetime.

This mechanism leading to high light-sensitivity in photoconducting NWs has been recently discussed in References 306–312. Because of the large surface-to-volume ratio, NWs own an extremely high density of surface states. Accordingly, due to the pinning of the Fermi energy at the surface, NWs form a depletion space charge layer which induces physical separation of electrons and holes and thus lead to significantly improved photocarrier lifetime (persistent photoconductivity). On account of the carrier distribution inside the NW being primarily dominated by the electric potential and Fermi energy pinning of the NW surface, which strongly rely on the shape of the NW, the dark- and photocurrents in NWs vary considerably with their different size. Theoretical calculations for InP NWs have shown that small diameters (i.e., minimal band bending) result in full-depletion of the NWs, thus minimizing the dark current which is the predominant source of noise in photodetectors, while large diameters (i.e., appreciable band-bending) increase the photoconductivity by obstructing photogenerated carrier recombination (308). Consistent results have been experimentally observed in GaN NWs with different diameters ranging from ~50 to 500 nm (307),

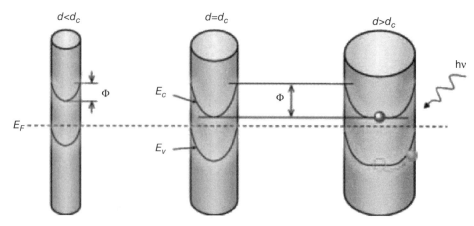

Figure 16.48   Schematic of the dependence of valence (E_V) and conduction (E_C) band profiles in NWs with different diameters d. Below the critical diameter d_c, the NW is fully depleted and band bending is minimal. For larger diameters, the recombination barrier Φ increases, thus the photocarrier lifetime is prolonged. The detail on the right shows the charge carrier separation mechanism upon photoexcitation. (*Source*: Reproduced from Reference 305, Copyright © 2010, American Scientific Publishers).

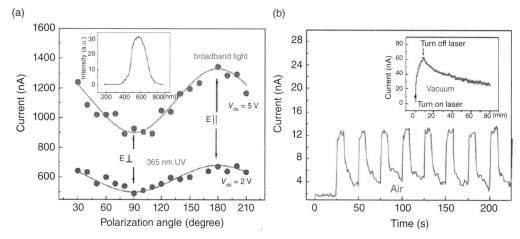

Figure 16.49   (a) Polarized photodetection of both UV (365 nm) and visible light (spectrum shown in the inset) show that NW conductance is maximized when the electric field component of the incident light is polarized parallel to the NW axis. (b) NW photoresponse to 633 nm laser in air compared to that in vacuum (inset). (*Source*: Reproduced from Reference 305, Copyright © 2004, American Institute of Physics).

as well as in SnO_2 NWs (313). A schematic of surface state effects on band bending and photogenerated carrier separation in n-type NWs with different diameters is illustrated in Figure 16.48.

Defect states related visible wavelength detection and polarized photodetection of ZnO NWs (314, 315), SnO_2 NWs (153) and In_2O_3 NWs have also been observed (316). Photoconductivity is found to be proportional to $\cos^2\theta$, where θ is the angle between the polarization of incident light and long axis of the NW. It is a maximum when the electric field component of the incident light is polarized parallel to the NW long axis (Fig. 16.49a). Because NW diameter is much smaller than the light wavelength, electric field component of light normal to NW axis is effectively attenuated inside the NW. This interesting property can lead to promising application for NW polarization-dependent photodetectors and optically gated switches. In the photoconductivity measurements of ZnO NWs, Fan et al. found that the environment has a crucial effect on the photoresponse (315, 317, 318). For example, surface chemical adsorption on the NWs could significantly expedite the photocurrent relaxation time. As shown in Figure 16.49b, the photocurrent relaxation time is around 8 s in air but hours in vacuum. This is because once upon illuminated in air, photogenerated holes discharge surface chemisorbed ions from NWs, while the photogenerated electrons significantly increase the conductivity. When illumination is switched off, oxidizing gas (primarily O_2) molecules in air re-adsorb onto the surface and reduce the conductivity. The sensitivity to surface chemisorption is obviously important for gas sensor application, which has been discussed in the preceding section.

The existence of wide bandgap in metal oxides has been extensively exploited for blue-UV range optoelectronic applications. However, some metal oxides also exhibit infrared (IR) photodetection due to indirect bandgaps. For example, CdO has a small indirect bandgap of

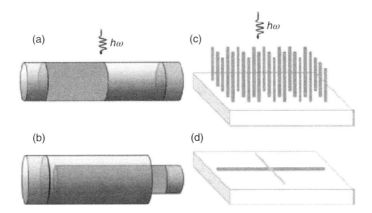

Figure 16.50 (a) Axial NW junction; (b) radial core/shell NW junction; (c) vertical NW array directly grown on the substrate as heterojunctions; (d) crossed NW junction.The segments in different colors represent either different doping or different materials for homogeneous and heterogeneous junctions, respectively. (*Source*: Reproduced from Reference 305, Copyright © 2010, American Scientific Publishers).

0.55 eV. Liu et al. had shown photoconductivity measurements of CdO-based field-effect transistor with IR light illumination. The IR detection on/off ratio is about 8.6 at 1.2 K and the relaxation time constant is estimated to be 8.6 s.

16.3.2.4.2 Photodiodes Nanowire Photodetectors NWs provide a lot of chances to construct photodiode devices, including Schottky metal–semiconductor junctions and homo- or heterojunction structures formed either axially along the NW (Fig. 16.50a), or radially by conformal NW coating as core–shell junctions, as shown in Figure 16.50b. Moreover, the possibility to form crossed (Fig. 16.50c) and branched NW junctions (319), or to directly grow vertical NW arrays on a variety of different substrates (Fig. 16.50d) largely raises the actualization of different photodetectors category by creating different device architectures and material combinations (320). It is worth to note that, depending whether illumination is perpendicular (Fig.16.50a and d) or parallel (Fig. 16.50b and c) to the junction plane, the boundary conditions for the continuity equation describing carrier diffusion across the junction at steady-state will vary considerably. Furthermore, in the case of light irradiation parallel to the junction plane, the effects of nonuniform absorption across the junction should also be taken into consideration.

HOMOGENEOUS AND HETEROGENEOUS JUNCTIONS Both single homo- and heterojunction NW photodiodes (Fig. 16.50a) have been achieved. For example, UV photodetectors based on indiviudal GaN NWs including axial p–n homojunctions have shown rectifying behavior, relatively fast photoresponse, and a photoconductivity increase of ~14 under 0.03 V reverse bias (321). Likewise, IR photodetectors based on individual NWs containing InAs/InAsP axial heterojunctions measured at 77 K showed very low dark current due to the conduction band offset formed at the NW heterointerface, strong polarization dependence, and a combined contribution to the photoconductive response from the InAs segment with onset ~0.5 eV and also from the InAsP segment with adjustable onset from 0.65 to 0.82 eV depending on the phosphorous proportion (322).

Similar design concepts for core–shell NW junctions (Fig. 16.50b) in photovoltaic applications could also be implemented for more efficient photodetector schemes that take advantage of the more effective charge carrier separation, and perhaps of the enhanced light absorption in growing radial NW junctions for vertical NW arrays (323). Till now, efficient photodetector with radial homojunctions has been demonstrated in both an indiviudal Si NW (324) and a vertical GaAs NWs array (325). Type-II core/shell heterostructure NW has also been proposed (326) and demonstrated (327) to stimulate charge carrier separation, enhance the photosensitivity, and simultaneously improve the spectral response.

Photodetectors based on heterogeneous junctions created by vertical NW arrays directly grown on the substrates (Fig. 16.50c) have also been demonstrated, where large NW densities are desirable to increase the photoresponse area and thus improve the photoconductivity. Particlly, due to owing a variety of fabrication methods, a large amount of work has been dedicated to heterojunction photodetectors by growing ZnO NWs on doped Si substrates. In these architectures, depending on the NW density, the top electrode can be either directly deposited onto the NW terminals layer or a transparent filling material such as spin-on-glass or an inert polymer which is usually used to reduce leakage current. Intrinsically doped, n-type ZnO NWs have been grown on both p-type or n-type Si substrates where the dependence of the dark current on applied bias is found to be similar to the ideal relationship for heterojunctions, with typical rectifying behavior as in the following (328):

$$I = I_s \left[exp \left(\frac{eV}{k_B T} \right) - 1 \right] \tag{16.7}$$

where I_s is the saturation current and the other symbols are in the usual meaning. In the view of the n-ZnO NWs/p-Si heterojunction, photoconductivity performance under UV light irradiationand reverse bias have exhibited responsivity of ~0.07 A/W at −20 V applied bias (329), while the fast and slow components of the photocarrier dynamics are in the order of ~300 ms and few minutes, respectively (330). It is interesting that, in the case of n-ZnO NWs/n-Si heterojunction, the spectral response can be adjusted from the visible to the UV spectral regions by applying forward or reverse bias to the device, through controlling the band-offset at the hetero interface and through selectively collecting photocarriers generated in the Si substrate, or in the ZnO NWs, respectively (331). In addition, recent progress in the direct heteroepitaxial growth of III–V NWs on Si substrate will also help foster new opportunities for vertical NW array photodetectors (332–334).

SCHOTTKY JUNCTIONS   Metal–semiconductor Schottky junctions can be used as photodiodes where both the photoexcited electrons from the metal and the photogenerated electron–hole pairs from the semiconductor can contribute to the photocurrent. One of the benefits of Schottky photodiodes is the fast response velocity, due to the high electric field which induces the short carrier transit time across the junction area under the reverse bias. The current in an ideal Schottky diode is still given by Equation 16.7, with the reverse bias saturation current given by (328)

$$I_s = A^* T^2 exp\left(-\frac{e\Phi_b}{kT}\right)$$
(16.8)

where A^* is Richardson's constant and Φ_b is the Schottky barrier height. If consider deviation from the ideal behavior, Equation 16.8 could be further adjusted as

$$I = I_s\left[exp\left(\frac{e\left(V - V_{th}\right)}{nk_BT}\right) - 1\right]$$
(16.9)

where V_{th} is the forward-bias threshold voltage and n is the ideality factor. When the current is determined by thermionic emission over the Schottky barrier $n = 1$.

The effects of Schottky barriers at the metal–semiconductor interface have been widely demonstrated in the literature of semiconductor NWs. Particulary, scanning photocurrent microscopy (SPCM) has provided an advantageous tool for the investigation of these effects in NWs photodetectors. In SPCM experiments conducted on Si (335), CdS (336, 337), and CdSe (338) NWs, photocurrent–voltage characteristics are typically asymmetric and, depending on the biasing conditions, photocurrent could be strongly localized near the metal electrode–NW contact. This technique is very effective to investigate the mechanisms of photoconduction at the nanoscale, for example, mapping the electronic band profile along the NW axis (335, 338) or determining the carrier mobility-lifetime products (336).

Although the good understanding of metal–semiconductor Schottky junctions in NWs devices, the intentional construction of Schottky photodiodes has rarely been traced in materials except for ZnO NWs. Nobel metals, for instance Au, Ag, and Pd, incline to form Schottky contacts with n-ZnO, thus two-terminal NWs devices with symmetric contacts made of these metals usually behave as back-to-back Schottky diodes and are responsive to UV irradiation (339–342). Ideality factors of ZnO NW Schottky photodiodes calculated by Equation 9 are often substantially larger than unity, because of the affect of both interface and surface states (343, 344). However, an almost perfect Pt/ZnO NW Schottky junction photodiode could still be achieved (345). In this study, the $I–V$ characteristic in the dark condition is well defined by the thermionic emission model (Eqs. 16.8 and 16.9), and the diode has an excellent ideality factor of $n = 1.1$ at room temperature and very low reverse current. It is interesting to note that under UV irradiation the device exhibited strong photoresponse, and the $I–V$ characteristic became almost linear. This transition from rectifying to ohmic behavior was ascribed to the reduction of the potential barrier between the Schottky contact and the ZnO NW uponirradiation. A similar behavior had been previously observed in ZnO NWs mesh contacted between two Au electrodes, where the transition from Schottky to ohmic behavior was observed exclusively under illumination with photon energy above the ZnO bandgap (318). The large difference in the photocurrent relaxation times observed in these two cases ($\tau < 33$ ms vs. $\tau > 10^4$) was attributed to bulk-dominated rather than surface-dominated transport.

16.3.2.4.3 Avalanche Nanowire Photodetectors   Avalanche photodiodes (APDs) are one of the very attractive photodetectors because they benefit from an internal gain due to the multiplication of the charge carriers generated from an internal gain due to the multiplication of the charge carriers generated by absorption of incident light. This occurs when these charge carriers cross a region of high electric field ($>10^5$ V/cm), thus acquiring enough energy to ionize the atoms of the crystal lattice, thereby creating new electron–hole pairs which are immediately separated, and can themselves create other electron–hole pairs: this leads step-by-step to an amplification of the primary photocurrent. This mechanism, known by term shock ionization or (band to band carrier) impact ionization, is also at the root of the reversible breakdown of reverse-biased p–n junctions. Carrier multiplication results in internal gain within the photodiode, which increases the effective responsivity of the device. The figure of merit for this process is the multiplication factor (or gain) M, which indicates the average number of carriers produced from the initial photocarriers (346):

$$M = \frac{1 - k}{e^{-\delta(1-k)} - k}, \quad k = \beta(E)/\alpha(E)$$
(16.10)

where δ is the average number of ionization events per electron transit, and α and β are the field-dependent ionization rates for electrons and for holes, respectively. So far, NW APDs have been demonstrated in two different configurations, namely, a crossed n-CdS/p-Si NW heterojunction (Fig. 16.48d) (347) and an axial p–i–n single Si NW homojunction (Fig. 16.50a) (348). In the case of the n-CdS/p-Si crossed avalanche photodiode, a photocurrent increase (I_{PC}/I_{dark}) of ~10^4 times higher than in individual n-CdS or p-Si NW photoconductors has been observed due to avalanche multiplication at the p–n crossed NW junction, with multiplication factors as high as $M = 7 \times 10^4$. Polarization dependence of the photoresponse has also been observed in the crossed structure, due to the predominant optical absorption in the CdS NW, as verified by spectral measurements. A detection limit of about 75 photons was estimated for these devices. Very similar results

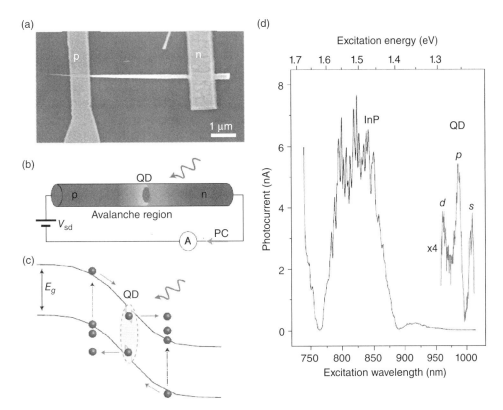

Figure 16.51 Single quantum dot in a NW APD. (a) SEM image of the NW photodiode. (b) A single quantum dot (QD) is located within the NW depletion region, where avalanche multiplication of the photocurrent (PC) is achieved under reverse voltage bias, V_{sd}. (c) Schematics of carrier multiplication starting from an exciton generated in the quantum dot, followed by tunnelling in the NW avalanche region. (d) Photocurrent spectroscopy at $V_{sd} = -2$ V with 1 μW and 20 mW excitation powers (black and red curves, respectively). Band-edge absorption in the NW is observed around 825 nm (black). Absorption in the quantum dot s, p, and d shells is observed at longer wavelengths (red). (*Source*: Reproduced from Reference 349, Copyright © 2012, Nature Publishing Group).

(namely, polarization sensitivity, high spatial resolution, and high sensitivity) were obtained in the case of the axial p–i–n Si NW APD, where complementary doping within a single NW was used instead of the assembly of two distinctly doped NWs. A maximum multiplication factor of $M = 40$ was derived in this case. Interestingly, multiplication factors for electron and hole injections could also be isolated, indicating that the multiplication factor for electrons ($M_n < 100$) was larger than that for holes ($M_p < 20$) due to larger electron ionization rate ($\alpha > \beta$) (348).

Recently, Bulgarini et al. (349) designed a device containing a single quantum dot embedded in a contacted InP NW, presented in Figure 16.51a. The InP NW was doped in situ during VLS growth to obtain a p–n junction. The depletion region of the p–n junction was used to multiply both electrons and holes as they gain enough energy to initiate the avalanche multiplication process. The operating principle of our NW APD is shown schematically in Figure 16.51b and c). A single photon incident on the device with a frequency equal to one of the quantum dot transitions is absorbed and creates a single exciton. Under reverse bias (V_{sd}, 0), the electron and hole separate and tunnel into the NW depletion region. Both the electron and hole then accelerate under the applied electric field, and once the carriers gain enough energy, additional electron–hole pairs are created by impact ionization. These additional electron–hole pairs can further trigger carrier multiplication and strongly enhance the photocurrent. The final result is that each exciton created in the quantum dot is multiplied into a macroscopic current. A unique feature of InP NWs is that the impact ionization energy is similar for both electrons and holes (1.84 eV and 1.65 eV, respectively) (350). Both carriers can thus contribute to the avalanche multiplication process and large gains can be achieved.

In Figure 16.51d, the quantum dot and NW absorption spectra were probed by tuning the laser excitation wavelength. The measured photocurrent shows a broad absorption peak around 825 nm originating from InP band-edge transitions, which suggests the presence of both wurtzite and zinc blende crystal structures, as confirmed by TEM (351). Three equally spaced photocurrent peaks are observed at higher excitation wavelengths (1007, 986, and 963 nm). By comparing these with typical photoluminescence spectroscopy of single quantum dots in intrinsic NWs, these three peaks were assigned to absorption in the quantum dot s, p, and d shells, respectively. From photoluminescence spectroscopy of the device at $V_{sd} = 0$ V, the peak at 1007 nm is confirmed due to absorption in the quantum dot ground state (s-shell). The observed shell separation of 26 meV corresponds to a diameter of ~27 nm according to calculations that assume an in-plane parabolic confinement in the quantum dot and is in agreement with the quantum dot size measured with TEM (352).

C.M. Lieber et al. also created nanoscale p–n diodes by crossing p- and n-type NWs (64, 245, 353–355). This concept was first demonstrated for p–n crossed InP NW junctions (64) and subsequently extended to crossed NW p–n diode junctions with p-Si/n-GaN (245),

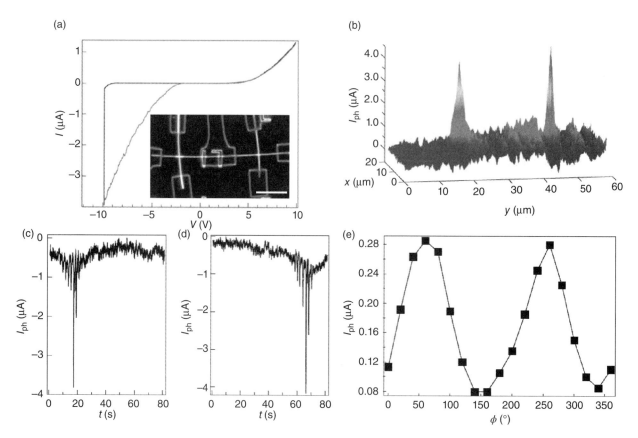

Figure 16.52 NW APD arrays. (a) *I–V* characteristic of the APD in dark (black line) and illuminated (red line) conditions; the device was illuminated with 500 nW of 488 nm light (red line). The inset shows optical micrograph of an array consisting of an n-CdS NW (horizontal) crossing two p-Si NWs (vertical); the larger rectangular features correspond to metal contacts (scale bar, 10 μm). (b) Spatially resolved photocurrent measured from the array in (a). Both devices were biased at −10 V and excited at 488 nm (200 nW, Ar+ -ion laser) with a scanning step size of 1 μm. (c and d) I_{ph} versus time traces for laser scanning measurements where individual devices were turned on (biased at −10 V); the left device is on in (c) and the right device on in (d). (e) I_{ph} versus polarization (*φ*) for a NW APD (see the Methods section); the device was biased at −5 V. (*Source*: Reproduced from Reference 347, Copyright © 2006, Nature Publishing Group).

p-GaN/n-GaN (353), and other systems. The AVPs were configured by crossing Si/CdS NWs p–n junctions, as shown in Figure 16.52a. These NW avalanche photodiodes (nanoAPDs) exhibit ultrahigh sensitivity with detection limits of less than 100 photons and subwavelength spatial resolution of 250 nm. Moreover, the elements in nanoAPD arrays can be addressed independently without electrical crosstalk (Fig. 16.52b).

16.3.2.4.4 Nanowire Phototransistor A phototransistor is a bipolar or unipolar transistor where light can hit the base, causing photo-generated carriers. This modulates the base-collector junction leading to an amplified current through transistor, which can thus further lead to much larger photo response. As discussed in the following, such structures have been successfully demonstrated in NW architectures. Typically, NW FETs have been fabricated by drop-casting NWs on a dielectric–semiconductor substrate (316, 356–359) or by patterning NWs through conventional photolithographic methods (360, 361). Subsequently, a gate bias is then applied by a lithographically patterned top gate, or a back gate. Sensitivity even down to a single charge carrier has been achieved using a double-gated Si NW phototransistor (361). This was realized by a very narrow gate which behaved as a trap for one carrier species. Once phototransistor is illuminated, only a few carriers are trapped under the gate, and their recombination could be detected through a sudden abrupt decrease in conductivity.

In the case of FET photodetectors, an electrical gate bias is utilized to modulate the lateral field across the NW. However, a similar effect is also present in NW photoconductors in which surface states lead to a radial electric field. As discussed earlier, this causes the separation of photogenerated carriers in the NW channel, which greatly improves the carrier recombination lifetime leading to a much greater sensitivity. Therefore, NW photoconductors can be considered as phototransistors where the internal field arising from the large density of surface states in conjunction with light irradiation taking the role of a photogate. For NWs, band bending can be caused by different surface effects, such as a strong surface electric (as in GaN NWs) (307, 310) or deep trap states (e.g., oxygenrelated hole-traps in ZnO NWs or surface states in Ge NWs) (306, 357). In order to explain these effects, photoconduction mechanism was studied in Si NWs, finding that the phototransistive gain mechanism is also trap related (362). It is experimentally discovered that p-type doped Si NW planar arrays fabricated by conventional photolithography and thermal oxidation (Fig. 16.53a) exhibited gain of $G > 3.5 \times 10^4$ at low light intensities (Fig. 16.53b).

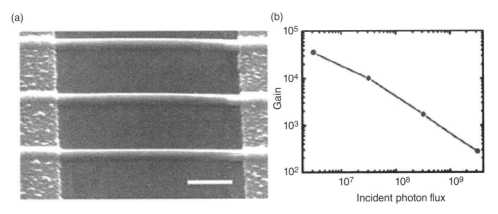

Figure 16.53 (a) SEM image of a planar-etched Si NW array fabricated from a silicon-on-insulator wafer. The scale bar is 2 μm. (b) Gain of a Si NW photodetector relative to the photon absorption rate extracted from photocurrent measurements at 0.5 V applied bias. (*Source*: Reproduced from Reference 362. American Institute of Physics).

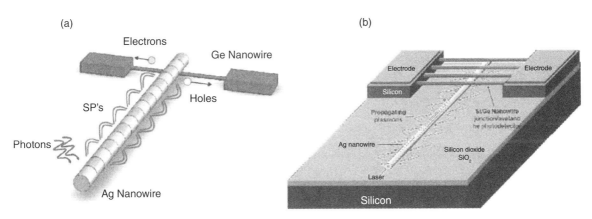

Figure 16.54 (a) Schematic representation of an Ag NW, acting as a launching pad for SPs and a crossing Ge NW, which supports an electrical current whenever electron–hole pairs are excited by the SP electric field at the Ag–Ge contact. (b) Schematics of a possible distributed photodetector based on Ag NW coupled into a series of NW photodetectors for efficient detection of SPPs. (*Source*: Reproduced from Reference 376, Copyright © 2011, Nature Publishing Group).

16.3.2.4.5 Plasmonic Nanowire Photodetectors An exciting new research area in nano-optoelectronics is represented by NW plasmonics that promises high integration density of nanophotonic devices. When free electrons on the surface of a metal NW are impinged by light of a specific wavelength, they generate a surface plasmon polariton (SPP), a surface wave resulting from collective electron oscillation. SPPs can propagate for long distances along metallic NWs, provided they do not encounter any defects in the structure (363–366). Ongoing efforts are directed toward the development of nanoscale photonic circuits based on integration of NWs with propagating SPPs that represent localized light below the diffraction limit. Controlling surface plasmons and guiding them through a specific path could lead to the optical equivalent of electronic circuits.

In the recent past, several methods for miniaturizing the size of optical waveguides with plasmonics have been reported including particle arrays (367, 368), thin metal films (369, 370), and metal NWs (371–373). Silver NWs have unique properties that make them particularly attractive for nanoscale confinement and guiding of light to nanoscale objects due to their smooth surface that contributed to lower propagation loss than metallic waveguides fabricated by electron-beam lithography (374). For practical applications in a system, plasmonic waveguides need to couple light to a nanoscale detector, and researchers proposed nanodots (375) and NWs (376–378) as potential candidates for plasmonic detection schemes. Falk et al. demonstrated a NW plasmonic PD that consists of a silver NW (~100 nm in diameter) that acts as a plasmonic waveguide and a crossing Ge NW that is connected to two metal pads, as shown in Figure 16.54a (376). When SPPs guided along the Ag NW to the Ag–Ge region excites electron–hole pairs in the Ge NW, it results in a detectable current between the electrodes. A calculated energy transfer of 23% is achieved using a single Ge NW photoconductor.

The current pick-and-place approach allows the coupling of one semiconductor NW to a single Ag nanorod. A process that allows the coupling of a single Ag NW to multiple semiconductor NWs can be developed based on the NW bridging method (379) as depicted in Figure 16.54b. To achieve this, a plasmonic waveguide via Ag NW or nanoribbon can be positioned using several methods including

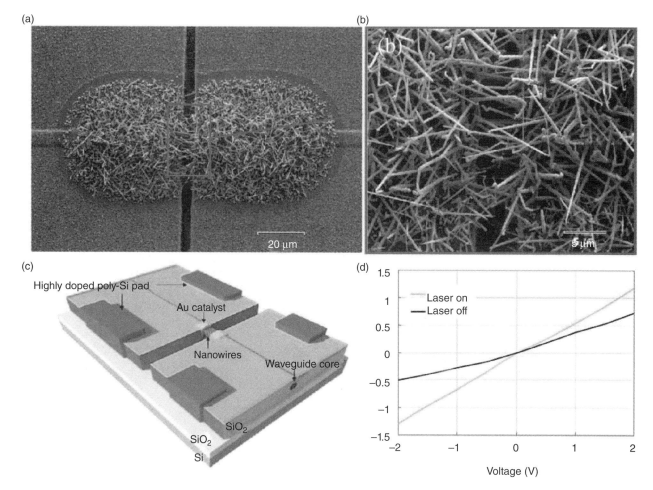

Figure 16.55 SEM images for NW-integrated waveguide device grown at 680 °C. (a) Top view of the microtrench with bridged NW and catalyst pads with NWs grown on the vertical sidewalls as well as on top of the upper cladding of the waveguide. (b) Close-up view of bridged NWs. (c) Schematic of the waveguide photodetector. (d) I–V characterization of waveguide-integrated NWs photoconductors across a 7 μm trench when illuminated with a 2 mW laser input of wavelength 780 nm. (*Source*: Reprinted from Reference 402, Copyright © 2011, IEEE).

DNA-templated metal NW synthesis (380, 381). Ag NWs can be physically positioned either below or on the surface of the detectors subsequent to the detector fabrication. For the characterization of the SPPs detectors, cylindrically symmetric SPP can be launched by illuminating the farthest end of the Ag NW with a tunable laser, while the detectors can be monitored for SPP-induced photocurrents. This device will be analogous to a distributed photodetector (382). This arrangement of distributed SPP detectors will allow an uncoupled and undetected SPPs in the first NW to be detected in subsequent NWs connected in series along the direction of the metallic NW waveguide. Multiple NWs along the axial direction are expected to enhance the efficiency to above 90% by absorbing uncoupled/undetected SPP in a series of NW detectors.

16.3.2.4.6 Waveguide-Integrated Nanowire Photodetectors Grego et al. recently reported the first monolithic waveguide-integrated photoconductors with Si NWs on an amorphous substrate (383, 384). This is significant because it will enable new building blocks for a self-contained CMOS-compatible photonic chip for both light guiding and detecting capability for high-speed optical interconnects (385–391). NWs have been shown to have the capability to waveguide photons (392–398) and plasmons (391, 399–401), and therefore, besides photodetection, they may also act as waveguide bridges between two waveguides integrated into optical interconnection devices.

As shown in Figure 16.55a–c, the photodetector comprises a passive optical waveguide designed for guiding 780-nm photons and laterally oriented photoconducting Si NWs. The waveguide is designed with a SiO_xN_y core layer and an amorphous SiO_2 cladding layer. Deep trenches were patterned in the waveguides by reactive ion etching (RIE). Subsequently, a highly p-doped polysilicon layer was deposited selectively on the vertical sidewalls of the trenches by a directional RIE-based etch back after a conformal thin film deposition. A selective deposition of Au film on the sidewalls at a tilt angle was done for catalyst-assisted NW growth. By employing a VLS method, undoped Si NWs were grown at 680 °C from the annealed Au thin-film catalyst. The NWs were grown without any globally specific orientation, since the growth template is polycrystalline silicon.

The device photoresponse was measured by probing the two electrodes bridged by the Si NWs while coupling the waveguide with a 2 mW laser input of 780 nm wavelength. The NW density that bridges the waveguide trenches was controlled by varying the diameter of the NW. Higher densities were achieved with NWs of an average diameter of ~100 nm as compared to lower densities with much thicker diameter NWs of ~900 nm. The devices with thicker NWs (low density) showed a much smaller increase in its photoresponse when normalized to the dark current at a bias of 5 V. This can be attributed to lower number of bridged photoconducting NWs and higher optical power loss by light scattering. On the other hand, devices with dense thin NWs showed much less reflectance and higher photosensitivity. The responsivity of the dense thin NW sample was about 0.03 A/W at 5 V bias for a waveguide illumination of 300 µW.

Considering the NW's doping concentration (nominally undoped) and the density of bridged connections (403), the dark current level of the devices seems much higher than that of a Si NW photodetector fabricated by RIE using Ni nanodots as a mask (404). This leakage current is most likely due to the surface states and defects on the NWs and uncatalyzed growth of thin films between the electrodes (e.g., amorphous and nanocrystalline Si on the NWs as well as on the substrates). Such an uncatalyzed film generates leakage paths among different terminals of a device as well as among adjacent devices in a wafer resulting in higher noise and crosstalk (405–408).

In order to reduce the surface states of Si NWs, many experimental efforts have focused on high-temperature annealing and hydrogen passivation (409, 410). The process of doping the NWs can directly contribute to higher carrier concentrations and possibly prevent complete depletion of the NW, and hence provide a stronger surface electric field for the effective separation of carriers at the surface (310). This procedure instead may increase the risk of leakage paths due to the uncatalyzed growth of polysilicon. By reducing the NW growth temperature, the uncatalyzed polysilicon deposition on the substrate and on the NW may be reduced as well. Moreover, the lower growth temperature results in a decrease of the average NW diameter, thus reducing the reflection of the incident photons (411). However, thinner NWs are more prone to being completely depleted (412). The trade-off between leakage current and photoconductivity should be considered for optimizing the performance of waveguide integrated NW PDs.

The photodetection results presented in Figure 16.55d demonstrate the feasibility of NW-integrated active optoelectronics devices fabricated with a scalable process. The photodetection performance can be improved in devices by using doped Si NWs, as opposed to the intrinsic Si used for this study, as well as the optimization process. The electrical probing of the devices is complicated by parasitic resistance at interfaces as demonstrated by the differences in two- and four-terminal measurements. Efforts are currently underway to improve the electrical contact with a new device design layout.

Further progress in this device requires an understanding of how passive waveguides and NWs interact with photons at the optical index transition, and how propagating optical modes from a micrometer-size waveguide switches to nanoscale semiconductor wires, and the impact of misorientation and nonuniformity in the distribution of the NWs. Investigation on the effects of light scattering, photon trapping, and absorption with the use of index-matching material may assist in improving the performance of these devices (411, 413–416). Photon propagation in a waveguide carries a specific polarization and as such NWs oriented along the direction of the waveguide will demonstrate varying coupling efficiency for different polarizations (315, 335, 417–420). Hence, it is also important to optimize the polarization preference of both the waveguide and the NWs for optimum performance.

16.3.2.5 Solar Cells

16.3.2.5.1 Fundamentals of Nanowire Photovotaics NW solar cells have some potential benefits over traditional wafer-based or thin-film devices related to optical, electrical, and strain relaxation effects; new charge separation mechanisms; and cost. Ordered arrays of vertical NWs with radial junctions take advantage of all these effects, although solar cells made using axial junctions or random arrays can still have some benefits over planar cells, as shown schematically in Figure 16.56. Functioning NW photovoltaics have been fabricated using a wide variety of materials including silicon, germanium, zinc oxide, zinc sulfide, cadmium telluride, cadmium selenide, copper oxide, titanium oxide, gallium nitride, indium gallium nitride, gallium arsenide, indium arsenide, and many polymer/NW combinations (422–436). Output efficiencies have steadily increased so that most material systems have now achieved efficiencies higher than 1%, with some close to 10%. In this regard, this section will provide a comprehensive review of recent progress on photovoltaic (PV) research based on single-NW and NW array.

Due to their unique 1D structure, semiconductor NWs have demonstrated remarkable electrical and optoelectronic properties, such as surface and environmental sensitive electrical transport (176), polarized photo-response, and so on. To understand the fundamental mechanism of these appealing behaviors, individual NWs are typical-fabricated into devices and characterized to decouple NW-to-NW interaction. Similarly, PV effect of individual NWs and nanorods were characterized and investigated firstly. Particularly, Si NWs were chosen as the model material due to the fact that Si is the most common PV material dominating the PV market for decades. Similar to thin film amorphous silicon p–i–n solar cells, axial p–i–n junction Si NW structures have been successfully grown with vapor phase catalytic method in conjunction with the *in-situ* doping modulation technique (324, 437), the intrinsic segments of Si NWs are photosensitive region which generates photo-carriers. The PV device was fabricated with electron-beam lithography (EBL) to contact p and n region, as shown in Figure 16.57a. Photoelectric characterization showed that an optimal device can produce open circuit voltage (V_{oc}) around 0.29 V (Fig. 16.57b), short circuit current density (J_{sc}) of 3.5 mA/cm^2, and a maximum conversion efficiency of 0.5% (437).

Though it has demonstrated the feasibility of building a PV device with a single NW, NWs with axial junctions horizontally lying on the substrate showed low efficiency mainly due to two reasons: (i) the direction of carrier transport is along the long axis of NWs, and large surface area of NWs results in significant surface recombination; (ii) photon incident angle is perpendicular to NW axis, resulting inefficient light absorption in conjunction with low absorption coefficient of Si (438). These limitations indicate that the structure design of PV devices has to be revolutionized in order to harvest solar energy efficiently while taking advantage of the unique properties of 1D nanomaterials. In this case, radial junction structure has been considered as a more promising structure for NW PV devices since the minority carrier collection/transport

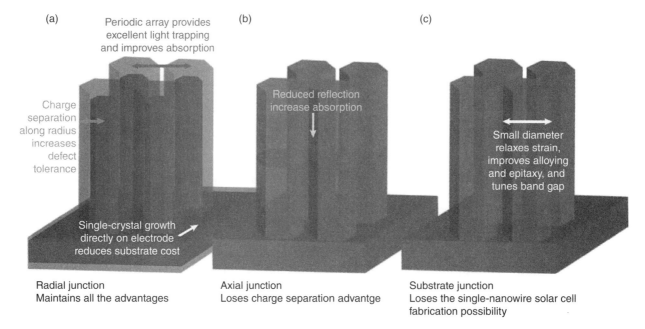

Figure 16.56 Benefits of the NW geometry. (a) Periodic arrays of NWs with radial junctions maintain all the advantages, including reduced reflection, extreme light trapping, radial charge separation, relaxed interfacial strain, and single-crystalline synthesis on nonepitaxial substrates. (b) Axial junctions lose the radial charge separation benefit but keep the others. (c) Substrate junctions lack the radial charge separation benefit and cannot be removed from the substrate to be tested as single NW solar cells. (*Source*: Reproduced from Reference 421, Copyright © 2011, Nature Publishing Group).

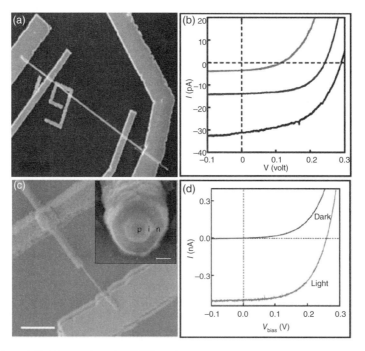

Figure 16.57 Single NW PV Device. (a) SEM image of a p–i–n Si NW device, scale bar is 4 μm. (b) Illuminated *I–V* characteristics for the *i*-length = 0, 2, and 4 μm a p–i–n devices under AM 1.5 G. (Reproduced from Reference 437, Copyright © 2008, American Chemical Society.) (c) SEM image of a core-shell NW photovoltaic device, scale bar is 1.5 μm, the inset shows the cross-section of a NW with p-type core, *i*-layer, and n-layer shell. (d) Dark and illuminated *I–V* curves of core-shell NW device. (*Source*: Reproduced from Reference 324, Copyright © 2008, Nature Publishing Group).

is along the radial direction, therefore greatly shortens the carrier travel and improves the collection efficiency (323, 324, 439–441). Such radial p–i–n NW structures have been fabricated with Si, GaAs, and so on (324, 431, 437, 442).

In a representative work, a Si single NW p–i–n structure was grown using an Au catalyzed vapor–liquid–solid (VLS) method. This core–shell NW consists of a p-type NW core with an intrinsic and n-type shell, as shown in the inset of Figure 16.57c. The p-type core was single crystalline, while the shells were polycrystalline. After etching shell, a PV device can be fabricated with EBL, as shown in Figure 16.57c. Photoelectric characterization in Figure 16.57d demonstrates the dark and light I–V curves measured for a corresponding core–shell NW PV device. As the result, such a core–shell NW PV device yields an open circuit voltage of 0.26 V, short circuit current of 0.503 nA, and fill factor of 55%, corresponding to a maximum power output of ~72 pW under 1 sun and device conversion efficiency ~3.4% after exclusion of the metal covered area, which is 7× improved from the axial p–i–n NW PV device. And further study at low temperature showed that the device conversion efficiency can be increased up to 6.6% at 80 K under 0.6-sun illumination, due to the reduced carrier recombination (324).

Core–shell structure clearly improves NW PV efficiency via shortening minority carrier travel distance; nevertheless, it does not overcome the light absorption limit as mentioned earlier, for a horizontal device, especially for Si which has relatively low optical absorption coefficient. On the other hand, GaAs has a much higher optical absorption coefficient and close to the ideal energy bandgap ($E_g = 1.45$ eV), therefore it is a more promising material for efficient NW solar cells. In fact, core-shell p–i–n GaAs NW PV devices have been successfully fabricated (431). In this work, GaAs NWs were grown with molecular beam epitaxy (MBE), and the PV device fabrication is similar to that of Si p–i–n devices. For a typical GaAs, core-shell p–i–n GaAs NW with thicknesses of the intrinsic and n-type shells of 15 and 50 nm, electrical characterization showed an efficiency of 4.5% and a fill factor of 0.65 under AM 1.5 illumination at room temperature, clearly improved from that of Si p–i–n NWs.

16.3.2.5.2 Performance Limiting Factors of Nanowire Solar Cells It is known that the performance of a PV device largely depends on minority carrier diffusion length in the material (438). However, it worth noting that the majority of NWs that have been utilized in PV investigation were grown by vapor phase metal catalytic approach, and metal incorporation in NWs leads to impurity states in bandgap. Particularly, in Si NWs growth, Au nanoparticles are often used as catalyst; however, it is well known that Au contamination in Si results in detrimental deep trap states, which can significantly reduce the minority carrier diffusion length in NWs (443). In fact, minority carrier diffusion lengths up to 4 μm have been reported for Au-catalyzed, VLS-grown Si NWs (444, 445). On the other hand, Cu-catalyzed Si NWs showed 10 μm minority carrier diffusion length as Cu is not as detrimental as Au to minority carrier lifetime in Si (443).

In addition to catalyst-induced impurity states in NWs, surface recombination also has to be taken into account when considering the limiting factors of NW solar cells as NWs usually have large surface-to-volume ratio. Although Si and GaAs NW arrays have been extensively explored for PV studies, they have relatively high surface recombination velocities, which degrade the cell conversion efficiency by the significant carrier surface recombination. Specifically, the reported surface recombination velocities of both non-passivated planar silicon and gallium arsenide structures have exceeded 10^6 cm/s making these materials nonideal for the NW solar cell (446–450). Therefore, in order to fabricate efficient PV devices with these materials, material dimension has to be relatively large to reduce surface-to-volume ratio, and surface states have to be well passivated. As an example, Kelzenberg et al. recently reported Si microwires with radial p–n junction demonstrating 9% PV efficiency, the microwire surfaces were passivated with amorphous Si and silicon nitride in this work (451).

On the other hand, there are a number of semiconductors that have relatively low surface recombination velocities, such as II–VI compound semiconductors. For example, typical cadmium sulfide (CdS) and cadmium telluride (CdTe) thin films have the untreated surface recombination velocities around 10^3 and 10^4 cm/s, which is much smaller than that of Si and GaAs. Therefore, they are more suitable material systems for solar cells (452, 453).

16.3.2.5.3 Property Investigation of Nanowire/Nanopillar Arrays Investigations on single NW PV devices demonstrated the feasibility of solar energy conversion with NWs; meanwhile, it has provided us with fundamental understanding of the nanoscale PV effect and indicated potential solutions for optimizing performance of NW solar cells, such as using core–shell device structure (324), choosing proper catalyst (443), applying surface passivation layer, and so on (454).

Single NW-based solar cells can certainly be used to drive nanoelectronic components (324). Nevertheless, scalable and cost-effective PV technologies are in more urgent demand. Therefore, fabricating PV devices made of ensembles of nanomaterials is of paramount importance and great interest. Note that this chapter will be focusing on 1D nanomaterials, primarily due to the fact that 1D geometry allows fast carrier extraction along axis direction. In conjunction with the their single crystalline nature as the result of bottom-up growth approach, 1D materials are believed to be highly promising for photovoltaics.

In addition, theoretical and experimental works have both revealed that well-organized array of 1D nanomaterials can significantly improve photo-carrier collection efficiency and light absorption, compared with the planar structure. Therefore, these interesting aspects will be further discussed in the following sections in order to facilitate our understanding on 1D nanomaterial–based solar cells at a larger scale.

IMPROVED OPTICAL ABSORPTION WITH NANOWIRE/NANOPILLAR ARRAYS Light absorption capability directly affect the performance of a solar cell, hence the commercial PV devices/panels usually have a layer of antireflection coating and/or texture on the surface. However, it is well known that broadband antireflection cannot be achieved with a single layer of antireflection coating, and multilayer antireflection coating can significantly increase PV cost. On the other hand, surface texturization requires chemical etching or laser ablation, which creates surface defects as well as increases the fabrication cost (455).

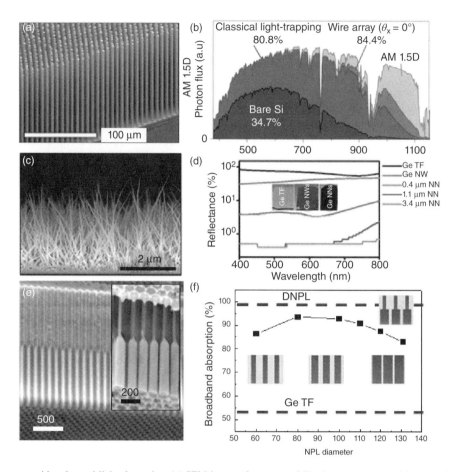

Figure 16.58 Nanostructures with enhanced light absorption. (a) SEM image of an array of Si wire array grown with vapor phase catalytic method; (b) Illustration of the normal-incidence, spectrally weighted absorption of the AM 1.5 D reference spectrum, corresponding to each of the three absorption cases: Si wire array which had an equivalent planar Si thickness of 2.8 μm, a 2.8 μm thick planar Si absorber, with an ideal back-reflector, assuming: bare, nontexturized surfaces and ideally light-trapping, randomly textured surfaces. (Reproduced from Reference 456, Copy right © 2010, Nature Publishing Group.) (c) SEM of a Ge nanoneedle array grown with catalytic chemical vapor deposition. (d) Surface reflectance spectra of Ge thin film, NW and nanoneedle arrays showing much lower optical reflectance for Ge nanoneedle arrays. (Reproduced from Reference 460, Copy right © 2010, American Chemical Society.) (e) Cross-sectional SEM images of a blank AAM with dual-diameter pores and the Ge DNPLs (inset) after the growth. (f) Average absorption efficiency over λ = 300–900nm for single-diameter NPLs as a function of diameter along with that of a DNPL array with D1 = 60 nm and D2 = 130 nm. (*Source*: Reproduced from Reference 459, Copy right © 2010, American Chemical Society).

Actually, the purpose of surface texturization is to create 3D structures on the surface to implement "light trapping" (455). This can be achieved by using 3D arrays of nano/microwires and pillars efficiently. There have been various 1D nanomaterials configured as 3D arrays, including Si (423, 456–458), Ge (459, 460), CdS (461), and so on. Extensive work has been done on their optical properties. An interesting series of experimental and theoretical works performed by Kelzenberg and Kosten et al. have shown that properly designed Si wire arrays have enhanced broadband light absorption exceeding ray-optics light trapping absorption limit of $2n^2$, which over-performs randomly textured planar Si with an equivalent volume over a broad range of incidence angles (456, 462). Figure 16.58a shows a Si wire array grown by a photo-lithographically patterned catalytic growth process in the work. As shown in Figure 16.58b, much improved absorption spectrum can be achieved on Si wire arrays. It was observed that with less than 5% areal fraction of wires, up to 96% peak absorption can be achieved leading to absorption of 85% of the day-integrated, above-bandgap direct sunlight (456).

Besides Si, 1D nanostructures based on Ge have also been explored for efficient light absorption. Chueh et al. reported "black" Ge nanoneedle arrays grown on plastic substrate with vapor phase catalytic growth, as shown in Figure 16.58c (460). The Ge nanoneedles have crystalline Ge core and amorphous shell. Due to the fact that they have gradually increased diameter from tip to root, the refractive index is gradually increased as the light is propagating into the nanoneedle array. Therefore, such a unique structure has demonstrated much lower reflectance as compared to a Ge thin film and Ge NW array, as shown in Figure 16.58d.

In another work, intead of using lithographic method and epitaxial wafer to obtain regular arrays of wire, Ge nanopillars (NPs) were assembled in the anodic alumina membrane (AAM) via the vapor phase catalytic growth method (461). By varying the diameter, pitch, and length of the pores, the shape of Ge nanopillars can be well controlled. Transmission electron microscopy investigation showed that the grown

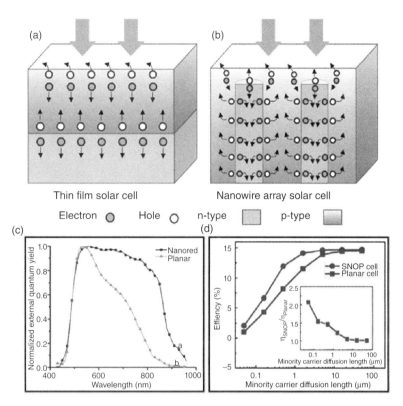

Figure 16.59 Carrier collection improvement: schematic of carrier collection in (a) a bulk/thin film solar cell and (b) a 3D NW/nanopillar solar cell. (c) spectral response of typical photo-etched planar and nanorod array photo-electrochemical cells with the external quantum yield normalized to its highest value. (Reproduced from Reference 463, Copyright © 2008, American Chemical Society.) (d) Conversion efficiencies of the nanopillar solar cell (SNOP cell) and planar cells versus the minority carrier (electron) diffusion length of the CdTe film. The inset shows their ratio, depicting the advantage of nanopillar solar cell, especially when the minority carrier life times are relatively low. (*Source*: Reproduced from Reference 461, Copyright © 2009, Nature Publishing Group).

NPLs are single crystalline which highly favors the carrier transport. Optical reflectance/transmission measurements showed that with the fixed pitch, nanopillar arrays with the large diameter has high reflectance but low transmission, and nanopillar arrays with small diameter has the opposite effect. As a result, there existed an optimal diameter which yielded a maximal 94% broadband light absorption, as shown in Figure 16.58e, as opposed to ~53% from a blank Ge film. To further improve light absorption, a unique dual-diameter nanopillar structure (Fig. 16.58f) was fabricated by simple two-step etching and anodization (459). This structure combines high transmission of small diameter structure and high absorption from large diameter structure, thus leading to ~99% broadband optical absorption (459).

CARRIER COLLECTION IMPROVEMENT   In addition to enhancing optical absorption, 1D nanostructures array can also significantly enhance the photo-carrier collection efficiency if the structure is properly designed. The principle is schematically shown in Figure 16.59a and b. It is known that photo-carrier collection and light absorption compete for planar structured solar cell, namely, efficient carrier collection requires thin material to shorten the minority carrier travel distance; however, efficient light absorption requires thick material for obvious reasons. This conflict may not be quite severe in single crystalline Si as it has hundreds of micrometer minority carrier diffusion length (455). However, it is fierce for polycrystalline or even nanocrystalline materials which have quite short carrier diffusion length. To greatly relax the competition between carrier collection and light absorption, it has been proposed and experimented to use 3D structure consisting of arrays of 1D nanomaterials for solar cells. As shown in Figure 16.59b, in such a structure p–n junction interface is parallel to light absorption direction, thus carrier collection occurs perpendicular to light absorption. This mechanism works for both core/shell type p–n junction NW array and nanopillar/thin film hybrid devices (456, 461, 463). In fact, a comparison between nanorod arrays and planar Cd (Se, Te) photoelectrodes has been performed (463). It was shown the fill factors of the nanorod array photoelectrodes were superior to those of the planar junction devices. More importantly, the spectral response of the nanorod array photoelectrodes exhibited better quantum yields for collection of near-IR photons relative to the collection of high-energy photons than the planar photoelectrodes, as shown in Figure 16.59c.

Furthermore, the performance benefit of orthogonalizing photon absorption and carrier collection is also demonstrated with the simulation of the CdS nanopillar/CdTe thin film hybrid solar cells (461). In this work, the conversion efficiency of a device structure consisting of CdS NPL arrays embedded in a CdTe thin film were compared to that of a planar CdS/CdTe cell. The nanopillar structure showed the improved conversion efficiency than its thin film counterpart, especially for small minority carrier diffusion lengths, as shown in Figure 16.59d. This result provides an important guideline for solar cell design using low cost and low-grade materials.

PHOTON MANAGEMENT PROPERTIES OF NANOWIRES   NWs and NPLs can both be categorized as quasi-one-dimensional materials, while NPLs normally refer to short and vertical standing NWs. In fact, NWs have been widely investigated and utilized for PV devices in the past few years, with various materials including Si (423, 456, 457, 464, 465), InAs (466), InP (467), ZnO (424), and so on. Nanostructure-based PV devices usually have large surface recombination rates compared to traditional Si solar cells and thin film solar cells. However, the carrier transport property of NWs and NPLs can be superior compared to other nanostructures, especially nanoparticles. Law et al. calculated that the electron diffusivity for single ZnO NWs is 0.05–0.5 cm^2/s, which is hundreds times larger than that of TiO$_2$ and ZnO nanoparticle films (424). Generally, there are two types of p–n junction configurations, that is, radial junctions and axial junctions (324, 421). Compared to axial junctions, the configuration of radial (core-shell) junction can further enhance the carrier collection efficiency, as long as the radii of NWs are much smaller than the minority diffusion length (324, 468, 469). NWs can also be embedded into thin films to form junctions, which was also proven to be an efficient way to improve carrier collection efficiency (461).

The photon management properties of NWs have been well studied (411, 442, 465, 467, 470–472). Theoretical and experimental works have shown that arrays of semiconductor NWs with well-defined diameter, length, and pitch have tunable reflectance, transmittance, and absorption. Figure 16.60a shows an SEM image of Si NW array on Si wafer reported by Garnett et al (423). The NW arrays were fabricated by deep reactive ion etching (DRIE) using a monolayer film of silica beads as a mask. The pitch and diameter of Si NWs can be controlled

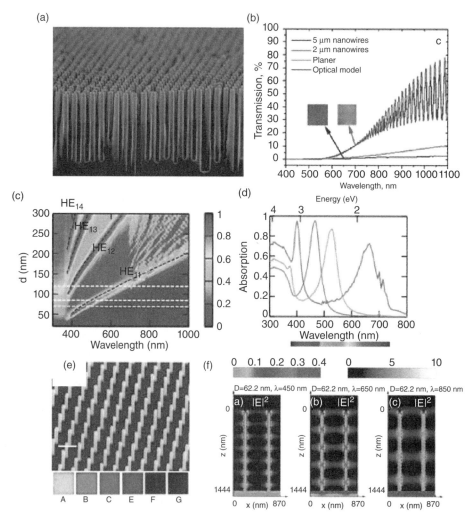

Figure 16.60   (a) SEM picture of an ordered silicon NW radial p–n junction array solar cells. (b) Transmittance spectra of silicon window structures with 5 μm (black), 2 μm (green) long NWs and with NWs (orange), with blue curve corresponding to optical model result. The insets are backlit color images of samples. (Reprinted from Reference 423, Copyright 2010 American Chemical Society.) (c) 2D contours of absorption as a function of NW diameter and wavelength for vertical Si NW arrays. Dash line corresponds to curves in (d). (d) Absorption curves of NW arrays with a diameter of 70, 85, and 120 nm. (Reprinted from Reference 442, Copyright 2012 The Optical Society.) (e) 30 degree tilted SEM picture of InAs NW arrays corresponding to sample A. Below are the color images of different samples. (f) Electric field intensity square distributions of NW with 62.2 nm diameter at different wavelengths. (*Source*: Reprinted from Reference 466, Copyright 2010 American Chemical Society.)

by silica beads, while the length is determined by etch time. Optical transmittance measurements showed that the absorption of Si thin films with NW arrays on the top are much higher than their planar counterpart with the same device thickness of about 8 μm, as shown in Figure 16.60b. For the planar device, because the absorption coefficient of Si at long wavelengths is much lower than that at short wavelengths, the transmittance is high for 700 nm and longer wavelength. The transmittance curve oscillates at this part due to the interference effect. While for sample with NW arrays, the transmittance is lower than 10% from 400 nm to 1000 nm wavelengths with 2 μm thick NWs layer. With a NW length of 5 μm, the transmittance is close to zero. Considering the fact that sample with an NW array has less material than a planar device, the light absorption performance of NW arrays is quite attractive.

In addition to photon absorption enhancement via scattering inside an NW array, it was found that even a single NW can demonstrate interesting photon management properties. Cao et al. showed that certain leaky mode resonances can occur in a single semiconductor NW, which can confine electromagnetic energy effectively (473). Wang et al. also calculated the leaky modes in silicon NW arrays (442). They demonstrated that due to symmetry matching requirements, incident light on vertical NWs can only couple to HE_{1m} leaky modes, as shown in Figure 16.60c. By varying the diameter of NWs, the wavelength associated to leaky modes and the resulting transverse resonance can be tuned. Accordingly, wavelength selective absorption can be realized by controlling the diameter, as shown in Figure 16.60d, which corresponds to the dash line cross-sections in Figure 16.60c, and the blue, green, and red curves are responsible for 70, 85, and 120 nm diameter NWs, respectively. The leaky modes resonance are caused by finite NW size and large refractive index contrast between semiconductor NWs and the surrounding, so this effect is applicable to other semiconductor materials, that is, Ge, amorphous Si, CdTe, and so on (474, 475). The aforementioned results are not only valuable for PV application, but also useful for other optoelectronic applications such as color-selective photodetectors.

Wu et al. also reported that optical absorption of InAs NWs can be tuned by geometrical tuning (466). Figure 16.60e shows the SEM image of InAs NWs grown in a high vacuum chemical beam epitaxy (CBE) unit by using electron-beam lithography (EBL) defined gold dots as seeds. With different diameter and length, the colors of NW sample array devices are different, which reveals different optical reflection spectrum. From the calculated electric field intensity distributions in Figure 16.60f, the vertical resonance can be clearly observed. Since the absorption coefficient of InAs decreases with increase of wavelength, optical absorption of 850 nm is not as strong as 450 nm wavelength. With a proper choice of diameter, length, and pitch, InAs NWs can also absorb either much more or much less light than a thin film counterpart, akin to Si NWs.

Besides NWs, NPL arrays have been extensively and successfully fabricated for efficient photon management and solar energy conversion with materials including CdS, Ge, Si, InP, SiO$_2$, and so forth (440, 441, 461, 476–479). In fact, NPLs can be more advantages than NWs for PV applications due to smaller surface area and less surface recombination which is one of the major issues for nanostructured solar cells (480, 481). While for single-diameter NPLs (Fig. 16.61a), which are similar to NWs discussed earlier, it was found that the increase of the light absorber material filling ratio leads to the increase of reflectance and the decrease of transmittance simultaneously, with the absorption showing a strong diameter dependency (459). The detailed relationship between absorption of an NPL array and the NPL diameter and pitch has been explored with finite difference time domain (FDTD) simulations with Ge as the model material, as shown in Figure 16.61b (475). It can be seen that optimal NPL structures can be identified with the best absorption ~80%.

To further enhance the broadband optical absorption capability, multi-diameter NPL (MNPL) structures have been studied, with smallest diameter tip for minimal reflectance and largest diameter base for maximal effective absorption coefficient (459, 475). Particularly, Fan et al. have presented ordered arrays of dual-diameter NPLs (DNPLs) with a small diameter tip and a large diameter base for an impressive absorption of ~99% of the incident light over wavelength range $\lambda = 300$–900 nm with a thickness of only 2 μm (459). Such a DNPL array was constructed via template-assisted vapor-liquid-solid (VLS) growth, utilizing dual-diameter anodic AAM as the template, which was achieved by a multistep anodization and etching process. Figure 16.61c shows an up-side-down cross-sectional SEM image of a blank AAM (i.e., before growth) with top and bottom pore diameters (D1 and D2) of ~40 and 110 nm, respectively. After the subsequent VLS growth, highly ordered Ge DNPLs embedded in the aforementioned AAM with $D1 \sim 60$ nm and $D2 \sim 130$ nm are formed (Fig. 16.61c, inset). The experimental absorption spectra of the obtained Ge DNPL arrays with equal lengths of about 1 μm for the two segments (total length of 2 μm), together with single-diameter NPL arrays with DNPL = 60 and 130 nm are plotted in Figure 16.61d. The Ge DNPL array exhibits 95–100% absorption for $\lambda = 900$–300 nm, which is a drastic improvement over single-diameter NPLs (Fig. 16.61d).

In order to further understand light coupling, propagation and absorption in NPL arrays, Hua et al. carried out a more systematic investigation on broadband solar spectrum absorption of MNPL arrays analyzed with FDTD simulations with Ge as the model material (475). The schematic of the Ge MNPL arrays is shown in Figure 16.61e, with the number of the segments $N = 3$. The lengths/height of the NPL arrays and the NCN (Nanocone) arrays here are all 2 μm. It was discovered that the broadband absorption of MNPLs was approaching that of NCNs when N increases, with $N = 7$ yielding the same light absorption level to NCNs, as demonstrated in Fig. 16.61f.

The above results have shown that by engineering shape of nanopillars, their optical absorption can be largely improved. On the other hand, PV performance of nanopillar based solar cells can still be limited by their relatively large surface area compared to thin film soalr cells. In this case, choice of materials is crucial, namely, materials systems such as CdS, CdTe, CIGS which have low surface recombination velocities are desirable as nanopillar materials (441, 480, 481). What's more, although materials such as Si and GaAs have high surface recombination velocity, decent device performance can also be achieved if proper surface passivation schemes can be applied (451, 482).

16.3.2.5.4 Photovoltaic Devices Based on Nanowire/Nanopillar Arrays Up to now, there have been a large number of works on NW-, nanorod-, nanopillar array-based PV devices using ZnO, Si, GaAs, CdS/CdTe, and so on as active materials (423, 431, 433, 456, 461, 483). In the following sections, the major research progress in this field will be discussed in detail.

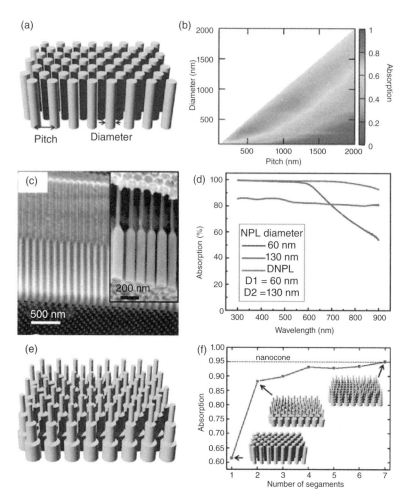

Figure 16.61 (a) Schematic of hexagonal Ge NPL arrays with single-diameter. (b) 2D contour of broadband absorption of Ge single-diameter NPL arrays. (Reprinted from Reference 475. Copyright 2013.) (c) Cross-sectional SEM images of a blank AAM with dual-diameter pores and the Ge DNPLs (inset) after the growth. (d) Experimental absorption spectra of a DNPL array with $D1 = 60\,nm$ and $D2 = 130\,nm$, and single-diameter NPL arrays with diameters of 60 and 130 nm. (Reprinted from Reference 459, Copyright 2010 American Chemical Society.) (e) Schematic of hexagonal Ge NPL arrays with multi-diameters. (f) Broadband-integrated absorption of 1000 nm pitch Ge MNPL arrays as a function of segment number. Dashed line represents the broadband-integrated absorption of 1000 nm pitch Ge NCN arrays. (*Source*: Reprinted from Reference 475, Copyright 2013).

ZNO NANOWIRE DYE-SENSITIZED SOLAR CELL   A pioneering work done by Law et al. in 2005 catalyzed research on 1D nanomaterial–based PVs (424). In this work, vertical arrays of ZnO NWs were grown on F:SnO$_2$ (FTO) substrates, as shown in Figure 16.62a. The electron diffusivity within the wires was estimated to be between 0.05 and 0.5 cm^2/s, much higher than that estimated for typically used ZnO nanoparticle films. PV measurements showed device characteristics of a short circuit current, $J_{sc} = 5.3$–5.85 mA/cm^2, open circuit voltage, $V_{oc} = 0.61$–0.71 V, fill factor, FF = 0.36–0.38 and the resulting efficiency, $\eta = 1.2$–1.5% as plotted in Figure 16.62b. The highest external quantum efficiency of the cell measured (Fig. 16.62b, inset) was 40% at 515nm wavelength, corresponding to the maximum absorption of the dye. The efficiency of the dye sensitized solar cell can be further improved by maximizing photon absorption at higher wavelengths in increasing the length of the cell, if electron diffusion length can be increased at the same time. In this regard, one simple method to increase electron diffusion lengths is through reducing the interfacial recombination rate by applying surface coatings on the nanocrystalline film (484–488).

For example, a conformal layer of TiO$_2$ was deposited on ZnO NWs via an atomic layer deposition system (424). This oxide shell is expected to suppress recombination by incorporating an energy barrier that physically separates the photo-generated electrons from the oxidized species within the electrolyte. The layer also acts to passivate the recombination centers on the surface of the NWs. The results showed an overall conversion efficiency of 2.25%, a significant improvement in performance of the cells.

SI NANOWIRE SOLAR CELL   As Si is still the dominant material for semiconductor industry as well as PV industry, Si-based nanostructured solar cells naturally attracted enormous attention. In fact, there are a number of works have been done on this aspect (423, 439, 456, 489–493). In these works, Si NW arrays were fabricated by vapor phase catalytic growth (456, 489), electroless chemical etching (439, 490–493) or DRIE, respectively (423). Among these three approaches, vapor phase catalytic growth has been discussed before, the main drawback being

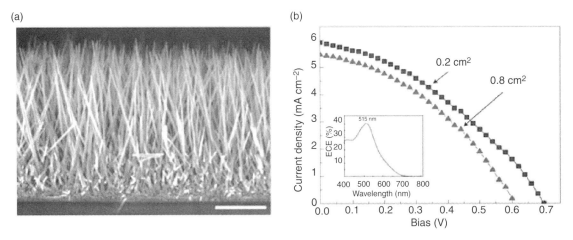

Figure 16.62   (a) SEM cross-section of a cleaved ZnO NW array on FTO, scalebar is 5 μm. (b) Traces of current density against voltage (*J–V*) for two cells with roughness factors of ~200. The small cell (active area: 0.2 cm^2) shows a higher V_{oc} and J_{sc} than the large cell (0.8 cm^2). The fill factor and efficiency are 0.37 and 1.51% and 0.38 and 1.26%, respectively. Inset shows the external quantum efficiency against wavelength for the large cell. (*Source*: Reproduced from Reference 424, Copyright © 2005, Nature Publishing Group).

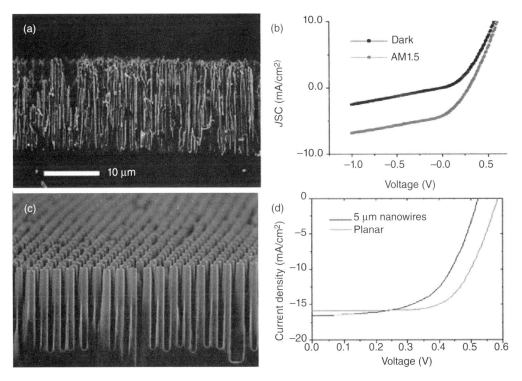

Figure 16.63   Si NW array solar cells. (a) SEM cross-section of a Si NW array prepared by electrochemical etching of a Si wafer; (b) dark and light *I–V* characteristics of the Si NW array solar cell 439. (c) SEM cross-section of a Si wire array fabricated with DRIE. (d) *I–V* characteristics of 5 μm Si wire array solar cell and planar Si solar cell. (*Source*: Reproduced from Reference 423, Copyright © 2010, American Chemical Society).

the metal incorporation in NWs to degrade minority carrier lifetime. To avoid this, Si NWs can be fabricated via electroless chemical etching (439). In this approach, n-type single crystalline Si wafer was simply etched inside mixed solution of silver nitrate and hydrofluoric acid. Figure 16.63a shows an SEM image of Si NWs array and thickness of it was controlled by etching time. After deposition of p-type a-Si, radial p–n junction NWs array was fabricated and its photovoltaic behavior was characterized using AM 1.5 illumination, as shown in Figure 16.63b. Take another work as example, DRIE was used to fabricate Si NW array. To fabricate Si NW array, Garnett et al. assembled a monolayer of silica spheres as DRIE etching mask (423). Figure 16.63c demonstrates a microscopy image of the resulting n-type Si NW array on the Si

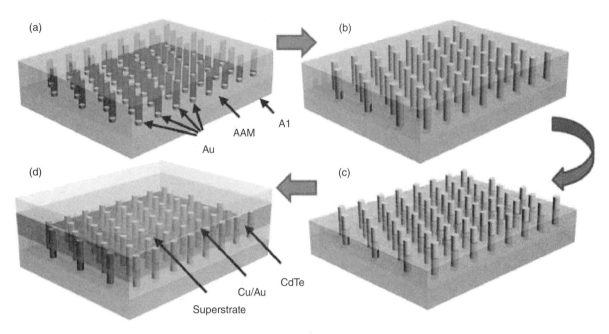

Figure 16.64 CdS NPLs/CdTe thin film cell fabrication scheme. (a) An AAM is grown on aluminum foil to form an array of pores in which Au catalyst is electrochemically deposited at the bottom. (b) Template-assisted VLS growth of CdS NPLs is then applied. (c) The AAM is partially etched to expose the top of the NPLs. (d) Thin film of CdTe is deposited on the exposed NPL array by CVD, followed by thermal evaporation of Cu/Au top contact. (*Source*: Reproduced from Reference 461, Copyright © 2009, Nature Publishing Group).

wafer. After DRIE, boron diffusion was performed on samples to produce radial p–n junctions. Figure 16.63c shows I–V characteristics of 5 μm length NWs arrays comparing with planar control sample. It was found that 4.83% energy conversion efficiency was achieved for 8 μm Si NW arrays which is about 20% performance improvement than the planar control sample, due to light trapping effect (423). Note that both electroless chemical and DRIE etching approaches were performed on Si wafer, which ensures that the resulting NWs are single crystalline without metallic contamination. However, fabricating Si into NWs array dramatically increases surface area, which can be detrimental since Si has high surface recombination velocity, although it greatly improved light absorption. Consequently, the obtained Si NW array solar cells have not overperformed the standard planar Si solar cells yet. In this case, materials with low surface recombination velocities are more promising candidates for nanostructure solar cells.

CDS/CDTE NANOPILLAR SOLAR CELL As mentioned earlier, II–VI compound semiconductors have much lower surface recombination velocities as compared to that of Si and GaAs. In addition, these materials are typically fabricated as polycrystalline films for solar cell applications. This fact suggests that bulk recombination, particularly at defective grain boundaries, can greatly affect PV performance. In this case, 3D structures are potentially beneficial in terms of carrier collection, as shown in Figure 16.64b. As an example, Fan et al. demonstrated low cost nanopillar array solar cells on Al foil with CdS/CdTe material system (461). A simple process schematic of the nanopillar solar cells is illustrated in Figure 16.64. In brief, about 2 μm thick anodic alumina membrane (AAM) with highly ordered pores was anodized on the aluminum foil. Then Au seeds were electrochemically deposited at the bottom of the pores as catalysts (Fig. 16.64) followed by CVD growth of CdS nanopillar (Fig. 16.64b). The processed AAM was partially and controllably etched in sodium hydroxide solution (Noah) at room temperature to expose the upper portion of the pillars to form the 3D structures (Fig. 16.64c). Importantly, this etching solution is highly selective and does not chemically react with and degrade the material quality of CdS nanopillars. Meanwhile, the exposed depth could be varied and controlled by adjusting the etching time to optimize the geometric configuration, where the device thickness comparable to the optical absorption depth and the bulk minority carrier diffusion length, for the enhanced conversion efficiency. After that, a p-type CdTe thin film with ~1 μm thickness was then deposited by CVD to serve as the photo-absorption layer due to its optimal bandgap (E_g = 1.5 eV) for the solar energy absorption (438). The top electrode was finally deposited by the thermal evaporation of Cu/Au (1 nm/13 nm) in order to achieve an acceptable transparency and to form an ohmic contact with the p-type CdTe film, as shown in Figure 16.64d. In such configuration, the backside electrical contact to the n-type CdS nanopillars was simply the aluminum support substrate while the entire solar cell device could then be bonded on the top to a glass slide by epoxy for the encapsulation.

In this work, CdS nanopillar arrays have been well engineered. Figure 16.65a shows an SEM image of the nanopillar array with pillar diameter of 200 nm and pitch of 500 nm. The perfect ordering was achieved using nanoimprint before anodization (461). Importantly, CdS nanopillars were confirmed to be single crystalline by the high-resolution TEM, as shown in Figure 16.65b. This ensured high electron mobility and efficient transport in the nanopillars. A photograph of a fully fabricated SNOP cell device is shown in Figure 16.66a. The cell performance was characterized under different illumination condition, and the I–V curves are shown in Figure 16.66b. Under AM 1.5G

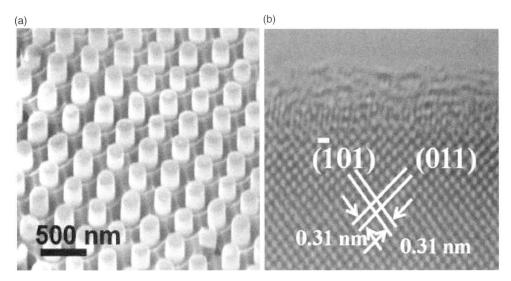

Figure 16.65 (a) SEM image of CdS NPL array partially embedded in AAM. (b) TEM of a CdS nanopillar shows its single crystallinity. (*Source*: Reproduced from Reference 461, Copyright © 2009, Nature Publishing Group).

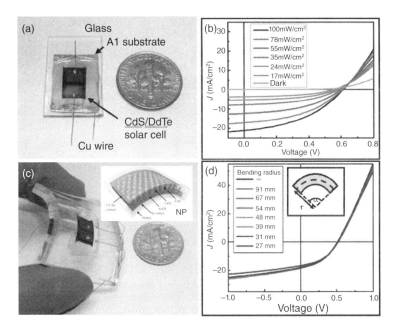

Figure 16.66 CdS NPL/CdTe thin film PVs. (a) An optical image of a fully fabricated cell bonded to a glass substrate. (b) *I–V* characteristics under various illumination conditions. (c) An optical image and schematic (insert) of a bendable module embedded in PDMS. (d) *I–V* characteristics of a flexible cell for various bending radii, showing minimal dependence on curvature. (*Source*: Reproduced from Reference 461, Copyright © 2009, Nature Publishing Group).

illumination ($100\,\mathrm{mW/cm^2}$), the cell produced a short circuit current density (J_{SC}) of ~$21\,\mathrm{mA/cm^2}$, an open circuit voltage (V_{OC}) of ~$0.62\,\mathrm{V}$ and a fill factor (FF) of ~43 %, yielding an efficiency (η) of ~6% and it was found that the efficiency was limited by the non-ideal metal-CdTe contact.

Besides using rigid substrates, nanopillar solar cell fabrication was also performed on plastics for flexible and high performance PVs, which are of particular interest for a number of technological applications (494–496). Simply put, a layer of polydimethylsiloxane (poly(dimethylsiloxane) (PDMS), ~$2\,\mathrm{mm}$ thick) is deposited and cured on the top surface following the top-contact metallization process. Then the aluminum substrate at the back is released by a wet chemical etching process. A layer of ~$200\,\mathrm{nm}$ thick indium film is deposited as the bottom contact to the n-CdS nanopillars and another ~$2\,\mathrm{mm}$ thick PDMS layer is put down and cured on the back side to complete the encapsulation. Figure 16.66c shows an optical image and device structural schematic. Notably, the entire SNOP cell structure is located in the neutral mechanical plane of the PDMS substrate which minimizes the strain on CdS nanopillars and CdTe film. Figure 16.66d demonstrates the *I–V* characteristics of a flexible nanopillar solar cell module under different bending conditions and confirms the negligible change in the cell performance such as the energy conversion efficiency.

Overall, the key advantage of this work is having combined a few low cost and scalable processes, including anodization, electrodeposition, and chemical vapor deposition. And the ability to directly grow single-crystalline active material structures on large aluminum sheets is highly attractive to minimize the material and processing cost. Further improving cell performance, by optimizing contacts and 3D geometry, including pillar size, pitch, and so on, can potentially lead to a cost-effective solution for next generation PV technology.

16.3.3 Nanowire Assembly and Integrated Circuitry

It is possible to grow NWs directly on the surface of some materials, avoiding the necessity to assemble them into desired structures. A few NWs can be synthesized with the required shape, dimensions, and composition straightforwardly within the device platform (497–501). These approaches are, however, limited to a few types of materials that can be assembled into a wire-like morphology. Separating the synthesis of NWs from their fabrication into a device has many advantages, including retaining the determined reaction conditions for NW growth and purgation without compromising the other materials within the ultimate device. Additionally, the synthesized NWs can be manipulated as individual structures and integrated into complex nanoarchitectures.

This section concentrates on techniques that have been developed to direct the assembly of many individual NWs over large areas for their incorporation into various technologies or materials. Our focus is limited to the post-synthetic assembly of anisotropic nanostructures with nanoscale diameters and a wire-like morphology. Directing the assembly of these structures into desired configurations presents a unique challenge. The assembly of NWs requires techniques that are able to appropriately manipulate and conduct the deposition of individual nanostructures. Ideally these processes would also be cost effective and able to assemble many NWs in parallel over large areas (several square centimeters).

NW materials synthesized mainly by bottom-up growth methods have shown crystalline nature enabling fast carrier transport for high-performance electronic applications (239, 254, 356, 409, 502–504). Concisely, individual NW devices are usually fabricated following such a process: NWs collected on a substrate are ultrasonicated and suspended in an organic solvent and then drop-casted onto an insulator substrate. Afterward, a photolithography or electron-beam lithography process followed by metallization and lift-off is used to pattern metallic electrodes to the NWs terminals (176, 237, 239, 505, 506). Typically, such drop-casting of random aligned NWs results in a very low device yield. However, it is still satisfactory to study their fundamental properties of single NW concept-proof devices. Obviously, this method is not suitable for low cost and large-scale integration of NW devices. The assembly of NWs with controlled orientations and interspacing over large areas is fundamental for the fabrication of complex logic circuits and high-performance electronic devices in which randomness of NW alignment would result in performance degradation, for example, deteriorated transistor gate electrostatic coupling (507). In the past, significant efforts have been put on investigating generic approaches to assemble NWs on various substrates, for example, flow-assisted alignment (64, 245, 355, 508, 509), Langmuir–Blodgett (510–515), bubble-blown techniques (516), electric field-directed assembly (63, 517–519), and contact/roll printing techniques (297, 504, 507, 520–525). The pros and cons of these methods are concisely summarized in Table 16.3 followed by the detailed discussions (532).

16.3.3.1 *Flow-Assisted Nanowire Alignment*
It is known that the motion of a fluid can create a shear force against a solid boundary (508, 526). This effect can be utilized to align NWs suspended in a solution. The orientations of NWs will be realigned along the flow direction of the fluid to minimize the fluid drag forces. Huang et al. (508) have further developed this technique to align NWs through confining the fluid in a microfluidic channel. In this flow-assisted assembly technique with the schematic picture shown in Figure 16.67a, a flat substrate is covered by a PDMS mold with a microchannel with width ranging from 50 to 500 μm and length from 6 to 20 mm. When flowing a suspension of NWs through the microchannel, the NWs can be assembled parallel to the flow direction. Figure 16.67b shows a SEM image of NWs assembled on substrate surface within microfuidic flow, representing that the NWs are aligned along the flow direction. This fluidic-directed assembly can extend over several hundreds of micrometers and is restricted only by the size of micro-channels. Huang et al. have also demonstrated high-performance p-Si nanowire transistors using this fluid-assisted alignment technique (508). Periodic aligned NWs arrays have also been achieved by Huang et al. by uniting a surface modification technique with the flow-assisted assembly technique (508). In this case, NH_2-terminated monolayers were patterned on the surface of a SiO_2/Si substrate in the shape of parallel stripes with a separation of a few micrometers. During the flow-assisted assembly process, the NWs are preferentially captured by the NH_2-terminated surface regions. The direction of the NWs is dominated by the Shear force created by the fluidic flow in the microchannels. Consequently, both the controlled location and orientation of the NWs are achieved.

Furthermore, this flow-directed technology can be utilized to align the NWs into more complex and crossed NWs structures, which are pivotal for constructing nanodevice arrays (64, 245, 355, 508, 526). As shown in Figure 16.67c, crossed NW arrays can be obtained by a layer-by-layer deposition process, that is to say, by interchanging the flow in perpendicular directions through a two-step assembly process. The crucial characteristic of this layer-by-layer assembly strategy is that each layer is independent with the others. Accordingly, various homo- and hetero-junctions can be constructed at the crossed points. NWs with different conduction types (e.g., p-Si and n-GaN NWs) in each step have been used to fabricate logic gates with computational functions and light emitting diodes from the assembled crossbar NW structures (Fig. 16.67d) (245, 355). Fancinatingly, this NWs assembly technique is compatible with various types of substrates, including plastics. Figure 16.67e and f demonstrate assembled NWs array TFTs and light-emitting diodes (LEDs) on plastic substrate. Further measurements have shown that NW array TFTs have >100 cm²/Vs FET, which is significantly better than that of a-Si and organic semiconductors for flexible

TABLE 16.3 A Summary of Nanowire Assembly Technologies

NW Assembly Technologies	Advantages	Disadvantages	Examples	References
Flow-assisted alignment in microchannels	1. Parallel and crossed NW arrays can be assembled 2. Compatible to both rigid and flexible substrates.	1. Area for NW assembly is limited by the size of fluidic microchannels 2. Difficult to achieve very high density of nanwoire arrays 3. NW suspension needs to be prepared first	Si, CdS, GaN, InP, GaP NWs	(64, 245, 355, 508, 526)
Bubble-blown technique	1. Area for NW assembly is large; 2. Compatible to both rigid and flexible substrates.	1. It is difficult to achieve high-density NW arrays 2. NW suspension needs to be prepared first	Si NWs	(516)
Contact printing	1. Area for NW assembly is large 2. High-density NW arrays can be achieved 3. Parallel and crossed NW arrays can be assembled 4. Direct transfer NW from growth substrate to receiver substrate 5. Compatible to both rigid and flexible substrates 6. NW assembly process is fast	1. Growth substratesneeds to be planar 2. The process works the best for long NWs	Si, Ge, InAs, CdSe, Ge/Si core/shell, ZnO, SnO$_2$ NWs	(297, 504, 507, 520–523, 525, 527–529)
Differential roll printing	1. Area for NW assembly is large 2. High-density NW arrays can be achieved 3. Direct transfer NW from growth substrate to receiver substrate 4. Compatible to both rigid and flexible substrates 5. NW assembly process is fast	1. Growth substratesneeds to becylindrical 2. The process works best for long NWs	Ge, ZnO NWs	(524, 530)
Langmuir– Blodgett technique	1. Area for NW assembly is large 2. High-density NW arrays can be achieved 3. Parallel and crossed NW arrays can be assembled 4. Compatible to both rigid and flexible substrates	1. NWs typically need to be functionalized with surfactant 2. The assembly process is slow, and has to be carefully controlled 3. NW suspension needs to be prepared first	Si, Ge, ZnSe, VO$_2$ NWs	(510–515)
Electric fields assisted orientation	1. NWs can be placed at specific location 2. Compatible to both rigid and flexible substrates 3. NW assembly process is fast	1. Patterned electrode arrays are needed 2. Area for NW assembly is limited by the electrode patterning 3. NW density is limited 4. It works the best for conductive NWs 5. NW suspension needs to be prepared first	Au, Ag, Si, SiC, InP NWs	(63, 64, 517–519, 531)

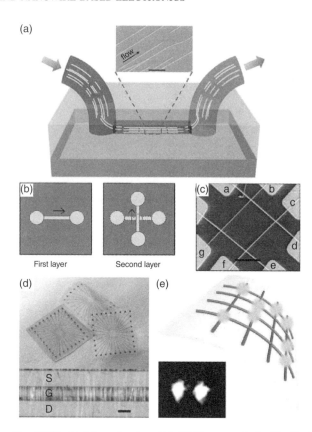

Figure 16.67 Fluidic flow-directed assembly of NWs. (a) Schematic of paralleled NWs arrays obtained by flowingNWs solution through a micro-channel on a substrate. The inset is the parallel NWs SEM image. (b) Schematic and (c) SEM image of crossed NWs matrix obtained by orthogonally changing the flow direction in a sequential flow alignment process. (Reprinted with permission from Reference 508. Copyright 2001American Association for the Advancement of Science.) (d) Upper photo is a picture of plastic devices with Si NW film. The plastic devices show high mechanical flexibility. Lower image is an optical micrograph of a locally gated NW-TFT. (Reprinted with permission from Reference 509. Copyright 2003 Nature Publishing Group.) Scale bar, 5 mm. (e) Schematic of a flexible light-emitting display consisting of a crossed-NW LED array on a flexible plastic substrate. Inset: electroluminescence images of localized emission from forward-biased Si-GaN junctions. (*Source*: Reprinted from Reference 533. Copyright 2003 American Chemical Society).

electronics (503, 507, 509, 534, 535). Notwithstanding, the main drawback of this process is that the area for NW alignment is restricted by the size of the fluidic microchannels. And it is rather difficult to form a uniform shear force in a large channel.

16.3.3.2   Bubble-Blown Nanowire Assembly on Plastic Substrate   The Shear force created by the expansion of a blow-bubble film can also be used for large-scale assembly of NW. Yu et al. developed a distinctive NWs assembly technique based on this principle and the process is schematically shown in Figure 16.68a (516). Briefly, a homogenous polymer suspension of NWs was firstly prepared then dispersed on a circular die followed by blow the die into a bubble at controlled pressure and expansion rate (516). During this process, the expansion caused shear force aligned more than 85% of the NWs along the upward expansion with only ±6° discrepancy (516). Then the NWs can be transferred to either rigid or flexible substrates by making them in contact with the bubble (516). Figure 16.68b and c demonstrate Si NW arrays assembled on plastic substrate and resulting FET devices. In experiments, Si NWs were transferred to flexible plastic sheets up to 225 by 300 mm^2, single-crystal wafers up to 200 mm in diameter. In principle, NWs density within the film can be adjusted by the concentration of the NWs in the polymer suspension, thus large NWFET arrays could also be fabricated by this method. However, it is challenging to achieve close packs of NWs due to the expansion of the bubble, and thus NWs density was only modest (~0.04 μm^{-2}), which can be seen from Figure 16.68c (516). Consequently, it is not an ideal technique for fabricating high current output devices with small dimension.

16.3.3.3   Contact/Roll Printing Techniques   Recently, a novel and convenient contact printing technique for wafer-scale assembly of aligned and high-density NW arrays has been reported (297, 520, 524). This contact printing process can directly align NWs into parallel arrays when transferring from the growth substrates to the receiver substrates under good control (297, 507, 520, 524, 530). As shown in Figure 16.69a, this process uses randomly grown NWs on donor substrate sliding over a receiver substrate with desired direction. During this contact sliding process, NWs are realigned by the sliding shear force, and are finally detached from the donor substrate when held by the van der Waals interactions with the contacted surface of the receiver substrate, leading to the direct transfer of paralleled NWs onto the receiver substrates (Fig. 16.69b). In order to minimize the undesirable NW-NWfriction and the uncontrolled breakage and discrete of NWs

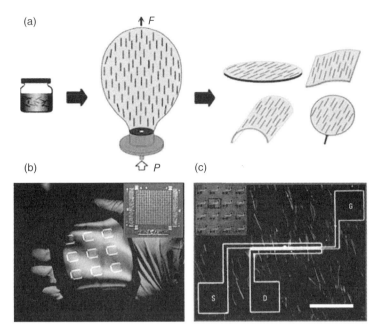

Figure 16.68 Bubble-blown assembly of NWs: (a) Schematic process flow of NWs assembly. (b) Plastic substrate with Si NWs device arrays. (c) Dark-field optical image of one top-gated Si NWFET device; the scale bar is 50 μm. (*Source*: Reproduced with permission from Reference 516, Copyright 2007 Nature Publishing Group).

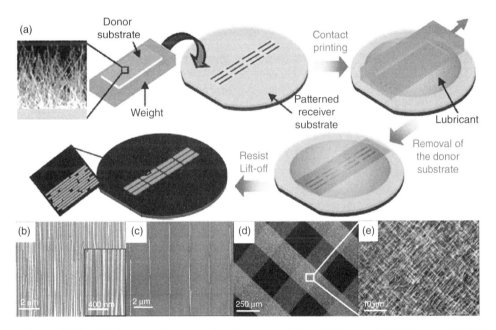

Figure 16.69 Contact printing of NWs. (a) Schematic of the process flow for contact printing of NW arrays. (b) SEM images of Ge NWs printed on a Si/SiO$_2$ substrate showing highly dense and aligned monolayer of NWs. (c) SEM image showing regular array assembly of single Ge NWs at predefined locations on a Si/SiO$_2$ substrate. (d and e) Optical images of double layer printing for SiNW crossed assembly. (*Source*: Reprinted from Reference 297, Copyright 2008 American Chemical Society).

during the sliding process, a mixture of octane and mineral oil is utilized as a spacing layer between the two substrates (297). By minimizing the interstitial mechanical friction, well-controlled transfer of NWs is enabled through chemical binding interactions between NWs and the surface of the receiver substrate (297). As a result, the printed NWs density is readily modulated and tuned by the appropriate surface chemical modification (297, 520). After NW printing, the patterned resist is removed by a standard lift-off process using acetone, leaving behind the NWs that were directly transferred to the patterned regions of the substrate. With this technique, ~7 NWs/μm high dense array of NWs (Fig. 16.69b) and streams of single NWs (Fig. 16.69c) have been assembled (297). In addition, NW crosses or controlled stacking of the

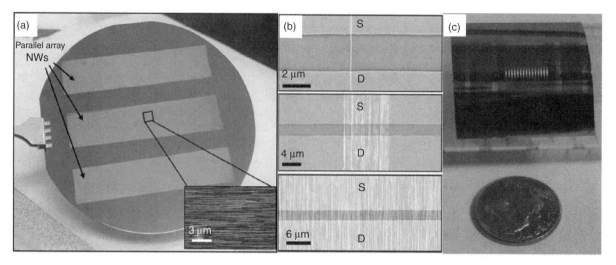

Figure 16.70 Wafer-scale NW printing and devices based on printed NW arrays. (a) Large area and highly uniform parallel arrays of aligned Ge NWs were assembled on a 4 inch Si/SiO$_2$ wafer by contact printing. (b) SEM images of parallel arrayed Ge/Si NW FETs with different channel widths (single NW, 10 μm, and 250 μm). (c) Optical image of a diode structure fabricated on parallel arrays of p-Si NWs on a flexible plastic substrate. Asymmetric Pd–Al contacts are used to obtain Schottky diodes. (*Source*: Reprinted from Reference 297, Copyright 2008 American Chemical Society).

NWs can be achieved by a two-step printing methodology (297). First, a layer of paralleled NW arrays were contact printed onto a receiver substrate. The sample was then spin-coated with a thin film of Poly (methyl methacrylate) (PMMA) to serve as a buffer layer, followed by a second printing step normal to the direction of the first layer. Finally, the polymer buffer was etched away by a moderate O$_2$ plasma process, resulting in the large arrays of NW crosses. Thus, large arrays of NW crosses were formed, as shown in Figure 16.69d.

The contact printing technique is not only versatile, more importantly, it is highly scalable. The researchers successfully assembled highly parallel and dense NWs arrays on a 4 inch Si wafer, shown in Figure 16.70a (297). Here, it is noted that the density of the NWs films that are obtained by this contact printing process is comparable to those attained by the Langmuir–Blodgett (LB) method, and is much larger than those obtained by the bubble-blown technique (∼0.3 NW/μm). A high density of NWs is desirable for achieving dense arrays of single NW devices, or high ON currents considering thin film transistors with parallel array NW channels. Figure 16.70b shows the SEM images of back-gated parallel arrayed NWFETs with different channel widths (single NW, 10 μm, and 250 μm). Furthermore, printed parallel array Si NW diodes can be also fabricated on flexible plastic Kapton substrates as shown in Figure 16.70c.

To further develop NW contact printing approach, the researchers have grown NWs on cylindrical substrates then transfer NWs to various types of substrate with a differential roll printing method (524). As shown in Figure 16.71a, this technique is based on the growth of NWs on a cylindrical substrate (i.e., roller) using the vapor-liquid-solid process, and the directional and aligned transfer of the as-grown NWs from the donor roller (e.g., glass or quartz tube) to a receiver substrate via differential roll printing. The receiver substrate can be patterned using photolithography to define the assembly regions. It was found that the NW assembly is relatively insensitive to the rolling speed (∼5 mm/min), and the printing outcome, however, highly depends on the roller-receiver substrate pressure. The optional pressure is ∼200 g/cm^2. At lower pressures, aligned transfer of NWs is not observed, and at higher pressures, mechanically induced damage to the NWs is observed, resulting in the assembly of short wires (<1 μm long). Following the NWs differential roll printing, the patterned resist is removed by a standard lift-off process, leaving behind assembled NWs at the predefined locations. Figure 16.71b shows the SEM image of the roller printed Ge NWs on a Si/SiO$_2$ substrate, clearly demonstrating the assembly of well aligned and dense (∼6 NW/μm) NW parallel arrays. Importantly, the roll printing process is compatible with various types of substrates, for example, rigid Si, glass, and flexible paper and plastic (Fig. 16.71c) (524). In addition, Figure 16.71d shows the roll printed NW FETs on a flexible Kapton substrate using Ge/Si core/shell parallel array NWs as the channel material, with NWdensity ∼6 μm^{-1}, demonstrating high ON current. It is worth noting that this roll printing method is highly promising for roll-to-roll production of flexible devices by nature. Combining single crystalline semiconductor NWs, high-performance flexible electronic device can be fabricated with high throughput using this technique (536).

16.3.3.4 *Other Approaches*

Besides the aforementioned NW assembly methods, other assembly approaches such as LB technique, electric field–assisted NW assembly (63, 64), and strain-release NW assembly (537), and so on have also been developed. These techniques will be briefly introduced in the following.

LB technique is usually utilized to deposit one or more organic monolayers from the surface of a liquid onto a solid substrate by immersing the substrate into the liquid to form extremely thin films with high degree of structure ordering (511–514, 532). It was found that this technique can also be used for assembly of nanomaterials including nanoparticles, nanorods, NWs, nanotubes, and nanosheets (513, 538–541). To perform assembly of these materials with LB technique, they are typically functionalized by surfactants in order to form stable suspensions in the organic solvents (510–515). The process of the LB technique can be briefly described as following, which is shown in

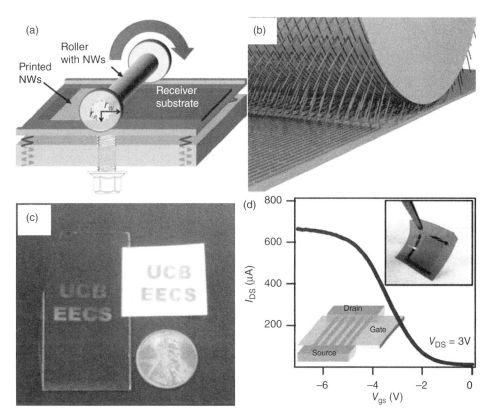

Figure 16.71 Differential roll printing. (a) Schematic of the differential roll printing setup. (b) Roll printing NWs onto the substrate. (c) Large area and aligned arrays of Ge NWs printed on glass and photography paper. The substrates were patterned using photolithography and functionalized with poly-L-lysine before printing to enable patterned assembly of NWs. (d) Parallel arrayed NW FETs on a plastic substrate. The bottom inset shows the schematic of the NW device structure and the top inset is a photograph of the substrate after NW device fabrication. (*Source*: Reprinted with permission from Reference 524, Copyright 2007 American Institute of Physics).

Figure 16.72. A NW-surfactant monolayer is firstly formed on a liquid surface in an LB trough. Then a barrier compresses the NW monolayer with an appropriate level of compression. The NWs with different orientations are realigned as parallel arrays with their longitudinal axes perpendicular to the compression direction to reduce the surface energy of the liquid. The vertical-dipping or horizontal-lifting techniques are then used to deposit the monolayer of the aligned NWs onto a substrate. The distance between the parallel NWs can be controlled by the lifting speed and compression pressure.

Up to date, many researchers have used LB technique to assemble various NWs in large scales, for example, silver NWs (511), Si NWs (510), Ge NWs (512), ZnSe NWs (513), and VO_2 NWs (514). In many works, the distance between the transferred NWs is adjusted from micrometer scale to well-ordered and close-packed structures by the compression process. The capability to assemble NWs with hierarchical structures makes the LB technique attractive for fabricating devices with complex structure. Hierarchical structures can be prepared by repeating the assembly process after altering the orientation of the substrate (510). Although LB technique has demonstrated its versatility for NW assembly, the disadvantages of this technique rest in several aspects, such as NWs have to be surface functionalized, LB process is slow and the condition control is rigid otherwise NW array films are not uniform, and so on, as summarized in Table 16.3.

Polarization of materials in electric field induces force parallel to the field which can be used to align semiconductor NWs (63, 64, 517–519, 531). Typically, such experiments are performed in NW suspension where NWs are free to move and reorient. Electric field can be created by the pair electrodes on the supporting substrate. In this case, multiple NWs could be aligned parallel along the electrical field direction, and individual NWs could also be fixed at specific positions with controlled directionality. Furthermore, the alignment of NWs can be accomplished in a layer-by-layer fashion to fabricate crossed NW junctions through altering electrical field direction of consecutive NWs solutions. (64)

Recent development of this technique combined dielectrophoretic force and fluidic NW alignment (63). In this work, laminar flow of NW suspension was established in a thin channel with a substrate patterned with pair electrodes serving as one side of the channel, as shown in Figure 16.73. In the stable flow, alternating electric field–induced dielectrophoretic force aligned NWs and trapped them in between the pair electrodes. By balancing NW–NW interactions, and dielectrophoretic, surface and hydrodynamic forces, the researchers demonstrated self-limiting assembly of single NWs with 98.5% yield on 16,000 electrode sites covering $400 \, mm^2$ (63).

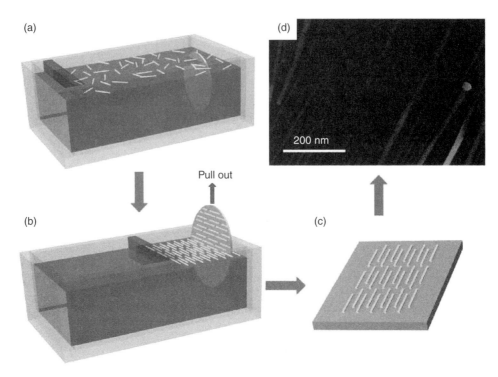

Figure 16.72   Langmuir–Blodgett assembly (532). (a) Random NWs suspended in the Langmuir–Blodgett trough. (b) A wafer being pulled vertically from the suspension in parallel with the lateral motion of the barrier. (c) A monolayer of parallel NWs array is assembled onto the substrate. (d) SEM image of paralleled Ge NWs film transferred to substrate by LB compression. (*Source*: Reprinted from Reference 512, Copyright 2005 American Chemical Society).

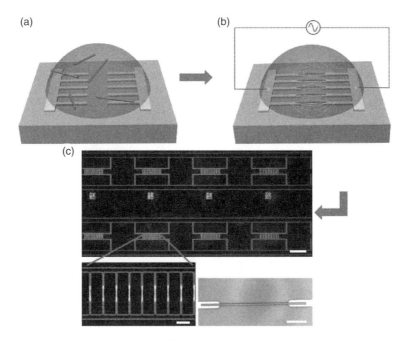

Figure 16.73   Electric field–assisted NW assembly 532 (a) Random NWs in the suspension on the electrodes before electric field–assisted assembly. (b) NWs are paralleled after electric field–assisted assembly. (c) Optical dark-field images of NWs assembled onto electrodes by electric field assisted–process. (*Source*: Reprinted with permission from Reference 63, Copyright 2010 Nature Publishing Group).

Electric field alignment/assembly of NWs is a technique of importance, particularly for the case that NWs are required to be positioned at specific locations precisely. However, the major disadvantage of this technique rests in the need for prefabricated microelectrode arrays to generate electric field on chip, which can increase the complexity of the process and limit the scalability of the process.

Recently, a strain-release assembly method was developed to align inorganic NWs on stretchable substrates (537). In this method, NWs on the as-grown substrate were first transferred to a stretched PDMS substrate by contact printing followed by release of strain afterward (537). During the strain releasing process, spacing between neighboring NWs was closed up resulting in parallel alignment of NW along the direction normal to stretching and increase of NW surface density (537). Thereafter, the aligned NW arrays could be transferred again to other rigid or flexible substrates for device fabrication. In this interesting assembly method, the large-strain elasticity of the substrate and the static friction between the NWs and the substrate are the two pivotal factors (537). In fact, besides transferring bottom-up grown NWs, the similar approach has also been utilized to transfer top-down fabricated Si, GaAs, InAs, and PZT nanomaterials for high-performance flexible electronics and energy harvesting (542–545).

16.3.3.5 Nanowire Integrated Circuits

16.3.3.5.1 Nanowire Parallel Arrays for Ultrahigh Frequency Transistors With the fast development of personal communication industry, ultrahigh frequency devices with light weight, low power consumption, and flexibility are in great demand. In this regard, high mobility inorganic semiconductor NWs are naturally ideal candidates. Recently, RF response of InAs NW array transistors on mechanically flexible substrates has been demonstrated by Takahashi et al. (503) In this work, InAs NWs (average diameter and length of ~30 nm and ~10 μm) were synthesized on Si/SiO_2 substrates by a physical vapor transport method. Subsequently, NWs are directly transferred from the grown substrate to the polyimide (PI) surface as parallel arrays by aforementioned contact printing technique. To achieve patterned assembly of NWs, the polyimide surface is first coated with a lithographically patterned resist layer, followed by NW printing and lift-off in a solvent. Nickel source (~50 nm) and drain electrodes were then formed, followed by atomic layer deposition of Al_2O_3 (~8 nm) as the gate dielectric. Finally, Al top-gate (G) electrodes (40 nm) were fabricated. All electrodes were defined by photolighography and lift-off process. The typical channel length (S/D electrode spacing) is $L \sim 1.5$ μm, and the channel width is 100–200 μm (503). The gate electrode length is $L_G \sim 1.4$ μm. The configuration of the electrical pads matches that of the conventional ground-signal-ground (GSG) microwave probes (150 μm pitch). Figure 16.74a shows the layered schematic of a NW array RF device. The printed NW density is ~4 NWs/μm as confirmed by scanning

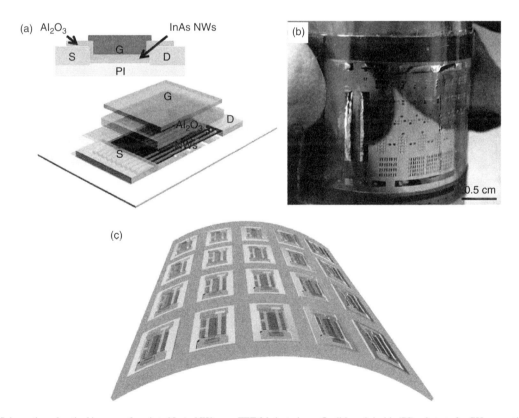

Figure 16.74 Schematic and optical images of a printed InAs NW array FET fabricated on a flexible polyimide (PI) substrate for GHz operation. (a) Schematic illustration of the NW parallel array FET, illustrating the various layers of the device. The cross-sectional image is shown in the top. (b) Photograph image of the fabricated NW device array on a bendable polyimide substrate. (c) Schematic of printed NW arrayfield-effect transistors integrated on aflexible substrate with high-frequency operation (532). (*Source*: Reprinted from Reference 503, Copyright 2010 American Chemical Society).

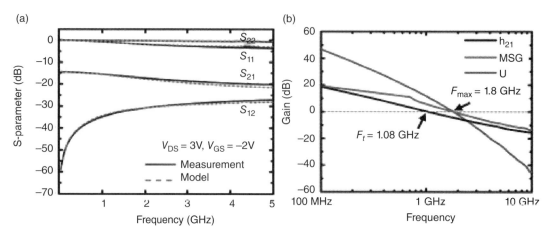

Figure 16.75 RF characterization of an InAs NW array FET. (a) Measured (black solid line) and modeled (red dashed line) scattering parameters, S11, S12, S21, and S22 of an InAs NW array FET with channel width ~200 μm after de-embedding for frequencies between 40 MHz and 5 GHz. (b) Current gain (h21), maximum stable gain (MSG), and unilateral power gain (U) extracted from measured S parameters as a function of frequency. The unity current gain frequency, ft, and unity power gain frequency, fmax, are ~1.08 and 1.8 GHz, respectively. (*Source*: Reprinted from Reference 503, Copyright 2010 American Chemical Society).

electron microscopy (503). After the completion of the fabrication process, the polyimide layer is peeled off from the rigid Si/SiO_2 handling wafer, resulting in mechanically flexible device arrays, as shown in Figure 16.74b.

To directly extract the high-frequency behavior, the two-port scattering parameters (S parameters) of InAs NW array FETs were measured in the common-source configuration using standard procedures with a vector network analyzer over a frequency range from 40 MHz to 10 GHz (503). The S parameters were then used to analyze the RF performance of the device (Fig. 16.75a). S_{11}, S_{22}, S_{21}, and S_{12} are, respectively, the reflection coefficient of the input, the reflection coefficient of the output, the forward transmission gain, and the reverse transmission gain. Figure 16.75b shows various radio frequency metrics of a representative device with channel width 200 μm, all derived from S parameters. Specifically, unity transit frequency of the current gain (h_{21}) of a transistor is called f_t and is an important factor for determining the high-frequency limit of the transistor for various analog/radio frequency and digital applications. To obtain f_t from the measurements, S parameters are first converted to hybrid parameters (*h* parameters). The parameter h_{21} is plotted as a function of frequency (Fig. 16.75b), and f_t occurs at the frequency where h_{21} equals 1, that is, 0 dB. As depicted in Figure 16.75b, InAs NW array FETs exhibit an impressive $f_t = 1.08$ GHz. Maximum stable gain (MSG) and maximum unilateral gain (U, Mason gain) are also extracted as a function of frequency. Mason gain is the maximum unilateral power gain the device can provide at a specific frequency of operation. The maximum frequency of oscillation, f_{max}, is determined by extracting Mason's gain at 0 dB. It is an important figure of merit, and at frequencies beyond f_{max}, a transistor cannot provide any power gain and turns into a passive component. As shown in Figure 16.75b, the NW array FET has $f_{max} = 1.8$ GHz (503). The high-frequency response of the device is due to the high saturation velocity of electrons in high-mobility InAs NWs (503). The results present a scalable platform for integration of ultrahigh frequency devices on flexible substrate, as schematically shown in Figure 16.74c, with potential applications on high-performance digital and analog circuitry.

16.3.3.5.2 Programmable Nanowire Circuits for Nanoprocessors A nanoprocessor constructed from intrinsically nanometer-scale building blocks is an essential component for controlling memory, nanosensors, and other functions proposed for nanosystems assembled from the bottom-up. Important steps toward this goal over the past 20 years include the realization of simple logic gates with individually assembled semiconductor NWs and carbon nanotubes (119, 245, 249, 270, 546, 547), but with only 16 devices or fewer and a single function for each circuit. Recently, logic circuits also have been demonstrated that use two or three elements of a one-dimensional memristor array (548), although such passive devices without gain are difficult to cascade. These circuits fall short of the requirements for a scalable, multifunctional nanoprocessor (549, 550) owing to challenges in materials, assembly, and architecture on the nanoscale. C.M. Lieber et al. (527) fabricated the programmable and scalable logic tiles for nanoprocessors that surmount these hurdles. The tiles were built from programmable, nonvolatile NW transistor arrays (PNNTAs), as shown in Figure 16.76. Ge/Si core/shell NWs coupled to designed dielectric shells yielded single-NW, nonvolatile field-effect transistors (FETs) with uniform, programmable threshold voltages and the capability to drive cascaded elements. Then an architecture was developed to integrate the programmable NW FETs and define a logic tile consisting of two interconnected arrays with 496 functional configurable FET nodes in an area of 960 mm². The logic tile was programmed and operated first as a full adder with a maximal voltage gain of 10 and input-output voltage matching, then the same logic tile be reprogrammed and used to demonstrate full-subtractor, multiplexer, demultiplexer, and clocked D-latch functions. These results represent a significant advance in the complexity and functionality of nanoelectronic circuits built from the bottom-up with a tiled architecture that could be cascaded to realize fully integrated nanoprocessors with computing, memory, and addressing capabilities.

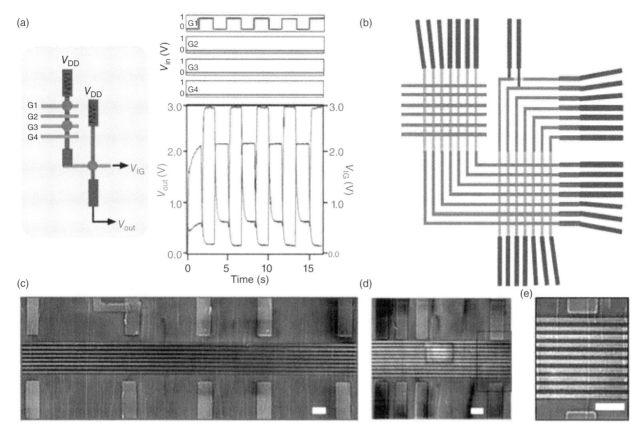

Figure 16.76 (a) Characterization of a NW–NW, coupled multigate device. Left: schematic of the device. Green dots indicate the gate nodes that were programmed as an active state. Top right: input signals to G1–G4. Bottom right: output signals from NW1 (V_{IG}, blue) and NW2 (V_{out}, red). (b) Design of the unit logic tile for integrated nanoprocessors containing two PNNTAs, block 1 (upper left) and block 2 (lower right), comprising charge-trapping NWs (pink) and metal gate electrodes (gray). The PNNTAs are connected to two sets of load devices (red). Lithographic-scale electrodes (blue) are integrated for input and output. Each PNNTA provides programmable logic functionality of up to approximately eight distinct logic gates. (c–e) SEM images of the different regions of the NW logic tiles. Scale bar, 1 µm (527).

16.3.3.5.3 All-Nanowire Integrated Image Sensor Circuitry The ability to fabricate a wide range of electronic and sensor devices with defined functionalities based on printed NW arrays enables the exploration of heterogeneous NW circuitry with on-chip integration. To examine this feasibility, proof-of-concept circuits were fabricated by Fan et al. that incorporate NW sensors and transistors to enable on-chip optical sensing and signal amplification (507). Figure 16.77a and b shows the schematic and circuit diagram of an all-NW photodetector based on ordered arrays of Ge/Si and CdSe NWs. Each individual circuit consists of three active device elements: (i) an optical nanosensor (NS) based on either a single or parallel array of CdSe NWs, (ii) a high-resistance FET (T_1) based on parallel arrays of 1–5 Ge/Si core/shell NWs, and (iii) a low-resistance buffer FET (T_2) with the channel consisting of parallel arrays of ∼2000 Ge/Si NWs. These two types of NWs were assembled with a two-step contact printing approach (507). The circuitry utilizes T_1 to match the output impedance of the nanosensor in a voltage divider configuration in such a way that once the illumination-dependent nanosensor current is translated into potential V_{G2}, the output current of T_2 is modulated according to its transfer characteristics, resulting in ∼5 orders of magnitude amplification of the nanosensor current signal. Figure 16.77c shows an optical microscopy image of a fabricated circuit and SEM images of each individual component, clearly demonstrating the highly ordered NW positioning and the on-chip integration. Time-resolved photoresponse measurements (operating bias $V_{DD} = -3$ V and $V_{G1} = 3$ V) were conducted for several illustration cycles, as shown in Figure 16.77d, showing average dark and light currents of ∼80 µA and ∼300 µA, respectively. As contact printing is a scalable technique for NW assembly, large-scale all-NW circuit arrays have been fabricated (507). Figure 16.78a shows the optical image of a large array (13 × 20) of the all-NW photodetection circuits, resembling an image sensor (507). Particularly, each individual circuit of the array functions as a single pixel to response to incident light. Furthermore, a halogen light source was focused and projected onto the center of the matrix (Fig. 16.78b) to illustrate the image-sensing function of the circuit array, and the photoresponse of each single circuit component was recorded. Eventually, photoresponse of each individual circuit pixel was organized into a gray-scale 2D matrix, reflecting the profile of the halogen light source (507). This work has clearly shown that with NW printing technique, all-NW integrated circuitry can be realized. As the printing technique has been demonstrated highly compatible to flexible substrate, such integrated circuitry can be realized for flexible electronics in the future.

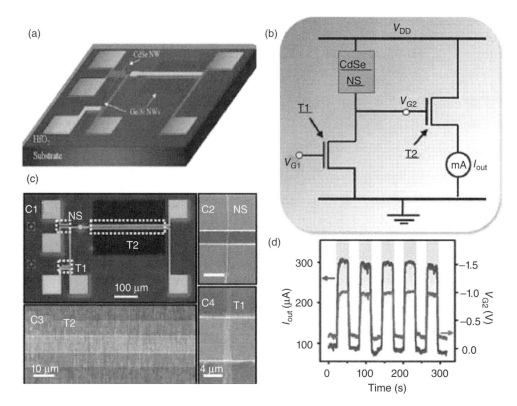

Figure 16.77 Heterogeneous integration of NWs for an optical sensor circuitry on Si substrates. (a) Schematic of the all-NW optical sensor circuit based on ordered arrays of Ge/Si and CdSe NWs. (b) Diagram of a proof-of-concept NW photosensor circuitry, consisting of a CdSe-NW light sensor (NS), an impedance-matching NW transistor (T_1), and a buffer transistor (T_2). (C1–C4) Optical and SEM images of an all-NW sensor circuitry with the aforementioned circuit elements. (d) The time-domain response of T_2 gate voltage, V_{G2}, and circuit output current, Iout, response to light illustration (4.4 mW/cm²). (*Source*: Reprinted with permission from Reference 507, Copyright 2008 The National Academy of Sciences of the USA).

Figure 16.78 All-NW image sensor on Si substrates. (a) Bright-field optical image of an array of all-NW photosensor circuitry composed of 20 × 13 individual circuits. (b) A perspective picture showing the imaging function of the circuit array. (c) An output response of the circuit array, imaging a circuit light spot. The contrast represents the normalized photocurrent, with the gray pixels representing the defective sites. (*Source*: Reprinted with permission from Reference 507, Copyright The National Academy of Sciences of the USA).

16.3.3.5.4 Nanowire Active-Matrix Circuitry for Artificial Skin Replacing organic semiconductor or a-Si with inorganic crystalline semiconductor NWs improves performance of flexible electronics. Meanwhile, it enables more functions particularly these need to operate in harsh environment. For example, an electronic skin is expected to be a multifunctional system with integrated sensors responsive to touch, temperature, chemical environment, and so on. Obviously sensors themselves and the supporting substrates are required to be flexible in reality. In this regard, NW active-matrix circuitry for low-voltage macroscale artificial skin has been demonstrated, and this type of skin can operate at low voltage due to high mobility of NW materials (551). Figure 16.79a shows the process scheme for the electronic skin

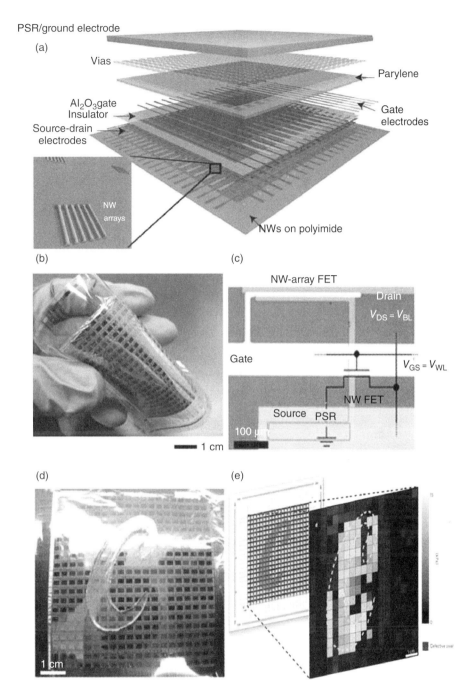

Figure 16.79 NW-based macroscale flexible devices. (a) Schematic of the passive and active layers of NW e-skin. (b) Optical photographs of a fully fabricated e-skin device (7×7 cm^2 with a 19×18 pixel array) under bending. (c) Optical-microscope image of a single sensor pixel in the array, depicting a Ge/Si NW-array FET (channel length $\sim$3 μm, channel width $\sim$250 μm) integrated with a PSR. The circuit structure for the pixel is also shown. (d) Photograph of a fabricated e-skin with a PDMS mould in the shape of "C" placed on the top for applying pressure and subsequent imaging. (e) The corresponding two-dimensional intensity profile obtained from experimental mapping of the pixel signals. The character "C," corresponding to the applied pressure profile, can be readily imaged by the e-skin. (*Source*: Reprinted with permission from Reference 551, Copyright 2010 Nature Publishing Group).

device with an integrated NW active-matrix backplane. Optical images of a fully fabricated electronic skin, consisting of a 19 × 18 pixel matrix with an active area of 7 × 7 cm², are shown in Figure 16.79b and c. The structure can easily be bent or rolled to a small radius of curvature, demonstrating the superb mechanical flexibility of the substrate and its integrated electronic components. The sensor array employs an active-matrix circuitry as shown in Figure 16.79c. Each pixel is connected to a NW-array FET that actively maintains the pixel state while other pixels are addressed. This presents an important advantage over passive-matrix circuitry where the sensor element itself must maintain its state without the use of an active switching device. Parallel arrays of NWs are used as the channel material of the active-matrix FETs, thereby reducing the stochastic device-to-device variation (551). A laminated pressure-sensitive rubber (PSR) is used as the sensing element. The source electrodes of NW-array FETs are connected to ground through the PSR. By applying an external pressure, the conductance of PSR changes, resulting in the modulation of NW FET characteristics and thus the pixel output signal. The gate (V_{GS}) and drain (V_{DS}) bias of FETs are used for addressing the word (i.e., row) and bit (i.e., column) lines of the matrix, respectively. By addressing and monitoring the conductance of each pixel in the active matrix, spatial mapping of the applied pressure can readily be attained.

To shed light on the response and relaxation time constants of the e-skin device, time-resolved measurements were carried out by applying an external pressure of ~15 kPa with a frequency of up to 5 Hz while measuring the electrical response of a pixel (551). A response and relaxation time of < ~0.1 s was observed. The devices are responsive without a significant signal degradation up to 5 Hz, which demonstrates the fast and deterministic response of the e-skin device. To demonstrate the functionality of the integrated e-skin, a piece of PDMS molded as the letter "C" with an area of ~3 cm² is placed on top of the sensor array, followed by the application of a normal pressure of ~15 kPa (Fig. 16.79d). The elastic PDMS mould enables the uniform distribution of the pressure over an area corresponding to that of the "C." The row and column line voltages of $V_{WL} = 5$ V and $V_{BL} = 0.5$ V are applied, respectively, to address each pixel individually (Fig. 16.79e). The output conductance for each individual pixel is measured and plotted as a two-dimensional intensity plot (Fig. 16.79e). The yield of functional active pixels was 84% with the defective pixels being mainly caused by the failure of the process integration steps and imperfect NW printing. As depicted in Figure 16.79e, the applied pressure profile can be spatially resolved by the integrated NW e-skin. The pixel resolution used in this work was set to ~2.5 mm to enable a manageable number of pixels without the need to implement system integration for the signal readout. In the future, the pixel size can be readily reduced down to the lithographic limits to enable imaging with higher spatial resolution. The artificial e-skin demonstrated here presents one example of the large-scale fabrication of low-cost NW sensor circuitry. The ability to controllably achieve functional electronics on truly macroscales using printed NW arrays presents a feasible route toward their implementation for practical applications.

REFERENCES

1. Avouris P, Chen Z, Perebeinos V. Carbon-based electronics. Nat Nanotechnol 2007;2:605–615.

2. Burghard M, Klauk H, Kern K. Carbon-based field-effect transistors for nanoelectronics. Adv Mater 2009;21:2586–2600.

3. Iijima S. Helical microtubules of graphitic carbon. Nature 1991;354:56–58.

4. Berger C, Song Z, Li T, Li X, Ogbazghi AY, Feng R, Dai Z, Marchenkov AN, Conrad EH, First PN. Ultrathin epitaxial graphite: 2D electron gas properties and a route toward graphene-based nanoelectronics. J Phys Chem B 2004;108:19912–19916.

5. Novoselov KS, Geim AK, Morozov SV, Jiang D, Zhang Y, Dubonos SV, Grigorieva IV, Firsov AA. Electric field effect in atomically thin carbon films. Science 2004;306:666–669.

6. Novoselov K, Geim AK, Morozov S, Jiang D, Grigorieva MKI, Dubonos S, Firsov A. Two-dimensional gas of massless Dirac fermions in graphene. Nature 2005;438:197–200.

7. Zhang Y, Tan Y, Stormer HL, Kim P. Experimental observation of the quantum Hall effect and Berry's phase in graphene. Nature 2005;438:201–204.

8. Hancock Y. The 2010 Nobel Prize in physics—ground-breaking experiments on graphene. J Phys D 2011;44:473001.

9. Kim K, Choi J, Kim T, Cho S, Chung H. A role for graphene in silicon-based semiconductor devices. Nature 2011;479:338–344.

10. Lin W, Lin-Hai T, Guo-Dong W, Feng-Mei G, Jin-Ju Z, Wei-You Y. Epitaxial growth of graphene and their applications in devices. J Inorg Mater 2011;26:1009–1019.

11. Neto AC, Guinea F, Peres N, Novoselov KS, Geim AK. The electronic properties of graphene. Rev Mod Phys 2009;81:109.

12. Liu M, Yin X, Ulin-Avila E, Geng B, Zentgraf T, Ju L, Wang F, Zhang X. A graphene-based broadband optical modulator. Nature 2011;474:64–67.

13. Geim AK, Novoselov KS. The rise of graphene. Nat Mater 2007;6:183–191.

14. Bunch JS, Yaish Y, Brink M, Bolotin K, McEuen PL. Coulomb oscillations and Hall effect in quasi-2D graphite quantum dots. Nano Lett 2005;5:287–290.

15. Geim AK. Graphene: status and prospects. Science 2009;324:1530–1534.

16. Bolotin K, Sikes K, Hone J, Stormer H, Kim P. Temperature-dependent transport in suspended graphene. Phys Rev Lett 2008;101:096802.

17. Bolotin KI, Sikes KJ, Jiang Z, Klima M, Fudenberg G, Hone J, Kim P, Stormer HL. Ultrahigh electron mobility in suspended graphene. Solid State Commun 2008;146:351–355.

18. Klein DL. STEP I STEP; 1969.

19. Morozov S, Novoselov K, Katsnelson M, Schedin F, Elias D, Jaszczak J, Geim A. Giant intrinsic carrier mobilities in graphene and its bilayer. Phys Rev Lett 2008;100:016602.

20. Shishir R, Ferry D. Intrinsic mobility in graphene. J Phys Condens Matter 2009;21:232204.

21. Stankovich S, Dikin DA, Dommett GH, Kohlhaas KM, Zimney EJ, Stach EA, Piner RD, Nguyen ST, Ruoff RS. Graphene-based composite materials. Nature 2006;442:282–286.

22. Balandin AA, Ghosh S, Bao W, Calizo I, Teweldebrhan D, Miao F, Lau CN. Superior thermal conductivity of single-layer graphene. Nano Lett 2008;8:902–907.

23. Liao L, Bai J, Cheng R, Lin Y, Jiang S, Qu Y, Huang Y, Duan X. Sub-100 nm channel length graphene transistors. Nano Lett 2010;10:3952–3956.

24. Frank I, Tanenbaum DM, Van der Zande A, McEuen PL. Mechanical properties of suspended graphene sheets. J Vac Sci Technol B 2007;25:2558–2561.

25. Liao L, Bai J, Qu Y, Huang Y, Duan X. Single-layer graphene on Al2O3/Si substrate: better contrast and higher performance of graphene transistors. Nanotechnology 2010;21:015705.

26. Reina A, Jia X, Ho J, Nezich D, Son H, Bulovic V, Dresselhaus MS, Kong J. Large area, few-layer graphene films on arbitrary substrates by chemical vapor deposition. Nano Lett 2008;9:30–35.

27. Emtsev KV, Bostwick A, Horn K, Jobst J, Kellogg GL, Ley L, McChesney JL, Ohta T, Reshanov SA, Röhrl J. Towards wafer-size graphene layers by atmospheric pressure graphitization of silicon carbide. Nat Mater 2009;8:203–207.

28. Bae S, Kim H, Lee Y, Xu X, Park J, Zheng Y, Balakrishnan J, Lei T, Kim HR, Song YI. Roll-to-roll production of 30-inch graphene films for transparent electrodes. Nat Nanotechnol 2010;5:574–578.

29. Chen C, Rosenblatt S, Bolotin KI, Kalb W, Kim P, Kymissis I, Stormer HL, Heinz TF, Hone J. Performance of monolayer graphene nanomechanical resonators with electrical readout. Nat Nanotechnol 2009;4:861–867.

30. Kim KS, Zhao Y, Jang H, Lee SY, Kim JM, Kim KS, Ahn J, Kim P, Choi J, Hong BH. Large-scale pattern growth of graphene films for stretchable transparent electrodes. Nature 2009;457:706–710.

31. Li X, Cai W, An J, Kim S, Nah J, Yang D, Piner R, Velamakanni A, Jung I, Tutuc E. Large-area synthesis of high-quality and uniform graphene films on copper foils. Science 2009;324:1312–1314.

32. Moon J, Curtis D, Hu M, Wong D, McGuire C, Campbell P, Jernigan G, Tedesco J, VanMil B, Myers-Ward R. Epitaxial-graphene RF field-effect transistors on Si-face 6H-SiC substrates. IEEE Electron Device Lett 2009;30:650–652.

33. Liu L, Zhou H, Cheng R, Chen Y, Lin Y, Qu Y, Bai J, Ivanov IA, Liu G, Huang Y. A systematic study of atmospheric pressure chemical vapor deposition growth of large-area monolayer graphene. J Mater Chem 2012;22:1498–1503.

34. Robinson JA, Wetherington M, Tedesco JL, Campbell PM, Weng X, Stitt J, Fanton MA, Frantz E, Snyder D, VanMil BL. Correlating Raman spectral signatures with carrier mobility in epitaxial graphene: a guide to achieving high mobility on the wafer scale. Nano Lett 2009;9:2873–2876.

35. Kedzierski J, Hsu P, Healey P, Wyatt PW, Keast CL, Sprinkle M, Berger C, de Heer WA. Epitaxial graphene transistors on SiC substrates. IEEE Trans Electron Devices 2008;55:2078–2085.

36. Li X, Zhang G, Bai X, Sun X, Wang X, Wang E, Dai H. Highly conducting graphene sheets and Langmuir–Blodgett films. Nat Nanotechnol 2008;3:538–542.

37. Stankovich S, Piner RD, Nguyen ST, Ruoff RS. Synthesis and exfoliation of isocyanate-treated graphene oxide nanoplatelets. Carbon 2006;44:3342–3347.

38. Abergel D, Russell A, Falko VI. Visibility of graphene flakes on a dielectric substrate. Appl Phys Lett 2007;91:063125–063125-3.

39. Blake P, Hill E, Neto AC, Novoselov K, Jiang D, Yang R, Booth T, Geim A. Making graphene visible. Appl Phys Lett 2007;91:063124.

40. Lin Y, Jenkins KA, Valdes-Garcia A, Small JP, Farmer DB, Avouris P. Operation of graphene transistors at gigahertz frequencies. Nano Lett 2008;9:422–426.

41. Lin Y, Chiu H, Jenkins KA, Farmer DB, Avouris P, Valdes-Garcia A. Dual-gate graphene FETs with of 50 GHz. IEEE Electron Device Lett 2010;31:68–70.

42. Lin Y-M, Dimitrakopoulos C, Jenkins KA, Farmer DB, Chiu H-Y, Grill A, Avouris P. 100-GHz transistors from wafer-scale epitaxial graphene. Science 2010;327:662–662.

43. Del Alamo JA. Nanometre-scale electronics with III-V compound semiconductors. Nature 2011;479:317–323.

44. He G, Sun Z, Li G, Zhang L. Review and perspective of Hf-based high-k gate dielectrics on silicon. Crit Rev Solid State Mater Sci 2012;37:131–157.

45. Likharev KK. CrossNets: neuromorphic hybrid CMOS/nanoelectronic networks. Sci Adv Mater 2011;3:322–331.

46. Wu Y, Lin Y, Bol AA, Jenkins KA, Xia F, Farmer DB, Zhu Y, Avouris P. High-frequency, scaled graphene transistors on diamond-like carbon. Nature 2011;472:74–78.

47. Lemme MC, Echtermeyer TJ, Baus M, Kurz H. A graphene field-effect device. IEEE Electron Device Lett 2007;28:282–284.

48. Meric I, Han MY, Young AF, Ozyilmaz B, Kim P, Shepard KL. Current saturation in zero-bandgap, top-gated graphene field-effect transistors. Nat Nanotechnol 2008;3:654–659.

49. Wang X, Tabakman SM, Dai H. Atomic layer deposition of metal oxides on pristine and functionalized graphene. J Am Chem Soc 2008;130:8152–8153.

50. Farmer DB, Chiu H, Lin Y, Jenkins KA, Xia F, Avouris P. Utilization of a buffered dielectric to achieve high field-effect carrier mobility in graphene transistors. Nano Lett 2009;9:4474–4478.

51. Kim S, Nah J, Jo I, Shahrjerdi D, Colombo L, Yao Z, Tutuc E, Banerjee SK. Realization of a high mobility dual-gated graphene field-effect transistor with Al_2O_3 dielectric. Appl Phys Lett 2009;94:062107–062107-3.

52. He G, Zhu L, Sun Z, Wan Q, Zhang L. Integrations and challenges of novel high-*k* gate stacks in advanced CMOS technology. Prog Mater Sci 2011;56:475–572.

53. Hollander MJ, LaBella M, Hughes ZR, Zhu M, Trumbull KA, Cavalero R, Snyder DW, Wang X, Hwang E, Datta S. Enhanced transport and transistor performance with oxide seeded high-*κ* gate dielectrics on wafer-scale epitaxial graphene. Nano Lett 2011;11:3601–3607.

54. Xu H, Zhang Z, Xu H, Wang Z, Wang S, Peng L. Top-gated graphene field-effect transistors with high normalized transconductance and designable Dirac point voltage. ACS Nano 2011;5:5031–5037.

55. Xu H, Zhang Z, Wang Z, Wang S, Liang X, Peng L. Quantum capacitance limited vertical scaling of graphene field-effect transistor. ACS Nano 2011;5:2340–2347.

56. Liao L, Duan X. Graphene–dielectric integration for graphene transistors. Mater Sci Eng R 2010;70:354–370.

57. Liao L, Bai J, Qu Y, Lin Y, Li Y, Huang Y, Duan X. High-*κ* oxide nanoribbons as gate dielectrics for high mobility top-gated graphene transistors. Proc Natl Acad Sci U S A 2010;107:6711–6715.

58. Qu Y, Liao L, Cheng R, Wang Y, Lin Y, Huang Y, Duan X. Rational design and synthesis of freestanding photoelectric nanodevices as highly efficient photocatalysts. Nano Lett 2010;10:1941–1949.

59. Liao L, Bai J, Cheng R, Zhou H, Liu L, Liu Y, Huang Y, Duan X. Scalable fabrication of self-aligned graphene transistors and circuits on glass. Nano Lett 2011;12:2653–2657.

60. Liao L, Lin Y, Bao M, Cheng R, Bai J, Liu Y, Qu Y, Wang KL, Huang Y, Duan X. High-speed graphene transistors with a self-aligned NW gate. Nature 2010;467:305–308.

61. Schwierz F. Electronics: industry-compatible graphene transistors. Nature 2011;472:41–42.

62. Kim D, Del Alamo JA. 30-nm InAs pseudomorphic HEMTs on an InP substrate with a current-gain cutoff frequency of 628 GHz. IEEE Electron Device Lett 2008;29:830–833.

63. Freer EM, Grachev O, Duan X, Martin S, Stumbo DP. High-yield self-limiting single-NW assembly with dielectrophoresis. Nat Nanotechnol 2010;5:525–530.

64. Duan XF, Huang Y, Cui Y, Wang JF, Lieber CM. Indium phosphide NWs as building blocks for nanoscale electronic and optoelectronic devices. Nature 2001;409:66–69.

65. Meric I, Baklitskaya N, Kim P, Shepard KL. RF performance of top-gated, zero-bandgap graphene field-effect transistors. *Electron Devices Meeting, 2008. IEDM 2008. IEEE International* Anonymous IEEE; 2008.

66. Chattopadhyay G, Schlecht E, Ward JS, Gill JJ, Javadi HH, Maiwald F, Mehdi I. An all-solid-state broad-band frequency multiplier chain at 1500 GHz. IEEE Trans Microwave Theory Tech 2004;52:1538–1547.

67. Maestrini A, Ward JS, Gill JJ, Javadi HS, Schlecht E, Tripon-Canseliet C, Chattopadhyay G, Mehdi I. A 540-640-GHz high-efficiency four-anode frequency tripler. IEEE Trans Microwave Theory Tech 2005;53:2835–2843.

68. Wang Z, Zhang Z, Xu H, Ding L, Wang S, Peng L. A high-performance top-gate graphene field-effect transistor based frequency doubler. Appl Phys Lett 2010;96:173104–173104-3.

69. Wang H, Nezich D, Kong J, Palacios T. Graphene frequency multipliers. IEEE Electron Device Lett 2009;30:547–549.

70. Wang H, Hsu A, Wu J, Kong J, Palacios T. Graphene-based ambipolar RF mixers. IEEE Electron Device Lett 2010;31:906–908.

71. Wang XD, Song JH, Liu J, Wang ZL. Direct-current nanogenerator driven by ultrasonic waves. Science 2007;316:102–105.

72. Lin Y, Valdes-Garcia A, Han S, Farmer DB, Meric I, Sun Y, Wu Y, Dimitrakopoulos C, Grill A, Avouris P. Wafer-scale graphene integrated circuit. Science 2011;332:1294–1297.

73. Odom TW, Huang J, Kim P, Lieber CM. Structure and electronic properties of carbon nanotubes. J Phys Chem B 2000;104:2794–2809.

74. Hamada N, Sawada S, Oshiyama A. New one-dimensional conductors: graphitic microtubules. Phys Rev Lett 1992;68:1579–1581.

75. Crespi VH, Cohen ML, Rubio A. In situ band gap engineering of carbon nanotubes. Phys Rev Lett 1997;79:2093–2096.

76. Yu Z, Brus LE. (n, m) structural assignments and chirality dependence in single-wall carbon nanotube Raman scattering. J Phys Chem B 2001;105:6831–6837.

77. Jorio A, Dresselhaus G, Dresselhaus M. *Carbon Nanotube*. 2008.

78. Dresselhaus M, Dresselhaus G, Jorio A, Souza Filho A, Pimenta M, Saito R. Single nanotube Raman spectroscopy. Acc Chem Res 2002;35:1070–1078.

79. Kataura H, Kumazawa Y, Maniwa Y, Umezu I, Suzuki S, Ohtsuka Y, Achiba Y. Optical properties of single-wall carbon nanotubes. Synth Met 1999;103:2555–2558.

80. Dresselhaus M, Saito R, Jorio A. Semiconducting carbon nanotubes. *AIP Conference Proceedings* Anonymous; 2005.

81. Treacy M, Ebbesen T, Gibson J. Exceptionally high Young's modulus observed for individual carbon nanotubes. Nature 1996;381:678–680.

82. Ebbesen T, Ajayan P. Large-scale synthesis of carbon nanotubes. Nature 1992;358:220–222.

83. Bethune D, Klang C, De Vries M, Gorman G, Savoy R, Vazquez J, Beyers R. Cobalt-catalysed growth of carbon nanotubes with single-atomic-layer walls. Nature 1993;363:605–607.

84. Journet C, Maser W, Bernier P, Loiseau A, Lamy de La Chapelle M, Lefrant DLS, Deniard P, Lee R, Fischer J. Large-scale production of single-walled carbon nanotubes by the electric-arc technique. Nature 1997;388:756–758.

85. Thess A, Lee R, Nikolaev P, Dai H, Petit P, Robert J, Xu C, Lee YH, Kim SG, Rinzler AG. Crystalline ropes of metallic carbon nanotubes. Sci AAAS Wkly Pap Ed 1996;273:483–487.

86. Liu J, Rinzler AG, Dai H, Hafner JH, Bradley RK, Boul PJ, Lu A, Iverson T, Shelimov K, Huffman CB. Fullerene pipes. Science 1998;280:1253–1256.

87. Cassell AM, Raymakers JA, Kong J, Dai H. Large scale CVD synthesis of single-walled carbon nanotubes. J Phys Chem B 1999;103:6484–6492.

88. Tibbetts GG. Why are carbon filaments tubular? J Cryst Growth 1984;66:632–638.

89. Tibbetts GG. Vapor-grown carbon fibers: status and prospects. Carbon 1989;27:745–747.

90. Tibbetts GG. Lengths of carbon fibers grown from iron catalyst particles in natural gas. J Cryst Growth 1985;73:431–438.

91. Tibbetts G, Devour M, Rodda E. An adsorption-diffusion isotherm and its application to the growth of carbon filaments on iron catalyst particles. Carbon 1987;25:367–375.

92. Baker R. Catalytic growth of carbon filaments. Carbon 1989;27:315–323.

93. Kumar M, Ando Y. Chemical vapor deposition of carbon nanotubes: a review on growth mechanism and mass production. J Nanosci Nanotechnol 2010;10:3739–3758.

94. Mamedov AA, Kotov NA, Prato M, Guldi DM, Wicksted JP, Hirsch A. Molecular design of strong single-wall carbon nanotube/polyelectrolyte multilayer composites. Nat Mater 2002;1:190–194.

95. Hiura H, Ebbesen T, Tanigaki K, Takahashi H. Raman studies of carbon nanotubes. Chem Phys Lett 1993;202:509–512.

96. Jorio A, Pimenta M, Souza Filho A, Saito R, Dresselhaus G, Dresselhaus M. Characterizing carbon nanotube samples with resonance Raman scattering. New J Phys 2003;5:139.

97. Jishi R, Venkataraman L, Dresselhaus M, Dresselhaus G. Phonon modes in carbon nanotubules. Chem Phys Lett 1993;209:77–82.

98. Kürti J, Kresse G, Kuzmany H. First-principles calculations of the radial breathing mode of single-wall carbon nanotubes. Phys Rev B 1998;58:R8869–R8872.

99. Sánchez-Portal D, Artacho E, Soler JM, Rubio A, Ordejón P. Ab initio structural, elastic, and vibrational properties of carbon nanotubes. Phys Rev B 1999;59:12678.

100. Dresselhaus M, Dresselhaus G, Jorio A, Souza Filho A, Saito R. Raman spectroscopy on isolated single wall carbon nanotubes. Carbon 2002;40:2043–2061.

101. Bandow S, Asaka S, Saito Y, Rao A, Grigorian L, Richter E, Eklund P. Effect of the growth temperature on the diameter distribution and chirality of single-wall carbon nanotubes. Phys Rev Lett 1998;80:3779–3782.

102. Milnera M, Kürti J, Hulman M, Kuzmany H. Periodic resonance excitation and intertube interaction from quasicontinuous distributed helicities in single-wall carbon nanotubes. Phys Rev Lett 2000;84:1324–1327.

103. Alvarez L, Righi A, Guillard T, Rols S, Anglaret E, Laplaze D, Sauvajol J. Resonant Raman study of the structure and electronic properties of single-wall carbon nanotubes. Chem Phys Lett 2000;316:186–190.

104. Bandow S, Chen G, Sumanasekera G, Gupta R, Yudasaka M, Iijima S, Eklund P. Diameter-selective resonant Raman scattering in double-wall carbon nanotubes. Phys Rev B 2002;66:075416.

105. Rao A, Richter E, Bandow S, Chase B, Eklund P, Williams K, Fang S, Subbaswamy K, Menon M, Thess A. Diameter-selective Raman scattering from vibrational modes in carbon nanotubes. Science 1997;275:187–191.

106. Strano MS, Doorn SK, Haroz EH, Kittrell C, Hauge RH, Smalley RE. Assignment of (n, m) Raman and optical features of metallic single-walled carbon nanotubes. Nano Lett 2003;3:1091–1096.

107. Venkateswaran U, Rao A, Richter E, Menon M, Rinzler A, Smalley R, Eklund P. Probing the single-wall carbon nanotube bundle: Raman scattering under high pressure. Phys Rev B 1999;59:10928.

108. Colomer J, Benoit J, Stephan C, Lefrant S, Van Tendeloo G, Nagy JB. Characterization of single-wall carbon nanotubes produced by CCVD method. Chem Phys Lett 2001;345:11–17.

109. Dai H. Carbon nanotubes: synthesis, integration, and properties. Acc Chem Res 2002;35:1035–1044.

110. Lee CJ, Park J. Growth model of bamboo-shaped carbon nanotubes by thermal chemical vapor deposition. Appl Phys Lett 2000;77:3397–3399.

111. Li W, Wen J, Ren Z. Effect of temperature on growth and structure of carbon nanotubes by chemical vapor deposition. Appl Phys A 2002;74:397–402.

112. Li W, Wen J, Tu Y, Ren Z. Effect of gas pressure on the growth and structure of carbon nanotubes by chemical vapor deposition. Appl Phys A 2001;73:259–264.

113. Colomer J, Stephan C, Lefrant S, Van Tendeloo G, Willems I, Konya Z, Fonseca A, Laurent C, Nagy J. Large-scale synthesis of single-wall carbon nanotubes by catalytic chemical vapor deposition (CCVD) method. Chem Phys Lett 2000;317:83–89.

114. Kuang M, Wang ZL, Bai X, Guo J, Wang E. Catalytically active nickel {110} surfaces in growth of carbon tubular structures. Appl Phys Lett 2000;76:1255–1257.

115. Bower C, Zhou O, Zhu W, Werder D, Jin S. Nucleation and growth of carbon nanotubes by microwave plasma chemical vapor deposition. Appl Phys Lett 2000;77:2767–2769.

116. Tans SJ, Verschueren AR, Dekker C. Room-temperature transistor based on a single carbon nanotube. Nature 1998;393:49–52.

117. Kong J, Franklin NR, Zhou C, Chapline MG, Peng S, Cho K, Dai H. Nanotube molecular wires as chemical sensors. Science 2000;287:622–625.

118. Misewich J, Martel R, Avouris P, Tsang J, Heinze S, Tersoff J. Electrically induced optical emission from a carbon nanotube FET. Science 2003;300:783–786.

119. Bachtold A, Hadley P, Nakanishi T, Dekker C. Logic circuits with carbon nanotube transistors. Science 2001;294:1317–1320.

120. Martel R, Schmidt T, Shea HR, Hertel T, Avouris P. Single- and multi-wall carbon nanotube field-effect transistors. Appl Phys Lett 1998;73:2447–2449.

121. Javey A, Guo J, Farmer DB, Wang Q, Wang D, Gordon RG, Lundstrom M, Dai H. Carbon nanotube field-effect transistors with integrated ohmic contacts and high-κ gate dielectrics. Nano Lett 2004;4:447–450.

122. Javey A, Guo J, Farmer DB, Wang Q, Yenilmez E, Gordon RG, Lundstrom M, Dai H. Self-aligned ballistic molecular transistors and electrically parallel nanotube arrays. Nano Lett 2004;4:1319–1322.

123. Franklin AD, Luisier M, Han S, Tulevski G, Breslin CM, Gignac L, Lundstrom MS, Haensch W. Sub-10 nm carbon nanotube transistor. Nano Lett 2012;12:758–762.

124. Qi P, Vermesh O, Grecu M, Javey A, Wang Q, Dai H, Peng S, Cho K. Toward large arrays of multiplex functionalized carbon nanotube sensors for highly sensitive and selective molecular detection. Nano Lett 2003;3:347–351.

125. Wang J, Musameh M. Carbon nanotube/teflon composite electrochemical sensors and biosensors. Anal Chem 2003;75:2075–2079.

126. Gao M, Dai L, Wallace GG. Biosensors based on aligned carbon nanotubes coated with inherently conducting polymers. Electroanalysis 2003;15:1089–1094.

127. Cai J, Jie J, Jiang P, Wu D, Xie C, Wu C, Wang Z, Yu Y, Wang L, Zhang X. Tuning the electrical transport properties of n-type CdS NWs via Ga doping and their nano-optoelectronic applications. Phys Chem Chem Phys 2011;13:14664–14668.

128. Wang S, Wang R, Sellin P, Zhang Q. DNA biosensors based on self-assembled carbon nanotubes. Biochem Biophys Res Commun 2004;325:1433–1437.

129. Kerman K, Morita Y, Takamura Y, Ozsoz M, Tamiya E. DNA-directed attachment of carbon nanotubes for enhanced label-free electrochemical detection of DNA hybridization. Electroanalysis 2004;16:1667–1672.

130. Snow E, Perkins F, Houser E, Badescu S, Reinecke T. Chemical detection with a single-walled carbon nanotube capacitor. Science 2005;307:1942–1945.

131. Shockley W, Queisser HJ. Detailed balance limit of efficiency of p-n junction solar cells. J Appl Phys 1961;32:510–519.

132. Wang S, Khafizov M, Tu X, Zheng M, Krauss TD. Multiple exciton generation in single-walled carbon nanotubes. Nano Lett 2010;10:2381–2386.

133. Landi BJ, Raffaelle RP, Castro SL, Bailey SG. Single-wall carbon nanotube–polymer solar cells. Prog Photovoltaics Res Appl 2005;13:165–172.

134. Wei L, Tezuka N, Umeyama T, Imahori H, Chen Y. Formation of single-walled carbon nanotube thin films enriched with semiconducting nanotubes and their application in photoelectrochemical devices. Nanoscale 2011;3:1845–1849.

135. Umeyama T, Tezuka N, Seki S, Matano Y, Nishi M, Hirao K, Lehtivuori H, Tkachenko NV, Lemmetyinen H, Nakao Y. Selective formation and efficient photocurrent generation of fullerene–single-walled carbon nanotube composites. Adv Mater 2010;22:1767–1770.

136. Tezuka N, Umeyama T, Matano Y, Shishido T, Yoshida K, Ogawa T, Isoda S, Stranius K, Chukharev V, Tkachenko NV. Photophysics and photoelectrochemical properties of nanohybrids consisting of fullerene-encapsulated single-walled carbon nanotubes and poly (3-hexylthiophene). Energy Environ Sci 2011;4:741–750.

137. D'Souza F, Das SK, Sandanayaka AS, Subbaiyan NK, Gollapalli DR, Zandler ME, Wakahara T, Ito O. Photoinduced charge separation in three-layer supramolecular nanohybrids: fullerene–porphyrin–SWCNT. Phys Chem Chem Phys 2012;14:2940–2950.

138. Umeyama T, Tezuka N, Kawashima F, Seki S, Matano Y, Nakao Y, Shishido T, Nishi M, Hirao K, Lehtivuori H. Carbon nanotube wiring of donor–acceptor nanograins by self-assembly and efficient charge transport. Angew Chem 2011;123:4711–4715.

139. Sgobba V, Rahman GA, Guldi DM, Jux N, Campidelli S, Prato M. Supramolecular assemblies of different carbon nanotubes for photoconversion processes. Adv Mater 2006;18:2264–2269.

140. Giancane G, Ruland A, Sgobba V, Manno D, Serra A, Farinola GM, Omar OH, Guldi DM, Valli L. Aligning single-walled carbon nanotubes by means of langmuir–blodgett film deposition: optical, morphological, and photo-electrochemical studies. Adv Funct Mater 2010;20:2481–2488.

141. Umeyama T, Kadota N, Tezuka N, Matano Y, Imahori H. Photoinduced energy transfer in composites of poly[(p-phenylene-1,2-vinylene)-co-(p-phenylene-1,1-vinylidene)] and single-walled carbon nanotubes. Chem Phys Lett 2007;444:263–267.

142. Campidelli S, Ballesteros B, Filoramo A, Díaz DD, de la Torre G, Torres T, Rahman GA, Ehli C, Kiessling D, Werner F. Facile decoration of functionalized single-wall carbon nanotubes with phthalocyanines via "click chemistry". J Am Chem Soc 2008;130:11503–11509.

143. Bissett MA, Köper I, Quinton JS, Shapter JG. Dendron growth from vertically aligned single-walled carbon nanotube thin layer arrays for photovoltaic devices. Phys Chem Chem Phys 2011;13:6059–6064.

144. Mountrichas G, Sandanayaka AS, Economopoulos SP, Pispas S, Ito O, Hasobe T, Tagmatarchis N. Photoinduced electron transfer in aqueous carbon nanotube/block copolymer/CdS hybrids: application in the construction of photoelectrochemical cells. J Mater Chem 2009;19:8990–8998.

145. Zhu H, Zeng H, Subramanian V, Masarapu C, Hung K, Wei B. Anthocyanin-sensitized solar cells using carbon nanotube films as counter electrodes. Nanotechnology 2008;19:465204.

146. Lee W, Lee J, Lee S, Chang J, Yi W, Han S. Improved Photocurrent in Ru (2, 2'-bipyridine-4, 4'-dicarboxylic acid) 2 (NCS) 2/Di (3-aminopropyl) viologen/single-walled carbon nanotubes/indium tin oxide system: suppression of recombination reaction by use of single-walled carbon nanotubes. J Phys Chem C 2007;111:9110–9115.

147. Chen T, Wang S, Yang Z, Feng Q, Sun X, Li L, Wang Z, Peng H. Flexible, light-weight, ultrastrong, and semiconductive carbon nanotube fibers for a highly efficient solar cell. Angew Chem Int Ed 2011;50:1815–1819.

148. Wadhwa P, Liu B, McCarthy MA, Wu Z, Rinzler AG. Electronic junction control in a nanotube-semiconductor Schottky junction solar cell. Nano Lett 2010;10:5001–5005.

149. Jia Y, Cao A, Bai X, Li Z, Zhang L, Guo N, Wei J, Wang K, Zhu H, Wu D. Achieving high efficiency silicon-carbon nanotube heterojunction solar cells by acid doping. Nano Lett 2011;11:1901–1905.

150. Wang C, Chien J, Takei K, Takahashi T, Nah J, Niknejad AM, Javey A. Extremely bendable, high-performance integrated circuits using semiconducting carbon nanotube networks for digital, analog, and radio-frequency applications. Nano Lett 2012;12:1527–1533.

151. Lu JG, Chang P, Fan Z. Quasi-one-dimensional metal oxide materials - synthesis, properties and applications. Mater Sci Eng R Rep 2006;52:49–91.

152. Dai ZR, Pan ZW, Wang ZL. Novel nanostructures of functional oxides synthesized by thermal evaporation. Adv Funct Mater 2003;13:9–24.

153. Liu ZQ, Zhang DH, Han S, Li C, Tang T, Jin W, Liu XL, Lei B, Zhou CW. Laser ablation synthesis and electron transport studies of tin oxide NWs. Adv Mater 2003;15:1754.

154. Park WI, Kim DH, Jung SW, Yi GC. Metalorganic vapor-phase epitaxial growth of vertically well-aligned ZnO nanorods. Appl Phys Lett 2002;80:4232–4234.

155. Zhao Q, Xu X, Zhang H, Chen Y, Xu J, Yu D. Catalyst-free growth of single-crystalline alumina NW arrays. Appl Phys A 2004;79:1721–1724.

156. Umar A, Kim SH, Lee YS, Nahm KS, Hahn YB. Catalyst-free large-quantity synthesis of ZnO nanorods by a vapor-solid growth mechanism: structural and optical properties. J Cryst Growth 2005;282:131–136.

157. Sekar A, Kim SH, Umar A, Hahn YB. Catalyst-free synthesis of ZnO NWs on Si by oxidation of Zn powders. J Cryst Growth 2005;277:471–478.

158. Cabrera N, Burton WK. Crystal growth and surface structure .2. Discuss Faraday Soc 1949:40–48.

159. Sears GW. A mechanism of whisker growth. Acta Metall 1955;3:367–369.

160. Blakely JM, Jackson KA. Growth of crystal whiskers. J Chem Phys 1962;37:428–430.

161. Wagner RS, Ellis WC. Vapor–liquid-solid mechanism of crystal growth. Appl Phys Lett 1964;4:89–90.

162. Wagner RS, Ellis WC, Arnold SM, Jackson KA. Study of filamentary growth of silicon crystals from vapor. J Appl Phys 1964;35:2993–3000.

163. Liu X, Wu XH, Cao H, Chang RPH. Growth mechanism and properties of ZnO nanorods synthesized by plasma-enhanced chemical vapor deposition. J Appl Phys 2004;95:3141–3147.

164. Wu YY, Yang PD. Direct observation of vapor–liquid-solid NW growth. J Am Chem Soc 2001;123:3165–3166.

165. Bae SY, Seo HW, Park JH. Vertically aligned sulfur-doped ZnO NWs synthesized via chemical vapor deposition. J Phys Chem B 2004;108:5206–5210.

166. Chen YJ, Li JB, Han YS, Yang XZ, Dai JH. The effect of Mg vapor source on the formation of MgO whiskers and sheets. J Cryst Growth 2002;245:163–170.

167. Liu X, Li C, Han S, Han J, Zhou CW. Synthesis and electronic transport studies of CdO nanoneedles. Appl Phys Lett 2003;82:1950–1952.

168. Wu JM, Shih HC, Wu WT, Tseng YK, Chen IC. Thermal evaporation growth and the luminescence property of TiO2 NWs. J Cryst Growth 2005;281:384–390.

169. Dai ZR, Gole JL, Stout JD, Wang ZL. Tin oxide NWs, nanoribbons, and nanotubes. J Phys Chem B 2002;106:1274–1279.

170. Dai L, Chen XL, Jian JK, He M, Zhou T, Hu BQ. Fabrication and characterization of In2O3 NWs. Appl Phys A 2002;75:687–689.

171. Chang PC, Fan ZY, Tseng WY, Rajagopal A, Lu JG. beta-Ga2O3 NWs: synthesis, characterization, and p-channel field-effect transistor. Appl Phys Lett 2005;87:222102.

172. Yang PD, Yan HQ, Mao S, Russo R, Johnson J, Saykally R, Morris N, Pham J, He RR, Choi HJ. Controlled growth of ZnO NWs and their optical properties. Adv Funct Mater 2002;12:323–331.

173. Wang XD, Summers CJ, Wang ZL. Large-scale hexagonal-patterned growth of aligned ZnO nanorods for nano-optoelectronics and nanosensor arrays. Nano Lett 2004;4:423–426.

174. Tseng YK, Lin IN, Liu KS, Lin TS, Chen IC. Low-temperature growth of ZnO NWs. J Mater Res 2003;18:714–718.

175. Huang HB, Yang SG, Gong JF, Liu HW, Duan JH, Zhao XN, Zhang R, Liu YL, Liu YC. Controllable assembly of aligned ZnO NWs/belts arrays. J Phys Chem B 2005;109:20746–20750.

176. Fan ZY, Wang DW, Chang PC, Tseng WY, Lu JG. ZnO NW field-effect transistor and oxygen sensing property. Appl Phys Lett 2004;85:5923–5925.

177. Picraux ST, Dayeh SA, Manandhar P, Perea DE, Choi SG. Silicon and germanium NWs: growth, properties, and integration. J Mater 2010;62:35–43.

178. Kolasinski KW. Catalytic growth of NWs: vapor–liquid–solid, vapor–solid–solid, solution–liquid–solid and solid–liquid–solid growth. Curr Opin Solid State Mater Sci 2006;10:182–191.

179. Lensch-Falk JL, Hemesath ER, Perea DE, Lauhon LJ. Alternative catalysts for VSS growth of silicon and germanium NWs. J Mater Chem 2009;19:849–857.

180. Hulteen JC, Martin CR. A general template-based method for the preparation of nanomaterials. J Mater Chem 1997;7:1075–1087.

181. Xue DS, Zhang LY, Gui AB, Xu XF. Fe3O4 NW arrays synthesized in AAO templates. Appl Phys A 2005;80:439–442.

182. Zheng MJ, Zhang LD, Li GH, Shen WZ. Fabrication and optical properties of large-scale uniform zinc oxide NW arrays by one-step electrochemical deposition technique. Chem Phys Lett 2002;363:123–128.

183. Shi KY, Xin BF, Chi YJ, Fu HG. Assembling porous Fe2O3 NW arrays by electrochemical deposition in mesoporous silica SBA-16 films. Acta Chim Sin 2004;62:1859–1861.

184. Liu XM, Zhou YC. Electrochemical deposition and characterization of Cu2O NWs. Appl Phys A 2005;81:685–689.

185. Mintz TS, Bhargava YV, Thorne SA, Chopdekar R, Radmilovic V, Suzuki Y, Devine TM. Electrochemical synthesis of functionalized nickel oxide NWs. Electrochem Solid-State Lett 2005;8:D26–D30.

186. Chen YH, Zhang XT, Xue ZH, Du ZL, Li TJ. Preparation of SnO2 NWs by AC electrodeposition in anodic alumina template and its deposition conditions. J Inorg Mater 2005;20:59–64.

187. Li Y, Cheng GS, Zhang LD. Fabrication of highly ordered ZnO NW arrays in anodic alumina membranes. J Mater Res 2000;15:2305–2308.

188. Chen YW, Liu YC, Lu SX, Xu CS, Shao CL, Wang C, Zhang JY, Lu YM, Shen DZ, Fan XW. Optical properties of ZnO and ZnO: in nanorods assembled by sol–gel method. J Chem Phys 2005;123:134701.

189. Wang XY, Wang XY, Huang WG, Sebastian PJ, Gamboa S. Sol–gel template synthesis of highly ordered MnO2 NW arrays. J Power Sources 2005;140:211–215.

190. Xu H, Qin DH, Yang Z, Li HL. Fabrication and characterization of highly ordered zirconia NW arrays by sol–gel template method. Mater Chem Phys 2003;80:524–528.

191. Zhang M, Bando Y, Wada K. Sol–gel template preparation of TiO2 nanotubes and nanorods. J Mater Sci Lett 2001;20:167–170.

192. Zhou YK, Li HL. Sol–gel template synthesis and structural properties of a highly ordered LiNi0.5Mn0.5O2 NW array. J Mater Chem 2002;12:681–686.

193. Yang Z, Huang Y, Dong B, Li HL. Fabrication and structural properties of LaFeO3 NWs by an ethanol-ammonia-based sol–gel template route. Appl Phys A 2005;81:453–457.

194. Xu CK, Xu GD, Liu YK, Wang GH. A simple and novel route for the preparation of ZnO nanorods. Solid State Commun 2002;122:175–179.

195. Xu CK, Zhao XL, Liu S, Wang GH. Large-scale synthesis of rutile SnO2 nanorods. Solid State Commun 2003;125:301–304.

196. Xu CK, Xu GD, Wang GH. Preparation and characterization of NiO nanorods by thermal decomposition of NiC2O4 precursor. J Mater Sci 2003;38:779–782.

197. Cao YL, Jia DZ, Liu L, Luo JM. Rapid synthesis of lead oxide nanorods by one-step solid-state chemical reaction at room temperature. Chin J Chem 2004;22:1288–1290.

198. Liang JH, Peng C, Wang X, Zheng X, Wang RJ, Qiu XPP, Nan CW, Li YD. Chromate nanorods/nanobelts: general synthesis, characterization, and properties. Inorg Chem 2005;44:9405–9415.

199. Vantomme A, Yuan ZY, Du GH, Su BL. Surfactant-assisted large-scale preparation of crystalline CeO2 nanorods. Langmuir 2005;21:1132–1135.

200. Thompson LH, Doraiswamy LK. Sonochemistry: science and engineering. Ind Eng Chem Res 1999;38:1215–1249.

201. Kumar RV, Koltypin Y, Xu XN, Yeshurun Y, Gedanken A, Felner I. Fabrication of magnetite nanorods by ultrasound irradiation. J Appl Phys 2001;89:6324–6328.

202. Hu XL, Zhu YJ, Wing SW. Sonochemical and microwave-assisted synthesis of linked single-crystalline ZnO rods. Mater Chem Phys 2004;88:421–426.

203. Gao T, Li QH, Wang TH. Sonochemical synthesis, optical properties, and electrical properties of core/shell-type ZnO nanorod/CdS nanoparticle composites. Chem Mater 2005;17:887–892.

204. Pol VG, Palchik O, Gedanken A, Felner I. Synthesis of europium oxide nanorods by ultrasound irradiation. J Phys Chem B 2002;106:9737–9743.

205. Miao JJ, Wang H, Li YR, Zhu JM, Zhu JJ. Ultrasonic-induced synthesis of CeO2 nanotubes. J Cryst Growth 2005;281:525–529.

206. Liu B, Zeng HC. Hydrothermal synthesis of ZnO nanorods in the diameter regime of 50 nm. J Am Chem Soc 2003;125:4430–4431.

207. Wang JM, Gao L. Wet chemical synthesis of ultralong and straight single-crystalline ZnO NWs and their excellent UV emission properties. J Mater Chem 2003;13:2551–2554.

208. Guo M, Diao P, Cai SM. Hydrothermal growth of well-aligned ZnO nanorod arrays: Dependence of morphology and alignment ordering upon preparing conditions. J Solid-State Chem 2005;178:1864–1873.

209. Cao MH, Wang YH, Guo CX, Qi YJ, Hu CW, Wang EB. A simple route towards CuO NWs and nanorods. J Nanosci Nanotechnol 2004;4:824–828.

210. Chen C, Zhuang ZB, Peng Q. Synthesis and characterization of cadmium orthosilicate NWs. Chem J Chin Univ Chin 2005;26:1220–1221.

211. Zhang J, Liu ZG, Lin CK, Lin J. A simple method to synthesize beta-Ga2O3 nanorods and their photoluminescence properties. J Cryst Growth 2005;280:99–106.

212. Zheng DS, Sun SX, Fan WL, Yu HY, Fan CH, Cao GX, Yin ZL, Song XY. One-step preparation of single-crystalline beta-MnO2 nanotubes. J Phys Chem B 2005;109:16439–16443.

213. Zhu DL, Zhu H, Zhang YH. Microstructure and magnetization of single-crystal perovskite manganites NWs prepared by hydrothermal method. J Cryst Growth 2003;249:172–175.

214. Zhou KB, Wang X, Sun XM, Peng Q, Li YD. Enhanced catalytic activity of ceria nanorods from well-defined reactive crystal planes. J Catal 2005;229:206–212.

215. Yuan ZY, Su BL. Titanium oxide nanotubes, nanofibers and NWs. Colloids Surf A 2004;241:173–183.

216. Martensson T, Carlberg P, Borgstrom M, Montelius L, Seifert W, Samuelson L. NW arrays defined by nanoimprint lithography. Nano Lett 2004;4:699–702.

217. Holmes JD, Johnston KP, Doty RC, Korgel BA. Control of thickness and orientation of solution-grown silicon NWs. Science 2000;287:1471–1473.

218. Heitsch AT, Akhavan VA, Korgel BA. Rapid SFLS synthesis of Si NWs using trisilane with in situ alkyl-amine passivation. Chem Mater 2011;23:2697–2699.

219. Hanrath T, Korgel BA. Supercritical fluid–liquid–solid (SFLS) synthesis of Si and Ge NWs seeded by colloidal metal nanocrystals. Adv Mater 2003;15:437–440.

220. Lu X, Hanrath T, Johnston KP, Korgel BA. Growth of single crystal silicon NWs in supercritical solution from tethered gold particles on a silicon substrate. Nano Lett 2003;3:93–99.

221. Tuan H, Ghezelbash A, Korgel BA. Silicon NWs and silica nanotubes seeded by copper nanoparticles in an organic solvent. Chem Mater 2008;20:2306–2313.

222. Heitsch AT, Fanfair DD, Tuan H, Korgel BA. Solution−liquid−solid (SLS) growth of silicon NWs. J Am Chem Soc 2008;130:5436–5437.

223. Reneker DH, Yarin AL. Electrospinning jets and polymer nanofibers. Polymer 2008;49:2387–2425.

224. Huang Z, Zhang Y, Kotaki M, Ramakrishna S. A review on polymer nanofibers by electrospinning and their applications in nanocomposites. Compos Sci Technol 2003;63:2223–2253.

225. Shao C, Kim H, Gong J, Lee D. A novel method for making silica nanofibres by using electrospun fibres of polyvinylalcohol/silica composite as precursor. Nanotechnology 2002;13:635.

226. Li D, Xia Y. Electrospinning of nanofibers: reinventing the wheel? Adv Mater 2004;16:1151–1170.

227. Choi S, Ankonina G, Youn D, Oh S, Hong J, Rothschild A, Kim I. Hollow ZnO nanofibers fabricated using electrospun polymer templates and their electronic transport properties. ACS Nano 2009;3:2623–2631.

228. Li D, Wang Y, Xia Y. Electrospinning of polymeric and ceramic nanofibers as uniaxially aligned arrays. Nano Lett 2003;3:1167–1171.

229. Sun D, Chang C, Li S, Lin L. Near-field electrospinning. Nano Lett 2006;6:839–842.

230. Bisht GS, Canton G, Mirsepassi A, Kulinsky L, Oh S, Dunn-Rankin D, Madou MJ. Controlled continuous patterning of polymeric nanofibers on three-dimensional substrates using low-voltage near-field electrospinning. Nano Lett 2011;11:1831–1837.

231. Chang C, Limkrailassiri K, Lin L. Continuous near-field electrospinning for large area deposition of orderly nanofiber patterns. Appl Phys Lett 2008;93:123111–123111-3.

232. Zheng G, Li W, Wang X, Wu D, Sun D, Lin L. Precision deposition of a nanofibre by near-field electrospinning. J Phys D 2010;43:415501.

233. Chang C, Tran VH, Wang J, Fuh Y, Lin L. Direct-write piezoelectric polymeric nanogenerator with high energy conversion efficiency. Nano Lett 2010;10:726–731.

234. Brown TD, Dalton PD, Hutmacher DW. Direct writing by way of melt electrospinning. Adv Mater 2011;23:5651–5657.

235. Pu J, Yan X, Jiang Y, Chang C, Lin L. Piezoelectric actuation of direct-write electrospun fibers. Sens Actuators A 2010;164:131–136.

236. Rinaldi M, Ruggieri F, Lozzi L, Santucci S. Well-aligned Ti O$_2$ nanofibers grown by near-field-electrospinning. J Vac Sci Technol B 2009;27:1829–1833.

237. Fan ZY, Wen XG, Yang SH, Lu JG. Controlled p- and n-type doping of Fe2O3 nanobelt field effect transistors. Appl Phys Lett 2005;87:013113.

238. Zhang DH, Li C, Han S, Liu XL, Tang T, Jin W, Zhou CW. Electronic transport studies of single-crystalline In2O3 NWs. Appl Phys Lett 2003;82:112–114.

239. Chang PC, Fan Z, Chien CJ, Stichtenoth D, Ronning C, Lu JG. High-performance ZnO NW field effect transistors. Appl Phys Lett 2006;89:133113.

240. Hossain FM, Nishii J, Takagi S, Sugihara T, Ohtomo A, Fukumura T, Koinuma H, Ohno H, Kawasaki M. Modeling of grain boundary barrier modulation in ZnO invisible thin film transistors. Physica E 2004;21:911–915.

241. Park WI, Kim JS, Yi GC, Bae MH, Lee HJ. Fabrication and electrical characteristics of high-performance ZnO nanorod field-effect transistors. Appl Phys Lett 2004;85:5052–5054.

242. Li Y, Qian F, Xiang J, Lieber CM. NW electronic and optoelectronic devices. Mater Today 2006;9:18–27.

243. Zheng GF, Lu W, Jin S, Lieber CM. Synthesis and fabrication of high-performance n-type silicon NW transistors. Adv Mater 2004;16:1890–1893.

244. Jin S, Whang DM, McAlpine MC, Friedman RS, Wu Y, Lieber CM. Scalable interconnection and integration of NW devices without registration. Nano Lett 2004;4:915–919.

245. Huang Y, Duan XF, Cui Y, Lauhon LJ, Kim KH, Lieber CM. Logic gates and computation from assembled NW building blocks. Science 2001;294:1313–1317.

246. Doh Y, van Dam JA, Roest AL, Bakkers EP, Kouwenhoven LP, De Franceschi S. Tunable supercurrent through semiconductor NWs. Science 2005;309:272–275.

247. Zhong Z, Fang Y, Lu W, Lieber CM. Coherent single charge transport in molecular-scale silicon NWs. Nano Lett 2005;5:1143–1146.

248. Tilke A, Blick R, Lorenz H, Kotthaus J. Single-electron tunneling in highly doped silicon NWs in a dual-gate configuration. J Appl Phys 2001;89:8159–8162.

249. Zhong ZH, Wang DL, Cui Y, Bockrath MW, Lieber CM. NW crossbar arrays as address decoders for integrated nanosystems. Science 2003;302:1377–1379.

250. Gudiksen MS, Lauhon LJ, Wang J, Smith DC, Lieber CM. Growth of NW superlattice structures for nanoscale photonics and electronics. Nature 2002;415:617–620.

251. Wu Y, Xiang J, Yang C, Lu W, Lieber CM. Single-crystal metallic NWs and metal/semiconductor NW heterostructures. Nature 2004;430:61–65.

252. Yang C, Zhong Z, Lieber CM. Encoding electronic properties by synthesis of axial modulation-doped silicon NWs. Science 2005;310:1304–1307.

253. Lu W, Xiang J, Timko BP, Wu Y, Lieber CM. One-dimensional hole gas in germanium/silicon NW heterostructures. Proc Natl Acad Sci U S A 2005;102:10046–10051.

254. Xiang J, Lu W, Hu YJ, Wu Y, Yan H, Lieber CM. Ge/Si NW heterostructures as high-performance field-effect transistors. Nature 2006;441:489–493.

255. Li Y, Xiang J, Qian F, Gradecak S, Wu Y, Yan H, Yan H, Blom DA, Lieber CM. Dopant-free GaN/AlN/AlGaN radial NW heterostructures as high electron mobility transistors. Nano Lett 2006;6:1468–1473.

256. Ng HT, Han J, Yamada T, Nguyen P, Chen YP, Meyyappan M. Single crystal NW vertical surround-gate field-effect transistor. Nano Lett 2004;4:1247–1252.

257. Nguyen P, Ng HT, Yamada T, Smith MK, Li J, Han J, Meyyappan M. Direct integration of metal oxide NW in vertical field-effect transistor. Nano Lett 2004;4:651–657.

258. Fan ZY, Lu JG. Chemical sensing with ZnO NW field-effect transistor. IEEE Trans Nano 2006;5:393.

259. Fan ZY, Lu JG. Gate-refreshable NW chemical sensors. Appl Phys Lett 2005;86:123510.

260. Demami F, Ni L, Rogel R, Salaun AC, Pichon L. Silicon NWs based resistors as gas sensors. Sens Actuators B 2011;170:158.

261. Zhang DH, Liu ZQ, Li C, Tang T, Liu XL, Han S, Lei B, Zhou CW. Detection of NO_2 down to ppb levels using individual and multiple In_2O_3NW devices. Nano Lett 2004;4:1919–1924.

262. Murray BJ, Walter EC, Penner RM. Amine vapor sensing with silver mesowires. Nano Lett 2004;4:665–670.

263. Liu Z, Searson PC. Single nanoporous gold NW sensors. J Phys Chem B 2006;110:4318–4322.

264. Im Y, Lee C, Vasquez RP, Bangar MA, Myung NV, Menke EJ, Penner RM, Yun M. Investigation of a single Pd NW for use as a hydrogen sensor. Small 2006;2:356–358.

265. Walter E, Favier F, Penner R. Palladium mesowire arrays for fast hydrogen sensors and hydrogen-actuated switches. Anal Chem 2002;74:1546–1553.

266. Favier F, Walter EC, Zach MP, Benter T, Penner RM. Hydrogen sensors and switches from electrodeposited palladium mesowire arrays. Science 2001;293:2227–2231.

267. Gu F, Zhang L, Yin X, Tong L. Polymer single-NW optical sensors. Nano Lett 2008;8:2757–2761.

268. Chen X, Yuan CA, Wong CK, Ye H, Leung SY, Zhang G. Molecular modeling of protonic acid doping of emeraldine base polyaniline for chemical sensors. Sens Actuators B 2012;174:210–216.

269. Chen X, Wong CK, Yuan CA, Zhang G. Impact of the functional group on the working range of polyaniline as carbon dioxide sensors. Sens Actuators B 2012;175:15–21.

270. Cui Y, Lieber CM. Functional nanoscale electronic devices assembled using silicon NW building blocks. Science 2001;291:851–853.

271. Cui Y, Wei QQ, Park HK, Lieber CM. NW nanosensors for highly sensitive and selective detection of biological and chemical species. Science 2001;293:1289–1292.

272. Shishiyanu ST, Shishiyanu TS, Lupan OI. Sensing characteristics of tin-doped ZnO thin films as NO2 gas sensor. Sens Actuators B 2005;107:379–386.

273. Ramamoorthy R, Dutta PK, Akbar SA. Oxygen sensors: materials, methods, designs and applications. J Mater Sci 2003;38:4271–4282.

274. Kolmakov A, Moskovits M. Chemical sensing and catalysis by one-dimensional metal-oxide nanostructures. Annu Rev Mater Res 2004;34:151–180.

275. Ramgir NS, Yang Y, Zacharias M. NW-based sensors. Small 2010;6:1705–1722.

276. Comini E. Metal oxide nano-crystals for gas sensing. Anal Chim Acta 2006;568:28–40.

277. Comini E, Sberveglieri G. Metal oxide NWs as chemical sensors. Mater Today 2010;13:36–44.

278. Nanto H, Minami T, Takata S. Zinc-oxide thin-film ammonia gas sensors with high-sensitivity and excellent selectivity. J Appl Phys 1986;60:482–484.

279. Faglia G, Nelli P, Sberveglieri G. Frequency effect on highly sensitive No2 sensors based on Rgto Sno2(al) thin-films. Sens Actuators B 1994;19:497–499.

280. Heilig A, Barsan N, Weimar U, Schweizer-Berberich M, Gardner JW, Gopel W. Gas identification by modulating temperatures of SnO2-based thick film sensors. Sens Actuators B 1997;43:45–51.

281. Zhang Y, Kolmakov A, Chretien S, Metiu H, Moskovits M. Control of catalytic reactions at the surface of a metal oxide NW by manipulating electron density inside it. Nano Lett 2004;4:403–407.

282. Azad AM, Akbar SA, Mhaisalkar SG, Birkefeld LD, Goto KS. Solid-state gas sensors - a review. J Electrochem Soc 1992;139:3690–3704.

283. Eranna G, Joshi BC, Runthala DP, Gupta RP. Oxide materials for development of integrated gas sensors - a comprehensive review. Crit Rev Solid State Mater Sci 2004;29:111–188.

284. Kang BS, Heo YW, Tien LC, Norton DP, Ren F, Gila BP, Pearton SJ. Hydrogen and ozone gas sensing using multiple ZnO nanorods. Appl Phys A Mater 2005;80:1029–1032.

285. Kim YS, Ha SC, Kim K, Yang H, Choi SY, Kim YT, Park JT, Lee CH, Choi J, Paek J, Lee K. Room-temperature semiconductor gas sensor based on nonstoichiometric tungsten oxide nanorod film. Appl Phys Lett 2005;86:213105.

286. Tien LC, Sadik PW, Norton DP, Voss LF, Pearton SJ, Wang HT, Kang BS, Ren F, Jun J, Lin J. Hydrogen sensing at room temperature with Pt-coated ZnO thin films and nanorods. Appl Phys Lett 2005;87:222106.

287. Wang HT, Kang BS, Ren F, Tien LC, Sadik PW, Norton DP, Pearton SJ, Lin J. Hydrogen-selective sensing at room temperature with ZnO nanorods. Appl Phys Lett 2005;86:243503.

288. Xu JQ, Chen YP, Li YD, Shen JN. Gas sensing properties of ZnO nanorods prepared by hydrothermal method. J Mater Sci 2005;40:2919–2921.

289. Wang CH, Chu XF, Wu MW. Detection of H2S down to ppb levels at room temperature using sensors based on ZnO nanorods. Sens Actuator B 2006;113:320–323.

290. Wan Q, Li QH, Chen YJ, Wang TH, He XL, Li JP, Lin CL. Fabrication and ethanol sensing characteristics of ZnO NW gas sensors. Appl Phys Lett 2004;84:3654–3656.

291. Wan Q, Li QH, Chen YJ, Wang TH, He XL, Gao XG, Li JP. Positive temperature coefficient resistance and humidity sensing properties of Cd-doped ZnO NWs. Appl Phys Lett 2004;84:3085–3087.

292. Chu XF, Wang CH, Jiang DL, Zheng CM. Ethanol sensor based on indium oxide NWs prepared by carbothermal reduction reaction. Chem Phys Lett 2004;399:461–464.

293. Li C, Zhang DH, Liu XL, Han S, Tang T, Han J, Zhou CW. In2O3 NWs as chemical sensors. Appl Phys Lett 2003;82:1613–1615.

294. Li C, Zhang DH, Lei B, Han S, Liu XL, Zhou CW. Surface treatment and doping dependence of In2O3 NWs as ammonia sensors. J Phys Chem B 2003;107:12451–12455.

295. Kolmakov A, Klenov DO, Lilach Y, Stemmer S, Moskovits M. Enhanced gas sensing by individual SnO2 NWs and nanobelts functionalized with Pd catalyst particles. Nano Lett 2005;5:667–673.

296. Sysoev VV, Button BK, Wepsiec K, Dmitriev S, Kolmakov A. Toward the nanoscopic "electronic nose": hydrogen vs carbon monoxide discrimination with an array of individual metal oxide nano- and mesowire sensors. Nano Lett 2006;6:1584–1588.

297. Fan ZY, Ho JC, Jacobson ZA, Yerushalmi R, Alley RL, Razavi H, Javey A. Wafer-scale assembly of highly ordered semiconductor NW arrays by contact printing. Nano Lett 2008;8:20–25.

298. Meyer BK, Alves H, Hofmann DM, Kriegseis W, Forster D, Bertram F, Christen J, Hoffmann A, Strassburg M, Dworzak M, Haboeck U, Rodina AV. Bound exciton and donor-acceptor pair recombinations in ZnO. Phys Status Solidi B 2004;241:231–260.

299. Yang YH, Chen XY, Feng Y, Yang GW. Physical mechanism of blue-shift of UV luminescence of a single pencil-like ZnO NW. Nano Lett 2007;7:3879–3883.

300. Park WI, Yi GC. Electroluminescence in n-ZnO nanorod arrays vertically grown on p-GaN. Adv Mater 2004;16:87–90.

301. Lyons DM, Ryan KM, Morris MA, Holmes JD. Tailoring the optical properties of silicon NW arrays through strain. Nano Lett 2002;2:811–816.

302. Chang PC, Chien CJ, Stichtenoth D, Ronning C, Lu JG. Finite size effect in ZnO NWs. Appl Phys Lett 2007;90:113101.

303. Shalish I, Temkin H, Narayanamurti V. Size-dependent surface luminescence in ZnO NWs. Phys Rev B 2004;69:245401.

304. Johnson JC, Yan H, Schaller RD, Haber LH, Saykally RJ, Yang P. Single NW lasers. J Phys Chem B 2001;105:11387–11390.

305. Soci C, Zhang A, Bao X, Kim H, Lo Y, Wang D. NW photodetectors. J Nanosci Nanotechnol 2010;10:1430–1449.

306. Soci C, Zhang A, Xiang B, Dayeh SA, Aplin DPR, Park J, Bao XY, Lo YH, Wang D. ZnO NW UV photodetectors with high internal gain. Nano Lett 2007;7:1003–1009.

307. Calarco R, Marso M, Richter T, Aykanat AI, Meijers R, Hart Avd, Stoica T, Lüth H. Size-dependent photoconductivity in MBE-grown GaN-NWs. Nano Lett 2005;5:981–984.

308. Wang L, Asbeck P. Analysis of photoelectronic response in semiconductor NWs. *Nanotechnology, 2006. IEEE-NANO 2006. 6th IEEE Conference on* Anonymous; IEEE; 2006.

309. Jie JS, Zhang WJ, Jiang Y, Meng XM, Li YQ, Lee ST. Photoconductive characteristics of single-crystal CdS nanoribbons. Nano Lett 2006;6:1887–1892.

310. Chen R, Chen H, Lu C, Chen K, Chen C, Chen L, Yang Y. Ultrahigh photocurrent gain in m−axial GaN NWs. Appl Phys Lett 2007;91:223106–223106-3.

311. Prades JD, Jimenez-Diaz R, Hernandez-Ramirez F, Fernandez-Romero L, Andreu T, Cirera A, Romano-Rodriguez A, Cornet A, Morante JR, Barth S. Toward a systematic understanding of photodetectors based on individual metal oxide NWs. J Phys Chem C 2008;112:14639–14644.

312. Prades JD, Hernández-Ramírez F, Jimenez-Diaz R, Manzanares M, Andreu T, Cirera A, Romano-Rodriguez A, Morante J. The effects of electron–hole separation on the photoconductivity of individual metal oxide NWs. Nanotechnology 2008;19:465501.

313. Mathur S, Barth S, Shen H, Pyun J, Werner U. Size-dependent photoconductance in SnO2 NWs. Small 2005;1:713–717.

314. Kind H, Yan HQ, Messer B, Law M, Yang PD. NW ultraviolet photodetectors and optical switches. Adv Mater 2002;14:158–160.

315. Fan ZY, Chang PC, Lu JG, Walter EC, Penner RM, Lin CH, Lee HP. Photoluminescence and polarized photodetection of single ZnO NWs. Appl Phys Lett 2004;85:6128–6130.

316. Zhang D, Li C, Han S, Liu X, Tang T, Jin W, Zhou C. Ultraviolet photodetection properties of indium oxide NWs. Appl Phys A 2003;77:163–166.

317. Heo YW, Tien LC, Norton DP, Kang BS, Ren F, Gila BP, Pearton SJ. Electrical transport properties of single ZnO nanorods. Appl Phys Lett 2004;85:2002–2004.

318. Keem K, Kim H, Kim GT, Lee JS, Min B, Cho K, Sung MY, Kim S. Photocurrent in ZnO NWs grown from Au electrodes. Appl Phys Lett 2004;84:4376–4378.

319. Wang D, Qian F, Yang C, Zhong Z, Lieber CM. Rational growth of branched and hyperbranched NW structures. Nano Lett 2004;4:871–874.

320. Agarwal R. Heterointerfaces in semiconductor NWs. Small 2008;4:1872–1893.

321. Son M, Im S, Park Y, Park C, Kang T, Yoo K. Ultraviolet photodetector based on single GaN nanorod p–n junctions. Mater Sci Eng C 2006;26:886–888.

322. Pettersson H, Trägårdh J, Persson AI, Landin L, Hessman D, Samuelson L. Infrared photodetectors in heterostructure NWs. Nano Lett 2006;6:229–232.

323. Kayes BM, Atwater HA, Lewis NS. Comparison of the device physics principles of planar and radial p-n junction nanorod solar cells. J Appl Phys 2005;97:114302.

324. Tian BZ, Zheng XL, Kempa TJ, Fang Y, Yu NF, Yu GH, Huang JL, Lieber CM. Coaxial silicon NWs as solar cells and nanoelectronic power sources. Nature 2007;449:885–889.

325. Czaban JA, Thompson DA, LaPierre RR. GaAs core− shell NWs for photovoltaic applications. Nano Lett 2008;9:148–154.

326. Zhang Y, Wang L, Mascarenhas A. "Quantum coaxial cables" for solar energy harvesting. Nano Lett 2007;7:1264–1269.

327. Wang K, Chen J, Zhou W, Zhang Y, Yan Y, Pern J, Mascarenhas A. Direct growth of highly mismatched type II ZnO/ZnSe core/shell NW arrays on transparent conducting oxide substrates for solar cell applications. Adv Mater 2008;20:3248–3253.

328. Sze SM, Ng KK. Physics of Semiconductor Devices. Wiley-Interscience; 2006.

329. Luo L, Zhang Y, Mao SS, Lin L. Fabrication and characterization of ZnO NWs based UV photodiodes. Sens Actuators A 2006;127:201–206.

330. Ghosh R, Basak D. Electrical and ultraviolet photoresponse properties of quasialigned ZnO NWs/p-Si heterojunction. Appl Phys Lett 2007;90:243106–243106-3.

331. Guo Z, Zhao D, Liu Y, Shen D, Zhang J, Li B. Visible and ultraviolet light alternative photodetector based on ZnO NW/n-Si heterojunction. Appl Phys Lett 2008;93:163501–163501-3.

332. Bao X, Soci C, Susac D, Bratvold J, Aplin DP, Wei W, Chen C, Dayeh SA, Kavanagh KL, Wang D. Heteroepitaxial growth of vertical GaAs NWs on Si (111) substrates by metal− organic chemical vapor deposition. Nano Lett 2008;8:3755–3760.

333. Martensson T, Svensson CPT, Wacaser BA, Larsson MW, Seifert W, Deppert K, Gustafsson A, Wallenberg LR, Samuelson L. Epitaxial III-V NWs on silicon. Nano Lett 2004;4:1987–1990.

334. Bakkers E, Borgström MT, Verheijen MA. Epitaxial growth of III-V NWs on group IV substrates. MRS Bull 2007;32:117–122.

335. Ahn Y, Dunning J, Park J. Scanning photocurrent imaging and electronic band studies in silicon NW field effect transistors. Nano Lett 2005;5:1367–1370.

336. Gu Y, Romankiewicz J, David J, Lensch J, Lauhon L, Kwak E, Odom T. Local photocurrent mapping as a probe of contact effects and charge carrier transport in semiconductor NW devices. J Vac Sci Technol B 2006;24:2172–2177.

337. Gu Y, Kwak E, Lensch J, Allen J, Odom TW, Lauhon LJ. Near-field scanning photocurrent microscopy of a NW photodetector. Appl Phys Lett 2005;87:043111–043111-3.

338. Doh Y, Maher KN, Ouyang L, Yu CL, Park H, Park J. Electrically driven light emission from individual CdSe NWs. Nano Lett 2008;8:4552–4556.

339. Zhang Z, Jin C, Liang X, Chen Q, Peng L. Current–voltage characteristics and parameter retrieval of semiconducting NWs. Appl Phys Lett 2006;88:073102–073102-3.

340. Liao Z, Liu K, Zhang J, Xu J, Yu D. Effect of surface states on electron transport in individual ZnO NWs. Phys Lett A 2007;367:207–210.

341. Liao Z, Xu J, Zhang J, Yu D. Photovoltaic effect and charge storage in single ZnO NWs. Appl Phys Lett 2008;93:023111–023111-3.

342. Cheng G, Li Z, Wang S, Gong H, Cheng K, Jiang X, Zhou S, Du Z, Cui T, Zou G. The unsaturated photocurrent controlled by two-dimensional barrier geometry of a single ZnO NW Schottky photodiode. Appl Phys Lett 2008;93:123103–123103-3.

343. Harnack O, Pacholski C, Weller H, Yasuda A, Wessels JM. Rectifying behavior of electrically aligned ZnO nanorods. Nano Lett 2003;3:1097–1101.

344. Cheng K, Cheng G, Wang S, Li L, Dai S, Zhang X, Zou B, Du Z. Surface states dominative Au Schottky contact on vertical aligned ZnO nanorod arrays synthesized by low-temperature growth. New J Phys 2007;9:214.

345. Heo Y, Tien L, Norton D, Pearton S, Kang B, Ren F, LaRoche J. Pt ZnONW Schottky diodes. Appl Phys Lett 2004;85:3107–3109.

346. Driggers RG. Encyclopedia of Optical Engineering: Las-Pho. CRC press; 2003. p 1025–2048.

347. Hayden O, Agarwal R, Lieber CM. Nanoscale avalanche photodiodes for highly sensitive and spatially resolved photon detection. Nat Mater 2006;5:352–356.

348. Yang C, Barrelet CJ, Capasso F, Lieber CM. Single p-type/intrinsic/n-type silicon NWs as nanoscale avalanche photodetectors. Nano Lett 2006;6:2929–2934.

349. Bulgarini G, Reimer ME, Hocevar M, Bakkers EP, Kouwenhoven LP, Zwiller V. Avalanche amplification of a single exciton in a semiconductor NW. Nat Photonics 2012;6:455–458.

350. Pearsall T. Threshold energies for impact ionization by electrons and holes in InP. Appl Phys Lett 1979;35:168–170.

351. Algra RE, Verheijen MA, Borgström MT, Feiner L, Immink G, van Enckevort WJ, Vlieg E, Bakkers EP. Twinning superlattices in indium phosphide NWs. Nature 2008;456:369–372.

352. Reimer ME, Bulgarini G, Akopian N, Hocevar M, Bavinck MB, Verheijen MA, Bakkers EP, Kouwenhoven LP, Zwiller V. Bright single-photon sources in bottom-up tailored NWs. Nat Commun 2012;3:737.

353. Zhong ZH, Qian F, Wang DL, Lieber CM. Synthesis of p-type gallium nitride NWs for electronic and photonic nanodevices. Nano Lett 2003;3:343–346.

354. McAlpine MC, Friedman RS, Lieber CM. Nanoimprint lithography for hybrid plastic electronics. Nano Lett 2003;3:443–445.

355. Huang Y, Duan XF, Lieber CM. NWs for integrated multicolor nanophotonics. Small 2005;1:142–147.

356. Han S, Jin W, Zhang DH, Tang T, Li C, Liu XL, Liu ZQ, Lei B, Zhou CW. Photoconduction studies on GaN NW transistors under UV and polarized UV illumination. Chem Phys Lett 2004;389:176–180.

357. Ahn Y, Park J. Efficient visible light detection using individual germanium NW field effect transistors. Appl Phys Lett 2007;91:162102–162102-3.

358. Wang ZL, Kong X, Zuo J. Induced growth of asymmetric nanocantilever arrays on polar surfaces. Phys Rev Lett 2003;91:185502.

359. Heo YW, Tien LC, Kwon Y, Norton DP, Pearton SJ, Kang BS, Ren F. Depletion-mode ZnO NW field-effect transistor. Appl Phys Lett 2004;85:2274–2276.

360. Francinelli A, Tonneau D, Clement N, Abed H, Jandard F, Nitsche S, Dallaporta H, Safarov V, Gautier J. Light-induced reversible conductivity changes in silicon-on-insulator NWs. Appl Phys Lett 2004;85:5272–5274.

361. Fujiwara A, Yamazaki K, Takahashi Y. Detection of single charges and their generation-recombination dynamics in Si NWs at room temperature. Appl Phys Lett 2002;80:4567–4569.

362. Zhang A, You S, Soci C, Liu Y, Wang D, Lo Y. Silicon NW detectors showing phototransistive gain. Appl Phys Lett 2008;93:121110–121110-3.

363. Barnes WL, Dereux A, Ebbesen TW. Surface plasmon subwavelength optics. Nature 2003;424:824–830.

364. Dragoman M, Dragoman D. Plasmonics: applications to nanoscale terahertz and optical devices. Prog Quantum Electron 2008;32:1–41.

365. Veronis G, Kocabas SE, Miller DA, Fan S. Modeling of plasmonic waveguide components and networks. J Comput Theor Nanosci 2009;6:1808–1826.

366. Veronis G, Yu Z, Kocabas SE, Miller DA, Brongersma ML, Fan S. Metal-dielectric-metal plasmonic waveguide devices for manipulating light at the nanoscale. Chin Opt Lett 2009;7:302–308.

367. Maier SA, Kik PG, Atwater HA. Optical pulse propagation in metal nanoparticle chain waveguides. Phys Rev B 2003;67:205402.

368. Maier SA, Brongersma ML, Kik PG, Meltzer S, Requicha AA, Atwater HA. Plasmonics—a route to nanoscale optical devices. Adv Mater 2001;13:1501–1505.

369. Devaux E, Ebbesen TW, Weeber J, Dereux A. Launching and decoupling surface plasmons via micro-gratings. Appl Phys Lett 2003;83:4936–4938.

370. Kocabas S, Veronis G, Miller D, Fan S. Transmission line and equivalent circuit models for plasmonic waveguide components. IEEE J Sel Top Quantum Electron 2008;14:1462–1472.

371. Takahara J, Yamagishi S, Taki H, Morimoto A, Kobayashi T. Guiding of a one-dimensional optical beam with nanometer diameter. Opt Lett 1997;22:475–477.

372. Collin S, Pardo F, Teissier R, Pelouard J. Efficient light absorption in metal–semiconductor–metal nanostructures. Appl Phys Lett 2004;85:194–196.

373. Collin S, Pardo F, Pelouard J. Resonant-cavity-enhanced subwavelength metal–semiconductor–metal photodetector. Appl Phys Lett 2003;83:1521–1523.

374. Pyayt AL, Wiley B, Xia Y, Chen A, Dalton L. Integration of photonic and silver NW plasmonic waveguides. Nat Nanotechnol 2008;3:660–665.

375. Hegg M, Lin LY. Near-field photodetection with high spatial resolution by nanocrystal quantum dots. Opt Express 2007;15:17163–17170.

376. Falk AL, Koppens FH, Chun LY, Kang K, de Leon Snapp N, Akimov AV, Jo M, Lukin MD, Park H. Near-field electrical detection of optical plasmons and single-plasmon sources. Nat Phys 2009;5:475–479.

377. Lou J, Tong L, Ye Z. Modeling of silica NWs for optical sensing. Opt Express 2005;13:2135–2140.

378. Rümke T, Sánchez Gil JA, Muskens O, Rümke M, Bakkers E, Gómez Rivas J. Local and anisotropic excitation of surface plasmon polaritons by semiconductor NWs. Opt Express 2008;16(7):5013–5021.

379. Islam MS, Sharma S, Kamins T, Williams RS. A novel interconnection technique for manufacturing NW devices. Appl Phys A 2005;80:1133–1140.

380. Yan H, Park SH, Finkelstein G, Reif JH, LaBean TH. DNA-templated self-assembly of protein arrays and highly conductive NWs. Science 2003;301:1882–1884.

381. Richter J, Mertig M, Pompe W, Monch I, Schackert HK. Construction of highly conductive NWs on a DNA template. Appl Phys Lett 2001;78:536–538.

382. Islam MS, Murthy S, Itoh T, Wu MC, Novak D, Waterhouse RB, Sivco DL, Cho AY. Velocity-matched distributed photodetectors and balanced photodetectors with pin photodiodes. IEEE Trans Microwave Theory Tech 2001;49:1914–1920.

383. Grego S, Gilchrist KH, Kim J, Kwon M, Islam MS. Waveguide-integrated NW photoconductors on a non-single crystal surface. Proceedings of SPIE Volume 7406. Nanoepitaxy: Homo-and Heterogeneous Synthesis, *Characterization* and *Device Integration* of *Nanomaterials* I; 2009. p 74060B-9.

384. Grego S, Gilchrist KH, Kim J, Kwon M, Islam MS. NW-based devices combining light guiding and photodetection. Appl Phys A 2011;105:311–316.

385. Wu F, Logeeswaran V, Islam MS, Horsley DA, Walmsley RG, Mathai S, Houng D, Tan MR, Wang S. Integrated receiver architectures for board-to-board free-space optical interconnects. Appl Phys A 2009;95:1079–1088.

386. Goodman JW, Leonberger FJ, Kung S, Athale RA. Optical interconnections for VLSI systems. Proc IEEE 1984;72:850–866.

387. Haurylau M, Chen G, Chen H, Zhang J, Nelson NA, Albonesi DH, Friedman EG, Fauchet PM. On-chip optical interconnect roadmap: challenges and critical directions. IEEE J Sel Top Quantum Electron 2006;12:1699–1705.

388. Hosako I, Hiromoto N. A novel wave-guide Ge: Ga photoconductor. Infrared and Millimeter Waves, 2004 and 12th International Conference on Terahertz Electronics. Conference Digest of the 2004 Joint 29th International Conference onAnonymous, IEEE; 2004.

389. Chandrasekhar S, Campbell J, Dentai A, Qua G. Monolithic integrated waveguide photodetector. Electron Lett 1987;23:501–502.

390. Hu X, Holzwarth CW, Masciarelli D, Dauler EA, Berggren KK. Efficiently coupling light to superconducting NW single-photon detectors. IEEE Trans Appl Supercond 2009;19:336–340.

391. Chang DE, Sørensen AS, Hemmer P, Lukin M. Strong coupling of single emitters to surface plasmons. Phys Rev B 2007;76:035420.

392. Zhang Z, Yuan H, Gao Y, Wang J, Liu D, Shen J, Liu L, Zhou W, Xie S, Wang X. Large-scale synthesis and optical behaviors of ZnO tetrapods. Appl Phys Lett 2007;90:153116–153116-3.

393. Yamada H, Shirane M, Chu T, Yokoyama H, Ishida S, Arakawa Y. Nonlinear-optic silicon-NW waveguides. Jpn J Appl Phys 2005;44:6541.

394. Wang X, Pan A, Liu D, Zou B, Zhu X. Comparison of the optical waveguide behaviors of Se-doped and undoped CdS nanoribbons by using near-field optical microscopy. J Nanosci Nanotechnol 2009;9:978–981.

395. Sirbuly DJ, Law M, Pauzauskie P, Yan H, Maslov AV, Knutsen K, Ning C, Saykally RJ, Yang P. Optical routing and sensing with NW assemblies. Proc Natl Acad Sci U S A 2005;102:7800–7805.

396. Daniel BA, Agrawal GP. Vectorial nonlinear propagation in silicon NW waveguides: polarization effects. J Opt Soc Am B 2010;27:956–965.

397. Gao S, Li Z, Zhang X. Power-attenuated optimization for four-wave mixing-based wavelength conversion in silicon NW waveguides. J Electromagn Waves Appl 2010;24:1255–1265.

398. Voss T, Svacha GT, Mazur E, Müller S, Ronning C. The influence of local heating by nonlinear pulsed laser excitation on the transmission characteristics of a ZnO NW waveguide. Nanotechnology 2009;20:095702.

399. Fedutik Y, Temnov V, Woggon U, Ustinovich E, Artemyev M. Exciton-plasmon interaction in a composite metal-insulator-semiconductor NW system. J Am Chem Soc 2007;129:14939–14945.

400. Oulton RF, Sorger VJ, Genov D, Pile D, Zhang X. A hybrid plasmonic waveguide for subwavelength confinement and long-range propagation. Nat Photonics 2008;2:496–500.

401. Li J, Engheta N. Core-shell NW optical antennas fed by slab waveguides. IEEE Trans Antennas Propag 2007;55:3018–3026.

402. Vj L, Oh J, Nayak AP, Katzenmeyer AM, Gilchrist KH, Grego S, Kobayashi NP, Wang S, Talin A, Dhar NK. A perspective on NW photodetectors: current status, future challenges, and opportunities. IEEE J Sel Top Quantum Electron 2011;17:1002–1032.

403. Chen R, Wang S, Lan Z, Tsai JT, Wu C, Chen L, Chen K, Huang Y, Chen C. On-chip fabrication of well-aligned and contact-barrier-free GaN nanobridge devices with ultrahigh photocurrent responsivity. Small 2008;4:925–929.

404. Kim H, Zhang A, Cheng J, Lo Y. High-sensitivity visible and IR (1550nm) Si NW photodetectors. *OPTO* Anonymous International Society for Optics and Photonics; 2010.

405. Sharma S, Kamins T, Islam MS, Williams RS, Marshall A. Structural characteristics and connection mechanism of gold-catalyzed bridging silicon NWs. J Cryst Growth 2005;280:562–568.

406. Tang Q, Kamins TI, Liu X, Grupp D, Harris JS. In situ p-n junctions and gated devices in titanium-silicide nucleated Si NWs. Electrochem Solid-State Lett 2005;8:G204–G208.

407. Garnett EC, Liang W, Yang P. Growth and electrical characteristics of platinum-nanoparticle-catalyzed silicon NWs. Adv Mater 2007;19:2946–2950.

408. Perea DE, Hemesath ER, Schwalbach EJ, Lensch-Falk JL, Voorhees PW, Lauhon LJ. Direct measurement of dopant distribution in an individual vapour–liquid–solid NW. Nat Nanotechnol 2009;4:315–319.

409. Cui Y, Zhong ZH, Wang DL, Wang WU, Lieber CM. High performance silicon NW field effect transistors. Nano Lett 2003;3:149–152.

410. Rivillon S, Amy F, Chabal YJ, Frank MM. Gas phase chlorination of hydrogen-passivated silicon surfaces. Appl Phys Lett 2004;85:2583–2585.

411. Hu L, Chen G. Analysis of optical absorption in silicon NW Arrays for photovoltaic applications. Nano Lett 2007;7:3249–3252.

412. Kimukin I, Islam MS, Williams RS. Surface depletion thickness of p-doped silicon NWs grown using metal-catalysed chemical vapour deposition. Nanotechnology 2006;17:S240.

413. Tsakalakos L, Balch J, Fronheiser J, Shih M, LeBoeuf SF, Pietrzykowski M, Codella PJ, Korevaar BA, Sulima OV, Rand J. Strong broadband optical absorption in silicon NW films. J Nanophotonics 2007;1:013552–013552-10.

414. Muskens OL, Rivas JG, Algra RE, Bakkers EPAM, Lagendijk A. Design of light scattering in NW materials for photovoltaic applications. Nano Lett 2008;8:2638–2642.

415. Muskens OL, Diedenhofen SL, Kaas BC, Algra RE, Bakkers EP, Gómez Rivas J, Lagendijk A. Large photonic strength of highly tunable resonant NW materials. Nano Lett 2009;9:930–934.

416. Muskens O, Borgstrom M, Bakkers E, Gómez Rivas J. Giant optical birefringence in ensembles of semiconductor NWs. Appl Phys Lett 2006;89:233117–233117-3.

417. Ruda H, Shik A. Polarization-sensitive optical phenomena in semiconducting and metallic NWs. Phys Rev B 2005;72:115308.

418. Wang JF, Gudiksen MS, Duan XF, Cui Y, Lieber CM. Highly polarized photoluminescence and photodetection from single indium phosphide NWs. Science 2001;293:1455–1457.

419. Ruda H, Shik A. Polarization-sensitive optical phenomena in thick semiconducting NWs. J Appl Phys 2006;100:024314–024314-6.

420. Lal S, Link S, Halas NJ. Nano-optics from sensing to waveguiding. Nat Photonics 2007;1:641–648.

421. Garnett EC, Brongersma ML, Cui Y, McGehee MD. NW solar cells. Annu Rev Mater Res 2011;41:269–295.

422. Cao L, Park J, Fan P, Clemens B, Brongersma ML. Resonant germanium nanoantenna photodetectors. Nano Lett 2010;10:1229–1233.

423. Garnett EC, Yang PD. Light trapping in Silicon NW solar cells. Nano Lett 2010;10:1082–1087.

424. Law M, Greene LE, Johnson JC, Saykally R, Yang PD. NW dye-sensitized solar cells. Nat Mater 2005;4:455–459.

425. Wang JX, Wu CML, Cheung WS, Luo LB, He ZB, Yuan GD, Zhang WJ, Lee CS, Lee ST. Synthesis of hierarchical porous ZnO disklike nanostructures for improved photovoltaic properties of dye-sensitized solar cells. J Phys Chem C 2010;114:13157–13161.

426. Fan Z, Razavi H, Do J, Moriwaki A, Ergen O, Chueh Y, Leu PW, Ho JC, Takahashi T, Reichertz LA. Three-dimensional nanopillar-array photovoltaics on low-cost and flexible substrates. Nat Mater 2009;8:648–653.

427. Lévy-Clément C, Tena-Zaera R, Ryan MA, Katty A, Hodes G. CdSe-sensitized p-CuSCN/NW n-ZnO heterojunctions. Adv Mater 2005;17:1512–1515.

428. Yuhas BD, Yang P. NW-based all-oxide solar cells. J Am Chem Soc 2009;131:3756–3761.

429. Varghese OK, Paulose M, Grimes CA. Long vertically aligned titania nanotubes on transparent conducting oxide for highly efficient solar cells. Nat Nanotechnol 2009;4:592–597.

430. Dong Y, Tian B, Kempa TJ, Lieber CM. Coaxial group III– nitride NW photovoltaics. Nano Lett 2009;9:2183–2187.

431. Colombo C, Heiss M, Gratzel M, Morral AFI. Gallium arsenide p-i-n radial structures for photovoltaic applications. Appl Phys Lett 2009;94:173108.

432. Briseno AL, Holcombe TW, Boukai AI, Garnett EC, Shelton SW, Fréchet JJ, Yang P. Oligo-and polythiophene/ZnO hybrid NW solar cells. Nano Lett 2009;10:334–340.

433. Greene LE, Law M, Yuhas BD, Yang PD. ZnO-TiO2 core-shell nanorod/P3HT solar cells. J Phys Chem C 2007;111:18451–18456.

434. Takanezawa K, Tajima K, Hashimoto K. Efficiency enhancement of polymer photovoltaic devices hybridized with ZnO nanorod arrays by the introduction of a vanadium oxide buffer layer. Appl Phys Lett 2008;93:063308.

435. Wei W, Bao X, Soci C, Ding Y, Wang Z, Wang D. Direct heteroepitaxy of vertical InAs NWs on Si substrates for broad band photovoltaics and photodetection. Nano Lett 2009;9:2926–2934.

436. Williams SS, Hampton MJ, Gowrishankar V, Ding I, Templeton JL, Samulski ET, DeSimone JM, McGehee MD. Nanostructured titania– polymer photovoltaic devices made using PFPE-based nanomolding techniques. Chem Mater 2008;20:5229–5234.

437. Kempa TJ, Tian B, Kim DR, Hu J, Zheng X, Lieber CM. Single and tandem axial *p-i-n* NW photovoltaic devices. Nano Lett 2008;8:3456–3460.

438. Fahrenbruch AL, Bube RH. Fundamentals of Solar Cells: Photovoltaic Solar Energy Conversion. New York: Academic Press, Inc.; 1983.

439. Garnett EC, Yang PD. Silicon NW radial p-n junction solar cells. J Am Chem Soc 2008;130:9224–9225.

440. Ergen O, Ruebusch DJ, Fang H, Rathore AA, Kapadia R, Fan Z, Takei K, Jamshidi A, Wu M, Javey A. Shape-controlled synthesis of single-crystalline nanopillar arrays by template-assisted vapor–liquid-solid process. J Am Chem Soc 2010;132:13972–13974.

441. Fan Z, Ruebusch DJ, Rathore AA, Kapadia R, Ergen O, Leu PW, Javey A. Challenges and prospects of nanopillar-based solar cells. Nano Res 2009;2:829–843.

442. Wang B, Leu PW. Tunable and selective resonant absorption in vertical NWs. Opt Lett 2012;37:3756–3758.

443. Putnam MC, Turner-Evans DB, Kelzenberg MD, Boettcher SW, Lewis NS, Atwater HA. 10 mu m minority-carrier diffusion lengths in Si wires synthesized by Cu-catalyzed vapor–liquid-solid growth. Appl Phys Lett 2009;95:163116.

444. Kelzenberg MD, Turner-Evans DB, Kayes BM, Filler MA, Putnam MC, Lewis NS, Atwater HA. Photovoltaic measurements in single-NW silicon solar cells. Nano Lett 2008;8:710–714.

445. Allen JE, Hemesath ER, Perea DE, Lensch-Falk JL, Li ZY, Yin F, Gass MH, Wang P, Bleloch AL, Palmer RE, Lauhon LJ. High-resolution detection of Au catalyst atoms in Si NWs. Nat Nanotechnol 2008;3:168–173.

446. Sharma AK, Agarwal SK, Singh SN. Determination of front surface recombination velocity of silicon solar cells using the short-wavelength spectral response. Sol Energy Mater Sol Cells 2007;91:1515–1520.

447. Sabbah AJ, Riffe DM. Measurement of silicon surface recombination velocity using ultrafast pump-probe reflectivity in the near infrared. J Appl Phys 2000;88:6954–6956.

448. Rowe MW, Liu HL, Williams GP, Williams RT. Picosecond photoelectron-spectroscopy of excited-states at Si(111) root-3 X root-3r 30-degrees-B, Si(111)7x7, Si(100)2x1, and laser-annealed Si(111)1x1 surfaces. Phys Rev B 1993;47:2048–2064.

449. Passlack M, Hong M, Mannaerts JP, Kwo JR, Tu LW. Recombination velocity at oxide-GaAs interfaces fabricated by in situ molecular beam epitaxy. Appl Phys Lett 1996;68:3605–3607.

450. Jastrzebski L, Lagowski J, Gatos HC. Application of scanning electron-microscopy to determination of surface recombination velocity - Gaas. Appl Phys Lett 1975;27:537–539.

451. Kelzenberg MD, Turner-Evans DB, Putnam MC, Boettcher SW, Briggs RM, Baek JY, Lewis NS, Atwater HA. High-performance Si microwire photovoltaics. Energy Environ Sci 2011;4:866–871.

452. Rosenwaks Y, Burstein L, Shapira Y, Huppert D. Effects of reactive versus unreactive metals on the surface recombination velocity at Cds and Cdse(1120) interfaces. Appl Phys Lett 1990;57:458–460.

453. Delgadillo I, Vargas M, CruzOrea A, Alvarado-Gil JJ, Baquero R, Sanchez-Sinencio F, Vargas H. Photoacoustic CdTe surface characterization. Appl Phys B 1997;64:97–101.

454. Gunawan O, Guha S. Characteristics of vapor–liquid-solid grown silicon NW solar cells. Sol Energy Mater Sol Cells 2009;93:1388–1393.

455. Nelson J. The physics of solar cells. In: *Anonymous*. Imperial College Press; 2003.

456. Kelzenberg MD, Boettcher SW, Petykiewicz JA, Turner-Evans DB, Putnam MC, Warren EL, Spurgeon JM, Briggs RM, Lewis NS, Atwater HA. Enhanced absorption and carrier collection in Si wire array for photovoltaic applications. Nat Mater 2010;9:239–244.

457. Zhu J, Yu ZF, Burkhard GF, Hsu CM, Connor ST, Xu YQ, Wang Q, McGehee M, Fan SH, Cui Y. Optical absorption enhancement in amorphous silicon NW and nanocone arrays. Nano Lett 2009;9:279–282.

458. Xie WQ, Oh JI, Shen WZ. Realization of effective light trapping and omnidirectional antireflection in smooth surface silicon NW arrays. Nanotechnology 2011;22:065704.

459. Fan ZY, Kapadia R, Leu PW, Zhang XB, Chueh YL, Takei K, Yu K, Jamshidi A, Rathore AA, Ruebusch DJ, Wu M, Javey A. Ordered arrays of dual-diameter nanopillars for maximized optical absorption. Nano Lett 2010;10:3823–3827.

460. Chueh YL, Fan ZY, Takei K, Ko H, Kapdia R, Rathore A, Miller N, Yu K, Wu M, Haller EE, Javey A. Black Ge based on crystalline/amorphous core/shell nanoneedle arrays. Nano Lett 2010;10:520–523.

461. Fan ZY, Razavi H, Do JW, Moriwaki A, Ergen O, Chueh YL, Leu PW, Ho JC, Takahashi T, Reichertz LA, Neale S, Yu K, Wu M, Ager JW, Javey A. Three dimensional nanopillar array photovoltaics on low cost and flexible substrate. Nat Mater 2009;8:648–653.

462. Kosten ED, Warren EL, Atwater HA. Ray optical light trapping in Silicon microwires: exceeding the $2n^2$ intensity limit. Opt Express 2011;19:3316–3331.

463. Spurgeon JM, Atwater HA, Lewis NS. A comparison between the behavior of nanorod array and planar Cd(Se, Te) photoelectrodes. J Phys Chem C 2008;112:6186–6193.

464. Seo K, Wober M, Steinvurzel P, Schonbrun E, Dan Y, Ellenbogen T, Crozier KB. Multicolored vertical silicon NWs. Nano Lett 2011;11:1851–1856.

465. Cao L, Fan P, Barnard ES, Brown AM, Brongersma ML. Tuning the color of silicon nanostructures. Nano Lett 2010;10:2649–2654.

466. Wu PM, Anttu N, Xu H, Samuelson L, Pistol ME. Colorful InAs NW arrays: from strong to weak absorption with geometrical tuning. Nano Lett 2012;12(4):1990–1995.

467. Diedenhofen SL, Janssen OT, Grzela G, Bakkers EP, Gómez Rivas J. Strong geometrical dependence of the absorption of light in arrays of semiconductor NWs. ACS Nano 2011;5:2316.

468. Tang J, Huo Z, Brittman S, Gao H, Yang P. Solution-processed core-shell NWs for efficient photovoltaic cells. Nat Nanotechnol 2011;6:568–572.

469. Tsai S, Chang H, Wang H, Chen S, Lin C, Chen S, Chueh Y, He J. Significant efficiency enhancement of hybrid solar cells using core–shell NW geometry for energy harvesting. ACS Nano 2011;5:9501–9510.

470. Cao L, Fan P, Brongersma ML. Optical coupling of deep-subwavelength semiconductor NWs. Nano Lett 2011;11:1463–1468.

471. Yu Y, Cao L. Coupled leaky mode theory for light absorption in 2D, 1D, and 0D semiconductor nanostructures. Opt Express 2012;20:13847–13856.

472. Li J, Yu H, Wong SM, Li X, Zhang G, Lo PG, Kwong D. Design guidelines of periodic Si NW arrays for solar cell application. Appl Phys Lett 2009;95:243113.

473. Cao LY, White JS, Park JS, Schuller JA, Clemens BM, Brongersma ML. Engineering light absorption in semiconductor NW devices. Nat Mater 2009;8:643–647.

474. Cao L, Fan P, Vasudev AP, White JS, Yu Z, Cai W, Schuller JA, Fan S, Brongersma ML. Semiconductor NW optical antenna solar absorbers. Nano Lett 2010;10:439–445.

475. Hua B, Wang B, Leu PW, Fan ZY. Rational geometrical design of multi-diameter nanopillars for efficient light harvesting. Nano Energy 2013. DOI: 10.1016/j.bbr.2011.03.031.

476. Cho K, Ruebusch DJ, Lee MH, Moon JH, Ford AC, Kapadia R, Takei K, Ergen O, Javey A. Molecular monolayers for conformal, nanoscale doping of InP nanopillar photovoltaics. Appl Phys Lett 2011;98:203101.

477. Lee MH, Takei K, Zhang J, Kapadia R, Zheng M, Chen Y, Nah J, Matthews TS, Chueh Y, Ager JW. p-type InP nanopillar photocathodes for efficient solar-driven hydrogen production. Angew Chem Int Ed 2012.

478. Ho C, Lin G, Fu P, Lin C, Yang P, Chan I, Lai K, He J. An efficient light-harvesting scheme using SiO_2 nanorods for InGaN multiple quantum well solar cells. Sol Energy Mater Sol Cells 2012;103:194–198.

479. Wang H, Tsai K, Lai K, Wei T, Wang Y, He J. Periodic Si nanopillar arrays by anodic aluminum oxide template and catalytic etching for broadband and omnidirectional light harvesting. Opt Express 2012;20:A94–A103.

480. Yu R, Lin Q, Leung SF, Fan Z. Nanomaterials and nanostructures for efficient light absorption and photovoltaics. Nano Energy 2011;1(1):57–72.

481. Kapadia R, Fan Z, Javey A. Design constraints and guidelines for CdS/CdTe nanopillar based photovoltaics. Appl Phys Lett 2010;96:103116.

482. Kim DR, Lee CH, Rao PM, Cho IS, Zheng XL. Hybrid Si microwire and planar solar cells: passivation and characterization. Nano Lett 2011;11:2704.

483. Law M, Greene LE, Radenovic A, Kuykendall T, Liphardt J, Yang PD. ZnO-Al2O3 and ZnO-TiO2 core-shell NW dye-sensitized solar cells. J Phys Chem B 2006;110:22652–22663.

484. Bandaranayake KMP, Senevirathna MKI, Weligamuwa PMGMP, Tennakone K. Dye-sensitized solar cells made from nanocrystalline TiO2 films coated with outer layers of different oxide materials. Coord Chem Rev 2004;248:1277–1281.

485. Diamant Y, Chappel S, Chen SG, Melamed O, Zaban A. Core-shell nanoporous electrode for dye sensitized solar cells: the effect of shell characteristics on the electronic properties of the electrode. Coord Chem Rev 2004;248:1271–1276.

486. Palomares E, Clifford JN, Haque SA, Lutz T, Durrant JR. Control of charge recombination dynamics in dye sensitized solar cells by the use of conformally deposited metal oxide blocking layers. J Am Chem Soc 2003;125:475–482.

487. K. Tennakone, J. Bandara, P. K. M. Bandaranayake, G. R. A. Kumara, and A. Konno, Enhanced efficiency of a dye-sensitized solar cell made from MgO-coated nanocrystalline SnO2, Jpn J Appl Phys Part 2 40, L732-L734 (2001).

488. Zaban A, Chen SG, Chappel S, Gregg BA. Bilayer nanoporous electrodes for dye sensitized solar cells. Chem Commun 2000:2231–2232.

489. Kuo CY, Gau C, Dai BT. Photovoltaic characteristics of silicon NW arrays synthesized by vapor–liquid-solid process. Sol Energy Mater Sol Cells 2011;95:154–157.

490. Wang X, Pey KL, Yip CH, Fitzgerald EA, Antoniadis DA. Vertically arrayed Si NW/nanorod-based core-shell p-n junction solar cells. J Appl Phys 2010;108:124303.

491. Chen C, Jia R, Yue H, Li H, Liu X, Wu D, Ding W, Ye T, Kasai S, Tamotsu H, Chu J, Wang S. Silicon NW-array-textured solar cells for photovoltaic application. J Appl Phys 2010;108:094318.

492. Kumar D, Srivastava SK, Singh PK, Husain M, Kumar V. Fabrication of silicon NW arrays based solar cell with improved performance. Sol Energy Mater Sol Cells 2011;95:215–218.

493. Jung J, Guo Z, Jee S, Um H, Park K, Hyun MS, Yang JM, Lee J. A waferscale Si wire solar cell using radial and bulk p-n junctions. Nanotechnology 2010;21:445303.

494. Yoon J, Baca AJ, Park S-I, Elvikis P, Geddes JB, Li L, Kim RH, Xiao J, Wang S, Kim T-H, Motala MJ, Ahn BY, Duoss EB, Lewis JA, Nuzzo RG, Ferreir PM, Huang Y, Rockett A, Rogers JA. Ultrathin silicon solar microcells for semitransparent, mechanically flexible and microconcentrator module designs. Nat Mater 2008;7:907–915.

495. Fan Z, Javey A. Solar cells on curtains. Nat Mater 2008;7:835.

496. Lungenschmied C, Dennler G, Neugebauer H, Sariciftci SN, Glatthaar M, Meyer T, Meyer A. Flexible, long-lived, large-area, organic solar cells. Sol Energy Mater Sol Cells 2007;91:379–384.

497. Avigal Y, Kalish R. Growth of aligned carbon nanotubes by biasing during growth. Appl Phys Lett 2001;78:2291–2293.

498. Shan Y, Fonash SJ. Self-assembling silicon NWs for device applications using the nanochannel-guided "grow-in-place" approach. ACS Nano 2008;2:429–434.

499. Chai J, Wang D, Fan X, Buriak JM. Assembly of aligned linear metallic patterns on silicon. Nat Nanotechnol 2007;2:500–506.

500. Messer B, Song JH, Yang P. Microchannel networks for NW patterning. J Am Chem Soc 2000;122:10232–10233.

501. Cheng C, Gonela RK, Gu Q, Haynie DT. Self-assembly of metallic NWs from aqueous solution. Nano Lett 2005;5:175–178.

502. Fortuna SA, Li X. GaAs MESFET with a high-mobility self-assembled planar NW channel. IEEE Electron Device Lett 2009;30:593–595.

503. Takahashi T, Takei K, Adabi E, Fan Z, Niknejad AM, Javey A. Parallel array InAs NW transistors for mechanically bendable, ultra-high frequency electronics. ACS Nano 2010;4:5855–5860.

504. Ford AC, Ho JC, Fan Z, Ergen O, Altoe V, Aloni S, Razavi H, Javey A. Synthesis, contact printing, and device characterization of Ni-catalyzed, crystalline InAs NWs. Nano Res 2008;1:32.

505. Fan ZY, Lu JG. Electrical properties of ZnO NW-field effect transistors characterized with scanning probes. Appl Phys Lett 2005;86:032111.

506. Fan ZY, Mo XL, Lou CF, Yao Y, Wang DW, Chen GR, Lu JG. Structures and electrical properties of Ag-tetracyanoquinodimethane organometallic NWs. IEEE Trans Nanotechnol 2005;4:238–241.

507. Fan ZY, Ho JC, Jacobson ZA, Razavi H, Javey A. Large-scale, heterogeneous integration of NW arrays for image sensor circuitry. Proc Natl Acad Sci U S A 2008;105:11066–11070.

508. Huang Y, Duan XF, Wei QQ, Lieber CM. Directed assembly of one-dimensional nanostructures into functional networks. Science 2001;291:630–633.

509. Duan XF, Niu CM, Sahi V, Chen J, Parce JW, Empedocles S, Goldman JL. High-performance thin-film transistors using semiconductor NWs and nanoribbons. Nature 2003;425:274–278.

510. Whang D, Jin S, Wu Y, Lieber CM. Large-scale hierarchical organization of NW arrays for integrated nanosystems. Nano Lett 2003;3:1255–1259.

511. Tao AR, Huang J, Yang P. Langmuir-blodgettry of nanocrystals and NWs. Acc Chem Res 2008;41:1662–1673.

512. Wang DW, Chang YL, Liu Z, Dai HJ. Oxidation resistant germanium NWs: bulk synthesis, long chain alkanethiol functionalization, and Langmuir-Blodgett assembly. J Am Chem Soc 2005;127:11871–11875.

513. S. Acharya, A. Panda, N. Belman, S. Efrima, and Y. Golan, A semiconductor-NW assembly of ultrahigh junction density by the Langmuir-Blodgett technique, Adv Mater 18, 210−213 (2006).

514. Mai L, Gu Y, Han C, Hu B, Chen W, Zhang P, Xu L, Guo W, Dai Y. Orientated Langmuir-Blodgett assembly of VO(2) NWs. Nano Lett 2009;9:826–830.

515. Panda AB, Acharya S, Efrima S, Golan Y. Synthesis, assembly, and optical properties of shape- and phase-controlled ZnSe nanostructures. Langmuir 2007;23:765–770.

516. Yu G, Cao A, Lieber CM. Large-area blown bubble films of aligned NWs and carbon nanotubes. Nat Nanotechnol 2007;2:372–377.

517. Fan DL, Zhu FQ, Cammarata RC, Chien CL. Efficiency of assembling of NWs in suspension by ac electric fields. Appl Phys Lett 2006;89:223115.

518. Liu Y, Chung J, Liu WK, Ruoff RS. Dielectrophoretic assembly of NWs RID B-7599-2009 RID B-7605-2009. J Phys Chem B 2006;110:14098–14106.

519. Smith PA, Nordquist CD, Jackson TN, Mayer TS, Martin BR, Mbindyo J, Mallouk TE. Electric-field assisted assembly and alignment of metallic NWs. Appl Phys Lett 2000;77:1399–1401.

520. Takahashi T, Takei K, Ho JC, Chueh YL, Fan Z, Javey A. Monolayer resist for patterned contact printing of aligned NW arrays. J Am Chem Soc 2009;131:2102–2103.

521. Min KW, Kim YK, Shin G, Jang S, Han M, Huh J, Kim GT, Ha JS. White-light emitting diode array of p(+)-Si/aligned n-SnO(2) NWs heterojunctions. Adv Funct Mater 2011;21:119–124.

522. Sun C, Mathews N, Zheng M, Sow CH, Wong LH, Mhaisalkar SG. Aligned tin oxide nanonets for high-performance transistors. J Phys Chem C 2010;114:1331–1336.

523. Yang J, Lee MS, Lee H, Kim H. Hybrid ZnO NW networked field-effect transistor with solution-processed InGaZnO film. Appl Phys Lett 2011;98:253106.

524. Yerushalmi R, Jacobson ZA, Ho JC, Fan Z, Javey A. Large scale, highly ordered assembly of NW parallel arrays by differential roll printing. Appl Phys Lett 2007;91:203104.

525. Javey A, Nam S, Friedman RS, Yan H, Lieber CM. Layer-by-layer assembly of NWs for three-dimensional, multifunctional electronics. Nano Lett 2007;7:773–777.

526. Wang MCP, Gates BD. Directed assembly of NWs. Mater Today 2009;12:34–43.

527. Yan H, Choe HS, Nam S, Hu Y, Das S, Klemic JF, Ellenbogen JC, Lieber CM. Programmable NW circuits for nanoprocessors RID D-5345-2011. Nature 2011;470:240–244.

528. Nam S, Jiang X, Xiong Q, Ham D, Lieber CM. Vertically integrated, three-dimensional NW complementary metal-oxide-semiconductor circuits. Proc Natl Acad Sci U S A 2009;106:21035–21038.

529. Skucha K, Fan Z, Jeon K, Javey A, Boser B. Palladium/silicon NW schottky barrier based hydrogen sensors. Sens Actuators B 2010;145:232–238.

530. Chang Y, Hong FC. The fabrication of ZnO NW field-effect transistors by roll-transfer printing. Nanotechnology 2009;20:195302.

531. Cao Y, Liu W, Sun JL, Han YP, Zhang JH, Liu S, Sun HS, Guo JH. A technique for controlling the alignment of silver NWs with an electric field. Nanotechnology 2006;17:2378–2380.

532. Liu X, Long Y, Liao L, Duan X, Fan Z. Large-scale integration of semiconductor NWs for high-performance flexible electronics. ACS Nano 2012;6:1888–1900.

533. McAlpine M, Friedman R, Jin S, Lin K, Wang W, Lieber C. High-performance NW electronics and photonics on glass and plastic substrates. Nano Lett 2003;3:1531–1535.

534. Mcalpine MC, Ahmad H, Wang D, Heath JR. Highly ordered NW arrays on plastic substrates for ultrasensitive flexible chemical sensors. Nat Mater 2007;6:379–384.

535. Duan XF. NW thin-film transistors: a new avenue to high-performance macroelectronics. IEEE Trans Electron Devices 2008;55:3056–3062.

536. Fan ZY, Ho JC, Takahashi T, Yerushalmi R, Takei K, Ford AC, Chueh YL, Javey A. Toward the development of printable NW electronics and sensors. Adv Mater 2009;21:3730–3743.

537. Xu F, Durham JW III, Wiley BJ, Zhu Y. Strain-release assembly of NWs on stretchable substrates RID D-1657-2010 RID F-5154-2011 RID A-7003-2008 RID C-4845-2008. ACS Nano 2011;5:1556–1563.

538. Li XL, Zhang L, Wang XR, Shimoyama I, Sun XM, Seo WS, Dai HJ. Langmuir-Blodgett assembly of densely aligned single-walled carbon nanotubes from bulk materials. J Am Chem Soc 2007;129:4890–4891.

539. Guo Q, Teng X, Rahman S, Yang H. Patterned Langmuir-Blodgett films of mondisperse nanoparticles of iron oxide using soft lithography. J Am Chem Soc 2003;125:630–631.

540. Kim F, Kwan S, Akana J, Yang P. Langmuir-Blodgett nanorod assembly. J Am Chem Soc 2001;123:4360–4361.

541. Yamaki T, Asai K. Alternate multilayer deposition from ammonium amphiphiles and titanium dioxide crystalline nanosheets using the Langmuir-Blodgett technique. Langmuir 2001;17:2564–2567.

542. Ko H, Takei K, Kapadia R, Chuang S, Fang H, Leu PW, Ganapathi K, Plis E, Kim HS, Chen S, Madsen M, Ford AC, Chueh Y, Krishna S, Salahuddin S, Javey A. Ultrathin compound semiconductor on insulator layers for high-performance nanoscale transistors. Nature 2010;468:286–289.

543. Baca AJ, Ahn JH, Sun Y, Meitl MA, Menard E, Kim HS, Choi WM, Kim DH, Huang Y, Rogers JA. Semiconductor wires and ribbons for high-performance flexible electronics. Angew Chem Int Ed 2008;47:5524–5542.

544. Qi Y, Kim J, Nguyen TD, Lisko B, Purohit PK, McAlpine MC. Enhanced piezoelectricity and stretchability in energy harvesting devices fabricated from buckled PZT ribbons. Nano Lett 2011;472:304–305.

545. Qi Y, Jafferis NT, Lyons K Jr, Lee CM, Ahmad H, McAlpine MC. Piezoelectric ribbons printed onto rubber for flexible energy conversion RID A-8025-2010. Nano Lett 2010;10:524–528.

546. Lu W, Lieber CM. Nanoelectronics from the bottom up. Nat Mater 2007;6:841–850.

547. Javey A, Kim H, Brink M, Wang Q, Ural A, Guo J, McIntyre P, McEuen P, Lundstrom M, Dai HJ. High-kappa dielectrics for advanced carbon-nanotube transistors and logic gates. Nat Mater 2002;1:241–246.

548. Borghetti J, Snider GS, Kuekes PJ, Yang JJ, Stewart DR, Williams RS. 'Memristive'switches enable 'stateful'logic operations via material implication. Nature 2010;464:873–876.

549. DeHon A. Array-based architecture for FET-based, nanoscale electronics. IEEE Trans Nanotechnol 2003;2:23–32.

550. Das S, Rose G, Ziegler MM, Picconatto CA, Ellenbogen JC. Architectures and simulations for nanoprocessor systems integrated on the molecular scale. In: Introducing Molecular Electronics, *Anonymous*. Springer; 2005. p 479–512.

551. Takei K, Takahashi T, Ho JC, Ko H, Gillies AG, Leu PW, Fearing RS, Javey A. NW active-matrix circuitry for low-voltage macroscale artificial skin. Nat Mater 2010;9:821–826.

17

NANOELECTRONICS BASED ON SINGLE-WALLED CARBON NANOTUBES

QING CAO AND SHU-JEN HAN

IBM T.J. Watson Research Center, Yorktown Heights, NY, USA

17.1 INTRODUCTION

Unceasingly scaling of solid-state electronic devices is the foundation of ongoing information revolution and will continue to bring new functions to our daily life in the future by providing more scaled devices with faster speed and lower power consumption to support emerging applications such as smarter mobile devices, cloud computing, and big data analytics. However, the conventional linear scaling of silicon devices has become more and more difficult (1), requiring the introduction of new materials and/or implementation of new device architectures.

The International Technology Roadmap for Semiconductors (ITRS) suggests that single-walled carbon nanotube (SWNT) could replace silicon in ultrascaled field-effect transistors (FETs) (2, 3), as both theoretical and experimental results indicate that SWNTs could offer several times better performance than what is possible with the most advanced silicon technologies (4), with their high mobility (5), small intrinsic capacitance (6), and atomically smooth ultrathin body (7). Great progress in carbon nanotube nanoelectronics has been made during the past decade, and currently, the major research interest has shifted from demonstrating proof-of-concept device operations toward performance benchmarks with other established and emerging technologies as well as practical matters of the manufacturability in a large scale. This chapter summarizes some most significant recent achievements in this field. The applications of SWNTs in other electronic systems, for example, low-cost thin-film electronics, conductive coatings, and chemical sensors, have been discussed in previous reviews and therefore will not be covered here (8–11). We will begin with highlighting the attractive electrical properties of carbon nanotubes and discussing associated device physics, with focus on works performed mainly on individual nanotubes. After that, the efforts on solving the manufacturability issues will be described in detail with emphasis on nanotube separation and assembly. Finally, higher level implementations in integrated circuits and functional systems are presented. The last section concludes with perspectives on future directions and remaining challenges in this rapid evolving field.

17.2 ELECTRICAL PROPERTIES OF NANOTUBES AND DEVICE PHYSICS

Understanding the electrical properties of nanotubes is the basis for their applications in nanoelectronics. Based on their intrinsic properties, compact device models and simulation platforms are established to help us predict the performance of devices built on SWNTs, which are then verified with experimental results. The contact properties between SWNT channel and source/drain electrodes are discussed in detail as they are critical for ultrascaled FETs. The polarity control of nanotube transistors can be realized in reliable fashions, which is important for constructing complementary metal–oxide–semiconductor (CMOS)-type circuits. These results clearly show the promise and benefit of replacing silicon with carbon nanotubes for high-performance nanoelectronics at the end of silicon scaling roadmap.

17.2.1 Electrical Properties of Carbon Nanotubes

The molecular structure of SWNTs can be described as a seamless cylinder created via rolling up a graphene sheet to a certain direction as designated by a pair of integers (n, m) as shown in Figure 17.1a. There are generally 10 ~ 40 carbon atoms along the circumference, which makes the diameter of a typical SWNT in the range of 0.5 ~ 2 nm. This atomically smooth ultrathin body of SWNT is a great advantage

Nanomaterials, Polymers and Devices: Materials Functionalization and Device Fabrication, First Edition. Edited by Eric S. W. Kong.
© 2015 John Wiley & Sons, Inc. Published 2015 by John Wiley & Sons, Inc.

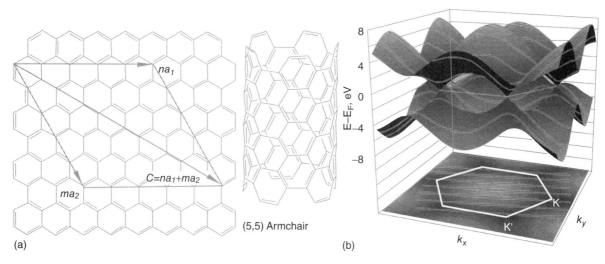

Figure 17.1   (a) A SWNT is formed by rolling up a graphene sheet along a chiral vector C to form a cylinder, such as the (5,5) vector shown here. (*Source*: Reproduced with permission from Reference 9. Copyright 2009, Wiley-VCH). (b) Three-dimensional plot of the band structure (top) and projected Brillouin zone (bottom) of a grapheme sheet. Solid lines represent the allowed wavevectors in the first Brillouin zone formed due to the quantization of circumferential momentum of a particular semiconducting nanotube. (*Source*: Reproduced with permission from Reference 13. Copyright 2007, Nature Publishing Group).

compared to conventional bulk semiconductor materials, whose mobility will be significantly reduced due to enhanced surface roughness scattering at a body thickness of around 1 nm (12). The band structure and the shape of the first Brillouin zone of a graphene sheet are plotted in Figure 17.1b (13). The valence band and the conduction band touch each other at six corners called K points, making graphene a semimetal. For nanotubes, electrons are confined around the circumference, which introduces a new quantization condition to split each band of graphene into a number of one-dimensional subbands. If these subbands pass through one of these K points, the nanotube is metallic; otherwise, it is semiconducting. It has been shown that the nanotube is metallic if n = m; the nanotube has a very small curvature-induced bandgap if n−m is divisible by 3; otherwise, the nanotube is semiconducting (14). As-synthesized SWNTs are typically composed of 66% semiconducting species. Separating semiconducting nanotubes from as-synthesized SWNTs is essential for most practical applications; however, it possesses a great challenge as semiconducting and metallic nanotubes share very similar structure and other properties, which will be discussed in Section 17.3. Semiconducting nanotubes demonstrate very high field-effect mobility and ballistic transport of electrons over several hundreds of nanometers at room temperature (5, 15), making them very attractive for applications in high-performance electronics. For an ultimately scaled SWNT FET with transparent electrical contacts, the device resistance will be limited by two resistance quantum ($R_Q = h/4e^2 = 6.5$ kΩ) (15).

17.2.2   Performance Projection and Benchmarking

The high carrier motilities, quasiballistic transport properties, and the intrinsic ultrathin body of SWNTs make them attractive to replace silicon in high-performance FETs. The rigorous nonequilibrium Green's function (NEGF) approach is adopted to simulate the performance of nanotube FETs, and the result suggests that they will show 3 ~ 4 times better performance over silicon technologies (16). Although NEGF approach is most accurate in physics, it is numerically too intensive for designing and predicting the performance of SWNT devices in a complex very large-scale integrated (VLSI) circuit. Recently, an analytical compact model for SWNT FETs based on ballistic transport with the inclusion of electrostatic capacitance, quantum capacitance, and source exhaustion effects due to the low density of states (DOS) in nanotubes has been developed (17) and implemented in a circuit performance optimizer to predict chip-level performance for a complex 4-core processor composed of 1.5 million logic gates (18). It is found that, under the optimized bias condition, gate length, and tube diameter, carbon nanotube FETs show over 5 times speed improvement over the partial depleted silicon-on-insulator technology projected at 11 nm node (17). The same group extends the circuit-level performance benchmarks for nanotube FETs and other emerging device technologies, such as III–V FETs and tunneling FETs (19). Through the energy–delay optimization by allowing different drive voltage and leakage current for different technologies, they show that carbon nanotube technology can outperform the 32 nm Si technology by 2–3.5-folds. When projecting into 11 nm node, based on their study, nanotube technology is 2–3 times better than ITRS target and expected to also significantly outperform III–V devices as shown in Figure 17.2.

17.2.3   Experimental Results on Nanodevices Based on SWNTs

The excellent performance of nanoscaled devices based on nanotubes predicted by simulation has also been verified by experimental results. There are two major arguments that bet against the adoption of SWNTs as a replacement of silicon in ultrascaled FETs. Firstly, SWNTs have much lower carrier effective mass. Such low effective mass could lead to more pronounced direct source–drain tunneling through the channel

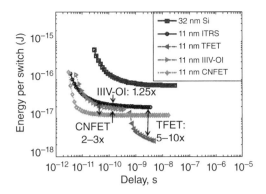

Figure 17.2 Simulation results of total energy per switch versus delay for 32 nm Si technology, projection based on state-of-the-art carbon nanotube technology (CNFET) and other emerging technologies at 11 nm node, including tunneling FET (TFET), III–V semiconductor on silicon (IIIV-OI), and the target for silicon devices predicted in ITRS. (*Source*: Reproduced with permission from Reference 19. Copyright 2011, IEEE).

energy barrier at the target channel length (L_{ch}) of nanotube transistors, that is, $L_{ch} < 10$ nm, and greatly decrease device power efficiency with a large subthreshold swing (*SS*) (20). Secondly, the ballistic transport of carrier within nanotubes could make it difficult to fully saturate the device output current (20, 21), and this current saturation is critically important for both logic and analog devices. To experimentally study the performance of nanotube FETs with a L_{ch} less than 10 nm, which will be the target gate length for 5 nm technology node, a nanotube transistor with 9 nm channel defined by electron-beam lithography is fabricated with a local bottom gate structure as shown in Figure 17.3a (22). The gate dielectric is 3 nm high κ HfO$_2$, whose equivalent oxide thickness (EOT) is around 0.65 nm. Such low EOT provides favorable electrostatics for such short-channel devices. Subthreshold characteristics of one device are shown in Figure 17.3b, demonstrating an *SS* of 94 mV/dec, which is comparable with that of most advanced Si devices at this dimension, including Si nanowire with gate-all-around configuration and extremely thin Si-on-insulator (ETSOI) devices (23–25). Clear current saturation behavior is also observed for such devices at a supply voltage less than 0.4 V, which is lower than the 0.64 V predicted by ITRS for 5 nm node devices. Both results verify the power benefit of nanotube FETs (26). At the same time, the width or pitch normalized device on current is several times higher than that of silicon devices, which verifies the performance benefit of nanotube FETs.

In the previous example, the nanotube FET is in the bottom gate configuration with significant overlap between gate and source/drain electrodes. Because of this overlapping and the associated high parasitic capacitance, the structure is not practical. It is important to demonstrate that SWNT FETs can still maintain their high performance in more technologically relevant device architectures. However, the conventional self-alignment process is difficult to be applied toward nanotube FETs. To solve this problem, in one example, a unique U-gate structure is utilized (27). This structure is formed relying on the profile control of photoresist and will be difficult to scale down to <10 nm gate length. In another example, SWNTs are first suspended on highly resistive single-crystalline silicon substrate with vertical trenches formed by anisotropic wet etching using source/drain electrodes as etch mask (28). Nanotubes are then coated with dielectric and TaN gate by atomic layer deposition (ALD) using aluminum oxynitride as adhesion layer to form a most favorable gate-all-around structure (Fig. 17.3c) (29, 30). Finally, a metal gate is deposited on top to connect with TaN gate. The depth of silicon trench allows us to subsequently remove this gate electrode from the source/drain region during liftoff, as shown in Figure 17.3c inset. Characteristics of such a device with L_{ch} of 30 nm are plotted in Figure 17.3d. The device demonstrates n-channel operation due to doping from fixed positive charges in gate dielectric. The doping from dielectric also helps to minimize the parasitic resistance arising from the existence of ungated region under spacers in this self-aligned structure, allowing the realization of high on-current and device transconductance. The superb electrostatics control from the gate-all-around structure makes the *SS* of this device less than 100 mV/dec despite an unoptimized dielectric bearing high trap density.

17.2.4 Electrical Contact to Carbon Nanotubes

The excellent electrical properties and the intrinsic ultrathin body of SWNTs are the most touted reasons for their advantage over silicon devices. However, for ultrascaled quasiballistic FETs, the metal–nanotube contacts are probably of more technological importance, since at this scale the parasitic contact resistance ($2R_C$), instead of channel resistance, limits device performance (31). We prefer to align the Fermi level in the metal electrodes either below the valence or above the conduction band edge of semiconducting nanotube to eliminate the Schottky barrier. For bulk materials, the Schottky barrier is generally determined by the metal-induced gap states, that is, interface states, at contacts (32). However, for nanotubes, such states have a much weaker impact on the band alignment due to electrostatics at such small dimension (33, 34), and therefore, the Schottky barrier (Φ_p) can be described as $\Phi_p = (\Phi_{SWNT} + 0.39\,\text{eV}/r(\text{nm})) - \Phi_m$, where r is the nanotube radius and Φ_{SWNT} and Φ_m are the work function of nanotube and metal contact, respectively. Pd forms ohmic contacts with nanotubes with r larger than 0.5 nm (15, 35). For ohmic Pd contacts, the exact $2R_C$ can be extracted based on the standard transmission line method (TLM), by making a series of FETs on the same nanotube with difference L_{ch}, as shown in Figure 17.4a (36). The I–V characteristics of devices spanning a wide range of L_{ch} are plotted in Figure 17.4b, showing how scaling affects the low-field slope and the device saturation current at the same device overdrive voltage. The extracted $2R_C$ is as small as 9 kΩ, as depicted in Figure 17.4c, with an intrinsic field-effect mobility

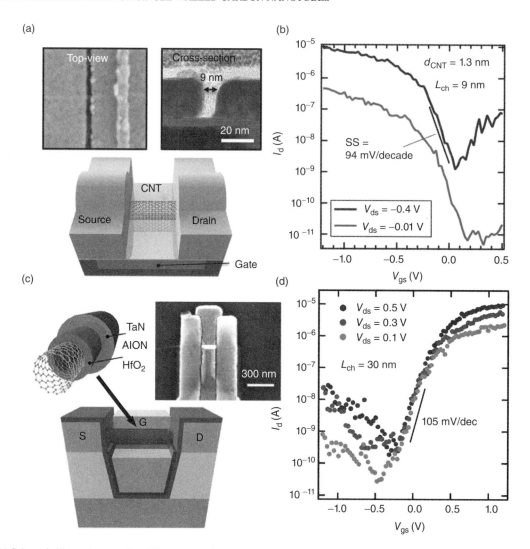

Figure 17.3   (a) Schematic illustration, top-view SEM, cross-sectional TEM, and (b) subthreshold transfer characteristics of an ultrascaled FET with an L_{ch} of approximately 9 nm, built on an SWNT with a diameter (d_{CNT}) of 1.3 nm. V_{gs}: gate bias. I_d: drain current. (*Source*: Reproduced with permission from Reference 22. Copyright 2012, American Chemical Society). (c) Schematic illustration, top-view SEM image (inset), and (d) subthreshold transfer characteristics of a gate-all-around nanotube FET, where an NO_2-functionalized nanotube is wrapped by ALD AlON/HfO_2 dielectric and TaN gate. (*Source*: Reproduced with permission from Reference 28. Copyright 2012, IEEE).

(μ_{FE}) > 1500 cm^2/(V s). Here, the $2R_C$ is mainly limited by the quantum resistance and is much lower than the parasitic resistance of state-of-the-art silicon technology, which is about 100 $\Omega \cdot \mu m$ or, in other words, >50 kΩ for approximately 2 nm device width (37).

The aforementioned "best-case" results are obtained from nanotubes grown by chemical vapor deposition (CVD) method. CVD tubes generally have better quality than solution-processed nanotubes, as evident from a smaller disorder D band in Raman spectrum shown in Figure 17.5a (38). They are very useful to study the physics or material properties of SWNTs. However, solution-processed nanotubes offer many process advantages, including capabilities of separating the nanotubes by electronic types and depositing them onto various substrates in the form of ultradensely aligned arrays at low temperature, which will be described in detail in Sections 17.3.3 and 17.3.4. Therefore, they are more attractive for real technological applications. As a result, it becomes necessary to benchmark the properties extracted from CVD tubes with those of solution tubes in a statistically important fashion. Figure 17.5b–5e compare the averaged $2R_C$ and μ_{FE} extracted via TLM from a large collection of CVD (frames b and c) and solution-processed (frames d and e) SWNTs (38, 39). They demonstrate similar average $2R_C$, both at around 40 kΩ. However, the measured μ_{FE} are dramatically different for CVD tubes, which is approximately 5700 cm^2/Vs, and solution tubes, which is only around 300 cm^2/Vs. These results, combined with defect density information extracted from Raman spectra, suggest that structural defects in solution-processed nanotubes, which might be caused by harsh chemical treatments and strong mechanical sonications applied during nanotube purification and/or suspension processes, act as electron scattering centers and therefore limit their electron transport capability. However, such defects do not affect the quality of electrical contacts between nanotube and metal electrodes, as evident from the fact that their extracted average $2R_C$ is comparable to that of CVD-grown nanotubes. Therefore, solution-processed nanotubes are expected

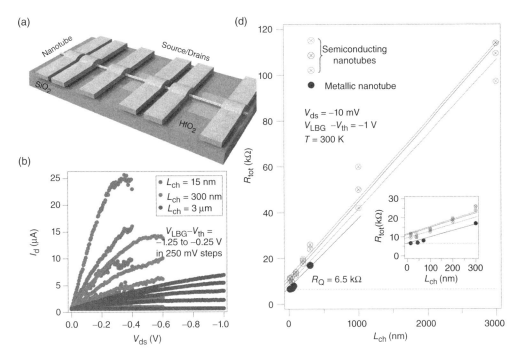

Figure 17.4 (a) Schematic illustration of a set of devices with different L_{ch} constructed on the same nanotube, which is transferred onto a local bottom metal gate covered by 10 nm HfO$_2$ gate dielectric. (b) Current–voltage characteristics for devices with L_{ch} of 3 μm, 300 nm, and 15 nm, respectively, from bottom to top. V_{ds}: drain-to-source bias. (c) Plot of total device resistance (R_{tot}) as a function of L_{ch}. Inset: magnified view for data corresponding with submicron devices. (*Source*: Reproduced with permission from Reference 36. Copyright 2010, Nature Publishing Group).

to provide performance similar to that of CVD-grown nanotubes in ultimately scaled FETs, where electron transport is largely limited by contacts rather than channel scattering. Such conclusion is further verified by building a 15 nm L_{ch} FET using a solution nanotube (40). This device demonstrates a low-field slope and saturation on current comparable with those of the best devices built on CVD tubes with identical L_{ch} and the same overdrive voltage, as shown in Figure 17.5d.

In addition to L_{ch} scaling, the contact length (L_c, length of source/drain) scaling is also important for a practical technology as together they limit the gate pitch scaling or the device packing density. L_c scaling behavior is heavily dependent on the contact geometries, which can be divided into two major categories: side-bonded contacts (Fig. 17.6a) and end-bonded contacts (Fig. 17.6b) (41). For end-bonded contacts, the nanotube ends at contact by forming direct chemical bond with the metal (42–44). In this configuration, both carbon pπ and pσ orbitals contribute to electron transmission, leading to a much higher transmission probability at the interface and therefore a potential to achieve more aggressively scaled L_c (45). However, such geometry is challenging to realize as it requires extreme conditions to form carbide bonds between SWNT and metal electrodes, for example, very high-temperature thermal annealing, which are generally not compatible with standard back-end-of-line processing. The side-bonded contacts are easier to produce, where nanotubes are embedded within metal directly deposited on top and linked through weak bondings, for example, van der Waals interactions (46). The L_c scaling behavior for Pd contact has been studied by constructing a series of ballistic short-channel FETs on the same nanotube as illustrated in Figure 17.6c (36). Smaller L_c significantly limits device on current, as shown in Figure 17.6d, due to the increase of $2R_C$. Furthermore, the dependence of $2R_C$ on L_c can be successfully described as $2R_C = R_Q + (\rho_c/L_c/r)$, where ρ_c is the specific contact resistivity, as plotted in Figure 17.6e. This result provides a lower limit for transfer length (L_T) at above 200 nm. Such high L_T value could be caused by either the weak coupling between metal and nanotube, or bad physical contact caused by poor wetting of metals on curvaceous nanotube surface (47). End-bonded contact, new contact material, or process need to be established to reduce L_T by at least 10 times for their successful adoption in scaled devices and represents a significant challenge for carbon nanotube electronics.

17.2.5 Polarity Control of Nanotube Transistors

FETs based on as-grown or as-deposited nanotubes typically exhibit unipolar p-channel behavior when built with high work function metals and exposed to oxygen (48, 49). Charge transfer doping with organic molecules or potassium atoms could realize n-channel device operation (50–53). However, such methods are not suitable for large-scale device integration due to their sensitivity to oxidation in ambient environment. SWNT FETs can also be converted from p- to n-channel mode via passivation with inorganic dielectrics, as shown in Figure 17.7a. The mechanism behind this process could involve (1) adjusting the doping level of SWNTs by eliminating oxygen molecules absorbed on the side of nanotubes in open air (49, 134), (2) modifying band alignment at contacts by removing oxygen attached at the SWNT/metal interface (48, 135), and (3) electrostatic doping of nanotubes by the deposited dielectric that has high density of fixed charges (123). In recent years,

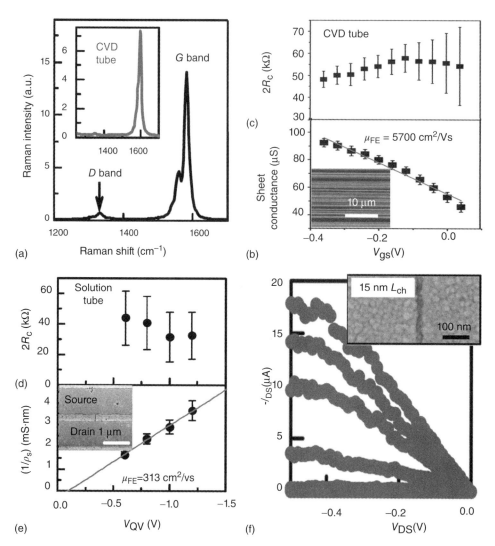

Figure 17.5 (a) Raman spectrum of solution-processed nanotubes, compared to CVD tubes (inset), showing the intensity of tangential G band and defect-induced D band. (b) Averaged $2R_C$ and (c) sheet conductance of aligned arrays of CVD nanotubes (SEM image shown as inset) extracted based on the TLM as a function of V_{gs}. (*Source*: Reproduced with permission from Reference 39. Copyright 2010, American Chemical Society). (d) Averaged $2R_C$ and (e) conductivity ($1/\rho_s$) of solution-processed nanotubes (SEM image of a device shown as inset) extracted based on the TLM as a function of gate overdrive voltage (V_{OV}: difference between V_{gs} and threshold voltage). (*Source*: Reproduced with permission from Reference 38. Copyright 2012, American Chemical Society). (e) Current–voltage characteristics for a device with L_{ch} of 15 nm constructed on a solution-processed nanotube. Device SEM image shown as inset. (*Source*: Reproduced with permission from Reference 40. Copyright 2013, American Chemical Society).

reliable and high-yield SWNT n-channel FETs (NFETs) are also demonstrated by solely adjusting the band alignment at contacts, with the adoption of rare earth metals with low enough work function as source/drain electrodes (57–60). Figure 17.7b plots the transfer characteristics of nanotube NFETs with erbium (Er) contact, benchmarked with p-channel FETs (PFETs) with Pd contact (57). Other low work function metals, including yttrium (Y) (58), scandium (Sc) (59), and gadolinium (Gd) (60), can also be utilized as n-channel contact for nanotube FETs. Control of device polarity with simple but reliable methods such as dielectric passivation or contact material modification is important for the construction of conventional CMOS-type circuits based on SWNT as the only semiconductor material (61).

17.3 MATERIALS AND PROCESSING

Extensive theoretical and experimental works have been done to demonstrate the advantages of carbon nanotubes over existing and other emerging technologies for high-performance, ultrascaled nanoelectronics, with the practical issues of contacts and polarity control in consideration. However, issues about manufacturability have to be solved to make it become a viable technology. Methods allowing the

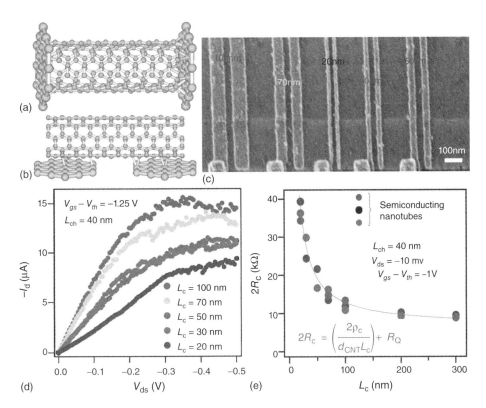

Figure 17.6 Schematic illustration of "end-bonded" (a) and "side-bonded" contacts between an SWNT and metal electrodes. (*Source*: Reproduced with permission from Reference 41. Copyright 2003, American Physical Society). (c) SEM image showing a test structure to study L_C scaling based on a single nanotube with output characteristics shown in frame (d). (e) Extracted $2R_C$ as a function of L_C for different set of devices with an empirical fitting. (*Source*: Reproduced with permission from Reference 36. Copyright 2010, Nature Publishing Group).

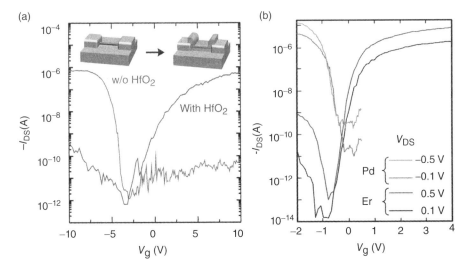

Figure 17.7 (a) Subthreshold plot of transfer characteristics of a nanotube transistor with L_{ch} of 0.7 μm under V_{DS} of −50 mV before (red) and after (blue) coating with 12 nm ALD HfO_2. (*Source*: Reproduced with permission from Reference (55). Copyright 2010, IOP Publishing). (b) Transfer characteristics of nanotube transistors with p-channel Pd contact and n-channel Er contact under different V_{DS}. (*Source*: Reproduced with permission from Reference 60. Copyright 2011, IEEE).

fabrication of SWNT wafers composed of high-purity semiconducting nanotubes aligned with tight pitch separation need to be established. Two major approaches are being explored: one is to synthesize aligned nanotubes on crystalline substrate and then remove metallic tubes afterward; the other one is to first separate nanotubes in solution, followed by assembling them into aligned arrays and depositing onto the target substrate for device fabrication. The latest progresses in both approaches are reviewed in this section.

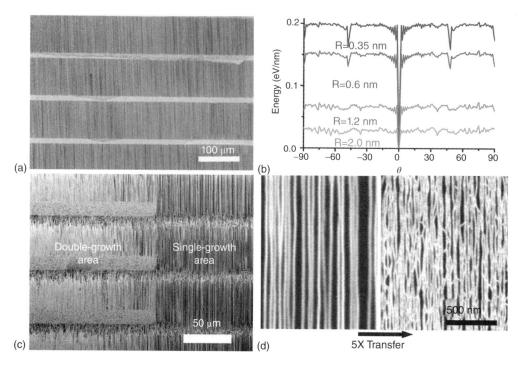

Figure 17.8   (a) Large area SEM image showing aligned arrays of SWNTs grown on single-crystalline quartz substrate. Bright horizontal lines correspond with prepatterned Fe catalyst strips. (*Source*: Reproduced with permission from Reference 68. Copyright 2007, American Chemical Society). (b) Oriental dependence of the van der Waals interaction energy per unit length of SWNT as a function of orientation angle (θ) for nanotubes with different radii (R). (*Source*: Reproduced with permission from Reference 69. Copyright 2009, American Chemical Society). (c) SEM image showing the density contrast between single-growth and double-growth area on the same substrate. (*Source*: Reproduced with permission from Reference 71. Copyright 2010, Wiley-VCH). (d) Five times consecutive transfer printing to increase the density of aligned arrays of nanotubes grown by CVD on quartz substrate from approximately 15 tubes/μm (left part) to approximately 55 tubes/μm on the final target substrate (right part). (*Source*: Reproduced with permission from Reference 72. Copyright 2010, Springer).

17.3.1   Aligned Growth of Nanotube Arrays

Aligned nanotube arrays can be directly grown on the substrate by CVD method, under the guidance of external electrical field or laminar flow of feeding gas (62, 63). However, such methods are difficult to achieve high tube density and linearity due to the thermal motion of SWNTs during high-temperature CVD growth and the fluctuation of gas flow. On the other hand, the wetting interaction between nanotubes and single-crystalline substrates can lead to the production of high-density arrays with nearly perfect alignment and linearity (64). The best result is obtained on ST-cut single-crystalline quartz substrate, with catalyst patterned into small strips to let the SWNTs grow primarily in regions free of catalyst particle contaminations, as shown in Figure 17.8a (65). SWNTs grown by such method demonstrate excellent alignment, with most nanotubes parallel with each other within 0.1 degree, and linearity, with maximum deviation less than 5 nm. Molecular mechanics simulation based on empirical interatomic potentials indicates that there is a global energy minimum when orientation angle θ of the nanotube, which is defined as the deviation of SWNT axis from [$2\bar{1}\bar{1}0$] orientation of quartz, equals zero, with the depth of the energy well decreases with increasing nanotube radius, as plotted in Figure 17.8b (66). This preferably orientation of as-grown SWNTs is determined by the surface atomic structure of quartz. At $\theta = 0$, the adjacent silicon atoms have the largest spacing, allowing SWNTs to react more effectively with the quartz substrate.

Although the substrate guidance approach produces SWNT arrays with high alignment quality, the nanotube density, measured as the number of tubes crossing per unit width, is limited to approximately 10 tubes/μm. The tube density is a very important topological parameter directly determining the electrical properties of nanotube arrays. First of all, the nanotube pitch separation must be smaller than the gate dielectric thickness to provide the highest possible gate capacitance per unit area for FETs (67). Moreover, for ultrascaled FETs, the increase of tube density will proportionally improve device on-current density by providing more transport pathways per unit width and reducing the effective $2R_C$ of the device, which dominates device operation at such geometry. A tube density above 100 tubes/μm is necessary to allow SWNTs to outperform Si technology. For CVD method, the tube density can be increased by improving the catalyst density or yield to certain limit, when higher tube density in the catalyst region will physically block the emergence of aligned nanotubes from the catalyst lines. Several approaches have been demonstrated to overcome this limitation. One simple route is to perform multiple growths on the same substrate (68). As shown in Figure 17.8c, after the first CVD growth, a new set of catalyst line is placed adjacent to the original ones, and a second CVD process yields additional nanotubes predominately from the new catalyst region and therefore increases the density of nanotube arrays, up to 20 ~ 30 tubes/μm range. In principle, such growth cycles can be repeated further for even higher tube density. Another route is to perform multiple transfer printings to multiply the tube densities as shown in Figure 17.8d (69, 70). In this process, SWNT arrays together with a metal film deposited on top are removed from the growth substrate by a stamp and then transferred onto the receiving substrate, followed by removing the metal film by wet etching (71). SWNTs grown on multiple substrates can be sequentially transferred to the same target substrate to increase the nanotube density. For example, a 4 times transfer increases the tube density from approximately 15 tubes/μm to approximately 55 tubes/μm (Fig. 17.8d). These two approaches can be combined, paving a route toward a tube density in the 100 tubes/μm range. Although these methods are effective in increasing the density of nanotube arrays, their generally low throughput and yield, together with other challenges including the damage of nanotube by hydrogen plasma during multiple growth process, the wiggling of nanotube after transfer, as well as the higher potential to form bundles with the increase of nanotube density, could limit their practical applications. Moreover, such CVD methods lack the capability of uniformly controlling nanotube density or nanotube pitch separation over large area. For current silicon VLSI chips, the tolerable device on-current variation is <15%. Without considering the influences from SWNT diameter distribution, the relative standard deviation for the nanotube pitch separation less than 25% is necessary to meet this requirement (72). Such level of pitch uniformity has not yet been demonstrated by CVD-grown SWNT arrays, likely due to (1) random positioning of deposited catalyst particles, (2) uncontrolled poisoning of catalyst before and during growth, and (3) random blocking of the emergence of SWNTs from catalyst lines.

17.3.2 Removal of Metallic Nanotubes after Growth

For nanoelectronic applications, the assembled nanotube arrays have to be electrically pure, and ideally with tight diameter distribution to minimize device performance variations. For CVD method, in most cases, the as-synthesized SWNT array is a mixture of 66% semiconducting nanotubes and 33% metallic ones. These metallic nanotubes have to be removed based on their different electrical, chemical, or optical properties (73). The most relevant difference between metallic and semiconducting SWNTs is their electrical properties. In a FET structure, increasing the source/drain bias with an applied gate voltage turning off semiconducting tubes can selectively breakdown metallic ones in an aligned array (74). Such method is effective in improving on/off ratio of nanotube FETs. However, it has poor manufacturability as it is difficult to apply this method to complex circuits, where independent access to each transistor might not be feasible. In addition, the nanotubes have to be exposed to oxygen during this process, which might not be possible for an integrated circuit requiring gate stack and metal interconnect formed on top. Finally, the heat dissipation from "burning" metallic tubes could also affect the structural integrity of the neighboring semiconducting ones (75).

Their difference in chemical reactivity could also be explored to remove metallic tubes after growth, as both experiments and simulations indicate that metallic nanotubes are more reactive with their finite DOS near the Fermi level, and these extra DOS might help to stabilize the reaction intermediate (Fig. 17.9a) (76). For example, diazonium salt can selectively react with metallic tubes as verified by Raman spectra shown in Figure 17.9b (77). After reaction, the intensity of the disorder D band for metallic tubes increases, suggesting the formation of covalent bonds on the sidewall of nanotubes and therefore the increase of sp^3 carbon. In the meantime, significantly less change occurs for semiconducting tubes under the same condition. This selectivity is also confirmed with *in situ* electrical measurements (Fig. 17.9c). The device on- and off-currents drop a similar amount, leading to a sharp increase of device on/off ratio without affecting too much device mobility. This observation is consistent with the selective elimination of metallic pathways. In another example, the selectivity is also observed for the gas-phase reaction with methane plasma (78). Atomic force microscopy (AFM) indicates the etching of metallic tubes into smaller segments and the sharp increase of device on/off ratio as shown in Figure 17.9d. This plasma-based selective reaction could also take place *in situ* during CVD growth, by providing an etching rate higher than the growth rate for metallic nanotubes, as demonstrated by introducing certain amount of water in the CVD chamber to promote the growth of enriched semiconducting nanotubes (Fig. 17.9f) (79). These approaches are promising. However, this chemical selectivity also depends on the nanotube diameter, as smaller tubes have higher curvature and therefore more exposed carbon π orbitals for reaction (Fig. 17.9e) (78). As a result, the reaction window (i.e., concentration, reaction time, temperature, etc.) that can completely remove metallic tubes without damaging semiconducting ones could be impractically small, especially considering the wide distribution of the diameter and chirality of as-synthesized nanotubes. More importantly, this synthesis-and-then-removal strategy will lead to significant device variation even in the perfect case where all metallic nanotubes are removed without affecting semiconducting ones, as this approach will randomly change the local tube density for each transistor inside an integrated circuit. These limitations make this strategy less attractive from a technology point of view.

17.3.3 Sorting of Nanotubes in Solution

Another strategy is to firstly suspend nanotubes (synthesized by one of several bulk methods) in solution using surfactant polymers or covalent sidewall functionalization, and then assemble/place them in the form of arrays on the substrate. Since charge transfer could occur between metallic SWNTs and certain polymers, the formed metallic nanotube/polymer complex might have lower linear charge density and/or high packing density compared to their semiconducting nanotube counterparts (80–83). Subsequent separation can be achieved by means such as ultracentrifugation, where nanotubes with different electronic types and diameters will be separated based on the difference of buoyant density and the viscous drag of nanotube/polymer complex, as shown in Figure 17.10a (82). Such separation scheme can be performed repeatedly and/or orthogonally for better purity and/or additional diameter as well as length sorting (84). Since semiconducting nanotubes form a more stable complex with some surfactants, this separation process can be combined with the suspension process. Under certain combination of appropriately selected surfactant, solvent, and suspension conditions, only semiconducting nanotubes will be suspended in solution with metallic ones removed by centrifugation as sediments, as verified by both optical methods and direct electrical measurements (Fig. 17.10b) (85, 86).

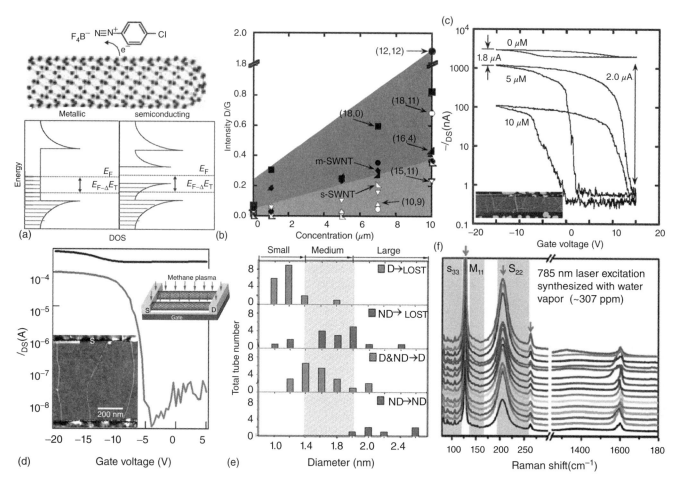

Figure 17.9 (a) Schematic illustration of the selective reaction between metallic nanotubes and diazonium salt, where selectivity comes from the difference in the density of states (DOS) of electrons near the Fermi level (E_F) for metallic and semiconducting nanotubes. (*Source*: Reproduced with permission from Reference 79. Copyright 2003 AAAS). (b) The ratio of the intensities of the disorder D mode to tangential G mode in Raman spectra (intensity D/G) of nanotubes with different chiral vectors after diazonium functionalization at different concentrations. Filled and open symbols refer to metallic (m-SWNT) and semiconducting nanotubes (s-SWNT), respectively. (c) Subthreshold transfer characteristics of a nanotube transistor before and after functionalization (V_{DS} = −0.1 V). Inset: AFM image of the channel. (*Source*: Reproduced with permission from Reference 80. Copyright 2005 American Chemical Society). (d) Subthreshold transfer characteristics of a nanotube transistor before and after selective plasma etching. Upper inset: schematic illustration. Lower inset: AFM image of device channel region after plasma etching, showing one nanotube severely damaged. (e) Diameter distribution of nanotubes with different responses toward plasma etching. (ND, nondepletable; D: depletable, LOST, electrically insulating.) (*Source*: Reproduced with permission from Reference 81, Copyright 2006, AAAS). (f) A collection of Raman spectra of nanotubes synthesized with the presence of water vapor. Transitions associated with metallic (M_{11}) and semiconducting (S_{22} and S_{33}) nanotubes are highlighted in grey shades. (*Source*: Reproduced with permission from Reference 82. Copyright 2012 American Chemical Society).

For practical applications, a suitable separation technique must be capable of easy iteration to enable multiple (i.e., 6–7) 9's purity, have high throughput, and can be scaled-up for manufacturing. Chromatography is a particularly promising method to meet these requirements, with its much higher throughput and easier scale-up compared to ultracentrifugation, and its capability of iteration compared to selective suspension (87–89). In this process, purified nanotube/sodium dodecyl sulfate (SDS) aqueous solution is loaded into a column for size-exclusion chromatography (89). The red band, which corresponds with enriched semiconducting nanotubes, moves more slowly than the dark-blue metallic band, because semiconducting nanotube/SDS complex has higher packing density and therefore is retained for longer time (Fig. 17.10c). This process can be repeated to get a better purity and narrower-diameter distribution of semiconducting nanotubes, which is shown by the change of characteristic absorption peaks in UV–Vis spectra. The electrical purity of separated nanotubes is further verified with direct electrical measurement. Figure 17.10d shows subthreshold transfer characteristics of one chiplet with approximately 150 devices, where no metallic tube is found. Measurement performed on a total of 941 devices using nanotube solution with two iterations gives a purity of 99.5 ± 0.2%. Further enhancement of semiconducting nanotube purity to a target of above 99.99% is possible with more refined engineering control over the separation processes. Currently, the biggest challenge is not necessarily the separation techniques themselves, but the lack of reliable and high throughput analytical techniques to detect the exact amount of residual metallic tubes and provide guidance for further

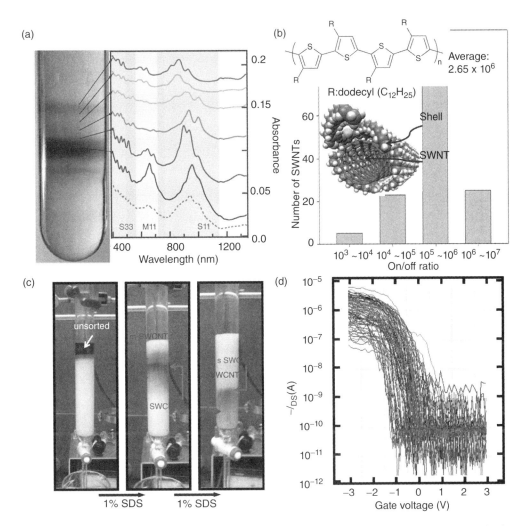

Figure 17.10 (a) Optical image and associated absorbance spectra for nanotubes enriched by diameter and electronic types, via ultracentrifugation. The second- and third-order semiconducting and first-order metallic optical transitions are labeled as S22, S33, and M11, respectively. (*Source*: Reproduced with permission from Reference 85. Copyright 2006 Nature Publishing Group). (b) Histogram of on/off ratios for individual nanotube transistors constructed on nanotubes sorted via selective polymer warping. (*Source*: Reproduced with permission from Reference 89. Copyright 2012 American Chemical Society). Inset: chemical structure of the polymer and schematic showing the warping of the polymer around a semiconducting nanotube. (*Source*: Reproduced with permission from Reference 88. Copyright 2011, Nature Publishing Group). (c) Optical images showing the separation of nanotubes based on size-exclusion chromatography, where m-SWNTs and s-SWNTs nanotubes separate into different bands after loading. (d) A collection of subthreshold plots of transistors fabricated on enriched semiconducting nanotubes. (*Source*: Reproduced with permission from Reference 92. Copyright 2013, American Chemical Society).

improving separation processes. Certain combinations of high-sensitivity optical techniques, for example, fluorescence and Raman, and direct electrical characterizations could offer a solution for conveniently detecting metallic nanotube purity at a much higher accuracy level.

17.3.4 Assembly of Nanotube Arrays

After sorting, semiconducting nanotubes need to be placed on the device substrate in the form of arrays aligned to the electron transport direction to provide the best performance (90). However, their nanometer size and high aspect ratio make the precise manipulation of nanotubes very challenging. Recently, it has been demonstrated that carefully designed deoxyribonucleic acid (DNA) oligomers can bind with multiple nanotubes and assemble them into bigger building blocks composed of several nanotubes with predefined pitch (91). As shown in Figure 17.11a, these DNA oligomers are composed of both single-stranded and double-stranded domains, where single-stranded domains bond with nanotubes and the more rigid double-stranded domains control the pitch. For a design where the double-stranded domain is composed of seven base pairs, a pitch of approximately 3 nm can be achieved (Fig. 17.11b). Such building blocks can be used together with other assembly techniques to achieve aligned nanotube arrays over large area.

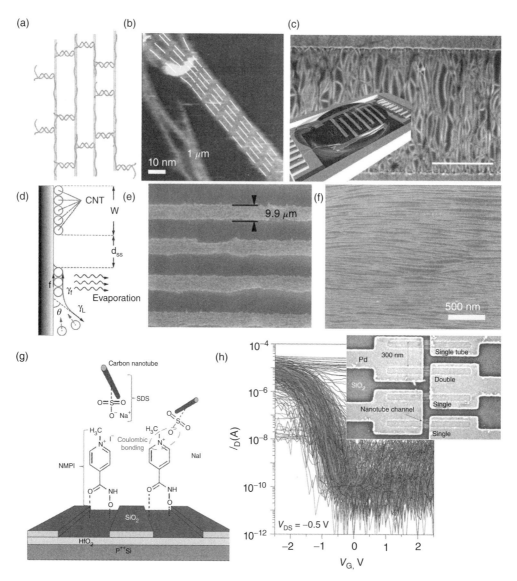

Figure 17.11 (a) Schematic showing the assembly of nanotubes with the help of noncovalently bonded DNA linkers. (b) AFM image showing an assembled DNA/nanotube array superstructure with the position of each nanotube highlighted by white dashed lines. (*Source*: Reproduced with permission from Reference 94. Copyright 2012, American Chemical Society). (c) SEM image showing semiconducting nanotube array with a density of 25 tubes/μm assembled via dielectrophoresis. (*Source*: Reproduced with permission from Reference 106. Copyright 2011, American Chemical Society). Inset: schematic illustration of the experimental setup. An alternating-current field applied through microelectrodes causes the deposition of aligned nanotubes. (*Source*: Reproduced with permission from Reference 101. Copyright 2003 AAAS). (d) Schematic showing the slip–stick mechanism to assemble nanotube arrays. (e) Low- and (f) high-resolution SEM images showing the nanotube array superstructure assembled via slip–stick mechanism. (*Source*: Reproduced with permission from Reference 107. Copyright 2008, American Chemical Society). (g) Schematic showing the assembly of nanotubes based on selective interactions between functionalized nanotubes and prepatterned self-assembled monolayer. (h) Subthreshold plot of a collection of transistors constructed on preplaced nanotubes. Inset: SEM image showing the device array structure with nanotubes inside the channel highlighted. (*Source*: Reproduced with permission from Reference 112. Copyright 2012, Nature Publishing Group).

There are several methods to align individual nanotubes or larger building blocks suspended in solution onto device substrate. One approach to produce this alignment is to rely on external forces including applied electrical or magnetic fields (92–94), and shear force (95–97). For example, alternating-current dielectrophoresis can be used to align nanotubes between pairs of prepatterned metal electrodes with an inhomogeneous electrical field created by applied voltage as shown in Figure 17.11c inset (98). Certain frequency for the alternating-current signal is necessary to prevent the formation of electrical double layer around electrodes, but the direction of the drag force exerted on nanotubes as a result of the interaction between applied electrical field and induced dipole within nanotube will remain the same with field reversal, since the induced dipole moment will be reversed as well (99, 100). This method will enrich metallic SWNTs in assembled arrays since metallic tubes have higher polarizability (98, 101), but it can also be applied with preenriched semiconducting nanotubes (102, 103). A linear tube density up to 25 tubes/µm has been demonstrated for 99% semiconducting nanotubes (Fig. 17.11c), with further increase of tube density limited by the formation of multilayers and bundles. Under the same principle, the shear force can also be utilized to guide the alignment. In one particular example, SWNTs are aligned along the liquid–solid–air contact line following a stick–slip mechanism as illustrated in Figure 17.11d (104). Convection flow of nanotubes during evaporation brings nanotubes to the contact line and they adhere to the substrate. Further evaporation of the solvent increases the capillary force γ_L, and when γ_L becomes larger than the combination of surface friction force f and liquid surface tension γ_f, the contact line will be depinned to another position and start depositing a new row of aligned nanotubes. This process forms the superstructure shown in Figure 17.11e with the row width W and spacing d_{ss} determined by many factors including the surface chemistry of the substrate and the concentration of nanotube solution. The degree of alignment is high with almost all nanotubes lying within 5° of one another, but the density of nanotube inside an array is still limited to 10~20 tubes/µm as measured by scanning electron microscopy (SEM) (Fig. 17.11f).

Another approach is to pattern the substrate with either top-down or bottom-up methods, and then precisely "place" nanotubes into these structures during deposition, mainly relying on the chemical recognition between nanotube/surfactant complex and chemical functionalized surface of the substrate (105–109). In one example, an SiO_2/HfO_2/Si substrate is patterned into many narrow trenches down to the HfO_2 surface. The exposed HfO_2 surface is then selectively functionalized by 4-(N-hydroxycarboxamido)-1-methylpyridinium iodide (NMPI) molecules at neutral pH, relying on the different isoelectric points for HfO_2 and SiO_2 (109). This NMPI molecule has a positively charged pyridinium cation, which will selectively bind with the negatively charged anionic SDS surfactant warping around SWNTs and therefore place the nanotube on the surface, as illustrated in Figure 17.11g. The nanotubes can be relatively aligned in the trench by making the trench width much smaller than the average tube length. A pitch separation as small as 200 nm or a nanotube density of 5 tubes/µm has been demonstrated, with the electrical properties of nanotubes deposited in each trench characterized in a FET structure. The image of the test structure is shown as inset of Figure 17.11h. Subthreshold plot of transfer curves for devices on a chiplet is also depicted (Fig. 17.11h), showing a device yield above 90%. Further scaling down the pitch separation of trenches can increase the tube density. To reach the target of >100 tubes/µm requires a patterning technique with at least 10 nm pitch resolution, which could possibly be achieved with directed self-assembly of diblock copolymers (110). Other challenges in addition to the patterning resolution include the increase of crossing defects with the reduction of SiO_2 barrier width under tighter pitch and poor surface wetting profile of the patterned hydrophobic monolayer at more scaled dimensions. The extreme requirement on surface cleanness for the chemistry to work may also limit the application of this approach in real fabrications.

The other approach to align the nanotubes relies on their liquid crystal nature (111, 112). With their cylinder structure, SWNTs favor a lateral alignment, especially with attractive intermolecular interactions among nanotubes, as explained in the Maier–Saupe mean field theory. To induce this nematic/smectic phase transition, nanotubes have to be confined to a two-dimensional surface with high nanotube density (113–117). In one example, a presorted 99% purity semiconducting nanotube solution in organic solvent is dispersed onto a water surface (117). The nanotubes are then spread out to cover the whole air–water interface, thanks to surface tension, and orient themselves randomly. Applied pressure is then aligning the nanotubes into regular arrays, which can be subsequently transferred to the device substrate using a horizontal dipping method (the Langmuir–Schaefer deposition) as illustrated in Figure 17.12a. SEM image shows the formation of such alignment structure over large area (Fig. 17.12b). Top-view and cross-sectional transmission electron microscopy (TEM) images indicate that such nanotube arrays have an almost full surface coverage and have a regular double layered structure. The distance between adjacent nanotubes within the arrays is less than 5 Å and is apparently limited only by the van der Waals separation of the nanotubes (Fig. 17.12c). The alignment is further confirmed with polarized Raman spectra (Fig. 17.12d), showing that most nanotubes are aligned within 17° with one another over the whole substrate. Misalignment defects are mainly caused by impurity particles present in the arrays and the looping of nanotubes during assembly, which could be minimized via adopting cleaner materials and more controlled experimental environments as well as further optimization of the assembly procedures. Such high nanotube density and degree of alignment lead to very large current densities for fabricated FETs. For devices with a L_{ch} of 120 nm, a drive current density above 100 µA/µm can be achieved, bringing the performance of nanotube array FETs to within a factor of ten of the drive currents of state-of-the-art silicon devices at the 22 nm technology node (118). Furthermore, respectable on-/off-current ratios of around 1000 can be maintained in these devices when they are devoid of metallic impurities and composed exclusively of semiconducting nanotubes (Fig. 17.12e). The performance is mainly limited by poor contacts between nanotubes and source/drain electrodes caused by the synergic effects from poor adhesion of metals on a full carbon surface, the limited contact area for each nanotube inside the array, and the lowering of nanotube valence band at the contact due to intertube Coulomb interactions (119). Converting to more efficient end-bonded contacts is expected to significantly improve device performance.

17.4   CIRCUIT- AND SYSTEM-LEVEL DEMONSTRATION OF NANOTUBE-BASED NANOELECTRONICS

The promising device performance demonstrated experimentally down to sub-10 nm channel length, along with the circuit-level simulation showing significant energy–delay benefit over silicon technology, encourages many groups to develop digital logic gates and basic circuits

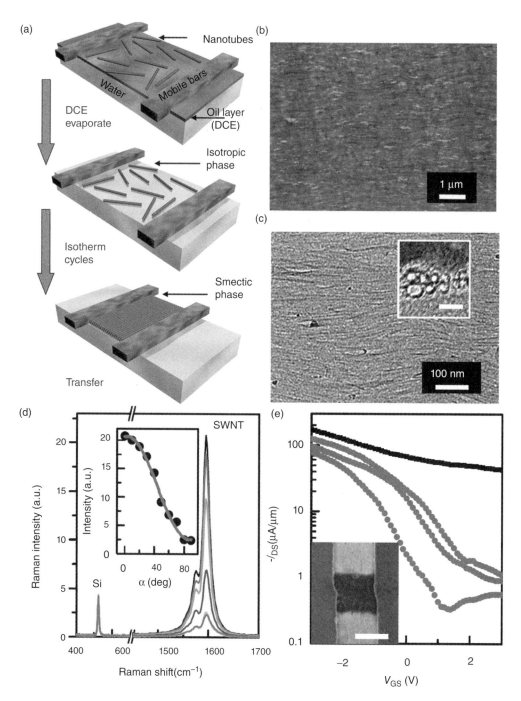

Figure 17.12   (a) Schematic showing the assembly of nanotube arrays based on Langmuir–Schaefer method. (b) SEM and (c) top-view TEM images showing the assembled full-coverage semiconducting nanotube arrays. Inset: high-resolution cross-sectional TEM image showing the nanotube pitch is self-limited by nanotube diameter plus van der Waals separation. Scale bar: 5 nm. (d) Polarized Raman spectra of Si and tangential G band of nanotube arrays for various angles α, 0, 20, 40, 60, 80, and 90 from top to bottom, between the polarization direction of the incident 532 nm wavelength laser and the nanotube overall alignment direction. Inset: angular dependence of the Raman intensity at 1594 cm^{-1}. The red solid line represents a fitting to cos$^2\alpha$ form. (e) Subthreshold plot of transfer characteristics of transistors constructed on purely semiconducting nanotube arrays (red) and array with metallic tube impurity (black). Inset: SEM image showing the device channel. Scale bar: 100 nm. (*Source*: Reproduced with permission from Reference 120. Copyright 2013, Nature Publishing Group).

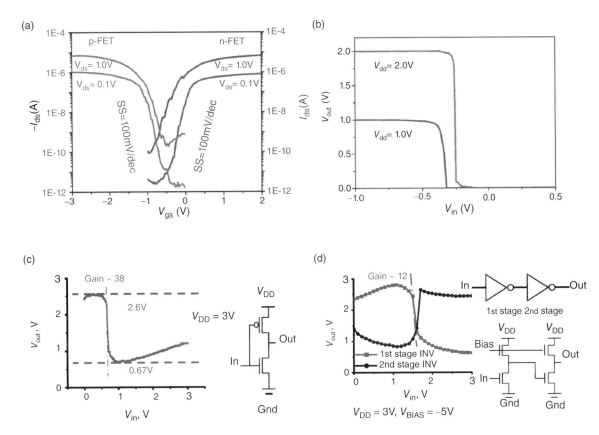

Figure 17.13 (a) Transfer characteristics of PFET (Pd contacts) and NFET (Sc contacts) showing nearly symmetric performance. (b) Voltage transfer characteristics showing high voltage gains. V_{in}: input voltage. V_{out}: output voltage. V_{dd}: drive voltage. (*Source*: Reproduced with permission from Reference 125. Copyright 2009, American Chemical Society). (c) Inverter voltage transfer curve from the cointegration of NFET (with Al layer) and PFET (without Al layer). (d) Two-stage inverter chain using NFETs only. V_{bias}: bias voltage. Gnd: ground. (*Source*: Reproduced with permission from Reference 56. Copyright 2011, IEEE).

using SWNTs. Individual gates demonstrated in the literatures, for example, inverter, NAND, and NOR, typically consist of only a few transistors. Although they are relatively simple structures, these experiments are important for showing the true complementary capability – a key feature for any VLSI technology. Most of the early circuit demonstrations are based on transistors consisting of a single nanotube in its channel. Moving forward to large-scale circuits with multiple tubes per transistor, the aforementioned material limitations, for example, alignment and placement, and the existing of metallic tubes have to be addressed. The approach to achieve robust carbon nanotube VLSI that is immune to nonidealities will also be reviewed.

17.4.1 Complementary Circuit Demonstration

As discussed in Section 17.2.5, the desorption of oxygen can convert p-channel carbon nanotube transistor to n-channel. When devices are measured under vacuum, high-bias current-induced Joule heating can cause oxygen desorption, and complementary logic gates including NOR, OR, NAND, and AND using up to six transistors have been built (120). To demonstrate air-stable circuits, NFETs based on Al contacts with large-diameter tubes (2–3 nm) are employed for the construction of the inverter (121). More recently, following the discovery of Sc electrodes as SWNT NFET contacts, symmetric NFET and PFET fabricated on a single CVD tube can be achieved, which provide a respectable inverter with voltage gain above 160 (Fig. 17.13a and 13b) (122). Electrical doping originated from trapped charges existing in naturally oxidized Al thin film deposited over carbon nanotubes also produces air-stable NFET (123). Cointegration of NFETs (with Al layer) and PFETs (without Al layer), both consisting of multiple CVD tubes, on the same wafer to form high-gain inverters and 2-stage inverter chains has been demonstrated, as shown in Figure 17.13c and 13d. On the other hand, the development of complementary alternating-current circuits, which are crucial for performance benchmarks with other technologies, has not been advanced since the demonstration of the ring oscillator built on a single tube (124).

17.4.2 Defect-Tolerant Nanotube Integrated Circuit

Improving the yield of the circuit blocks built by tube arrays assembled from the solution (described in Section 17.3.4) is more similar to the yield improvement in conventional semiconductor technologies – the "starting" substrate has to be nearly perfect before the device and circuit fabrication. In the case of carbon nanotubes, it means that semiconducting tubes with an extremely high purity (>5 9's) have to be precisely placed onto designed locations with a very fine tube–tube pitch. In spite of more stringent requirements for the starting substrate, this approach provides a better chance of satisfying the yield requirement in modern VLSI technologies by continuous advances in material purification and assembly schemes.

As described in Section 17.3.1, transferring aligned tubes grown by CVD from quartz substrates onto Si wafers followed with selective removal of metallic tubes is another popular method to prepare large-scale SWNT substrates for circuit fabrication. However, a very different approach has to be developed to utilize this kind of substrate due to the difficulty of purifying and realigning CVD tubes once they are transferred. A commonly practiced approach in literatures is to accept the inherent imperfections in starting nanotube substrates, and try to overcome these imperfections by performing post substrate processing and/or employing special design methodologies. It is known that the existence of metallic tubes degrades the device off current and can introduce incorrect circuit function if the metallic concentration is too high. To remove metallic nanotubes on the substrate, the group at Stanford University expended the simple electrical breakdown method to VLSI scale by firstly patterning interdigitated electrodes at minimum metal pitch (these electrodes also serve as final device contacts, as shown in Fig. 17.14a), turning off semiconducting tubes by applying a global back-gate bias, and then performing breakdown by applying high voltage across electrodes (125). The claimed area penalty and delay penalty at the chip level using this method are 2% and 3%, respectively. The major drawback of this method is the large device-to-device variations due to the randomness of the presence of metallic nanotubes. In addition, the requirement of forming metal contacts first is not necessary compatible to the device fabrication flow, and the practicability of applying this method to build billions of transistors remains questionable. The exemplary XNOR circuit based on this technique is shown in Figure 17.14b (126). Using this technique, an application-specific circuit completely built with carbon nanotube transistors (PFET only) has recently been demonstrated (127). The circuit consists of two 9-stage oscillators and a D-latch and serves as a simple capacitive sensor interface (Fig. 17.14c).

Another technique to eliminate the impact of metallic CVD tubes without using electrical breakdown was also proposed and demonstrated by the same group. The technique utilizes the asymmetric correlation properties of CVD tubes, in which devices made along the direction of tube growth share correlated tubes and devices spaced perpendicular to the growth direction contain uncorrelated tubes (128). Therefore, by simply connecting multiple uncorrelated devices in series, the probability that one device along the chain contains no metallic tubes increases and eventually allows high on/off ratio to be restored. Multiple chains also need to be connected in parallel along the tube growth direction in order to compensate the drive current penalty from the series connection of devices. The concept is illustrated in Figure 17.14d. Although this technique can be directly applied to build devices on unsorted carbon nanotubes, the drawback is quite obvious as it incurs significant overheads due to many transistors used to construct a single functional device. It is also found that, among various sources of variations, the variation originated from the nanotube density per device has the largest impact on circuit performance (129). Aside from the improvement of carbon nanotube material itself, the performance penalty due to variations can also be reduced through circuit design techniques such as device upsizing and aligned-active layout (129).

17.5 SUMMARY AND PERSPECTIVE

The scaling of transistor technology is expected to continue for the next 15 years, with the adoption of novel device structures, for example, 3D FinFET and ETSOI devices, and the adoption of new materials, for example, SiGe/SiC stressors, $HfSiON/La_2O_3$ gate dielectrics, TiSi contacts, and SiBCN spacers (130). The ultimate challenge is to replace the silicon channel with new materials such as SWNT. Experimental results on devices based on individual nanotubes directly demonstrate a superior performance, that is, capability to deliver higher current and requiring lower operating voltage, than most advanced silicon technologies at an extreme dimension of 9 nm (26). Simulation indicates that such performance advantage will translate to a microprocessor chip that runs three to five times faster at system level (17). In recent years, more technology relevant progress has been made by researchers in both academia and top chip manufacturers like IBM. The contacts between nanotube and metals have been carefully studied in the 10~20 nm geometry range (36). The integration of SWNTs in an electrostatically most favorable self-aligned gate-all-around structure has also been demonstrated (28). The polarity of nanotube FETs can be readily controlled in a manner compatible with CMOS manufacturing (57).

With the performance advantage of nanotube over existing technologies clearly demonstrated in both simulations and experiments, the manufacturability becomes the central issue. Procedures have been established to successfully sort nanotubes according to their electronic type, chirality, and length. Methods that are capable of producing semiconducting nanotubes with purity above 99% have been developed by several research groups (82, 85, 89). Those separated nanotubes can then be assembled into arrays and deterministically placed onto device substrate (109), with the distance between neighboring nanotubes scaled down to the van der Waals separation (117). Challenges for perfecting the separation and placement of nanotubes to meet manufacturing target, that is, improving the nanotube purity to 99.999% and aligning arrays covering the whole wafer with low defect density at tight pitch, mainly come from the requirement of extreme engineering control rather than intrinsic limitations of the material or processes.

At circuit and system level, design methodology that can tolerate certain degree of material imperfections has been established. Designing tools based on device compact model for nanotube FETs and fault-tolerant design methodology have been established. Complex integrated unipolar digital circuits composed of up to 44 SWNT FETs (127) and low-power CMOS circuits with a supply voltage as small as 0.4 V have

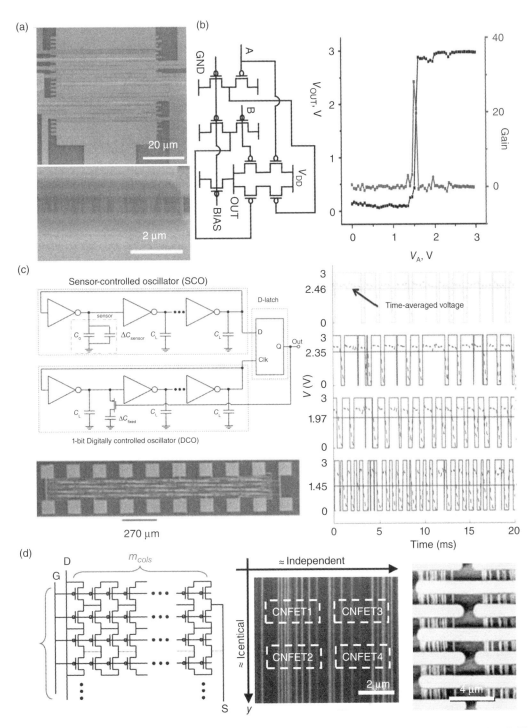

Figure 17.14   (a) SEM showing arrays of electrodes and aligned SWNTs compatible with metallic tube removal and device integration. (b) The circuit diagram and output characteristics of an XNOR function from carbon nanotube half-adder sum generator circuit. (*Source*: Reproduced with permission from Reference 128. Copyright 2011, IEEE). (c) The circuit diagram of carbon nanotube sensor interface circuit and SEM images of the full circuit. The digitized circuit output shows clear responses to the capacitance change of the sensor. (*Source*: Reproduced with permission from Reference 129. Copyright 2011, IEEE). (d) The concept of building metallic tube immune FET using the asymmetric correlation properties of CVD tubes. Each final FET is made by connecting several independent transistors in series (to eliminate the probability of the presence of metallic tubes) and multiple transistor chains in parallel (to restore the drive current). (*Source*: Reproduced with permission from Reference 130. Copyright 2009, IEEE).

been demonstrated (131). High-frequency performance of SWNT FETs has also been explored with devices that demonstrate an intrinsic cutoff frequency up to 100 GHz regime (132). Since semiconducting nanotubes have a finite bandgap, nanotube RF devices are more practical compared to graphene devices with its ability to achieve much higher power gain (133).

In spite of this progress, some obstacles remain. First, and perhaps the most important, techniques to form high-quality contact for nanotubes with $L_T < 10$ nm must be developed to allow the scale of gate pitch. Second, the device-to-device variation must be greatly improved for such surface sensitive ultrathin body semiconductor. The diameter and chirality distribution of nanotubes need to be further tightened, and the interface traps have to be minimized via suitable passivation to eliminate device hysteresis and reduce the variations of device threshold voltage and on current. Finally, the compatibility of SWNT technology with established CMOS processes and equipment need to be demonstrated, as the cost structure will ultimately determine the success of nanotubes in replacing silicon in nanoelectronics.

REFERENCES

1. Ieong M, Doris B, Kedzierski J, Rim K, Yang M. Science 2004;306:2057.

2. ITRS Web-Site: http://public.itrs.net/

3. Graham AP, Duesberg GS, Hoenlein W, Kreupl F, Liebau M, Martin R, Rajasekharan B, Pamler W, Seidel R, Steinhoegl W, Unger E. Appl Phys A-Mater Sci Process 2005;80:1141.

4. De Volder MFL, Tawfick SH, Baughman RH, Hart AJ. Science 2013;339:535.

5. Zhou XJ, Park JY, Huang SM, Liu J, McEuen PL. Phys Rev Lett 2005;95:146805.

6. Ilani S, Donev LAK, Kindermann M, McEuen PL. Nat Phys 2006;2:687.

7. Appenzeller J. Proc IEEE 2008;96:201.

8. Cao Q, Rogers J. Nano Res 2008;1:259.

9. Cao Q, Rogers JA. Adv Mater 2009;21:29.

10. Hu L, Hecht DS, Gruner G. Chem Rev 2010;110:5790.

11. Park S, Vosguerichian M, Bao Z. Nanoscale 2013;5:1727.

12. Uchida K, Watanabe H, Kinoshita A, Koga J, Numata T, Takagi S IEDM Technical Digest; 2002. p 47.

13. Avouris P, Chen ZH, Perebeinos V. Nat Nanotechnol 2007;2:605.

14. Avouris P. Acc Chem Res 2002;35:1026.

15. Javey A, Guo J, Wang Q, Lundstrom M, Dai HJ. Nature 2003;424:654.

16. Guo J, Javey A, Dai H, Lundstrom M. IEDM Technical Digest; 2004. p 703.

17. Wei L, Frank DJ, Chang L, Wong HSP. IEDM Technical Digest; 2009. p 37.1.1.

18. Frank DJ, Haensch W, Shahidi G, Dokumaci OH. IBM J Res Develop 2006;50:419.

19. Wei L, Oh S, Wong HSP. IEEE Trans Electron Devices 2011;58:2430.

20. Leonard F, Stewart DA. Nanotechnology 2006;17:4699.

21. Guo J, Datta S, Lundstrom M. IEEE Trans Electron Devices 2004;51:172.

22. Franklin AD, Luisier M, Han S-J, Tulevski G, Breslin CM, Gignac L, Lundstrom MS, Haensch W. Nano Lett 2012;12:758.

23. Ming L, Kyoung Hwan Y, Sung Dae S, Yun-young Y, Dong-Won K, Tae Young C, Kyung-Seok O, Lee W-S. VLSI technology. 2009 Symposium on; 2009. p 94.

24. Yu B, Chang L, Ahmed S, Wang H, Bell S, Yang C-Y, Tabery C, Ho C, Xiang Q, King T-J, Bokor J, Hu C, Lin M-R, Kyser D. IEDM Technical Digest; 2002. p 251.

25. Doris B, Ieong M., Zhu T., Zhang Y, Steen M, Natzle W., Callegari S, Narayanan V, Cai J, Ku SH, Jamison P, Li Y, Ren Z, Ku V, Boyd T, Kanarsky T, D'Emic C, Newport M, Dobuzinsky D, Deshpande S, Petrus J, Jammy R, Haensch W. IEDM Technical Digest; 2003. p 27.3.1.

26. Kreupl F. Nature 2012;484:321.

27. Ding L, Wang Z, Pei T, Zhang Z, Wang S, Xu H, Peng F, Li Y, Peng L-M. ACS Nano 2011;5:2512.

28. Franklin, AD, Koswatta, S, Farmer, D, Tulevski, GS, Smith, JT, Miyazoe, H, Haensch, W IEDM Technical Digest; 2012. p 4.5.1.

29. Farmer DB, Gordon RG. Nano Lett 2006;6:699.

30. Chen ZH, Farmer D, Xu S, Gordon R, Avouris P, Appenzeller J. IEEE Electron Device Lett 2008;29:183.

31. Leonard F, Talin AA. Nat Nanotechnol 2011;6:773.

32. Tersoff J. Phys Rev Lett 1984;52:465.

33. Leonard F, Tersoff J. Phys Rev Lett 2000;84:4693.

34. Leonard F, Talin AA. Phys Rev Lett 2006;97:026804.

35. Chen ZH, Appenzeller J, Knoch J, Lin YM, Avouris P. Nano Lett 2005;5:1497.

36. Franklin AD, Chen Z. Nat Nanotechnol 2010;5:858.

37. Zhang Z, Pagette F, D'Emic C, Yang B, Lavoie C, Zhu Y, Hopstaken M, Maurer S, Murray C, Guillorn M, Klaus D, Bucchignano J, Bruley J, Ott J, Pyzyna A, Newbury J, Song W, Chhabra V, Zuo G, Lee KL, Ozcan A, Silverman J, Ouyang Q, Park DG, Haensch W, Solomon PM. IEEE Electron Device Lett 2010;31:731.

38. Cao Q, Han S-J, Tulevski GS, Franklin AD, Haensch W. ACS Nano 2012;6:6471.

39. Ho XN, Ye LN, Rotkin SV, Cao Q, Unarunotai S, Salamat S, Alam MA, Rogers JA. Nano Lett 2010;10:499.

40. Choi S-J, Bennett P, Takei K, Wang C, Lo CC, Javey A, Bokor J. ACS Nano 2013;7:798.

41. Palacios JJ, Pérez-Jiménez AJ, Louis E, SanFabián E, Vergés JA. Phys Rev Lett 2003;90:106801.

42. Martel R, Derycke V, Lavoie C, Appenzeller J, Chan KK, Tersoff J, Avouris P. Phys Rev Lett 2001;87:256805.

43. Rodriguez-Manzo JA, Janowska I, Pham-Huu C, Tolvanen A, Krasheninnikov AV, Nordlund K, Banhart F. Small 2009;5:2710.

44. Huang L, Chor EF, Wu Y, Guo Z. Nanotechnology 2010;21:095201.

45. Matsuda Y, Deng W-Q, Goddard WA. J Phys Chem C 2010;114:17845.

46. Banhart F. Nanoscale 2009;1:201.

47. Xia F, Perebeinos V, Lin Y-m, Wu Y, Avouris P. Nat Nanotechnol 2011;6:179.

48. Heinze S, Tersoff J, Martel R, Derycke V, Appenzeller J, Avouris P. Phys Rev Lett 2002;89:106801.

49. Collins PG, Bradley K, Ishigami M, Zettl A. Science 2000;287:1801.

50. Klinke C, Chen J, Afzali A, Avouris P. Nano Lett 2005;5:555.

51. Shim M, Ozel T, Gaur A, Wang CJ. J Am Chem Soc 2006;128:7522.

52. Javey A, Tu R, Farmer DB, Guo J, Gordon RG, Dai H. Nano Lett 2005;5:345.

53. Cao Q, Xia MG, Shim M, Rogers JA. Adv Funct Mater 2006;16:2355.

54. Sorescu DC, Jordan KD, Avouris P. J Phys Chem B 2001;105:11227.

55. Derycke V, Martel R, Appenzeller J, Avouris P. Appl Phys Lett 2002;80:2773.

56. Sreenivasan R, McIntyre PC, Kim H, Saraswat KC. Appl Phys Lett 2006;89:112903.

57. Shahrjerdi D, Franklin AD, Oida S, Tulevski GS, Han S-J, Hannon JB, Haensch W. IEDM Technical Digest; 2011. p 23.3 .1.

58. Ding L, Wang S, Zhang ZY, Zeng QS, Wang ZX, Pei T, Yang LJ, Liang XL, Shen J, Chen Q, Cui RL, Li Y, Peng LM. Nano Lett 2009;9:4209.

59. Zhang ZY, Liang XL, Wang S, Yao K, Hu YF, Zhu YZ, Chen Q, Zhou WW, Li Y, Yao YG, Zhang J, Peng LM. Nano Lett 2007;7:3603.

60. Wang C, Ryu K, Badmaev A, Zhang J, Zhou C. ACS Nano 2011;5:1147.

61. Ding L, Zhang ZY, Liang SB, Pei T, Wang S, Li Y, Zhou WW, Liu J, Peng LM. Nat Commun 2012;3:677.

62. Zhang YG, Chang AL, Cao J, Wang Q, Kim W, Li YM, Morris N, Yenilmez E, Kong J, Dai HJ. Appl Phys Lett 2001;79:3155.

63. Jin Z, Chu HB, Wang JY, Hong JX, Tan WC, Li Y. Nano Lett 2007;7:2073.

64. Kang SJ, Kocabas C, Ozel T, Shim M, Pimparkar N, Alam MA, Rotkin SV, Rogers JA. Nat Nanotechnol 2007;2:230.

65. Kocabas C, Kang SJ, Ozel T, Shim M, Rogers JA. J Phys Chem C 2007;111:17879.

66. Xiao J, Dunham S, Liu P, Zhang Y, Kocabas C, Moh L, Huang Y, Hwang K-C, Lu C, Huang W, Rogers JA. Nano Lett 2009;9:4311.

67. Cao Q, Xia MG, Kocabas C, Shim M, Rogers JA, Rotkin SV. Appl Phys Lett 2007;90:023516.

68. Hong SW, Banks T, Rogers JA. Adv Mater 2010;22:1826.

69. Wang CA, Ryu KM, De Arco LG, Badmaev A, Zhang JL, Lin X, Che YC, Zhou CW. Nano Res 2010;3:831.

70. Shulaker MM, Wei H, Patil N, Provine J, Chen HY, Wong HSP, Mitra S. Nano Lett 2011;11:1881.

71. Kang SJ, Kocabas C, Kim H-S, Cao Q, Meitl MA, Khang D-Y, Rogers JA. Nano Lett 2007;7:3343.

72. Jie Z, Patil N, Hazeghi A, Mitra S. Des Aut Con 2009:71.

73. Hersam MC. Nat Nanotechnol 2008;3:387.

74. Collins PC, Arnold MS, Avouris P. Science 2001;292:706.

75. Liao A, Alizadegan R, Ong ZY, Dutta S, Xiong F, Hsia KJ, Pop E. Phys Rev B 2010;82:205406.

76. Strano MS, Dyke CA, Usrey ML, Barone PW, Allen MJ, Shan H, Kittrell C, Hauge RH, Tour JM, Smalley RE. Science 2003;301:1519.

77. Wang CJ, Cao Q, Ozel T, Gaur A, Rogers JA, Shim M. J Am Chem Soc 2005;127:11460.

78. Zhang GY, Qi PF, Wang XR, Lu YR, Li XL, Tu R, Bangsaruntip S, Mann D, Zhang L, Dai HJ. Science 2006;314:974.

79. Zhou W, Zhan S, Ding L, Liu J. J Am Chem Soc 2012;134:14019.

80. Strano MS, Zheng M, Jagota A, Onoa GB, Heller DA, Barone PW, Usrey ML. Nano Lett 2004;4:543.

81. Tu X, Manohar S, Jagota A, Zheng M. Nature 2009;460:250.

82. Arnold MS, Green AA, Hulvat JF, Stupp SI, Hersam MC. Nat Nanotechnol 2006;1:60.

83. Ju SY, Doll J, Sharma I, Papadimitrakopoulos F. Nat Nanotechnol 2008;3:356.

84. Green AA, Hersam MC. Adv Mater 2011;23:2185.

85. Lee HW, Yoon Y, Park S, Oh JH, Hong S, Liyanage LS, Wang H, Morishita S, Patil N, Park YJ, Park JJ, Spakowitz A, Galli G, Gygi F, Wong PHS, Tok JBH, Kim JM, Bao Z. Nat Commun 2011;2:541.

86. Park S, Lee HW, Wang H, Selvarasah S, Dokmeci MR, Park YJ, Cha SN, Kim JM, Bao Z. ACS Nano 2012;6:2487.

87. Liu HP, Nishide D, Tanaka T, Kataura H. Nat Commun 2011;2:309.

88. Moshammer K, Hennrich F, Kappes MM. Nano Res 2009;2:599.

89. Tulevski GS, Franklin AD, Afzali A. ACS Nano 2013;7:2971.

90. Pimparkar N, Kocabas C, Kang SJ, Rogers JA, Alam MA. IEEE Electron Device Lett 2007;28:593.

91. Han SP, Maune HT, Barish RD, Bockrath M, Goddard WA. Nano Lett 2012;12:1129.

92. Diehl MR, Yaliraki SN, Beckman RA, Barahona M, Heath JR. Angew Chem Int Ed 2001;41:353.

93. Krupke R, Linden S, Rapp M, Hennrich F. Adv Mater 2006;18:1468.

94. Long DP, Lazorcik JL, Shashidhar R. Adv Mater 2004;16:814.

95. Xiong X, Jaberansari L, Hahm MG, Busnaina A, Jung YJ. Small 2007;3:2006.

96. Yu GH, Cao AY, Lieber CM. Nat Nanotechnol 2007;2:372.

97. LeMieux MC, Roberts M, Barman S, Jin YW, Kim JM, Bao Z. Science 2008;321:101.

98. Krupke R, Hennrich F, von Lohneysen H, Kappes MM. Science 2003;301:344.

99. Dimaki M, Boggild P. Nanotechnology 2004;15:1095.

100. Dimaki M, Boggild P. Nanotechnology 2005;16:759.

101. Shekhar S, Stokes P, Khondaker SI. ACS Nano 2011;5:1739.

102. Krupke R, Hennrich F, Kappes MM, Löhneysen HV. Nano Lett 2004;4:1395.

103. Sarker BK, Shekhar S, Khondaker SI. ACS Nano 2011;5:6297.

104. Engel M, Small JP, Steiner M, Freitag M, Green AA, Hersam MC, Avouris P. ACS Nano 2008;2:2445.

105. Hannon JB, Afzali A, Klinke C, Avouris P. Langmuir 2005;21:8569.

106. Lee M, Im J, Lee BY, Myung S, Kang J, Huang L, Kwon YK, Hong S. Nat Nanotechnol 2006;1:66.

107. Bardecker JA, Afzali A, Tulevski GS, Graham T, Hannon JB, Jen AKY. J Am Chem Soc 2008;130:7226.

108. Tulevski GS, Hannon J, Afzali A, Chen Z, Avouris P, Kagan CR. J Am Chem Soc 2007;129:11964.

109. Park H, Afzali A, Han S-J, Tulevski GS, Franklin AD, Tersoff J, Hannon JB, Haensch W. Nat Nanotechnol 2012;7:787.

110. Bates CM, Seshimo T, Maher MJ, Durand WJ, Cushen JD, Dean LM, Blachut G, Ellison CJ, Willson CG. Science 2012;338:775.

111. Song W, Kinloch IA, Windle AH. Science 2003;302:1363.

112. Puech N, Blanc C, Grelet E, Zamora-Ledezma C, Maugey M, Zakri C, Anglaret E, Poulin P. J Phys Chem C 2011;115:3272.

113. Lynch MD, Patrick DL. Nano Lett 2002;2:1197.

114. Ko H, Tsukruk VV. Nano Lett 2006;6:1443.

115. Lu L, Chen W. ACS Nano 2010;4:1042.

116. Li X, Zhang L, Wang X, Shimoyama I, Sun X, Seo W-S, Dai H. J Am Chem Soc 2007;129:4890.

117. Cao Q, Han S-j, Tulevski GS, Zhu Y, Lu DD, Haensch W. Nat Nanotechnol 2013;8:180.

118. Klauk H. Nat Nanotechnol 2013;8:158.

119. Leonard F. Nanotechnology 2006;17:2381.

120. Javey A, Wang Q, Ural A, Li Y, Dai H. Nano Lett 2002;2:929.

121. Javey A, Qian W, Woong K, Dai H. IEDM Technical Digest; 2003, p. 31.2.1.

122. Zhang Z, Wang S, Wang Z, Ding L, Pei T, Hu Z, Liang X, Chen Q, Li Y, Peng L-M. ACS Nano 2009;3:3781.

123. Wei H, Chen HY, Liyanage L., Wong H.S.P., Mitra S. IEDM Technical Digest; 2011. p 23.2.1.

124. Chen ZH, Appenzeller J, Lin Y-M, Sippel-Oakley J, Rinzler AG, Tang J, Wind SJ, Solomon PM, Avouris P. Science 2006;311:1735.

125. Patil N, Lin A, Zhang J, Wei H, Anderson K, Wong HSP, Mitra S. IEDM Technical Digest; 2009. p 23.4.1.

126. Patil N, Lin A, Zhang J, Wei H, Anderson K, Wong HSP, Mitra S. IEEE Trans Nanotech 2011;10:744.

127. Shulaker M, Rethy JV, Hills G, Chen H-Y, Gielen G, Wong HSP, Mitra S ISSCC Technical Digest; 2013. p 112.

128. Lin A, Zhang J, Patil N, Wei H, Mitra S, Wong HSP. IEEE Trans Electron Devices 2009;57:2284.

129. Zhang J, Patil N, Wong HSP, Mitra S. IEDM Technical Digest; 2011. p 4.6.1.

130. Service RF. Science 2009;323:1000.

131. Ding L, Liang S, Pei T, Zhang Z, Wang S, Zhou W, Liu J, Peng L-M. Appl Phys Lett 2012;100:263116.

132. Steiner M, Engel M, Lin Y-M, Wu Y, Jenkins K, Farmer DB, Humes JJ, Yoder NL, Seo J-WT, Green AA, Hersam MC, Krupke R, Avouris P. Appl Phys Lett 2012;101:053123.

133. Koswatta SO, Valdes-Garcia A, Steiner MB, Yu-Ming L, Avouris P. IEEE Trans Microwave Theory Tech 2011;59:2739.

134. Kaminishi D, Ozaki H, Ohno Y, Maehashi K, Inoue K, Matsumoto K, Seri Y, Masuda A, Matsumura H. Appl Phys Lett 2005;86:113115.

135. Moriyama N, Ohno Y, Kitamura T, Kishimoto S, Mizutani T. Nanotechnology 2010;21:165201.

18

MONOLITHIC GRAPHENE–GRAPHITE INTEGRATED ELECTRONICS

MICHAEL CAI WANG, JONGHYUN CHOI, JAEHOON BANG, SUNGGYU CHUN, BRANDON SMITH, AND SUNGWOO NAM

Department of Mechanical Science and Engineering, Department of Materials Science and Engineering, University of Illinois, Urbana-Champaign, USA

18.1 INTRODUCTION

Graphene (1), an atomically thin sp^2 carbon, has garnered substantial interest from the areas of fundamental low-dimensional physics and surface science/chemistry, and several potential application areas have been recently explored, including flexible electronic devices, high-frequency transistors, optoelectronic devices, and thermal management of nanoelectronic devices. In this article, we review controlled synthesis of graphene and graphite and discuss potential applications to flexible and unconventional nanoelectronic devices. We will begin our discussions with a review of the recent developments of graphene synthesis and applications, then turn to our recent work on the controlled synthesis of monolithic graphene–graphite nanostructures (2), and further extend the discussion to potential future directions. We expect that numerous opportunities and novel uses of monolithic graphene–graphite nanostructures will present themselves in the areas of nanoelectromechanical devices as well as biointegrated systems in the near future.

18.2 GRAPHENE SYNTHESIS AND APPLICATIONS

18.2.1 Graphene Synthesis

To realize the full potential of graphene's superior electrical, mechanical, and thermal properties, significant progress must be made toward developing manufacturing techniques of laboratory-grade quality in an efficient and scalable fashion, in addition to achieving material properties suited to each of the myriad of applications. The quality specifications of graphene and graphitic materials range from graphitic flakes used as bulk filler and binder additives for composite materials and coatings on the low end, to that of pristine monolayer graphene, which is demanded by high-performance electronic devices on the high end. While there are dozens of methods to create graphene and graphitic materials of various sizes, morphology, crystallinity, and quality, emphasis is placed here on exploring those that are applicable to nanoscale high-performance-based applications and those synthesis pathways that are projected to be scalable (large sheet synthesis) and of commercial interest as opposed to bulk methods. Specifically, large-area single- and few-layer graphene is of interest. In such a case, it is important to evaluate the relative quality yielded via the various methods on the basis of the material properties, defect control, substrate compatibility, and postprocessing requirements.

18.2.1.1 Exfoliation of Graphitic Crystals Graphene, in its modern "macroscopic" reincarnation, was isolated by Geim and Novoselov in 2004 through mechanical exfoliation of graphite using adhesive tape (1). Similarly, solvent or aqueous surfactant-based liquid-phase exfoliation of larger graphite crystals was achieved, whereby the favored state of surface tension increases the overall total amount of exposed surface area to yield monolayer graphene, often with the aid of sonication. A derivative method involves prior oxidation of graphite, followed

Nanomaterials, Polymers and Devices: Materials Functionalization and Device Fabrication, First Edition. Edited by Eric S. W. Kong.
© 2015 John Wiley & Sons, Inc. Published 2015 by John Wiley & Sons, Inc.

by subsequent aqueous exfoliation, deposition as thin film onto a substrate, and finally reduction back to graphene (4–6). Such methods to delaminate graphene from its stacked graphitic configuration may also be assisted by intercalation where the combination of small molecule intercalation and thermal processing can assist isolation of the individual constituent graphene sheets.

18.2.1.2 Silicon Carbide Substrate Precursor

Graphene surfaces can also be created via sublimation of silicon carbide (SiC) surfaces, thus yielding high-quality graphene crystal domains in the order of several micrometers (7–11). However, SiC is expensive in comparison to other metallic substrate materials and the high processing temperature (>1000 °C) required for sublimation is incompatible with existing electronic device fabrication processes. Meanwhile, techniques are being developed to manufacture atomically thin SiC layers on cheaper plain Si substrates and to control undesired doping at the high processing temperatures.

18.2.1.3 Chemical Vapor Deposition

The standard approach for producing "large-area" uniform graphene sheets is via chemical vapor deposition (CVD) using various precursors and on different (catalyst) substrates, with copper and nickel substrates being the most commonly used (12). Li et al. first demonstrated centimeter-scale graphene synthesis using CVD with gaseous methane and hydrogen as precursors on copper foil substrates with reproducible monolayer growth (13–15). Bae et al. extended this concept to produce mostly monolayer graphene up to 30 inches in size in a roll-to-roll processing (16). In general, CVD-synthesized graphene approaches the quality of exfoliated graphene. Aside from the requisite high processing temperatures and high vacuum, other issues concerning current CVD approaches include other variables such as control of grain size, rippling, layer count, and relative crystallographic orientations of an individual layer. Improvements to the transfer process subsequent to CVD synthesis could be further optimized to minimize graphene damage and also facilitate removal and/or recovery of catalyst metals. Efforts are being made to achieve CVD growth directly on arbitrary surfaces and under low temperatures in order to forego the chemical intensive post-synthesis transfer process and facilitate manufacturing compatibility with temperature-sensitive materials, respectively. The transition from high-vacuum to ambient or "higher"-pressure processes would also contribute to the competitiveness of CVD-based approaches.

18.2.1.4 CVD Synthesis Using Different Catalysts and Precursors

Aside from the predominantly used Cu, Ni, and Co catalysts (12, 17–21) – which are covered thoroughly in subsequent sections – there exist other metallic materials that have been used as catalysts for graphene synthesis. While some of these methodologies are alloyed derivatives (such as palladium–cobalt, gold–nickel, copper–nickel, and nickel–molybdenum), there also exist several metals from groups VII to XI that can support graphene synthesis in polycrystalline metallic form (12, 22). Specifically, these include iron and ruthenium in group VIII, rhodium and iridium in group IX, platinum in group X, and gold in group XI (23, 24). However, in some cases, special synthesis configurations such as ambient pressure or plasma-enhanced CVD and/or different gaseous precursors such as ethylene are required (25). On the other hand, there also exist nonmetallic catalyzed CVD methods on dielectric substrates such as silicon dioxide, silicon nitride (Si_3N_4), and sapphire (26–29).

18.2.2 Graphene Integrated Electromechanical Devices

Due to the outstanding electrical properties of graphene (i.e., electron mobility of ~250,000 $cm^2/V \cdot s$) (30), researchers in the electronic device community have started to explore graphene as a part of electronic devices (31). In particular, field-effect transistors (FETs) employing graphene as an active channel material have been developed rapidly with different applications such as radio-frequency (RF) transistors and ultrafast photodetectors (32–35). Wu et al. investigated high-frequency graphene transistors based on top-gated CVD-graphene RF transistors with gate lengths down to 40 nm (32). Cutoff frequencies as high as 155 GHz have been obtained for the 40 nm transistors, and the cutoff frequency was found to scale inversely to the gate length. Xia et al. proposed a novel application using graphene-based FET device with single- and few-layer graphene transistor working as a high-speed photodetector (34). They observed that the device could be operated at a frequency up to 40 GHz, responding to incoming photons as photocurrent. Vicarelli et al. exploited similar photodetector devices at even higher frequencies (0.3 THz) working at room temperatures (35).

Although the graphene-based FET devices show outstanding performance, they still have inherent limitations of contact materials. Graphene-based transistors, although employing graphene as a part of the device, use metals for the source and drain. The contact resistance between metal and graphene could limit the device performance at short channel length. Moreover, the transistors can only be used on a hard and flat surface where the excellent mechanical flexibility of graphene could not be fully utilized.

Owing to its unique mechanical properties, graphene can also be utilized in various applications such as mechanical switches, actuators, or 3D sensor devices (36–39). Conley et al. investigated the temperature-dependent deflection of graphene/substrate bimetallic cantilevers (36). They have demonstrated a thermally controllable microactuator made of graphene layers on a silicon nitride (Si_3N_4) substrate. Bao et al. proposed that the morphology of suspended graphene can be modulated with electrostatic or thermal control (37). With an applied gate voltage, a flat graphene membrane can change its shape into a parabolic profile. There exists a proportional correlation between the amount of deflection and the applied gate voltage, and almost no hysteresis was observed. These investigations suggest that graphene can be an excellent platform for various types of nanoelectromechanical system (NEMS) devices, although a deeper understanding of electrical, mechanical, and thermal characteristics as well as strain engineering of graphene devices are required.

18.3 MONOLITHIC GRAPHENE–GRAPHITE INTEGRATED ELECTRONICS

18.3.1 Monolithic Graphene–Graphite Synthesis Using Heterogeneously Patterned Catalysts

Metals such as Cu, Ni, and Co can be used as catalysts to dissociate hydrocarbon gas molecules into carbon and hydrogen. These catalyst metals each have varying levels of carbon solubility that are different from each other; Ni and Co have higher carbon solubility than what is optimally required to synthesize graphene (12, 14). Therefore graphite, which is thicker than graphene, can be synthesized by CVD onto these catalyst metals. Cu, on the other hand, has almost negligible carbon solubility, so only 1–3 layers of graphene can be synthesized on Cu catalyst surface (13, 40, 41).

An experimental process to demonstrate and exploit the difference in carbon solubility among the catalyst metals is presented in Figure 18.1. Figure 18.1a and b show the spatially patterned heterogeneous catalyst structure prepared by photolithography and thermal evaporation, which are composed of Cu, Ni, and Co thin film on a 285 nm-thick SiO_2 on Si substrate. The background is covered with Cu (700 nm), and a thin Ni protection layer (5 nm) is coated over the Cu layer to prevent the Cu from any potential oxidation that may be caused by water and organic solvents during photolithography processes. The consonant and vowel characters are covered with Co (400 nm) and Ni (20 nm), respectively, to form the word "graphene." When the temperature reaches up to 1000 °C in the CVD process, the Co and Ni catalysts layers are diffused locally into the Cu layer and form alloys. Figure 18.1c and d shows the atomic force microscopy (AFM) measurements at the interface between Co and Cu catalysts. Before the CVD synthesis (Fig. 18.1c), there is a clear difference in height between the two catalysts, but after the CVD (Fig. 18.1d), the height difference at the interface becomes less pronounced. This demonstrates that Co diffuses into the underlying Cu layer at the high synthesis temperature (1000 °C). Before the AFM characterization, oxygen plasma is applied to etch off the graphene–graphite layers in order to clearly show the height change after the CVD synthesis (Fig. 18.1d).

Synthesis of graphene and graphite are based on different growth mechanisms (41). Graphene is synthesized on Cu catalyst that has very low carbon solubility, and the synthesis is the result of physical adsorption of carbon atoms on the Cu layer. On the other hand, Co and Ni catalysts are diffused into the underlying Cu layer at high synthesis temperatures, and the carbon atoms are dissolved into the Co/Cu (or Ni/Cu) alloys. When the synthesis is completed and the cooling down process starts, the dissolved carbon is precipitated again onto the catalyst layers to create a relatively thicker graphite layer.

CVD is conducted with the flow of CH_4 and H_2 at atmospheric pressure, following the initial heating process. When the center temperature of the CVD furnace reaches 1000 °C, the growth substrate is put into the center of the furnace using load-lock system for the purpose of rapid heating. By doing rapid heating, undesirable lateral diffusion of heterogeneous metal catalysts during the heating stage could be avoided. When the synthesis step is finished, the chamber is cooled down in Ar atmosphere. After the synthesis, the metal catalyst layers are removed by an etchant solution of $FeCl_3$ and HCl, and the monolithic graphene–graphite structure is then transferred to a 285 nm-thick SiO_2 on a Si substrate (Fig. 18.1b). The three different areas (Cu, Ni, and Co) show different colors and contrasts because the reflectance varies at the air/graphene multilayers and graphene multilayers/SiO_2 interfaces depending on the number of graphene layers (42).

Figure 18.1e shows the Raman spectra of the monolithic graphene–graphite structure. Raman characterization is recorded with a 532 nm laser as an excitation source using 100× objective lens (spot size ~1 μm). Typically, D band (~1350 cm^{-1}) may appear because of the residue oxygen, water molecules, or excessive carbon atoms during the synthesis in atmospheric pressure. With G (~1590 cm^{-1}) and 2D (~2680 cm^{-1}) bands, one can distinguish the existence of graphene or graphite and estimate the number of graphene layers. The peak intensity ratio between the G and 2D bands allows one to infer the number of graphene layers present. In Figure 18.1e, the number of graphene layers (Cu area) is smallest at the background, followed by the vowel (Ni) and consonant (Co) area. This result confirms that Co has the highest carbon solubility, followed by Ni and Cu. The Raman result suggests that the number of graphene layers on the background region is 2–4 layers (14, 43). In addition, the full width at half maximum (FWHM) of the 2D band peak is another tool to estimate the number of graphene layers: the FWHM increases as the number of graphene layers increases (44). For example, the FWHM of a single-layer graphene is around 24 cm^{-1}, and that for bilayer graphene is 46 cm^{-1}. In Figure 18.1e, FWHM is smallest at the background (Cu) and has larger values in the case of Ni and Co catalysts (the 2D peaks in these area are also nonsymmetric, dispersive, and broader, with a slight blueshift due to interlayer binding). Figure 18.1f shows the two-dimensional Raman mapping of the square area in Figure 18.1a, and G and 2D bands show clear intensity contrast compared to the background Cu area. Furthermore, Raman maps of the graphite layers (characters) are continuous with the 2–4 graphene layers in the background, which further confirms the monolithic synthesis of graphene–graphite structure with heterogeneous catalyst pattern.

AFM measurement further confirms that the thickness of graphite is about ~40–190 nm at the consonant area (Co), and at the vowel area (Ni), there is a much thinner graphite layer with a thickness of ~2 ± 0.94 nm (~6–8 layers). AFM measurement of the background (Cu) also demonstrated layer variations at the graphene area (<1.7 nm, 2–4 layers) (15, 16).

Figure 18.2a shows the transmission electron microscope (TEM) image at the graphene–graphite interface and verifies the monolithic structure, demonstrating the continuous interface between the graphene and graphite; one can see the number of layers gradually changing at the interface of graphite and graphene. The thickness of graphite is intentionally prepared to be thinner using Ni catalyst. The hexagonal electron diffraction patterns of graphene and graphite are shown in Figure 18.2b and c, respectively, to further confirm that graphene and graphite are successfully synthesized.

To verify that the metal catalysts are completely removed after the etching and transferring process, the energy dispersive X-ray analysis (EDAX) and X-ray photoelectron spectroscopy (XPS) analyses are performed with a monolithic graphene–graphite transferred onto a

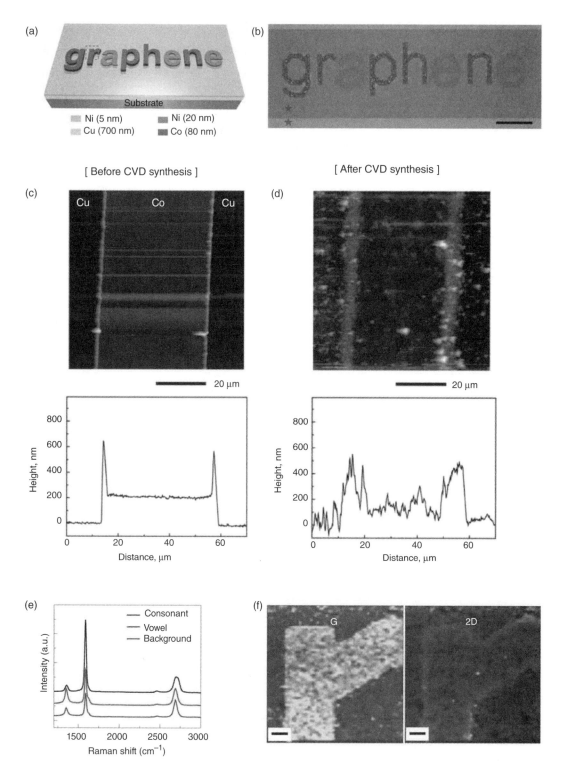

Figure 18.1 (a) Schematic illustration of the heterogeneous metal catalyst pattern. (b) Optical micrograph image of graphene–graphite synthesized by CVD from the heterogeneous catalyst structure. (c) and (d) AFM images at the interface of the Co and Cu catalysts, showing the height difference between the two different catalysts, before and after the CVD. (e) Raman spectra on the three different catalyst regions (Co, Ni, and Cu). (f) Raman mapping on the dashed area in (a). (*Source*: Reprinted with permission from Reference 2).

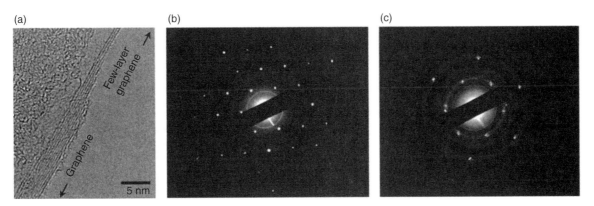

Figure 18.2 TEM images of graphene–graphite interface (side view) (a) and the hexagonal electron diffraction pattern of graphene (b) and graphite (c). (*Source*: Reprinted with permission from Reference 2).

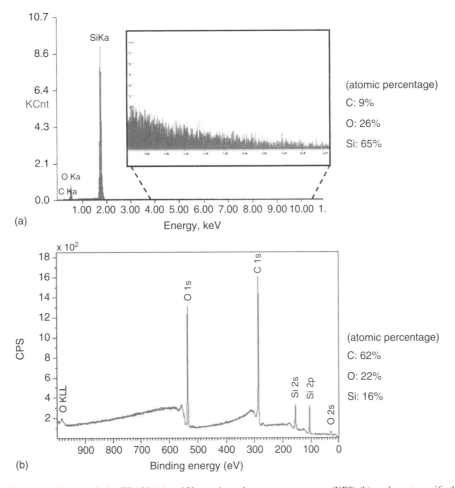

Figure 18.3 The energy dispersive X-ray analysis (EDAX) (a) and X-ray photoelectron spectroscopy (XPS) (b) analyses to verify that all metal catalysts are etched away after the transfer process. (*Source*: Reprinted with permission from Reference 2).

Si wafer with a 285 nm-thick SiO_2 layer after the removal of metal catalysts (Fig. 18.3a and b). In this analysis, no metal peaks are observed.

The size of the graphite grains increases as the thickness of the Co catalyst increases. Figure 18.4a shows the Co catalyst patterns exhibiting different thicknesses and a synthesized graphene–graphite structure transferred on a 285 nm-thick SiO_2/Si wafer. Figure 18.4b shows the AFM characterizations demonstrating that the size of the graphite grain increases with the thickness of the Co catalyst.

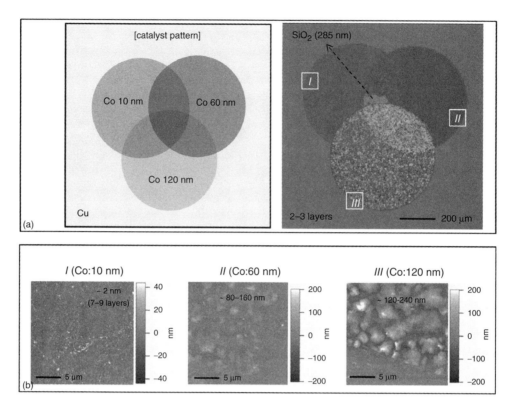

Figure 18.4   The size of graphite grains increases as the thickness of Co catalyst increases. Co catalysts prepared with different thickness followed by the CVD synthesis (a) and the size of graphite grains observed by AFM (b). (*Source*: Reprinted with permission from Reference 2).

18.3.2   Flexible Transistors and Sensors Based on Monolithic Graphene–Graphite Nanostructure

18.3.2.1   Graphene–Graphite Field-Effect Transistor   The electrical properties of graphene and graphite, such as sheet resistance and field-effect response, are controllable by changing the number of graphene layers (n). By increasing or decreasing the number of graphene layers, the sheet resistance could be modulated by more than an order of magnitude. Figure 18.5 shows the correlation between the sheet resistance and the number of layers measured by a four-point probe setup. As the number of layers increases from ~2–4 to ~850 layers, the sheet resistance is reduced by a factor of about 25 (from 2463 ± 1037 Ω/sq down to 98 ± 46 Ω/sq). The deviation of the sheet resistance is attributed to the local variation in the number of layers of graphene.

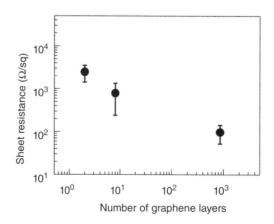

Figure 18.5   Sheet resistance as a function of the number of graphene layers (n). (*Source*: Reprinted with permission from Reference 2).

In addition, the field-effect responses of 2–4 layers of graphene and approximately 850 layers of graphite were investigated (Fig. 18.6). Source and drain electrodes consisted of Au with a thin Cr adhesion layer, and degenerated doped Si was used as a back gate. The current versus back-gate characteristics showed much stronger modulation in 2–4-layer graphene, compared with a negligible change in the thick graphite (~850 layers) owing to the stronger screening effect as the number of layers increases (45). The low sheet resistance and field-effect response of graphite are advantageous for applications in conductive films or electrodes. In contrast, the superior transconductance level of the 2–4-layer graphene is appropriate for the active channels of FETs.

Figure 18.7a and b shows the schematic illustration of the all-carbon monolithic graphene–graphite field-effect transistor structure and a Raman map (G band), respectively. Graphene channel is synthesized with Cu/Ni (700 nm/5 nm) catalyst layers, and graphite electrodes (source and drain) are synthesized with additional Co layer (400 nm) upon the Cu/Ni layers. After the synthesis, the patterned graphene–graphite FET structure is coated with the poly(methyl methacrylate) (PMMA) supporting layer and then transferred onto a degenerately doped Si wafer

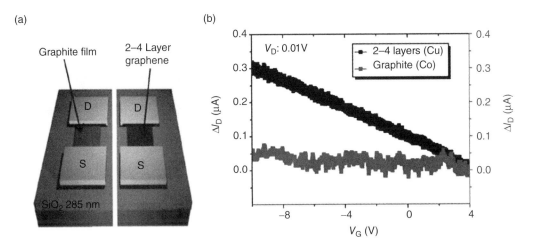

Figure 18.6 Schematic illustration of the graphene–graphite back-gate field-effect transistor (a) and I_D–V_G characteristic of the FETs with channels of graphene and graphite (b). (*Source*: Reprinted with permission from Reference 2).

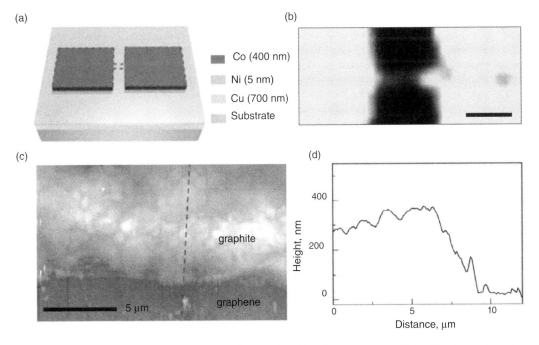

Figure 18.7 Schematic illustration of the graphene–graphite field-effect transistor (a) and the Raman map (G band) at the graphene channel area with the graphitic source/drain (b). AFM image of the interface between the graphite electrode and graphene channel (c) and the section profile of the dashed line (d). (*Source*: Reprinted with permission from Reference 2).

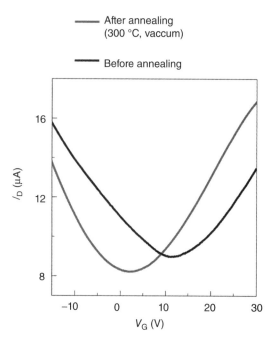

Figure 18.8   S/D current (I_D) versus back-gate bias (V_G) characterization of the graphene FETs before and after the annealing process. (*Source*: Reprinted with permission from Reference 2).

with a 285 nm-thick thermal oxide layer. PMMA layer is then removed with acetone to yield monolithic graphene–graphite back-gate FETs. Thermal annealing treatment (~300 °C, vacuum) of monolithic graphene–graphite FET devices is performed for approximately 3 h to remove polymer residues on graphene/graphite surface. AFM scan of the interface between the graphite electrode and graphene channel is shown in Figure 18.7c and d.

Figure 18.8 shows the correlation between the source (S)/drain (D) current (I_D) and back-gate bias (V_G) at room temperature. Before and after the annealing, the device shows a clear ambipolar behavior, which is consistent with our expectation from the semimetallic characteristics of graphene (14, 46, 47). After the annealing, the hole (electron) mobility increased to ~1800 (1400) cm²/V·s (calculated by the standard metal–oxide–semiconductor FET model) because the resist residues are removed during the annealing process.

The contact properties between the graphite electrode and the graphene channel are compared to the contact properties of traditional Cr/Au electrodes and the graphene channel (Fig. 18.9). Both graphite and Cr/Au electrodes show ohmic contact behavior, a linear I_D–V_D characteristics at room temperature. The contact resistances of both cases are estimated using the transfer length method at the room temperature (Fig. 18.10). Compared with the Cr/Au electrodes (~1100 Ω μm), covalent graphite contacts to the graphene channel showed a

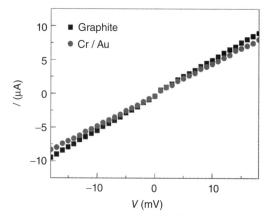

Figure 18.9   Comparison of contact properties between the monolithic graphene channel and the graphite electrodes (square) and between graphene and the Cr/Au electrodes (circle). (*Source*: Reprinted with permission from Reference 2).

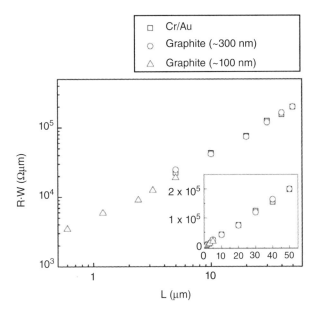

Figure 18.10   Estimation of contact resistances by transfer length method (TLM). The contact resistance values estimated from y-intercepts (at $L = 0$) of this graph are 1110, 880, and 790 Ω μm for the three cases of Cr/Au, 100 nm graphite, and 300 nm graphite, respectively. (*Source*: Reprinted with permission from Reference 2).

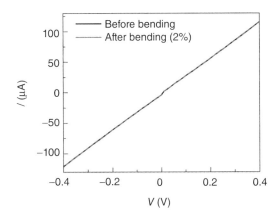

Figure 18.11   *I–V* characteristic before and after bending (2% strain) the graphite electrodes. (*Source*: Reprinted with permission from Reference 2).

relatively lower contact resistance ($\sim$700–900 Ω μm). Moreover, varying the thickness of the graphite contacts does not significantly change the contact resistance.

Another advantageous property of the graphite electrode is its mechanical flexibility. Up to 2% of strain, the *I–V* characteristic remains almost the same, as in Figure 18.11. The mechanical deflection is performed with a motor-driven micromanipulator to precisely control the position and movement. Euler–Bernoulli beam theory is used to estimate the strain resulting from mechanical deflection.

18.3.2.2   *Chemical Sensing Capability of Graphene Field-Effect Sensors*

The one-step synthesis method with heterogeneous catalyst enables the creation of a large-scale field-effect sensor arrays. Figure 18.12a shows the schematic illustration of the graphene–graphite field-effect sensors. Each block of sensors consists of nine of 2–4-layer graphene field-effect sensors with a single common graphite source and independent drains. The sensor chip has four of these blocks, so a total of 36 field-effect sensors are fabricated, as illustrated in Figure 18.12a. The graphite electrodes (source and drain) are covered with a passivation layer, leaving the 2–4-layer graphene channel area between the graphite electrodes.

Figure 18.12b describes the correlation between the source/drain current and Ag/AgCl water-gate voltage of the graphene field-effect sensor. Multichannel pH sensing is performed by monitoring the conductance with a small AC bias (5 mV, 79 Hz) and DC bias set to 0 V, with custom-designed variable gain amplifier to amplify the drain current from multiple devices, and filtered using computer-based virtual lock-in amplifier, with time constant set to 300 ms. Time-variant conductance changes are recorded while different pH solutions are delivered through a microfluidic channel. As pH increases, the charge neutrality point shifts positively with the sensitivity of approximately 17 mV/pH.

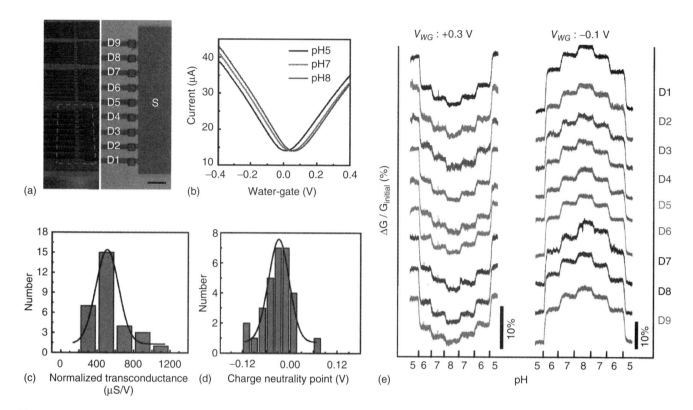

Figure 18.12   (a) Schematic illustration of the graphene–graphite field-effect chemical sensing arrays, composed of four blocks consisting of nine individual FETs for each block, yielding a total of 36 sensor devices. (b) I_D/V_G characteristic with different pH levels of Ag/AgCl solution. (c) and (d) Statistical distribution of normalized transconductance and charge neutrality point at pH 7. (e) Real-time, complementary pH sensing using nine field-effect sensor arrays in n-type regime (left) and p-type regime (right). (*Source*: Reprinted with permission from Reference 2).

Also, the water-gate response shows an ambipolar characteristic, meaning that field-effect sensors can be used as both p-type and n-type sensors. These electrical properties are similar to that fabricated by conventional CVD or exfoliated graphene (48, 49). Figure 18.12c and d shows the statistical distribution of the normalized transconductance and charge neutrality point for the field-effect sensor device array at pH 7. According to the Gaussian fits of the two device parameters, the center values of the normalized transconductance and charge neutrality point are found to be 540 ± 199 μS/V and -0.03 ± 0.038 V, respectively. Figure 18.12e shows real-time multiplexed pH sensing using the nine field-effect sensor arrays. In the case of the p-type regime, the conductance increases (decreases) as the pH of the solution increases (decreases) and vice versa in the n-type regime. This complementary sensing capability of graphene field-effect sensors has advantages over other unipolar field-effect sensors because it enables the discrimination of potential electrical cross talk and/or false-positive signals (49).

18.3.3   Graphene–Graphite Devices as Highly Flexible and Transferable Electronics

As previously demonstrated, graphene–graphite devices have unparalleled mechanical properties that allow for highly flexible and transferable electronic applications. Figure 18.13 depicts key properties (a) and applications (b) of graphene–graphite devices. The high elastic modulus and hydrophobicity of the graphene–graphite devices allow the entire monolithic device arrays to float on water in free-standing form (Fig. 18.13a). Additionally, the flexibility of the graphene–graphite devices enables the integrated device structures of monolithic graphene–graphite to be transferred on various nonplanar substrates. As examples, Figure 18.13b depicts monolithic devices wrapped on the outside surfaces of a thin cylindrical glass tube (outer diameter: ~1.5 mm), an eye contact lens (soft galyfilcon A polymer), a glove finger (latex), a coin, and three different body parts of an insect (*Odontolabis sarasinorum* specimen).

A principal application area that can take advantage of monolithically integrated graphene–graphite devices is flexible electronics. Vertical, 3D integration by assembling layers of the monolithic graphene–graphite device components on polymeric films is demonstrated in Figures 18.14 and 18.15. To realize this fabrication, monolithically integrated graphene–graphite structures are transferred onto a polyether ether ketone film (PEEK), followed by deposition of a SiO_2 dielectric layer. Flexible monolithic FET arrays are created by assembling a graphitic top-gate layer.

Along with mechanical flexibility, optical transparency is another principal characteristic of monolithically integrated graphene–graphite electronics. Figure 18.15b depicts the semitransparency of the top-gated devices that are positioned on top of the printed logo. The transmittance of single-layer graphene is approximately 97% at 550 nm wavelength and decreases as the number of layers *n* increases (16). Consequently, the overall transparency of monolithic graphene–graphite devices can be adjusted by varying *n* for each device component.

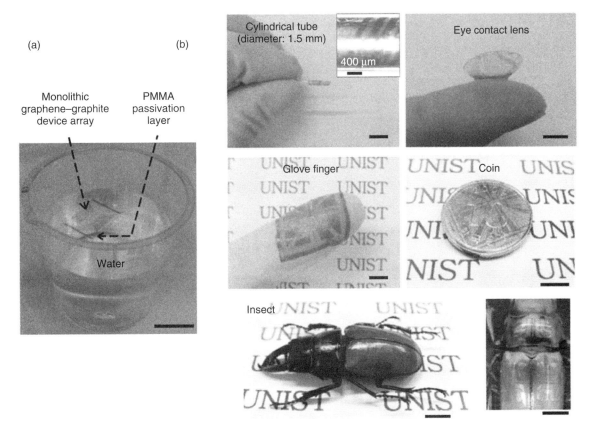

Figure 18.13 (a) Photograph of free-standing, monolithic graphene–graphite integrated sensor networks floating on water. Scale bar, 20 mm. (b) Photographs of monolithic device structures transferred onto various nonplanar substrates, including cylindrical glass tube (outer diameter: ~1.5 mm), eye contact lens (soft galyfilcon A polymer), glove finger, coin, and the epidermis of an insect (*Odontolabis sarasinorum* specimen) with a magnified top-view image of the insect. Scale bars: 5 mm. (*Source*: Reprinted with permission from Reference 2).

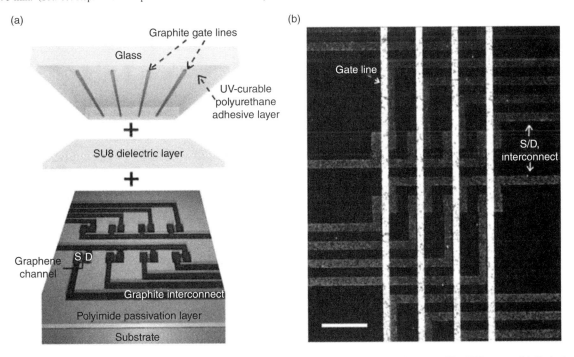

Figure 18.14 (a) Schematic illustration of the assembly process to fabricate top-gate monolithic graphene–graphite FET arrays. (b) Optical micrograph (dark-field) of the top-gate FET arrays after removing the top glass substrate using HF. Scale bar, 200 μm. (*Source*: Reprinted with permission from Reference 2).

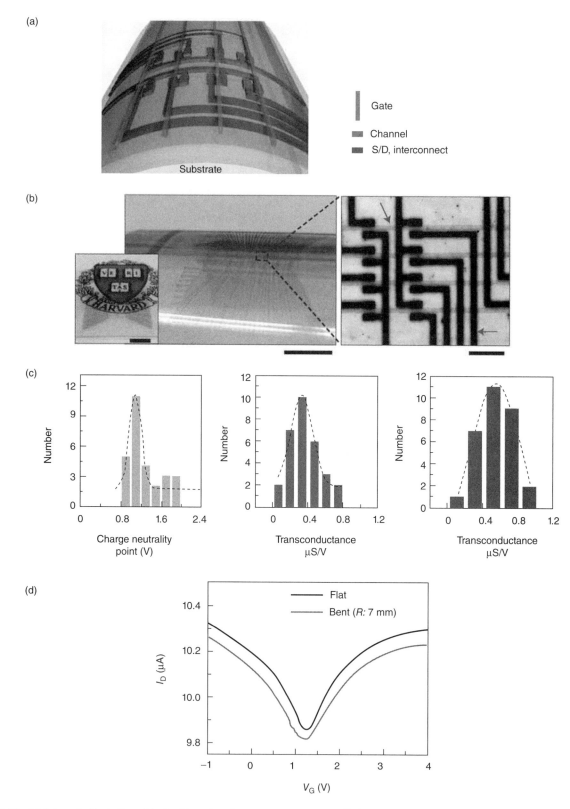

Figure 18.15 (a) Schematic illustration of the flexible and semitransparent top-gate monolithic graphene–graphite FET arrays. (b) Photograph (main panel) of the devices wrapped on a cylindrical glass (radius of curvature: 1.2 cm). The device is rested on a paper printed with a logo to demonstrate the device's semitransparent characteristics (left inset). Scale bars, 4 mm. Optical micrograph of the top-gate FET arrays (right image). Scale bar, 200 μm. (c) Statistical distributions of the charge neutrality point (left panel) and transconductance in the n-type (center) and p-type (right panel) regimes. (d) I_D–V_G curves of the top-gate FET measured when the substrate is flat and when it is bent (radius of curvature: 0.7 cm). (*Source:* Reprinted with permission from Reference 2).

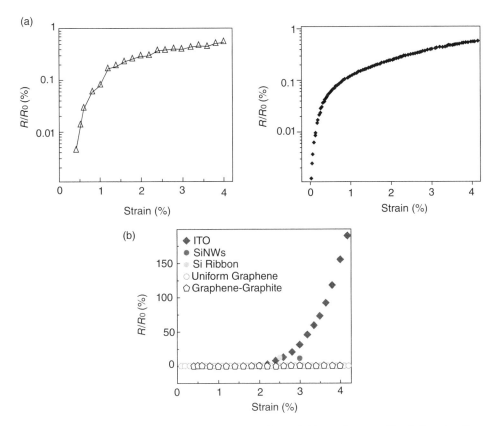

Figure 18.16 (a) Log-scaled resistance variations (R/R_0) versus strain plot of monolithic graphene–graphite (left) and uniform graphene (right), (b) comparison with the R/R_0 changes in other materials (16, 50, 51). (*Source*: Reprinted with permission from Reference 2).

The statistical distributions fitting Gaussian profiles of the charge neutrality point and transconductance of the top-gate FET arrays (average mobility: 675 $cm^2/V{\cdot}s$) are provided in Figure 18.15c. In comparison to the back-gate FETs in Figure 18.8, the charge neutrality point shifts closer to zero, most likely due to oxygen desorption from graphene in the SiO_2 evaporation step. The mechanical flexibility of monolithic graphene–graphite top-gate FET devices is explored in Figure 18.15d, which presents current versus top-gate voltage curves of the FET when the substrate is flat versus when it is bent (radius of curvature: 0.7 cm). From the plot, bending to a radius of curvature of 0.7 cm (estimated bending-induced strain: ~0.6%) causes no significant change in the electric response (i.e., mobility values remained constant). Furthermore, the electrical properties of monolithic graphene–graphite devices remain almost constant (less than 1% change) when applying a maximum strain of approximately 4%, demonstrating the unique mechanical flexibility of monolithic graphene–graphite integrated electronics (Fig. 18.16).

18.4 FUTURE DIRECTIONS: GRAPHENE BIOELECTRONICS

The intersection between nanomaterials and biology and life sciences can be understood in many ways, perhaps the most obvious of which is by reference to the size and organization of common structures (52, 53). The similarity in size of synthetic and natural nanostructures to the relevant molecule or organism makes nanomaterials an optimal candidate for sensors and other tools of biological detection. Nanomaterials enable detection in powerful new ways, by increasing the spatial and temporal resolution of detection. Furthermore, the large surface-to-volume ratio of nanostructures enables superior sensitivity of signal transduction. Moreover, as the size and thickness of nanostructures are reduced, flexural rigidity becomes many orders of magnitude smaller than that of bulk materials, offering mechanical flexibility and noninvasive, three-dimensional conformal interfacing capabilities (54).

Superb electromechanical properties of monolithic graphene–graphite, where large elastic deformation is achievable without perturbation of its electrical properties, offer great promise for advanced electromechanical platforms to biological interfaces. Unlike conventional integrated electronics that have mechanically fragile heterogeneous metal–semiconductor interfaces, the monolithic interface of graphene–graphite structure offers mechanical robustness and flexibility and is suitable for applications to flexible and conformal bioelectronics (55).

One of the strengths of the ultrathin monolithic graphene–graphite structure for bioelectronic applications is that it can be integrated onto various nonplanar 3D substrates. As shown in Figure 18.17, the whole monolithic device is integrated on a human eye model, demonstrating

Figure 18.17   Flexible and conformal monolithic graphene–graphite bioelectronics. Monolithic graphene–graphite bioelectronics sensor arrays are integrated on a human eye model. The monolithic graphene–graphite integrated electronics suggests substantial promise toward flexible, wearable electronics and implantable biosensor devices.

the unique potential for application as an artificial retina. The softness of the material used for the flexible bioelectronic device further enables the device to be conformally interfaced in accordance with the corresponding mechanical properties of biological systems.

To summarize, this article reviews a unique one-step synthesis of monolithic graphene–graphite nanostructures that can be transferred to various substrates and surfaces. The monolithic integration enables high-sensitivity field-effect detection of chemical and biological signals with potential for mechanically flexible and robust interface with biological systems. We believe the capability to synthesize monolithic graphene–graphite integrated electronics provides a promising strategy toward flexible, wearable electronics and implantable biosensor devices and also suggests substantial promise toward chemical and biological detection and conformal interface with biological systems in the future.

REFERENCES

1. Novoselov KS, Geim AK, Morozov SV, Jiang D, Zhang Y, Dubonos SV, Grigorieva IV, Firsov AA. Electric field effect in atomically thin carbon films. Science (New York, NY) 2004;306:666–669.

2. Park J-U, Nam S, Lee M-S, Lieber CM. Synthesis of monolithic graphene-graphite integrated electronics. Nat Mater 2012;11:120–125.

3. Stankovich S, Dikin DA, Piner RD, Kohlhaas KA, Kleinhammes A, Jia Y, Wu Y, Nguyen ST, Ruoff RS. Synthesis of graphene-based nanosheets via chemical reduction of exfoliated graphite oxide. Carbon 2007;45:1558–1565.

4. Schniepp HC, Li J-L, McAllister MJ, Sai H, Herrera-Alonso M, Adamson DH, Prud'homme RK, Car R, Saville DA, Aksay IA. Functionalized single graphene sheets derived from splitting graphite oxide. J Phys Chem B 2006;110:8535–8539.

5. Potts JR, Lee SH, Alam TM, An J, Stoller MD, Piner RD, Ruoff RS. Thermomechanical properties of chemically modified graphene/poly(methyl Methacrylate) composites made by in situ polymerization. Carbon 2011;49:2615–2623.

6. Forbeaux I, Themlin J-M, Debever J-M. Heteroepitaxial graphite on 6H-SiC(0001): interface formation through conduction-band electronic structure. Phys Rev B 1998;58:16396–16406.

7. Ohta T, Bostwick A, Seyller T, Horn K, Rotenberg E. Controlling the electronic structure of bilayer graphene. Science (New York, NY) 2006;313:951–954.

8. Virojanadara C, Syväjarvi M, Yakimova R, Johansson L, Zakharov A, Balasubramanian T. Homogeneous large-area graphene layer growth on 6H-SiC(0001). Phys Rev B 2008;78:245403.

9. Emtsev KV, Bostwick A, Horn K, Jobst J, Kellogg GL, Ley L, McChesney JL, Ohta T, Reshanov SA, Röhrl J, Rotenberg E, Schmid AK, Waldmann D, Weber HB, Seyller T. Towards wafer-size graphene layers by atmospheric pressure graphitization of silicon carbide. Nat Mater 2009;8:203–207.

10. Peng T, Lv H, He D, Pan M, Mu S. Direct transformation of amorphous silicon carbide into graphene under low temperature and ambient pressure. Sci Rep 2013;3:1148.

11. Edwards RS, Coleman KS. Graphene film growth on polycrystalline metals. Acc Chem Res 2013;46:23–30.

12. Li X, Cai W, An J, Kim S, Nah J, Yang D, Piner R, Velamakanni A, Jung I, Tutuc E, Banerjee SK, Colombo L, Ruoff RS. Large-area synthesis of high-quality and uniform graphene films on copper foils. Science (New York, NY) 2009;324:1312–1314.

13. Kim KS, Zhao Y, Jang H, Lee SY, Kim JM, Kim KS, Ahn J-H, Kim P, Choi J-Y, Hong BH. Large-scale pattern growth of graphene films for stretchable transparent electrodes. Nature 2009;457:706–710.

14. Bhaviripudi S, Jia X, Dresselhaus MS, Kong J. Role of kinetic factors in chemical vapor deposition synthesis of uniform large area graphene using copper catalyst. Nano Lett 2010;10:4128–4133.

15. Bae S, Kim H, Lee Y, Xu X, Park J-S, Zheng Y, Balakrishnan J, Lei T, Kim HR, Song YI, Kim Y-J, Kim KS, Ozyilmaz B, Ahn J-H, Hong BH, Iijima S. Roll-to-roll production of 30-inch graphene films for transparent electrodes. Nat Nanotechnol 2010;5:574–578.

16. Celebi K, Cole MT, Choi JW, Wyczisk F, Legagneux P, Rupesinghe N, Robertson J, Teo KBK, Park HG. Evolutionary kinetics of graphene formation on copper. Nano Lett 2013;13:967–974.

17. Ago H, Ogawa Y, Tsuji M, Mizuno S, Hibino H. Catalytic growth of graphene: toward large-area single-crystalline graphene. J Phys Chem Lett 2012;3:2228–2236.

18. Losurdo M, Giangregorio MM, Capezzuto P, Bruno G. Graphene CVD growth on copper and nickel: role of hydrogen in kinetics and structure. Phys Chem Chem Phys 2011;13:20836–20843.

19. Huang L, Chang QH, Guo GL, Liu Y, Xie YQ, Wang T, Ling B, Yang HF. Synthesis of high-quality graphene films on nickel foils by rapid thermal chemical vapor deposition. Carbon 2012;50:551–556.

20. Reina A, Jia X, Ho J, Nezich D, Son H, Bulovic V, Dresselhaus MS, Kong J. Large area, few-layer graphene films on arbitrary substrates by chemical vapor deposition. Nano Lett 2009;9:30–35.

21. Kim ES, Shin H-J, Yoon S-M, Han GH, Chae SJ, Bae JJ, Gunes F, Choi J-Y, Lee YH. Low-temperature graphene growth using epochal catalyst of PdCo alloy. Appl Phys Lett 2011;99:223102.

22. McCarty KF, Feibelman PJ, Loginova E, Bartelt NC. Kinetics and thermodynamics of carbon segregation and graphene growth on Ru(0001). Carbon 2009;47:1806–1813.

23. Oznuluer T, Pince E, Polat EO, Balci O, Salihoglu O, Kocabas C. Synthesis of graphene on gold. Appl Phys Lett 2011;98:183101.

24. Vo-Van C, Kimouche A, Reserbat-Plantey A, Fruchart O, Bayle-Guillemaud P, Bendiab N, Coraux J. Epitaxial graphene prepared by chemical vapor deposition on single crystal thin iridium films on sapphire. Appl Phys Lett 2011;98:181903.

25. Chen J, Wen Y, Guo Y, Wu B, Huang L, Xue Y, Geng D, Wang D, Yu G, Liu Y. Oxygen-aided synthesis of polycrystalline graphene on silicon dioxide substrates. J Am Chem Soc 2011;133:17548–17551.

26. Chen J, Guo Y, Wen Y, Huang L, Xue Y, Geng D, Wu B, Luo B, Yu G, Liu Y. Graphene: two-stage metal-catalyst-free growth of high-quality polycrystalline graphene films on silicon nitride substrates (Adv. Mater. 7/2013). Adv Mater 2013;25:938.

27. Fanton MA, Robinson JA, Puls C, Liu Y, Hollander MJ, Weiland BE, Labella M, Trumbull K, Kasarda R, Howsare C, Stitt J, Snyder DW. Characterization of graphene films and transistors grown on sapphire by metal-free chemical vapor deposition. ACS Nano 2011;5:8062–8069.

28. Hwang J, Kim M, Campbell D, Alsalman HA, Kwak JY, Shivaraman S, Woll AR, Singh AK, Hennig RG, Gorantla S, Rummeli MH, Spencer MG. Van Der Waals epitaxial growth of graphene on sapphire by chemical vapor deposition without a metal catalyst. ACS Nano 2013;7:385–395.

29. Geim AK. Graphene: status and prospects. Science (New York, NY) 2009;324:1530–1534.

30. Schwierz F. Graphene transistors. Nat Nanotechnol 2010;5:487–496.

31. Wu Y, Lin Y, Bol AA, Jenkins KA, Xia F, Farmer DB, Zhu Y, Avouris P. High-frequency, scaled graphene transistors on diamond-like carbon. Nature 2011;472:747–748.

32. Sui Y, Appenzeller J. Screening and interlayer coupling in multilayer graphene field-effect transistors. Nano Lett 2009;9:2973–2977.

33. Xia F, Mueller T, Lin Y-M, Valdes-Garcia A, Avouris P. Ultrafast graphene photodetector. Nat Nanotechnol 2009;4:839–843.

34. Vicarelli L, Vitiello MS, Coquillat D, Lombardo A, Ferrari AC, Knap W, Polini M, Pellegrini V, Tredicucci A. Graphene field-effect transistors as room-temperature terahertz detectors. Nat Mater 2012;11:865–871.

35. Conley H, Lavrik NV, Prasai D, Bolotin KI. Graphene bimetallic-like cantilevers: probing graphene/substrate interactions. Nano Lett 2011;11:4748–4752.

36. Bao W, Myhro K, Zhao Z, Chen Z, Jang W, Jing L, Miao F, Zhang H, Dames C, Lau CN. In situ observation of electrostatic and thermal manipulation of suspended graphene membranes. Nano Lett 2012;12:5470–5474.

37. Li P, You Z, Cui T. Graphene cantilever beams for nano switches. Appl Phys Lett 2012;101:093111.

38. Zhu S-E, Shabani R, Rho J, Kim Y, Hong BH, Ahn J-H, Cho HJ. Graphene-based bimorph microactuators. Nano Lett 2011;11:977–981.

39. Jauregui LA, Cao H, Wu W, Yu Q, Chen YP. Electronic properties of grains and grain boundaries in graphene grown by chemical vapor deposition. Solid State Commun 2011;151:1100–1104.

40. Li X, Cai W, Colombo L, Ruoff RS. Evolution of graphene growth on Ni and Cu by carbon isotope labeling. Nano Lett 2009;9:4268–4272.

41. Blake P, Hill EW, Castro Neto AH, Novoselov KS, Jiang D, Yang R, Booth TJ, Geim AK. Making graphene visible. Appl Phys Lett 2007;91:063124.

42. Dresselhaus MS, Jorio A, Hofmann M, Dresselhaus G, Saito R. Perspectives on carbon nanotubes and graphene Raman spectroscopy. Nano Lett 2010;10:751–758.

43. Lee S, Lee K, Zhong Z. Wafer scale homogeneous bilayer graphene films by chemical vapor deposition. Nano Lett 2010;10:4702–4707.

44. Zhang Y, Small JP, Pontius WV, Kim P. Fabrication and electric-field-dependent transport measurements of mesoscopic graphite devices. Appl Phys Lett 2005;86:073104.

45. Novoselov KS, Geim AK, Morozov SV, Jiang D, Katsnelson MI, Grigorieva IV, Dubonos SV, Firsov AA. Two-dimensional gas of massless dirac fermions in graphene. Nature 2005;438:197–200.

46. Zhang Y, Tan Y-W, Stormer HL, Kim P. Experimental observation of the quantum hall effect and Berry's Phase in graphene. Nature 2005;438:201–204.

47. Ohno Y, Maehashi K, Yamashiro Y, Matsumoto K. Electrolyte-gated graphene field-effect transistors for detecting pH and protein adsorption. Nano Lett 2009;9:3318–3322.

48. Cohen-Karni T, Qing Q, Li Q, Fang Y, Lieber CM. Graphene and nanowire transistors for cellular interfaces and electrical recording. Nano Lett 2010;10:1098–1102.

49. Ryu SY, Xiao J, Park WI, Son KS, Huang YY, Paik U, Rogers JA. Lateral buckling mechanics in silicon nanowires on elastomeric substrates. Nano Lett 2009;9:3214–3219.

50. Timko BP, Cohen-Karni T, Qing Q, Tian B, Lieber CM. Design and implementation of functional nanoelectronic interfaces with biomolecules, cells, and tissue using nanowire device arrays. IEEE Trans Nanotechnol 2010;9:269–280.

51. Rogers JA, Someya T, Huang Y. Materials and mechanics for stretchable electronics. Science (New York, NY) 2010;327:1603–1607.

52. Lee C, Wei X, Kysar JW, Hone J. Measurement of the elastic properties and intrinsic strength of monolayer graphene. Science (New York, NY) 2008;321:385–388.

53. Chabot V, Kim B, Sloper B, Tzoganakis C, Yu A. High yield production and purification of few layer graphene by gum arabic assisted physical sonication. Sci Rep 2013;3:1378.

54. Khang D-Y, Jiang H, Huang Y, Rogers JA. A stretchable form of single-crystal silicon for high-performance electronics on rubber substrates. Science (New York, NY) 2006;311:208–212.

55. Dvir T, Timko BP, Kohane DS, Langer R. Nanotechnological strategies for engineering complex tissues. Nat Nanotechnol 2011;6:13–22.

19

THIN-FILM TRANSISTORS BASED ON TRANSITION METAL DICHALCOGENIDES

WOONG CHOI[1] AND SUNKOOK KIM[2]

[1]*School of Advanced Materials Engineering, Kookmin University, Seoul, South Korea*
[2]*Department of Electronic and Radio Engineering, Kyung Hee University, Yongin, South Korea*

19.1 INTRODUCTION

The discovery of graphene opened the door to the exotic electronic, optical, and mechanical properties of two-dimensional (2D) nanomaterials (1). Graphene has a conical Dirac spectrum of energy states without a bandgap and a linear dispersion. While these properties are the root of much of the novel electronic and optical phenomena of graphene, the gapless band structure also makes it unsuitable for conventional transistors for electronic switching. Although there are a number of techniques to open a bandgap in graphene, such as graphene nanoribbons or graphene bilayer, they are some way from being suitable for use in real applications (2).

Similar to the storyline of the graphite and graphene family, transition metal dichalcogenides of the form MX_2, where M = transition metal and X = S, Se, or Te, are emerging as highly attractive candidates for the study of fundamental physics in 2D and in layered (thin-film) structures. These materials form layered structures, where layers of covalently bonded X–M–X atoms are held together by van der Waals interactions. But because of the broken symmetry in the atomic basis, they can have bandgaps of ~1 eV. Among these layered materials, recently, special emphasis has been given to MoS_2 due to its intriguing electrical and optical properties. While bulk MoS_2 is usually an n-type semiconductor with an indirect bandgap (~1.3 eV) (3) and carrier mobility in the 50–200 cm²/(V s) range at room temperature (4), single-layer MoS_2 is found to have a direct bandgap of ~1.8 eV (5, 6). Field-effect transistors (FETs) using single-layer MoS_2 exhibited high on/off ratios (~10^8) and low subthreshold swing (SS, ~70 mV/decade) (7). The electron mobility of single-layer MoS_2 FETs varied from ~1 cm²/(V s) (in air/MoS_2/SiO_2 structures) to ~200 cm²/(V s) (in HfO_2/MoS_2/SiO_2 structures) depending on dielectric environment. Thus, in addition to fundamental scientific interest, MoS_2 FETs can be an attractive alternative for electronic switches in the form of TFTs for high-resolution liquid crystal displays (LCDs) and organic light-emitting diode (OLED) displays.

While MoS_2 has been mainly studied for decades as a solid lubricant or a catalyst, not much is known in its electrical or electronic properties. Therefore, in this chapter, we first review materials properties of MoS_2, followed by recent literatures on MoS_2 transistors and technological challenges for the real application of MoS_2 transistors.

19.2 MATERIALS PROPERTIES OF MoS_2

19.2.1 Crystal Structures

Transition metal dichalcogenides are a class of materials of the form MX_2 where M is a transition metal element and X is a chalcogen element (S, Se, or Te). These materials form layered structures, where layers of covalently bonded X–M–X atoms are held together by van der Waals interactions (Fig. 19.1). They display a wide range of physical properties including superconductivity, half-metallic magnetism, semiconductors, and charge density wave depending on the coordination and oxidation state of the metal atoms (8). The stable metal dichalcogenides that can form stable 2D monolayer structure are shown in Figure 19.2. Transition metal atoms indicated by M are divided into $3d$, $4d$, and $5d$ groups. MX_2 compounds partially highlighted indicate that only some of the metal dichalcogenides form layered structures. The resulting structures can be semimetallic, metallic, or semiconducting with direct or indirect bandgaps.

Nanomaterials, Polymers and Devices: Materials Functionalization and Device Fabrication, First Edition. Edited by Eric S. W. Kong.
© 2015 John Wiley & Sons, Inc. Published 2015 by John Wiley & Sons, Inc.

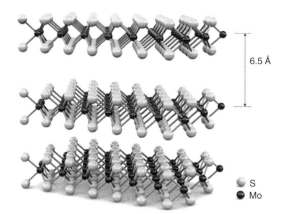

Figure 19.1   Three-dimensional depiction of MoS_2 crystal structure. (*Source*: Reprinted with permission from Reference (7), © 2011 Macmillan Publishers Ltd).

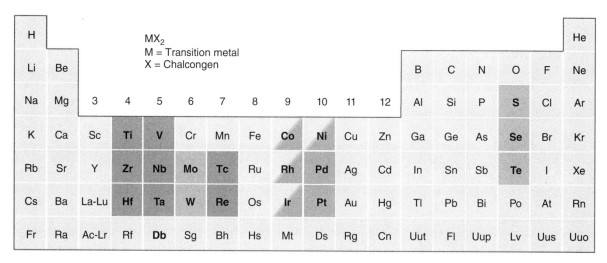

Figure 19.2   The transition metal elements and the three chalcogen elements that predominantly crystallize in layered structure (highlighted). Partial highlights indicate that only some of the dichalcogenides form layered structures. (*Source*: Reprinted with permission from Reference (8), © 2013 Macmillan Publishers Ltd).

For MoS_2, within a single layer, a Mo atom is coordinated with six S atoms either in an octahedral or a trigonal prismatic symmetry. The planar S–Mo–S sandwich layer is formed by face-sharing octahedrons or edge-sharing trigonal prisms. Due to the variation in stacking sequence, MoS_2 has three polytype crystal structures (Fig. 19.3) (9). For the trigonal prismatic crystal structures, the unit cell contains either two (2H; hexagonal symmetry) or three (3R; rhombohedral symmetry) repeating layers. 2H MoS_2 has an AbA–BaB stacking sequence, while 3R MoS_2 has an AbA–BcB–CaC stacking sequence. The lower and upper case letters represent Mo and S layers, respectively. Both 2H and 3R polytypes exhibit semiconducting behavior. For the octahedral coordination, the unit cell contains only one AbC (1 T; tetragonal symmetry) repeating layer. The Mo–S coordination for 1 T MoS_2 changes from trigonal prismatic to octahedral through simultaneous glide motion by the molybdenum and sulfur planes. The 1 T MoS_2 polytype exhibits metallic behavior because there is no bandgap (10). The structural transition significantly reduces the crystal symmetry from $P6_3/mmc$ to $Pm1$.

MoS_2 occurs naturally in appreciable quantities, and the naturally occurring MoS_2 has been found in 3R as well as with more common 2H polytype (11). However, 1 T MoS_2 is known to be metastable. The lattice constants are in the range of 3.1–3.7 Å for different materials and the interlayer spacing is ~6.5 Å.

19.2.2   Electronic Band Structures

The electronic properties of transition metal dichalcogenides exhibit a wide range: superconductors, charge density waves, metals, and semiconductors. This is due to the combination of crystal structures and splitting of hybridization between metal d orbitals and chalcogen p orbitals. Trigonal prismatic MoS_2(2H and 3R) has completely filled valence band and is semiconducting. But this can be modified by inserting foreign compounds into the interlayer spaces. For example, intercalating alkali or alkaline earth metals such as $Rb_{0.3}MoS_2$ induces superconductivity (12).

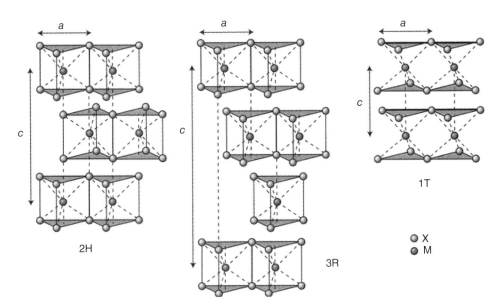

Figure 19.3 Schematics of MoS$_2$ polytypes: 2H (hexagonal symmetry), 3R (rhombohedral symmetry), and 1 T (tetragonal symmetry). (*Source*: Reprinted with permission from Reference (9), © 2012 Macmillan Publishers Ltd).

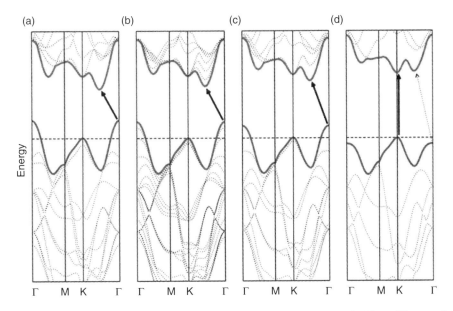

Figure 19.4 Calculated band structures of (a) bulk, (b) quadrilayer, (c) bilayer, and (d) monolayer MoS$_2$. The solid arrows indicate the lowest energy transitions. (*Source*: Reprinted with permission from Reference (5), © 2010 American Chemical Society).

The bulk MoS$_2$ crystal is an indirect bandgap semiconductor with a bandgap of ~1.3 eV. The bandgap increases with decreasing crystal thickness <100 nm due to quantum confinement. Moreover, when bulk MoS$_2$ is thinned to a monolayer, the nature of bandgap as well as its width changes from indirect to direct bandgap. Figure 19.4 shows the calculated band structures of bulk, quadrilayer, bilayer, and monolayer MoS$_2$ (5). The direct excitonic transition energy at K point remains relatively unchanged as the number of MoS$_2$ layers decreases. Interestingly, the indirect transition energy increases as the MoS$_2$ is thinned. It becomes even higher (~1.9 eV) than the direct transition energy for monolayer MoS$_2$ rendering monolayer MoS$_2$ a 2D direct bandgap semiconductor. Figure 19.5 shows the bandgap of MoS$_2$ with respect to the number of layers (6).

The quantum confinement and the resulting change in hybridization between p_z orbitals of S atoms and d orbitals of Mo atoms cause the change in the band structure with respect to layer number. According to density functional theory (DFT) calculations, localized d orbitals of Mo atoms mainly comprise the conduction band states at the K point (5). But the conduction band states near Γ point are due to combinations

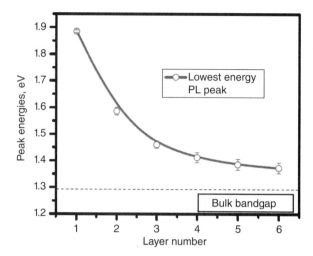

Figure 19.5 Bandgap energy of MoS_2 thin layers. The dashed line indicates the indirect bandgap energy of bulk MoS_2. (*Source*: Reprinted with permission from Reference (6), © 2010 American Physical Society).

of the antibonding p_z orbitals of S atoms and the d orbitals of Mo atoms and have a strong interlayer coupling effect. Hence, as the layer number decreases, the direct transition at K point is almost unchanged, but the transition at Γ point changes significantly. The transition from an indirect bandgap to a direct bandgap provides important implications for photonics, optoelectronics, and sensing.

19.2.3 Electrical Properties of Bulk MoS_2

The conduction of MoS_2 can be either p- or n-type although n-type conduction is common for naturally occurring MoS_2 single crystals (11). The electrical conductivity values parallel and perpendicular to the layers show a ratio of $\sim 10^3$: it is ~ 0.1–$1/\Omega \cdot cm$ in parallel with c-axis and $\sim 10^{-4}/\Omega \cdot cm$ perpendicular to c-axis (13). The two values show similar temperature behavior. The Hall effect measurements on MoS_2 indicate mobility of 50–200 $cm^2/(V\ s)$ at room temperature (4). Typical self-consistent values at room temperature are as follows (11):

Carrier concentration $(n) = 10^{16}$–$10^{17}\ cm^{-3}$,

Conductivity $(\sigma_{\perp c}) = 0.1$–$1\ \Omega \cdot cm^{-1}$,

Hall mobility $(\mu_e \approx \mu_h) \approx 100\ cm^2/(V\ s)$.

Although all the electrical properties of MoS_2 are very anisotropic, electrical measurement on MoS_2 single crystals are often difficult to obtain satisfactorily due to the layered structure (11). Because of the exceedingly soft crystal edges, any damage on a microscopic level leads to extensive screening of the anisotropic properties.

19.2.4 Optical and Vibrational Properties

As single-layer MoS_2 is a direct bandgap semiconductor, it is of great interest for applications in optoelectronics, which can generate, detect, interact with, or control light. Optoelectronic devices that are flexible and transparent are expected to become important in photovoltaic panels, wearable, and transparent electronics.

The changes in the nature and the width of MoS_2 bandgap are observed in changes in photoconductivity, absorption spectra, and photoluminescence. For instance, photoluminescence increases up to 10^4 times as MoS_2 changes from bulk to single layer (6). But the quantum yield of single-layer MoS_2 ($\sim 10^{-3}$) is much lower than that expected from a typical direct bandgap semiconductor (~ 1). As suspended MoS_2 and MoS_2 on hexagonal BN substrates show higher quantum yields (14), more work is needed to increase the quantum yield in the future.

The photoluminescence peak of suspended monolayer MoS_2 consists of a single narrow one centered at 1.9 eV (Fig. 19.6) (6). However, multilayer MoS_2 displays multiple peaks. Similarly, the optical absorption spectrum of bulk MoS_2 shows two main peaks corresponding to direct transitions at K point of Brillouin zone as well as one peak from indirect transition (Fig. 19.7) (6).

The calculated phonon dispersions of MoS_2 have been correlated with experimental Raman spectra showing anisotropic character of the vibrational modes. The main Raman peaks are from in-plane E^1_{2g} and E_{1u} phonon modes and the out-of-plane A_{1g} mode (Fig. 19.8) (15, 16). The frequencies, widths, and intensities of the Raman E^1_{2g} and A_{1g} peaks are strongly influenced by the thicknesses of the ultrathin flakes. As thickness decreases, the A_{1g} mode decreases in frequency, whereas the E^1_{2g} mode increases. But unlike the frequencies of E^1_{2g} and A_{1g} peaks, the widths and intensities vary arbitrarily (17). Hence, MoS_2 layer thickness can be identified based on these peak shifts in Raman spectra. The origins of the shifts are the neighboring layers affecting on the effective restoring forces on atoms and the increase of dielectric screening of long-range Coulomb interactions.

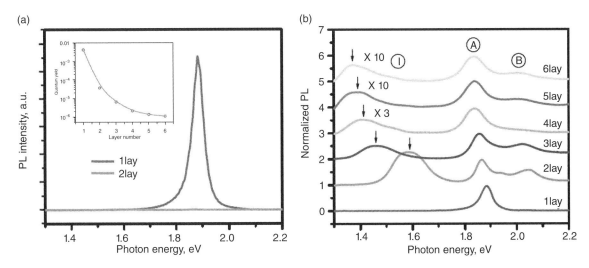

Figure 19.6   (a) Photoluminescence spectra for mono- and bilayer MoS$_2$. Inset: photoluminescence quantum yield of thin layers for N = 1–6. (b) Normalized photoluminescence spectra by the intensity peak A of thin layers of MoS$_2$. (*Source*: Reprinted with permission from Reference (6), © 2010 American Physical Society).

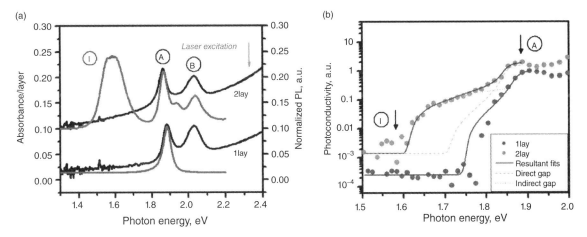

Figure 19.7   (a) Absorption spectra (left axis) and the corresponding photoluminescence spectra (right axis). (b) Photoconductivity spectra for mono- (red dots) and bilayer (green dots) MoS$_2$. (*Source*: Reprinted with permission from Reference (6), © 2010 American Physical Society).

19.3   PERFORMANCE OF MoS$_2$ THIN-FILM TRANSISTORS

19.3.1   Single-Layer Transistors

In 2005, Novoselov et al. obtained single-layer MoS$_2$ by mechanical exfoliation method and fabricated the first TFT based on single-layer MoS$_2$ (1). The field-effect mobility of the single-layer MoS$_2$ transistor was rather low (0.5–3 cm^2/(V s)) compared to the mobility of bulk MoS$_2$. However, in 2011, Radisavljevic et al. reported that the field-effect mobility of single-layer MoS$_2$ transistors could be increased up to ∼200 cm^2/(V s) if a high-k HfO$_2$ layer is deposited on top of the MoS$_2$ channel (Fig. 19.9) (7). They also reported outstanding transistor performances including high on/off current ratios (∼10^8) and low SS (∼70 mV/decade). Interestingly, they too obtained low field-effect mobility (0.1–10 cm^2/(V s)) without the high-k HfO$_2$ capping layer. They attributed the increase of the mobility to the suppression of Coulombic impurity scattering due to the high-k dielectric environment and modification of phonon dispersion in MoS$_2$ monolayers (Fig. 19.10) (18). Since then, the field-effect mobility of single-layer MoS$_2$ transistors has been reported up to ∼700 cm^2/(V s) (19, 20).

Regarding the origins of mobility boost, there have been contradictory reports. The calculated room temperature mobility of ∼410 cm^2/(V s) was found to be dominated by optical phonon scattering and was weakly dependent on the carrier density (21). In contrast, Yun et al. suggested that mobility in the single- or multilayer MoS$_2$ sheets was reduced mainly due to the decrease in carrier concentration and the increase in effective mass of the carriers on lowering the number of layers (22). Hence, a deeper understanding is needed on the microscopic picture of the transport in single- or multilayer MoS$_2$ devices.

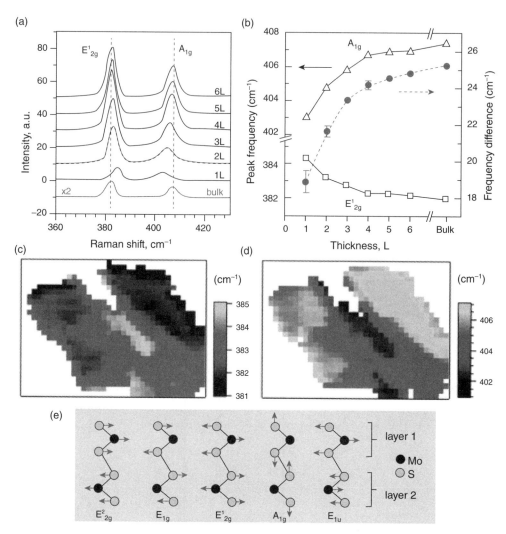

Figure 19.8 (a) Raman spectra of thin layer and bulk MoS_2. (b) Frequencies of E^1_{2g} and A_{1g} Raman modes (left axis) and their difference (right axis) as a function of layer thickness. (c, d) Spatial maps (23×10 μm) of the Raman frequency of E^1_{2g} (c) and A_{1g} (d) modes. (e) Four Raman- and an IR-(E_{1u}) vibrational modes of bulk MoS_2 crystals. (*Source*: Reprinted with permission from Reference (16), © 2010 American Chemical Society).

Another ongoing debate on the single-layer MoS_2 transistors is the estimation of the field-effect mobility using a double-gated structure. Fuhrer and Hone suggested that the consideration of the coupling between top-gated and back-gated capacitances in the measurement of Radisavljevic et al. (7) reduced the mobility in the range of 2–7 cm²/(V s) (Fig. 19.11) (23). Radisavljevic and Kis argued that the mobility of their devices became in the range of 15–55 cm²/(V s) even after considering the capacitance coupling (24). They claimed their original conclusion that the high-k HfO$_2$ layer enhances the mobility was still valid.

There have been projections on the ultimate performance limit of single-layer MoS_2 transistors by theoretical calculations. Yoon et al. reported that MoS_2 transistors may not be ideal for high-performance applications due to heavier electron effective mass ($m^* = 0.45\ m_0$) and a lower mobility than most of the III–V semiconductors (25). However, they also reported that MoS_2 transistors can be an attractive alternative for low standby and operating power applications as MoS_2 has a large bandgap and excellent electrostatic integrity inherent in a 2D system.

Ghatak et al. explored the nature of disorder and that of the electronic states in single- or multilayer MoS_2 transistors (Fig. 19.12) (26). Based on low-temperature electrical transport of MoS_2 transistors fabricated on SiO$_2$/Si substrates, they suggested that the transport occurs through variable range hopping rather than direct excitation to conduction band minima or mobility edge from the Fermi energy. They also suggested that the Coulomb potential from trapped charges in the substrate is the dominant source of disorder in single- or multilayer MoS_2 FETs, which also leads to carrier localization. Yet, understanding on the transport of single-layer MoS_2 transistors needs more work.

19.3.2 Multilayer Transistors

FETs using single-layer MoS_2 exhibited high on/off ratios ($\sim 10^8$) and low SS (~ 70 mV/decade). The electron mobility of single-layer MoS_2 FETs varied from ~ 1 cm²/(V s) (in air/MoS_2/SiO$_2$ structures) to ~ 200 cm²/(V s) (in HfO$_2$/MoS_2/SiO$_2$ structures) depending on dielectric

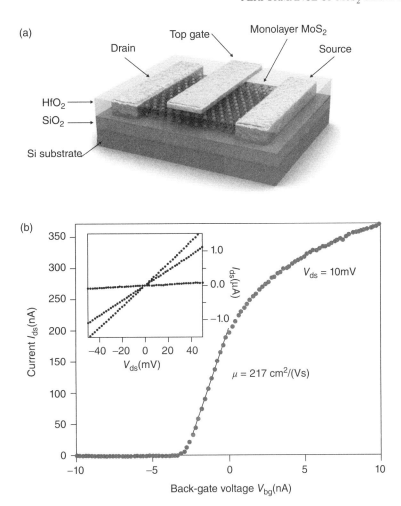

Figure 19.9 (a) Schematic view of single-layer MoS₂ transistor fabricated on top of an oxidized Si wafer. (b) Room temperature transfer characteristic of single-layer MoS₂ transistor with 10 mV applied bias voltage V_{ds}. Back-gated voltage V_{bg} is applied to the substrate and the top gate is disconnected. Inset: I_{ds}–V_{ds} curve acquired for V_{bg} values of 0, 1, and 5 V. (*Source*: Reprinted with permission from Reference (7), © 2011 Macmillan Publishers Ltd).

environment (7). Thus, in addition to fundamental scientific interest, single-layer MoS₂ FETs can be an attractive alternative for electronic switches in the form of TFTs for high-resolution LCDs and OLED displays. These devices have a critical need for high field-effect mobility (>30 cm²/(V s)), high on/off ratio (>10³), steep subthreshold slopes for low power consumption, and electrical and optical reliabilities (27). But the synthesis of single-layer MoS₂ followed by a deposition of an additional high-*k* dielectric layer may not be well suited for commercial fabrication processes. Based on the physics of MoS₂, there are a number of reasons why multilayer MoS₂ can be more attractive for TFT application than single-layer MoS₂. The density of states (DOS) of multilayer MoS₂ is three times that of single-layer MoS₂, which will lead to considerably high drive currents in the ballistic limit. In long-channel TFTs, multiple conducting channels can be created by field effect in multilayer MoS₂ for boosting the current drive of TFTs, similar to silicon-on-insulator metal-oxide-semiconductor field-effect transistors (MOSFETs).

However, multilayer MoS₂ and corresponding dichalcogenide semiconductors have been even less studied for use in electronics than single-layer semiconductors. Moreover, the characteristics in the few early reports are not vastly competitive with current TFT technologies. For example, in 2005, Ayari et al. reported field-effect mobility of multilayer MoS₂ (8–40 nm) ranged from 10 to 50 cm²/(V s) and an on/off current ratio higher than 10⁵ (28). The case of multilayer MoS₂ FETs was further explored in the previous publications by the authors, which showed that it offers a compelling case for applications in TFTs (29, 30). Hence, in this section, we reiterate our previous publications on the multilayer MoS₂ transistors.

19.3.2.1   Electrical Performance of Multilayer Transistors

19.3.2.1.1   Device Fabrication   TFTs based on multilayer MoS₂ were fabricated with the architecture shown in Figure 19.13a. After multilayer MoS₂ flakes were mechanically exfoliated from bulk MoS₂ crystals and transferred on atomic layer deposition (ALD)-Al₂O₃-covered Si substrates, electrical contacts of Au/Ti were patterned on top of MoS₂ flakes as described in reference (29). Optical micrograph of a completed

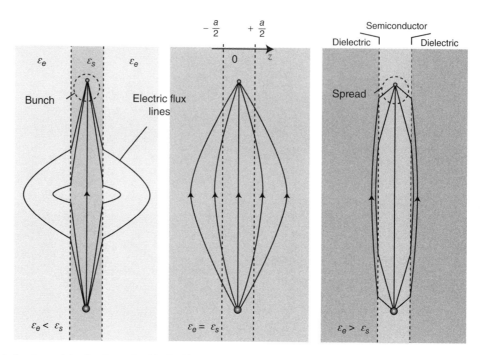

Figure 19.10 Electric flux lines originating from a fixed ionized impurity and terminating on a mobile electron, and the effect of the dielectric environment. The flux lines bunch closer inside the semiconductor layer if $\varepsilon_e < \varepsilon_s$, and spread farther apart if $\varepsilon_e < \varepsilon_s$, thus enhancing Coulomb interaction in the former case and damping it in the latter. (*Source*: Reprinted with permission from Reference (18), © 2007 American Physical Society).

device is shown is Figure 19.13b. The thickness of MoS_2 channels measured by atomic force microscope (AFM) was in the range of 20–80 nm. Between the two stacking polytypes that are observed in bulk MoS_2 crystals (hexagonal 2H and rhombohedral 3R), crystal orientation mapping by electron backscatter diffraction indicated only 2H MoS_2 within the measured devices as shown in Figure 19.13c. Subsequent inverse pole figure in Figure 19.13c confirmed that the MoS_2 channels are [0001]-oriented single crystals. Raman spectra of MoS_2 channels were almost identical with those of bulk single crystals as shown in Figure 19.13d, and no noticeable difference in Raman shifts of MoS_2 channels were found among measured devices, indicating minimal structural modifications.

19.3.2.1.2 Electronic Properties Figure 19.14 shows the measured device characteristics of a multilayer MoS_2 TFT. The thickness of the MoS_2 layer is $t_{ch} \sim 30$ nm, and the back-gated oxide thickness is $t_{ox} \sim 50$ nm (Fig. 19.14a). Figure 19.14b shows the major features observed: the *n*-type nature of the MoS_2 layer as indicated by it turning on at positive gate biases due to accumulation of electrons, and a window of gate biases where the device stays off (depletion). This feature was observed in *all* measured devices, but a fraction of the devices also showed a recovery of the current at large negative gate biases as in Figure 19.14b, which is a clear indication of an inversion channel. We note that the gate capacitance in this geometry is ~ 20 times higher than a recent report (28). As a result, while exhibiting the high on/off ratio expected of a semiconductor with a bandgap of 1.3 eV, a sharp SS (~ 80 mV/decade for Fig. 19.14; ~ 70 mV/decade for some devices) is also measured at room temperature in deep depletion. In a typical field-effect geometry, the subthreshold slope is given by (31) $SS = \left(1 + \frac{C_S + C_{it}}{C_{ox}}\right)\frac{kT}{q}ln10$, where C_S is the capacitance in the MoS_2 conducting channel, $C_{it} = qD_{it}$ is the capacitance due to interface traps of density D_{it}, and $C_{ox} = \varepsilon_{ox}/t_{ox}$ is the oxide capacitance. Based on this model, $(C_S + C_{it}) \sim C_{ox}/3$. Since the semiconductor capacitance is negligible in the deep subthreshold region, the interface trap density is $D_{it} \sim 2.6 \times 10^{11} (eV \cdot cm^2)^{-1}$, a very low value indeed (32). The SS value was measured to be ~ 24 mV/decade at 77 K, indicating D_{it} does not vary with temperature. We note that similar to the Radisavljevic et al.'s report (7) for monolayer MoS_2, the subthreshold slope (and D_{it}) of multilayer MoS_2 TFTs is also exceptionally low, comparable to state-of-the-art silicon transistors. But it is obtained without the decades of processing improvement that was necessary to reduce interface trap densities between the dielectric insulator and silicon. Currently III–V semiconductors are facing the same challenge (33). This is because of the layered nature of the semiconductor – for 2D crystals and their stacks, there are no out-of-plane broken bonds, and thus the interface traps are expected to be in the dielectrics and materials placed in contact with them, not in the semiconductor itself. This is a major advantage of MoS_2 for TFT applications.

19.3.2.1.3 Current Saturation The second major boost is seen in Figure 19.14c. The drain current is observed to saturate at high drain biases for all gate voltages. The saturation occurs over a wide drain voltage window (unlike graphene). Current saturation in transistors is an important feature toward real applications as the TFTs in OLED displays are operated in the saturation region. Like in a long-channel transistor made of a covalent semiconductor, the saturation of current occurs in the MoS_2 TFT due to pinch-off of the conducting channel

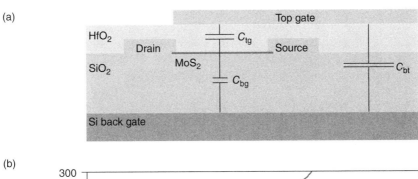

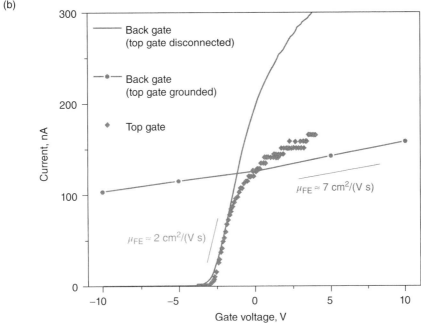

Figure 19.11   Field-effect mobility measurements in MoS$_2$ dual-gated transistors. (a) Device schematic showing capacitances. Drawing is not to scale. (b) Drain current versus gate voltage for a single-layer MoS$_2$ transistor with gate voltage applied to the back gate with the top gate disconnected, back gate with the top gate grounded, or top gate. Drain voltage is 10 mV in all cases. (*Source*: Reprinted with permission from Reference (23), © 2013 Macmillan Publishers Ltd).

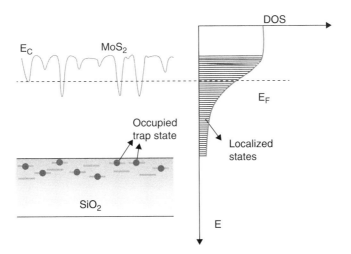

Figure 19.12   Schematic representation of the fluctuations in the conduction band of MoS$_2$ arising due to the proximity of the trapped charges at the SiO$_2$/MoS$_2$ interface (left) leading to the band tail and localized states (right). (*Source*: Reprinted with permission from Reference 27, © 2011 American Chemical Society).

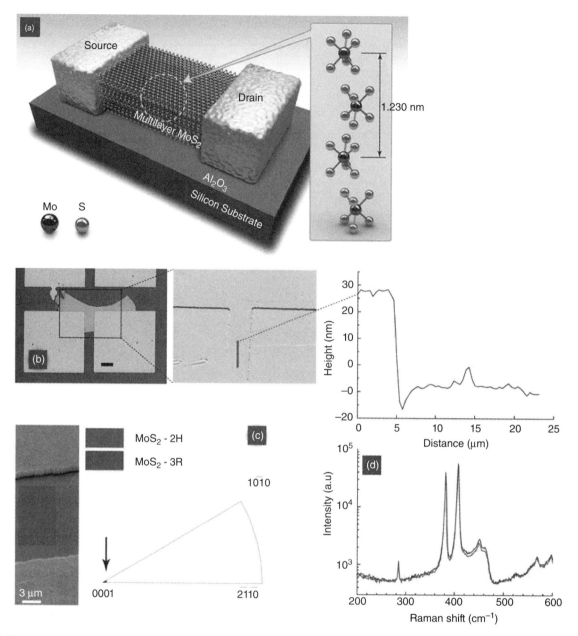

Figure 19.13 (a) Schematic perspective view of a MoS$_2$ TFT with a multilayer MoS$_2$ crystal. (b) Optical and AFM image of a device deposited on top of a silicon substrate with a 50-nm-thick Al$_2$O$_3$ layer. Also shown is a cross-sectional plot along the red line in AFM image. (c) Scanning electron microscope image of the MoS$_2$ channel with a crystal orientation mapping. Also displayed is a partial inverse pole figure indicating a <0001>-oriented single crystal. (d) Raman spectroscopy measurements on a bulk single crystal and a transistor channel. (29)

at the drain side as the gate-drain diode becomes reverse biased at high V_{DS}. Because graphene has zero bandgap, instead of pinch-off, the drain side of the conducting channel becomes *p*-type at high drain bias, restricting current saturation *and* current modulation to a very small window, if at all. The bandgap of MoS$_2$ makes both current modulation and saturation robust, as borne out by Figure 19.14b and c.

The saturation of current observed here is quantitatively understood based on a long-channel device model based on surface potential. Poisson equation is solved to determine the surface potential at the MoS$_2$/ALD oxide interface as a function of the gate and the drain bias voltages. Then, the Shockley model of transistor performance is used to calculate the current. The only unknown input parameters to the model are – the carrier mobility, the doping density, and the contact resistance. At high positive gate biases, the channel is flooded with accumulated carriers and is highly conductive, and the contact resistances limit the current. This helps us to deduce the contact resistance to be ~17 kΩ as described in the supplementary section of Reference 30. This value is rather high, and reducing it by an order of magnitude is necessary for the future.

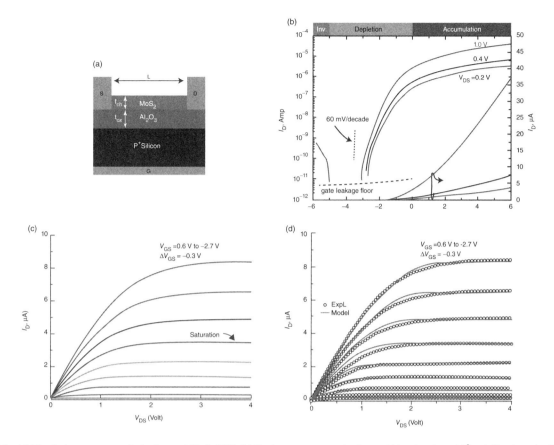

Figure 19.14   (a) The device geometry of a back-gated MoS$_2$ TFT. (b) Drain current versus back-gated bias showing ~10^6 on/off ratio and ~80 mV/decade subthreshold slope. (c) Drain current versus drain bias showing current saturation. (d) Same as (c), including a long-channel model (red lines) showing excellent agreement between the TFT model and measured device behavior. (29)

Thus, the mobility and the doping density remain as unknown parameters. In Figure 19.14d, using a mobility of 100 cm^2/(V s) and a doping density of $N_D \sim 10^{16}$cm^{-3}, an excellent fit to the measured device characteristics is obtained. We assert that these values are not fortuitous, as borne out by complementary capacitance–voltage and field-effect mobility measurements, which are described next. The device model for TFTs provides insight into the performance of 2D-layered semiconductor devices and thus is a powerful tool to extract physical parameters of the material.

19.3.2.1.4 Accumulation and Inversion Channels   At large negative gate biases, the drain current recovers as seen in Figure 19.14b, indicating the formation of an inversion channel (formation of a hole gas in a *n*-type semiconductor). However, the source/drain contacts are formed to the conduction band, and therefore there exists a large barrier for conduction through a *p*-type inversion channel. To explore these features quantitatively, capacitance–voltage measurements were performed, and the measured device characteristics were compared to energy-band diagram-based models.

Figure 19.15a–c shows the calculated energy-band diagrams for the MoS$_2$ TFT structure for various gate bias conditions. For the calculation, we self-consistently solve Poisson and Schrödinger equations in the effective-mass approximation. On the one hand, the bandgap of single-layer MoS$_2$ occurs at the K-points in the hexagonal k-space lattice (6), implying a valley degeneracy of $g_v = 2$, similar to graphene. On the other hand, the conduction band minimum of multilayer MoS$_2$ moves to a lower symmetry point in the k-space along the T-K line. This results in a higher valley degeneracy ($g_v = 6$ for the T-K line) than single-layer MoS$_2$, effectively *tripling* the DOS, implying higher carrier densities and higher currents in the ballistic limit (34). The net drive current for a given voltage is a product of the carrier density and the velocity. Thus, in addition to high velocity, a high DOS is equally attractive for attaining high speed. Driven by higher valley degeneracy, multilayer MoS$_2$ has the potential for considerably higher current drives than single-layer MoS$_2$ in the ultrascaled limit, and high charge densities have recently been reported (35). Even in the long-channel structure, thin-film MoS$_2$ can take advantage of its multilayer nature. It can provide multiple conducting channels for boosting the current drive by using double gates, similar to silicon-on-insulator MOSFETs. The semiconducting materials properties of multilayer MoS$_2$ such as the conduction and valence band offsets with Si and ALD Al$_2$O$_3$, the valley degeneracy, band-edge effective masses, dielectric constant, and bandgap were used in the calculation (29). The model indicates that due to the work-function difference between MoS$_2$ and p^+-Si, the thin MoS$_2$ layer is initially depleted of mobile carriers. Upon application

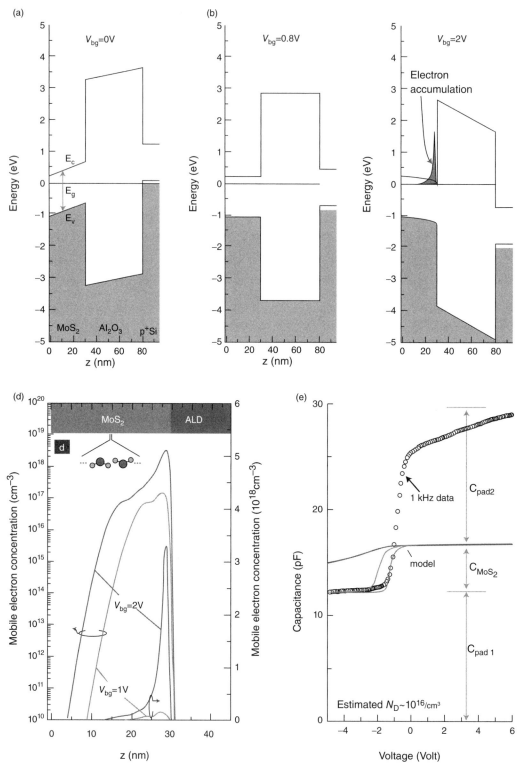

Figure 19.15   (a–c) Energy-band diagrams of the $MoS_2/Al_2O_3/p^+Si$ device under various bias conditions. The band offsets and physical parameters relevant for the calculation are described in the supplementary information. The self-consistent Schrödinger-Poisson calculation shows that at large positive gate bias, a 2D electron gas is formed at the MoS_2/Al_2O_3 interface. (d) The 2D electron gas that forms the conductive channel is shown in an enlarged scale. Most conduction occurs by electrons accumulated in a few layers at the MoS_2/Al_2O_3 interface. With increasing bias the centroid of the electron distribution shifts closer to the interface. This indicates that the "quantum" capacitance in the semiconductor increases with positive gate bias. (e) Measured capacitance–voltage curves of the MoS_2/Al_2O_3/back-gated capacitors (circles). The solid lines show the calculated capacitance for three different doping densities, not including parasitic pad capacitances. The slope indicates a doping density close to 10^{16} cm^{-3}. (29)

of positive bias on the Si gate, a 2D electron gas in the MoS$_2$ layers closest to the ALD Al$_2$O$_3$ forms. This accumulation channel conducts current between the source and the drain. Figure 19.15d shows the accumulation carrier density profile in more detail. Most of the carriers are electrostatically confined close to the MoS$_2$/ALD interface, similar to the case in a silicon-on-insulator MOSFET. Thus, the "quantum capacitance," which dictates the voltage drop in the semiconductor to sustain the conducting charge $C_S \sim \varepsilon_S / \langle z \rangle$, is large since the centroid of the charge distribution $\langle z \rangle$ is ~1–2 nm from the interface.

19.3.2.1.5 Capacitance Measurements   Figure 19.15e shows the measured two-terminal capacitance as a function of the voltage between an ohmic contact pad to the MoS$_2$ and the back gate. Several interesting features are evident from the measurement. Since the pads are large as seen in Figure 19.13b, they form a parasitic pad capacitance, which sets the floor of the measured value (~12 pF). When the MoS$_2$ layer is depleted of mobile carriers, the measured capacitance is this pad capacitance. As positive gate biases are applied, the formation of an electron accumulation layer in the MoS$_2$ results in an increase in the capacitance. The electron accumulation layer also electrically connects the two ohmic pads, effectively doubling the parasitic pad capacitance – this is exactly what is measured. The capacitance of the MoS$_2$ layer alone (C_{MoS_2}) is dependent on its doping density. The calculated values of C_{MoS_2} for three different doping densities are shown in Figure 19.15e, from which it is concluded that the unintentional doping density in the measured MoS$_2$ layer is of the order of $N_D \sim 10^{16}$ cm^{-3}. We note that this is a low doping level, and can vary between naturally occurring samples not grown by controlled means. The extracted doping density has a direct impact on impurity scattering and carrier mobility, which is discussed in the next section.

19.3.2.1.6 Charge Transport and Scattering   The field-effect mobility of MoS$_2$ TFTs was extracted from the I_D–V_{GS} curves of Figure 19.14b, and the corresponding measurements for temperature 77–300 K. The mobility values are extracted in the linear region at $V_{GS} = 2$ V or equivalently at a carrier density $n = C_{ox} (V_{GS}-V_T) \approx 1.6 \times 10^{12}$ cm^{-2}. The intrinsic carrier mobility (mobility without the effect of the contact resistance R_c) was calculated using an equivalent circuit as shown in Figure 19.16c and are shown as blank circles in Figure 19.16a. Also shown are data reported from Fivaz and Mooser's work (36) from 1967. The values are very similar. We note here that

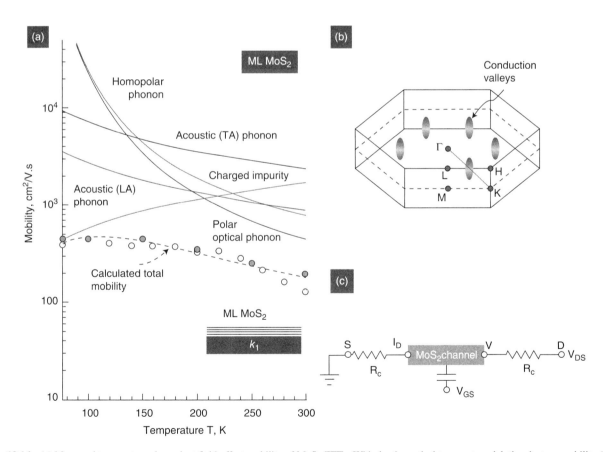

Figure 19.16   (a) Measured temperature-dependent field-effect mobility of MoS$_2$ TFTs. With the theoretical transport model, the electron mobility (dashed line) is limited by impurity scattering at low temperatures. At room temperature, the mobility is limited by the combined effect of the homopolar (out-of-plane) phonon and the polar optical phonon scattering. Details of these scattering mechanisms are described in the supplementary information. (b) The hexagonal Brillouin zone of multilayer MoS$_2$ with the high symmetry points and six equivalent conduction valleys. (c) An equivalent circuit model for the MoS$_2$ TFT including the effect of the contact resistance R_c. (29).

Fivaz and Mooser used MoS_2 crystals grown by transport reaction as opposed to the naturally occurring MoS_2 used in this work, which bodes well for large-area growth for practical TFT applications. The decrease in mobility with temperature is a typical signature of diffusive band transport, as opposed to activated (hopping) transport. If there were significant Schottky barrier heights, the mobility would appear to increase with temperature. This can lead to erroneous conclusions on the nature of charge transport (i.e., activated vs band transport). To avoid such confusion, samples that exhibited ohmic contacts over the entire temperature range were carefully chosen for mobility extraction.

To explain the temperature-dependent carrier mobility, the semiclassical Boltzmann transport equation under the relaxation time approximation is used. A typical characteristic of layered structures (such as MoS_2) is that the carriers move independently in each layer. Neglecting the vanishing interlayer interaction, the energy dispersion of carriers becomes (37) $E(k) = \hbar^2(k_x^2 + k_y^2)/2m^*$, where $k = (k_x, k_y)$ is the 2D wave vector of carriers and m^* is the in-plane effective mass. The corresponding 2D DOS is $g_{2D} = g_v m^*/\pi\hbar^2$, where $\hbar$ is the reduced Planck constant and g_v is the valley degeneracy. Recent experiments and models indicate the conduction band minima to be along the T-K line (38, 39) of the Brillouin zone, as indicated in Figure 19.16b. The sixfold symmetry of this point leads to a valley degeneracy of $g_v = 6$. We have used this value for transport calculations, consistent with the energy-band diagram and capacitance calculations of Figure 19.15. Carrier scattering from (a) ionized impurities, (b) acoustic phonons, (c) in-plane polar optical phonons, and (d) out-of-plane lattice vibrations (homopolar phonons) are taken into account to explain the transport measurements. A 2D ionized impurity scattering model is invoked where carriers scatter from a sheet of impurities located at the surface of the MoS_2 layers. Coupling of carrier with both longitudinal acoustic (LA) phonons and transverse acoustic (TA) phonons are taken into account under the deformation potential approximation. The energies of optical phonons in bulk MoS_2 are in the frequency range of 400–500 cm^{-1} (50–60 meV) (40) as also measured in Figure 19.13d. Electron-polar optical phonon scattering is described by the Fröhlich interaction (41) with a static dielectric constant $\varepsilon_0 = 7.6$ and optical dielectric constant (42) $\varepsilon_\infty = 7.0$ with phonon energy $\hbar\omega_{op} = 49$ meV for the E_{2g}^1 mode. For homopolar phonon modes, the sulfur atoms of opposite planes vibrate out of phase (A_{1g} mode) and the corresponding phonon energy is $\hbar\omega_{hp} = 52$ meV (16, 40).

The resultant mobility is calculated using Matthiessen's rule $\mu^{-1} = \mu_{imp}^{-1} + \mu_{LA}^{-1} + \mu_{TA}^{-1} + \mu_{op}^{-1} + \mu_{hp}^{-1}$. The calculated mobility associated with the individual scattering mechanisms as well as the resultant mobility is shown in Figure 19.16a along with the measured data. At low temperatures, the mobility is limited by ionized impurity scattering. At room temperature, the mobility decreases by enhanced optical phonon and acoustic phonon scattering. A reasonably good match is found between calculated and experimentally measured mobility (at low temperatures) for a choice of impurity sheet density $n_{imp} \sim 1.8 \times 10^{10}$ cm^{-2}, which corresponds to a volume density of $\sim 3 \times 10^{15}$ cm^{-3}. This value is comparable to the background doping density for unintentionally doped bulk MoS_2, and consistent with the value estimated from the capacitance–voltage measurements in Figure 19.15e.

A large electron effective mass and the strong optical phonon (out-of-plane and polar phonon) scattering set an upper bound on the mobility in multilayer MoS_2. We note here that the accuracy of the calculation is subject to the uncertainty in the electron–phonon coupling coefficients and the band structure parameters used in the model. However, these parameters are not expected to be vastly different from those assumed. Higher mobilities could potentially be achieved by intercalation of MoS_2 layers (similar to recently demonstrated encapsulations of silicon carbide crystals in graphite enclosure (43)). By sandwiching single-layer MoS_2 between two dielectric layers, the out-of-plane phonon vibrations can be suppressed. If the homopolar phonon mode is damped, ionized impurity scattering and in-plane polar optical phonon scattering determine the charge transport. Strain effects can also potentially be used to deform the band structure, leading to lowering of the electron effective mass and improvement in mobility. Electron mobility in MoS_2 can be further expected to improve as the growth and processing methods improve, leading to fewer impurities. Moreover, charged impurity scattering in these devices can be damped using high-k dielectrics (dielectric engineering) (18, 19, 44). Effects such as remote phonon scattering could limit this improvement (45). Since the current experimental values are far below limits expected of remote phonon scattering, there is ample room for improvement.

19.3.2.1.7 Discussion Nonetheless, the mobility measured for multilayer MoS_2 already exceeds most competing semiconductor materials for large-area TFTs by orders of magnitude. The values may be further improved by proper dielectric choices to near the intrinsic phonon limits. When combined with the large current modulation, the low subthreshold slope, and robust current saturation, multilayer MoS_2 makes a compelling case for TFT applications. All these properties are achieved in a back-gated structure, without the need for an additional dielectric layer on top, which is highly attractive for TFT implementation. The first demonstration of these attractive properties combined with the comprehensive modeling of the behavior is expected to move multilayer MoS_2 toward real applications. The multilayer structure is easier to achieve over large areas by chemical vapor deposition or allied techniques, which are well suited for large-area applications. Looking beyond MoS_2, other transition metal dichalcogenides can offer competitive or complementary features. In addition to technological applications, conduction band electron states in these layered semiconductors have contributions from d orbitals, quite unlike traditional group IV and III–V semiconductors and carbon nanomaterials, where chemical bonding is restricted to s and p orbitals. Thus, a rich range of physical phenomena that depend on d orbitals, such as magnetism, correlated-electron effects, and superconductivity, can be expected in these materials. Such features can possibly be integrated with the semiconducting properties demonstrated here seamlessly, since there are no out-of-plane bonds to be broken.

19.3.2.2 Optoelectronic Performance of Multilayer Transistors

19.3.2.2.1 Optical Absorption Before fabricating the MoS_2 phototransistors, we first measured the optical absorption in multilayer MoS_2 flakes with different thicknesses across the visible and near-infrared (NIR) spectral ranges (Fig. 19.17a). The thicknesses of MoS_2 flakes measured by a tapping mode atomic force microscopy (AFM) are ~ 40 nm, ~ 4 nm, and ~ 1 nm, in sequence. Regardless of their thicknesses, all of the MoS_2 flakes show two excitonic absorption peaks – "A" and "B" – between 600 and 700 nm arising from the K point of the Brillouin zone (5, 6). Their energy differences are due to the spin-orbital splitting of the valence band, as indicated in Figure 19.17b. For the thick MoS_2

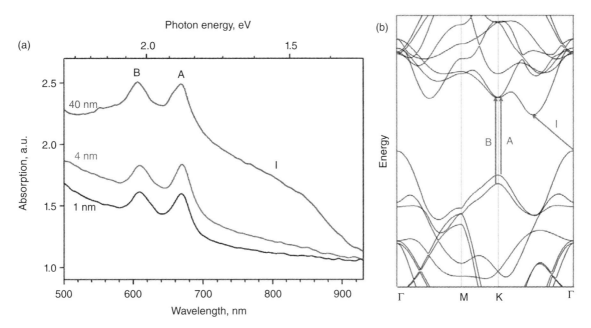

Figure 19.17 (a) Absorption spectra of MoS₂ crystals with three different thicknesses. Absorption peaks A and B correspond to the direct band transitions with the energy split from the valence band spin-orbital coupling. Broad absorption tail "I" corresponds to the indirect band transition. As the MoS₂ crystal becomes thinner, this tail becomes weaker. (b) Band structure of bulk MoS₂. Direct band transitions A and B occur at K point. Indirect band transition I occurs between the valence band maximum at Γ point and the conduction band minimum. (*Source*: Reprinted with permission from Reference (30), © 2012 Wiley).

flake (40 nm), an optical absorption tail − labeled "I" − is observed through the indirect band transition at a wavelength longer than ∼700 nm. However, as the thickness of MoS₂ flakes approached 1 nm, this absorption tail becomes weaker. This observation is consistent with the fact that single-layer MoS₂ is a direct bandgap semiconductor where the lowest energy interband transition occurs at the K point of the Brillouin zone (5, 6).

19.3.2.2.2 Electrical Transport Properties Multilayer MoS₂ TFTs are fabricated as shown in Figure 19.18a and their electrical transport properties are measured in a back-gated structure at room temperature. The thicknesses of MoS₂ flakes are in the range of 10–60 nm. Figure 19.18c and d shows the transistor characteristics for multilayer MoS₂ TFTs (gate length ∼3 μm, width ∼7 μm, and thickness of MoS₂ channel ∼20 nm) with a 50-nm-thick ALD Al₂O₃ gate insulator. The integration of high-κ dielectrics allows lower power consumption than SiO₂ (κ = 3.9) due to increased gate capacitance and dielectric screening of Coulomb scattering (18). The measured current–voltage (I–V) behavior shows good agreement with a conventional long-channel NMOS transistor, exhibiting a linear triode regime at low drain voltages (V_{ds}) and a saturation regime at high V_{ds}. Unlike the reported 2D-crystal-based electronics, such as pristine graphene FETs (46, 47) and single-layer MoS₂ TFTs (48), our transistors exhibit a current saturation at high V_{ds}. Saturation current at the "pinched-off" condition is independent of V_{ds} and is operated by the gate voltage according to the CMOS square-law model, in which the high output resistance in the saturation regime is a key factor to achieve a high voltage gain and to isolate output from the input signal in digital circuits. The representative multilayer MoS₂ TFT exhibits a maximum transconductance ($g_m = dI_d/dV_{gs}|_{Vds = 0.2\ V}$) of 3.12 μS, and an I_{on}/I_{off} of ∼1 × 10⁶. Note that the leakage current through the back gate in the operating regime is lower than the drain current by at least eight orders of magnitude. Based on the standard model of MOSFETs and a parallel plate model of gate capacitance, the field-effect mobility ($\mu_{eff} = Lg_m/(WC_{ox}V_{ds})$) extracted from our experimental transfer curves is >70 cm²/(V s) in the linear regime (V_{ds}= 0.2 V). Interestingly, our multilayer MoS₂ transistors provide a higher μ_{eff} than those reported in conventional TFTs that are based on amorphous Si, low-temperature poly-Si, or amorphous oxide semiconductors (49).

19.3.2.2.3 Optoelectronic Properties Figure 19.19 shows the optoelectronic behavior of our multilayer MoS₂ phototransistors in the dark and under incident light with a schematic energy-band diagram illustrating the photogeneration process of the electron–hole pairs. When a tungsten lamp with λ = 630 nm and an intensity of 50 mW/cm² is illuminated on the MoS₂ channel at V_{ds}= 1 V, we observe a 10³-fold increase in I_d in the "OFF-state"; yet, the accumulation current in the "ON-state" is independent of the incident light (Fig. 19.19a). Such opposing behavior can be explained by combining the dominant carrier transport mechanisms in the two distinct regimes: (i) photogenerated current, which dominates the depletion regime, and (ii) thermionic emission and tunneling current, which dominate the accumulation regime. Figure 19.19b shows an energy-band diagram of a multilayer MoS₂ phototransistor with a Schottky barrier. Under equilibrium conditions, the Schottky barrier (Φ_B) between the Ti/Au electrodes and the n-type semiconducting MoS₂ channel can be expressed as $\Phi_B = \Phi_M - \chi$, where Φ_M is the Ti/Au metal work function and χ is the electron affinity of MoS₂. The schematic band diagram in the OFF-state under light

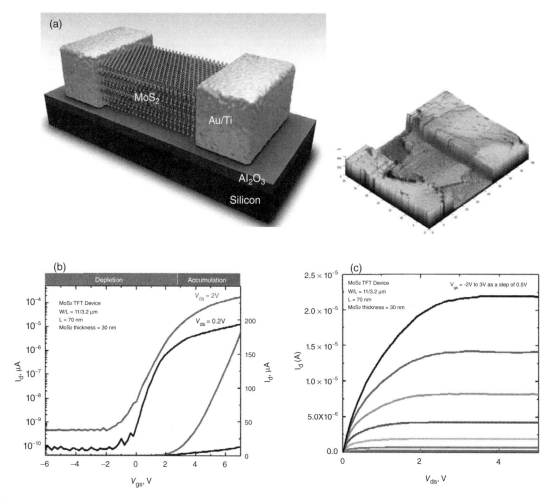

Figure 19.18 (a) Cross-sectional view and atomic force microscopy of multilayer MoS_2 TFTs consisting of an ALD Al_2O_3 gate insulator (50 nm), patterned Au electrodes (300 nm), and multilayer MoS_2 (thickness ~60 nm) as an active channel. (b) I–V characteristics of the multilayer MoS_2 (thickness ~35 nm) transistor with a gate length of 3.2 μm and MoS_2 width of 11 μm. The I_d–V_{gs} curves were measured under V_{ds} = 200 mV and 2 V. (c) I_d–V_{ds} curves recorded for various back-gated voltages with a step of 0.5 V. (*Source*: Reprinted with permission from Reference (30), © 2012 Wiley).

illumination depicts the photogeneration of electron–hole pairs by the absorption of light inside the MoS_2. However, the schematic band diagram in the ON-state during accumulation ($V_{gs} > 0$) under light illumination shows the dominating effects of the thermionic and tunneling currents, and the negligible contribution of the photogenerated current.

Figure 19.20a schematically shows the carrier profile along the MoS_2 channel in the saturation regime of the transistor. The drain voltage primarily controls the carrier profile close to the drain region, which pinches off the channel at high source/drain voltages. This process leads to current saturation, as shown in Figure 19.18c. When light is illuminated on the MoS_2 channel, carriers (both electrons and holes) are generated due to the band-to-band transition in addition to the electrons accumulated by the gate voltage. These photogenerated carriers modify the carrier profile along the channel particularly at the drain side where the carrier density is vanishingly small before illumination, as shown in Figure 19.20b. The photogenerated electrons and holes move in opposite directions under the high source/drain electric field, leading to a generation current (I_G) in addition to the dark current. Hence, if P_{in} is the light power incident on the surface of the MoS_2 film, the residual power at a distance x from the surface is given by $P(x) = P_{in}e^{-\alpha x}$, where α is the absorption coefficient of the MoS_2 film at the incident photon energy. The amount of power absorbed by a slab of MoS_2 with thickness Δx at a distance x from the surface is $dR_a = -(dP/dx)\Delta x$. Then, the total power absorbed by the MoS_2 film of thickness d is $R_a = P_{in}(1-e^{-\alpha d})$. For $\alpha d1$, the absorbed power can be written as $R_a = P_{in}\alpha d$. Note that for a MoS_2 film with a thickness of 30 nm and an absorption coefficient of $\alpha = 2 \times 10^5$ cm^{-1} (42), only 60% of the incident power is absorbed. If $h\nu$ is the energy of an incident photon, the number of electron–hole pairs generated per unit time per unit area is $G = R_a/h\nu$, where it is assumed that a single photon generates only one electron–hole pair. Defining τ as the carrier lifetime, the number of excess electron and hole generated per unit area is $\Delta n = \Delta p = G\tau$. Thus, the current due to these photogenerated carriers is $I_G = 2\Delta ne\mu(W/L)V_{ds}$, where e is the electronic charge, μ is the carrier mobility (assuming an identical value for electrons and holes), W is the device width, L is the device length, and V_{ds} is the applied source/drain voltage. The total drain current under illumination is therefore

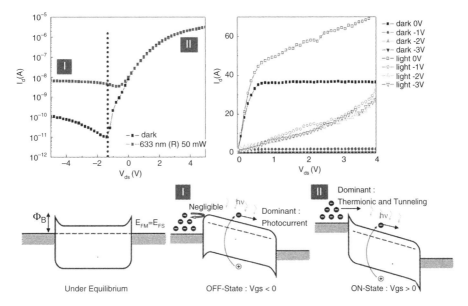

Figure 19.19 (a) Comparison of the I–V characteristics of an MoS$_2$ phototransistor under dark and light illumination conditions (λ_{ex} = 630 nm and power ~50 mW/cm^2). b) Energy-band diagram of a multilayer MoS$_2$ phototransistor with a Schottky barrier: Under equilibrium conditions, a Schottky barrier (Φ_B) between Ti/Au electrodes and an *n*-type semiconducting MoS$_2$ channel can be expressed as $\Phi_B = \Phi_M - \chi$, where χ is the electron affinity of MoS$_2$ and Φ_M is the Ti/Au metal work function. (i) Schematic OFF-state band diagram under light illumination, depicting the photogeneration of electron–hole pairs by the absorption of light inside MoS$_2$. (ii) Schematic ON-state band diagram in accumulation ($V_{gs} > 0$) with light. Photocurrent generated by light is negligible as thermionic and tunneling currents dominate channel current in the accumulation regime. (*Source*: Reprinted with permission from Reference (30), © 2012 Wiley).

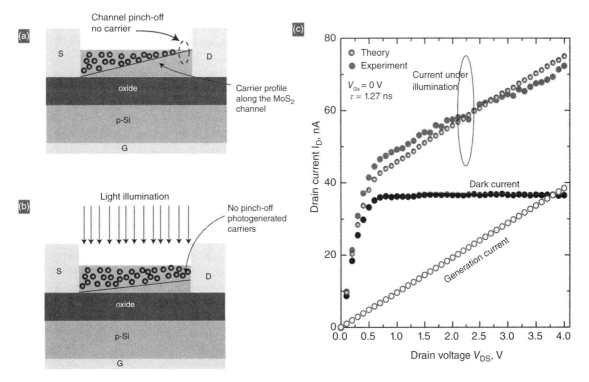

Figure 19.20 Schematic illustration of the carrier profile along the channel MoS$_2$ in the saturation regime (a) before and (b) after light illumination. Note that, due to photogenerated carriers after illumination, the channel does not pinch off at the drain side. (c) Comparison between the theoretical model and the experimental measurements for the drain current under illumination. (*Source*: Reprinted with permission from Reference (30), © 2012 Wiley).

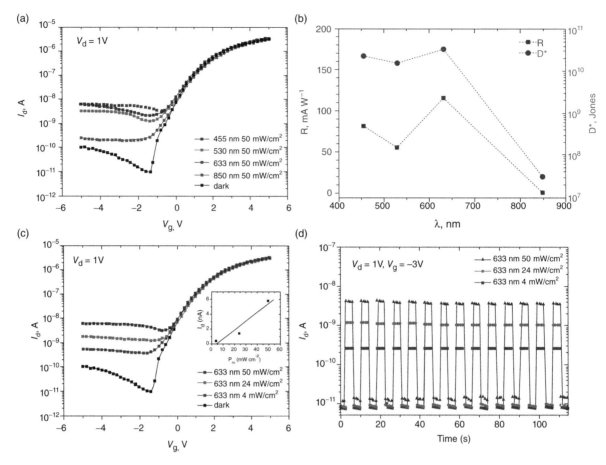

Figure 19.21 (a) Transfer characteristics of the phototransistor at different wavelengths. (b) Calculated responsivity and specific detectivity at different wavelengths. (c) Transfer characteristics of the phototransistor under visible light (633 nm) for different light intensities. Inset shows photocurrent response to light illumination (633 nm) for different light intensities. (d) Photocurrent as a function of light intensity at a wavelength of 633 nm. (*Source*: Reprinted with permission from Reference (30), © 2012 Wiley).

$I_D = I_D(\text{dark}) + I_G$. By comparing the experimentally measured current under illumination with the theory, we determine the carrier lifetime τ to be 1.27 ns, as shown in Figure 19.20c. Note that the photogenerated current is not only proportional to V_{ds} (as in Fig. 19.20c), but it also varies linearly with the incident power, which agrees well with the experimental results (see inset of Fig. 19.21c).

As a next step, we also measure the dark currents and photocurrents of the MoS$_2$ phototransistor across a wide range of wavelengths and powers (Fig. 19.21). In Figure 19.21a, illuminating the phototransistor with monochromatic visible light (455 nm, 530 nm, and 633 nm) at a power density of 50 mW/cm^2 increases the current up to almost three orders of magnitudes at an OFF-state gate bias. Under an infrared light (850 nm), a significantly higher power density (2.3 W/cm^2) is needed to increase the current by an order of magnitude at the same gate bias. This low sensitivity for infrared light is related to the weak absorption tail of the indirect bandgap semiconductor MoS$_2$ at the wavelength of 850 nm. The performance of the phototransistor as a photodetector can be evaluated by its figures of merit such as responsivity (R) and specific detectivity (D^*) (50). Responsivity is a measure of the electrical response to light and is given by $R = I_{ph}/P_{in}$, where I_{ph} is the photocurrent flowing in a detector and P_{in} is the incident optical power. Specific detectivity is a measure of detector sensitivity and, assuming that shot noise from dark current is the major contributor to the total noise, it is given by $D^* = RA^{1/2}/(2eI_d)^{1/2}$, where R is the responsivity, A is the area of the detector, e is the unit charge, and I_d is dark current (51). Figure 19.21b shows the calculated R and D^* of the phototransistor at different wavelengths. For visible light, R and D^* exist in the range of 50–120 mA/W and 10^{10}–10^{11} Jones, respectively. However, the R and D^* of infrared are significantly reduced to 9×10^{-2} mA/W and 5×10^7 Jones, respectively. Although our MoS$_2$ phototransistors show much inferior performances to silicon photodiodes ($R \sim 300$ A/W and $D^* \sim 10^{13}$ Jones) (52, 53), their performance is better than phototransistors based on graphene ($R \sim 1$ mA/W at $V_g = 60$ V) (54) or single-layer MoS$_2$ ($R \leq 7.5$ mA/W at $V_g = 50$ V) (48). Future work involving optimizing the device architecture and processing will greatly enhance the performance of our MoS$_2$ phototransistors.

To further characterize our phototransistors, the illumination intensity-dependence of the transfer curves is measured under a visible light (633 nm). As shown in Figure 19.21c, as the illumination intensity increases from 4 mW/cm^2 to 50 mW/cm^2, the photocurrent also increases. Since the linear device response to the incident light intensity is important, a plot of photocurrent as a function of illumination intensity is shown in the inset of Figure 19.21c at $V_{ds} = 1$ V and $V_{gs} = -3$ V. The good linear output between the photocurrents and the illumination intensity indicates that photocurrent is determined by the number of photogenerated carriers under illumination. From the slope of the linear fit, a responsivity of ~12 mA/W is obtained, which is consistent with the result in Figure 19.21b. In addition, the time-resolved photoresponse is measured for multiple illumination cycles, as depicted in Figure 19.21d. Although an accurate response time is not measurable within our experimental setup, a nearly identical response was observed for multiple cycles, which demonstrates the robustness and reproducibility of our phototransistors.

19.4 ISSUES IN MoS₂ THIN-FILM TRANSISTORS

19.4.1 Large-Area Synthesis

Reliable production of MoS₂ thin films with uniform properties is essential for device applications. Thin flakes of MoS₂ can be obtained by mechanical exfoliation of MoS₂ single crystals using the same techniques that were developed for graphene. While these flakes by mechanical exfoliation have high purity and cleanliness that are suitable for fundamental studies, this method is not scalable and does not allow systematic control of thickness and size. Recently, a focused laser beam was used to thin down MoS₂ to single layer by thermal ablation (55). Yet, the requirement for laser scanning is also challenging to scale up this method. In this section, we review several methods for large-area synthesis of MoS₂.

19.4.1.1 Exfoliation in Liquid It has been known that the intercalation of transition metal dichalcogenides by ionic species (such as Li) allows the layers to be exfoliated in liquid (56, 57). The typical procedure involves submerging MoS₂ powder in a solution of a Li-containing compound such as *n*-butyllithium for more than a day to allow Li ions to intercalate, followed by exposing the intercalated material to water. The water vigorously reacts with the Li between the layers to evolve H₂ gas, which rapidly separates the layers (56, 58). Such chemical exfoliation methods can produce gram quantities of sub-μm-sized single layers (59). But the crystal structure and electronic properties of resulting exfoliated material are different from those of the bulk material. The Li intercalation process changes both crystal structure (2H → 1 T) and electronic properties (semiconducting → metallic) (58, 60–62). Recently, it was reported that annealing at 300 °C can cause a phase transformation of MoS₂ from 1 T to 2H, restoring the semiconducting properties of the pristine material (58).

MoS₂ can also be exfoliated by ultrasonication in appropriate liquids, including organic solvents, aqueous surfactant solutions, or solutions of polymers in solvents (63). The ultrasonication typically results in exfoliated layers with a few hundred nanometers in size. While the main advantage of ion exfoliation is the high yield of single layers, the process needs to be conducted under inert gas due to the flammability of Li compounds under ambient conditions. Moreover, as Li becomes increasingly expensive material, alternative intercalants are needed. Exfoliation by ultrasonication is not sensitive to ambient conditions, but its yield of single layers is relatively low.

19.4.1.2 Chemical Vapor Deposition (CVD) Several approaches of CVD methods have been demonstrated for the synthesis of large-area MoS₂ layers. A two-step thermolysis process was reported for the deposition of MoS₂ by dip-coating in (NH₄)₂MoS₄ and converting to MoS₂ by annealing at 500 °C followed by sulfurization at 1000 °C in sulfur vapor (64). The FETs based on this material showed *n*-type conduction with the on/off current ratio ~10⁵ and mobility up to 6 cm²/(V s). Deposition of single-layer MoS₂ by sulfurization of Mo metal thin films was also reported (65). Alternatively, gas-phase reaction of vaporized MoO₃ and S powders was used to deposit MoS₂ (66). Figure 19.22 shows several CVD methods. In many of these CVD methods, the final MoS₂ film thickness depends on the concentration of thickness of initial precursor. However, in spite of initial reports showing promise, precise control of the thickness over a large area has not yet been reported and remains a challenge.

19.4.1.3 Physical Vapor Deposition (PVD) Sputtering has been commonly used to deposit MoS₂ thin films for lubricant coatings. The deposited MoS₂ thin films by sputtering are either amorphous or polycrystalline columnar depending on deposition temperatures. Very low Hall mobility (<10 cm²/(V s)) was observed in MoS₂ thin films deposited by sputtering (67). Alternatively, MoS₂ thin films can be obtained by pulsed laser deposition or ion-beam-assisted deposition (68, 69).

19.4.2 Metal Contacts and Doping

Electrical contacts often play a more crucial role in nanoscale devices than the semiconductor itself. Compared to the carrier mobility of bulk MoS₂ (4), the observed mobility in single-layer MoS₂ is very low without HfO₂ capping layer (1, 7). Although currently Au is the most common contact metal for MoS₂ devices, the existence of Schottky barrier between Au and MoS₂ is reported by both computation and experiments (70, 20). The nonideal electric contacts formed on MoS₂ can fundamentally hamper any attempts to improve transistor performance.

To reduce contact resistivity between metal electrodes and MoS₂, the use of thermal annealing, highly doped interface, or scandium electrodes has been suggested (Fig. 19.23) (7, 20, 71). However, as the conventional thermal annealing affects entire components, it can also induce damages to the components where the annealing should be avoided. In addition, the doping stability of highly doped interface significantly limits commercial feasibility as the chemical doping effect is gradually reduced over time. The use of scandium metals offers low commercial feasibility because of the low availability and the difficulties in the preparation of metallic scandium.

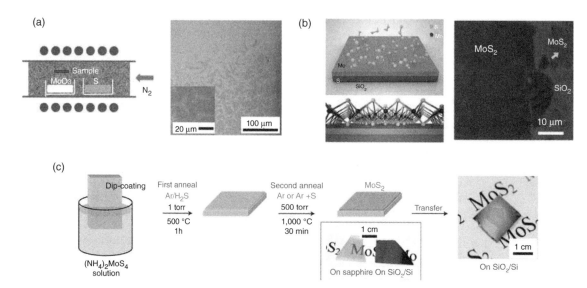

Figure 19.22   Several CVD methods for MoS$_2$ layers. (a) CVD from solid S and MoO$_3$ precursors (left) and resulting MoS$_2$ films on SiO$_2$ (right). (b) CVD from a solid layer of Mo on SiO$_2$ exposed to S vapor (top left) resulting in MoS$_2$ layers that are visible in optical microscopy (right). (c) CVD from a dip-coated precursor on the substrate and growth in the presence of Ar gas and S vapor. (*Source*: Reprinted with permission from (a) Reference (64), © 2011 Wiley; (b) Reference (65), © 2012 Wiley; (c) Reference (66), © 2012 American Chemical Society).

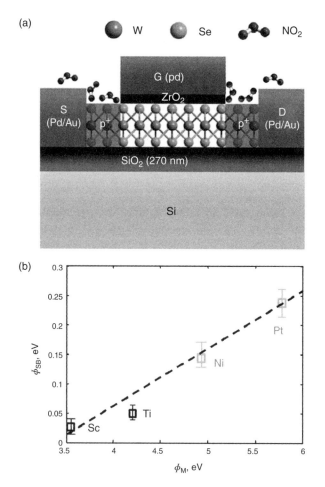

Figure 19.23   (a) Schematic of a top-gated WSe$_2$ transistor with chemically p-doped source/drain contacts by NO$_2$ exposure. (b) Extracted Schottky barrier height for Sc, Ti, Ni, and Pt. The dotted line is a guide to eyes. (*Source*: Reprinted with permission from (a) Reference (71), © 2012 American Chemical Society; (b) Reference (20), © 2013 American Chemical Society).

Zhang et al. demonstrated ambipolar transport in a MoS_2 electric double-layer transistor (an FET with an ionic liquid as gate) using a thin flake of MoS_2 (>10 nm) (72). They reported on/off ratio >10^2 for both hole and electron transport, which is much lower than that of single-layer MoS_2 transistors with a HfO_2 capping layer. Hall mobilities were measured to be 44 and 86 $cm^2/(V s)$ for electrons and holes, respectively. The demonstration of both p- and n-type transports will be useful for applications such as complementary metal-oxide-semiconductor (CMOS) logic devices and optoelectronic devices based on p–n junctions.

19.4.3 High-k Deposition

A field-effect mobility of 100−500 $cm^2/(V s)$ and a remarkable SS as low as ~70 mV/decade have been reported in single or multilayer MoS_2 FETs, in which high-dielectric-constant (high-k) materials were formed by an ALD technique and adopted as a top- or bottom-gated dielectric layer. In these MoS_2 FET structures, engineering the high-k gate dielectric that covers the MoS_2 surface plays a critical role in the enhancement of carrier mobility in 2D-layered MoS_2. It was suggested that the dielectric engineering can strongly dampen the Coulombic scattering of charge carriers in the 2D-layered nanostructure through the dielectric constant mismatch effect between the nanoscale semiconducting material and the high-k dielectric (18). For example, the field-effect mobility drastically drops down to 1−20 $cm^2/(V s)$ without an ALD high-k dielectric film on MoS_2 (1, 7). Furthermore, as an additional benefit, the high-k layer on MoS_2 may reduce the hysteresis of FETs by preventing the moisture absorption from ambient air (73).

However, due to the absence of dangling bonds or functional groups (e.g., hydroxyl groups) on the MoS_2 surface to react with ALD precursors, especially metal-organic precursors, Liu et al. claimed that complete coverage of Al_2O_3 films could not be achieved at an elevated ALD temperature over 200 °C (74). They attributed it to the physically adsorbing nature of ALD precursors on MoS_2 with a much weaker binding energy than the chemical adsorption, and solved this problem by lowering the ALD temperature down to 200 °C (19, 74). Lowering the ALD temperature could be a facile way for acquiring a uniform and continuous high-k film on MoS_2. However, the electrical properties of the high-k film can be degraded with decreasing ALD temperature, because a large amount of carbon impurities released from metal-organic precursors can remain in the high-k film (75). In addition, structural defects such as pinholes may not be completely removed, even at the optimal ALD temperature, which may vary depending on the precursor chemistry and hardware configuration.

As there is also interest in using MoS_2 transistors in flexible electronics, several groups reported the fabrication of MoS_2 transistors on flexible substrates (76–78). The best transistor performance was obtained from MoS_2 transistors with high-k gate dielectrics, where experimental investigation showed that crack formation in the dielectric such as Al_2O_3 or HfO_2 is responsible for the dielectric failure (78). Therefore, the mechanical properties of the high-k dielectric layer are also critical for the realization of flexible electronics based on MoS_2.

19.5 CONCLUSION

The TFTs based on transition metal dichalcogenides that we described previously provide promising device performance metrics including high carrier mobility, high on/off current ratio, and low SS. They also exhibit intriguing optoelectronic properties such as optical stability, wider spectral response, and higher photoresponsivity than graphene. The TFTs based on transition metal dichalcogenides can be attractive for a variety of applications, including OLED displays, touch sensor panels, image sensors, solar cells, and communication devices. However, to realize their technological potentials in real applications, several challenges must be overcome, including uniform synthesis on large-area substrates, formation of ohmic contacts, and interface control between high-k capping layers and semiconductors. Nonetheless, as the properties and applications of transition metal dichalcogenides are rapidly expanding area of research, the rapid progress in this field will enable a wide variety of applications as well as better scientific understanding.

REFERENCES

1. Novoselov KS, Jiang D, Schedin F, Booth TJ, Khotkevich VV, Morozov SV, Geim AK. Proc Natl Acad Sci U S A 2005;102:10451–10453.

2. Schwierz F. Nat Nanotechnol 2010;5:487–496.

3. Frey GL, Elani S, Homyonfer M, Feldman Y, Tenne R. Phys Rev B 1998;57:6666–6671.

4. Fivaz R, Mooser E. Phys Rev 1967;163:743–755.

5. Splendiani A, Sun L, Zhang Y, Li T, Kim J, Chim C, Galli G, Wang F. Nano Lett 2010;10:1271–1275.

6. Mak KF, Lee C, Hone J, Shan J, Heinz TF. Phys Rev Lett 2010;105:136805.

7. Radisavljevic B, Radenovic A, Brivio J, Giacometti V, Kis A. Nat Nanotechnol 2011;6:147–150.

8. Chhowalla M, Shin HS, Eda G, Li L-J, Loh KP, Zhang H. Nat Chem 2013;5:263–275.

9. Wang QH, Kalantar-Zadeh K, Kis A, Coleman JN, Strano MS. Nat Nanotechnol 2012;7:699–712.

10. Kertesz M, Hoffmann R. J Am Chem Soc 1984;106:3453–3460.

11. Wilson JA, Yoffe AD. Adv Phys 1969;18:193–335.

12. Woollam JA, Somoano RB. Phys Rev B 1976;13:3843–3853.

13. Evans BL, Young PA. Proc R Soc A 1965;284:402–422.

14. Mak KF, He K, Shan J, Heinz T. Nat Nanotechnol 2012;7:494–498.

15. Ramana CV, Becker U, Shutthanandan V, Julien CM. Geochem Trans 2008;9:8.

16. Lee C, Yan H, Brus LE, Heinz TF, Hone J, Ryu S. ACS Nano 2010;4:2695–2700.

17. Li H, Zhang Q, Yap CCR, Tay BK, Edwin THT, Olivier A, Baillargeat D. Adv Funct Mater 2012;22:1385–1390.

18. Jena D, Kona A. Phys Rev Lett 2007;98:136805.

19. Liu H, Ye PD. IEEE Electron Device Lett 2011;33:546–548.

20. Das S, Chen H-Y, Penumatcha AV, Appenzeller J. Nano Lett 2013;13:100–105.

21. Kaasbjerg K, Thygesen KS, Jacobsen KW. Phys Rev B 2012;85:115317.

22. Yun WS, Han SW, Hong SC, Kim IG, Lee JD. Phys Rev B 2012;85:033305.

23. Fuhrer MS, Hone J. Nat Nanotechnol 2012;8:146–147.

24. Radisavljevic B, Kis A. Nat Nanotechnol 2012;8:147–148.

25. Yoon Y, Ganapathi K, Salahuddin S. Nano Lett 2011;11:3768–3773.

26. Ghatak S, Pal AN, Ghosh A. ACS Nano 2011;5:7707–7712.

27. Kamiya T, Nomura K, Hosono H. Sci Technol Adv Mater 2010;11:044305.

28. Ayari A, Cobas E, Ogundadegbe Q, Fuhrer MS. J Appl Phys 2007;101:014507.

29. Kim S, Konar A, Hwang W-S, Lee JH, Lee J, Yang J, Jung C, Kim H, Yoo J-B, Choi J-Y, Jin YW, Lee SY, Jena D, Choi W, Kim K. Nat Commun 2012;3:1011.

30. Choi W, Cho MY, Konar A, Lee JH, Cha G-B, Hong SC, Kim S, Kim J, Jena D, Joo J, Kim S. Adv Mater 2012;24:5832–5836.

31. Sze SM. *Physics of Semiconductor Devices*. New York: John Wiley & Sons, Inc; 1981.

32. Park D-G, Cho H-J, Lim K-Y, Lim C, Yeo I-S, Roh J-S, Park JW. J Appl Phys 2001;89:6275–6280.

33. Wallace RM, McIntyre PC, Kim J, Nishi Y. MRS Bull 2009;34:493–503.

34. Natori K. J Appl Phys 1994;76:4879–4890.

35. Zhang Y, Ye J, Matsuhashi Y, Iwasa Y. Nano Lett 2012;12:1136–1140.

36. Fivaz R, Mooser E. Phys Rev 1967;163:743–755.

37. Fivaz R. J Phys Chem Solids 1967;28:839–845.

38. Han SW, Kwon H, Kim SK, Ryu S, Yun WS, Kim DH, Hwang JH, Kang J-S, Baik J, Shin HJ, Hong SC. Phys Rev B 2011;84:045409.

39. Coehoorn R, Haas C, Dijkstra J, Flipse CJF, de Groot RA, Wold A. Phys Rev B 1987;35:6195–6202.

40. Verble JL, Wieting TJ. Phys Rev Lett 1970;25:362–365.

41. Gelmont BL, Shur M. J Appl Phys 1995;77:657–660.

42. Frindt RF, Yoffe AD. Proc R Soc A 1963;273:69–83.

43. de Heer WA, Berger C, Ruan M, Sprinkle M, Li X, Hu Y, Zhang B, Hankinson J, Conrad E. Proc Natl Acad Sci U S A 2011;108:16900–16905.

44. Konar A, Jena D. J Appl Phys 2007;102:123705.

45. Konar A, Fang T, Jena D. Phys Rev B 2010;82:115452.

46. Kedzierski J, Hsu P-L, Healey P, Wyatt PW, Keast CL, Sprinkle M, Berger C, de Heer WA. IEEE Trans Electron Dev 2008;55:2078–2085.

47. Li X, Wang X, Zhang L, Lee S, Dai H. Science 2008;319:1229–1232.

48. Yin Z, Li H, Li H, Jiang L, Shi Y, Sun Y, Lu G, Zhang Q, Chen X, Zhang H. ACS Nano 2012;6:74–80.

49. Jeon S, Ahn S-E, Song I, Kim C-J, Chung U-I, Lee E, Yoo I, Nathan A, Lee S, Robertson J, Kim K. Nat Mater 2012;11:301–305.

50. Konstantatos G, Sargent EH. Nat Nanotechnol 2010;5:391–400.

51. Jha AR. *Infrared Technology*. New York: John Wiley & Sons, Inc; 2000.

52. Gong X, Tong M, Xia Y, Cai W, Moon JS, Cao Y, Yu G, Shieh C-L, Nilsson B, Heeger AJ. Science 2009;325:1665–1667.

53. Guo Y, Yu G, Liu Y. Adv Mater 2010;22:4427–4447.

54. Xia F, Mueller T, Lin Y, Valdes-Garcia A, Avouris P. Nat Nanotechnol 2009;4:839–843.

55. Castellanos-Gomez A, Barkelid M, Goossens AM, Calado HS, van der Zant HSJ, Steele GA. Nano Lett 2012;12:3187–3192.

56. Joensen P, Frindt RF, Morrison SR. Mater Res Bull 1986;21:457–461.

57. Dines MB. Mater Res Bull 1975;10:287–291.

58. Eda G, Yamaguchi H, Voiry D, Fujita T, Chen M, Chhowalla M. Nano Lett 2011;11:5111–5116.

59. Tsai H-L, Heising J, Schindler JL, Kannerwurf CR, Kanatzidis MG. Chem Mater 1997;9:879–882.

60. Frey GL, Reynolds KJ, Friend RH, Cohen H, Feldman Y. J Am Chem Soc 2003;125:5998–6007.

61. Bissessur R, Kanatzidis MG, Schindler JL, Kannerwurf CR. J Chem Soc, Chem Commun 1993:29:1582–1585.

62. Gordon RA, Yang D, Crozier ED, Jiang DT, Frindt RF. Phys Rev B 2002;65:125407.

63. Coleman JN, Lotya M, O'Neill A, Bergin SD, King PJ, Khan U, Young K, Gaucher A, De S, Smith RJ, Shvets IV, Arora SK, Stanton G, Kim H-Y, Lee K, Kim GT, Duesberg GS, Hallam T, Boland JJ, Wang JJ, Donegan JF, Grunlan JC, Moriarty G, Shmeliov A, Nicholls RJ, Perkins JM, Grieveson EM, Theuwissen K, McComb DW, Nellist PD, Nicolosi V. Science 2011;331:568–571.

64. Liu K-K, Zhang W, Lee Y-H, Lin Y-C, Chang M-T, Su C-Y, Chang C-S, Li H, Shi Y, Zhang H, Lai C-S, Li L-J. Nano Lett 2012;12:1538–1544.

65. Zhan Y, Liu Z, Najmaei S, Ajayan PM, Lou J. Small 2012;8:966–971.

66. Lee Y-H, Zhang X-Q, Zhang W, Chang M-T, Lin C-T, Chang K-D, Yu Y-C, Wang JT-W, Chang C-S, Li L-J, Lin T-W. Adv Mater 2012;24:2320–2325.

67. Bichsel R, Levy F. J Phys D 1986;19:1809–1819.

68. Donley MS, Murray PT, Barber SA, Haas TW. Surf Coat Technol 1988;36:339–340.

69. Seitzman LE, Bolster RN, Singer IL. Thin Solid Films 1995;260:143–147.

70. Popov I, Seifert G, Tomanek D. Phys Rev Lett 2012;108:156802.

71. Zhang Y, Ye J, Matsuhashi Y, Iwasa Y. Nano Lett 2012;12:1136–1140.

72. Fang H, Chuang S, Chang TC, Take K, Takahashi T, Javey A. Nano Lett 2012;12:3788–3792.

73. Late DJ, Liu B, Matte HSSR, Dravid VP, Rao CNR. ACS Nano 2012;6:5635–5641.

74. Liu H, Xu K, Zhang X, Ye PD. Appl Phys Lett 2012;100:152115.

75. Swerts J, Peys N, Nyns L, Delabie A, Franquet A, Maes JW, Elshocht SV, Gendt SD. J Electrochem Soc 2010;157:G26–G31.

76. Pu J, Yomogida Y, Liu KK, Li LJ, Iwasa Y, Takenobu T. Nano Lett 2012;12:4013–4017.

77. Yoon J, Park W, Bae G-Y, Kim Y, Jang HS, Hyun Y, Lim SK, Kahng YH, Hong W-K, Lee BH, Ko HC. Small 2013. DOI: 10.1002/smll.201300134.

78. Chang H-Y, Yang S, Lee J, Tao L, Hwang W-S, Jena D, Lu N, Akinwande D. ACS Nano 2013. DOI: 10.1021/nn401429w.

INDEX

Nanomaterials, Polymers and Devices: Materials Functionalization and Device Fabrication, First Edition. Edited by Eric S. W. Kong.
© 2015 John Wiley & Sons, Inc. Published 2015 by John Wiley & Sons, Inc.